Methods in Enzymology

Volume 218
RECOMBINANT DNA

Part I

METHODS IN ENZYMOLOGY

EDITORS-IN-CHIEF

John N. Abelson   Melvin I. Simon

DIVISION OF BIOLOGY
CALIFORNIA INSTITUTE OF TECHNOLOGY
PASADENA, CALIFORNIA

FOUNDING EDITORS

Sidney P. Colowick and Nathan O. Kaplan

Methods in Enzymology

Volume 218

Recombinant DNA

Part I

EDITED BY

Ray Wu

CORNELL UNIVERSITY
ITHACA, NEW YORK

ACADEMIC PRESS, INC.

HARCOURT BRACE & COMPANY

San Diego   New York   Boston
London   Sydney   Tokyo   Toronto

Copyright © 1993 by ACADEMIC PRESS, INC.

All Rights Reserved.

No part of this publication may be reproduced or transmitted in any form or by any means, electronic or mechanical, including photocopy, recording, or any information storage and retrieval system, without permission in writing from the publisher.

Academic Press, Inc.
1250 Sixth Avenue, San Diego, California 92101-4311

United Kingdom Edition published by
Academic Press Limited
24–28 Oval Road, London NW1 7DX

Library of Congress Catalog Number: 54-9110

International Standard Book Number: 0-12-182119-6

PRINTED IN THE UNITED STATES OF AMERICA
93 94 95 96 97 98   E B   9 8 7 6 5 4 3 2 1

Table of Contents

Section I. Methods for Sequencing DNA

Section II. Polymerase Chain Reaction for Amplifying and Manipulating DNA

Section III. Methods for Detecting DNA–Protein Interaction

Section IV. Other Methods

Contributors to Volume 218

Article numbers are in parentheses following the names of contributors.
Affiliations listed are current.

MARIE ALLEN (1), *Department of Medical Genetics, University of Uppsala Biomedical Center, S-751 23 Uppsala, Sweden*

FRANCISCO JOSÉ AYALA (21), *Department of Organismic and Evolutionary Biology, Harvard University, Cambridge, Massachusetts 02138*

ALAN T. BANKIER (13), *Medical Research Council Laboratory of Molecular Biology, Cambridge CB2 2QH, England*

BARCLAY G. BARRELL (13), *Medical Research Council Laboratory of Molecular Biology, Cambridge CB2 2QH, England*

STEVEN R. BAUER (33), *Laboratory of Molecular Immunology, Department of Health and Human Services, Food and Drug Administration, Bethesda, Maryland 20892*

PETER B. BECKER (40), *Gene Expression Program, European Molecular Biology Laboratory, D-6900 Heidelberg, Germany*

MICHAEL BECKER-ANDRÉ (32), *GLAXO Institute for Molecular Biology, 1228 Plan-les-Ouates, Geneva, Switzerland*

CLAIRE M. BERG (19, 20), *Departments of Molecular and Cell Biology, University of Connecticut, Storrs, Connecticut 06269*

DOUGLAS E. BERG (19, 20), *Departments of Molecular Microbiology and Genetics, Washington University School of Medicine, St. Louis, Missouri 63110*

CYNTHIA D. K. BOTTEMA (29), *Department of Biochemistry and Molecular Biology, Mayo Clinic/Foundation, Rochester, Minnesota 55905*

SYDNEY BRENNER (18), *Department of Medicine, Cambridge University, Cambridge CB2 2QH, England, and The Scripps Research Institute, La Jolla, California 92121*

IGOR BRIKUN (19), *Department of Molecular Microbiology, Washington University School of Medicine, St. Louis, Missouri 63110*

CAROL M. BROWN (13), *Medical Research Council Laboratory of Molecular Biology, Cambridge CB2 2QH, England*

MICHAEL BULL (22), *Department of Immunology, Mayo Clinic, Rochester, Minnesota 55905*

GLADYS I. CASSAB (48), *Plant Molecular Biology and Biotechnology, Institute of Biotechnology, National Autonomous University of Mexico, Cuernavaca 62271, Mexico*

JOSLYN D. CASSADY (29), *Department of Biochemistry and Molecular Biology, Mayo Clinic/Foundation, Rochester, Minnesota 55905*

MARK S. CHEE (13), *Affymax Research Institute, Palo Alto, California 94304*

CATHIE T. CHUNG (43), *Hepatitis Viruses Section, Laboratory of Infectious Diseases, National Institute of Allergy and Infectious Diseases, National Institutes of Health, Bethesda, Maryland 20892*

GEORGE M. CHURCH (14), *Department of Genetics, Howard Hughes Medical Institute, Harvard Medical School, Boston, Massachusetts 02115*

JOHN A. CIDLOWSKI (38), *Department of Physiology, University of North Carolina at Chapel Hill, Chapel Hill, North Carolina 27599*

MOLLY CRAXTON (13), *Medical Research Council Laboratory of Molecular Biology, Cambridge CB2 2QH, England*

PETER B. DERVAN (15), *Arnold and Mabel Beckman Laboratories of Chemical Synthesis, Division of Chemistry and Chemical Engineering, California Institute of Technology, Pasadena, California 91125*

CRAIG A. DIONNE (30), *Cephalon, Inc., West Chester, Pennsylvania 19380*

DAVID M. DORFMAN (23), *Department of Pathology, Brigham and Women's Hospital, Boston, Massachusetts 02115, and Harvard Medical School, Harvard University, Cambridge, Massachusetts 02138*

ROBERT L. DORIT (4), *Department of Biology, Yale University, New Haven, Connecticut 06511*

HOWARD DROSSMAN (12), *Department of Chemistry, Colorado College, Colorado Springs, Colorado 80903*

ZIJIN DU (10), *Department of Genetics, Washington University School of Medicine, St. Louis, Missouri 63110*

CHARYL M. DUTTON (29), *Department of Biochemistry and Molecular Biology, Mayo Clinic/Foundation, Rochester, Minnesota 55905*

DAVID D. ECKELS (22), *Immunogenetics Research Section, Blood Research Institute, The Blood Center of Southeastern Wisconsin, Milwaukee, Wisconsin 53233*

FRITZ ECKSTEIN (8), *Abteilung Chemie, Max-Planck-Institut für Experimentelle Medizin, D-3400 Göttingen, Germany*

HENRY ERLICH (27), *Department of Human Genetics, Roche Molecular Systems, Alameda, California 94501*

JAMES A. FEE (50), *Spectroscopy and Biochemistry Group, Isotope and Nuclear Chemistry Division, Los Alamos National Laboratory, Los Alamos, New Mexico 87545*

MICHAEL A. FROHMAN (24), *Department of Pharmacological Sciences, State University of New York at Stony Brook, Stony Brook, New York 11794*

ODD S. GABRIELSEN (36), *Department of Biochemistry, University of Oslo, N-0316 Oslo, Norway*

MELISSA A. GEE (49), *Worcester Foundation for Experimental Biology, Shrewsbury, Massachusetts 01545*

MARY JANE GEIGER (22), *Department of Medicine, Duke University Medical Center, Durham, North Carolina 27710*

WALTER GILBERT (4), *Department of Cellular and Developmental Biology, Harvard University, Cambridge, Massachusetts 02138*

JACK GORSKI (22), *Immunogenetics Research Section, Blood Research Institute, The Blood Center of Southeastern Wisconsin, Milwaukee, Wisconsin 53233*

MICHAEL M. GOTTESMAN (45), *Laboratory of Cell Biology, National Cancer Institute, National Institutes of Health, Bethesda, Maryland 20892*

RICHARD W. GROSS (17), *Division of Bioorganic Chemistry and Molecular Pharmacology, Washington University School of Medicine, St. Louis, Missouri 63110*

TOM J. GUILFOYLE (49), *Department of Biochemistry, University of Missouri, Columbia, Missouri 65211*

ULF B. GYLLENSTEN (1), *Department of Medical Genetics, University of Uppsala Biomedical Center, S-751 23 Uppsala, Sweden*

PERRY B. HACKETT (5), *Department of Genetics and Cell Biology, University of Minnesota, St. Paul, Minnesota 55108*

GRETCHEN HAGEN (49), *Department of Biochemistry, University of Missouri, Columbia, Missouri 65211*

MICHAEL K. HANAFEY (51), *Agricultural Products Department, E. I. DuPont de Nemours & Company, Wilmington, Delaware 19880*

DANIEL L. HARTL (3, 21), *Department of Organismic and Evolutionary Biology, Harvard University, Cambridge, Massachusetts 02138*

CHERYL HEINER (11), *Applied Biosystems, Inc., Foster City, California 94404*

LEROY HOOD (10, 11), *Department of Molecular Biotechnology, School of Medicine, University of Washington, Seattle, Washington 98195*

BRUCE H. HOWARD (45), *Laboratory of Molecular Growth Regulation, National Institute of Child Health and Human Development, National Institutes of Health, Bethesda, Maryland 20892*

TAZUKO HOWARD (45), *Laboratory of Cell Biology, National Cancer Institute, National Institutes of Health, Bethesda, Maryland 20892*

HENRY V. HUANG (19), *Department of Molecular Microbiology, Washington University School of Medicine, St. Louis, Missouri 63110*

JANINE HUET (36), *Service de Biochimi et Génétique Moléculaire, Centre d'Etudes de Saclay, 91191 Gif-sur-Yvette, France*

TIM HUNKAPILLER (11), *Department of Molecular Biotechnology, School of Medicine, University of Washington, Seattle, Washington 98195*

NORIO ICHIKAWA (46), *Department of Biochemistry, School of Hygiene and Public Health, The Johns Hopkins University, Baltimore, Maryland 21205*

SETSUKO II (29), *Department of Biochemistry and Molecular Biology, Mayo Clinic/ Foundation, Rochester, Minnesota 55905*

BRENT L. IVERSON (15), *Arnold and Mabel Beckman Laboratories of Chemical Synthesis, Division of Chemistry and Chemical Engineering, California Institute of Technology, Pasadena, California 91125*

MICHAEL JAYE (30), *Department of Molecular Biology, Rhône-Poulenc Rorer Central Research, Collegeville, Pennsylvania 19426*

D. S. C. JONES (9), *Medical Research Council, Molecular Genetics Unit, Cambridge CB2 2QH, England*

MICHAEL D. JONES (31), *Department Virology, Royal Postgraduate Medical School, Hammersmith Hospital, University of London, London W12 0NN, England*

VINCENT JUNG (25), *Cold Spring Harbor Laboratories, Cold Spring Harbor, New York 11724*

ROBERT KAISER (11), *Department of Molecular Biotechnology, School of Medicine, University of Washington, Seattle, Washington 98195*

ERNEST KAWASAKI (27), *Procept, Inc., Cambridge, Massachusetts 02139*

J. ANDREW KEIGHTLEY (50), *Spectroscopy and Biochemistry Group, Isotope and Nuclear Chemistry Division, Los Alamos National Laboratory, Los Alamos, New Mexico 87545*

DAVID J. KEMP (37), *Menzies School of Health Research, Casuarina, Northern Territory 0811, Australia*

DANGERUTA KERSULYTE (19), *Department of Molecular Microbiology, Washington University School of Medicine, St. Louis, Missouri 63110*

BRUCE C. KLINE (26), *Department of Biochemistry and Molecular Biology, Mayo Clinic/Foundation, Rochester, Minnesota 55905*

TONY KOSTICHKA (12), *Cary, North Carolina 27511*

JAN P. KRAUS (16), *Department of Pediatrics, University of Colorado School of Medicine, Denver, Colorado 80262*

MARTIN KREITMAN (2), *Department of Ecology and Evolution, University of Chicago, Chicago, Illinois 60637*

KEITH A. KRETZ (7), *Department of Neurosciences and Center for Molecular Genetics, School of Medicine, University of California, San Diego, La Jolla, California 92093*

B. RAJENDRA KRISHNAN (19), *Department of Medicine, Washington University School of Medicine, St. Louis, Missouri 63110*

LAURA F. LANDWEBER (2), *Department of Cellular and Developmental Biology, Biological Laboratories, Harvard University, Cambridge, Massachusetts 02138*

JEFFREY G. LAWRENCE (3), *Department of Biology, University of Utah, Salt Lake City, Utah 84112*

JEAN-CLAUDE LELONG (42), *Institut d'Oncologie Cellulaire et Moléculaire Humaine, Université de Paris Nord, 93000 Paris, France*

ANDREW M. LEW (37), *Burnet Clinical Research Unit, The Walter and Eliza Hall Institute of Medical Research, Royal Melbourne Hospital, Parkville, Victoria 3050, Australia*

ZHANJIANG LIU (5), *Institute of Human Genetics, University of Minnesota, St. Paul, Minnesota 55108*

KENNETH J. LIVAK (18), *DuPont Merck Pharmaceutical Company, Wilmington, Delaware 19880*

MATTHEW J. LONGLEY (41), *Department of Biochemistry, Duke University Medical Center, Durham, North Carolina 27710*

JOHN A. LUCKEY (12), *Department of Chemistry, University of Wisconsin-Madison, Madison, Wisconsin 53706*

VIKKI M. MARSHALL (37), *Immunoparasitology Unit, The Walter and Eliza Hall Institute of Medical Research, Royal Melbourne Hospital, Parkville, Victoria 3050, Australia*

MICHAEL W. MATHER (50), *Department of Biochemistry and Molecular Biology, Oklahoma State University, Stillwater, Oklahoma 74078*

BRUCE A. MCCLURE (49), *Department of Biochemistry, University of Missouri, Columbia, Missouri 65211*

TERRI L. MCGUIGAN (18), *DuPont Merck Pharmaceutical Company, Wilmington, Delaware 19880*

ROGER H. MILLER (43), *Hepatitis Viruses Section, National Institute of Allergy and Infectious Diseases, National Institutes of Health, Bethesda, Maryland 20892*

DALE W. MOSBAUGH (41), *Departments of Agricultural Chemistry, Biochemistry, and Biophysics, Oregon State University, Corvallis, Oregon 97331*

JOHN S. O'BRIEN (7), *Department of Neurosciences and Center for Molecular Genetics, School of Medicine, University of California, San Diego, La Jolla, California 92093*

HOWARD OCHMAN (3, 21), *Department of Biology, University of Rochester, Rochester, New York 14627*

OSAMU OHARA (4), *Shinogi Research Laboratories, Osaka, Japan*

DAVID B. OLSEN (8), *Merck Sharp and Dohme Research Laboratories, West Point, Pennsylvania 19486*

R. PADMANABHAN (45), *Department of Biochemistry and Molecular Biology, University of Kansas Medical Center, Kansas City, Kansas 66103*

RAJI PADMANABHAN (45), *Department of Health and Human Services, National Institutes of Health, Bethesda, Maryland 20892*

SIDNEY PESTKA (25), *Department of Molecular Genetics & Microbiology, University of Medicine and Dentistry of New Jersey, Robert Wood Johnson Medical School, Piscataway, New Jersey 08854*

STEVEN B. PESTKA (25), *North Caldwell, New Jersey 07006*

MICHAEL GREGORY PETERSON (35), *Tularik, Inc., South San Francisco, California 94080*

JAMES W. PRECUP (26), *Department of Biochemistry and Molecular Biology, Mayo Clinic/Foundation, Rochester, Minnesota 55905*

J. ANTONI RAFALSKI (51), *Agricultural Products Department, E. I. DuPont de Nemours & Company, Wilmington, Delaware 19880*

WILLIAM D. RAWLINSON (13), *Medical Research Council Laboratory of Molecular Biology, Cambridge CB2 2QH, England*

PETER RICHTERICH (14), *Department of Human Genetics and Molecular Biology, Collaborative Research, Inc., Waltham, Massachusetts 02154*

RANDALL SAIKI (27), *Department of Human Genetics, Roche Molecular Systems, Alameda, Calfornia 94501*

GURPREET S. SANDHU (26), *Department of Biochemistry and Molecular Biology, Mayo Clinic/Foundation, Rochester, Minnesota 55905*

GOBINDA SARKAR (28, 29), *Department of Biochemistry and Molecular Biology, Mayo Clinic/Foundation, Rochester, Minnesota 55905*

RICHARD H. SCHEUERMANN (33), *Department of Pathology, University of Texas Southwestern Medical Center, Dallas, Texas 75235*

J. P. SCHOFIELD (9), *Medical Research Council, Molecular Genetics Unit, Cambridge CB2 2QH, England*

GÜNTHER SCHÜTZ (40), *Institute of Cell and Tumor Biology, German Cancer Research Center, D-6900 Heidelberg, Germany*

WENYAN SHEN (6), *Whitehead Institute, Cambridge, Massachusetts 02142*

HARINDER SINGH (39), *Department of Molecular Genetics and Cell Biology, Howard Hughes Medical Institute, University of Chicago, Chicago, Illinois 60637*

LLOYD M. SMITH (12), *Department of Chemistry, University of Wisconsin-Madison, Madison, Wisconsin 53706*

VICTORIA SMITH (13), *Department of Genetics, Stanford University, Stanford, California 94305*

HANS SÖDERLUND (34), *Biotechnical Laboratory, Technical Research Centre of Finland, 02150 Espoo, Finland*

STEVE S. SOMMER (28, 29), *Department of Biochemistry and Molecular Biology, Mayo Clinic/Foundation, Rochester, Minnesota 55905*

YAH-RU SONG (47), *Department of Plant Physiology, Institute of Botany, Academia Sinica, Beijing 10044, China*

DAVID L. STEFFENS (17), *Research and Development, Li-Cor, Inc., Lincoln, Nebraska 68504*

LINDA D. STRAUSBAUGH (20), *Department of Molecular and Cell Biology, University of Connecticut, Storrs, Connecticut 06269*

ANN-CHRISTINE SYVÄNEN (34), *Department of Human Molecular Genetics, National Public Health Institute, 00300 Helsinki, Finland*

TAKAHIRO TAHARA (16), *Department of Pediatrics, National Okura Hospital, Tokyo 157, Japan*

SCOTT V. TINGEY (51), *Agricultural Products Department, E. I. DuPont de Nemours & Company, Wilmington, Delaware 19880*

ROBERT TJIAN (35), *Howard Hughes Medical Institute, Department of Molecular and Cell Biology, University of California, Berkeley, Berkeley, California 94720*

PAUL O. P. TS'O (46), *Department of Biochemistry, School of Hygiene and Public Health, The Johns Hopkins University, Baltimore, Maryland 21205*

DOUGLAS B. TULLY (38), *Department of Physiology, University of North Carolina at Chapel Hill, Chapel Hill, North Carolina 27599*

ANGELA UY (8), *Abteilung Medizinische Mikrobiologie des Zentrums für Hygiene und Humangenetik der Universität, D-3400 Göttingen, Germany*

JOSEPH E. VARNER (47), *Department of Biology, Washington University, St. Louis, Missouri 63130*

M. VAUDIN (9), *Medical Research Council, Molecular Genetics Unit, Cambridge CB2 2QH, England*

GAN WANG (20), *Department of Molecular and Cell Biology, University of Connecticut, Storrs, Connecticut 06269*

MARY M. Y. WAYE (6), *Department of Biochemistry, The Chinese University of Hong Kong, Shatin, New Territories, Hong Kong*

FALK WEIH (40), *Department of Molecular Biology, Bristol Myers Squibb Pharmaceutical Research Co., Princeton, New Jersey 08543*

PAUL A. WHITTAKER (44), *Clinical Biochemistry, University of Southampton, and South Laboratory and Pathology Block, Southampton General Hospital, Southampton S09 4XY, England*

JOHN G. K. WILLIAMS (51), *Data Management Department, Pioneer Hi-Bred International, Johnston, Iowa 50131*

RICHARD K. WILSON (10), *Department of Genetics, Washington University School of Medicine, St. Louis, Missouri 63110*

GERD WUNDERLICH (8), *Abteilung Medizinische Mikrobiologie des Zentrums für Hygiene und Humangenetik der Universität, D-3400 Göttingen, Germany*

ZHENG-HUA YE (47), *Department of Biology, Washington University, St. Louis, Missouri 63130*

MING YI (46), *Department of Biochemistry, School of Hygiene and Public Health, The Johns Hopkins University, Baltimore, Maryland 21205*

Preface

Recombinant DNA methods are powerful, revolutionary techniques for at least two reasons. First, they allow the isolation of single genes in large amounts from a pool of thousands or millions of genes. Second, the isolated genes from any source or their regulatory regions can be modified at will and reintroduced into a wide variety of cells by transformation. The cells expressing the introduced gene can be measured at the RNA level or protein level. These advantages allow us to solve complex biological problems, including medical and genetic problems, and to gain deeper understandings at the molecular level. In addition, new recombinant DNA methods are essential tools in the production of novel or better products in the areas of health, agriculture, and industry.

The new Volumes 216, 217, and 218 supplement Volumes 153, 154, and 155 of *Methods in Enzymology*. During the past few years, many new or improved recombinant DNA methods have appeared, and a number of them are included in these new volumes. Volume 216 covers methods related to isolation and detection of DNA and RNA, enzymes for manipulating DNA, reporter genes, and new vectors for cloning genes. Volume 217 includes vectors for expressing cloned genes, mutagenesis, identifying and mapping genes, and methods for transforming animal and plant cells. Volume 218 includes methods for sequencing DNA, PCR for amplifying and manipulating DNA, methods for detecting DNA–protein interactions, and other useful methods.

Areas or specific topics covered extensively in the following recent volumes of *Methods in Enzymology* are not included in these three volumes: "Guide to Protein Purification," Volume 182, edited by M. P. Deutscher; "Gene Expression Technology," Volume 185, edited by D. V. Goeddel; and "Guide to Yeast Genetics and Molecular Biology," Volume 194, edited by C. Guthrie and G. R. Fink.

RAY WU

METHODS IN ENZYMOLOGY

VOLUME 74. Immunochemical Techniques (Part C)
Edited by JOHN J. LANGONE AND HELEN VAN VUNAKIS

VOLUME 75. Cumulative Subject Index Volumes XXXI, XXXII, XXXIV–LX
Edited by EDWARD A. DENNIS AND MARTHA G. DENNIS

VOLUME 76. Hemoglobins
Edited by ERALDO ANTONINI, LUIGI ROSSI-BERNARDI, AND EMILIA CHIAN-
CONE

VOLUME 77. Detoxication and Drug Metabolism
Edited by WILLIAM B. JAKOBY

VOLUME 78. Interferons (Part A)
Edited by SIDNEY PESTKA

VOLUME 79. Interferons (Part B)
Edited by SIDNEY PESTKA

VOLUME 80. Proteolytic Enzymes (Part C)
Edited by LASZLO LORAND

VOLUME 81. Biomembranes (Part H: Visual Pigments and Purple Membranes, I)
Edited by LESTER PACKER

VOLUME 82. Structural and Contractile Proteins (Part A: Extracellular Matrix)
Edited by LEON W. CUNNINGHAM AND DIXIE W. FREDERIKSEN

VOLUME 83. Complex Carbohydrates (Part D)
Edited by VICTOR GINSBURG

VOLUME 84. Immunochemical Techniques (Part D: Selected Immunoassays)
Edited by JOHN J. LANGONE AND HELEN VAN VUNAKIS

VOLUME 85. Structural and Contractile Proteins (Part B: The Contractile Apparatus and the Cytoskeleton)
Edited by DIXIE W. FREDERIKSEN AND LEON W. CUNNINGHAM

VOLUME 86. Prostaglandins and Arachidonate Metabolites
Edited by WILLIAM E. M. LANDS AND WILLIAM L. SMITH

VOLUME 87. Enzyme Kinetics and Mechanism (Part C: Intermediates, Stereo-
chemistry, and Rate Studies)
Edited by DANIEL L. PURICH

VOLUME 88. Biomembranes (Part I: Visual Pigments and Purple Membranes, II)
Edited by LESTER PACKER

VOLUME 89. Carbohydrate Metabolism (Part D)
Edited by WILLIS A. WOOD

VOLUME 90. Carbohydrate Metabolism (Part E)
Edited by WILLIS A. WOOD

VOLUME 91. Enzyme Structure (Part I)
Edited by C. H. W. HIRS AND SERGE N. TIMASHEFF

[1] Sequencing of *in Vitro* Amplified DNA

By ULF B. GYLLENSTEN and MARIE ALLEN

Introduction

The polymerase chain reaction (PCR)[1,2] method for *in vitro* amplification of specific DNA fragments has opened up a number of fields in molecular biology that were previously intangible because of lack of sufficiently sensitive analytical methods. The PCR is based on the use of two oligonucleotides to prime DNA polymerase-catalyzed synthesis from opposite strands across a region flanked by the priming sites of the two oligonucleotides. By repeated cycles of DNA denaturation, annealing of oligonucleotide primers, and primer extension an exponential increase in copy number of a discrete DNA fragment can be achieved. Many applications of PCR, including diagnosis of heritable disorders, screening for susceptibility to disease, and identification of bacterial and viral pathogens, require determination of the nucleotide sequence of amplified DNA fragments. In this chapter we review alternate methods for the generation of sequencing templates from amplified DNA and sequencing by the method of Sanger.[3]

Generation of Sequencing Template for Direct Sequencing

Traditionally, templates for DNA sequencing have been generated by inserting the target DNA into bacterial or viral vectors for multiplication of the inserts in bacterial host cells. These cloning methods have been simplified, but are still subject to inherent problems associated with the maintenance and use of systems dependent on living cells, such as *de novo* mutations in vector and host cell genomes. By using PCR, templates for sequencing can be generated more efficiently than with cell-dependent methods either from genomic targets or from DNA inserts cloned into vectors. Amplification of cloned inserts of unknown sequence can be achieved using oligonucleotides that are priming inside, or close to, the polylinker of the cloning vector.[2]

Sequencing the PCR products directly has two advantages over se-

---

[1] K. B. Mullis and F. Faloona, this series, Vol. 155, p. 335.

[2] R. K. Saiki, D. H. Gelfand, S. Stoffel, S. J. Scharf, R. Higuchi, G. T. Horn, K. B. Mullis, and H. A. Erlich, *Science* **239**, 487 (1988).

[3] F. Sanger, S. Nicklen, and A. R. Coulson, *Proc. Natl. Acad. Sci. U.S.A.* **74**, 5463 (1979).

METHODS IN ENZYMOLOGY, VOL. 218

quencing of cloned PCR products. First, it is readily standardized because it is a simple enzymatic process that does not depend on the use of living cells. Second, only a single sequence needs to be determined for each sample (for each allele). By contrast, when PCR products are cloned, a consensus sequence based on several cloned PCR products must be determined for each sample, in order to distinguish mutations present in the original genomic sequence from random misincorporated nucleotides introduced by the *Taq* polymerase during PCR.

Optimization of Polymerase Chain Reaction Conditions for Direct Sequencing

The ease with which clear and reliable sequences can be obtained by direct sequencing depends on the ability of the PCR primers to amplify *only* the target sequence (usually called the specificity of the PCR), and the method used to obtain a template suitable for sequencing. The specificity of the PCR is to a large extent determined by the sequence of the oligonucleotides used to prime the reaction. For an individual pair of primers the specificity of the PCR can be optimized by changing the ramp conditions, the annealing temperature, and the $MgCl_2$ concentration in the PCR buffer. A titration, in 0.2 mM increments, of $MgCl_2$ concentrations from 1.0 to 3.0 mM in the final reaction is advised if the standard 1.5 mM concentration fails to produce the necessary specificity of the PCR.

In cases in which optimization of PCR conditions fails to produce the desired priming specificity, either new oligonucleotides are required or the different PCR products can be separated by gel electrophoresis and reamplified individually for sequencing.

When the PCR primers amplify several related sequences of the same length, for example, the same exon from several recently duplicated genes, or repetitive or conserved signal sequences, electrophoretic separation of the different products can be achieved either by the use of restriction enzymes that cut only certain templates and subsequent gel purification of the intact PCR products, or by the use of an electrophoretic system (denaturing gradient gel electrophoresis, temperature gradient gel electrophoresis) for separation that will differentiate between the products based on their nucleotide sequence difference.[4,5]

[4] R. M. Myers, V. C. Shemeld, and D. R. Cox, *in* "Genome Analysis—A Practical Approach" (K. E. Davies, ed.), p. 95. IRL Press, Oxford, 1988.

[5] V. C. Shemeld, D. R. Cox, L. S. Lerman, and R. M. Myers, *Proc. Natl. Acad. Sci. U.S.A.* **86,** 232 (1989).

Double-Stranded DNA Templates

Many of the problems associated with direct sequencing of PCR products are not due to lack of specificity, but result from the ability of the two strands of the linear amplified product to reassociate rapidly after denaturation, thereby either blocking the primer–template complex from extending or preventing the sequencing oligonucleotide from annealing efficiently.[6] This problem is more severe for longer PCR products. To circumvent the strand reassociation of double-stranded DNA (dsDNA), a number of alternate methods have been developed.

Precipitation of Denatured DNA

Denature the template in 0.2 M NaOH for 5 min at room temperature, transfer the tube to ice, neutralize the reaction by adding 0.4 vol of 5 M ammonium acetate (pH 7.5), and immediately precipitate the DNA with 4 vol of ethanol. Resuspend the DNA in sequencing buffer and primer at the desired annealing temperature.[7]

Snap-Cooling of Template DNA

Denature the template by heating (95°) for 5 min. Quickly freeze the tube by putting it in a dry ice–ethanol bath to slow down the reassociation of strands. Add sequencing primer either prior to or after denaturation and bring the reaction to the proper temperature.[8]

Cycling of Polymerase Chain Reactions

A third method for generating enough sequencing template is to cycle the sequencing reaction, using *Taq* polymerase as the enzyme for both amplification and sequencing. Even though only a small fraction of the templates will be utilized in each round of extension–termination, the amount of specific terminations will accumulate with the number of cycles.[8–10]

[6] U. B. Gyllensten, and H. A. Erlich, *Proc. Natl. Acad. Sci. U.S.A.* **85,** 7652 (1988).
[7] L. A. Wrischnik, R. G. Higuchi, M. Stoneking, H. A. Erlich, N. Arnhein, and A. C. Wilson, *Nucleic Acids Res.* **15,** 529 (1987).
[8] N. Kusukawa, T. Uemori, K. Asada, and I. Kato, *Biotechniques* **9,** 66 (1990).
[9] M. Craxton, *Methods: Companion Methods Enzymol.* **3,** 20 (1991).
[10] J.-S. Lee, *DNA* **10,** 67 (1991).

Single-Stranded DNA Templates

Sequencing problems derived from strand reassociation can be avoided by preparing single-stranded DNA (ssDNA) templates by any of the following number of methods.

Strand-Separating Gels

Agarose strand-separating gels may be successfully employed to obtain ssDNA of fragments of more than about 500 bp.[11] This method is suitable primarily for long products, or where other methods may not give sufficient yields of ssDNA.

Blocking Primer Polymerase Chain Reaction

An alternative way of generating ssDNA in the PCR, without the inherent lower efficiency achieved using an asymmetric PCR, is to use blocking primer PCR. In this method, an excess of a third primer that is complementary to one of the PCR primers is added during the PCR (after about 15–20 cycles). The third oligonucleotide will outcompete the newly synthesized target molecules in each cycle as priming sites for the PCR primer and thereby prevent synthesis of one of the DNA strands. The PCR is thereby transformed at any suitable stage into a primer-extension reaction.

Solid-State Sequencing

In this procedure, one of the oligonucleotide primers is labeled with biotin prior to the PCR. After a balanced synthesis of dsDNA, the strands are denatured and put through a streptavidin–agarose column,[12] or mixed with magnetic beads to which streptavidin has been attached.[13] The strand labeled through the incorporated PCR primer will be bound to the solid support, and the unbound strand can be removed. The bound ssDNA is subsequently eluted for direct sequencing, or sequencing is performed with the templates still bound to the matrix. The magnetic beads do not interfere with the sequencing reagents, and can even be loaded on the sequencing gel without distorting the migration of termination products. The benefit of this method is that the reaction will be cleaned up for sequencing, at the same time as the ssDNA template is generated.

[11] T. Maniatis, E. F. Fritsch, and J. Sambrook, "Molecular Cloning: A Laboratory Manual," p. 179. Cold Spring Harbor Press, Cold Spring Harbor, New York, 1982.
[12] L. G. Mitchell and C. R. Merill, *Anal. Biochem.* **178**, 239 (1989).
[13] J. Wahlberg, J. Lundberg, T. Hultman, and M. Uhlen, *Proc. Natl. Acad. Sci. U.S.A.* **87**, 6569 (1990).

λ Exonuclease-Generated Single-Stranded DNA

In this procedure one of the oligonucleotide primers is treated with polynucleotide kinase to introduce a 5'-phosphate prior to the PCR. After a symmetric PCR, the products are exposed to λ 5' → 3'-exonuclease, and the strand containing a 5'-phosphatased primer will be digested. The ssDNA is then purified from the reaction mix and used for sequencing.[14] The efficiency of this method in generating ssDNA depends to a large extent on the proportion of primers that have been successfully kinased.

Transcript Sequencing

A radically different approach for template generation is to combine PCR with reverse transcription, using a phage promotor sequence attached to one of the PCR primers.[15] A standard PCR is performed initially to generate dsDNA. The PCR product is subsequently used in a transcription reaction that will yield a further increase in copy number of the desired single-stranded (RNA) template. This transcript is then sequenced using reverse transcriptase. Either a thermolabile reverse transcriptase, with a temperature range of 37–45°, or a thermostable recombinant reverse transcriptase (rTh; Perkin-Elmer Cetus, Norwalk, CT) with a temperature optimum of 75°, is available for the sequencing.

Asymmetric Polymerase Chain Reaction

In this procedure an asymmetric, or unequal, ratio of the two amplification primers is used in the PCR[6] (Fig. 1). During the first 20–25 cycles dsDNA is generated, but when the limiting primer is exhausted ssDNA is produced for the next 5–10 cycles by primer extension. The accumulation of dsDNA and ssDNA during a typical amplification of a genomic sequence, using an initial ratio of 50 pmol of one primer to 0.5 pmol of the other primer in a 100-μl PCR, is shown schematically in Fig. 2. The amount of dsDNA accumulates exponentially to the point at which the primer is almost exhausted, and thereafter essentially stops. The ssDNA generation starts at about cycle 25, the point at which the limiting primer is almost depleted. Following a short (one or two cycles) initial phase of rapid increase, the ssDNA accumulates linearly as expected when only one primer is present (primer extension). In general, a ratio of 50 pmol : 1–5 pmol for a 100-μl PCR reaction will result in about 1–3 pmol of ssDNA after 30 cycles of PCR. The yield of ssDNA can be estimated by adding 0.1 μl of [α-^{32}P]dCTP (3000 Ci/mmol) to the PCR, and examining the

[14] R. G. Higuchi and H. Ochman, *Nucleic Acids Res.* **17**, 5865 (1989).
[15] E. S. Stoflet, D. D. Koeberl, G. Sarkar, and S. S. Sommer, *Science* **239**, 491 (1988).

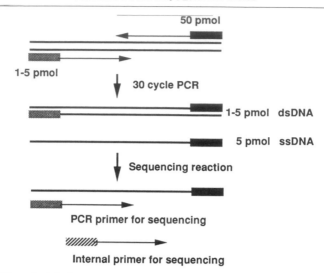

FIG. 1. The principle for asymmetric PCR. When the primer in limited concentration is exhausted, ssDNA is produced. The ssDNA produced can be sequenced either using the limiting PCR primer or an internal primer complementary to the ssDNA.

reaction products on a gel. The ssDNA yield cannot be consistently quantified from staining with ethidium bromide, because the tendency of ssDNA to form secondary structures may vary between templates. However, we routinely obtain a qualitative estimate by assaying 10 μl on a 3% (w/v) NuSieve (FMC, Rockland, ME), 1% (w/v) regular agarose gel. The ssDNA is visible after the bromphenol blue has migrated about 2 cm as a discrete fraction migrating ahead of the dsDNA. If a ssDNA fraction is visible by ethidium staining, the asymmetric PCR contains enough material for one to four sequencing reactions.

The overall efficiency of amplification is lower when an asymmetric primer ratio is used compared to when both are present in vast excess. This can usually be compensated for by increasing the number of PCR cycles. In addition, titrations may be needed to find the optimal primer ratio for each strand. An example of such a titration is shown in Fig. 3. In this case the most asymmetric ratios did not produce sufficient amounts of ssDNA. Instead, large amounts of high molecular weight, nonspecific PCR products were obtained. The optimal ratios for this primer pair were found to be 50:5 for one strand and 5:50 for the other. Low yields of ssDNA using the asymmetric PCR may reflect either too little of the limiting primer, preventing the accumulation of enough dsDNA as a template for the primer–extension reaction, or too high amounts of the limiting primer, saturating the reaction with dsDNA before any ssDNA is produced.

The ssDNA generated can then be sequenced using either the PCR primer that is limiting or an internal primer and applying conventional protocols for incorporation sequencing or labeled primer sequencing.[16] The population of ssDNA strands produced should have discrete 5' ends but may be truncated at various points close to the 3' end due to premature termination of extension. However, for any primer used in the sequencing reaction, only full-length ssDNA can be recruited as template.

The ssDNA of choice can be generated either directly in the original PCR, by using an asymmetric molar ratio of the two oligonucleotide primers, or in a second PCR reaction with an excess of one PCR primer, using a gel-purified fragment from an initial regular (symmetric) PCR as a target, or a 1/100 dilution of a previous symmetric PCR.[6,17] The asymmetric PCR has the advantage that, because the limiting primer is exhausted, there is no need to remove excess primers prior to initiating the sequencing reaction.

Protocol for Generation of Templates by Asymmetric Polymerase Chain Reaction. This protocol is suitable for generation of templates from a previous successful symmetric PCR.

1. Mix 80 μl distilled H_2O, 10 μl 10× PCR buffer (500 mM KCl, 100 mM Tris, pH 8.3, 15 mM $MgCl_2$), 5 μl premixed primers, with 50 pmol of one primer and 1–5 pmol of the other primer in a total of 5 μl, 5μl mix of nonionic detergents [10% (v/v) each of Nonidet P-40 (NP-40) and Tween 20], 0.8 μl deoxynucleoside triphosphate (dNTP) mix (25 mM with respect to each dNTP), 2.5 units *Taq* polymerase, and 2 drops of mineral oil.

2. Dilute the previous symmetric PCR 1/100.

3. Add 1 μl of diluted PCR to the asymmetric PCR mix and cap the tubes.

4. Run 40 PCR cycles.

5. After completion of PCR, assay for the presence of single-stranded DNA by running out 10 μl of the reaction on a 3% NuSieve, 1% regular agarose gel. Run the bromphenol blue about 2 cm into the gel before examining the fluorescence. A successful reaction should have two bands, the ssDNA migrating slightly ahead of the dsDNA.

6. If ssDNA can be seen, remove the oil from the rest of the PCR by a single chloroform extraction.

[16] U. Gyllensten, *in* "PCR Technology: Principles and Applications for DNA Amplification" (H. A. Erlich, ed.), p. 45. Stockton Press, New York, 1989.

[17] T. D. Kocher, W. K. Thomas, A. Meyer, S. V. Edwards, S. Pääbo, F. X. Villablanca, and A. C. Wilson, *Proc. Natl. Acad. Sci. U.S.A.* **86,** 6196 (1989).

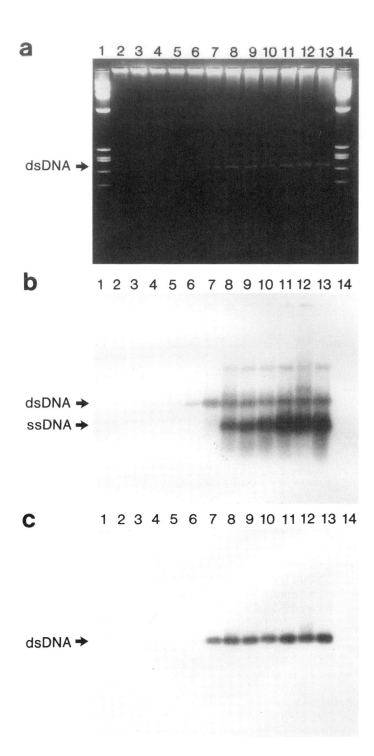

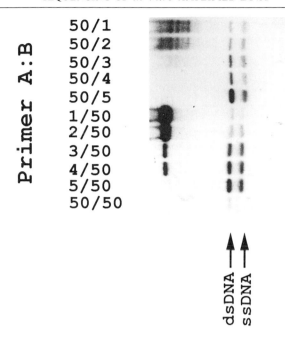

Primer A:B

50/1
50/2
50/3
50/4
50/5
1/50
2/50
3/50
4/50
5/50
50/50

dsDNA →
ssDNA →

FIG. 3. Titration of optimal primer concentrations in the asymmetric PCR. Exon 13 of the human CFTR gene [J. R. Riordan, J. M. Rommens, B.-S. Kerem, N. Alon, R. Rozmahel, Z. Grzelczak, J. Zielenski, S. Lok, N. Plavsic, J.-L. Chou, M. L. Drumm, M. C. Ianuzzi, F. S. Collins, and L.-C. Tsui, *Science* **245**, 1066 (1989)] was amplified using primer A (5'-CTGTGTCTGTAAACTGATGGCTA-3') and primer B (5'-GTCTTCTTCGTTAATTTCTT-CAC-3'). The PCR mix included 0.1 μl [α-^{32}P]dCTP (3000 Ci/mmol); the reaction products were separated on a 3% NuSieve, 1% regular agarose gel, and the gel was dried and autoradiographed.

7. Remove the buffer components and residual dNTPs from the ssDNA templates using centrifuge-driven dialysis [either Centricon 30 (Amicon, Danvers, MA) or Millipore (Bedford, MA)]. Collect the retentate (40 μl).

8. Use 10–25 μl for the sequencing reaction. [As an alternative to dialysis, precipitate the DNA in 4 M ammonium acetate to remove excess dNTPs and buffer components. Combine 100 μl PCR reaction and 100 μl

FIG. 2. The accumulation of PCR products during an asymmetric PCR. A 242-bp product from the second exon of the HLA-DQA1 gene was amplified using primers GH26 and GH27.[6] Lanes 1 and 14 contain the size standard ϕx174 cut with *Hae*III. Lanes 2–13 contain samples amplified for 5, 10, 13, 16, 19, 25, 28, 31, 34, 37, 40, and 43 cycles, respectively. (a) Genomic DNA was amplified with 50 pmol of primer GH26 and 0.5 pmol of primer GH27. (b) Southern blot of the agarose gel hybridized with an oligonucleotide complementary to both the dsDNA and ssDNA. (c) Same blot reprobed with an oligonucleotide with the same sequence as the ssDNA generated.

4 M ammonium acetate and mix. Add 200 μl 2-propanol, mix, leave at room temperature for 10 min, and then spin for 10 min. Remove the supernatant and wash the pellet carefully with propanol, mix, leave at room temperature for 10 min, and then spin for 10 min. Remove the supernatant and wash the pellet carefully with 500 μl 70% (v/v) ethanol. Dry down the pellet and dissolve in 10 μl TE (10 mM Tris-HCl, pH 7.5, 0.5 mM EDTA) buffer.]

Direct Sequencing with T7 DNA Polymerase

The sequencing protocol consists of two steps: labeling and termination.

1. Use 20–60% of the PCR reaction (purified) in a total volume of 7 μl.

2. Add 2 μl 5× sequencing buffer (1×: 40 mM Tris-HCl, pH 7.5, 20 mM MgCl$_2$, 50 mM NaCl).

3. Add 1 μl (1–10 pmol) sequencing primer (in an asymmetric PCR use either the limiting primer or an internal primer complementary to the ssDNA generated).

4. Heat the primer–template mix to 65°, leave for 4 min, and then allow it to cool to 30° over a period of 5 min.

5. Mix 2 μl labeling mix with 50 μl distilled water. When the yield of ssDNA template is low the labeling mix can be diluted to 1 : 100. [*Note:* The undiluted labeling mix is 750 μM (with respect to dTTP, dCTP, and dGTP) and lacks dATP.]

6. Add 1 μl of 0.1 M dithiothreitol (DTT) to the primer–template mix.

7. Add 2 μl of diluted labeling mix to the primer–template mix.

8. Add 0.5 μl of [α-^{35}S]thio-dATP (>1000 mCi/mmol).

9. Dilute T7 DNA polymerase to 1.6 units/μl in 7μl enzyme dilution buffer [enzyme dilution buffer: 10mM Tris-HCl, pH 7.5, 5 mM DTT, 0.5 mg/ml bovine serum albumin (BSA)].

10. Add 2.0 μl of diluted T7 DNA polymerase (3.2 units).

11. Incubate the mixture at room temperature for 5 min.

12. Add 3.5 μl of the labeling reaction to each of the four tubes, or a microtiter plate, with 2.5 μl of each termination mix [each containing 80 μM concentrations of each dNTP and an 8 μM concentration of the appropriate dideoxynucleoside phosphate (ddNTP)], and incubate the reaction at 37° for 5 min.

13. Stop the reaction by adding 4 μl formamide–dye stop solution [90% (v/v) formamide, 20 mM ethylenediaminetetraacetic acid (EDTA), pH 8.0, and 0.05% (v/v) each of the dyes xylene cyanol and bromphenol blue].

14. Store the reaction at −20° until loading onto a sequencing gel.

Direct Sequencing with *Taq* Polymerase

Taq polymerase is an ideal enzyme for DNA sequencing because it has high processivity and an absence of detectable $3' \rightarrow 5'$-exonuclease activity, which help to avoid false terminations.[18] In addition to these properties, which it shares with the thermolabile T7 DNA polymerase, it permits reaction temperatures between 55 and 85°, which will melt the secondary structure of most templates.

Protocol for Sequencing of Amplified DNA Using Taq Polymerase

1. In a 0.5-ml microfuge tube, prepare one labeling reaction mixture per sample by adding in the following order: 4 μl distilled H_2O, 1 μl sequencing primer (1 pmol/μl), 1 μl [α-^{35}S]thio-dATP (>1000 mCi/mmol), 4 μl labeling mix (the labeling mix contains 0.57 units/μl *Taq* DNA polymerase, 0.86 μM dGTP, 0.86 μM dCTP, 0.86 μM dTTP, 143 mM Tris-HCl, pH 8.8, 20 mM MgCl$_2$), and 10 μl DNA template.
2. Cap the tube and mix.
3. Incubate the tube for 5 min at 45°.
4. Dispense 4 μl of the labeling reaction into each of four tubes, or one microtiter plate, with 4 μl of the four termination mixes A, T, C, and G (G termination mix: 20 μM dGTP, 20 μM dATP, 20 μM dTTP, 20 μM dCTP, 60 μM ddGTP; A termination mix: 20 μM dGTP, 20 μM dATP, 20 μM dTTP, 20 μM dCTP, 800 μM ddATP; T termination mix: 20 μM dGTP, 20 μM dATP, 20 μM dTTP, 20 μM dCTP, 1200 μM ddTTP; C termination mix: 20 μM dGTP, 20 μM dATP, 20 μM dTTP, 20 μM dCTP, 400 μM ddCTP).
5. Cap the tubes and incubate at 72° for 5 min.
6. Remove the plate or tubes and add 4 μl stop solution (see above) to all samples.
7. Cover the plate or cap the tubes. If the samples cannot be analyzed immediately, they can be stored up to 1 week at $-20°$.

Sequencing of Regions with Strong Secondary Structure

Regions of DNA with strong secondary structure may give rise to two problems: (1) low efficiency of the PCR, due to a high frequency of templates that are not being fully extended by the *Taq* polymerase, and (2) compression of the DNA sequences in the sequencing reactions. It appears that the high reaction temperature of PCR using *Taq* polymerase

---

[18] M. A. Innis, K. B. Myambo, D. H. Gelfand, and M. A. D. Brow, *Proc. Natl. Acad. Sci. U.S.A.* **85,** 9436 (1988).

(50–75°) should be sufficient to resolve most short secondary structures. However, strong inhibition of more complex regions has been observed, and efficient PCR of these can be achieved only after the addition of the base analog c^7dGTP in the appropriate ratio relative to dGTP.[19] Similarly, base analogs may have to be used in the sequencing reactions to avoid compression problems. *Taq* polymerase will incorporate c^7dGTP but not inosine efficiently.[18]

Direct Sequencing of Heterozygous Individuals

When two alleles differ by a single point mutation, direct sequencing using a PCR primer will display the heterozygote position. However, when the allelic templates differ by more than one mutation direct sequencing will not resolve the phase of the mutations. In addition, the presence of short insertions or deletions in one of the alleles will generate compound sequencing ladders. There are four ways to resolve the phase of point mutations and obtain sequences of individual alleles from heterozygotes: (1) separating the alleles by cloning, (2) separating the different templates on the basis of their nucleotide sequence prior to sequencing, using a gradient gel electrophoretic system, (3) priming only one allele in the sequencing reaction, and (4) amplifying only one allele at a time.[6] Approaches 3 and 4 are applicable only to loci where the sequence of some of the alleles is known. In the sequencing reaction, oligonucleotides made to known allele-specific regions are used to selectively prime only one of the two allelic templates in a heterozygote.

Errors Involved in Sequencing of Polymerase Chain Reaction Products

Individual PCR products can differ from the sequence to be amplified by point mutations (Fig. 4) and by events of *in vitro* recombination in the PCR. Based on a fidelity assay for phage M13, the frequency of base substitution errors (1/10,000) and frameshift errors (1/40,000) of *Taq* polymerase was found to be considerably higher than for Klenow polymerase (1/29,000 base substitution errors, 1/65,000 frameshift errors) and T4 DNA polymerase (1/160,000 base substitution errors, 1/280,000 frameshift errors).[20] These assays were not performed under the same conditions as a standard PCR, and because the processivity and rate of synthesis by DNA polymerase are affected by $MgCl_2$ and dNTP concentration, buffer components, and the temperature profile of the cycle, these absolute

[19] L. McConologue, M. A. D. Brow, and M. A. Innis, *Nucleic Acids Res.* **16**, 9869 (1988).
[20] K. R. Tindall and T. A. Kunkel, *Biochemistry* **27**, 6008 (1988).

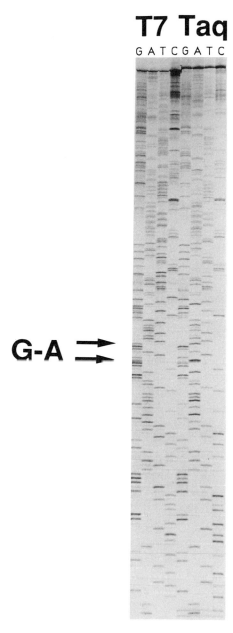

T7 Taq

G A T C G A T C

G-A ⇉

FIG. 4. Comparison of the sequencing ladders obtained by sequencing of asymmetric PCR templates by either T7 DNA polymerase (left four lanes) and *Taq* DNA polymerase (right four lanes). A portion (450 bp) of the human mitochondrial D loop [S. Anderson, A. T. Bankier, B. G. Barrell, M. H. L. de Bruijn, A. R. Coulson, J. Drouin, I. C. Eperon, D. P. Nierlich, A. Roe, F. Sanger, P. H. Schreier, A. J. H. Smith, R. Staden, and I. G. Young, *Nature* (*London*) **290,** 457 (1981)] was amplified using the primers UG142 (5′-GGTCTATCACCCTATTAACCAC-3′) and UG143 (5′-CTGTTAAAAGTGCATACCGC-CA-3′) and sequenced using UG142. The arrows indicate the location of point mutational differences between the two individuals.

numbers may not apply directly to PCR. The error rate in the PCR, estimated by sequencing of individual PCR products after 30 cycles (starting with 100–1000 ng of genomic target DNA), suggested that two random PCR products may be expected to differ once every 400–4000 bp.[2]

The mosaic, or *in vitro* recombinant, PCR products are the result of partially extended DNA strands that can act as primers on other allelic templates in later cycles. Both of these artifact products are likely to accumulate primarily at the end point of PCR because of insufficient enzyme to extend all available templates and an abundance of DNA strands for annealing. These artifact products have been seen primarily in studies of highly degraded DNA, or in studies of archaeological remains.[21,22] In PCR analyses of high molecular weight samples, these products are likely to constitute less than 1% of all templates.

Both these types of errors must be considered when PCR products are cloned and allelic sequences inferred from individual PCR products. In direct sequencing, by contrast, these artifact PCR products will not be visible against the consensus sequence on the gel. Even when starting from a single DNA copy, such as that found in a single sperm, a misincorporation that arises in the first PCR cycle will appear only with, at the most, 25% of the intensity of the consensus nucleotide, given that all templates have an equal probability of being replicated.[6] Thus, direct sequencing is to be preferred, unless the primer sequences do not allow sufficient specificity to amplify only a single target, or the individual allelic sequences cannot be determined due to genetic polymorphism at multiple positions between the primers. The relatively high error rate of *Taq* polymerase may, however, create problems when individual products are to be used for expression studies, or analysis of mutation frequencies. Unless a population of linear PCR products can be used in the expression system, several molecules must be cloned and sequenced to identify the unmodified clones.

Acknowledgments

U.B.G. was supported by a Fellowship from the Knut and Alice Wallenberg Foundation and a grant from the Swedish Natural Science Research Council.

[21] S. Pääbo, J. A. Gifford, and A. C. Wilson, *Nucleic Acids Res.* **16,** 9775 (1988).
[22] S. Pääbo, *Proc. Natl. Acad. Sci. U.S.A.* **86,** 1939 (1989).

[2] Producing Single-Stranded DNA in Polymerase Chain Reaction for Direct Genomic Sequencing

By LAURA F. LANDWEBER and MARTIN KREITMAN

Introduction

Direct sequencing of polymerase chain reaction (PCR)-amplified DNA is a powerful tool for analyzing DNA sequences, because it eliminates the need for constructing genomic DNA libraries or cloning the PCR product.[1] One important application of this technique is direct sequence analysis of variation among individuals. The ability to sequence the PCR product directly from genomic DNA relies on having an efficient method for sequencing the amplified DNA.

Several approaches have been taken for sequencing PCR-amplified DNA. The ease and reproducibility of the method are the main factors in deciding which strategy to follow. Double-stranded DNA has been compared to single-stranded DNA as a template for dideoxy sequencing. Double-stranded templates are easier to prepare, because the PCR product is generally a linear double-stranded molecule; however, single-stranded templates tend to produce better sequencing ladders. Protocols for sequencing double-stranded amplified DNA (dsDNA) by the dideoxy method usually involve a DNA denaturation step followed by a rapid annealing to a specific oligonucleotide primer.[1-3] The most successful results have been obtained with a snap-cooling approach[4] that minimizes reannealing of linear template DNA. Another way of improving the quality of double-stranded PCR sequences is to incorporate a nonradioactive or ^{32}P label onto the primer and perform multiple rounds of primer annealing and chain extension.[1-3,5,5a] This requires additional steps and obviates the use of ^{35}S and its superior base ladder resolution.

We described a strategy for obtaining single-stranded DNA (ssDNA)

[1] R. K. Saiki, D. H. Gelfand, S. Stoffel, S. J. Scharf, R. Higuchi, G. T. Horn, K. B. Mullis, and H. A. Erlich, *Science* **239**, 487 (1988).

[2] D. R. Engelke, P. A. Hoener, and F. S. Collins, *Proc. Natl. Acad. Sci. U.S.A.* **85**, 544 (1987).

[3] C. Wong, C. E. Dowling, R. K. Saiki, R. G. Higuchi, H. A. Erlich, and H. H. Kazazian, Jr., *Nature (London)* **330**, 384 (1987).

[4] J.-L. Cassanova, C. Pannetier, C. Jaulin, and P. Kourilsky, *Nucleic Acids Res.* **18**, 4028 (1990).

[5] M. Hunkapiller, *Nature (London)* **333**, 478 (1988).

[5a] S. M. Adams and R. Blakesley, *Focus (Life Technol.)* **13**, 56 (1991).

METHODS IN ENZYMOLOGY, VOL. 218

directly from the PCR amplification reaction.[6,7] We investigated two methods for synthesizing single-stranded DNA, both of which produce templates for standard ^{32}P, ^{35}S, or nonradioactive dideoxy sequencing protocols.[8,9] Both methods are based on an initial geometric amplification of approximately 1 pmol of double-stranded DNA, followed by a linear amplification of only one strand by one primer. In the first method, the two amplification primers, which remain in excess after geometric amplification, are removed by a selective ethanol precipitation. A single primer is then added and additional rounds of the PCR, which is now a primer extension reaction, produce an excess of one DNA strand. The second method combines geometric and linear amplification in a single PCR reaction. It uses a limiting amount of one primer (approximately 1 pmol) and an excess of the second primer (approximately 15 pmol). As the amplification proceeds for 10 to 20 rounds beyond the depletion of the limiting primer, the nonlimiting primer continues to direct the synthesis of one DNA strand. Both methods produce a stoichiometric excess of one DNA strand, which serves as a template for dideoxy sequencing. We also review two additional methods for producing single-stranded DNA from double-stranded amplification products, as well as other applications of limiting-primer and "linear" PCR.

Materials and Methods

Organisms, Loci, and Primers

We investigated PCR amplification in different gene regions of two organisms, the *Adh-dup* (duplicate) gene locus of *Drosophila melanogaster*[10] and the *Tcp-1 29x* region of *Mus spretus*.[11,12] All oligonucleotide primers are 20 bases and are selected to have duplex stability, $\Delta H°$, in the range of 160–170 kcal/mol (calculated according to Breslauer *et al.*[13]). Primers are named according to the 5' starting positions on sequences and according to the strand (+ or −). Primers can be purified on acrylamide

[6] M. Kreitman and L. F. Landweber, *Gene Anal. Tech.* **6,** 84 (1989).

[7] U. B. Gyllensten and H. A. Erlich, *Proc. Natl. Acad. Sci. U.S.A.* **85,** 7652 (1988).

[8] M. D. Biggins, T. J. Gibson, and G. F. Hong, *Proc. Natl. Acad. Sci. U.S.A.* **80,** 3963 (1983).

[9] S. Beck, T. O'Keeffe, J. M. Coull, and H. Koster, *Nucleic Acids Res.* **17,** 5115 (1989).

[10] S. W. Shaeffer and C. F. Aquadro, *Genetics* **117,** 61 (1987).

[11] K. R. Willison, K. Dudley, and J. Potter, *Cell* **44,** 727 (1986).

[12] K. Willison, unpublished results, 1988.

[13] K. J. Breslauer, R. Frank, H. Blocker, and L. A. Marky, *Proc. Natl. Acad. Sci. U.S.A.* **83,** 3746 (1986).

gels, Du Pont (Wilmington, DL) NENSORB columns, or on thin-layer chromatography plates [$60F_{254}$ 20 × 20 cm (Merck, Rahway, NJ); running buffer: 55% *n*-propanol, 35% NH_4OH in water; visualized with short-wavelength ultraviolet (UV) light; eluted with water].

Double-Stranded Polymerase Chain Reaction Amplification

Amplifications are usually performed in 100-μl reaction volumes containing 100 ng to 1 μg of genomic DNA in 50 mM KCl, 10 mM Tris (pH 8.3), 2 mM $MgCl_2$, 0.01% (w/v) gelatin (optional), 0.2 μM solutions of each primer, 200–250 μM solutions of each dNTP (dATP, dCTP, dTTP, and dGTP), and 2 units of AmpliTaq DNA polymerase (Perkin-Elmer Cetus, Norwalk, CT). Overlay the samples with approximately 50 μl of paraffin oil to prevent evaporation and then amplify for 25 to 30 rounds (20 sec to 1 min denaturing at 94°, 1–2 min annealing at 50–65°, and 2 to 3 min synthesis at 72°). After the last cycle, allow the samples to cool slowly to room temperature and add 2 μl 100 mM ethylenediaminetetraacetic acid (EDTA) (for general storage of the amplified DNA). Examine an 8- to 10-μl aliquot on a 2% (w/v) NuSieve: 1% (w/v) LE (FMC, Rockland, ME) agarose gel for small products or on a 1% LE agarose gel for larger products, as shown in Fig. 1.

Removal of Amplification Primers and Single-Stranded DNA Synthesis

The double-stranded PCR product is selectively ethanol precipitated in 2.5 M ammonium acetate and 1 vol ethanol. For a 100-μl PCR reaction, add 50 μl 7.5 M ammonium acetate and 150 μl 100% ethanol ($-20°$). Wait 5 min at room temperature, then spin the sample for 15 min at high speed in a microfuge at room temperature. Wash the pellet in 70% (v/v) ethanol, dry under vacuum, and resuspend in 20 μl 1 mM Tris, 0.1 mM EDTA. Under these conditions the oligonucleotide primers do not precipitate.

Reamplify approximately 0.2 pmol of double-stranded amplified DNA (4 to 10 μl of the selectively ethanol-precipitated DNA) for 15 to 20 rounds with 10 to 20 pmol of one primer in a 100 μl reaction volume. The PCR reaction is set up in the same way as the double stranded PCR reaction, but without the genomic DNA and with one of the primers missing. Either one of the original PCR primers or an internal primer can be chosen to synthesize the ssDNA. An example of ssDNA produced with an internal primer from both strands of a double-stranded PCR fragment is shown in Fig. 2. The ssDNA was labeled with a small amount of ^{32}P-kinased primer in the PCR for resolution on a 6% polyacrylamide sequencing gel (Fig. 2) and quantification.

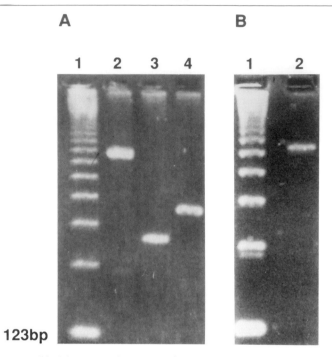

FIG. 1. Amplified fragments from genomic DNA visualized on a 3.5% NuSieve agarose gel. (A) *Drosophila Adh-dup* gene locus. Lane 1, 123-bp ladder; lane 2, the amplified fragment from bases 93–922; lane 3, bases 133–408; lane 4, bases 93–538. (B) Mouse genomic DNA. Lane 1, 123-bp ladder; lane 2, the amplified fragment from bases 385–1045 of *Tcp-1*. Reprinted by permission from Ref. 6. Copyright 1989 by Elsevier Science Publishing.

Synthesis of Single-Stranded DNA by Limiting Primer Method

Set up a 100-μl PCR reaction as described above for double-stranded PCR from 1 μg genomic DNA but with the following modifications: Use only 1 pmol of the limiting primer (0.01 μM, approximately 5 ng for a 20′-mer) and amplify for a total of 40 to 45 rounds. When the PCR is finished, add 2 μl 100 mM EDTA, and examine an 8- to 10-μl aliquot on an agarose gel to determine the yield of dsDNA. Single-stranded DNA sometimes migrates ahead of the dsDNA, but it does not stain well with ethidium bromide.

To determine the effect of limiting the amount of one primer on the yield of dsDNA and ssDNA, we conducted the experiment shown in Fig. 3. Different amounts of one primer (50, 20, 15, 10, 5, and 2.5 ng) were added to otherwise identical reaction mixtures containing an excess of the second primer (100 ng). The yield of dsDNA after 25 rounds of the PCR decreased only in the 5- and 2.5-ng limiting primer reactions (Fig. 3A).

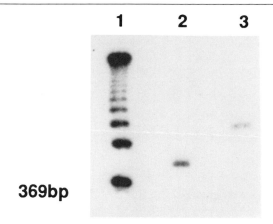

FIG. 2. Single-stranded DNA produced from the double-stranded amplified fragment shown in Fig. 1B, lane 2. Lane 1, 123-bp ladder; lane 2, the plus strand synthesized with *Tcp-1* primer 609+; lane 3, the minus strand synthesized with primer 1026−. Reprinted by permission from Ref. 6. Copyright 1989 by Elsevier Science Publishing.

This suggests that approximately 5 ng, or 1 pmol, of each primer is incorporated into the amplification product, with a corresponding yield of approximately 1 pmol of dsDNA in a standard PCR reaction.

Single-stranded DNA and dsDNA produced in each of the limiting primer amplifications were autoradiographically visualized on 6% (w/v) polyacrylamide denaturing gels by ^{32}P end-labeling a small amount of the nonlimiting primer prior to amplification (Fig. 3B and C). Digestion with the restriction enzyme *Cla*I cleaves the end-labeled 446-bp double-stranded product to 374 bp, whereas the single-stranded DNA remains 446 bases long. After 25 rounds of amplification, ssDNA was observed in the reactions containing 2.5 and 5 ng, the only two amplifications in which essentially all of the limiting primer was incorporated into double-stranded product (see above), whereas little or no ssDNA was observed in the 10- to 50-ng amplifications. On further amplification (5, 10, or 15 rounds) additional ssDNA was produced in the 2.5- and 5-ng limiting primer PCR amplifications (Fig. 3C).

It is possible to estimate the amount of ssDNA produced by either method by measuring the radioactivity in a gel slice containing the ssDNA. However, we generally assess ssDNA yield directly in sequencing reactions.

DNA Sequencing

Purify the ssDNA prepared by either method for sequencing. Phenol and chloroform extract the amplified DNA to remove the *Taq* DNA poly-

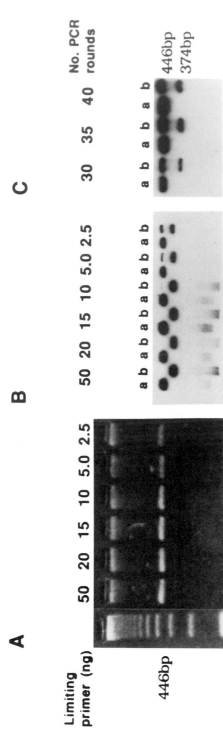

FIG. 3. Effect of primer amount on amplification yield. One microgram of *Drosophila* genomic DNA was amplified with primers 93+ and 538−. Expected PCR product length is 446 bp. On completion of the PCR the reactions were slowly cooled to allow complete reannealing. (A and B) Six identical 100-μl PCR reactions containing 100 ng of ^{32}P end-labeled 538− primer and 50, 20, 15, 10, 5, or 2.5 ng of 93+ primer were amplified for 25 rounds. (A) Reaction mix (10 μl) was electrophoresed on a 3.5% NuSieve agarose gel and ethidium bromide stained. (B) After phenol and chloroform extraction and ethanol precipitation DNA was resuspended in 20 μl water. One microliter was removed for electrophoretic analysis (*a* columns) and the remaining sample was digested in 30 μl with *Cla*I. Complete digestion was confirmed by agarose gel electrophoresis and ethidium bromide staining (not shown); 1.5 μl was removed for electrophoretic analysis (*b* columns). Equal volumes of formamide loading buffer were added to *a* and *b*, respectively, and samples were heat denatured and electrophoresed on a 6% polyacrylamide sequencing gel. The gel was autoradiographed for 10 hr on Kodak XAR X-ray film. The 446-base band in the *b* columns contains the excess unreannealed ssDNA; the 374-bp bands contain reannealed *Cla*I-cut dsDNA. (C) Same as (B) (5 ng limiting primer) but after 30, 35, and 40 rounds of amplification. A majority of the amplified DNA is single-stranded. Reprinted by permission from Ref. 6. Copyright 1989 by Elsevier Science Publishing.

merase. (This step is optional but produces cleaner results.) Ethanol precipitate the ssDNA in 2.5 M ammonium acetate and 1 vol ethanol as described earlier for removing the amplification primers. It is essential to remove the unused nucleotides completely before sequencing, but it is not always necessary to remove the amplification primers at this step because the primer that is present in excess (the single-strand synthesis primer) should not interfere with the sequencing reactions. However, incorporation of a labeled nucleotide, rather than labeled primers, often produces more background when the PCR primers are still present.

Resuspend the ssDNA in 10 to 20 μl 1 mM Tris, 0.1 mM EDTA. Sequence 2–7 μl of ssDNA (approximately 0.2 to 1 pmol ssDNA) with 1 pmol of primer according to standard dideoxy sequencing protocols for single-stranded templates, such as the Sequenase protocol (U.S. Biochemical Corp., Cleveland, OH). DNA is electrophoresed on 6% Tris–borate–EDTA gradient gels.[8] Autoradiographic exposures on Kodak (Rochester, NY) XAR X-ray film vary from 10 to 48 hr with ^{35}S or from 2 to 30 min with a biotinylated primer and chemiluminescent detection.[9] Sample sequencing ladders obtained with ^{35}S]dATP (1200 Ci/mmol, 10 mCi/ml; Amersham, Arlington Heights, IL) and modified T7 DNA polymerase (Sequenase; U.S. Biochemical Corp.) are shown in Fig. 4.

Conclusions and Discussion

Polymerase chain reaction amplification relies on the geometric principle that each strand is copied once in a single round of amplification. Factors favoring the synthesis of one strand relative to the other would be expected to reduce the overall rate of amplification rather than result in a differential expansion of the favored strand. Therefore, any strategy for synthesizing ssDNA with the PCR should involve the efficient depletion or removal of one primer after the production of a suitable amount of double-stranded template. This can be accomplished either by removal of the PCR primers followed by linear amplification or by limiting-primer PCR.

These methods allow a flexible primer strategy to be used. For amplifications of single-copy DNA that yield an individual product as visualized on an agarose gel (Fig. 1) satisfactory sequence ladders can be obtained with the original primers (Fig. 4B, C, E, and F). Long amplification products, on the other hand, can be conveniently sequenced with primers that produce overlapping sequences. Background due to nonspecific annealing can also be reduced by using an internal sequencing primer or by using a nested, or internal, primer to generate ssDNA by linear PCR. Alternatively, if there is more than one amplified band present, the target band

A   B   C   D   E F

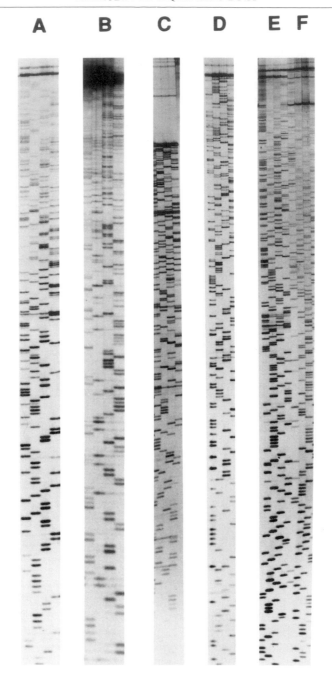

can be gel purified after electrophoresis. This effectively removes the primers as well, and the purified fragment can either be subjected to reamplification with one primer, or an aliquot diluted at least 1 : 100 can be reamplified by limiting primer PCR.

At the time we evaluated several schemes for purifying the PCR product from excess primers, only the ethanol precipitation in 2.5 M ammonium acetate and 1 vol of ethanol efficiently removed the oligonucleotides and retained the ssDNA and dsDNA. However, several products have recently been introduced on the market, including Centricon-100 spin dialysis (Amicon, Danvers, MA) and Millipore (Bedford, MA) Ultrafree-MC 30,000 NMWL, which we have used successfully, and other similar devices and gel filtration columns. Most of these remove the primers with equal efficiency, but at a much higher cost than ethanol precipitation.

Two methods have been reported for producing ssDNA efficiently from double-stranded PCR products. One method uses λ-exonuclease to digest the phosphorylated strand of a dsDNA fragment.[14] This requires that one primer is phosphorylated before amplification. We find that this method works well and does not require product purification before exonuclease treatment. The ssDNA can be purified from excess primers and nucleotides by selective ethanol precipitation (as described in Materials and Methods) before sequencing. A second method for conversion of dsDNA into ssDNA uses a biotinylated primer and alkaline denaturation to separate the two strands of a double-stranded molecule.[15] This method works quite well but requires more expensive oligonucleotide primers. Limiting primer and linear PCR both have the advantage that they use unmodified oligonucleotides. Limiting primer PCR is a simple one-step

[14] R. G. Higuchi and H. Ochman, *Nucleic Acids Res.* **17**, 5865 (1989).
[15] T. Hultman, S. Bergh, T. Moks, and M. Uhlen, *BioTechniques* **10**, 84 (1991).

FIG. 4. Sample sequencing ladders obtained from PCR-amplified ssDNA. (A–C) Amplifications from *Mus spretus* genomic DNA; (D–F) amplifications from *Drosophila melanogaster* genomic DNA. (A and D) Single stranded DNA was produced by selective ethanol precipitation and reamplification with one primer. (A) Single-stranded DNA amplification product shown in Fig. 2, lane 3, sequenced with 609+ primer. (B) Double-stranded DNA amplification product shown in Fig. 1B; ssDNA synthesized with 696+ primer and sequenced with 1045−. (C–F) Single-stranded DNA produced by limiting primer method. DNA from 45 rounds of amplification was extracted once each with phenol and chloroform, ethanol precipitated, and resuspended in 20 μl water. Either 3 or 7 μl was used for sequencing. (C) Mouse DNA amplified with 696+ (100 ng) and 1045− (5 ng), sequenced with 1045− primer. (D) *Drosophila* DNA amplified with 93+ (5 ng) and 538− (100 ng), sequenced with 133+. (E) *Drosophila* DNA amplified with 93+ (5 ng) and 922− (100 ng), sequenced with 93+. (F) *Drosophila* DNA amplified with 93+ (100 ng) and 922− (5 ng), sequenced with 922−.

reaction, but linear PCR can be more selective, because it has an added step in which nested primers can be used to generate ssDNA. The method that works most effectively for any template will vary, however, and needs to be determined experimentally.

Limiting primer and linear PCR have found many applications, including the generation of ssDNA probes[16] and cycle sequencing with *Taq* polymerase.[5a] The development of nonradioactive detection systems that combine biotin[9] and digoxigenin[17] labeled sequencing reactions will increase the thoroughput of direct PCR sequencing, allowing a label-multiplexing[18] approach to be used.

Acknowledgments

Laura Landweber is a Howard Hughes Medical Institute Predoctoral Fellow. This research was supported by NIH Grant GM39355 to M.K. The figures in this chapter are reprinted by permission of the publisher from "A Strategy for Producing Single-Stranded DNA in the Polymerase Chain Reaction: A Direct Method for Genomic Sequencing" by Martin Kreitman and Laura Landweber, *Gene Analysis Techniques* **6,** 84–88. Copyright 1989 by Elsevier Science Publishing Co., Inc.

[16] U. Gyllensten, *in* "PCR Technology: Principles and Applications for DNA Amplification" (H. A. Erlich, ed.), p. 55. Stockton Press, New York, 1989.
[17] P. Richterich and G. M. Church, this volume [14].

[3] Sequencing Products of Polymerase Chain Reaction

By Jeffrey G. Lawrence, Daniel L. Hartl, and Howard Ochman

Introduction

The polymerase chain reaction (PCR) is a method for amplifying specific DNA sequences from complex genomes and has many applications to molecular biology, medicine, genetics, evolution, and forensics.[1,2] The speed and specificity of the PCR has led to its utilization as an alternative to traditional cloning methods for generating templates for nucleotide sequence analysis. Although the PCR is rapid and effective in its production of relatively short double-stranded DNA (dsDNA) products, deter-

[1] H. A. Erlich, ed. "PCR Technology: Principles and Applications for DNA Amplification." Stockton Press, New York, 1989.
[2] M. A. Innis, D. H. Gelfand, J. J. Sninsky, and T. J. White, "PCR Protocols: A Guide to Methods and Applications." Academic Press, San Diego, 1990.

mining the nucleotide sequence of these products generally requires the isolation and purification of sequencing templates from reaction mixtures. Most of the existing procedures for sequencing products of the PCR utilize the Sanger sequencing methodology, which employs oligonucleotide primers for the initiation of chain elongation and incorporates dideoxynucleotide (ddNTP) terminators in four parallel reactions.[3] Because this method is based on the annealing of specific primers and the utilization of chain terminators, the following factors must be considered when sequencing amplification products:

1. During sequencing reactions, the original amplification primers can serve to prime DNA synthesis of both strands of the PCR-generated template.
2. High concentrations of dNTPs will disrupt dNTP/ddNTP ratios as well as reduce incorporation of labeled nucleotides.
3. The PCR is executed in a buffer that is inappropriate for some enzymes utilized for DNA sequencing.
4. Complementary strands of dsDNA templates can reassociate when annealing the sequencing primers.
5. Multiple templates are often generated in a single PCR reaction mixture.

To circumvent problems associated with determining the nucleotide sequence of PCR products, numerous investigators have resorted to subcloning these fragments into conventional sequencing vectors,[4] an approach that requires the subsequent sequencing of at least three clones to resolve any errors accumulated during amplification. Direct sequencing of PCR products without subcloning avoids this problem because any misincorporation, even one occurring in the first round of amplification, will represent a minor fraction of the total product. In this chapter we review several procedures developed to create templates directly from PCR reaction mixtures that are suitable for DNA sequencing with the Sanger methodology (the Maxam and Gilbert sequencing protocol, which relies on chemical degradation of DNA at specific nucleotides,[5] may also be applied to amplification products[6]). Although individual methods are often effective in sequencing certain amplification products, they may be problematic when applied to other classes of PCR products. We detail an approach that circumvents the problems inherent to sequencing PCR

[3] F. Sanger, S. Miklen, and A. R. Coulson, *Proc. Natl. Acad. Sci. U.S.A.* **74,** 5463 (1977).
[4] S. J. Scharf, G. T. Horn, and H. A. Erlich, *Science* **233,** 1076 (1986).
[5] A. M. Maxam and W. Gilbert, this series, Vol. 65, p. 499.
[6] T. Tahara, J. P. Kraus, and L. E. Rosenberg, *BioTechniques* **8,** 366 (1990).

products and has produced superior results with a variety of templates and primers.

Sequencing Products of Polymerase Chain Reaction

Wrishnik *et al.*[7] were among the first to report a nucleotide sequence of a PCR product. Their protocol required the removal of salts and dNTPs from the reaction mixture and the amplification product was sequenced using radiolabeled primers annealing internal to the original amplification primers. Similar procedures were employed to examined globin gene polymorphisms.[8] This "third-primer" method avoids problems resulting from extension products of competing DNA templates and residual primers. However, this technique requires the preparation of an additional primer, distinct dNTP/ddNTP mixes, and lengthy exposures of the resultant autoradiograms.

Template Purification Procedures

Several procedures can be used to remove excess primers and dNTPs but are useful only if a single amplification product is obtained. These include purification through Sepharose CL-6B[9] or microfiltration columns [e.g., Centricon (Amicon, Danvers, MA), Ultrafree (Millipore, Bedford, MA), and Qiagen (Chatsworth, CA)],[10] selective precipitation of large DNA fragments by the addition of polyethylene glycol,[11] removal of oligonucleotides by treatment with glass powder, or optimization of the PCR so that no primers or dNTPs remain following amplification.[12]

Other methods have been applied to reactions yielding multiple amplification products, including high-performance liquid chromatography (HPLC) purification,[13] sequencing directly in low melting temperature agarose,[14] and standard purification methods for extracting DNA fragments from agarose and acrylamide gels (e.g., glass powder, DEAE membranes, standard or electrophoretic elution). Although providing a high

---

[7] L. A. Wrishnik, R. H. Higuchi, M. Stoneking, H. A. Erlich, M. Arnheim, and A. C. Wilson, *Nucleic Acids Res.* **15**, 529 (1977).
[8] D. R. Engelke, P. A. Hoener, and F. S. Collins, *Proc. Natl. Acad. Sci. U.S.A.* **85**, 544 (1988).
[9] R. F. DuBose and D. L. Hartl, *BioTechniques* **8**, 271 (1990).
[10] M. Mihovilovic and J. E. Lee, *BioTechniques* **7**, 14 (1989).
[11] N. Kusukawa, T. Uemori, K. Asada, and I. Kato, *BioTechniques* **9**, 66 (1990).
[12] S. J. Meltzer, S. M. Mane, P. K. Wood, L. Johnson, and S. W. Needleman, *BioTechniques* **8**, 142 (1990).
[13] W. Warren and J. Doninger, *BioTechniques* **10**, 216 (1991).
[14] K. A. Kretz, G. S. Carson, and J. S. O'Brien, *Nucleic Acids Res.* **17**, 5864 (1989).

degree of purity, size fractionation and purification on gels are time consuming, and these techniques yield variable amounts of template.

Production of Single-Stranded Templates

Single-stranded sequencing templates can be generated directly in the
PCR by the addition of unequal molar ratios of the amplification primers.
After several cycles of amplification, the limiting primer is exhausted, and
the remaining cycles result in a linear increase in the number of single-
stranded products. This technique eliminates the need to remove excess
primers because none of the limiting primer remains after the PCR. Although standard PCR may reliably produce double-stranded products,
simply altering the initial concentrations of the primers rarely yields sufficient quantities of single-stranded templates. The successful production
of single strands can be enhanced using a purified double-stranded fragment as the template in a second reaction with an excess of one primer.[10,15]
Alternatively, ssDNA templates may be created by the addition of complementary ssDNA to dsDNA templates generated from a single-stranded
bacteriophage clone.[16] This method proves useful if the nucleotide sequences of many variants of the same template are to be determined.

Other techniques, based on chemical or enzymatic modification of only
one of the amplification primers, have been developed to produce single-
stranded sequencing templates from products of the PCR. These methods
avoid not only the problems associated with sequencing dsDNA but eliminate the need for further purification from oligonucleotides. Stoflet et al.[17]
incorporated the sequence of an RNA polymerase promoter into one of the
amplification primers. RNA templates are transcribed and subsequently
sequenced with reverse transcriptase. Higuchi and Ochman[18] utilized one
kinased oligonucleotide in their amplification reactions. Single-stranded
sequencing templates were then produced by treating the PCR products
with λ-exonuclease, a $5' \rightarrow 3'$ nuclease that specifically attacks double-
stranded DNA bearing a $5'$ terminal phosphate. Mitchell and Merrill[19]
incorporated biotin into one of the amplification primers, and the PCR
products were passed over a column containing strepavidin-agarose,
which binds the strand synthesized from the biotinylated primer yielding
single-stranded templates. Ward et al.[20] protected one strand of the double

[15] U. B. Gyllensten and H. A. Erlich, *Proc. Natl. Acad. Sci. U.S.A.* **85,** 7652 (1988).
[16] S. Gal and B. Hohn, *Nucleic Acids Res.* **18,** 1076 (1990).
[17] E. S. Stoflet, D. D. Koeberl, G. Sarker, and S. S. Sommer, *Science* **239,** 491 (1988).
[18] R. G. Higuchi and H. Ochman, *Nucleic Acids Res.* **17,** 5865 (1989).
[19] L. G. Mitchell and C. R. Merrill, *Anal. Biochem.* **178,** 239 (1989).
[20] M. A. Ward, A. Skandalis, B. W. Glickman, and A. J. Grosovsky, *Nucleic Acids Res.* **17,**
8394 (1989).

stranded PCR product from exonuclease III digestion by filling a restriction site included in one of the primers with thiotriphosphates.

Direct Sequencing with Taq Polymerase

Innis et al.[21] described a method that allows the direct sequencing of PCR-amplified templates using Taq polymerase, eliminating the need to purify samples. Krishnan et al.[22] applied the method of Craxton[23] for the concurrent linear amplification and dideoxy sequencing of DNA cloned in λ.

Principle of Method

Each of the methods described above is applicable to the sequencing of certain templates. Ideally, a method for isolating and sequencing products of the PCR should meet the following standards:

1. Applicable to amplification reactions that yield multiple products
2. Applicable to dsDNA templates to avoid extra manipulations and enzymatic steps
3. Isolation of template results in substantial yields of highly purified nucleic acids suitable not only for DNA sequencing but also for labeling for Southern blot or in situ hybridization, microinjection, and long-term storage
4. Remains cost effective and avoids elaborate materials or instrumentation

Many of the existing methods for preparing sequencing templates by the PCR require a single amplified product. If multiple products are present, gel purification of the desired fragment is the preferred, albeit time-consuming, method of isolation and purification. However, one typically performs gel electrophoresis to determine if a PCR has been successful and to verify the presence of a single product prior to executing a more rapid approach to template purification. Therefore it is possible to simultaneously verify the success of the PCR as well as to isolate individual sequencing templates from PCR reactions by recovering the size-fractionated DNA fragments from the same agararose gels. Moreover, in this manner it is possible to recover DNA fragments that would be unavailable if the PCR were optimized to yield a single band.

[21] M. A. Innis, K. B. Myambo, D. H. Gelfand, and M. A. D. Brow, *Proc. Natl. Acad. Sci. U.S.A.* **85,** 9546 (1988).
[22] B. R. Krishnan, R. W. Blakesley, and D. E. Berg, *Nucleic Acids Res.* **19,** 1153 (1991).
[23] M. Craxton, *Methods* **3,** 20 (1991).

The method described below fulfills the standards outlined above and has allowed sequencing of PCR products unsuitable for use with other methods. Following electrophoresis, template DNA is transferred to a DEAE membrane and recovered by ethanol precipitation. In addition to providing high yields of purified template (>90% recovery), this method allows visual confirmation during isolation and purification of the template DNA that is unavailable in most other methods (e.g., column purification).

Materials and Methods

Polymerase Chain Reaction

The PCR is performed in a buffer containing 50 mM KCl, 10 mM Tris (pH 8.4), 2.5 mM MgCl$_2$, 0.01% (w/v) gelatin, 800 μM dNTP, 2.5 U/ml *Taq* polymerase (Perkin-Elmer Cetus, Norwalk, CT), 2 ng/μl of each oligonucleotide primer (0.5 pmol), and an appropriate concentration of template DNA (10–100 ng, depending on the complexity of the target genome). Typical reaction protocols comprise 25 cycles of denaturation at 94° for 30 sec, primer annealing at an appropriate temperature for 40 sec, and elongation at 72° for 1 min/kbp. We find a reaction volume between 50 and 200 μl convenient for amplification and purification. Because the DNA isolated from 10 μl of a PCR mixture is utilized in one DNA sequencing reaction, this volume yields sufficient DNA to completely sequence both strands of typical amplification products. If necessary, the samples are reduced in volume in a vacuum desiccator prior to gel purification.

Template Preparation

The DNA fragments are size fractionated on 0.8% (w/v) agarose gels—cast with 10 μg/μl ethidium bromide and electrophoresed in 0.5× TBE [45 mM Tris-borate (pH 8.0), 5 mM ethylenediaminetetraacetic acid (EDTA)] and visualized under ultraviolet (UV) light. With a razor or scalpel, incisions are made and Schleicher & Schuell (Keene, NH) NA45 DEAE membranes, prepared and sized according to the instructions of the manufacturer, are inserted into the gel on both sides of the desired fragment. Electrophoresis is continued until the band has migrated onto the cathodal membrane. The membrane placed behind the DNA fragment of interest prevents copurification of larger molecular weight DNA species. The membrane containing the fragment of interest is removed and the mobilization of DNA from the gel onto the membrane

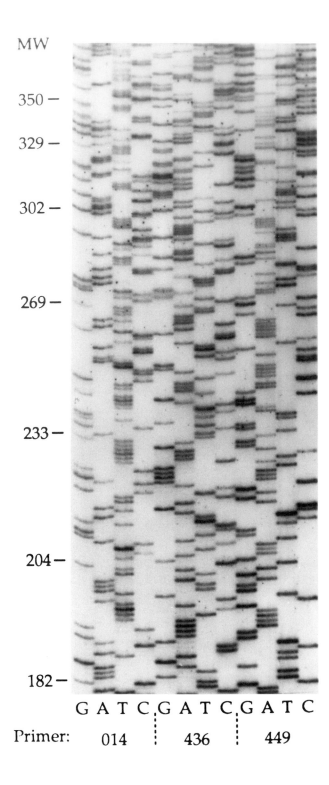

is verified under UV light. The membrane-bound DNA is briefly washed (1 min) in a microtiter plate containing 3–5 ml of 150 mM NaCl, 50 mM Tris (pH 8.0), 10 mM EDTA at room temperature to cleanse the membrane of agarose debris. The membrane is transferred to a 1.5-ml microcentrifuge tube and the DNA is eluted in 150 μl of 1 M NaCl, 50 mM Tris (pH 8.0), 10 mM EDTA at 68° for 30 min. The eluant is removed and the membrane is incubated in fresh buffer for an additional 15 min. The elution of the DNA from the membrane is verified under UV light. The eluants are pooled (300-μl total volume) and mixed well with an equal volume of phenol saturated with 3% (w/v) NaCl (pH 8.0). The mixture is centrifuged for 15 min at 4° to remove trace amounts of DEAE membrane. The aqueous phase is carefully removed and extracted with an equal volume of 24 : 1 (v/v) chloroform/isoamyl alcohol. Following the addition of 3 μl of 1% linear polyacrylamide carrier,[24] the DNA is precipitated from the aqueous phase by the addition of 2.5 vol of 100% ethanol and incubated at $-20°$ for 30 min. DNA is recovered by centrifugation for 15 min at room temperature. After removal of the supernatant, the pellet is rinsed with 70% (v/v) ethanol to remove salts and centrifuged for 2 min. The supernatant is again removed and the pellet is dried under vacuum for 3 min. The DNA is resuspended in a volume of 10 mM Tris, 1 mM EDTA equivalent to that of the original PCR mixture, yielding concentrations of DNA on the order of 10 ng/μl for typical reactions.

DNA Sequencing

One hundred nanograms of each template DNA is denatured in 0.2 M NaOH for 5 min at room temperature, neutralized by the addition of one-third volume 3 M potassium acetate (pH 4.6), and precipitated with 2.5 vol of 100% ethanol in the presence of 1 μl of additional polyacrylamide carrier. A typical reaction would be as follows:

DNA	10 μl (100 ng)
Distilled H$_2$O	8 μl
NaOH (2 N)	2 μl

[24] C. Galliard and F. Strauss, *Nucleic Acids Res.* **18**, 378 (1990).

FIG. 1. An internal portion of the transposon IS*30* was amplified by PCR, isolated on a DEAE membrane, and sequenced with three oligonucleotide primers. Molecular weight markers indicate number of nucleotides from the primer.

Incubate 5 min at room temperature. Then add

Potassium acetate (3 M), pH 4.6	7 μl
n-Acrylamide (1%)	1 μl
Ethanol (100%)	70 μl

Following incubation at $-20°$ for 30 min, DNA is recovered by centrifugation, washed with 70% ethanol, and dried under vacuum. Following precipitation, the DNA may be stored for several weeks. The templates are sequenced with Sequenase v2.0 (U.S. Biochemicals, Cleveland, OH) according to the instructions of the manufacturer, with the following minor modifications. To initiate sequencing reactions, the pellets are quickly resuspended in 6 μl water, 2 μl (2 pmol) oligonucleotide primer, and 2 μl Sequenase buffer. This mixture is incubated at 39° for 8–10 min to anneal the primers. The labeling reaction is executed as per the instructions of the manufacturer, except the dNTP-containing labeling mix is diluted 1 : 100 prior to use and the reaction is allowed to proceed 2–3 min prior to the addition of the termination mixes. The termination reactions are allowed to incubate at 39–42° for 7 min prior to the addition of stop buffer (U.S. Biochemicals) containing EDTA and NaOH. Terminated reactions are heated to 85° for 2 min and chilled on ice prior to electrophoresis on 6% (w/v) acrylamide, 7.5 M urea gels cast and run in 0.5× TBE.

Experimental Results

We have utilized this method to determine the sequences of a variety of PCR products, including transposons and single-copy genes amplified from plasmid, phage, prokaryotic, and eukaryotic genomes. Sequences internal to IS30 were amplified from a strain of *Escherichia coli* and isolated on DEAE membranes as described above. The template DNA was sequenced using the amplification primers as well as internally annealing oligonucleotides. As shown in Fig. 1, when template DNA is purified by this method one can easily resolve over 350 bp elongated from a single primer. Because these templates were gel purified, there was no need to optimize the PCR to yield a single product. In this manner, we have isolated an IS30 from a strain of *E. coli* into which IS3411 has inserted.[25] Had the conditions of amplification been optimized to yield a single product, this template would not have been isolated. Moreover, the presence of the inverted terminal repeats of IS3411 did not interfere with the sequencing of the double-stranded IS30 :: IS3411 template. Such templates proved difficult to sequence with other methods, notably those involving

[25] J. G. Lawrence and D. L. Hartl, unpublished results.

heat denaturation of dsDNA templates or the generation of ssDNA templates.

Discussion

The method described above allows the isolation, purification, and sequencing of dsDNA templates generated by the polymerase chain reaction. In addition, it is applicable to the sequencing of other dsDNA templates, such as restriction endonuclease fragments or plasmid DNA. In practice, we have found purification of DNA on DEAE membranes to be rapid, efficient, and effective for the simultaneous isolation of many templates. In addition, visualization of DNA under UV light allows verification of successful purification unavailable in most other methods. Although isolation of DNA on DEAE membranes proves inefficient for fragments greater than 10 kbp in length, typical PCR products do not achieve this length. The utilization of DEAE membranes is cost effective, requires no additional enzymatic treatment of oligonucleotides or templates, and utilizes standard laboratory technology. Templates prepared in this manner are stable at 4° for several months.

In performing DNA sequencing, we have found denaturation by alkali an effective method for the simultaneous processing of multiple templates. In addition, denaturation by alkali has proved useful for sequencing templates containing repetitive sequences. Methods involving heat denaturation or the production of ssDNA proved problematic in sequencing such templates. Moreover, following ethanol precipitation, centrifugation, and desiccation, individual templates may be stored indefinitely prior to the initiation of DNA sequencing reactions. Because additional nucleic acid species (i.e., additional templates, oligonucleotides, and dNTPs) have been removed, templates prepared in this manner are suitable not only for DNA sequencing but for Southern blot hybridizations, *in situ* hybridization, and microinjection.

[4] One-Sided Anchored Polymerase Chain Reaction for Amplification and Sequencing of Complementary DNA

By ROBERT L. DORIT, OSAMU OHARA, and WALTER GILBERT

Introduction

Conventional approaches to the characterization and sequencing of specific mRNA molecules generally involve reverse transcription of pooled mRNA, preparation of a cDNA library, screening of the library with specific probes, and subsequent isolation of single cDNA clones from within the library. Although this methodology has proved useful overall, success depends critically on the quality of the cDNA library. Frequently, cDNA libraries are vitiated by prematurely terminated cDNA molecules that do not reflect full-length message. In addition, rare messages may frequently be lost altogether in the preparation of cDNA libraries.

The polymerase chain reaction (PCR) offers an opportunity to select, amplify, and isolate a message of interest directly. The speed and convenience of PCR has led to its rapid adoption as a method for mRNA characterization. Conventional PCR, however, does have certain limitations, including the need for prior knowledge of the sequences flanking the region of interest in order to design the PCR primers. The advent of PCR has now spawned a number of procedures that do not require full *a priori* knowledge of the sequence being amplified.[1-5] Many of these modified PCR methods assume that only a small region of sequence is known. The PCR amplifications are consequently anchored by a primer derived from the region of known sequence; the second primer is nonspecific and is targeted to a general feature of the sequence, such as the poly(A) tail present in the plus strand of the majority of complementary DNAs (cDNAs) derived from nature mRNA. Alternatively, the target sequence can be enzymatically modified to provide an annealing site for the second primer. Such modifications include the addition (via terminal transferase) of a homopolymeric tail, restriction cleavage and subsequent circulariza-

[1] M. A. Frohman, M. K. Dush, and G. R. Martin, *Proc. Natl. Acad. Sci. U.S.A.* **85,** 8998 (1988).
[2] E. Y. Loh, J. F. Elliott, S. Cwirla, L. L. Lanier, and M. M. Davis, *Science* **243,** 217 (1989).
[3] P. R. Mueller and B. Wold, *Science* **246,** 780 (1989).
[4] H. Ochman, A. S. Gerber, and D. L. Hartl, *Genetics* **120,** 621 (1988).
[5] O. Ohara, R. L. Dorit, and W. Gilbert, *Proc. Natl. Acad. Sci. U.S.A.* **86,** 5673 (1989).

tion of the target molecule, or the ligation of a sequence "splint" at the end of the target molecule.

We present a modified PCR protocol, anchored PCR, that allows amplification of a specific full-length mRNA when only a small amount of sequence information is available. Unlike the amplification of RNA (or of cDNA) by conventional PCR, anchored PCR assumes that only a small region of sequence lying within the target mRNA is known in advance. In theory, only a single specific primer (17 bp or longer) is required to perform anchored PCR. In practice, amplification of a single homogeneous product will require a second round of PCR amplification, using a second nested specific primer (see Fig. 1). Both the original amplification and the reamplification use as a second flanking primer an oligo(dT)$_{20}$ complementary either to the poly(A) tail of the mature mRNA (3' to the known sequence) or, when amplifying 5' to the known sequence, to an enzymatically synthesized homopolymer tail added to the target cDNA following first-strand synthesis.

The two rounds of anchored PCR amplification result in a single homogeneous amplified product that can then be directly sequenced or cloned into an appropriate vector for further analysis. We have employed this protocol to obtain the DNA sequence directly of full-length cDNA for skeletal muscle α-tropomyosin from both zebrafish (*Brachydanio rerio*) and European frog (*Rana temporaria*),[5] as well as to determine the sequence of rat membrane-associated phospholipase A$_2$.[6] A modification of this method has also been used to detect transcription initiation sites in the phospholipase A$_2$ gene.[7] The sensitivity of this method for detecting transcription initiation may be 10-fold greater than that of conventional primer extension approaches, making it particularly useful for the analysis of rare mRNA species.

Principle

We describe an anchored PCR protocol for the amplification of cDNA targets. We have subdivided the discussion of this protocol into two sections. The first of these deals with the amplification of mRNA targets for which downstream (3') sequences are unknown. The amplification into the unknown downstream (3') region capitalizes on the poly(A) tail present on most mRNAs. It therefore will not succeed with poly(A)$^-$ mRNAs

---

[6] J. Ishizaki, O. Ohara, E. Nakamura, M. Tamaki, T. Ono, A. Kanda, N. Yoshida, H. Teraoka, H. Tojo, and M. Okamoto, *Biochem. Biophys. Res. Commun.* **162**, 1030 (1989).
[7] O. Ohara, J. Ishizaki, T. Nakano, H. Arita, and H. Teraoka, *Nucleic Acids Res.* **18**, 6997 (1990).

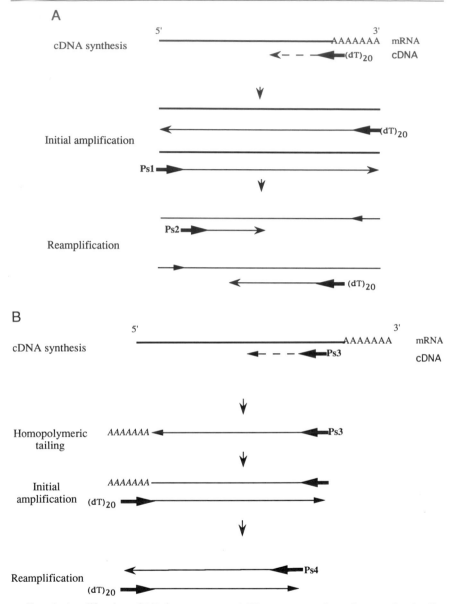

FIG. 1. Amplification of (A) downstream and (B) upstream regions. See text for details.

without poly(A) tails, such as those encoding histones. The region of interest is amplified by annealing an oligo(dT)$_{20}$ to this poly(A) tail, and the sequence-specific primer to the appropriate complementary region of the cDNA.

The second section presents a method for the amplification of mRNA target molecules when the upstream (5') sequence is unknown. The amplification into the unknown upstream (5') region makes use of an artificial poly(A) tail that is enzymatically added to the 3' end of an initial, specific cDNA product.

Materials and Reagents

MoMLV (Moloney murine leukemia virus) reverse transcriptase buffer (5×): 250 mM Tris-HCl (pH 8.3), 375 mM KCl, 50 mM dithiothreitol (DTT), 15 mM MgCl$_2$

PCR buffer (10×): 500 mM KCl, 100 mM Tris-HCl (pH 8.8), 15 mM MgCl$_2$, 30 mM DTT, 1 mg/ml bovine serum albumin (BSA)

dNTP mix: 10 mM dATP, 10 mM dCTP, 10 mM dGTP, 10 mM dTTP

PCR dNTP mix: 2.5 mM dATP, 2.5 mM dCTP, 2.5 mM dGTP, 2.5 mM dTTP

TdT buffer (5×): 1 M potassium cacodylate, 125 mM Tris-HCl (pH 7.4), 1.25 μg/μl BSA

Poly(A) RNA (>100 ng/μl)

Actinomycin D (500 ng/μl)

BSA (5 μg/μl)

MoMLV reverse transcriptase (>200 units/μl) (Boehringer Mannheim Biochemicals, Indianapolis, IN)

Taq polymerase (2.5 units/μl) (Perkin-Elmer Cetus, Norwalk, CT)

NaCl (1 M)

Sodium acetate (3 M)

Tris-HCl (pH 7.5) (200 mM)

Ethylenediaminetetraacetic acid (EDTA) (25 mM)

Terminal transferase (25 units/μl)

CoCl$_2$ (15 mM)

dATP (1 mM)

Sequence-specific primer (Ps1) [100 pmol/μl (670 ng/μl)]

Sequence-specific primer (Ps2) [100 pmol/μl (670 ng/μl)]

Sequence-specific primer (Ps3) [1 fmol/μl (6.7 pg/μl)]

Sequence-specific primer (Ps4) [100 pmol/μl (670 ng/μl)]

Oligo(dT)$_{20}$ [15 pmol/μl (100 ng/μl)]

Methods

Amplification of Regions Downstream (3') of Known Sequence

This method can be used to characterize messages of interest from any source of poly(A) RNA. Poly(A) RNA can be prepared following conventional protocols. While the amount of poly(A) RNA required for this procedure will depend on the genome complexity of organism being studied and on the relative abundance of the targeted mRNA, in general, 100–300 ng of total poly(A) RNA prepared from vertebrate tissue is sufficient for this protocol. Under usual circumstances, three PCR primers are required: a sequence-specific anchoring primer (Ps1), a second nested (internal) sequence-specific primer (Ps2) for reamplification, and an oligo(dT)$_{20}$ complementary to the 3' poly(A) tail of the mRNA molecule.

Synthesis of Generalized cDNA Pool

1. Prepare 50–100 ng of poly(A) RNA, at a concentration >100 ng/μl. This high concentration of mRNA is necessary to keep the final volume of the reverse transcriptase reaction to a minimum, preferably <10 μl. Larger amounts of poly(A) RNA can be used if readily available (up to 1 μg). Starting mRNA amounts of less than 100 ng may not include the full complement of rare mRNAs.

2. Incubate the poly(A) RNA at 65° for 2 min. Spin briefly and place immediately on ice. This incubation melts the secondary structure of the mRNA, removing hairpins and loops that interfere with the synthesis of cDNA.

3. Synthesize cDNA pool by reverse transcription of poly(A) RNA, using MoMLV reverse transcriptase primed with the oligo(dT)$_{20}$. The following reverse transcription mix should be prepared and kept on ice:

MoMLV reverse transcriptase buffer (5×)	2 μl
BSA (5 μg/μl)	1 μl
Poly(A) RNA (>100 ng/μl)	1 μl
dNTP mix	1 μl
Actinomycin D (500 ng/μl)	1 μl
MoMLV reverse transcriptase (>200 units/μl)	1 μl
Oligo(dT)$_{20}$ (100 ng/μl)	1 μl
Sterile distilled H$_2$O	2 μl

4. Mix gently, spin briefly in a microcentrifuge, and incubate for 1 hr at 37°. These general incubation conditions result in the annealing of the oligo(dT)$_{20}$ to the poly(A) tails of the mature mRNAs and the synthesis of a complete cDNA pool.

5. Following the incubation, bring the reaction volume up to 50 μl by the addition of Tris-EDTA buffer (10 mM Tris, pH 8.0, 1 mM EDTA).

Specific Amplification of cDNA Target Molecule

6. Remove a 1-μl aliquot of the newly synthesized cDNA pool. Carry out 30–40 cycles of conventional PCR in a mix containing:

cDNA template (from step 4)	
PCR buffer (10×)	10 μl
Oligo(dT)$_{20}$ (100 pmol/μl)	1 μl
Sequence-specific primer (Ps1; 100 pmol/μl)	1 μl
PCR dNTP mix	6 μl
Taq polymerase (2.5 units/μl)	1 μl
Sterile distilled H$_2$O	80 μl

Initial PCR amplifications should be carried out under generalized conditions: 1.5 mM MgCl$_2$ (final concentration), annealing temperatures 37–42°, and 100 pmol of each primer. The magnesium concentration and annealing temperatures and times can be subsequently modified and optimized for the specific template used, and primer concentrations reduced to 30 pmol of each primer.

The results of this amplification should be checked on an agarose gel. Seldom will enough specificity be conferred by a single sequence-specific oligonucleotide; the result of this first PCR is usually a smear around the expected size range. The presence of the desired product can be confirmed by probing a Southern blot of the gel containing the PCR products with a sequence-specific oligomer (Ps2) internal to the amplified sequence. Additional specificity is achieved by reamplification using a second sequence-specific internal primer.

7. Remove a 1-μl aliquot of the previous amplification to serve as the template for reamplification. Carry out a new round of PCR (30–40 cycles), as described in step 6, using 40–100 pmol each of the oligo(dT)$_{20}$ and a second internal, sequence-specific primer (Ps2).

This second sequence specific internal primer can be immediately adjacent (3′) to, or even partially overlap with, Ps1 (see Fig. 1). Thus, knowledge of 40 nucleotides within the mRNA of interest is usually sufficient for a successful two-stage amplification.

Once again, the product of the reamplification should be checked on an agarose gel. The amplified product, stretching from the target site of the sequence-specific primer (Ps2) to the 3′ end of the cDNA, should now appear as a single band on an agarose gel.

Under certain circumstances, the products of this second amplification may still appear as a smear when visualized by ethidium bromide staining of an agarose gel. This smear results in part from the multiple positions that the oligo(dT)$_{20}$ primer can occupy along the length of the poly(A) tail. In such cases, the largest amplified product can be size selected and reamplified again using oligo(dT)$_{20}$ and Ps2 primers. Alternatively, the imprecise annealing of the oligo(dT)$_{20}$ can be reduced by synthesizing an oligo(dT)$_{20}$ with an additional specific sequence at the 5' end, as is done in the RACE protocol.[1] Subsequent reamplifications will then rely on a primer directed to this synthetic sequence. As a third option, the oligo(dT)$_{20}$ can be synthesized with a degenerate (A, G, or C) 3'-most position, resulting in more precise annealing at the poly(A) tail boundary of the mRNA.

The amplified product obtained after two cycles of amplification is now suitable for subsequent cloning into an appropriate vector and/or for additional characterization by direct sequencing.

Amplification of Regions Upstream (5') of Known Sequence

In contrast to the previous protocol, here the sequence-specific primer is used to initiate the synthesis of a specific cDNA strand. This cDNA is then modified by the addition of a poly(A) "tail." Polymerase chain reaction amplifications, mediated by two adjacent sequence-specific primers and a oligo(dT)$_{20}$ complementary to the newly synthesized tail, yield the desired unique product.

Synthesis of Specific cDNA Target

1. Prepare, by conventional methods, at least 100 ng of poly(A) RNA, at a concentration >100 ng/μl.

2. Prepare a 5-μl annealing mix containing:

Sequence-specific primer (Ps3; 1 fmol/μl)	1 μl
NaCl (1 M)	1 μl
Tris-HCl (pH 7.5, 200 mM)	1 μl
EDTA (25 mM)	1 μl
Poly(A) RNA (>100 ng/μl)	1 μl

The amount of primer to be used for sequence-specific targeted cDNA synthesis depends in part on the abundance of the target mRNA. As a general rule, a 5- to 10-fold molar excess of oligomer primer relative to target template yields the best results.

3. Incubate the annealing mix at 65° for 3 min, spin briefly, and place immediately on ice. This incubation melts the secondary structure of the

mRNA, removing hairpins and loops that may interfere with the synthesis of cDNA.

4. Incubate at 40° for 3–4 hr.

The long incubation time, relatively high annealing temperature, and low primer concentration are designed to increase the specificity of the primer–template annealing reaction. Shorter annealing times may be used, but can result in insufficient primer–template annealing. Similarly, higher primer concentrations will speed the reaction, but may do so at the expense of specificity.

5. Add 15 μl of cold 100% ethanol, place in a dry ice–ethanol bath for 10 min, and spin at 4° in a microcentrifuge for 10 min.

6. Decant the supernatant. Add 50 μl of 70% (v/v) ethanol, and gently invert the tube several times to rinse and desalt the pellet (do not vortex!). Spin for 2 min in a microcentrifuge, remove the supernatant, and dry briefly under vacuum.

7. Resuspend the pellet in 10 μl of sterile, distilled H₂O.

8. On ice, prepare the reverse transcriptase mix, containing:

MoMLV reverse transcriptase buffer (5×)	5 μl
BSA (5 μg/μl)	2.5 μl
dNTP mix	2.5 μl
Actinomycin D (500 ng/μl)	2.5 μl
Poly(A) mRNA/Ps3 mix (from step 7)	10 μl
Sterile distilled H₂O	1.5 μl

9. Add 1 μl (>200 units/μl) of MoMLV reverse transcriptase. Incubate at 37° for 1 hr.

10. Carry out the phenol extraction, following standard protocols, to remove reverse transcriptase.

11. Following phase separation by centrifugation, carefully transfer the supernatant to a new microcentrifuge tube. Add 2.5 μl of 3 M sodium acetate (yielding a final sodium acetate concentration of 0.3 M) and 75 μl of 100% ice-cold ethanol. Place in a dry ice–ethanol bath for 5 min, spin in a microcentrifuge at 4° for 20 min, and remove and discard supernatant. In cases in which small amounts of poly(A) RNA are being used, 1–5 μg of carrier tRNA may be added to this step to facilitate precipitation.

12. Resuspend the pellet in 25 μl of TE (10 mM Tris-HCl, pH 8.0, 1 mM EDTA) and repeat the ethanol precipitation as described in the previous step. Rinse the pellet with 70% ethanol, spin for 5 min in a microcentrifuge, remove the supernatant, and dry the pellet briefly under vacuum.

Addition of Synthetic Homopolymeric Tail to 3' End of cDNA Molecule

13. Resuspend the pellet in 5.0 μl of water, and boil for 2 min to denature the cDNA/RNA hybrid molecules prior to tailing the cDNA molecule with terminal transferase. Spin briefly and place immediately on ice.

14. Prepare terminal transferase mix, on ice:

TdT buffer (5×)	2 μl
CoCl$_2$ (15 mM; final concentration of CoCl$_2$, 1.5 mM)	1 μl
dATP (1 mM; final concentration, 100 μmol)	1 μl

Add 1 μl terminal transferase (25 units/μl); incubate for 30 min at 37°.

15. Following the tailing reaction, inactivate the enzyme by heating at 65° for 2 min.

16. Ethanol precipitate the tailed cDNA in the presence of 0.3 M sodium acetate, as described in step 11. As before, in cases in which small amounts of poly(A) RNA are being used, 1–5 μg of carrier tRNA may be added at this step to facilitate the precipitation. Wash and dry the pellet.

Polymerase Chain Reaction Amplification of Tailed Target cDNA

17. Resuspend the pellet in 10 μl of water. Add 40–100 pmol each of sequence specific primer (Ps3) and oligo(dT)$_{20}$ and all necessary components for PCR amplification in a final volume of 50–100 μl (see step 6 under Amplification of Specific cDNA Target Molecule, above). Carry out 35–40 cycles of amplification in the presence of 1.5 mM MgCl$_2$. The annealing temperature should be in the 40–42° range. These conditions, as well as primer concentrations, can subsequently be optimized for the particular template being used.

The products of this round of PCR may well not be visible by ethidium bromide staining of an agarose gel, but will certainly be detectable when a Southern blot of the gel is probed with a sequence-specific oligomer internal to the amplified region (Ps4). The use of a single sequence-specific primer—in this case to prime the synthesis of the cDNA strand—usually does not result in the precise amplification of a single homogeneous product. A subsequent round of PCR amplification using a second nested primer resolves this problem.

18. Remove a 1-μl aliquot of the amplification to serve as template for a new round of PCR amplification in the presence of 40–100 pmol each of a second, internal sequence-specific primer (Ps4) and the nonspecific oligo(dT)$_{20}$. Carry out 35 cycles of PCR reamplification under standard conditions.

The second sequence-specific primer can be immediately adjacent to, or even partially overlap, the first specific primer. Because the anchored PCR procedures are carried out independently in the 3′ and 5′ directions, sequence-specific primers Ps3 and Ps4 can be the complements of primers Ps1 and Ps2 used in the previous protocol (see Fig. 1).

This reamplified product, stretching from the target site of the sequence-specific primer (Ps4) to the 3′ end of the cDNA, should now appear as a single band on an ethidium bromide-stained agarose gel. If a smear is still present, size-select a small aliquot of product at the upper range of the smear (see Concluding Remarks and Troubleshooting, below) and reamplify 20–25 cycles using Ps4 and oligo(dT)$_{20}$.

The two cycles of PCR amplification yield a single product that can now be cloned into an appropriate vector and/or further characterized by direct sequencing.

Concluding Remarks and Troubleshooting

The protocols presented here are specifically designed for the amplification of full-length cDNA, when only a fragment of the cDNA sequence is known. The amplification into the unknown downstream (3′) region capitalizes on the poly(A) tail present on most mRNAs. The region of interest is amplified by annealing the oligo(dT)$_{20}$ to this poly(A) tail, and the sequence specific primer to the appropriate region of the cDNA.

The amplification into the unknown upstream (5′) region makes use of an artificial poly(A) tail that is enzymatically added to the 3′ end of an initial, specific cDNA product.

In theory, both upstream and downstream amplifications can succeed with any mRNA for which sufficient sequence (20–40 bp) to design a single sequence-specific oligonucleotide is known. In practice, however, a single sequence-specific primer does not always confer sufficient selectivity in the amplification. This primer may bind to inappropriate or partially complementary sequences, resulting in the amplification of more than one sequence. This problem is further compounded in situations in which the only sequence available at the outset comes from the partial sequencing of the protein produced by the mRNA of interest, or from identifying conserved regions in homologous mRNAs extracted from other organisms. Such cases require the use of degenerate (heterologous) sequence-specific primers, further reducing the specificity of the anchored PCR amplifications. The consequences of using degenerate primers will depend on the template being amplified, the degree of degeneracy, and the amplification conditions. We strongly recommend the use of fully sequence-specific primers for the protocols described here.[5]

It is important to keep in mind that the initial stages of anchored PCR will seldom result in a fully specific amplification, and the heterogeneous amplification products appear as a smear when visualized on an agarose gel. Polymerase chain reaction products produced by this protocol should be monitored by visualizing them on a Southern blot probed with a target-specific internal oligomer. Such monitoring is particularly important in'the initial phases of this protocol, in order to confirm that the desired product is in fact being amplified.

A number of steps may be taken to increase the target specificity of anchored PCR. A reamplification carried out using a second internal (nested) specific sequence-specific primer greatly reduces spurious amplification products. This second primer can lie immediately adjacent to the first specific primer (see Fig. 1). Thus, 40–50 nucleotides of known sequence are sufficient to fully anchor this two-step procedure. Specificity can also be increased if the first-round amplification products are size-selected by extracting an aliquot of the main PCR band within the "smear" (or of the "correct" target band, previously determined by Southern hybridization with an internal specific oligomer). These size-selected products can now serve as templates for a second round of PCR amplification. If the first round products are run out on a low-melting agarose gel, the desired target can be size-selected simply by piercing the band on the gel with a Pasteur pipette. This small, low-melting gel core sample can then be placed directly into an appropriate PCR mix and reamplified without any further purification.

A second problem that may interfere with the production of a single, distinct PCR product when using anchored PCR stems from the length of either the naturally occurring or the enzymatically synthesized poly(A) tail. A long poly(A) tail provides a large number of potential pairing sites for the nonspecific oligo(dT)$_{20}$ primer. The resultant PCR products may consequently be of variable length. As previously discussed, a number of possible solutions to this problem are available. These include the synthesis of a hybrid oligomer containing a poly(dT) stretch at the 3' end and a nonhomopolymeric specific sequence at the 5' end, as in the RACE procedure,[1] or ligation of a specific oligonucleotide at the 3' ends of specific cDNAs.[8] Subsequent PCR amplifications then capitalize on this nonhomopolymeric region. Similar results can also be achieved by designing an oligomer with a degenerate (A, G, or C) 3'-most position, thus ensuring the annealing of the oligomer at the boundary of the poly(A) tract.

We have found that certain DNA targets cannot be efficiently amplified using standard PCR conditions. In such cases, the low-yield PCR products

[8] J. Edwards, J. Delort, and J. Mallet, *Nucleic Acids Res.* **19,** 5227 (1991).

can be readily cloned into an appropriate vector[9,10] and subsequently screened by colony hybridization to isolate the appropriate clones. Given the relatively high error rate of *Taq* polymerase, sequence obtained from cloned PCR fragments should be confirmed by isolating and sequencing multiple clones.

As with any method for the analysis of mRNA, the success of these protocols depends critically on the quality and integrity of the mRNA being used and on the successful synthesis of cDNA templates. Appropriate precautions to ensure a high yield of intact mRNA and full-length cDNA should be taken. In addition, because several steps in these protocols involve single-stranded nucleic acids (mRNA and first-strand cDNA), we recommend the use of siliconized microcentrifuge tubes and pipette tips.

[9] T. A. Holton and M. W. Graham, *Nucleic Acids Res.* **19,** 1156 (1991).
[10] D. Marchuck, M. Drumm, A. Saulino, and F. S. Collins, *Nucleic Acids Res.* **19,** 1154 (1991).

[5] Reverse Cloning Procedure for Generation of Subclones for DNA Sequencing

By ZHANJIANG LIU and PERRY B. HACKETT

Molecular biology has entered the era of genomic mapping, and genes are being isolated and characterized at an exponentially increasing rate. The cloning of megabase sequences from a wide variety of organisms, coupled with the availability of automated DNA sequencers, should lead to rapid sequencing of extended stretches of DNA. However, there is a bottleneck in the procedures that develops during the isolation of long DNA segments and subsequent sequence analysis. Rapid DNA sequencing procedures provide sequences of 400–800[1–3] nucleotides. Longer sequences must be built up from many overlapping short sequences obtained from subclones of the original segments. Three subcloning procedures are generally used to generate overlapping fragments for extended sequencing of the target DNA segment: (1) the *shotgun* procedure,[4–6] wherein the

[1] A. M. Maxam and W. Gilbert, *Proc. Natl. Acad. Sci. U.S.A.* **74,** 560 (1977).
[2] F. Sanger, S. Nicklen, and A. R. Coulson, *Proc. Natl. Acad. Sci. U.S.A.* **74,** 5463 (1977).
[3] F. Sanger and A. R. Coulson, *FEBS Lett.* **87,** 107 (1978).
[4] J. Messing, this series, Vol. 101, p. 20.
[5] S. Anderson, *Nucleic Acids Res.* **9,** 3015 (1981).
[6] P. L. Deininger, *Anal. Biochem.* **129,** 216 (1983).

METHODS IN ENZYMOLOGY, VOL. 218

target DNA is reduced in size by either restriction endonucleases, DNases, or sonication, (2) the *sequential digestion* procedures (see Table I), and (3) the lesser used *transposon-mediated, in vivo deletion* procedure.[7–9] With the shotgun procedure, overlapping sequences comprising 80–90% of the long DNA segment can be easily accumulated. However, obtaining missing portions of the initial DNA segment can be frustrating due to extended searches through large numbers of redundant subclones. The frustration can be further aggravated when repetitive elements exist in the long target DNA. For this reason sequential digestion procedures were developed to avoid the repetitive sequencing of related sequences.

The sequential digestion procedures, which produce ordered, extended deletions of the target DNA, require a single endonucleolytic cleavage of the cloned, double-stranded DNA followed by progressive exonucleolytic digestion. The single cleavage may be achieved by (1) restriction endonuclease digestion at a unique site, (2) nicking with DNase I in the presence of ethidium bromide, or (3) cleavage of a single-stranded plasmid containing a limited double-stranded region that contains a unique restriction enzyme site.[10] The cleaved DNA can be progressively shortened by enzymatic digestion with either *Bal*31,[11,12] exonuclease III in combination with nuclease S1 or exonuclease VII,[13–15] or the exonuclease activity of T4 DNA polymerase.[10] DNase I can be used to cut the target DNA randomly and a restriction endonuclease can be used to cut in the polylinker region.[16–20] To obtain clones that have ordered deletions, only a single cut in the target DNA is assumed. To avoid problems with some DNA clones in which a single cut was difficult to achieve, Dale *et al.*[10] developed an efficient single-stranded cloning strategy. However, this procedure, which requires a special primer, often produces a broad range of size deletions

[7] A. Ahmed, *J. Mol. Biol.* **178,** 941 (1984).
[8] A. Ahmed, *Gene* **39,** 305 (1985).
[9] L. Peng and R. Wu, *Gene* **45,** 247 (1986).
[10] R. M. K. Dale, B. A. McClure, and J. P. Houchins, *Plasmid* **13,** 31 (1985).
[11] M. Poncz, D. Solowiejczyk, M. Ballantine, E. Schwartz, and S. Surrey, *Proc. Natl. Acad. Sci. U.S.A.* **79,** 4298 (1982).
[12] T. K. Misra, *Gene* **34,** 263 (1985).
[13] L. Guo, R. C. A. Yang, and R. Wu, *Nucleic Acids Res.* **11,** 5521 (1983).
[14] S. Henikoff, *Gene* **28,** 351 (1984).
[15] E. Ozkaynak and S. D. Putney, *BioTechniques* **5,** 770 (1987).
[16] W. M. Barnes and M. Bevan, *Nucleic Acids Res.* **11,** 349 (1983).
[17] A. M. Frichauf, H. Garoff, and H. Lehrach, *Nucleic Acids Res.* **8,** 5541 (1980).
[18] G. F. Hong, *J. Mol. Biol.* **158,** 539 (1982).
[19] A. Laughon and M. P. Scott, *Nature (London)* **310,** 25 (1984).
[20] Q. Li and G. Wu, *Gene* **56,** 245 (1987).

TABLE I
SEQUENTIAL CLONING STRATEGIES[a]

Initial cleavage	Deletion enzymes	Limitations
RE	ExoIII, S1, or *Bal*31	Target DNA must be smaller than vector[b]
RE	ExoIII, S1 Klenow, ExoIII, ExoVII	Two appropriate restriction sites on either side, but not within, target DNA are required.[c,d]
RE	*Bal*31, RE	Target DNA must be smaller than vector. Two proximal restriction sites must be close to, but not inside, the target DNA[e,f]
RE	T4 DNA polymerase	Specific primers are required. T4 DNA polymerase digestion is asynchronous[g]
RE DNase I DNase I with EtBr	DNase I RE ExoIII, *Bal*31, RE	Target DNA must be cut only once. DNA must be fractionated.[h,i,j,k,l] Linkers are required.[i,l]

[a] Unless otherwise specified, each limitation applies to all of the variations in each category. RE, Restriction endonuclease; ExoIII, exonuclease III; ExoVII, exonuclease VII; S1, nuclease S1; EtBr, ethidium bromide.

[b] L. Guo, R. C. A. Yang, and R. Wu, *Nucleic Acids Res.* **11,** 5521 (1983).

[c] S. Henikoff, *Gene* **28,** 351 (1984).

[d] E. Ozkaynak and S. D. Putney, *BioTechniques* **5,** 770 (1987).

[e] M. Poncz, D. Solowiejczyk, M. Ballantine, E. Schwartz, and S. Surrey, *Proc. Natl. Acad. Sci. U.S.A.* **79,** 4298 (1982).

[f] T. K. Misra, *Gene* **34,** 263 (1985).

[g] R. M. K. Dale, B. A. McClure, and J. P. Houchins, *Plasmid* **13,** 31 (1985).

[h] Q. Li and G. Wu, *Gene* **56,** 245 (1987).

[i] A. M. Frichauf, H. Garoff, and H. Lehrach, *Nucleic Acids Res.* **8,** 5541 (1980).

[j] G. F. Hong, *J. Mol. Biol.* **158,** 539 (1982).

[k] A. Laughon and M. P. Scott, *Nature (London)* **310,** 25 (1984).

[l] W. M. Barnes and M. Bevan, *Nucleic Acids Res.* **11,** 349 (1983).

due to the asynchronous behavior of the nuclease activity of T4 DNA polymerase.[15]

All of the sequential deletion protocols have problems (Table I). A major problem for all of the procedures is that the "deleting" enzymes are all template dependent, such that the rate of DNA digestion varies as a function of the nucleotide (nt) sequence. Consequently, the time required for generation of subclones from different templates can vary more than

10-fold. Therefore, we developed the *reverse cloning* procedure for generating subclones for DNA sequencing[21] The procedure takes advantage of two properties of modified bacteriophage T7 DNA polymerase: its extremely rapid polymerization rate and its lack of exonuclease activity.[22] The modified polymerase is composed of the 84-kDa gene 5 protein of bacteriophage T7 and the 12-kDa thioredoxin protein of *Escherichia coli* associated in a one-to-one stoichiometry.[21] The reverse cloning procedure is similar to the polymerase time course method developed by Burton *et al.*,[23,24] which uses the *E. coli* DNA polymerase Klenow fragment. We have used the reverse cloning procedure to sequence the v-*src* gene from the Pr-A strain of Rous sarcoma virus and three actin genes from fish.[25–27]

Experimental Rationale

Materials

Sequenase kits are obtained from United States Biochemical (Cleveland, OH). Restriction endonucleases, ligase, and mung bean nuclease are purchased from Bethesda Research Laboratories (Gaithersburg, MD). Nucleotides and T4 polynucleotide kinase are purchased from Pharmacia (Piscataway, NJ) [γ-^{32}P] ATP (>3000 Ci/mmol) is purchased from New England Nuclear (Boston, MA) and [α-^{35}S]dATP (>1000 Ci/mmol) from Amersham (Arlington Heights, IL).

Insertion of Target DNA into M13 Origin Plasmids

The strategy for reverse cloning using the complementary, single-stranded/double-stranded plasmid vectors pUC118 and pUC119 is shown in Fig. 1.[28] The target DNA fragments should be inserted either (1) in both orientations into the polycloning site of either pUC118 or pUC119 when a single restriction enzyme is used for cloning, or (2) into the polycloning sites of pUC118 and pUC119 if two restriction enzymes are used for force

[21] S. Tabor and C. C. Richardson, *Proc. Natl. Acad. Sci. U.S.A.* **84**, 4767 (1987).

[22] Z. Liu and P. B. Hackett, *BioTechniques* **7**, 722 (1989).

[23] F. H. Burton, D. D. Loeb, C. A. Hutchinson, III, and M. H. Edgell, *DNA* **5**, 239 (1986).

[24] F. H. Burton, D. D. Loeb, R. A. McGraw, M. H. Edgell, and C. A. Hutchinson, *Gene* **67**, 159 (1988).

[25] Z. Liu and P. B. Hackett, *Nucleic Acids Res.* **17**, 5850 (1990).

[26] Z. Liu, Z. Zhu, K. Roberg, A. Faras, K. Guise, A. Kapuscinski, and P. B. Hackett, *Nucleic Acids Res.* **17**, 3986 (1990).

[27] Z. Liu, Z. Zhu, K. Roberg, A. Faras, K. Guise, A. Kapuscinski, and P. B. Hackett, *DNA Sequence* **1**, 125 (1990).

[28] J. Vieira and J. Messing, this series, Vol. 153, p. 3.

cloning so that each strand of the DNA can be sequenced separately. In case 1, the orientations of complementary clones can be determined by the C test.[4,29] The F^+ bacterial strain JM101, transformed[10] by the appropriate recombinant pUC plasmids, was then infected with the helper phage M13KO7 to induce single-stranded plasmids, which were prepared by the polyethylene glycol separation method.[30]

Annealing Primer and Sequenase Reaction

The universal primer (17-mer) was annealed to the template according to the instructions of the manufacturer. The primer was extended with modified bacteriophage T7 DNA polymerase (Sequenase),[21] and terminated by the limited availability of deoxyribonucleotides (dNTPs) to produce molecules of different lengths of double-stranded DNA. To establish conditions for extensions of different length, the concentrations of dNTPs were varied, while the period of DNA synthesis and temperature were kept constant. Under these conditions, about 50% of the added dNTPs were incorporated into DNA (Table II). For best results, the reaction should be conducted at room temperature for 5 min without subsequently inactivating the enzyme. To increase the range of random clone sites, when concentrations of dNTPs of 10 μM or greater are used, it is best to remove aliquots of the reaction mix every minute into a prewarmed (65°) Eppendorf tube to stop DNA synthesis. Incubation at 65° for 10 min will inactivate polymerase activity.

Digestion of Single-Stranded DNA

Digestion of single-stranded regions with mung bean nuclease generated a population of blunt-ended, double-stranded DNA fragments of different lengths with restriction enzyme sites in the polycloning region proximal to the primer. Three microliters of 10× mung bean nuclease buffer (pH 4.5)[31] and 1–2 units of mung bean nuclease were added to the reaction mixture, and incubated at 30° for 10 min. The solution of double-stranded DNAs was diluted with 30 μl water, extracted with phenol and chloroform, and precipitated with ethanol. In practice both mung bean nuclease and S1 nuclease often left molecules with single-stranded exten-

[29] P. B. Hackett, J. A. Fuchs, and J. W. Messing, in "An Introduction to Recombinant DNA Techniques: Basic Experiments in Gene Manipulation," 2nd ed. Benjamin/Cummings, Menlo Park, California, 1988.

[30] K. R. Yamomoto, B. M. Alberts, R. Benzinger, L. Lawhorne, and G. Treiber, *Virology* **40**, 731 (1970).

[31] T. Maniatis, E. F. Fritsch, and J. Sambrook, "Molecular Cloning: A Laboratory Manual." Cold Spring Harbor Press, Cold Spring Harbor, New York, 1982.

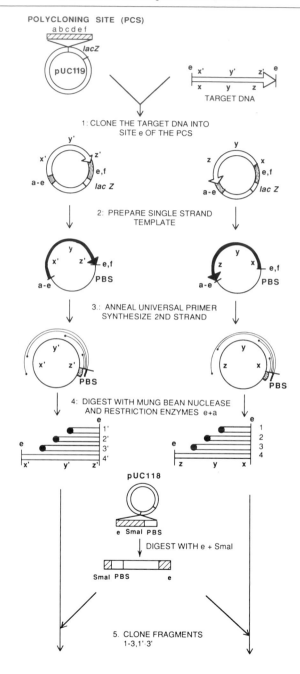

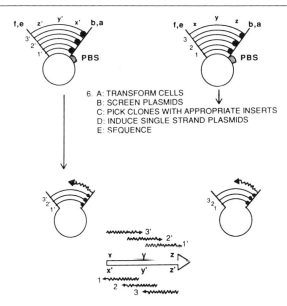

FIG. 1. Reverse cloning strategy. (1) Target DNA with marker sites x, y, and z and their complements x', y', and z' is cloned into one (e.g., pUC119) of a pair of complementary plasmid vectors (open circle) in both orientations (for force-cloned target DNA, both of the complementary vectors are used) in the polycloning site (PCS; letters a–f represent various restriction enzyme cleavage sites). The orientations of the target DNAs are indicated by the arrows. Two parallel reactions are conducted to obtain subclones for both strands. (2) Single-stranded copies of the plasmids containing the target DNA are induced by infection of the host with M13KO7 and the single-stranded plasmids are prepared. (3) Universal primer is annealed to the primer binding site (PBS) and variable lengths of single-stranded plasmids are synthesized, using modified T7 DNA polymerase, by random termination due to limited availability of nucleotide substrate. (4) The resulting duplexes are digested with mung bean nuclease and appropriate restriction endonucleases (e.g., a plus e). (5) The resulting fragments are cloned into the compatible sites of either a complementary vector (pUC118) or the same vector (not shown; see text) to reverse the orientation of the inserts. (6) After sizing of the inserts (1–3, 1'–3'), clones are picked for sequencing. The bottom part of the figure shows the relationship of the initial target DNA (large open arrow) and two complementary sets of three overlapping sequences (wavey arrows with the direction of sequencing indicated by the arrowheads).

sions at their ends. To increase the proportion of molecules with blunt ends, the ethanol-precipitated DNA fragments were resuspended in 10 mM Tris (pH 7.4), 1 mM ethylenediaminetetraacetic acid (EDTA) (Tris-EDTA) at a concentration of about 50 ng/μl, and treated with the Klenow fragment of E. $coli$ DNA polymerase I in the presence of all four dNTPs.[31] The products were phenol/chloroform extracted, ethanol precipitated, and resuspended in Tris-EDTA (pH 7.4).

TABLE II
EXTENT OF SEQUENASE REACTIONS[a]

Time (min)	Trichloroacetic acid-precipitated counts (cpm)	
	dC, dG, dT (= 3 pmol)	dC, dG, dT (= 15 pmol)
1	20,737	104,436
2	27,938	101,755
3	21,023	94,448
5	19,292	113,391
10	19,585	106,693
Average counts:	21,715	104,144
Radioactivity incorporated:	0.9%	4.1%
dNTPs incorporated	51%	47%

[a] [^{32}P]dATP (3.3 pmol; >3000 Ci/mmol) was added to the standard polymerase reactions carried out at room temperature in the presence of 168 pmol of cold dATP in a total volume of 20 μl. At each time point, 1.5-μl samples were diluted into 1 ml of 10 mM EDTA to stop the modified T7 DNA polymerase and precipitated with 1 ml 20% (w/v) trichloroacetic acid (TCA) to determine incorporation of dNTPs into DNA. The background counts (2549 cpm) from reactions lacking the T4 DNA polymerase were subtracted from the counts given.

Preparation of Target DNAs for Reverse Cloning

The blunt-ended fragments of different lengths were cleaved with an appropriate restriction endonuclease, generally the same as that used for cloning the target DNA. Other enzymes that cleave in the polycloning region proximal to the primer site, and that do not cleave the target DNA, can be used. The fragments could be readily cloned into the complementary vector cleaved with the same restriction enzyme plus SmaI or HincII, either of which leaves blunt ends. SmaI or HincII should be used such that the blunt end will be proximal to the sequencing primer. In some cases, the double-stranded fragments extended beyond the target DNA into the initial cloning vector sequence. To avoid including vector sequences in the subclones, the double-stranded DNA fragments were cleaved with a restriction endonuclease at the polycloning site downstream of the target DNA; obviously an enzyme that does not cleave the target DNA must be used. Consequently, the reverse cloning procedure can eliminate the introduction of vector sequences into the second round of clones used for sequencing (steps 4 and 5 of Fig. 1).

Reverse Cloning

The double-stranded DNA fragments from one of the complementary vectors (e.g., pUC119) were force cloned into the other member of the

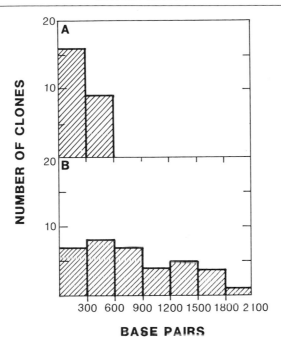

FIG. 2. Size distribution of randomly selected clones as a function of the concentration of dNTPs. Twenty-four clones obtained from reactions containing 3 pmol/20 μl of dNTPs; (B) 36 clones obtained from reactions containing 15 pmol/20 μl of dNTPs.

vector pair (e.g., pUC118) when the original cleavage site is between the primer-binding site (PBS) and SmaI or HincII blunt end cleavage sites; otherwise, the fragments were force cloned (reverse cloned) into the same vector (pUC119) when the original cleavage was distal to the PBS/SmaI–HincII region. As a result, the portion of each extended fragment distal to the initial primer site originally was proximal to the primer site in the complementary vector. Plasmids from 2 ml of a minipreparation of cells were analyzed on 1% (w/v) agarose gels to determine the sizes of the inserts. Clones with size differences of 300–400 bp were used for DNA sequencing by the dideoxy-termination procedure.[2]

Results

Clones Longer than Calculated Average Size

Modified bacteriophage T7 DNA polymerase extends DNA at the rapid rate of about 300 nucleotides/sec.[22] Accordingly, the length of a DNA extension is dependent on (1) the relative time at which the primer, template, and enzyme complex to initiate DNA replication and (2) the availability of nucleotide substrates. Figure 2 shows the size distribution of clones made from 1 μg of single-stranded DNA template, when either 3 or 15 pmol of each dNTP was provided. The data shown in Table II indicate that the polymerization reaction was finished within 1 min and that about 50% of the dNTPs were incorporated into DNA. The linear relationship of DNA synthesized/[dNTPs] can serve as the basis for adjusting the levels of dNTPs when working with target DNA of different sizes. Consequently, the average lengths of the extended DNAs in Fig. 2A and B should be 15–20 and 80–100 nucleotides, respectively. The data in Fig. 2 show much longer extensions than average and a relatively uniform distribution of clone sizes. With 3 pmol of dNTPs, most clones had inserts of 100–300 bp, although some inserts were as long as 600 bp, about 40 times longer than the calculated average size. Similarly, inserts as long as 2.1 kb were obtained with 15 pmol of dNTPs in the reaction, about 30 times longer than the calculated average size. The results shown in Fig. 2 can serve as a general guide for working with any target DNA. After size analysis on agarose gels, a series of clones with incremental size differences of about 300 bp can be picked for DNA sequencing.

Randomness of Clone Sizes

Although the data shown in Fig. 2 indicated random termination of DNA polymerization, analysis of short extensions of the primer demonstrated that some preferential termination occurs, presumably at polymerization pause sites.[22] There appears to be a slight phasing of preferential termination with a periodicity of roughly 10 bp, or integral turns of the B-form DNA double helix.[22] We do not understand the cause of this periodicity.

Reverse Cloning for Linker Scanning Clones

Reverse cloning can be used for generating subclones of fragments for uses other than DNA sequencing. For instance, reverse cloning can be employed to generate linker scanning mutants.[32] A deletion library at the 5' and 3' ends can be made with the standard procedure to generate randomly terminated subclones (Fig. 1). The desirable 5' and 3' deletion clones can be ligated after restriction enzyme digestion at the adjacent polylinker region, thus introducing a linker 10 bp in length. If longer stretches of the substitution are desired, thereby permitting a greater latitude for scanning, cleavage at a restriction site distal to the initial

[32] S. L. McKnight and R. Kingsbury, *Science* **217**, 316 (1982).

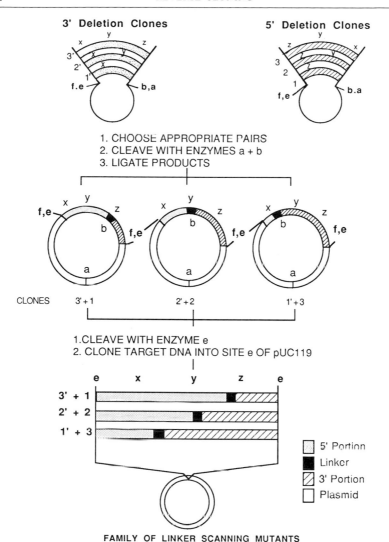

FIG. 3. Application of reverse cloning procedure to linker scanning. Deletion libraries (5' and 3') are made by reverse cloning (see Fig. 1). Pairwise deletion clones (1 and 3', 2 and 2', 3 and 1') are digested with appropriate restriction endonucleases (a plus b) and the products are ligated, giving the intermediate products in which a linker of 10 bp is introduced between the 5' and 3' deletion clones. Note that the original fragments in opposite orientation are now juxtaposed in the same orientation. These intermediates are cleaved with another restriction endonuclease (e), and cloned into pUC119 cleaved with the same restriction endonuclease (e).

cloning site can be used. The basic strategy for using the reverse cloning procedure in linker scanning is illustrated in Fig. 3. The reverse cloning procedure permits construction of subclones of all sizes, which is particularly important for mapping resulting domains of promoters, and enhancers, as well as functional and structural domains of proteins.

Conclusion

We have described an easy, efficient, and reliable method for preparing subclones of long DNA fragments for sequencing. Reverse cloning is simple; it avoids gel purification or fractionation of DNA fragments, the use of linkers, and requires only a standard primer conventionally used for DNA sequencing. Repeated sequencing is minimized, thereby considerably shortening the time of sequencing projects.

A potential problem with the reverse cloning procedure is the accuracy of the sequences generated by the modified T7 DNA polymerase. The reverse cloning procedure depends on the rapid speed of the enzyme. However, a consequence of the high speed of polymerization is the lack of proofreading by the DNA polymerase. To date we, and others who have used the procedure, have not detected any errors due to improper copying of the single-stranded template. We have sequenced thousands of bases, including the Pr-A pp60^{v-src} gene and three actin genes from fish without any indications of problems. Of course, the sequencing of both strands of a specific region of DNA, from independently generated clones, provides the necessary check on the fidelity of the overall DNA sequence. The separate generation of complementary subclones should identify any low-probability errors in subclone sequences.

[6] Unidirectional Deletion Mutagenesis for DNA Sequencing and Protein Engineering

By WENYAN SHEN and MARY M. Y. WAYE

Introduction

Reducing the length of target DNA sequences inserted into M13 or plasmid vectors has become a routine step for dideoxynucleotide sequence analysis. The commonly used methods involve deletion with exonucleases

METHODS IN ENZYMOLOGY, VOL. 218

or restriction enzymes.[1-4] Most of these methods require extensive purification of double-stranded DNA (dsDNA) or single-stranded DNA (ssDNA), and sufficient knowledge of the restriction map of the target DNA to allow selection and use of a unique restriction site that is absent inside the target DNA. In this chapter, we describe an efficient deletion method that uses M13 ssDNA templates that are prepared the same way as for deoxynucleotide sequencing.[5] It is based on the use of a mixture of oligodeoxynucleotide primers that have fixed sequences at 5′ ends, defining the end point of the deletion, and variable sequences at 3′ ends, composed of mixtures of all four nucleotides at six positions. The 5′ ends of the primers are hybridized to a fixed location of the single-stranded M13K11RX vectors, which by design contain four copies of *Eco*K restriction enzyme recognition sites, and the 3′ ends are hybridized randomly to the target DNA to be analyzed (Figs. 1 and 2). Such primers when extended with DNA polymerase can direct deletions of intervening parts of the single-stranded DNA that harbor the four copies of *Eco*K sites as well as part of the target DNA. These deletion products are selected in a host strain with the *Eco*K restriction system (e.g., JM101) and only those phages that have deletions will be able to grow. By using this method, the template DNA does not require extensive purification and previous knowledge of the restriction map of the target DNA is not necessary. Therefore, it is an efficient way of generating a nested set of unidirectional deletion mutants useful for dideoxy sequencing. This is also a particularly useful method when making a nested set of truncated proteins.

Principle and Advantage of Method

The strategy of our method was based on the fact that a mixture of oligodeoxynucleotides (16-mers) with a random nucleotide sequence at six positions of the 3′ portion (5′-GGATCCCCTANNNNNN-3′) can prime randomly at different positions on a DNA template (Fig. 1). The underlined triplet indicates the position of the stop codon of the complementary strand. The unidirectional nature of the deletion was achieved because the 5′ portion of the oligodeoxynucleotide was designed to hybridize specifically to a site upstream from the universal priming site (Figs. 1 and 2), whereas the 3′ portion was designed to prime randomly on the target DNA. A schematic diagram of the method is shown in Fig. 1. This method

[1] C. Yanisch-Perron, J. Vieira, and J. Messing, *Gene* **33,** 103 (1985).
[2] S. Henikoff, *Gene* **28,** 351 (1984).
[3] C. J. Barcak and R. E. Wolf, Jr., *Gene* **49,** 119 (1986).
[4] R. M. K. Dale, B. A. McClure, and J. P. Houchins, *Plasmid* **13,** 31 (1985).
[5] W. Shen and M. M. Y. Waye, *Gene* **70,** 205 (1988).

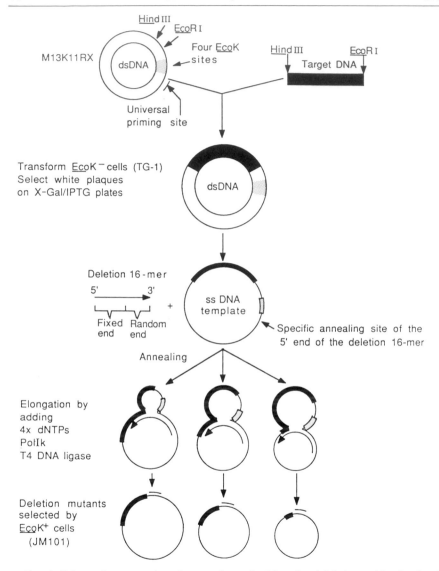

FIG. 1. Scheme for preparation of a nested set of unidirectional deletions with a fixed end point. The M13K11RX double-stranded replicative form (RF) DNA [M. M. Y. Waye, M. E. Verhoeyen, P. J. Jones, and G. Winter, *Nucleic Acids Res.* **13,** 8561 (1985)] was prepared from TG1 cells, which are *Eco*K⁻ (T. J. Gibson, Ph.D. Thesis, Cambridge Univ., 1984). The target DNA was cloned between the *Hin*dIII and *Eco*RI sites of M13K11RX. The recombinant containing the target DNA can be selected by picking white plaques on X-Gal/IPTG plates. DNA from the M13 recombinant phage was then prepared by growing in TG1 cells. Phosphorylated mutagenic oligodeoxynucleotides (10 pmol) were then annealed to 1 μg of single-stranded template DNA in 10 m*M* Tris, pH 8, and 10 m*M* MgCl₂ (from 85° to room temperature

is highly efficient because it selects against the parental M13K11 or M13K11RX recombinant template (with cloned target DNA insert and four copies of *Eco*K sites) when the mutagenized DNA is transfected into JM101, the host strain, which has the restriction enzyme *Eco*K.

One specific advantage of the system is that the ssDNA that is used as the template can be prepared rapidly. There is no need to purify dsDNA (replicative form) as in (1) the exonuclease III method,[6] (2) the exonuclease III plus exonuclease VII method,[1] (3) the BAL 31 method,[7] or (4) the exonuclease III plus S1 method.[2] There is no need to purify ssDNA completely free of (dNTPs) as in the RD20 oligodeoxynucleotide-directed mutagenesis method.[4]

Another specific advantage is that only two enzymes are needed: *Escherichia coli* DNA polymerase I Klenow fragment and T4 DNA ligase. Furthermore, the optimal amount of enzymes required does not have to be titrated exactly. This is quite different from other methods; for example, the amounts of exonuclease III and VII,[1,6] S1,[2] DNase I,[8] or T4 polymerase[4] must be titrated carefully.

A further specific advantage is that the end point of the deletion can be designed precisely using a mixture of oligodeoxynucleotides with fixed 5' as well as variable 3' ends. Thus the method will provide a means not only to facilitate DNA sequencing, but also to encode for unfused proteins when the test DNA to be truncated represents a translatable gene sequence. This is due to the fact that a translational termination codon is present in the M13K11RX sequence between the universal priming site and the four copies of *Eco*K sites, which is complementary to the predetermined portion of the primer (underlined triplet illustrated above). In this

[6] L. H. Guo and R. Wu, *Nucleic Acids Res.* **10,** 2065 (1982).
[7] M. Poncz, K. Solowiejczyk, M. Ballantine, E. Schwartz, and S. Surrey, *Proc. Natl. Acad. Sci. U.S.A.* **79,** 4298 (1982).
[8] G. F. Hong, *Biosci. Rep.* **1,** 243 (1981).

---

in 30 min). The heteroduplex was then put on ice, and four dNTPs, rATP, DTT (final concentrations of 0.5 m*M* each, 0.5 m*M*, and 10 m*M*, respectively), 1 unit Klenow fragment of DNA polymerase, and 10 units T4 DNA ligase were added to transform 0.2 ml of JM101 competent cells, which were prepared by the Hanahan method [D. Hanahan, *J. Mol. Biol.* **166,** 557 (1983)], to select against the parental clones with the four copies of *Eco*K sites. The solid black box represents the target DNA of interest. The shaded box represents the four copies of *Eco*K sites used for the biological selection. The three curved lines with arrows represent the deletion oligonucleotides. The three short lines above the three deletion mutants shown at the bottom indicate the positions of the universal priming sites. Reprinted by permission from Ref. 5.

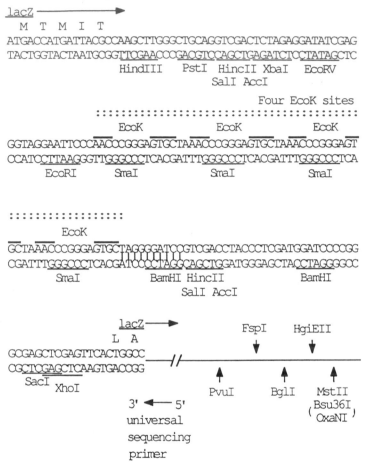

FIG. 2. Nucleotide sequence of the polylinker region of M13K11 and M13K11RX. The colons above the sequence indicate the location of the four copies of *Eco*K recognition sites, AACNNNNNNGTGC. Short vertical lines indicate the location of base pairing between the 5′ end of the mutagenic oligodeoxynucleotide and the template DNA. The first five and two amino acids of the LacZ protein (flanking the polylinker region) are also shown. For the position of the universal sequencing primer, see Fig. 1. Reprinted by permission from Ref. 5.

manner, the test DNA, when cloned in the appropriate orientation in relation to the termination codon, after truncation, will terminate in one out of three instances at this termination codon. This will lead to the encoding of a truncated form of the unfused protein with a shortened carboxyl terminus. Conversely, when cloned in the reverse orientation, an initiation codon, in place of the termination codon, in the same construct should be used. In this manner, a shortened unfused protein truncated from the amino terminus will be obtained. It is also possible to introduce any additional amino acids extended from the truncated proteins by using primers that have the appropriate codons inserted between the predetermined portion and the random portion of the primer.

Materials and Reagents

Enzymes, Chemicals, and Radiochemicals

Klenow (large) fragment of *E. coli* DNA polymerase I, dideoxynucleotide triphosphates, deoxynucleotide triphosphates, and T4 ligase are purchased from Pharmacia-PL Biochemicals (Milwaukee, WI), and [α-^{32}P] dATP for sequencing is obtained from New England Nuclear (Boston, MA). Oligodeoxynucleotides (a mixture of 16-mers with the following sequences: 5'-GGATCCCCTANNNNNN-3') are synthesized by Allelix, Inc. (Mississauga, Ontario, Canada).

M13 Vectors, M13 Recombinants, and Host Strains

The constructions of M13K11RX and M13K11, which have four copies of *Eco*K restriction enzyme recognition sites (Fig. 2), have been described previously.[9] The *tyrS* gene from *Bacillus stearothermophilus* (1.7-kb *Hind*III–*Eco*RI fragment of *tyrS* gene) is cloned between the *Hind*III and *Eco*RI sites of M13K11RX[10]; the *H1* gene from *Xenopus laevis* (1.3-kb *Taq*I fragment of pXLHW7) is cloned into the *Acc*I site of M13K11[11]; the *H4* gene from *X. laevis* (2.3-kb *Pst*I fragment of pXLHW7) is cloned into the *Pst*I site of M13K11.[12]

Escherichia coli TG1 [K12 strain: Δ(*lac-pro*) *supE thi hsd*Δ5 (r⁻, m⁻) (F' *traD36 proA⁺B⁺ lacIq lacZ*ΔM15)] is obtained from T. Gibson.[13] *Escherichia coli* TG1 has an inactive *Eco*K restriction system that supports growth of phages such as M13K11 and M13K11RX containing *Eco*K sites, whereas *E. coli* JM101 has an active *Eco*K restriction system that restricts propagation of such phages.[14] The presence of multiple *Eco*K sites makes

[9] M. M. Y. Waye, M. E. Verhoeyen, P. J. Jones, and G. Winter, *Nucleic Acids Res.* **13**, 8561 (1985).

[10] M. M. Y. Waye and G. Winter, *Eur. J. Biochem.* **158**, 505 (1986).

[11] P. C. Turner, T. C. Aldridge, H. R. Woodland, and R. W. Old, *Nucleic Acids Res.* **11**, 4093 (1983).

[12] A. F. M. Moorman, P. A. J. De Boer, R. T. M. De Laaf, W. M. A. M. Van Dongen, and O. H. J. Destree, *FEBS Lett.* **136**, 45 (1981).

[13] R. J. Gibson, Ph.D. Thesis, Cambridge Univ., 1984.

[14] J. Messing, B. Gronenborn, B. Muller-Hill, and P. H. Hofschneider, *Proc. Natl. Acad. Sci. U.S.A.* **75**, 3642 (1977).

the selection more efficient. A kit based on this method is supplied by Bio 101, Inc. (San Diego, CA).

Competent Cells

TG1 competent cells, made fresh using the standard Ca^{2+} treatment method,[15] are used for transforming the vectors as well as cloned recombinants. To transform deletion mutants at high efficiency, JM101 competent cells are made fresh using the standard Hanahan method.[16]

Transformation of M13 DNA

1. Add single- or double-stranded M13 DNA (for cloning, use up to 40 ng DNA; for deletion mutants, use half of the ligation mixture) to 200 μl of the competent cells, which are prechilled in 17 × 100 mm polypropylene tubes. (Falcon 2059); Becton Dickinson, Oxnard, CA).

2. Mix it gently by swirling and store on ice for 30 min.

3. Add 0.2 ml of log-phase cell culture (this cell strain should be the same as in the competent cell) into the above tube and then add 25 μl 5-bromo-4-chloro-3-indolyl-β-D-galactoside (X-Gal) (25 mg/ml in dimethylformamide) and 25 μl isopropyl-β-D-thiogalactopyranoside (IPTG) (25 mg/ml in H_2O).

4. Add 3 ml of melted top agar kept at 42° [per liter: 8 g Bacto-tryptone (Difco, Detroit, MI) 5 g Bacto-yeast extract, 5 g NaCl, and 6 g Bacto-agar] into the tube, cap the tube, and mix by inverting the tube three times.

5. Pour quickly onto a prewarmed (37°) 2× YT (per liter: 10 g Bacto-tryptone, 10 g Bacto-yeast extract, and 5 g NaCl) plate, and swirl to spread the top agar evenly.

6. Let the agar harden at room temperature, invert the plate, and incubate at 37°. Plaques should appear in about 10 hr.

Protocol

Replicative Form of M13K11 or M13K11RX DNA Preparation

1. Inoculate 1 ml of TG1 cells (1/100 dilution of the overnight culture in 2× YT medium) by toothpicking one well-isolated plaque from the above transformation.

2. Shake 5–6 hr at 37°.

3. Spin in the Eppendorf centrifuge at 22° for 5 min.

[15] T. Maniatis, E. F. Fritsch, and J. Sambrook, "Molecular Cloning: A Laboratory Manual," p. 92. Cold Spring Harbor Press, Cold Spring Harbor, New York, 1982.
[16] D. Hanahan, *J. Mol. Biol.* **166,** 557 (1983).

4. Collect the supernatant. The supernatant contains the phages, which remain viable for more than 1 year when stored at 4°.

5. Pour 1 ml of the above supernatant into 100 ml 2× YT medium that contains 1 ml of overnight culture of the host cells.

6. Shake 4–5 hr at 37°.

7. Centrifuge at 5000 g for 5 min at 4° to precipitate cells.

8. Make replicative form (RF) DNA using any of the procedures that work for plasmid DNA preparation.[17] The expected yield for RF DNA is 40–50 μg per 100 ml of culture.

Cloning Target DNA

The target DNA is eluted from an agarose gel and cloned into the appropriate restriction site(s) in M13K11 or M13K11RX. The recombinant DNA clones are selected by picking white plaques after transfecting TG1 cells in the presence of IPTG and X-Gal.[14] To facilitate sequencing through the deletion junction, it is necessary to clone the target DNA such that the EcoK sites are in between the target DNA and the universal priming site [i.e., into the HindIII, PstI, XbaI, EcoRI, or EcoRV site(s)]. After the four copies of EcoK sites are deleted, the new junction created by the oligodeoxynucleotides can be sequenced from the universal priming site (Figs. 1 and 2). The genes are cloned into M13 in an orientation that depends on whether a nested set of 5′ or 3′ deletion mutants are required.

Preparation of Template for Mutagenesis

The oligodeoxynucleotide-directed deletion procedure is modified from the procedure described by Carter et al.[18]

Day Zero

Pick a single TG1 colony and grow in 5 ml of 2× YT medium at 37° overnight.

Day 1

1. Inoculate 6 ml 2× YT medium with 60 μl of a saturated fresh overnight culture of TG1 cells (described above).

2. Toothpick a single M13K11RX recombinant plaque (transfer about 10^7 phages) into the medium and incubate with shaking at 37° for 5–6 hr.

[17] P. M. G. F. Van Wezenbeck, T. J. M. Hulsoebos, and J. G. G. Schoenmakers, Gene 11, 129 (1980).

[18] P. Carter, H. Bedouelle, and G. Winter, Nucleic Acids Res. 13, 4431 (1985).

3. Pour into 5 × 1.5 ml Eppendorf tubes, each containing 1 ml of the above culture.

4. Centrifuge for 5 min in a microfuge to bring down the cells.

5. Pour the supernatant into five tubes, each containing 200 μl 2.5 M NaCl, 20% (w/v) polyethylene glycol (PEG) 6000. Make no effort to transfer the supernatant completely, as it is important to avoid carrying across any cells.

6. Mix briefly and then leave at room temperature for 15 min (do not put samples on ice, because RNA may be precipitated on ice). Centrifuge for 5 min at 22° in an Eppendorf microfuge.

7. Pour off the supernatant with a flick of the wrist, and respin for a few seconds to bring liquid off the walls of the tube. Carefully pipette off all the remaining traces of PEG supernatant with a drawn-out capillary. It is important to remove all the PEG for good results. The PEG pellet should be visible. The PEG pellet can be stored at −20° if necessary.

8. Add to each tube 100 μl of TE buffer (10 mM Tris, 0.1 mM ethylenediaminetetraacetic acid (EDTA), pH 8.0) and vortex for 30 sec to resuspend the pellet. [*Optional step:* Add 5 μl (10 mg/ml) RNase; the RNase should have been boiled in 0.3 M sodium acetate, pH 5.5, for 15 min to remove traces of DNase, and is kept frozen at −20°. Incubate at 37° for 30 min. This should digest any RNA, which can give extensive self-priming activity.]

9. Extract with 50 μl phenol neutralized with 0.5 M Tris, 0.1 M EDTA, PH 8.0. Vortex for 30 sec and spin for 2 min in a microfuge to separate the phases. Pipette off the upper aqueous layer and transfer to another Eppendorf tube.

10. Remove traces of phenol by one extraction with 0.5 ml diethyl ether or chloroform neutralized with TE buffer, and precipitate DNA by adding 10 μl 3 M sodium acetate, pH 5.5, 250 μl ethanol to the lower aqueous phase. Vortex and quick freeze in pellets of dry ice (or crushed dry ice) for 20 min.

11. Spin for 5 min in a microfuge at 22° to bring down the DNA, remove the supernatant, and then wash the DNA pellet with 1 ml of 95% (v/v) ethanol. Pour off the supernatant and respin for a few seconds to bring traces of ethanol off the walls of the tubes. Carefully remove all the remaining ethanol with a drawn-out capillary. Resuspend all five tubes of DNA in a total of 100 μl. Measure the optical density (OD) at 260 nm by adding a 1-μl aliquot to 1 ml TE in a quartz cuvette: 1 OD/ml of ssDNA corresponds to 40 μg/ml. The total yield should be about 15 μg (1 ml of culture giving approximately 3 μg of DNA). Ethanol precipitate again and make up to 1 μg/μl (optional).

Kinasing Oligonucleotide for Mutagenesis

Kinase enough oligonucleotides for 10 experiments:

Primer (10 pmol/μl), 10 μl (100 pmol total)
Kinase buffer (10×) (500 mM Tris-HCl, pH 8.0, 100 mM MgCl$_2$), 2 μl
Dithiothreitol (DTT; 100 mM), 1 μl
Adenosine 5' triphosphate (rATP; 10 mM), 2 μl
Water, 4 μl

Add polynucleotide kinase [e.g., 0.5 μl of polynucleotide kinase (10 U/μl) from Pharmacia]. Incubate in an Eppendorf tube for 30 min at 37°, and then incubate for 10 min at 70° to denature the kinase, then freeze at −20° for storage if necessary. (*Note:* Frozen ssDNA template and frozen kinased oligonucleotide sometimes give poor results.)

Hybridization and Extension

1. Anneal primer and template together in an Eppendorf tube:

Kinased primer (5 pmol/μl)	2 μl
Template (1 μg/μl)	1 μl
TM buffer (10×) (100 mM Tris-HCl, pH 8.0, 100 mM MgCl$_2$)	1 μl
Water	6 μl

Place the tube containing the sample in a small beaker of hot water (80°), and let cool to room temperature. This will take about 30 min.

2. Extension/ligation: Add to the annealing mix:

TM buffer (10×)	1 μl
rATP (5 mM)	1 μl
dNTPs (5 mM)	1 μl
DTT (100 mM)	1 μl
Water	4 μl

Place on ice and then add T4 DNA ligase (10 U; Pharmacia) and Klenow fragment of DNA polymerase (1 U; Pharmacia). Then incubate 12–20 hr at 12–15°.

Transformation

Day 2

The above reaction mixture is transformed into JM101 using the method of Hanahan.[16] Only those phages that have lost all the *Eco*K sites

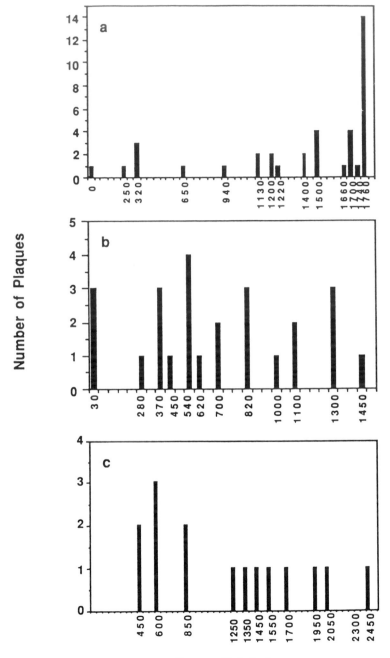

are able to transfect. Approximately 100 plaques are obtained from 1 μg of ssDNA template.

Selection of Deletion Mutants

Day 3

Twenty to 60 phage lysates are prepared and electrophoresed on a 0.7% (w/v) agarose gel.[19] Phage supernatants are prepared according to the procedure described above for the preparation of RF DNA (steps 1–4) and 25 μl of phage supernatant is added to 7.5 μl of a 6% (w/v) sodium dodecyl sulfate (SDS), 30% (v/v) glycerol, 0.3% (v/v) bromphenol blue solution and heated to 65° over 15 min. The entire mixture is loaded onto the 0.7% agarose gel and then fractionated by electrophoresis. Clones containing the appropriate deletions are then sequenced using the universal primer and the dideoxy-sequencing method. Only one of the four tracks (bases) is sequenced as a preliminary screen for the retention of the universal priming site. The appropriate clones (normally more than 50%) that retain the universal priming site are then sequenced.[20]

Illustrative Examples

To demonstrate the usefulness of this method, we have obtained three different sets of unidirectional deletion mutants.[5] As shown in Fig. 3, the distribution of deletion sizes is quite random. In Fig. 4, ~50% of the plaques analyzed on a 0.7% agarose gel had detectable deletions. When plaques were picked randomly and sequenced, 35% (6 of 17 clones) had lost the universal priming site. However, these clones could be eliminated readily by analyzing the sequence of one of the four nucleotides. Ten of

[19] J. Messing, this series, Vol. 101, p. 20.
[20] F. Sanger, S. Nicklen, and A. R. Coulson, *Proc. Natl. Acad. Sci. U.S.A.* **74**, 5463 (1977).

FIG. 3. A histogram showing the number of clones recovered with different size deletions after oligonucleotide-directed deletion mutagenesis. The parental M13 clone used has the following inserts: (a) *tyrS* (1.7 kb, tyrosyl-tRNA synthetase gene of *Bacillus stearothermophilus*) [M. M. Y. Waye and G. Winter, *Eur. J. Biochem.* **158**, 505 (1986)]; (b) *H4* (2.3 kb, histone H4 gene of *Xenopus laevis*) [A. F. M. Moorman, P. A. J. De Boer, R. T. M. De Laaf, W. M. A. M. Van Dongen, and O. H. J. Destree, *FEBS Lett.* **136**, 45 (1981)]; (c) *H1* (1.3 kb, histone H1 gene of *X. laevis* [P. C. Turner, T. C. Aldridge, H. R. Woodland, and R. W. Old, *Nucleic Acids Res.* **11**, 4093 (1983)]. The sizes of the inserts were estimated by their mobility on 0.7% agarose gels. Sequencing data suggest that the mobility of M13 ssDNA is linear over the range of 0–2.3 kb, and this method could resolve M13 clones that differed by 100 bp. Reprinted by permission from Ref. 5.

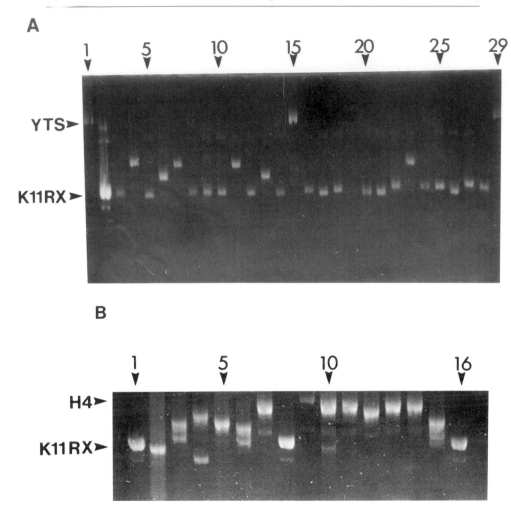

FIG. 4. Agarose gel electrophoresis of M13 phage DNA after oligonucleotide-directed deletion mutagenesis. Phage supernatants (25 μl) were added to 7.5 μl of a 6% SDS, 30% glycerol, 0.3% bromphenol blue solution and heated at 65° for 15 min. The entire mixture was loaded onto the 0.7% agarose gel and then fractionated by electrophoresis. The direction of migration was from top to bottom. M13 recombinants with different DNA fragments in the polylinker region were used as the parental clones. (A) M13*tyrS* (M13K11RX with a 1.7-kb *tyrS* gene fragment) was used as the parental clone (lanes 1, 15, and 29). Lane 2 is the control DNA, that is, M13K11RX with no insert; lanes 3–14 and 16–28 are ssDNA of various deletion clones using the M13*tyrS* clone as the parental clone. (B) M13*H4* (M13K11 with a 2.3-kb *H4* gene fragment) was used as the parental clone (lane 9). Lanes 1, 8, and 16 are control DNA, that is, M13K11RX with no insert; lanes 3–7 and 10–15 are various deletion clones using M13*H4* clone as the parental clone. Reprinted by permission from Ref. 5

11 (90%) of these sequenced mutants that retained the universal sequencing site had the correct junction, as specified by the 10 nucleotides of the 5' end of the oligodeoxynucleotide. Thus the overall percentage of deletion clones that were indeed unidirectional was ~60% (65% × 90%). We have completely sequenced two of the deletion mutants containing the truncated *tyrS* gene using three sequencing primers separately along the gene by a method described by Wilkinson *et al.*[21] No spurious mutation in the truncated gene was observed. To increase the low molar ratio of any particular sequence in the mixed oligodeoxynucleotides, six random nucleotides were designed at the 3' end. A longer stretch of random nucleotide would dramatically reduce the molar ratio. The use of another mixture of oligodeoxynucleotides that were longer (with 10 random nucleotides: 5'-GGAT-CCCCTANNNNNNNNNN-3', instead of 6 random positions at the 3' end) did not offer any improvement.

Limitation of Method

This method is not applicable if the target DNA contains any *Eco*K site. *Eco*K sites [AAC(N)$_6$GTGC] are present on the average once every 4^7 bp = 16,384 bp, assuming all 4 nucleotides are represented randomly at equal frequencies. This can be circumvented by using another vector that has four copies of *Eco*B restriction enzyme recognition sites instead of *Eco*K sites.[22,23] Although this method is applicable for deletions in targets ranging from 1.3 to 2.3 kb, as illustrated, a shorter target of 500 bp has also been used successfully.[24] However, with an insert of 500 bp, the number of plaques obtained after mutagenizing 1 μg of template DNA was less (approximately 30 plaques) than those obtained with larger inserts, presumably due to the high proportion of lethal deletion mutants of M13 vector itself.

Conclusion

We have developed a method that allows efficient generation of nested sets of deletion mutants. This method utilizes only one set of oligodeoxynucleotides and does not need a unique restriction site for deletion mutagenesis.

[21] A. J. Wilkinson, A. R. Fersht, D. M. Blow, P. Carter, and G. Winter, *Nature (London)* **307**, 187 (1984).
[22] M. M. Y. Waye, F. Mui, and K. Wong, *Technique* **1**, 188 (1990).
[23] M. M. Y. Waye, this series, Vol. 217 [16].
[24] M. M. Y. Waye and A. K. Gupta, unpublished results, 1988.

[7] Direct Sequencing of Polymerase Chain Reaction Products from Low Melting Temperature Agarose

By KEITH A. KRETZ and JOHN S. O'BRIEN

Introduction

The polymerase chain reaction (PCR) technique[1,2] for the amplification of specific segments of DNA has provided researchers with a powerful tool for the characterization of disease-causing mutations. Small quantities of genomic DNA or cDNA derived from RNA may be used as substrate for PCR amplification, yielding up to microgram quantities of specific DNA product for analysis. Many techniques have been employed to screen PCR products for known mutations, the most common of which include restriction enzyme digestion[3,4] and allele-specific oligonucleotide hybridization.[5,6] These techniques are useful but each has limitations. Screening by restriction enzyme digestion limits the user to only those mutations that either create new restriction sites or destroy existing sites. Although allele-specific oligonucleotide probe hybridization can be used to screen for any known mutation, it cannot be used to identify new mutations. To screen for these new mutations it is necessary to sequence the amplified DNA and to this end many methods have been developed.

Originally, the products of PCR amplification were cloned into vectors for traditional single- or double-stranded sequencing. Although this is facilitated by the addition of restriction sites within the primers, multiple clones must be sequenced because of the possible errors introduced by the *Taq* polymerase used in the PCR. To overcome this problem, direct sequencing of PCR products is preferred and many methods have been

[1] K. B. Mullis and F. A. Faloona, this series, Vol. 155, p. 335.
[2] R. K. Saiki, D. H. Gelfand, S. Stoffel, S. J. Scharf, R. Higuchi, G. T. Horn, K. B. Mullis, and H. A. Erlich, *Science* **239**, 487 (1988).
[3] R. K. Saiki, S. Scharf, F. Faloona, K. B. Mullis, G. T. Horn, H. A. Erlich, and N. A. Arnheim, *Science* **230**, 1350 (1985).
[4] K. A. Kretz, J. K. Darby, P. J. Willems, and J. S. O'Brien, *J. Mol. Neurosci.* **1**, 177 (1989).
[5] R. K. Saiki, T. L. Bugawan, G. T. Horn, K. B. Mullis, and H. A. Erlich, *Nature (London)* **324**, 163 (1986).
[6] M. Verlaan-de Vries, M. E. Bogaard, H. van den Elst, J. H. van Boom, A. J. van der Eb, and J. L. Bos, *Gene* **50**, 313 (1986).

described, including (1) ^{32}P end-labeled internal primers,[7,8] (2) ^{32}P end-labeled primers used after ultrafiltration of PCR product to remove the original primers,[9] (3) genomic amplification with transcript sequencing,[10] (4) generation of excess single-stranded DNA (ssDNA) using unequal amounts of PCR primers,[11] (5) PCR using one biotinylated primer, which can be used to affinity purify one strand of the PCR product for ssDNA sequencing,[12] and (6) production of ssDNA templates by exonuclease digestion following PCR.[13] These methods are useful, but often require time-consuming purification of the PCR product, end labeling primers with ^{32}P, and/or synthesis of a third primer specifically for sequencing. A method was described in which PCR products were sequenced directly using *Taq* polymerase.[14] Here, we describe a method in which PCR products purified in low melting temperature agarose can be sequenced directly using commercially available sequencing kits with little or no variation from the directions of the manufacturer. Therefore, the PCR product can be prepared for sequencing in a single step and sequenced under the same conditions as a normal double-stranded template using techniques already available in most laboratories.

Materials and Methods

Low melting temperature agarose is purchased from FMC BioProducts (Rockland, ME) and the Sequenase version 2.0 sequencing kit is purchased from United States Biochemical Corporation (Cleveland, OH). The AmpliTaq sequencing kit, GeneAmp λ control PCR reagents, and AmpliTaq recombinant *Taq* DNA polymerase are kindly provided by Perkin-Elmer Cetus (Norwalk, CT).

Polymerase Chain Reaction Reactions

Two different PCR products are synthesized for these sequencing studies. The first template is the 500-base pair (bp) λ control product supplied

[7] L. A. Wrischnik, R. G. Higuchi, M. Stoneking, H. A. Erlich, N. Arnheim, and A. C. Wilson, *Nucleic Acids Res.* **15**, 529 (1987).

[8] D. R. Engelke, P. A. Hoener, and F. S. Collins, *Proc. Natl. Acad. Sci. U.S.A.* **85**, 544 (1988).

[9] C. Wong, C. E. Dowling, R. K. Saiki, R. G. Higuchi, H. A. Erlich, and H. H. Kazazian, *Nature (London)* **330**, 384 (1987).

[10] E. S. Stoflet, D. D. Koeberl, G. Sarkar, and S. S. Sommer, *Science* **239**, 491 (1988).

[11] U. B. Gyllensten and H. A. Erlich, *Proc. Natl. Acad. Sci. U.S.A.* **85**, 7652 (1988).

[12] L. G. Mitchell and C. R. Merril, *Anal. Biochem.* **178**, 239 (1989).

[13] R. G. Higuchi and H. Ochman, *Nucleic Acids Res.* **17**, 5865 (1989).

[14] K. B. Gorman and R. B. Steinberg, *BioTechniques* **7**, 326 (1989).

by Perkin-Elmer Cetus in the GeneAmp λ control reagent kit. The second template is a 541-bp fragment of the human fucosidase gene spanning the last two exons.[15] Both templates are amplified using the following conditions: 10 mM Tris-HCl (pH 8.4), 50 mM KCl, 1.5 mM MgCl$_2$, dNTPs (100 μM each), primers (0.25 μM each), 2 units Taq polymerase, and either 1 ng λ DNA (control) or 2 μg genomic DNA (fucosidase gene fragment) in a total reaction volume of 100 μl. The reactions are overlaid with 75 μl of mineral oil. Amplification is performed using a temperature profile of 1 min at 94°, 1 min at 55°, and 1 min at 72° for 30 cycles in a Perkin-Elmer Cetus DNA thermal cycler.

Low Melting Temperature Gel Electrophoresis

For these templates we use either 1% (w/v) SeaPlaque GTG agarose or 1% (w/v) NuSieve GTG agarose (FMC Bioproducts). In previous studies we have used agarose concentrations ranging from 0.5 to 2.0% agarose without any difference in subsequent sequencing reactions. The manufacturer of these low melting temperature agaroses recommends SeaPlaque agarose for separation of bands >1000 bp and NuSieve for bands <1000 bp and this has worked well. (*Note*: NuSieve gels are quite fragile and should not be poured at concentrations of <1%.) The gels are prepared and run in 1× Tris–borate–ethylenediaminetetraacetic acid (EDTA) (TBE) buffer[16] containing 0.5 mg/ml EDTA.

1. Mix 20-μl aliquots of the PCR reaction product with 4 μl gel loading dye.
2. Load these mixes into individual lanes of the agarose gel.
3. Run the gels long enough to yield well-resolved bands.
4. Excise the bands from the gels using a scalpel and long-wavelength ultraviolet (UV) light source.

It is imperative that all excess agarose be trimmed away from the product band. This will increase the concentration of the DNA in the gel slice after it is melted, allowing the maximum amount of DNA to be added to the sequencing reactions. If the PCR reaction does not yield a strong band, the reaction product may be ethanol precipitated and resuspended in a volume such that the entire reaction product may be loaded on the agarose gel to increase the amount of template present in the gel slice recovered.

[15] K. A. Kretz, D. Cripe, G. S. Carson, H. Fukushima, and J. S. O'Brien, *Genomics* **12**, 276 (1992).
[16] T. Maniatis, E. F. Fritsch, and J. Sambrook, "Molecular Cloning: A Laboratory Manual," p. 454. Cold Spring Harbor Press, Cold Spring Harbor, New York, 1982.

Gel electrophoresis serves to remove excess primers, nucleotides, and other components of the PCR reaction as well as assuring that nonspecific PCR products are not present to contaminate the sequencing reactions. The presence of nonspecific amplification products would require the use of a new, internal primer made specifically for sequencing to avoid background contamination of the sequence data.

Sequencing with Sequenase Sequencing Kit

The Sequenase reactions are performed according to the instructions of the manufacturer except that the labeling reactions are performed at 37° to prevent the agarose from solidifying.

1. Melt the DNA-containing gel slice by heating for 5 min at 68°.
2. Mix 10 μl of DNA with 2 μl of sequencing primer (5 μM) and 3 μl of 5× Sequenase buffer and denature by heating to 95° (or placing in boiling water) for 10 min.
3. Quickly centrifuge (10–15 sec) and incubate at 37° for 5 min to anneal.
4. Prepare the labeling reaction mixture and equilibrate to 37°.
5. Add 5-μl aliquots of this labeling reaction mixture to the annealed primer/template mixture.
6. Incubate at 37° for 3 min.
7. Add 4-μl aliquots of the labeling reaction from step 5 to each of the four ddNTP termination mixtures previously aliquotted (2.5 μl) and equilibrated at 37°.
8. Incubate at 37° for 3 min.
9. Add 4 μl of formamide stop mix to each reaction.

Sequencing with AmpliTaq Sequencing Kit

The instructions provided by the manufacturer are followed, except that only 50% of the recommended amount of [α-^{35}S]dATP is used.

1. Melt the DNA-containing gel slice by heating for 5 min at 68°.
2. Mix 14 μl of melted gel with 2 μl of sequencing primer (5 μM) and denature by heating to 95° (or placing in boiling water) for 10 min.
3. Quickly centrifuge (10–15 sec) and add 0.5 μl of [α-^{35}S]dATP and 4 μl of labeling mix.
4. Incubate for 3 min at 45°.
5. Add 4-μl aliquots of the labeling reaction to each of the four ddNTP termination mixtures previously aliquotted (4 μl).
8. Incubate at 72° for 3 min.
9. Add 4 μl of stop mix.

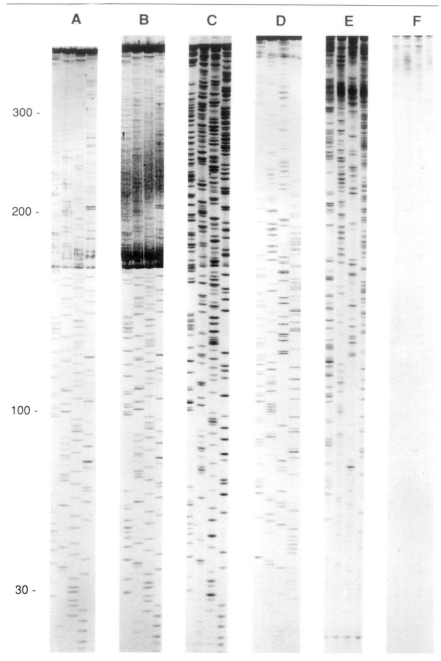

Sequencing Gel Electrophoresis

Denaturing polyacrylamide gel electrophoresis is performed on 6% polyacrylamide gels containing 8 M urea in 1× TBE. Gels are fixed for 1 hr in 15% (v/v) methanol and 5% (v/v) acetic acid and dried. Although overnight exposure is often sufficient for ^{35}S-labeled autoradiographs, longer exposures are sometimes necessary.

Results and Discussion

As can be seen in Fig. 1, both the Sequenase and AmpliTaq sequencing kits yield easily readable sequence from both the λ control product and the fucosidase gene product under the standard conditions (Fig. 1, lanes B–E). We can generally obtain sequence within 20 bp of the sequencing primer and using longer sequencing gels we have obtained sequence more than 400 bp from the primer. To date we have obtained direct sequence from templates ranging in size from 200 to 2000 bp. In addition, we have seen no difference in the quality of sequence obtained when using internal sequencing primers as compared to the original PCR primer.

Since our initial reports concerning this technique[17,18] we have investigated the effects of a number of different parameters on the quality of sequence obtained using this procedure. The manufacturer (FMC BioProducts) of the low melting temperature agarose has suggested in their literature that when preparing DNA for use in DNA synthesis reactions (i.e., nick translation, etc.) the agarose gel should be prepared and run in TBE containing only 10% the normal amount of EDTA. We have investigated this suggestion and found that in most instances the amount of EDTA in the preparative agarose gel has no significant effect on sequence quality. However, using the Sequenase kit with the λ control template under the low-EDTA conditions reduced the obstruction seen in the middle of the control sequence (Fig. 1, lanes A and B). In addition, we observed a

[17] K. A. Kretz, G. S. Carson, and J. S. O'Brien, *Nucleic Acids Res.* **17**, 5864 (1989).
[18] K. A. Kretz, G. S. Carson, and J. S. O'Brien, *Comments* (U.S. Biochem. Newsl.) **16**, 4 (1990).

Fig. 1. Sequences obtained using the direct-sequencing protocols described in the text. Lane A, λ control product, Sequenase, and low EDTA; lane B, λ control product and Sequenase; lane C, λ control product and AmpliTaq; lane D, human fucosidase gene product and Sequenase; lane E, human fucosidase gene product and AmpliTaq; lane F, human fucosidase gene product, AmpliTaq, and low EDTA. Exposure was for 14 hr at room temperature with Kodak (Rochester, NY) X-Omat AR film.

drastic reduction in synthesis of DNA under the low-EDTA conditions using the AmpliTaq sequencing kit and the fucosidase gene product (Fig. 1, lanes E and F). These two examples are the only instances, among many different templates investigated, in which we have seen any differences when comparing sequence obtained from templates prepared in gels with differing amounts of EDTA.

Another interesting parameter investigated was the use of the Mn^{2+} buffer supplied with the Sequenase kit. Adding Mn^{2+} to the Sequenase reactions causes the polymerase to incorporate the ddNTPs more efficiently. This means more sequences are terminated sooner. This generates more signal in the bands close to the primer, allowing this sequence to be read more easily. This has been of benefit only when trying to sequence suboptimal amounts of DNA, that is, when the PCR band is faint in the gel. On many occasions we have been able to read within 10 bp of the primer without changing the original sequencing protocol. Therefore, the Mn^{2+} buffer is helpful only when the quantity of DNA is limiting in the sequencing reaction.

The results presented here demonstrate that this method is fast, simple, and generates easily readable sequence comparable in quality to that obtained by using highly purified plasmid DNA. Polymerase chain reaction primers may be used without end labeling and the PCR reaction products need not be processed prior to low melting temperature agarose gel electrophoresis. We have investigated a number of different parameters that have the potential to affect the quality of sequence obtained. We have determined that templates of any size may be used to obtain reliable sequence information using either the original PCR primers or internal primers as sequencing primers. In rare cases in which this technique fails to yield good sequences, we have noted several parameters that may be changed to improve the results (i.e., EDTA concentrations, sequencing kit used).

We previously demonstrated the utility of this method in identifying disease-causing mutations in cDNA made from patient mRNA.[19] In cases in which the organization of a suspected mutant gene is known, additional mutations can be detected using this method. In this instance, intron-specific primers can be made to flank each of the exons and the upstream regulatory region. Direct sequencing of these products could detect mutations affecting the regulation of gene expression or intron/exon boundary mutations that affect mRNA processing. This substantially increases the chances of finding disease-causing mutations. This method also provides

[19] K. A. Kretz, G. S. Carson, S. Morimoto, Y. Kishimoto, A. L. Fluharty, and J. S. O'Brien, *Proc. Natl. Acad. Sci. U.S.A.* **87,** 2541 (1990).

the ability to analyze multiple products from a single amplification, which can arise in the case of tissue-specific processing of mRNAs. In addition, we have found that degenerate oligonucleotide mixtures can be used as sequencing primers in cases in which they were also used as primers for the PCR reaction. This may be useful when using PCR amplification to identify members of conserved families of proteins. We feel that this technique for directly sequencing PCR products is widely applicable and can be a useful technique in many molecular biology laboratories.

Acknowledgments

This work has been supported in part by NIH Grants NS08682 and HD18983 to Dr. J. S. O'Brien.

[8] Direct Sequencing of Polymerase Chain Reaction Products

By DAVID B. OLSEN, GERD WUNDERLICH, ANGELA UY, and FRITZ ECKSTEIN

Introduction

In vitro amplification of DNA by the polymerase chain reaction (PCR) method has provided an invaluable tool for molecular biologists. The necessary equipment needed to perform such reactions can be found in almost every laboratory that utilizes recombinant DNA techniques. Although there are some potential misuses of this powerful procedure,[1] an undisputed advantage is the ability of the research scientist to obtain direct sequence information from as little as one copy of target DNA.[2,3] This was considered an impossible task before the advent of PCR.

The major advantage of sequencing genomic or complex mixtures of DNAs directly using PCR, versus sequencing after amplification and cloning, is that only a single sequence need be determined. The data obtained will represent an average of the sequences of the target DNAs in solution. Alternatively, when "PCR clones" are sequenced, a minimum of four or five different data sets should be determined to distinguish the

[1] P. Karlovsky, *TIBS* **15**, 419 (1990).
[2] K. Mullis and F. A. Faloona, this series, Vol. 155, p. 335.
[3] H. Li, U. B. Gyllensten, X. Cui, R. K. Saiki, H. A. Erlich, and N. Arnheim, *Nature* (*London*) **335**, 414 (1988).

positions of point mutations[4–8] as well as to determine if recombinant "phase disruption"[9] had occurred during cycling. (Several reports have linked the production of PCR-mediated recombinants,[10] also referred to as "shuffle clones"[11] or "phase disruptions,"[9] to the production of prematurely terminated PCR products[12].)

To date a large number of different PCR sequencing methodologies are available that claim to be "direct." However, almost without exception these methods require gel purification, additional "nested primers," column chromatography, asymmetric amplification reactions, precipitations, and/or removal of excess primers and nucleotides. In contrast, the sequencing method that we have developed is a direct and novel approach that avoids all of these time-consuming manipulations.

Overview

In 1988 Nakamaye *et al.*[13] developed a reliable PCR sequencing method by the chemical degradation of phosphorothioate-containing PCR fragments. This method required four amplification reactions, each with one dNTP replaced by the appropriate dATPαS analog. The DNA fragments were specifically cleaved at the positions of the phosphorothioate groups using iodoethanol or epoxipropanol. However, the PCR fragments first required gel purification prior to this chemical cleavage step.[14]

In an attempt to understand the cause of interference with direct sequencing of crude PCR products, a sample was analyzed by polyacrylamide gel electrophoresis (PAGE) following amplification. We found that the expected product was accompanied by a large number of shorter fragments even though further analysis by agarose gel electrophoresis had revealed a single band.[12] The yield of these prematurely terminated PCR products was found to be dependent on the number of cycles. Presumably, it is the presence of these smaller fragments that also interferes with typical

[4] S. Pääbo and A. C. Wilson, *Nature (London)* **334**, 387 (1988).

[5] P. Keohavong, A. G. Kat, N. F. Cariello, and W. G. Thilly, *DNA* **7**, 63 (1988).

[6] A. Belyavsky, T. Vinogradova, and K. Rajewsky, *Nucleic Acids Res.* **8**, 2919 (1989).

[7] K. R. Tindall and T. A. Kunkel, *Biochemistry* **27**, 6008 (1988).

[8] K. A. Eckert and T. A. Kunkel, *Nucleic Acids Res.* **18**, 3739 (1990).

[9] R. Jansen and F. D. Ledley, *Nucleic Acids Res.* **18**, 5153 (1990).

[10] A. R. Shuldiner, A. Nirula, and J. Roth, *Nucleic Acids Res.* **17**, 4409 (1989).

[11] U. Gyllensten, *in* "PCR Technology: Principles and Applications for DNA Amplification" (H. A. Erlich, ed.), Chap. 5. Stockton Press, New York, 1989.

[12] D. B. Olsen and F. Eckstein, *Nucleic Acids Res.* **17**, 9613 (1989).

[13] K. L. Nakamaye, G. Gish, F. Eckstein, and H. P. Vosberg, *Nucleic Acids Res.* **16**, 9947 (1988).

[14] G. Gish and F. Eckstein, *Science* **240**, 1520 (1988).

dideoxy (dd) sequencing methods. Further experiments showed that a similar distribution and intensity of these prematurely terminated PCR products after DNA amplification using dNTPαS residues in the polymerization mix. We reasoned that these multiple-length fragments could yield sequencing information if they could be base-specifically digested in a 3' → 5' direction to the next phosphorothioate group (see Scheme I). Two exonucleases, snake venom phosphodiesterase and exonuclease III, were tried. These enzymes are 3' → 5' specific and cleave phosphorothioate internucleotidic linkages over 100 times more slowly than normal phosphate linkages.[15–18] Both enzymes were able to utilize the complex mixture of substrates and yielded the desired sequence information. Therefore, the unwanted by-products that disrupt normal PCR sequencing analysis become the basis for the phosphorothioate-based PCR sequencing protocol. Figure 1 shows the results from the direct sequencing of a small quantity of serum-isolated hepatitis B genomic DNA.

Experimental

Materials

Sep-Pak C_{18} columns are from Waters Associates (Millipore, Milford, MA). The thermocycler used for DNA amplification is supplied by Perkin-Elmer Cetus (Norwalk, CT). Amplification reaction tubes (0.5 ml) are from Sarstedt (Germany).

Enzymes

Snake venom phosphodiesterase (*Crotalus durissus,* 1 mg/0.5 ml, 1.5 U/mg) is supplied by Boehringer Mannheim (Mannheim, Germany) and *Taq* polymerase is purchased from Amersham (Amersham, England) (4.2 U/μl) or Perkin-Elmer Cetus (5 units/μl). Exonuclease III (100 units/μl) is obtained from New England BioLabs (Beverly, MA) and polynucleotide kinase (30 units/μl) is from United States Biochemicals (Cleveland, OH).

Nucleotides and Oligonucleotides

The *Sp*-diastereomer of dNTPαS analogs are either prepared as described[19] or purchased from Amersham and dNTPs are from Boehringer

[15] P. M. J. Burgers and F. Eckstein, *Biochemistry* **18,** 592 (1979).
[16] A. P. Gupta, P. A. Benkovic, and S. J. Benkovic, *Nucleic Acids Res.* **12,** 5897 (1984).
[17] S. Labeit, H. Lehrach, and R. S. Goody, *DNA* **5,** 173 (1986).
[18] S. Labeit, H. Lehrach, and R. S. Goody, this series, Vol. 155, p. 166.
[19] J. Ludwig and F. Eckstein, *J. Org. Chem.* **54,** 631 (1989).

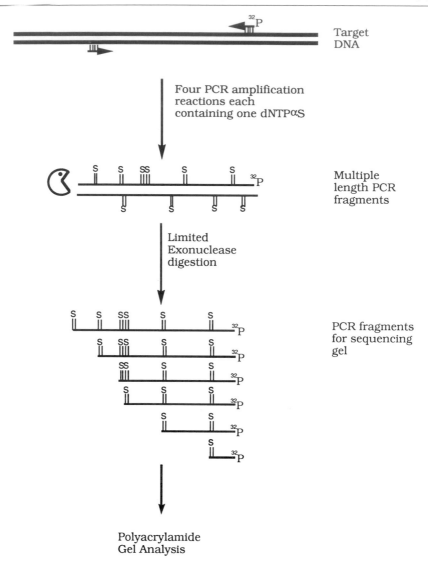

SCHEME 1. Schematic representation of the Olsen and Eckstein[12] phosphorothioate-based PCR sequencing method.

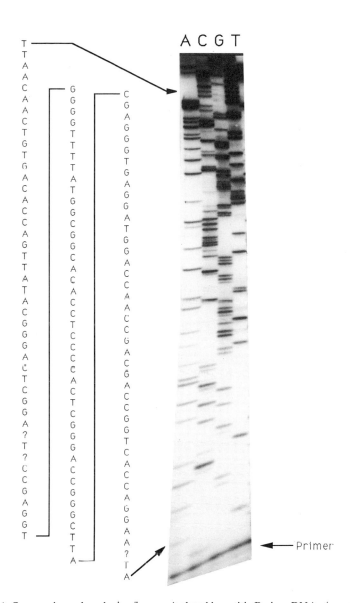

FIG. 1. Sequencing gel analysis of serum-isolated hepatitis B virus DNA. Approximately 50,000 copies of native hepatitis B virus genomic DNA isolated using the procedure of Kaneko *et al.*[31] from a chronically infected person were amplified using 40 cycles of PCR. The buffer conditions and the cycling program are described under Methods. Amplification reactions were performed with 50% dNTPαS mixes and digested with exonuclease III as described. The primers used in these experiments had the sequence 5'-AATCCAGATTGG-GACTTCAAC-3' (annealing from nucleotide 2991 to 2971 in the published HBV sequence) and 5'-CCGATTGGCGGAGGCAGGAGGA-3' (annealing from nucleotide 3150 to 3129). The question marks found in the sequence to the left represent blanks on the gel corresponding to deoxycytosine residues as confirmed by sequencing the other strand.

Mannheim. Radioactively labeled [γ-^{32}P]ATP (10 mCi/ml in water) is obtained from Amersham.

Oligonucleotide primers are prepared by the phosphoramidite method using an Applied Biosystems (Foster City, CA) 380B DNA synthesizer[20] and are purified by ethanol precipitation. Priming sites are checked by sequence alignment to all published hepatitis B virus sequences, using the computer program Microgenie.[21]

Reagents

End-labeling buffer (5×): 20 μl 100 mM MgCl$_2$, 20 μl 500 mM Tris-HCl (pH 8), 20 μl 100 mM 2-mercaptoethanol

Oligonucleotide Sep-Pak Purification Solutions

Solution A: 2% aqueous acetonitrile/0.1 M triethylammonium bicarbonate

Solution B: 40% aqueous acetonitrile/0.1 M triethylammonium bicarbonate

Amplification buffer (5×): 12 μl 100 mM MgCl$_2$, 8 μl 500 mM Tris-HCl (minimum pH, 8.5[8,22]), 10 μl 1% (v/v) Nonidet P-40 (Sigma, St. Louis, MO), 10 μl 1% (v/v) Tween 20 (Bio-Rad, Richmond, CA), 40 μl water

Preparation of dNTPαS Mixes and dNTP4 Mix

Stock solutions	dATPαS mix (μl)	dCTPαS mix (μl)	dGTPαS mix (μl)	dTTPαS mix (μl)	dNTP4 mix (μl)
5 mM dATP	5	10	10	10	10
5 mM dATPαS	5	—	—	—	—
5 mM dCTP	10	5	10	10	10
5 mM dCTPαS	—	5	—	—	—
5 mM dGTP	10	10	5	10	10
5 mM dGTPαS	—	—	5	—	—
5 mM dTTP	10	10	10	5	10
5 mM dTTPαS	—	—	—	5	—

[20] T. Brown and D. J. S. Brown, *in* "Oligonucleotides and Their Analogues" (F. Eckstein, ed.), p. 1. IRL Press, Oxford, 1991.

[21] C. Queen and L. J. Korn, *Nucleic Acids Res.* **12**, 581 (1984).

[22] U. Linz, U. Delling, and H. Ruebsamen-Waigmann, *J. Clin. Chem. Clin. Biochem.* **28**, 5 (1990).

Methods

Determination of Optimal Polymerase Chain Reaction Conditions

In most cases the researcher will have already established PCR conditions that yield reasonable quantities of a desired product. In this case, continue with the section on the incorporation of phosphorothioate residues (below).

Choice of Primer Pairs: Specificity and Compatibility

For the novice, the most important consideration for the successful amplification of DNA is the choice of the oligonucleotides to be used in the reaction. Ideally, the primers should anneal within 400 bp of each other on the target DNA/RNA, be approximately 18–22 nucleotides in length, and have a G : C to-A : T base content ratio of about 1 : 1. It is of the utmost importance that the oligonucleotides do not have 3' ends that are self-complementary. This is because the primers can be elongated by the polymerase using another oligonucleotide as a template (because they are in such high concentration during an amplification reaction this is not a minor reaction), thus destroying their specificity for the intended target sequence. Several computer programs are available that check for inter- and intramolecular secondary structures for a given set of PCR primers.[23,24]

Trial Polymerase Chain Reactions

It is usually necessary to manipulate the PCR conditions to optimize the yield of product. It is beyond the scope of this chapter to discuss the large number of variables that can influence an amplification reaction. Therefore, we refer the reader to books on PCR or to two publications[22,23] to obtain more detailed information. We have supplied conditions that typically yield significant amounts of PCR products. These should be taken as the starting points for the optimization of a standard reaction using a dNTP4 mix described under Reagents. [Note: We have determined that the presence of KCl disrupts the exonuclease activity required in the "clean-up" step (Fig. 2). Therefore, use of this salt should be avoided if at all possible.]

Incorporation of Phosphorothioate Residues

After conditions have been established for the PCR reaction we recommend that identical conditions be employed for the production of the

[23] D. Y. Wu, Ugozzoli, B. K. Pal, J. Qian, and R. B. Wallace, *DNA Cell Biol.* **10**, 233 (1991).
[24] W. Rychlik and R. E. Rhoads, *Nucleic Acids Res.* **17**, 8543 (1989).

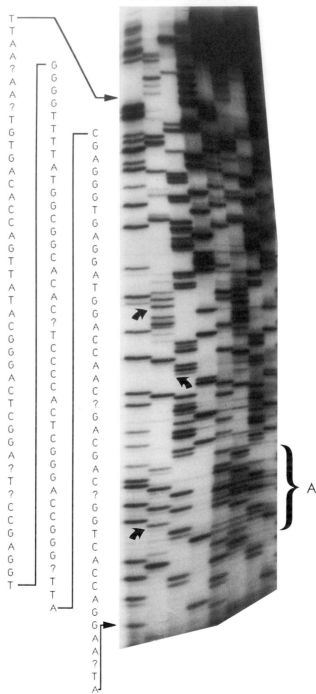

phosphorothioate-containing PCR fragments. Four reactions should be set up, each with a different dNTPαS mix described above.

Amplification Reactions

Add the following reagents to four 500-µl microcentrifuge tubes labeled with the letters A, C, G, and T:

Amplification buffer (5×)	10 µl
Primer A (~5 pmol) (stock of A_{260} = 0.1 OD/ml)	5 µl
Primer B (~5 pmol) (stock of A_{260} = 0.1 OD/ml)	5 µl
Plasmid DNA to be amplified (or 50,000 copies of viral DNA as in the experiments described in this chapter; the number of copies of DNA to be amplified will vary depending on the source)	5 ng
Water	to 45 µl (total volume)

Anneal the primers to the target DNA by incubation of each tube at 96° for 5 min and then cool to 37°. This step is important because it denatures the DNA and, in addition, the high temperature destroys any contaminating enzyme activity that would interfere with amplification.

To start the polymerization reactions, add 5 µl of each dATPαS, dCTPαS, dGTPαS, dTTPαS mix to the appropriate tubes (i.e., dATPαS to be labeled A, etc.) and 0.5 µl *Taq* polymerase (~2 units) for a total volume of 50.5 µl.

Vortex the tube and spin briefly in a microcentrifuge. Then overlay each reaction with 70 µl paraffin oil (for reproducible results add the same amount of oil using a pipettor[25]).

[25] L. M. Mazei, *Amplifications—Forum PCR Users (Perkin–Elmer Cetus)* **11** (1990).

FIG. 2. Comparison of the sequences obtained using PCR fragments prepared by polymerization with 25% dNTPαS mixes with (four lanes to the right) and without salt. The details are the same as for Fig. 1 except that 25% dNTPαS mixes were used in the amplification reactions and one of the sets (right) was carried out using 50 m*M* KCl. Arrows indicate some of the positions where deoxycytosine bands are weak or missing. In one area of the gel (bracket A) the sequence data are ambiguous due to the effect of salt on the exonuclease digestion reaction.

Typical Polymerase Chain Reaction Thermal Cycling Program

The PCR program given below yields generous amounts of PCR product using the Perkin-Elmer Cetus heating block described under Materials. However, a different set of cycling parameters is required when using a thermal cycler (programmable Dri-Block PHC-1; Techne, Princeton, NJ). Therefore, it seems that there is a lack of reproducibility from one PCR machine to another.[26,27] There is also concern about temperature variability from one sample position to the next within the same heating block.[26,27] It is therefore important to be consistent. Optimization of a reaction should be performed on a single instrument using the same sample wells each time and other variables (such as the source of dNTPs as well as *Taq* polymerase) should be carefully controlled.

Thermal cycling conditions (30–40 cycles)
    Polymerization: 1.0 min at 72°
    Denaturation: 15 sec at 96°
    Annealing: 15 sec at 55°

After the last cycle incubate at 71° for 10 min. This program has a short cycle time and is used successfully in combination with the thermal cycler supplied by Perkin-Elmer Cetus.

The following program is employed for amplification using the Techne heating block mentioned above.

Polymerization: 1 min at 72°
Denaturation: 1 min at 95°
Annealing: 1 min at 55°

After cycling is complete, a small (10 μl) sample can be taken from the reaction and checked for the production of a fragment with the desired length by 2% agarose gel electrophoresis.[28] A suitable DNA length standard, such as a *Hae*III digest of ϕX174 (New England BioLabs) or a *Dde*I digest of pUC19, should accompany every analysis of PCR products.

If all four reactions yield significant and equivalent amounts of product then the PCR reactions should be repeated using a ^{32}P end-labeled primer. The polynucleotide kinase oligonucleotide labeling procedure and subsequent purification steps are given below. Obviously, both oligonucleotides can be labeled at the same time, but in separate reactions, so that both strands of the DNA can be sequenced.

[26] R. Hoelzel, *TIG* **6,** 237 (1990).
[27] U. Linz, *BioTechniques* **9,** 286 (1990).
[28] J. Sambrook, E. F. Fritsch, and T. Maniatis, "Molecular Cloning: A Laboratory Manual," 2nd Ed., p. 6.3. Cold Spring Harbor Press, Cold Spring Harbor, New York, 1989.

End Labeling with [γ-³²P]ATP and Purification of the Oligonucleotide

Established safety guidelines should always be followed when working with radioactive materials.
To a 1.5-ml microcentrifuge tube add the following:

End-labeling buffer (5×):	7 µl
[γ-³²P]ATP (10 µCi/µl in water)	10 µl
Polynucleotide kinase (30 units/µl)	1 µl
Sequencing grade oligonucleotide	~5 pmol
Water	to 35 µl (total volume)

Incubate at 37° for 30 min, and heat inactivate at 70° for 15 min.

Sep-Pak Purification of End-Labeled Oligonucleotide

Wash Sep Pak C_{18} cartridge using a 20-ml syringe with

Acetonitrile	20 ml
Double-distilled sterile water	20 ml
Solution B	10 ml
Solution A	20 ml

Carefully apply the radioactive sample from the end-labeling reaction, using a pipette. Wash with 30 ml solution A and direct the flow-through into a radioactive waste container (*Note*: This initial wash contains a large amount of unreacted radiolabeled ATP and should be handled carefully.) Place 20 sterile microcentrifuge tubes into an empty rack. Elute the oligonucleotide from the Sep Pak with solution A (~3 ml) and carefully collect 4-drop aliquots in each microcentrifuge tube. Pool the peak of radioactivity into an appropriately labeled 1.5-ml microcentrifuge tube (simply measure ³²P using a Geiger counter) and Speed-Vac to dryness. Add 100 µl water, vortex and quick freeze with liquid nitrogen, or use immediately.

Exonuclease-Mediated Sequencing Reactions

After the PCR cycling is complete, 5-µl aliquots from the aqueous bottom layer of each tube are placed into 1.5-ml microcentrifuge tubes labeled A, C, G, and T and stored on ice. It is not necessary to remove the oil overlay before taking the 5-µl samples. The digestion reactions are initiated by the addition of 1 µl of exonuclease III [previously diluted to 10 units/µl using the recommended dilution buffer (see Reagents)]. The tube is vortexed briefly, spun in a microcentrifuge for 3 min at 10,000 rpm at room temperature, and then placed into a 16° water bath for 30 min.

The reaction is stopped by the addition of 2 μl stop mix (see Reagents) and heated to 80–95° for 3 min before being placed on ice.

The samples are ready for loading onto a 6% polyacrylamide sequencing gel.[29] If desired, 5-μl samples can be taken from the PCR reactions and loaded without exonuclease treatment. This will provide a control for the amplification reactions as well as the exonuclease digestions. Figure 1 shows a typical result of the phosphorothioate-based PCR sequencing method of a 179-bp region amplified from hepatitis B genomic DNA isolated from blood serum.

Reproducibility

To a large degree the lack of reproducibility of a PCR experiment is the result of inconsistency of the PCR user. We have already indicated some of the variables that can adversely affect an amplification reaction, including oligonucleotide sequence, PCR machine, amount of oil overlay, pH of the reaction, as well as the supplier of the polymerase and dNTPs. It is also worthwhile to point out that the magnesium concentration can be an important variable. It is important to keep to a minimum the components of the reaction that can effectively decrease the amount of Mg^{2+} available to the polymerase. These variables include concentration of the oligonucleotides, dNTPs, target DNA, and the use of ethylenediaminetetraacetic acid (EDTA) in any of the stock solutions. In addition, Table I lists other troubleshooting hints.

Factors Influencing Choice of Exonuclease

The method that we have described would not yield sequence data if the exonuclease were to dissociate from the DNA at random positions before gapping to the next phosphorothioate group. This problem has been encountered occasionally when we tried to determine the DNA sequence that was very close to the 5' end of the PCR fragments. To obtain the sequence information in this region, the DNA must still be a substrate for the exonuclease. However, as the enzyme gaps very close to the 5' end of the DNA fragment (near the end-labeled oligonucleotide), the exonuclease dissociates from the substrate more frequently.[30] When using exonuclease III the presence of salt such as KCl decreases the processivity of the

---

[29] J. Sambrook, E. F. Fritsch, and T. Maniatis, "Molecular Cloning: A Laboratory Manual," 2nd Ed., p. 11.23. Cold Spring Harbor Press, Cold Spring Harbor, New York, 1989.

[30] S. G. Rogers and B. Weiss, this series, Vol. 65, p. 201. (1980).

TABLE I

TROUBLESHOOTING GUIDE FOR SEQUENCING PHOSPHOROTHIOATE-CONTAINING
POLYMERASE CHAIN REACTION FRAGMENTS

Problem	Possible cause	Remedy
No sequencing bands in one lane	Overdigestion	Use less exonuclease
	No amplification; Check for product by PAGE before digestion	Repeat PCR reaction with new dNTPαS mix
Sequence ambiguities		
Bands in more than one lane	^{32}P-Labeled primer has multiple annealing sites; KCl in PCR?	Increase annealing temperature or choose new primer sequence; Remove KCl
Weak band intensities at specific positions (C lane)	Over digestion by exonuclease	Increase percentage of dNTPαS for PCR or end label other primer and sequence other strand
More than one PCR band obtained after amplification	Low specificity of primers or heterogeneous target DNA sequence	Increase annealing temperature; or each band can be isolated by PAGE and sequenced individually according to Ref. 13

enzyme even further,[30] which leads to ghost bands on the sequencing gel as seen in Fig. 2 (brace A).

Another problem was encountered with the snake venom phosphodiesterase, which is dependent on the degree of phosphorothioate group substitution of the substrate PCR fragments. Initially, we reported that the best sequence information was obtained using snake venom phosphodiesterase as the exonuclease for the "clean-up" reaction.[12] The PCR fragments that were prepared for digestion in this case were synthesized by polymerization using 100% dNTPαS mixes (i.e., for the dGTPαs mix no dGTP was added). However, a problem of aberrant migration of the bands on the sequencing gel became noticeable when we tried to read beyond 150 bp. This migrational abnormality disappeared if the proportion of dNTPαS in the mix used to prepare the PCR fragments was reduced to 50% or below.

Interestingly, if the percentage of dNMPs incorporated into the PCR fragments was decreased below 100%, we found that we could not obtain a sequencing pattern by digestion with snake venom exonuclease. Fortunately, successful digestion of PCR fragments containing 50% phosphorothioate group substitution can be accomplished using exonuclease III (Fig.

1). This enzyme is able to yield sequencing patterns with DNA containing as little as 10% of the modified internucleotidic linkages. However, one of the difficulties that we encountered in using exonuclease III and very low percentage dNMPS-containing fragments is that some bands on the sequencing gel were very weak or missing (Fig. 2). These positions most often correspond to the position of a deoxycytosine. Missing bands are less of a problem if the PCR fragments are made with 50% dNTPαS mixes, as seen from the results in Fig. 1.[31] Sequence ambiguities can be verified simply by labeling the other primer and sequencing the other strand.

Scope of Procedure

The PCR sequencing procedure that we have described is attractive because of its simplicity, speed, and the relative ease with which the analysis can be carried out. Because all sample purification steps have been eliminated, this methodology should easily lend itself to automation. In addition, most automated DNA sequencing machines utilize primers that contain a fluorescent label on the 5' end. Therefore, nonspecific exonucleases could potentially be employed for the "clean-up" reaction in addition to those mentioned here.

Acknowledgment

We wish to thank Fritz Benseler for supplying the oligonucleotides, Annette Fahrenholz for expert technical assistance, and Dr. R. Mackin for critical reading of the manuscript.

[31] S. Kaneko, R. H. Miller, S. M. Feinstone, M. Unoura, K. Kobazashi, N. Hattori, and R. H. Purcell, *Proc. Natl. Acad. Sci. U.S.A.* **86,** 312 (1989).

[9] Fluorescent and Radioactive Solid-Phase Dideoxy Sequencing of Polymerase Chain Reaction Products in Microtiter Plates

By J. P. SCHOFIELD, D. S. C. JONES, and M. VAUDIN

Introduction

Following the introduction of the polymerase chain reaction (PCR)[1] several methods of directly sequencing the amplified products have been described. The fact that several have been proposed is indicative of the difficulty involved in designing a universally applicable technique. Each has its own particular limitations, usually with respect to the quality of sequence, the length of the read, and the maximum PCR product length to be sequenced. For example, asymmetric PCR[2] is widely used on templates less than 500 bp in length, but several workers have found poor results from longer templates, as well as variability in sequencing both strands of the amplified product. In our experience direct sequencing of PCR products bound to a solid support is rapid, reliable, and provides high-quality sequencing results on short (<500 bp) and longer (>1.5 kb) templates.

Principles of Solid-Phase Sequencing

Solid-phase techniques play an important role in DNA synthesis, peptide synthesis, and sequencing. A single strand of DNA, suitable as a template for dideoxy sequencing, can be immobilized on a solid support prior to sequencing.[3] This immobilization can now be readily achieved by using streptavidin-coated magnetic beads and terminally biotinylated DNA.[4] This facilitates template purification, and is convenient for rapid removal of the complementary strand, PCR primers, and excess nucleotides. During target amplification one of the DNA strands incorporates a 5'-biotinylated primer (Fig. 1). Following PCR, streptavidin beads are added directly to the reaction tube, and DNA is allowed to bind via the biotin group. Alkali is used to denature the double-stranded product, and the nonbiotinylated strand removed, together with the excess primers,

---

[1] R. K. Saiki, S. Scharf, F. Faloona, K. B. Mullis, G. T. Horn, H. A. Erlich, and N. Arnheim, *Science* **230**, 1350 (1985).

[2] U. B. Gyllensten and H. A. Erlich, *Proc. Natl. Acad. Sci. U.S.A.* **85**, 7652 (1988).

[3] A. Rosenthal, R. Jung, and H.-D. Hunger, this series, Vol. 155, p. 301.

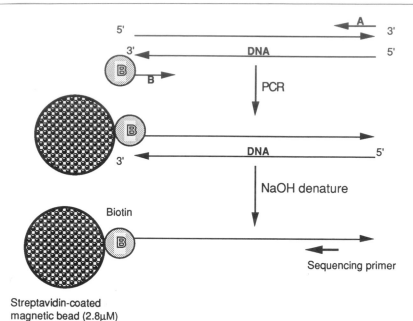

FIG. 1. Polymerase chain reaction and solid-phase sequencing. Either of the primer pairs (A or B) may be biotinylated prior to PCR. Following PCR, the biotinylated strands are bound to the streptavidin-coated beads and then rendered single-stranded by alkali denaturation and washing of the beads. Optimum sequence information is obtained by using a sequencing primer internal to the original nonbiotinylated PCR primer. Oligo walking is easily applicable, the only limitation being the length of the amplified PCR product.

deoxynucleotides, and buffer by aspiration from the pelleted magnetic beads. (The biotin–streptavidin bond is stable in dilute alkali). The single-stranded DNA remains attached to the beads and is a suitable template for dideoxy sequencing. A sequencing primer is annealed to the DNA template, and chain termination proceeds toward the solid support. We have modified the original method of Hultman *et al.*[4] by reducing the amplification volume, performing PCR directly from bacterial colonies or phage plaques, and using [α-^{35}S]dATP instead of an end-labeled primer. Hultman has described solid-phase sequencing using a fluorescently labeled primer and the Pharmacia (Piscataway, NJ) ALF (Automated Laser Fluorescence). Here we describe the use of fluorescent dideoxy terminators in a simple one-tube reaction in conjunction with the Du Pont (Wilmington, DE) Genesis 2000,[5] and dye-labeled primers with an Applied

[4] T. Hultman, S. Stahl, E. Hornes, and M. Uhlen, *Nucleic Acids Res.* **17,** 4937 (1989).
[5] D. S. C. Jones, J. P. Schofield, and M. Vaudin, *DNA Sequence* **1,** 279 (1991).

Biosystems (Foster City, CA) 373A DNA sequencer. We have also scaled up the number of templates that can be simultaneously handled by performing all reactions in a microtiter plate format together with a robotic workstation and purpose-built magnetic shaker/separator device.[6]

Materials

The polymerase chain reaction is performed in thermostable polycarbonate plates (Hi-temp 96; Techne, Inc., Princeton, NJ) on a purpose-built Dri-plate thermocycler (MW-1; Techne). All reagents are of the highest quality molecular biology grade. *Taq* polymerase (5 units/μl) is purchased from Perkin-Elmer Cetus (Norwalk, CT). Sterile neutralized deoxynucleotide triphosphates are from Pharmacia, and light mineral oil is from Sigma (St. Louis, MO). Oligonucleotide primers are synthesized on an Applied Biosystems 380B oligonucleotide synthesizer. Streptavidin-coated magnetic beads are Dynabead-M280 (Dynal, Oslo, Norway). A DNA sequencing reagent kit, including Sequenase 2.0 enzyme, is from U.S. Biochemical (Cleveland, OH). Fluorescent dye-terminator sequencing reagents are supplied by Du Pont and fluorescently labeled primers are from Applied Biosystems. Automated DNA sequencing is performed on the Genesis 2000 (Du Pont) or the Applied Biosystems 373A as appropriate. The magnetic shaker–separator was developed with Techne (UK) Ltd. (Cambridge, England) (patent pending).

Methods

Oligonucleotide Primers

All oligonucleotide primers are synthesized using an Applied Biosystems 380B DNA synthesizer, according to the instructions of the manufacturer. Following cleavage from the column support and deprotection at 55°, overnight in NH$_4$OH, the oligonucleotides are precipitated by the addition of 0.1 vol of 3 M potassium acetate, pH 4.8, and 3 vol of ethanol, vortexed, then placed at −70° for 10 min. Following microcentrifugation for 10 min at room temperature (13,000 rpm), the pellets are drained and washed in 2 ml of 80% (v/v) ethanol, dried, and resuspended in a total of 100 μl of TE (10 mM Tris-HCl, pH 7.4, 1 mM EDTA). This procedure is rapid and convenient and has been used for oligonucleotides as short as 11-mers.

Biotinylation is readily achieved using a two-step method. The oligonucleotide is synthesized with an amine group at its 5' terminus using Amino-

[6] J. P. Schofield, M. Vaudin, S. Kettle, and D. S. C. Jones, *Nucleic Acids Res.* **17,** 9498 (1989).

link II (Applied Biosystems) according to the recommendations of the manufacturer and ethanol precipitated (*vide supra*). A portion (15 μl) of this material is resuspended in 55 μl of sterile water and the following reagents added in order: 10 μl 1 M sodium bicarbonate buffer, pH 9.0, and 20 μl 100 mM biotin N-hydroxysuccinimide ester (Sigma) in dry dimethylformamide (DMF). (The quality of the DMF used here is of great importance.) This is incubated overnight and the oligonucleotide recovered by ethanol precipitation. The biotinylated oligonucleotide must then be purified free of unincorporated biotin and nonbiotinylated oligonucleotide. This can be achieved most effectively by reversed-phase high-performance liquid chromatography (HPLC) or preparative polyacrylamide gel electrophoresis. Using either of these methods the biotinylated oligonucleotide is retarded in comparison to the parent compound, a portion of which should be analyzed in parallel. To purify the biotinylated oligonucleotide by preparative polyacrylamide gel electrophoresis a 1 mm $\times$ 20 cm $\times$ 40 cm gel is suitable. A 15% polyacrylamide gel (19 : 1, acrylamide : bisacrylamide) in 1$\times$ TBE (89 mM Tris-borate, 89 mM boric acid, 2 mM EDTA, pH 8.3 at room temperature) *without* urea is made with a 1-cm toothed comb. The gel is prerun at 65 W constant power for 30 min. The precipitated oligonucleotide is resuspended in 20 μl of 50% (v/v) formamide and loaded into one well. In an adjacent lane formamide containing bromphenol blue and xylene cyanol is loaded to monitor the gel run. The gel is run at 30 W constant power until the bromphenol blue is approximately 15 cm from the base. The oligonucleotide is visualized by ultraviolet (UV) shadowing. [The gel is placed on a piece of Saran wrap on top of a thin-layer chromatography (TLC) plate and illuminated from above with long-wavelength UV light. The oligonucleotide casts a shadow on the fluorescing plate.] The biotinylated oligonucleotide is excised from the gel and eluted in sterile H$_2$O by incubation at 65° for 3–12 hr twice. To maximize DNA recovery the exposure of the gel to UV light should be minimized. The pooled eluates are filtered (0.2-μm pore size) and the oligonucleotide concentration determined spectrophotometrically (OD$_{260\ nm}$).

Polymerase Chain Reaction

A variety of DNA templates can be used for amplification. If phage plaques or bacterial colonies are used, they are lightly "stabbed" with a sterile toothpick and then stirred into the prealiquotted PCR mixture. Frozen bacterial cultures stored in 5% (v/v) glycerol in a microtiter plate can be transferred directly using a 96-pin replica-plating device. All these templates use 20-μl PCR volumes with 35 cycles of amplification to generate sufficient template for sequencing. However, if diluted (1 : 100) minipreparation DNA is to be amplified, then 20 cycles are performed. Amplification is performed in the following mixture: 10 mM Tris-HCl, pH 8.3, at 25°, 1.5 mM MgCl$_2$, 100 μM dNTP, 0.5 μM forward primer, 0.5 μM 5'-biotinylated reverse primer, 0.25 units Taq polymerase and then overlaid with light mineral oil. Cycles routinely used are 95° for 90 sec; (95°, 30 sec; 55°, 30 sec; 72°, 30 sec) for 35 cycles; 72°, 3 min. Following all amplifications, one-tenth of the reaction is checked for purity and yield of DNA by agarose gel electrophoresis before proceeding to direct sequencing.

The method described can also be adapted for oligo walking along an insert. A recombinant pUC19 plasmid minipreparation containing a 2.4-kb insert was amplified with M13 universal primers for 20 cycles using a cycling profile similar to that above but with a 72° extension segment of 1 min. Due to the length of the template and the need to perform many separate sequencing reactions, four 50-μl reactions were used to achieve sufficient template. A single product was amplified and the reaction divided into seven 20-μl aliquots in microtiter wells prior to the addition of beads. The template produced was then sequenced with seven walking primers (Fig. 2).

Radioactive DNA Sequencing

For each amplified template, 30 μl of washed (in TE/0.1 M NaCl) beads is added directly to the reaction tube or microplate well (including the oil overlay).

The biotinylated PCR products are allowed to bind to the streptavidin-coated beads by incubation at room temperature for 10 min with gentle stirring. Beads are pelleted toward the side of the well (via a magnet), and the aqueous phase removed, taking care not to aspirate any beads. The nonbiotinylated strand of the PCR product attached to the beads is removed by resuspension in 20 μl of 0.15 M NaOH and incubation for 5 min at room temperature. Again the beads are pelleted, the supernatant removed, and the beads washed in 20 μl 0.15 M NaOH, followed by three washes with 40 μl of distilled H$_2$O. This effectively removes excess primers, buffer, and dNTPs prior to sequencing.

The beads with attached single-stranded template are resuspended in a final volume of 7.5 μl distilled H$_2$O, 2 μl 5× Sequenase reaction buffer (200 mM Tris-HCl, pH 7.5, 100 mM MgCl$_2$, 250 mM NaCl), and 3 pmol of sequencing primer (e.g., the nonbiotinylated PCR primer). Primer annealing is carried out by heating the reaction to 65° for 2 min and then placing on ice. The labeling reaction is initiated by the addition of 5.5 μl labeling mix comprising 1 μl 0.1 M dithiothreitol (DTT), 2 μl of a 1:5 dilution labeling mix (1.5 μM concentrations of dTTP, dCTP, and dGTP),

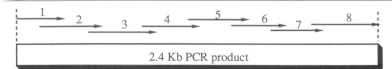

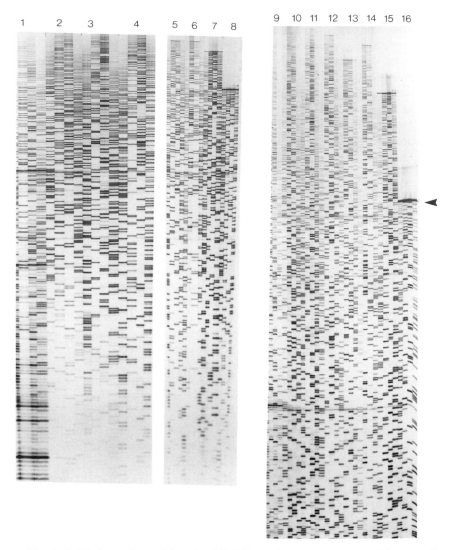

FIG. 2. Solid-phase oligo walking. A 1.4-kb PCR product was sequenced. This autoradiograph illustrates the typically inferior quality obtained when using the nonbiotinylated PCR primer as a sequencing primer (walk 1, lanes 1 and 9). The remaining tracks were oligo walks

0.5 μl [α-^{35}S]dATP (1000 Ci/mmol), and 2 μl of a 1 : 8 dilution of Sequenase 2.0 enzyme (U.S. Biochemical). During the 5-min labeling reaction at room temperature the mixture, including beads, is equally divided (3.5 μl) between four microplate wells labeled T, C, G, and A. Onto the lip of each of these wells 2.5 μl of appropriate dideoxy (dd) chain terminator mix (80 μM dNTPs, 8 μM ddNTP, 50 mM NaCl) is carefully dispensed. At the end of 5 min, the microplate is carefully placed in a microplate centrifuge, and the termination reaction simultaneously begun by a brief pulse centrifugation (<1000 rpm, room temperature, 5 sec). The beads pellet by centrifugation and are quickly resuspended by agitation prior to incubating at 37° for 5 min. Finally, the reaction is stopped by the addition of 4 μl of dye/formamide/ethylenediaminetetraacetic acid (EDTA) to each well [95% (v/v) formamide, 20 mM EDTA, 0.05% (w/v) bromphenol blue, 0.05% (w/v) xylene cyanol FF].

Immediately prior to loading onto a denaturing 6% (w/v) polyacrylamide sequencing gel, DNA is denatured by heating the samples at 80° for 10 min in the uncovered microplate, then placing the samples on ice. Approximately 2–3 μl of sample is loaded per gel track, and it does not matter if beads are loaded as well, as they remain trapped in the well and do not interfere with the subsequent gel running. The gel is run at constant power, fixed in 10% acetic acid/10% methanol, dried down, and autoradiographed for 2–14 hr at room temperature.

Fluorescent DNA Sequencing with Dye-Labeled Terminators

When using fluorescently labeled dideoxynucleoside triphosphate, the growing DNA strand is simultaneously labeled and specifically terminated in a single step. Consequently this method is simpler to perform than the radioactive protocol.

The initial denaturation and preparation of single-stranded DNA bound to the beads is performed exactly as for the radioactive method described above. Following washing, the beads are resuspended in a final volume of 22.5 μl by the addition of 12 μl of double-distilled water (e.g., Milli-Q; Millipore, Bedford, MA), 6 μl of Sequenase buffer, and 3 pmol of sequencing primer. The primer is annealed at 37° for 10 min and then placed on ice. There is no separate labeling step. The sample is simultaneously

---

from internal primers. Lanes 2–8 and 9–16 are the short and long sequencing ladders, respectively, obtained by double-loading of a 6% polyacrylamide gel. The quality of these ladders is high, with little background and few compressions. The final walk (lane 16) ends in a stop in all 4 lanes as the bead is reached (arrowhead).

labeled and terminated at 37° for 5 min by addition of 2.5 μl 0.1 M DTT, 3 μl dNTPs (75 μM each), 1 μl fluorescent ddNTPs (125 μM each), and 1 μl (3 units) of Sequenase 2.0 enzyme. The reaction is stopped by the addition of 40 μl water, and the beads pelleted with a magnetic source. Excess fluorescent dideoxynucleotides are removed by a further three washes with 40 μl water, before resuspending the beads in 3 μl formamide dye mix (95% formamide, 20 mM EDTA, pH 8.0, 2 mg/ml Crystal Violet). The labeled DNA is denatured and eluted from the immobilized template by heating at 80° for 10 min immediately prior to loading onto a polyacrylamide gel placed in the Genesis 2000 sequencing machine. Each of the four fluorescent dideoxynucleotides has a distinct emission spectrum following excitation from an argon laser-generated light source. This is detected by a pair of photomultiplier tubes within the sequencing machine, and the output is acquired and base-called by a dedicated Apple Macintosh microcomputer. This explains the requirement for a one-tube reaction, and for one-lane sample gel loading.

A size-fractionated library (0–2 kb) made from human genomic DNA digested with HpaII has been cloned into pBluescribe plasmid. Following storage of colonies as glycerol stocks in microtiter plates, individual clones have been amplified by PCR and sequenced using the Genesis 2000 fluorescent method. About 200 kb of sequence has been generated, a testimony to the reproducibility of the method (Fig. 3A).[7] The average sequence data for these templates was ~300 bp with 98% accuracy.

Fluorescent Sequencing with Dye-Labeled Primers

An alternative to using dye-labeled dideoxynucleotide triphosphates for fluorescent DNA sequencing is to attach a fluorescent dye directly to the sequencing primer. This is the approach used by the Pharmacia ALF DNA sequencer, which uses a single dye (and therefore requires four separate lanes per clone for electrophoresis). The Applied Biosystems 373A sequencer utilizes four different dyes, which allows electrophoresis in a single lane, although like the ALF sequencer the four sequencing reactions must be performed separately. Here we present a protocol using a solid-phase single-stranded DNA template in conjunction with the Applied Biosystems 373A using four dye-labeled primers and Sequenase enzyme, according to the instructions of the manufacturer.

The template is prepared as already described and resuspended in 16 μl of H_2O, prior to mixing with dye–primer and buffer as shown in the tabulation below. The primer in this instance was M13 forward labeled

---

[7] M. Vaudin, personal communication.

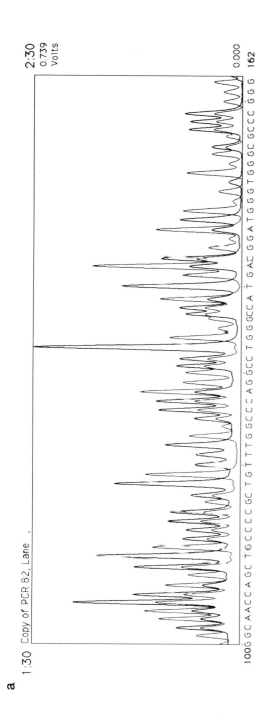

2:30
0.739
Volts

0.000
162

100 G C A A C C A G C T G C C C C G C T G T T T G G C C C A G G C C T G G G C C A T G A C G G A T G G G T G G G C G C C C G G G

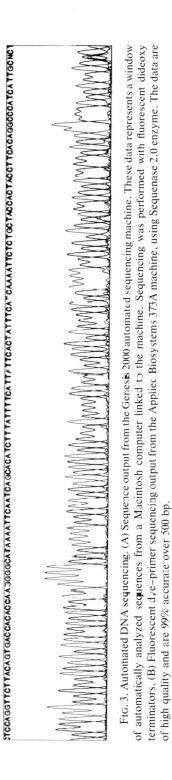

FIG. 3. Automated DNA sequencing. (A) Sequence output from the Genesis 2000 automated sequencing machine. These data represents a window of automatically analyzed sequences from a Macintosh computer linked to the machine. Sequencing was performed with fluorescent dideoxy terminators. (B) Fluorescent dye-primer sequencing output from the Applied Biosystems 373A machine, using Sequenase 2.0 enzyme. The data are of high quality and are 99% accurate over 500 bp.

Component	A (μl)	C (μl)	G (μl)	T (μl)
Template	2	2	6	6
Buffer (5×)	1	1	3	3
Joe	0.5			
Fam		0.5		
Tamra			1	
Rox				1
H$_2$O	0.5	0.5	1	1
Total	4	4	11	11

with each of the four dyes: JOE, FAM, TAMRA, and ROX (supplied by Applied Biosystems). Annealing is carried out by heating briefly to 70° (2 min) and cooling on the bench for 3 min. Nucleotides (both dNTP and the appropriate ddNTP) and diluted Sequenase are then added as follows:

Component	A (μl)	C (μl)	G (μl)	T (μl)
ddNTP/dNTP	1	1	2	2
Diluted enzyme	1	1	2	2

The reactions are incubated at 37° for 30 min and stopped by pooling into 50 mM EDTA. The beads with the labeled product still attached are recovered using a magnet, and resuspended in 5 μl of 80% formamide. The fluorescently labeled DNA is eluted from the beads by heating at 95° for 2 min, chilled on ice, and then loaded into a single lane of the Applied Biosystems 373A DNA sequencer. Data acquisition and base calling are all performed automatically by a Macintosh computer. As can be seen in Fig. 3B this protocol yields high-quality DNA sequence information with few ambiguities (~99% accuracy over 500 bp).

Conclusions

The reactions presented here are easy to perform and are reproducible, using various different sources of DNA template. Both short and long PCR products can be readily sequenced and oligo walking along the solid-phase bound template has been demonstrated.[8] A further advantage of solid-phase sequencing is the high quality of the sequence data produced.

The main limitation was the limited availability of a wide selection of 5'-

[8] G. Elgar and J. P. Schofield, *J. DNA Sequencing Mapp.* **2,** 219 (1992).

biotinylated primers. Several companies have now responded by offering biotinylated phosphoramidites for automated synthesis; this resource, together with custom synthesis services, should greatly alleviate this problem. We and others have noted the generally poor quality of sequence obtained when using a primer at the very 3' end of a template produced by PCR [e.g., if the original PCR primer is used as the sequencing primer (Fig. 2, lane 1)]. This is easily overcome by using a specific sequencing primer just internal to the 3' end (Fig. 1). Hultman et al.[9] have sequenced the eluted strand as well as the immobilized strand, but the sequence quality is inferior.

It quickly became apparent during the development of solid-phase sequencing that automation would facilitate the simultaneous handling of numerous templates. The use of a robot to dispense PCR reaction mix into a microtiter plate was effective, but required expensive equipment and was time consuming to set up and maintain. Also, contamination must be avoided at all costs during PCR, and this was a major concern when using the robotic workstation. We have therefore abandoned the robot, and have concentrated on the repetitive steps of bead washing and magnetic pelleting. A simple portable unit has been developed to facilitate all the mixing and washing steps as well as the magnetic separation in a microtiter plate format. Following PCR amplification, the microtiter plate is placed on the miniworkstation (~15 × 30 cm footprint) and the majority of the sequencing, including template denaturation and bead preparation, performed in situ. This is a great advantage over performing manipulations using hand-held or plate magnets.

In summary, we have developed a robust technique for directly sequencing terminally biotinylated PCR products by selective immobilization to streptavidin-coated magnetic beads. This technique can be applied with different labeling approaches (either [α-35S]dATP, fluorescently labeled dideoxynucleotide triphosphates, or fluorescently labeled sequencing primers). It has been used to generate DNA sequences amounting in excess of 200 kb in total. The method is applicable to short and long templates, and semiautomation of the technique has been described.

---

[9] T. Hultman, S. Bergh, T. Moks, and M. Uhlen, *BioTechniques* **10**, 84 (1991).

[10] Automated Fluorescent DNA Sequencing of Polymerase Chain Reaction Products

By Zijin Du, Leroy Hood and Richard K. Wilson

In designing strategies for both large- and small-scale DNA sequence analysis projects, some use of the polymerase chain reaction[1,2] (PCR) may be considered. Polymerase chain reaction-based approaches can be used to prepare template DNA for direct sequence analysis, facilitate the amplification and cloning of regions which are underrepresented in recombinant DNA libraries as well as other regions of interest, and provide a means of checking the assembly and linearity of the final nucleotide sequence. An extraordinary number of DNA sequencing methods utilizing templates produced by PCR have been described.[3-22] Although some of these methods have appeared promising and useful in some laboratories, they often have been difficult to reproduce in others. Accordingly, the large number of PCR-related sequencing manuscripts that have appeared in the literature is a testament to the problems associated with this general approach. In

[1] R. K. Saiki, S. Scharf, F. Faloona, K. B. Mullis, G. T. Horn, H. A. Erlich, and N. Arnheim, *Science* **230**, 1350 (1985).
[2] K. B. Mullis and F. Faloona, this series, Vol. 155, p. 335 (1987).
[3] U. B. Gyllensten and H. A. Erlich, *Proc. Natl. Acad. Sci. U.S.A.* **85**, 7652 (1988).
[4] R. K. Wilson, C. Chen, and L. Hood, *BioTechniques* **8**, 184 (1990).
[5] R. A. Gibbs, P. N. Nguyen, L. J. McBride, S. M. Koepf, and C. T. Caskey, *Proc. Natl. Acad. Sci. U.S.A.* **86**, 1919 (1989).
[6] U. B. Gyllensten, *BioTechniques* **7**, 700 (1989).
[7] R. Higuchi and H. Ochman, *Nucleic Acids Res.* **17**, 5865 (1989).
[8] T. Hultman, S. Stahl, E. Hornes, and M. Uhlen, *Nucleic Acids Res.* **17**, 4937 (1989).
[9] E. Hornes, T. Hultman, T. Moks, and M. Uhlen, *BioTechniques* **9**, 730 (1990).
[10] L. G. Mitchell and C. R. Merril, *Anal. Biochem.* **178**, 239 (1989).
[11] G. Ruano and K. K. Kidd, *Proc. Natl. Acad. Sci. U.S.A.* **88**, 2815 (1991).
[12] K. A. Kretz, G. S. Carson, and J. S. O'Brien, *Nucleic Acids Res.* **17**, 5864 (1989).
[13] K. B. Gorman and R. A. Steinberg, *BioTechniques* **7**, 326 (1989).
[14] P. R. Winship, *Nucleic Acids Res.* **17**, 1266 (1989).
[15] D. P. Smith, E. M. Johnstone, S. P. Little, and H. M. Hsiung, *BioTechniques* **9**, 48 (1990).
[16] N. Kusukawa, T. T. Vemori, K. Asada, and I. Kato, *BioTechniques* **9**, 66 (1990).
[17] S. Gal and B. Hohn, *Nucleic Acids Res.* **18**, 1076 (1990).
[18] V. Murray, *Nucleic Acids Res.* **17**, 8889 (1989).
[19] J. L. Casanova, C. Pannetier, C. Javlin, and P. Kourilisky, *Nucleic Acids Res.* **18**, 4028 (1990).
[20] D. B. Olsen and F. Eckstein, *Nucleic Acids Res.* **17**, 9613 (1989).
[21] M. Mihovilovic and J. E. Lee, *BioTechniques* **7**, 14 (1989).
[22] M. W. Berchtold, *Nucleic Acids Res.* **17**, 453 (1989).

METHODS IN ENZYMOLOGY, VOL. 218

this chapter, we describe four methods for using PCR templates to perform automated fluorescent DNA sequence analysis. Asymmetric PCR (APCR) is a method that has worked well for some investigators,[3-5] although others have experienced difficulties in obtaining satisfactory results and maintaining acceptable reproducibility from template to template. Because our experience with APCR and fluorescent DNA sequencing has produced both good and bad results using two different cloning systems, we shall describe the methods we have employed and discuss some of the pitfalls of the technique. We also describe two useful methods for direct sequencing of the products of symmetric PCR, including purification of double-stranded template by polyethylene glycol (PEG) precipitation and enzymatic conversion to single-stranded template (outlined in Fig. 1). We also provide solutions to some of the common problems of directly sequencing PCR products.

Coupled Polymerase Chain Reaction and DNA Sequencing Methods

The advent of PCR has yielded new approaches to molecular characterization of a particular region, gene, or single base pair within a complex genome.[23] For DNA sequence analysis, PCR amplification can be used to prepare templates from purified genomic DNA, a single bacteriophage λ or M13 plaque, a bacterial colony, tissue culture cells, blood, or individual sperm cells. Many investigators have developed methods for direct DNA sequencing of PCR-produced templates. Some of the methods have utilized unpurified PCR product; many have utilized material purified by agarose gel electrophoresis,[12-14] and eluted by phenol extraction, glass affinity, and electroelution, or used directly within molten agarose. Other methods for DNA sequencing of PCR products have employed ultrafiltration,[3] affinity or ion-exchange chromatography, high-performance liquid chromatography (HPLC), enzymatic removal of one DNA strand,[7] and biotin-mediated strand separation and purification.[8-10] Another alternative, APCR, results in the production of an excess of one of the two template strands.[3-6] In addition to the main problem of unreliability, there are specific disadvantages associated with each of these methods. For example, only a small number of PCR templates can be reasonably purified using preparative agarose gel electrophoresis or HPLC. Depending on the actual technique used, such methods can provide DNA of suitable quality that is free from deoxynucleotides and amplification primers, and should consistently provide sufficient DNA for the linear amplification sequencing method described below. However, if a large number of samples must be

[23] H. A. Erlich, D. Gelfand, and J. J. Sninsky, *Science* **252**, 1643 (1991).

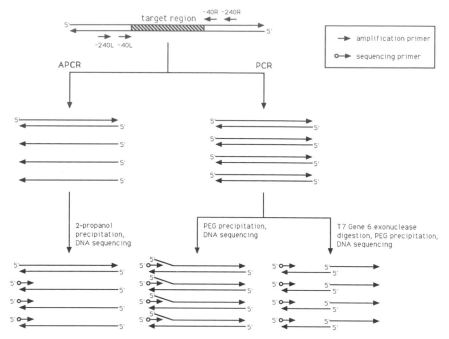

FIG. 1. Methods for coupled PCR amplification and DNA sequence analysis. In this chapter, four methods are described. The first two methods use APCR to produce an excess of single-stranded DNA from the target region, while the third and fourth methods utilize a standard PCR amplification. For M13 subclones, the −240 amplification primers are utilized; for fragments cloned in the pUC family of vectors, −40 amplification primers are used. In the third method, double-stranded PCR product is purified by PEG precipitation. In the fourth method, double-stranded PCR product is converted to single-stranded DNA with T7 gene 6 exonuclease. DNA produced from all four methods can be sequenced using *Taq* DNA polymerase with a linear amplification sequencing method, while that produced using the APCR and T7 gene 6 exonuclease methods can be sequenced using the modified T7 DNA polymerase.

sequenced, preparative agarose gel electrophoresis is not an attractive method for template purification. Likewise, purification methods that utilize HPLC or disposable columns or cartridges are time consuming as well as expensive when applied to a large number of PCR samples.

APCR produces single-stranded template that is much more suitable for DNA sequencing using standard dideoxynucleotide methods than templates produced by symmetric PCR (see Fig. 1). Because APCR amplification does not proceed exponentially, the yield of template is significantly lower than with standard PCR.[3] This reduced yield has often hampered efforts to implement APCR methods in conjunction with automated fluorescent DNA sequencing. However, problems of insufficient template yield usually can be overcome by slight adjustments to the thermal cycling program. When sufficient DNA is produced by APCR, the single-stranded template may be sequenced directly or following a simple 2-propanol precipitation step. Thus, for automated DNA sequencing, APCR amplification can be an inexpensive and high-throughput method for template production.

Direct sequencing of unpurified PCR products would seem to offer the most rapid and inexpensive approach; however, the quality of the DNA sequence data typically produced is significantly lower than that of other methods. These problems are due mainly to the difficulties of sequencing linear double-stranded DNA using the standard dideoxynucleotide method.[24] The most successful procedures for sequencing the products of standard PCR have utilized physical or enzymatic methods of DNA strand separation followed by sequencing of the resulting single-stranded template.[7-10] In one physical approach, a biotin molecule is attached to one of the amplification primers.[8-10] After PCR, the biotin-labeled double-stranded DNA is bound to streptavidin-coated beads (paramagnetic or agarose). The beads are treated with alkali to denature the double-stranded DNA, washed to remove the unlabeled DNA strand, and captured and used as single-stranded template in a solid-phase DNA sequencing reaction. The unlabeled strand also may be collected by ethanol purification and used for sequencing. This method has worked well with small fragments (1–2 kbp), but the streptavidin-coated beads can be expensive and a pure preparation of biotin-labeled primer is required for effective recovery of single-stranded template. Another approach utilizes the selective removal of one strand of PCR products by bacteriophage λ exonuclease.[7] In this technique, one of the amplification primers is phosphorylated prior to PCR. After amplification, the phosphorylated strand is degraded to mononucleotides by the exonuclease. A second enzymatic method, described below, utilizes T7 gene 6 exonuclease to degrade PCR products into single-stranded DNA. However, unlike λ exonuclease, T7 gene 6 exonuclease does not require 5' phosphorylation and allows DNA sequencing to proceed from either end of the PCR products (see Fig. 1). A similar method using T7 gene 6 exonuclease to produce single-stranded template DNA from restriction fragments has been described.[25]

A variety of methods have been utilized for DNA sequence analysis of PCR and APCR products.[3-22] Most commonly, PCR products have been sequenced using the chemically modified T7 DNA polymerase (Sequenase;

[24] E. Y. Chen and P. H. Seeburg, *DNA* **4**, 165 (1985).
[25] N. A. Strauss and R. J. Zagursky, *BioTechniques* **10**, 376 (1991).

U.S. Biochemical, Cleveland, OH) and [α-^{35}S]dATP. This DNA sequencing chemistry is available in kit form and works well with single-stranded templates.[26] For sequence analysis of double-stranded PCR products, the use of an end-labeled sequencing primer is necessary, and the template first must be denatured by boiling or alkaline treatment. Alternatively, double-stranded template may be sequenced using *Taq* DNA polymerase and primer extension temperatures of 70–80°.[27] This sequencing chemistry also is available in kit form, although the quality of the resulting data generally is not as good as that produced with the T7 DNA polymerase. A significant improvement in data quality can be produced when *Taq* DNA polymerase is utilized with the linear amplification sequencing method.[28] Here, a small amount of single- or double-stranded template DNA is used in a single-primer amplification reaction that contains one of the four dideoxynucleotides. As with PCR, the reaction temperature is automatically cycled to allow DNA denaturation, primer–template annealing, and primer extension. At the conclusion of thermal cycling, sufficient DNA sequencing reaction products are typically formed and can be analyzed by gel electrophoresis and autoradiography or fluorescence detection.

The DNA sequencing approaches described in this chapter utilize the Sanger dideoxynucleotide method[29,30] with four-color fluorescent dye–primer chemistry[31] and subsequent analysis on a model 373A automated DNA sequencer (Applied Biosystems, Inc., Foster City, CA). This approach has the advantages of sequencing with labeled primers to minimize artifacts, high-throughput processing of multiple samples, real-time nonradioactive detection, and automatic sequence reading. Use of the linear amplification sequencing method solves many of the problems routinely experienced with automated fluorescent DNA sequencing. Most notably, less template DNA is required for the sequencing reactions, and problematic data analysis due to a low signal-to-noise ratio is minimized.[28] Additionally, preliminary data indicate that single-stranded templates produced using either the APCR or T7 gene 6 exonuclease methods can be readily sequenced using modified T7 DNA polymerase and a set of

[26] S. Tabor and C. C. Richardson, *Proc. Natl. Acad. Sci. U.S.A.* **84**, 4767 (1987).
[27] M. D. Brow, *in* "PCR Protocols: A Guide to Methods and Applications" (M. A. Innis, D. H. Gelfand, J. J. Sninsky, and T. J. White, eds.), p. 189. Academic Press, San Diego, 1990.
[28] M. Craxton, *Methods: Companion Methods Enzymol.* **3**, 20 (1991).
[29] F. Sanger, S. Nicklen, and A. R. Coulson, *Proc. Natl. Acad. Sci. U.S.A.* **74**, 5463 (1977).
[30] F. Sanger, A. Coulson, B. Barrell, A. Smith, and B. Roe, *J. Mol. Biol.* **143**, 161 (1980).
[31] L. M. Smith, J. Z. Sanders, R. J. Kaiser, P. Hughes, C. Dodd, C. R. Connell, C. Heiner, S. B. H. Kent, and L. E. Hood, *Nature (London)* **321**, 674 (1986).

fluorescent dye labeled-dideoxynucleotides.[32] Because each of the four dideoxynucleotides is labeled with a different fluorescent dye, this chemistry has the additional advantages of single-tube reactions and the ability to use any sequencing primer.

Asymmetric Polymerase Chain Reaction Methods

The first APCR procedure described below can be used to prepare sequencing templates directly from M13 plaques. This procedure has been used to sequence a 9.6-kbp region of the murine T cell receptor α/δ chain locus.[33] There is no need to pretreat the phage with phenol or detergent because sufficient release of phage DNA occurs during the thermal cycling program. Sequencing artifacts are minimized by use of a pair of amplification primers that leave at least 200 bp between the 3' terminus of the template DNA and the site of sequencing primer annealing.[4] Following PCR, the bulk of unincorporated primers and deoxynucleotides is removed by a room-temperature 2-propanol precipitation. Fluorescent DNA sequence analysis may be performed using either the modified T7 DNA polymerase or linear amplification sequencing methods (described below). Furthermore, because the intermediate product of APCR is double-stranded DNA, either template strand can be produced in excess and used in the sequencing reaction.

The second APCR method described below was developed specifically for use with recombinant plasmid subclones. In our laboratory, random flush-ended DNA fragments are subcloned into the HincII (SalI) site of pUC118 or 119.[34] For DNA sequence analysis of the resulting recombinant subclones, the M13 universal primer, which anneals adjacent to the cloning site, may be used. Similar to the first APCR method, amplification is performed directly from bacterial colonies without prior lysis of the cells. For the pUC-based subclones, we use amplification primers that anneal at sites approximately 40 bp from either side of the multiple cloning site. Templates produced using this APCR procedure have been directly sequenced using the linear amplification sequencing method without the 2-propanol precipitation step described for the first procedure.

Standard Polymerase Chain Reaction Methods

Two approaches for direct sequencing of symmetric PCR products using plasmids, bacteriophage λ clones, and genomic DNA as starting

[32] R. K. Wilson, unpublished data, 1991.

[33] R. K. Wilson, B. F. Koop, C. Chen, N. Halloran, R. Sciammis, and L. Hood, *Genomics* **13**, 1198 (1992).

[34] J. Vieira and J. Messing, this series, Vol. 153, p. 3 (1987).

material are described below. Both of these procedures have worked well in our experience. In the first procedure, PCR-amplified DNA is purified by precipitation with PEG and used as template for linear amplification sequencing reactions. Polyethylene glycol precipitation appears to yield cleaner sequencing template than alcohol precipitation, presumably because low molecular weight DNA fragments, primers, primer–dimers, and mononucleotides are not precipitated efficiently.[35] In the second procedure, the DNA produced by PCR is converted to single-stranded template using T7 gene 6 exonuclease. This method is required if T7 DNA polymerase chemistry is to be used to sequence a standard PCR product.

Materials and Reagents

Several instruments are currently available for automated thermal cycling. All of the methods described in this chapter were developed using either a DNA thermal cycler or a 9600 thermal cycler, both manufactured by Perkin-Elmer Cetus (Norwalk, CT). If other thermal cycling equipment is used, amplification reaction conditions should be carefully optimized. Oligonucleotide primers were synthesized on an Applied Biosystems 380B DNA synthesizer. Following cleavage from the synthesis column, oligonucleotides were deprotected by addition of 1 ml of fresh concentrated NH$_4$OH and incubation at 55° for 4–16 hr. Oligonucleotides were evaporated to dryness under vacuum, dissolved in TE buffer [10 mM Tris-HCl (pH 8.0), 0.1 mM ethylenediaminetetraacetic acid (EDTA)], and quantitated by ultraviolet spectrophotometry. *Thermus aquaticus* DNA polymerase was purchased from Perkin-Elmer Cetus. Best results were obtained using either native *Taq* DNA polymerase, or the cloned AmpliTaq preparation. Deoxy- and dideoxynucleotide triphosphates for PCR and DNA sequencing were purchased as fast protein liquid chromatography (FPLC)-purified solutions from Pharmacia-LKB (Piscataway, NJ) T7 gene 6 exonuclease is available from U.S. Biochemical.

All of the sequencing methodologies described have been developed for use with an Applied Biosystems 373A automated DNA sequencer. Communication of results from other investigators indicates that these methods also are applicable to DNA sequencing methods utilizing ^{32}P-labeled primers. Fluorescent dye-labeled primers used for sequence analysis include a −21 M13 universal primer (5'-TGT-AAA-ACG-ACG-GCC-AGT-3'), a −32 M13 reverse primer (5'-CAG-GAA-ACA-GCT-ATG-ACC-3'), and an SP6 promoter-specific primer (5'-ATT-TAG-GTG-ACA-CTA-TAG-3'). All of these sequencing primers are available from Applied

[35] K. R. Paithankar and K. S. N. Prasad, *Nucleic Acids Res.* **19,** 1346 (1991).

Biosystems. Modified T7 DNA polymerase (Sequenase) is available from U.S. Biochemical.

Asymmetric Polymerase Chain Reaction and DNA Sequencing of Recombinant M13 Clones

The following protocol has been described previously for amplification and DNA sequence analysis of fragments cloned in the M13 family of vectors.[4] Resulting templates have been sequenced using the modified T7 DNA polymerase chemistry, although they also should be suitable for the linear amplification sequencing method described below.

Reagents

PCR buffer (10×): 500 mM KCl, 100 mM Tris-HCl (pH 8.3), 15 mM MgCl$_2$, 0.1% (w/v) gelatin (store at 4°)

dNTP mix (1.25 mM): 12.5 μl each of 100 mM dATP, dCTP, dGTP, and dTTP in 10 mM Tris-HCl (pH 8.0), 0.1 mM EDTA to a final volume of 1.0 ml (store at −20°)

M13 amplification primers: −240L (universal) primer: 5'-GGA-CGA-CGA-CCG-TAT-CGG-3'; −240R (reverse) primer: 5'-GAA-AAA-CCA-CCC-TGG-CGC-3' (prepare 10 and 0.2 μM dilutions, store at −20°)

Hind/DTT (sequencing) buffer (5×): 50 mM Tris-HCl (pH 7.5), 300 mM NaCl, 5 mM dithiothreitol (DTT)

MnCl$_2$ (1 M): Dissolve 20.33 g in 100 ml distilled water (immediately before sequencing, add 3 μl of 1 M MnCl$_2$ to 197 μl of 5× Hind/DTT buffer)

dNTP mix (8 mM): 100 μl each of 100 mM dATP, dCTP, dGTP, and dTTP in 10 mM Tris-HCl (pH 8.0), 0.1 mM EDTA to a final volume of 1.25 ml (store at −20°)

ddXTP mixes (50 μM): ddA: 20 μl of 5 mM ddATP, 180 μl of 10 mM Tris-HCl (pH 8.0), 0.1 mM EDTA; ddC: 20 μl of 5 mM dCTP, 180 μl of 10 mM Tris-HCl (pH 8.0), 0.1 mM EDTA; ddG: 40 μl of 5 mM dGTP, 360 μl of 10 mM Tris-HCl (pH 8.0), 0.1 mM EDTA; ddT: 40 μl of 5 mM dTTP, 360 μl of 10 mM Tris-HCl (pH 8.0), 0.1 mM EDTA (store all four mixes at −20°)

8 mM dNTPs + 50 μM ddXTP mixes: Combine equal volumes of the 8 mM dNTP mix and one of the four 50 μM ddXTP mixes

mT7 DNA polymerase: Stock enzyme is 12.5 units/μl, dilute to 1.5 units/μl with 10 mM Tris-HCl (pH 8.0), 0.1 mM EDTA immediately before using

loading solution: Mix 5 vol deionized formamide with 1 vol 0.05 *M* EDTA

Amplification Procedure

M13 bacteriophage plaques are cored from an agar plate using the small end of a sterile Pasteur pipette, and the plug is transferred to a 1.5-ml microcentrifuge tube containing 50 μl of sterile water. Each sample is vortexed for a few seconds to release the phage particles. These phage stocks may be stored at 4° for several days. For sequencing reactions using the M13 universal primer, 10 μM of -240R primer and 0.2 μM of -240L primer should be used in the amplification reaction. If reverse primer is to be used for sequencing, 10 μM of -240L and 0.2 μM of -240R primers should be used for amplification. For ease of dealing with multiple samples, all of the reagents except the DNA template may be combined and added to reaction tubes in a single pipetting step. The APCR amplification should be set up as follows:

Distilled water	59 μl
PCR buffer (10×)	10 μl
dNTP mix (1.25 m*M*)	16 μl
Primer 1 (10 μM)	5 μl
Primer 2 (0.2 μM)	5 μl
Phage stock (template DNA)	5 μl
Taq DNA polymerase (5 units/μl)	0.2 μl
Total volume	100 μl

The reaction is mixed gently and overlaid with 100 μl of light mineral oil. Thermal cycling is performed for 35 cycles (a typical cycle for the Perkin-Elmer Cetus DNA thermal cycler is 30 sec at 94°, 30 sec at 55°, 90 sec at 72°, with the minimum time interval between temperature segments). At the conclusion of thermal cycling, samples should be maintained at 4°. The reaction mixture is carefully removed from under the mineral oil and transferred to a clean 1.5-ml microcentrifuge tube. Two microliters may be analyzed by agarose gel electrophoresis. The remainder of the reaction mixture is precipitated by the addition of 10 μl of 3 *M* sodium acetate (pH 5.2) and 100 μl of 2-propanol. The samples are mixed briefly and allowed to stand at room temperature for at least 30 min. The DNA is pelleted by centrifugation at 13,000 *g* for 15 min at room temperature. The DNA pellet is washed once with 350 μl of 70% (v/v) ethanol (room temperature), and dried for a few minutes under vacuum. The DNA then is dissolved in 25 μl of TE buffer and stored at 4°.

DNA Sequencing Procedure

In this procedure, reactions are performed using the modified T7 DNA polymerase chemistry[26,36] as adapted for sequencing with fluorescent dye–primers.[4] The reactions are conveniently performed in 96-well U-bottom plates (Falcon No. 3911; Becton Dickinson, Oxnard, CA), either by hand or using an automated pipetting station.[37] Because the fluorescent dyes used in the G and T reactions produce a weaker signal, the reaction volumes are doubled.

Sequencing annealing reactions should be set up as follows:

	A (μl)	C (μl)	G (μl)	T (μl)
Hind/DTT (5×) (+Mn; make fresh)	1	1	2	2
Template DNA (from above APCR)	3	3	6	6
Dye–primer	1	1	2	2

The reactions are heated to 55° for 3 to 5 min, then cooled slowly to room temperature for 10–15 min. If the reactions are performed in a 96-well plate, a dry block heater may be modified to heat all 96 wells effectively.[38,39] To each annealing reaction is added:

	A (μl)	C (μl)	G (μl)	T (μl)
8 mM dNTPs + 50 μM ddXTP mix	2	2	4	4
Modified T7 DNA polymerase (1.5 units/μl)	1.5	1.5	3	3

The reactions are incubated at 37° for 5 to 10 min. At the conclusion of extension and termination, the four reactions must be stopped before they may be combined and concentrated for electrophoresis. A simple method for stopping the reactions is to place 6 μl of 5 M ammonium acetate (pH 7.4) and 120 μl of 95% ethanol in 1.5-ml microcentrifuge tubes (one tube for each template). For each template set, 8 μl of the A reaction is transferred to the tube and two quick cycles of up-and down pipetting are used to rinse the tip and mix the sample with the ethanol solution. Without changing the pipette tip, 8 μl of the C reaction and 16 μl of the G and T reactions are transferred to the tube, with up-and-down pipetting following each addition. The ethanol solution effectively stops further enzymatic activity. This procedure is repeated for each template set. The combined

[36] S. Tabor and C. C. Richardson, *Proc. Natl. Acad. Sci. U.S.A.* **86**, 4076 (1989).
[37] R. K. Wilson, C. Chen, N. Avdalovic, J. Burns, and L. Hood, *Genomics* **6**, 626 (1990).
[38] B. F. Koop, R. K. Wilson, C. Chen, N. Halloran, R. Sciammis, and L. Hood, *BioTechniques* **9**, 32 (1990).
[39] D. Seto, *Nucleic Acids Res.* **19**, 2506 (1991).

reactions are cooled at −70° for 15 min to precipitate the DNA products. The DNA is pelleted by centrifugation at 13,000 g for 15 min at room temperature, washed once with 300 μl of 70% ethanol (room temperature), and dried briefly under vacuum. The dried sample may be stored at −20° for several days. Immediately before the samples are loaded on the sequencing gel, the DNA pellet is dissolved in 4 μl of the formamide–EDTA loading solution, heated at 100° for 3 to 5 min, and placed on ice. The entire sample is loaded into single wells on the Applied Biosystems instrument.

Asymmetric Polymerase Chain Reaction and DNA Sequencing of Recombinant Plasmid Clones

For APCR production of DNA cloned in the pUC family of vectors, a slightly different protocol is used. Here, we use amplification primers that more closely flank the insert DNA and a modified thermal cycling program that ensures template production from subclones containing inserts as large as 4000 bp. In addition, we describe the linear amplification sequencing procedure that can be used for sequencing APCR-produced templates from either M13 or pUC subclones.

Reagents

pUC amplification primers: −40L (universal) primer: 5′-GTT-TTC-CCA-GTC-ACG-AC-3′; −40R (reverse) primer: 5′-GGA-TAA-CAA-TTT-CAC-ACA-GG-3′ (prepare 10 and 0.2 μM dilutions, store at −20°)

LASR buffer (5×): 400 mM Tris-HCl (pH 8.9), 100 mM $(NH_4)_2SO_4$, 25 mM $MgCl_2$

Amplification Procedure

Using a sterile toothpick, single recombinant colonies are picked from agar plates and resuspended in 5 μl of sterile water. The APCR amplification is set up as follows:

Distilled water	59 μl
PCR buffer (10×)	10 μl
dNTP mix (1.25 mM)	20 μl
Primer 1 (10 μM)	5 μl
Primer 2 (0.25 μM)	5 μl
Taq DNA polymerase (5 units/μl)	1 μl
Total volume	100 μl

The reaction is mixed gently and overlaid with 100 μl of light mineral oil. Thermal cycling is performed for 35 cycles (a typical cycle for the

Perkin-Elmer Cetus DNA thermal cycler is 40 sec at 94°, 60 sec at 55°, 4 min at 72°, with the minimum time interval between temperature segments). At the conclusion of thermal cycling, an additional 10-min incubation at 72° is performed, and the samples are then maintained at 4°. The reaction mixture is carefully removed from under the mineral oil and transferred to a clean 1.5-ml microcentrifuge tube. The samples may be stored at −20° or used directly for sequencing as follows.

DNA Sequencing Procedure

In this procedure, linear amplification sequencing reactions are performed.[28] Although the linear amplification method does not produce the even peak intensities characteristic of the modified T7 DNA polymerase, this procedure is less sensitive to template concentration and thereby results in a better overall success rate. Furthermore, this method typically produces better data than one-step extension sequencing protocols using *Taq* DNA polymerase. Here, fluorescent sequencing reactions are set up much the same as for PCR, either in 0.5-ml microcentrifuge tubes or in 0.2-ml MicroAmp tubes for the Perkin-Elmer Cetus 9600 thermal cycler. Sequencing reactions should contain the following components:

	A (μl)	C (μl)	G (μl)	T (μl)
LASR buffer (5×)	1.1	1.1	2.2	2.2
dNTP/ddXTP mix	1	1	2	2
Dye–primer (0.4 pmol/μl)	1	1	2	2
APCR product	1.5	1.5	3	3
Taq DNA polymerase (0.7 units/μl)	1	1	2	2
Total volume	5.6	5.6	11.2	11.2

The reactions are mixed gently and overlaid with 100 μl of light mineral oil (if necessary). Thermal cycling is performed as follows, depending on the type of instrument available:

DNA thermal cycler	9600 thermal cycler
For 15 cycles:	For 15 cycles:
95° for 30 sec	95° for 4 sec
55° for 30 sec	55° for 10 sec
70° for 60 sec	70° for 60 sec
For an additional 15 cycles:	For an additional 15 cycles:
95° for 30 sec	95° for 4 sec
70° for 60 sec	70° for 60 sec

Note: If the fluorescent dye-labeled SP6 sequencing primer is used, the annealing temperature should be reduced from 55 to 50°. Similarly, for other sequencing primers, the melting temperature (T_m) should be calculated, and the thermal cycling conditions adjusted accordingly.

At the conclusion of thermal cycling, the samples are maintained at 4° until ethanol precipitation. As described for the modified T7 DNA polymerase chemistry, the four sequencing reactions must be stopped before they may be combined and concentrated for electrophoresis. Again, a simple method for stopping the reactions is to transfer each sample to a 1.5-ml microcentrifuge tube containing the ammonium acetate/ethanol solution. After ethanol precipitation, the dried samples may be stored at −20° for several days. Sample preparation for electrophoresis on the Applied Biosystems sequencer should be performed as described for the modified T7 DNA polymerase sequencing method.

Standard Polymerase Chain Reaction and DNA Sequencing

Two strategies for direct sequencing of PCR products are provided. In the first, PCR-amplified DNA is precipitated with PEG. In the second, the DNA produced by PCR amplification is converted to single-stranded template using T7 gene 6 exonuclease and then precipitated with PEG. In our experience, sequencing template produced with the second strategy gives fewer artifacts and a better overall success rate than that with the first method.

Reagents

PEG-8000 (40%, w/v), 10 mM MgCl$_2$
Sodium acetate (3 M), (pH 4.8)
T7-G6 exo buffer (10×): 500 mM Tris-HCl (pH 8.1), 50 mM MgCl$_2$, 200 mM KCI, 50 mM 2-mercaptoethanol

Amplification Procedure

Polymerase chain reaction is performed with bacteriophage lysate, DNA (0.01–1.0 μg), or bacterial colonies or bacteriophage plaques that have been transferred to a small amount of water or TE buffer (as described above). The reactions should contain the following components:

Distilled water	x μl
PCR buffer (10×)	5 μl
dNTPs (1.25 mM)	8 μl
Primer 1 (20 μM)	2.5 μl
Primer 2 (20 μM)	2.5 μl
Template	y μl
Taq DNA polymerase (5 units/μl)	0.2 μl
Total volume	50 μl

If necessary, each reaction should be overlaid with 80 μl of light mineral oil. The thermal cycler is preheated to 95°, and the samples are placed in the preheated block and incubated at 95° for 2 min. Thermal cycling then is performed using the following parameters:

DNA thermal cycler	9600 thermal cycler
For 35 cycles:	For 35 cycles:
94° for 30 sec	92° for 10 sec
55° for 30 sec	55° for 60 sec
72° for 60 sec	72° for 60 sec

At the conclusion of thermal cycling, the samples are maintained at 4°.

Purification by Polyethylene Glycol Precipitation

The samples are removed from the reaction tubes and transferred to 1.5-ml microcentrifuge tubes containing 8 μl of 3 M sodium acetate (pH 4.8) and 20 μl of 40% (w/v) PEG-8000, 10 mM MgCl$_2$. If the samples were covered with mineral oil, it is important to avoid transferring any of the oil. The samples are mixed by vortexing and allowed to stand at room temperature for 10 min. The DNA is pelleted by centrifugation at 13,000 g for 15 min at room temperature. All of the supernatant is carefully and completely removed by aspiration. The DNA pellets are washed twice with 250 μl of 100% ethanol (room temperature) and dried briefly under vacuum. Each sample is dissolved in 20 μl of 10 mM Tris-HCl (pH 8.0), 0.1 mM EDTA. One or 2 μl of each sample may be analyzed on an agarose gel. For DNA sequence analysis using the linear amplification sequencing method described above, 1 μl of the PCR-prepared DNA is used in the A and C reactions and 2 μl in the G and T reactions.

Conversion to Single-Stranded Template with T7 Gene 6 Exonuclease

Add to each 50-μl PCR sample:

Distilled water	38.5 μl
T7-G6 exo buffer (10×)	10 μl
T7 gene 6 exonuclease (80 units/μl)	1.5 μl

The reactions are incubated at 37° for 30 min. Forty microliters of phenol : chloroform (1 : 1) is then added to each sample, and the mixtures are vortexed vigorously for a few seconds and centrifuged at 13,000 g for 5 min. The aqueous phases are transferred to 1.5-ml microcentrifuge tubes, and the single-stranded DNA is precipitated from the aqueous phase using the PEG method described above. DNA pellets are dissolved in 20 μl of 10 mM Tris-HCl (pH 8.0), 0.1 mM EDTA and may be used for either sequencing method described above.

Troubleshooting

Problems experienced with sequencing PCR products can be divided into two groups: (1) insufficient or multiple amplification products, and (2) insufficient or impure sequencing template. Here we describe some of the causes of these problems and offer some suggestions for solving them.

In troubleshooting problems with any of the methods for sequencing PCR products, the first step should be assessment of the quality of PCR amplification of the target DNA. Agarose or polyacrylamide gel electrophoresis should be employed to approximate DNA concentration and to detect the presence of extraneous fragments. Sufficient DNA should be produced by PCR to allow gel analysis using 1/25 of the total sample. If this is not possible, PCR conditions should be optimized to provide sufficient template. Alternatively, and certainly less desirable, the products of several reactions may be combined. To increase the yield of a PCR amplification, several different parameters may be altered, either singularly or in combination. The most common solution is simply to increase the number of cycles. We routinely utilize 35 cycles for amplification experiments that start with plaques, colonies, or lysates, although ample DNA may be produced using 25 to 30 cycles. Additionally, increasing the concentration of *Taq* DNA polymerase, deoxynucleotides, or amplification primers can help increase product yield. Also, increasing the incubation time at 72° may help to maximize yields, especially for larger products. Another parameter that may affect yield is magnesium concentration. In our standard PCR conditions, the final magnesium concentration is 1.5 mM, although a range of 0.5 to 4 mM can produce improved amplification for some primer–template combinations. The amplification primers also can be altered. In our experience, primers that are 17–22 bp in length and with a T_m of 52 to 56° usually work well and may be used with an annealing temperature of 55°. If the T_m of an amplification primer is lower than 52°, the annealing temperature of the PCR may be lowered. However, at temperatures below 50°, extraneous products will be more common. For some primer–template combinations, other changes in thermal cycling, such as denaturation temperature and time, may be necessary. Interestingly, we have observed significant differences in the yields of APCR products when identical experiments were performed on two different instruments from the same manufacturer. In this case, the production of sufficient template for fluorescent DNA sequencing by APCR necessitated adjustment of denaturation and annealing temperatures by a few degrees in one of the instruments. Therefore, regardless of the instrument used for PCR or APCR, cycle times and temperatures should be carefully optimized.

The second problem often observed with the PCR amplification is the presence of more than one product. This is especially troublesome when the same oligonucleotide primer is to be used for both PCR and DNA sequencing. Often, one or more extra fragments are produced when the stringency of primer annealing is low, allowing nonspecific hybridization of a primer(s) to nontarget regions. This problem can be solved by increasing the annealing temperature or by modifying the sequence of the amplification primer to increase specificity. An excessive number of cycles also can result in extraneous products, some of which may not be obvious on an agarose gel. Often, the DNA sequence data will contain a significant number of artifacts from a sample that appears homogeneous by agarose gel electrophoresis. Decreasing the number of amplification cycles often is a good first step toward solving this problem. Increasing the amount of target DNA in the PCR also may help reduce such nonspecific products, as can reducing certain reaction components such as *Taq* DNA polymerase, deoxynucleotides, or amplification primers. Obviously, in troublesome experiments, the concentrations of all components should be titrated carefully to maximize product yield while minimizing nonspecific amplification. In cases where more than one amplification product is present and cannot be eliminated by adjustment of the experimental conditions, the products must be purified by preparative gel electrophoresis or HPLC. Note that for APCR amplifications, agarose gel electrophoresis will show two bands; a double-stranded product and a faster migrating single-stranded product.

A major problem for fluorescent DNA sequencing of APCR product is insufficient quantity of single-stranded template DNA. This problem is especially acute when the modified T7 DNA polymerase is utilized. Because single-stranded template production is limited by linear amplification, less-than-optimal APCR conditions will not produce the relatively large amount of template required for sequencing with this enzyme. Such a problem usually can be solved by carefully adjusting amplification conditions to increase product yield. An alternate solution would be to switch to linear amplification sequencing reactions or to utilize the methods for sequencing symmetric PCR products.

For DNA sequence analysis of both APCR and PCR products, the presence of contaminants in the DNA template is a second major problem. Contaminants may include unincorporated deoxynucleotides or primers from the amplification reaction, extraneous DNAs produced in the amplification reaction, short amplification products that result when the *Taq* enzyme pauses or dissociates from the template strand, and agarose, salt, PEG, or other chemicals used for purification of the template DNA. Deoxynucleotides remaining from the PCR will cause a significant decrease in the signal of the smaller products of the sequencing reactions. Contaminating amplification primer can result in overall lower signal, especially if the amplification and sequencing primers share all or part of the annealing site. Both of these problems can be minimized by reducing the amounts present in the reaction. Alternatively, 2-propanol or PEG precipitation can be used to remove unincorporated deoxynucleotides and primers. Other purification methods such as preparative gel electrophoresis, HPLC, affinity chromatography, and ultrafiltration also will remove these contaminants. The template DNA itself can be a "contaminant" when used in excess. In this case, strong signals will be observed for the smaller products of the sequencing reaction, followed by a marked reduction in signal as dideoxynucleotides are consumed. Extraneous DNA products resulting from nonspecific annealing of amplification primers can be troublesome if the same primer(s) is used for DNA sequencing. Here, overlapping sequence patterns will be observed. This problem should be solved by adjusting amplification primers or conditions to minimize the production of nontarget DNAs. Alternatively, the use of a specific internal sequencing primer may correct the problem. Another common problem that has been observed with many sequencing methods is a "stop" or "cross-band," characterized by a large, four-color peak (or a dark band across all four lanes of autoradiograph). This type of artifact can be produced with any of the sequencing chemistries described here and most likely is the result of truncated amplification products or gaps and nicks in some fraction of the full-length product. Interestingly, such artifacts often seem to be location specific within a large number of different sequences, appearing at constant distances from the priming site. Many explanations have been offered for these artifacts; however, there is no universal cure. To minimize "stops," we have adjusted amplification primers (different annealing site or T_m), decreased the amount of Taq DNA polymerase in the PCR, and utilized fewer amplification cycles. These modifications have worked effectively for some primer–template combinations.

Summary

    The methods described in this chapter provide some useful approaches for DNA sequencing of templates produced by PCR. These procedures have been employed successfully for large-scale DNA sequencing of cosmid fragments subcloned in plasmid or M13 vectors, and for sequence analysis of cDNAs cloned in bacteriophage λ vectors. In addition, the method describing direct sequencing from PEG-precipitated PCR product has been used successfully for analysis of *Caenorhabditis elegans* genomic and cDNA sequences.[40] It is important to reiterate that for every combination of amplification primer pair and target DNA, there is an optimal method for PCR amplification; the ability to sequence the products of any PCR experiment directly will also vary. A coupled PCR/DNA sequencing method that works well for one experimental system may work quite poorly with others. Hence, a few days or hours spent optimizing PCR amplification conditions and selecting the best DNA sequencing method for the target DNA of interest will be time well spent.

Acknowledgments

    We thank Dr. Eric Green for critical review of this chapter, Ms. Molly Craxton for providing PEG precipitation conditions, and Dr. Robert Waterston and John Sulston for support and advice.

[40] R. Waterston, C. Martin, M. Craxton, C. Huynh, A. Coulson, L. Hillier, R. Durbin, P. Green, R. Shownkeen, N. Halloran, T. Hawkins, R. Wilson, M. Berks, Z. Du, K. Thomas, J. Thierry-Mieg, and J. Sulston, *Nature Genetics* **1,** 114 (1992).

[11] Specific Primer-Directed DNA Sequence Analysis Using Automated Fluorescence Detection and Labeled Primers

By ROBERT KAISER, TIM HUNKAPILLER, CHERYL HEINER, and LEROY HOOD

Introduction

DNA sequence analysis, the determination of the linear order of the four deoxyribonucleotides A, G, C, and T comprising a given molecule of DNA, is a fundamental tool of the modern molecular biologist. The enzymatic chain-terminating (often referred to as dideoxy) method pioneered by Sanger and co-workers[1] has become a widely used and highly reliable method for rapidly and easily obtaining DNA sequence information. The method relies on the ability of the enzyme, DNA polymerase, to synthesize a complementary copy of a single-stranded template from an annealed primer oligonucleotide. The inclusion of analogs of the normal deoxynucleotides, dideoxynucleotides, in the reaction mix results in the premature termination of the new strand synthesis, and thus the production of a nested set of fragments, all with a common start point (the primer) but some ending at each base in the template. Vigorous efforts at both the academic and industrial levels have resulted in the development and availability of new sequencing enzymes, new cloning vectors, ultrahigh-purity sequencing reagents, and specialized sequencing equipment, such that high-quality DNA sequencing runs of considerable length (>400 bases) from either single- or double-stranded templates are routinely accessible to the average molecular biology laboratory.

The evolution of DNA sequencing technology has advanced to the point whereby molecular biologists can now begin the investigation of complex biological systems, that is, those whose genes span extensive chromosomal regions or even complete genomes themselves. The massive volume of DNA sequence information that will be required to understand and describe these systems has triggered the development of several new approaches (chemistries and instrumentation) for the automation of DNA sequencing.[2-5]

[1] F. Sanger, S. Nicklen, and A. R. Coulson, *Proc. Natl. Acad. Sci. U.S.A.* **74,** 5463 (1977).
[2] L. M. Smith, J. Z. Sanders, R. J. Kaiser, P. Hughes, C. Dodd, C. R. Connell, C. Heiner, S. B. H. Kent, and L. E. Hood, *Nature (London)* **321,** 674 (1986).

One powerful approach to automating the data acquisition and analysis steps of the dideoxy sequencing procedure has been the replacement of the more conventional radioisotopic labels used to visualize the products of the enzymatic reactions with a set of four spectrally discriminable, highly sensitive fluorochromes (fluorescent dyes).[2–4] A different fluorochrome is used to color code each of the four enzymatic sequencing reactions. The products of the four reactions are combined and electrophoretically separated by size in a single lane of a polyacrylamide gel. The labeled fragments move through the gel into a region near its bottom, where the fluorochromes are excited to fluorescence by a laser. The resulting emitted light is collected and a four-point spectrum obtained by a high-sensitivity detector in real time. This spectrum is used to automatically identify which dye, and therefore which terminal base, is present in each band of DNA fragments. The sequence is automatically computed from the temporal order in which the different colored bands pass the detector. A prototype instrument based on these principles has been described in an earlier volume of this series.[6] A second-generation automated DNA sequencer (model 373A), vastly improved from the original prototype system, is currently available from Applied Biosystems, Inc. (Foster City, CA). The instrument is capable of routinely producing 450 bases of quality sequence data from each of the 24 lanes of the gel, or approximately 10,000 total bases/day.

Automated DNA sequencing technologies such as this offer a great potential for the rapid accumulation of DNA sequence. However, it is generally the case that a single sequencing run will be of insufficient length to provide the entire sequence of a nucleic acid of interest. A variety of strategies have been devised to obtain the complete sequence of large DNAs. In the random, or "shotgun," strategy, a large DNA is randomly fragmented into small pieces, which are then subcloned into an appropriate sequencing vector. Clones are picked randomly, and are sequenced using the same vector-specific primer. The finished sequence is assembled based on local overlapping sequence similarity among the random sequences. This strategy is simple to implement and generates a great deal of data

[3] C. Connell, S. Fung, C. Heiner, J. Bridgham, V. Chakerian, E. Heron, B. Jones, S. Menchen, W. Mordan, M. Raff, M. Recknor, L. Smith, J. Springer, S. Woo, and M. Hunkapiller, *BioTechniques* **5**, 342 (1987).

[4] J. M. Prober, G. L. Trainor, R. J. Dam, F. W. Hobbs, C. W. Robertson, R. J. Zagursky, A. J. Cocuzza, M. A. Jensen, and K. Baumeister, *Science* **238**, 336 (1987).

[5] W. Ansorge, B. Sproat, J. Stegemann, C. Schwager, and M. Zenke, *Nucleic Acids Res.* **15**, 4593 (1987).

[6] L. M. Smith, R. J. Kaiser, J. Z. Sanders, and L. E. Hood, this series, Vol. 155, p. 260.

rapidly. However, it does require substantial subcloning, clone preparation, and repetitive sequencing to obtain the entire sequence of interest. The final sequence is obtained only after significant effort, because the position of any random fragment relative to the whole is initially unknown. Alternatively, directed strategies have the advantage of requiring less actual cloning and sequencing, and of knowing the location of a particular fragment relative to the entire sequence.

This chapter will present the current chemistries involved in the synthesis of dye–primer sets for use with the Applied Biosystems 373A automated DNA sequencer, as well as reliable protocols for their utilization in enzymatic DNA sequencing. It will also present some selection rules for assessing candidate primer sequences. We believe that there are three primary applications of specific primers: (1) the sequencing of nucleic acids flanked by the same primer sequence, such as cloning vector-specific (so-called "universal") primers or primers for the routine repetitive sequence analysis of polyallelic genetic regions, such as in the determination of HLA haplotypes, (2) the filling of occasional small gaps in extended sequence obtained by random sequencing methods, and (3) classic specific primer-directed sequencing (SPDS), often termed gene walking, wherein the entire sequence of a large DNA is obtained by walking along its length using information obtained from a previous round of sequence analysis to select a new primer for a subsequent round.

General Principles of Specific Primer-Directed Sequencing

The following discussion is predicated on the premise that the DNA to be sequenced has been cloned into an appropriate host vector using standard procedures.[7–10] This provides a stable, pure source of the DNA in sufficient quantity for DNA sequence analysis.

Use of Specific Primers for Analysis of Multiple Sequences

The partial DNA sequences of the multiple DNAs cloned in the same vector can be obtained using a "universal" primer. The sequence of this primer is selected to be complementary to a known region of the cloning vector near the multiple cloning site. This primer is therefore universal in the sense that it can be used to obtain sequence information from any

[7] A.-M. Frischauf, this series, Vol. 152, p. 190.
[8] A. G. DiLella and S. L. C. Woo, this series, Vol. 152, p. 199.
[9] H. Miller, this series, Vol. 152, p. 145.
[10] J. Messing, this series, Vol. 101, p. 20.

unknown insert that has been cloned into the particular vector from which the primer was derived. Alternatively, when studying polyallelic genes, a previously obtained sequence from a nonpolymorphic region of the gene may be used to provide the universal primer. Enzymatic sequencing reactions are performed with this primer, the products are separated by gel electrophoresis, and some sequence of the unknown is obtained. In this case, the same set of specific primers is used many times.

Filling of Small Gaps

Specific primers can be used to fill gaps between contiguous stretches of extended sequence (contigs) obtained using random sequencing methods. In general, random sequencing will provide about 90% of the complete sequence of a large clone before the effort needed to further extend the contigs reaches an unacceptably high level. The remaining 10% usually consists of several small gaps between the contigs. These gap sequences may be obtained by choosing a specific primer from the end of a known contig sequence and using it to obtain sequence across the immediately adjacent gap region. By sequentially applying this strategy a limited number of times, the entire gap sequence can be obtained. Closure is realized when this process produces end sequence from the adjacent contig.

Gene Walking

Extended sequence of a particular cloned DNA can be obtained using sequence information generated from a universal primer to select a new insert-specific primer for a further round of sequence analysis. Successive cycles of sequencing and new primer generation yield the complete sequence of interest. The advantage of the approach lies in the large amount of sequence information that may be obtained from a single clone. However, the speed with which the entire insert sequence is acquired is a function of several factors: the rate at which new specific primers can be selected, synthesized, and purified, the rate of new sequence acquisition for each primer; the amount of sequence information sufficiently accurate to pick a succeeding primer obtained per run; and the percentage of selected primers giving rise to interpretable sequence. For sizeable clones, a large number of primers is required and the gene-walking process can be quite slow and expensive, often prohibitively so.

The three applications of fluorescence-labeled primers described above are diagrammed in Fig. 1.

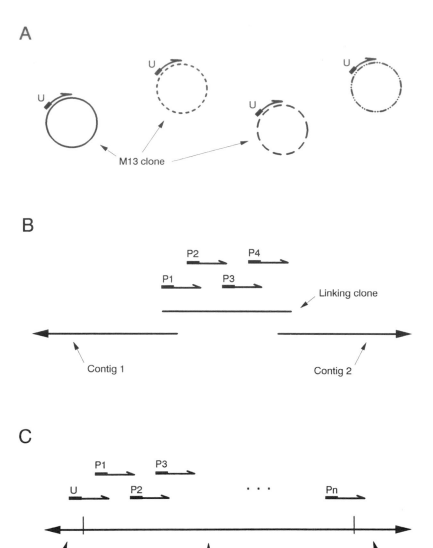

FIG. 1. Three applications of specific primers for DNA sequence analysis. (A) Use as universal primer for the partial sequencing of multiple M13 clones. In this case, a single primer (U) can be used to obtain DNA sequence from a large number of unknowns. (B) Use as specific primers in conjunction with a linking clone to fill gaps between contigs. A limited number of primers (P1, P2, P3, and P4) would be required for this application. (C) Use as specific primers for gene walking. Initial sequence from a large clone is obtained using the universal primer (U). This sequence is then used to generate a new insert-specific primer (P1), from which additional sequence is obtained. Successive cycles of sequencing and new primer generation (P2, P3, P4, . . . , Pn) yield the complete sequence of interest. For sizeable clones, the value of n can be very large.

*Specific Primer-Directed Sequencing and 373A Automated
  DNA Sequencer*

The current automated sequencer consists of two linked items of hardware: (1) a module in which the gel for the electrophoretic separation of the sequencing reaction products is run, which houses the electrophoresis hardware, the optical components (laser and detector), and the microprocessor that controls these elements; and (2) a computer (Apple Macintosh II) that collects the raw fluorescence data and carries out the automated sequence determination. The instrument uses a slab gel presently capable of handling 24 different samples, each sample being the combined set of 4 enzymatic extension reactions for a particular clone and/or primer. Approximately 600 bases of raw data is obtained per gel lane in a 14-hr period, from which more than 400 bases of analyzed sequence can be obtained at an accuracy of at least 95%.

As mentioned previously, the Applied Biosystems 373A automated DNA sequencer uses a set of four fluorescent dyes to color code the four reactions in enzymatic DNA sequencing. The instrument is capable of utilizing either dye-labeled primers or dye-labeled dideoxy analogs, either of which can be employed for SPDS. Currently, however, the best protocols (those giving the longest and most accurate sequence runs) are those involving dye–primers. Thus, we shall concentrate on the use of labeled primers for SPDS with the automated sequencer here.

The use of dye–primers in SPDS requires the synthesis and purification of a set of four fluorochrome-labeled oligonucleotides per primer sequence. Labeling is accomplished in two steps. First, the desired oligonucleotide sequence is synthesized on an automated DNA synthesizer using standard procedures.[11] The final reaction cycle of the synthesis incorporates a protected aliphatic amino alcohol at the 5' terminus.[3] The deprotected crude aminooligonucleotide is then reacted with the appropriate fluorescent dyes as amino-reactive derivatives, and the products of the reactions are purified. Clearly, to rapidly accrue sequence information using the SPDS approach in conjunction with dye–primers, a minimum amount of time should be spent in the above synthesis and purification.

Materials and Methods

Overview

An attractive feature of the automated sequencer is the capability of employing a wide variety of sequencing chemistries in conjunction with it.

[11] T. Atkinson and M. Smith, *in* "Oligonucleotide Synthesis: A Practical Approach" (M. J. Gait, ed.), p. 35. IRL Press, Oxford, 1984.

In particular, any of the DNA polymerases common to modern sequencing procedures (Klenow fragment of *Escherichia coli* DNA polymerase I,[1] modified T7 DNA polymerase,[12] DNA polymerase from *Thermus aquaticus* [*Taq*][13]) may be used, albeit with somewhat different results. From experience, we prefer to use *Taq* polymerase, because it exhibits several features desirable for automated fluorescence-based DNA sequencing. Due to its thermostability and high temperature of optimal activity (70–72°), problems associated with primer annealing at secondary template binding sites are reduced, as are problems related to template secondary structure. The enzyme is compatible with commonly used nucleotide analogs such as 7-deazadeoxyguanosine (d-c^{7}GTP). Extension of the primer is fast, processive, and efficient, giving rise to strong fluorescence signals. The lack of exonuclease activity minimizes the occurrence of false terminations. Additionally, it is possible to utilize *Taq* polymerase in conjunction with a DNA thermal cycler in a useful procedure termed "cycle sequencing," which will be described in detail below (see *Taq* DNA Polymerase Cycle Sequencing Protocol).

In this chapter, methods and illustrative data will be provided for the sequencing of single-stranded M13 templates with *Taq* DNA polymerase. Applied Biosystems provides users of the 373A with complete, detailed sequencing protocols for use with a variety of templates and enzymes, as well as continuing User Bulletins of updated information and in-house technical assistance.

The following procedures are modifications of generally accepted methods such as those found elsewhere in this series. All reagents are those commonly used in biological laboratory preparations, and must be of the highest purity possible. Manufacturers listed for many of the reagents reflect those routinely utilized in our laboratory with good success; comparable reagents purchased from other sources may be suitable as well.

Preparation of Template DNA

The success of any DNA sequencing procedure is directly linked to the purity of the template DNA to be sequenced. The saying "garbage in, garbage out" is especially applicable to automated DNA sequencing.

Reagents

Escherichia coli strain JM 101 competent cell culture (fresh)
M13 cloned DNA phage culture

---

[12] S. Tabor and C. C. Richardson, *Proc. Natl. Acad. Sci. U.S.A.* **84,** 4767 (1987).
[13] M. A. Innis, K. B. Myambo, D. H. Gelfand, and M. D. Brow, *Proc. Natl. Acad. Sci. U.S.A.* **85,** 9436 (1988).

YT medium (2×): 16.0 g Bacto-tryptone (Difco, Detroit, MI), 5.0 g
   yeast extract (Difco), 5.0 g NaCl per liter of distilled, deionized water,
   pH 7.2–7.4; sterilize by autoclaving
Polyethylene glycol 8000 (PEG) (20%, w/v), 2.5 M NaCl: Prepare daily
   by mixing equal volumes of stock 40% (w/v) PEG and 5 M NaCl)
TE buffer: 10 mM Tris base, 1 mM Na$_2$EDTA, titrated to pH 8.0 with
   HCl
TE-saturated phenol
TE-saturated phenol : chloroform : isoamyl alcohol (25 : 24 : 1, v/v/v)
Distilled deionized H$_2$O–saturated diethyl ether
Ammonium-acetate (5 M)
Absolute ethanol
Cold 70% (v/v) ethanol
Tris buffer (10 mM), titrated to pH 8.0 with HCl

Procedure. Inoculate 2× YT medium (10 ml) with a single colony of
JM 101, and incubate with shaking for 12–18 hr at 37°. Add an aliquot (2
ml) of this culture to 2× YT (150 ml) and incubate a further hour at 37°
with shaking. Transfer aliquots (10 ml) of this culture into sterile culture
tubes. With a sterile Pasteur pipette or sterile toothpick, remove a desired
M13 clone plaque and transfer it to the tube containing the bacterial
culture. Shake the inoculated culture for about 6 hr at 37°.

Centrifuge the cultures at 4° and 2500 rpm for 20–30 min to pellet the
bacteria. Transfer the cleared supernatant to sterile centrifuge tubes that
will withstand 10,000 rpm. Add PEG/NaCl (2 ml) solution and mix well.
Allow the mixture to stand at 4° for at least 30 min (if more convenient,
overnight is acceptable). Centrifuge the resulting suspension for 15 min at
10,000 rpm and 4°. Carefully aspirate the supernatant completely (residual
PEG and NaCl will inhibit polymerase activity), leaving the translucent
phage pellet. Suspend the pellet in TE (600 μl) and transfer to a microcentrifuge tube.

Isolate the single-stranded DNA by extracting (use a fume hood for all
manipulations of organic solvents) the aqueous suspension once with TE-
saturated phenol (600 μl) and once with TE-saturated phenol : chloroform :
isoamyl alcohol (600 μl). For each extraction, add the organic to the
suspension, cap the tube, vortex, and spin in a microcentrifuge to separate
the layers cleanly. Remove the upper aqueous layer to a fresh tube each
time, taking care not to transfer any of the insoluble material that may be
present at the interface. Finally, extract twice with distilled, deionized
H$_2$O–saturated ether until the aqueous (bottom) phase is clear. Remove
residual ether either by briefly drying in a vacuum centrifuge (e.g., Speed-
Vac; Savant Instruments, Farmingdale, NY), or by placing the open microfuge tubes in a heating block at 65° for about 15 min.

Precipitate the template DNA by adding one-twelfth volume of 5 M ammonium acetate and 2 vol of absolute ethanol. Mix by inverting the tube several times, and allow to precipitate overnight at $-20°$. Pellet the precipitated DNA at maximum speed in a microcentrifuge at 4°. Remove the supernatant and wash the pellet with cold 70% ethanol (100 μl). Collect the precipitate by centrifugation as before, remove the supernatant, and dissolve the DNA in TE (30 μl). Determine the concentration of the DNA solution by measuring the absorbance (A) at 260 nm; 1 A_{260} unit = 33 μg/ml of single-stranded DNA in a 1-cm cuvette. The average yield from this procedure is about 10 μg, sufficient for several sequencing runs. Also, assess the purity of the preparation by calculating the A_{260}/A_{280} ratio; a value of less than 1.6 indicates possible contamination by protein or phenol, and the preparation will probably not yield quality data on the automated sequencer.

Preparation of 6% Polyacrylamide Sequencing Gel

Special care should be taken in the preparation of gels for use with the automated sequencer in order to maximize resolution, especially for long runs. It is critical that the glass plates used be clean and free of dust, scratches, and markings of any kind that would alter their fluorescent or light-scattering properties and increase the background noise detected by the 373A optical system. Similarly, all solutions must be carefully filtered through 0.45-μm membrane filters to remove any suspended particulate matter.

Reagents

Acrylamide solution (40%, w/v): 20% (w/v) T, 5% (w/v) C, 38 g acrylamide (ultrapure; Schwarz/Mann Biotech, Orangeburg, NY), 2 g N-methylenebisacrylamide (ultrapure; Bethesda Research Laboratories, Gaithersburg, MD) per 100 ml distilled, deionized H_2O; make fresh every 2–3 weeks, and store at 4°

Urea (ultrapure; Schwarz/Mann Biotech)

N,N',N'-Tetramethylethylenediamine (TEMED) (ultrapure; Bethesda Research Laboratories

Ammonium persulfate (ultrapure; Bethesda Research Laboratories)

TBE (10×): 108 g Tris base, 55 g boric acid, 8.3 g Na_2EDTA per liter of distilled, deionized H_2O; filter through a 0.45-μm membrane; pH should be 8.3–8.5; make fresh weekly

Mixed-bed ion-exchange resin such as Amberlite MB-1 (Sigma, St. Louis, MO)

Apparatus

Optically clear glass plates, spacers, and combs for use with the 373A
    sequencer: Available from Applied Biosystems (Foster City, CA)
Permacel laboratory tape
Clamps
Cup membrane filters (0.45 μm) (Corning, Corning, NY)

Procedure

Assembly of Glass Plates. Clean both sides of the glass plates well with
warm water and a laboratory detergent that will not leave a residue, such
as Alconox (Fisher Scientific, Pittsburgh, PA). Be careful not to use any
abrasives that would damage the glass surfaces. Rinse the plates well with
warm water, distilled, deionized H_2O, and 95% (v/v) ethanol, and stand
upright to drain and air dry; do not wipe, because residual lint on the glass
surface will affect results. Wash the combs and spacers in a similar manner.
It is useful to etch a small mark into each plate near the top, indicating the
outside of the plate, that is, the side not in contact with the gel. Having
the same side of each plate in contact with the gel from run to run will
help ensure consistent results

Place the unnotched plate horizontally on a protected benchtop (ab-
sorbant paper and paper towels) with the outside surface facing down.
Arrange the spacers on the long edges of the plate, and carefully lay the
notched plate on top, outside surface facing up. Align the plates and
spacers, and clamp one side in place to hold the assembly temporarily in
position. Carefully and tightly tape the other side and bottom (edge without
the notch) of the assembly with wide Permacel tape, removing any air
pockets between the tape and the plates to prevent leakage. Remove the
clamps and finish taping the last edge. The top (edge with the notch) is left
untaped. During this procedure, wear laboratory gloves and refrain from
touching the front surfaces of the plates; handle by the edges as much as
possible.

Pouring Gel. Mix urea (50 g) and 40% acrylamide solution (15 ml), and
adjust the volume to 85–90 ml with distilled, deionized H_2O. Add mixed-
bed ion-exchange resin (1–2 g), and stir until all of the urea has dissolved
(15–30 min). Filter the mixture through a 0.45-μm membrane and transfer
to a 100-ml graduated cylinder. Add 10× TBE solution (10 ml), adjust the
volume to 100 ml with distilled, deionized H_2O, and mix gently, avoiding
entrapment of air bubbles in the viscous solution. Add 10% (w/v) ammonium persulfate in distilled, deionized H_2O (500 μl; make fresh daily),
followed by TEMED (45 μl). Mix thoroughly, once again avoiding air
bubbles. Quickly and carefully pour the acrylamide solution into the plate

assembly. Insert the flat-edged temporary comb and allow the gel to polymerize for at least 60 min in a horizontal position.

Sequencing Reactions Using Universal Primer

General Considerations. As mentioned previously, *Taq* polymerase is quite effective and reliable when used for automated fluorescence-based DNA sequence analysis. There are several general considerations that, if carefully observed, will help ensure consistently good results.

Dye–primers are obtained as lyophilized solids, and should be stored as such at $-20°$ until needed. They should then be dissolved in sterile TE buffer to a concentration of 0.8 pmol/μl and stored at $-20°$. The dye–primers tend to dissolve slowly, so that frequent vortexing and long (20–30 min) incubations on ice are required to assure complete dissolution. The fluorochrome labels are somewhat light sensitive, so that all manipulations should be done under subdued lighting or in the dark.

Taq polymerase is quite sensitive to the concentrations of nucleotide triphosphates in its environment. If starved for a particular nucleotide, it will begin to misincorporate other nucleotides, giving rise to sequencing errors. Thus, it is important to use high-purity deoxynucleotide triphosphate (dNTP) and dideoxynucleotide triphosphate (ddNTP) mixes in *Taq* sequencing protocols. Degradation due to repeated cycles of freezing and thawing of these mixtures is a common source of problems in this procedure, so it is best to divide stock solutions into small aliquots suitable for a day's worth of reactions. These aliquots are stored at $-20°$ until needed, and are kept on ice while assembling the sequencing reactions.

We recommend the use of d-c^{7}GTP in place of dGTP in DNA sequencing reactions. This analog is well tolerated by *Taq* polymerase and greatly reduces the incidence of band compression in the gel.[14]

Excellent results have been obtained using *Taq* polymerase from Perkin-Elmer Cetus (Norwalk, CT). However, enzyme from alternative reputable suppliers may give equally high quality data. It should be noted, however, that the use of a preparation of *Taq* polymerase that gives good amplification results in the polymerase chain reaction (PCR) does not always guarantee the realization of good results in DNA sequencing.

Applied Biosystems markets kits for both the standard *Taq* protocol and the cycling protocol containing sufficient reagents [enzyme, buffer(s), dNTP/ddNTP mixes, and dye–primers] for about 100 sequences. Alternatively, reagents (except dye–primers) may be obtained from any reputable supplier of chemicals and enzymes for molecular biology.

[14] S. Mizusawa, S. Nishimura, and F. Seela, *Nucleic Acids Res.* **14**, 1319 (1986).

Standard Taq DNA Polymerase Sequencing Protocol

The standard protocol for the sequencing of single-stranded DNA templates with *Taq* polymerase is simple and rapid. It does require a significant number of manipulations, however. The procedure has been optimized for use with a total of 0.4 to 0.5 pmol of template, corresponding to a few micrograms of most M13 clones.

Reagents

Taq sequencing buffer (5 ×): 50 mM Tris base, 50 mM MgCl$_2$, 250 mM
    NaCl, pH to 8.5 at room temperature with HCl
Taq dilution buffer (10 ×): 500 mM KCl, 100 mM Tris base, 15 mM
    MgCl$_2$, 0.1% (w/v) gelatin, pH to 8.3 at room temperature with HCl
TE buffer
M13 cloned DNA template (0.6 pmol in distilled, deionized H$_2$O or TE
    at about 0.1 pmol/μl)
Universal M13(-21) dye–primer set (FAM, JOE, TAMRA, and ROX
    primers) (Applied Biosystems): Dilute to 0.2 pmol/μl in TE
dNTP/ddNTP stock mixes (see below)
Taq DNA polymerase (AmpliTaq, or AmpliTaq for sequencing; Perkin-
    Elmer Cetus)
Ethanol (95%, v/v)
Sodium acetate (3 M), pH 5.2
Ethanol (70%, v/v)

dNTP/ddNTP stock mixes

dNTP mix	Concentration			
	dATP (μM)	dCTP (μM)	d-c^{7}GTP (μM)	dTTP (μM)
dA	125	500	750	500
dC	500	125	750	500
dG	500	500	188	500
dT	500	500	750	125

ddNTP solutions

ddNTP stock	Concentration
ddA	3.0 mM ddATP
ddC	1.5 mM ddCTP
ddG	0.25 mM ddGTP
ddT	2.5 mM ddTTP

Prepare a 1 : 1 (v/v) mixture of each ddNTP stock solution and dNTP mix (dA/ddATP, dC/ddCTP, dg/ddGTP, and dT/ddTTP); vortex and keep these working solutions on ice.

Apparatus

Heating blocks at 55–65, 70, and 37°, capable of holding microcentrifuge tubes (0.5 or 1.5 ml)

Procedure. The following describes the procedure for the sequencing of a single template. For multiple templates, make working stock solutions by multiplying the given volumes by the number of templates plus an excess of 5% to account for pipetting errors.

Dilute *Taq* DNA polymerase (10 units; volume will depend on concentration of stock) with $10 \times$ *Taq* dilution buffer (2 μl) and sufficient distilled, deionized H_2O to give a total volume of 20 μl. Mix by flicking the tube with a finger (do not vortex) and keep on ice.

For purposes of the following description, A and C reactions are each $1 \times$; G and T are $2 \times$ (that is, double the quantities and volumes needed for the $1 \times$ reaction). For each $1 \times$ reaction, mix $5 \times$ *Taq* sequencing buffer (1.8 μl), template DNA (0.1 pmol), and dye–primer solution (1.0 μl; FAM–primer for C, JOE–primer for A, TAMRA–primer for G, and ROX–primer for T). Dilute to 6.0 μl with distilled, deionized H_2O. Incubate each tube at 55–65° for 5 to 10 min. Cool slowly (20–30 min) to 4–8° to anneal the primer to the template. Spin the tubes briefly in a microcentrifuge to collect the condensate.

Add the appropriate dN mix/ddNTP working solution (1.0 μl) to each annealed $1 \times$ mixture (that is, dA mix/ddATP to the JOE–primer-containing solution, etc.). Add the diluted enzyme (2.0 μl) to each $1 \times$ reaction, and incubate at 70° for 5 to 10 min. Centrifuge briefly.

Cool the tubes to 37°, and add an additional aliquot (1.0 μl) of the diluted *Taq* polymerase solution as a chase. Incubate 5 to 10 min. Spin briefly, then incubate at 70° for 5 to 10 min.

Precipitate the products of the four reactions by sequentially adding each to a microcentrifuge tube containing 3 M sodium-acetate (6 μl) and cold 95% (v/v) ethanol (144 μl). Incubate the mixture on ice for at least 10 min. Centrifuge the precipitate for 30 min at 4°. Carefully remove the supernatant and wash the pellet with 70% ethanol (200 μl). Centrifuge for 5 min at 4° and remove the supernatant. Dry the pellet *briefly* (1 to 3 min) in a vacuum centrifuge (e.g., SpeedVac) to remove residual ethanol, taking care not to overdry.

At this point, if electrophoresis cannot be run immediately, the reaction products are stable in the dried form for several months if stored in the dark at $-20°$. However, if the separation is to be run within about 48 hr of performing the reactions, they may be allowed to sit at $-20°$ in the original sodium acetate/ethanol mixture during the interim. The precipitation can then be completed immediately prior to electrophoresis.

Taq DNA Polymerase Cycle Sequencing Protocol

This procedure is based on a technique reported by Carothers et al.,[15] and takes advantage of the thermal stability of Taq polymerase. The reactions are carried out in a DNA thermal cycler of the sort commonly used for DNA amplification by the polymerase chain reaction. The sequencing mixtures (enzyme, template, primer, and deoxy- and dideoxynucleotide triphosphates) are subjected to repeated cycles of denaturation, annealing, and extension, resulting in a linear amplification of the terminated products. This has several benefits. The amplification inherent in the procedure means that significantly less starting template is required to obtain the same signal. Thus, excellent sequencing results can be obtained from a few hundred nanograms of M13-cloned DNA. This means that a variety of procedures for rapidly isolating small quantities of template, quantities too small to use reliably with the standard protocol, can now be employed for the production of templates for use with the cycling procedure. The quantity of enzyme used, a significant sequencing expense, is less than for the standard protocol. Repeated denaturation yields improved signal and length of read from double-stranded templates. Operationally, the procedure is extremely simple, with many fewer manipulations than the standard protocol because most of the work is performed by the thermal cycler. The disadvantage to the method is that the thermal cycling is a more lengthy process than the standard protocol. We have been using cycle sequencing almost exclusively now for about 6 months, and routinely have obtained excellent results with both single-stranded M13 clones, and with double-stranded plasmid and cosmid clones.

Reagents

Taq cycle sequencing buffer ($5\times$): 400 mM Tris base, 100 mM $(NH_4)_2SO_4$, 25 mM $MgCl_2$, pH to 8.9 at room temperature with HCl TE buffer

M13-cloned DNA template (500–750 ng in distilled, deionized H_2O or TE at about 100 ng/μl

Universal M13(-21) dye–primer set (Applied Biosystems): dilute to 0.4 pmol/μl in TE

dNTP/ddNTP stock mixes (see below)

---

[15] A. M. Carothers, G. Urlaub, J. Mucha, D. Grunberger, and L. A. Chasin, BioTechniques 7, 494 (1989).

Taq DNA polymerase (AmpliTaq, or AmpliTaq for sequencing; Perkin-Elmer Cetus)
Ethanol (95%, v/v)
Sodium acetate (3 *M*), pH 5.2
Ethanol (70%, v/v)
Light mineral oil

dNTP/ddNTP stock mixes: These mixes are the same as those used in the standard protocol above. Make the same 1:1 mixtures as working solutions and keep on ice during assembly of the sequencing reactions.

Apparatus. The protocol was developed using the DNA thermal cycler from Perkin-Elmer Cetus Instruments. We have been routinely using a temperature cycler (TwinBlock System) from Ericomp (San Diego, CA) in our laboratory. Temperature cycles from other manufacturers may be equally effective.

Procedure. As above, the following describes the procedure for the sequencing of a single template. For multiple templates, make working stock solutions by multiplying the given volumes by the number of templates plus an excess of 5% to account for pipetting errors.

Dilute *Taq* DNA polymerase (6 units) with 5× cycle sequencing buffer (1.0 μl) and sufficient distilled, deionized H_2O to give 7.0 μl. Mix by flicking the tube with a finger (do not vortex) and keep on ice.

For purposes of the following description, A and C reactions are each 1×; G and T are 2× (i.e., double the quantities and volumes needed for the 1× reaction). For each 1× reaction, in a 0.5-ml microcentrifuge tube or other vessel compatible with the temperature cycler to be used, mix d/ddNTP working solution (1.0 μl), the appropriate dye–primer solution (1.0 μl; JOE–primer for the dA/ddA-containing reaction, FAM-primer for dC/ddC, TAMRA–primer for dG/ddG, and ROX–primer for dT/ddT), 5× cycle sequencing buffer (1.0 μl), DNA template solution (1.0 μl), and diluted enzyme solution (1.0 μl). Overlay each reaction with light mineral oil (about 20 μl), and spin briefly in a microcentrifuge.

Place the tubes in the temperature cycler preheated to 95°. Immediately begin cycling according to the following program:

1. 95°, 30 sec
2. 55°, 30 sec
3. 70°, 1 min

Repeat this cycle 15 times. Continue with

1. 95°, 30 sec
2. 70°, 60 sec

Repeat this cycle 15 times. All transition times between temperature plateaus should be as fast as possible (maximum temperature ramping). End the program with a rapid decrease to room temperature or below.

Ethanol precipitation of the reaction products is carried out essentially as described for the standard protocol, taking into account the smaller volumes associated with the cycling procedure. Mix 95% ethanol (80 μl) with 3 M sodium-acetate (1.5 μl). Pipette each of the four reaction mixtures sequentially into this solution, taking care to transfer as little oil as possible. This is best accomplished by immersing the end of the pipette tip just below the oil–water interface and then carefully removing the bottom aqueous layer. Mix thoroughly and incubate the mixture on ice for at least 10 min. Centrifuge the precipitate for 30 min at 4°. Carefully remove the supernatant and wash the pellet with 70% ethanol (200 μl). Centrifuge for 5 min at 4° and remove the supernatant. Dry the pellet *briefly* (1 to 3 min) in a vacuum centrifuge to remove residual ethanol, taking care not to overdry.

Procedural Options. Removing the small-volume (5 μl) A and C reactions from under the oil can be difficult. There are two options one can use to alleviate this difficulty. First, try adding distilled, deionized H_2O (5 μl) to both the A and C reactions. Spin briefly, and remove the aqueous layer (now 10 μl). If this option is used, increase the volume of 95% ethanol used in the precipitation to 100 μl, and the volume of 3 M sodium acetate to 2 μl. Alternatively, it is possible to perform all four reactions at double all amounts given above (i.e., A and C reactions become 2×, G and T reactions become 4×). Realize that this option provides double amounts of sequencing reaction products for gel loadings. It should be noted that doubling the reaction volume alone does *not* give good results.

Occasionally, analyzed sequence generated by the cycling procedure exhibits local regions of small or apparently absent signal due to sequence positions at which *Taq* polymerase has poorly incorporated the dideoxy analog. This appears to be a reproducible feature of the enzyme, and seems to be dependent on the nearby surrounding sequence. Often this problem can be corrected by increasing the $MgCl_2$ concentration in the reactions. It is important to use this only for problem sequences, however, as increasing $MgCl_2$ shortens the length of readable sequence.

Electrophoresis

Reagents

TBE (1×): Dilute 200 ml of the same 10× TBE used in the preparation of the gel with distilled, deionized H_2O to 2 liters

EDTA (50 mM), pH 8.0

Formamide (molecular biology grade, deionized with mixed-bed ion-exchange resin; International Biotechnologies, New Haven, CT)

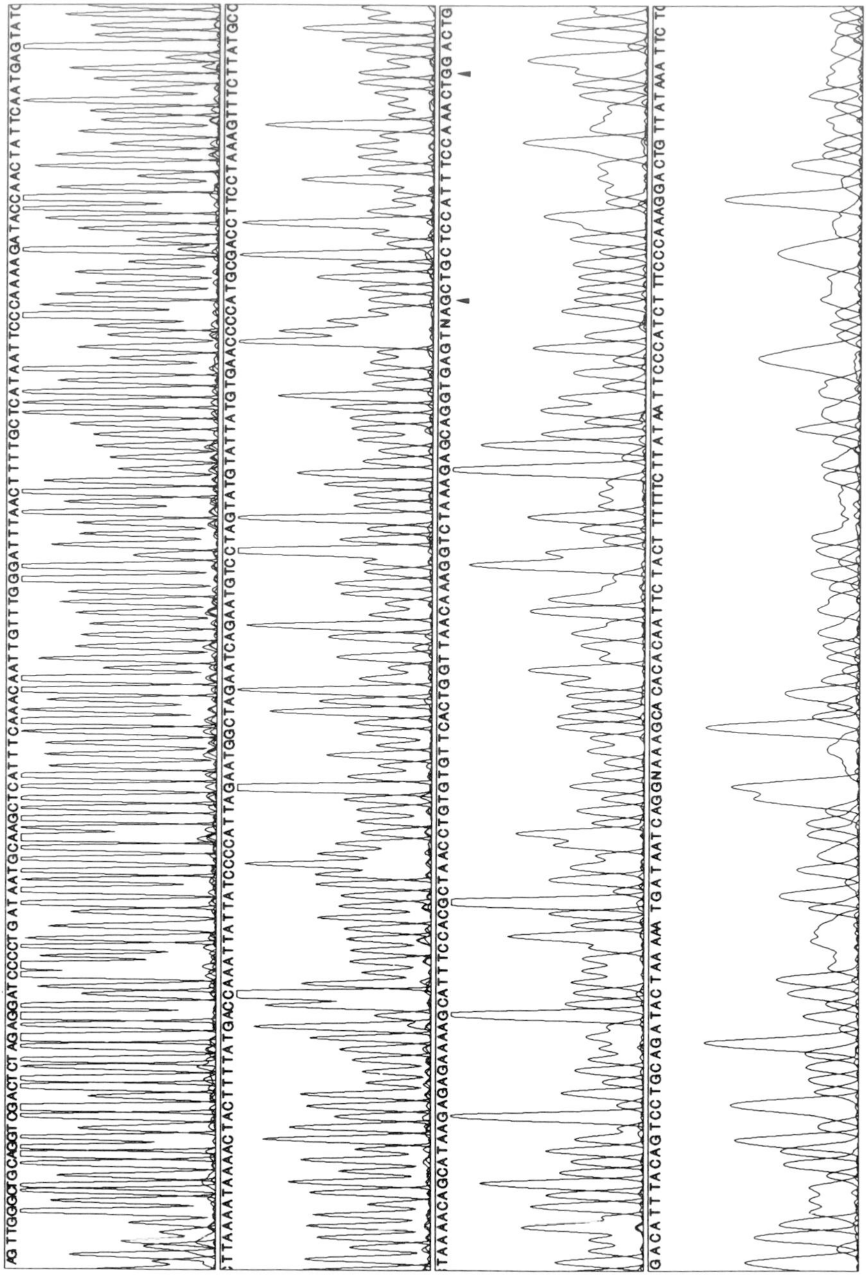

Procedure. Install the polymerized gel/plate assembly, buffer chambers, and shark's-tooth comb in the 373A as instructed by the operator manual. Fill the buffer chambers with 1 × TBE. Test the plates for fluorescence background using the "plate check" option. The four colored traces on the computer monitor should be essentially flat; spikes indicate some perturbation of the fluorescence background such as contamination on the glass surfaces or bubbles in the gel. If such spikes are observed in regions where sequencing reactions will be run, reclean the plates or repour the gel as necessary. Set the electrophoresis parameters from the instrument softkey pad. We generally electrophorese at slightly lower power (25 W) than that recommended by Applied Biosystems (30 W). This somewhat reduces the amount of raw data obtained per lane, but maintains good band resolution further into the run. All other parameters (voltage, current, temperature, and laser power) are the same as recommended by the manufacturer. Preelectrophorese the gel for 15–60 min prior to loding to stabilize the electrophoresis. During this time, add 50 mM EDTA (1 μl) and deionized formamide (5 μl) to each tube containing the dried, precipitated reaction products. Vortex to dissolve, which may take several minutes. Spin briefly to collect the solution in the bottom of the tube. Place the samples in a heating block at 90° for 2 min to denature. Transfer immediately to ice. Clean the wells of the gel with buffer to remove urea, and *carefully* load the samples. Begin electrophoresis and data collection as instructed in the operator manual.

Automated Data Analysis

At the conclusion of the run (usually 14 hr after loading), the data collected by the computer is automatically analyzed, the sequence is determined and stored as a separate data file, and a hard copy of the analyzed data from each lane is printed. This hard copy is an electropherogram (fluorescence signal versus time) in four colors of each lane of data with the sequence determined by the computer appended above it. The

---

FIG. 2. Representative analyzed sequence data from the 373A automated sequencer. The template is a fragment from the human T cell receptor α gene locus cloned into M13mp10. Depicted are 450 bases (from the end of the primer) of automatically analyzed sequence obtained using the cycle sequencing protocol and universal M13(-21) primer (5'-TGTAAAACGACGGCCAGT-3') (see text). "N" in the sequence printout indicates a position at which the fluorescence data was too ambiguous for the analysis software to determine the identity of the base accurately. Although this figure is reproduced in black and white, the actual data presentation from the automated sequencer is in four colors: C, blue; A, green; G, black; T, red.

user may then view the raw and analyzed data, and edit or manipulate it in a variety of ways. Figure 2 is a black-and-white version of typical data from an M13 clone obtained using the cycle sequencing protocol with M13(-21) primer as described.

Sequencing Reactions Using Insert-Specific Primer

The following section describes the synthesis and use of custom-labeled primers in automated DNA sequencing on the 373A, as well as illustrates the strategy of gene walking in this context. Once initial sequence information about a cloned insert is obtained through the use of a universal primer, a new insert-specific primer sequence can be generated for the next round of enzymatic sequencing. The sequence of a specific primer should be chosen to maximize specificity and hybrid stability under the conditions of the sequencing procedure to be used. Once a candidate sequence has been chosen, the set of four fluorochrome-labeled oligonucleotides must be synthesized and purified. The new dye–primer set can then be used to generate additional sequence information about the cloned insert.

Selection of Specific Primer Sequences

Selection Rules. While the selection of a candidate primer sequence is conceptually straightforward, observation of some general criteria will improve the chances of obtaining a long run of high-quality sequence data. The following set of rules is compiled primarily from three literature sources[16-18] and our own experiences with the automated sequencer.

1. The length of the primer should be sufficient to have a high probability of forming a unique hybrid with the template. The melting temperature (T_m) of the primer–template hybrid should be high enough to assure formation of a significant population of stable hybrids under the conditions of the sequencing reactions. This is an especially important consideration when using *Taq* DNA polymerase due to its high temperature of optimal activity. Primers of base length 17 to 20 and 40% to 60% G/C composition generally fulfill these two criteria satisfactorily. Primers with too great an A/T composition will generally form hybrids of poor stability (due to low T_m).

2. Attempt to select a sequence having as random a linear distribution of the four bases as possible. Homopolymeric regions should be avoided,

[16] W. M. Barnes, this series, Vol. 152, p. 538.
[17] E. C. Strauss, J. A. Kobori, G. Siu, and L. E. Hood, *Anal. Biochem.* **154,** 353 (1986).
[18] K. J. Breslauer, R. Frank, H. Blocker, and L. A. Marky, *Proc. Natl. Acad. Sci. U.S.A.* **83,** 3746 (1986).

especially if they occur at the 3' end. Deleterious secondary structure formation within the primer itself can be minimized by avoiding inverted repeats (dyad symmetries). Likewise, direct repeats can produce less specific priming. While a stable T_m is critical, too great a G/C composition will frequently suffer from significant self-complementation and nonspecific template complementation. The 3' nucleotide is particularly important in strand extension by *Taq* polymerase. Experience suggests choosing a primer sequence whose 3' terminus ends in G and whose 3' penultimate base is a pyrimidine (C or T), if possible.

3. Candidate primer sequences should come from highly reliable regions of the fragment sequence. Insertions, deletions, or incorrect base identifications will severely degrade the ability of the oligonucleotide to function as a primer, especially if the errors occur near the 3' end. With fluorescence data from the automated sequencer, this translates to regions of good, uniform signal strength and resolution that have been completely analyzed by the instrument without the inclusion of ambiguous assignments (represented by "N" in Fig. 2). The electropherographic data giving rise to a particular candidate primer sequence should be checked for homopolymeric stretches of low resolution, as bases may often be added or deleted in these regions by the computer. This is particularly true in the region greater than about 400 bases from the priming site. In general, we perform three sequencing runs per primer, preferably on different gels, and look for sequences that are unambiguous in all three runs.

4. A computer-assisted search of the known vector sequence as well as of any previously obtained insert sequence for complementarity to the candidate primer is useful in establishing a high probability of uniqueness. Pay particular attention to significant complementarity at the 3' end of the primer sequence. Any incidence of multiple site priming by the oligonucleotide can significantly affect the data from the sequencer, resulting in low accuracy in automated sequence determination and severely attenuated length of run.[3] Additionally, compare the candidate sequence against all known repeat sequences such as *Alu* to avoid multiple-site priming from this source.

5. Primers should be of a length and composition that can be cleanly and rapidly assembled on the automated DNA synthesizer. In the case of fluorochrome-labeled oligonucleotides, purification of the final products is most easily accomplished by reversed-phase high-performance liquid chromatography (RP-HPLC). This requires that the finished oligonucleotide be no longer than about 25 bases, in order to maintain good chromatographic separation and resolution.

There are several computer-assisted tools available to aid in the definition and choice of primer sequences for a particular template. We have utilized the program Oligo (version 3.4) marketed by National Biosciences (Hamel, MN). This is an expanded version of a program for determining hybrid stabilities reported by Rychlik and Rhoads,[19] and runs on an IBM-PC or compatible computer. The program calculates T_m values for oligonucleotides complementary to a user-specified sequence under user-defined conditions of length, start point, and salt concentrations using nearest neighbor free energy calculations. Potentially interesting sequences may also be searched for self-complementarity, potential for dimer formation, and complementarity to other sequence regions.

Compatibility with 373A Analysis. As discussed previously, four fluorescent dyes are used as labels with the automated sequencer, two fluorescein derivatives (FAM and JOE), and two rhodamine derivatives (TAMRA and ROX). Each of these dyes has a different mass and/or charge. Thus, they impart an altered electrophoretic mobility to the terminated fragments to which they are attached. These mobility differences are unique to each dye, different and reproducible. However, these differences can be compensated for through the use of calibration factors during data analysis. Unfortunately, the magnitude of the mobility shift depends not only on the chemical nature of each dye, but also on the local DNA sequence of the 5′ end of the primer. Thus, a new set of mobility factors would be required for every specific primer. The process for generating these factors is not difficult, but it is time consuming and thus unsatisfactory for primer sets that will be employed only a few times. We have discovered a much more simply implemented alternative.[20] By adding 5 bases from the 5′ end of a primer whose mobility correction factors are known to the 5′ end of the new specific primer, the extant factors can be utilized for the new, fused sequence. We routinely use the end sequence for Applied Biosystems commercial M13 reverse primer (5′-CAGGA-3′), analyzing the resulting raw sequence data with the existing mobility file "Dye Primer {M13RP1}." This simple procedure has worked quite well for the specific-primer sets we have used in our laboratory.

Synthesis of Dye–Primer Sets

The procedure involved in the synthesis of dye primers is diagrammed in Fig. 3 and summarized in Table I. The oligonucleotide corresponding to the primer sequence is synthesized using standard automated phosphoramidite chemistry.[11] The final addition to the 5′ end of the primer incorporates a protected primary amine on a six-carbon linker. After cleavage and

[19] W. Rychlik and R. E. Rhoads, *Nucleic Acids Res.* **17**, 8543 (1989).
[20] R. J. Kaiser, S. L. MacKellar, R. S. Vinayak, J. Z. Sanders, R. A. Saavedra, and L. E. Hood, *Nucleic Acids Res.* **17**, 6087 (1989).

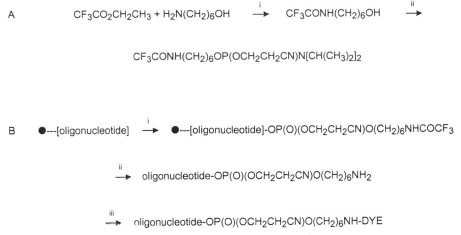

A   CF$_3$CO$_2$CH$_2$CH$_3$ + H$_2$N(CH$_2$)$_6$OH $\xrightarrow{\text{i}}$ CF$_3$CONH(CH$_2$)$_6$OH $\xrightarrow{\text{ii}}$

CF$_3$CONH(CH$_2$)$_6$OP(OCH$_2$CH$_2$CN)N[CH(CH$_3$)$_2$]$_2$

B   ●---[oligonucleotide] $\xrightarrow{\text{i}}$ ●---[oligonucleotide]-OP(O)(OCH$_2$CH$_2$CN)O(CH$_2$)$_6$NHCOCF$_3$

$\xrightarrow{\text{ii}}$ oligonucleotide-OP(O)(OCH$_2$CH$_2$CN)O(CH$_2$)$_6$NH$_2$

$\xrightarrow{\text{iii}}$ oligonucleotide-OP(O)(OCH$_2$CH$_2$CN)O(CH$_2$)$_6$NH-DYE

FIG. 3. (A) Scheme for the synthesis of the N-trifluoroacetyl-6-aminohexan-1-ol phosphor-amidite linker employed in the labeling of oligonucleotides for use as dye–primers: (i) absolute methanol, reflux, overnight; (ii) 2-cyanoethyl N,N-diisopropylaminochlorophosphoramidite, N,N-diisopropylethylamine, dichloromethane, room temperature, 30 min. (B) Scheme for the synthesis of dye-labeled oligonucleotides: (i) standard automated DNA synthesis using the protected amino linker amidite from (A) above; (ii) concentrated ammonium hydroxide, 55°, 6 hr; (iii) dye–NHS ester in aqueous sodium bicarbonate, pH 9, room temperature, 2 hr.

deprotection, the crude amino oligonucleotide is derivatized with activated N-hydroxysuccinimide esters of the carboxyfluorochromes, and the products are purified, first using gel filtration to remove excess unreacted dye, and finally by RP-HPLC. The following procedure is a simplified version of a similar procedure reported by us in an earlier volume of this series.[6]

Protected six-carbon amino alcohol linker phosphoramidites suitable for use in the synthesis of dye–primers are currently available from several commercial sources. We generally use "Aminolink 2" from Applied Bio-systems. The amino group of this amidite is protected with the base-labile trifluoroacetyl (TFA) moiety, which is automatically removed during the standard cleavage and deprotection of the oligonucleotide. Appropriate TFA-protected amidites are also available from Pharmacia-LKB (Piscataway, NJ) and Clontech (Palo Alto, CA). Alternative linker amidites protected with the acid-labile monomethoxytrityl (MMT) group are also available from Clontech and Glen Research (Herndon, VA), and can be used with minor modifications to this procedure. As an example, a synthesis of the 2-cyanoethyl-N,N-diisopropylaminophosphoramidite of N-TFA-6-aminohexan-1-ol is presented here.

TABLE I
SYNTHESIS OF FLUOROCHROME-LABELED OLIGONUCLEOTIDES

Step	Time (hr)	Relative cost[a]
Oligonucleotide synthesis	0.1 per cycle[b]	1×
Cleavage and deprotection	8[c]	—
Lyophilization	3	—
Precipitation	1	—
Fluorochrome labeling	2	20×
Purification		
Gel filtration	1 (4 × 0.25)	7.5×
RP-HPLC	5 (4 × 1.25)	2.5×

[a] Cost estimate based on comparative price of materials.

[b] Cycle time is based on the use of an Applied Biosystems 380B three-column DNA synthesizer with the fast cycle hardware modification, 0.2-μmol scale, and standard Applied Biosystems small-scale fast cycle.

[c] Time required for cleavage and deprotection using standard amidite protecting groups. The use of amidites with more labile protecting groups [see, e.g., J. C. Schulhof, D. Molko, and R. Teoule, *Nucleic Acids Res.* **15**, 397 (1987)] can significantly reduce this time. Such amidites are currently available from Applied Biosystems, Sigma, and Pharmacia-LKB.

Synthesis of Amino Linker Phosphoramidite

N-Trifluoroacetyl-6-aminohexan-1-ol

Reagents

Ethyl trifluoroacetate and 6-aminohexan-1-ol (Aldrich, Milwaukee, WI)
AG50-X8 cation-exchange resin (H⁺ form) (Bio-Rad, Richmond, CA)
Silica gel 60 F254 thin-layer chromatography (TLC) plates, 5 × 10 cm (Whatman, Clifton, NJ)
Reagent-grade absolute methanol and chloroform

Procedure. 6-Aminohexan-1-ol (11.7 g, 100 mol) is dissolved in absolute methanol (100 ml), and ethyl trifluoroacetate (71.0 g, 500 mmol) is added. The reaction mixture is heated under reflux overnight. The solution is then cooled to room temperature, and AG50-X8 cation-exchange resin (10 g dry weight) is added. After stirring for 1 hr, the resin is filtered and washed with methanol (50 ml). The combined methanol solutions are evaporated to give a pale brown liquid, which is distilled under high vacuum (oil pump) to give a clear viscous liquid that solidifies on standing at room temperature. The yield is 18.1 g (85%), and the molecular weight is 213. Thin-layer chromatography on silica gel in 9 : 1 (v/v) chloroform :

methanol and visualization using iodine vapor shows a single species, R_f 0.48.

N-Trifluoroacetyl-6-aminohexan-1-ol 2-Cyanoethyl-N,N-diisopropylaminophosphoramidite

Reagents

2-Cyanoethyl N,N-diisopropylaminochlorophosphoramidite, anhydrous dichloromethane, and N,N-diisopropylethylamine (Aldrich)
Silica gel for flash chromatography (Baker, Phillipsburg, NJ)
Reagent-grade hexanes, ethyl acetate, anhydrous diethyl ether, and triethylamine

Procedure. N-Trifluoroacetyl-6-aminohexan-1-ol (1.06 g, 5 mmol) is dissolved in anhydrous dichloromethane (25 ml) in an oven-dried flask fitted with a rubber septum, and the solution is stirred under a dry argon atmosphere. N,N-Diisopropylethylamine (1.3 ml, 25 mmol) is added via syringe, followed by 2-cyanoethyl N,N-diisopropylaminochlorophosphine (2.2 ml, 10 mmol) dropwise. The reaction is stirred for 30 min at room temperature. The mixture is diluted with anhydrous diethyl ether (200 ml) and is extracted once with cold 5% (v/v) aqueous sodium bicarbonate (100 ml) and once with cold saturated aqueous sodium chloride (100 ml). The solution is dried with anhydrous magnesium sulfate, filtered, and evaporated to a thick, pale yellow oil. This oil is purified by flash chromatography[21] on a 6 × 1 in. column of silica gel using hexanes : ethyl acetate : triethylamine (60 : 30 : 10, v/v/v) as the eluant. Fractions are assayed for product by spotting on a TLC plate, spraying with a 0.25% (w/v) solution of ninhydrin in 1-butanol, and warming gently. The spray reacts with the product amidite, liberating N,N-diisopropylamine, which produces a yellow color on reaction with ninhydrin. Fractions containing product are pooled and evaporated to a clear, almost colorless oil. The yield is 1.18 g (55%), and the molecular weight is 413. Thin-layer chromatography on silica gel in 70 : 25 : 5 hexanes : ethyl acetate : triethylamine showed a single species (ninhydrin spray), R_f 0.42. The product is stored under dry argon in a desiccator at −20°.

Synthesis of 5'-Aminooligonucleotide

The synthesis of the aminooligonucleotide proceeds according to well-documented and widely used procedures. Any automated DNA synthesizer and attendant reagents from a reputable manufacturer may be used.

[21] W. C. Still, M. Khan, and A. Mitra, *J. Org. Chem.* **43**, 2923 (1978).

The crude dried product is precipitated from LiCl with an ethanol/acetone mixture to remove residual ammonium ions from the cleavage and deprotection that would deleteriously compete with the 5′-amino group during subsequent dye labeling.[22]

Reagents

Standard reagents for automated DNA synthesis by the solid-phase phosphoramidite method

N-Trifluoroacetyl-6-aminohexan-1-ol amidite (0.2 M) in anhydrous acetonitrile

LiCl (4 M)

Ice-cold absolute ethanol : acetone (1 : 1, v/v)

Procedure. Assemble the oligonucleotide sequence as usual. Because little primer DNA is required for use in the sequencing reactions, perform the synthesis on the smallest scale possible, commonly 0.1 to 0.2 μmol. The amino linker amidite is added as the final coupling, using a cycle identical to that for the standard nucleotide amidites. End the procedure using the "5′ trityl on" option. Cleave and deprotect the modified DNA under standard conditions (concentrated ammonium hydroxide, 55°, six or more hours). Evaporate the oligonucleotide-containing solution to dryness in a vacuum centrifuge. Completely dissolve the crude DNA pellet in 4 M LiCl (200 μl). Add the cold ethanol/acetone mixture (1 ml); a flocculent white precipitate should form almost immediately. Allow to stand at $-20°$ for at least 1 hr. Collect the precipitate by centrifugation. Decant the solution from the precipitate carefully, and dissolve the white pellet in distilled, deionized H_2O (200 μl).

Synthesis of Fluorochrome-Labeled Oligonucleotides

The following procedure is quite ubiquitous, and has been used successfully in our laboratory to modify aminooligonucleotides with a variety of moieties, such as biotin, fluorochromes, chelating agents, and cross-linkers.

Reagents

Crude aminooligonucleotide solution [about 0.2 μmol in distilled, deionized H_2O (200 μl)]

Aqueous sodium bicarbonate (1 M), adjusted to pH 9 with NaOH

---

[22] C. Levenson and C. Chang, *in* "PCR Protocols: A Guide to Methods and Applications" (M. A. Innis, D. H. Gelfand, J. J. Sninsky, and T. J. White, eds.), p. 99. Academic Press, San Diego, 1990.

Activated dye solutions (FAM–NHS, JOE–NHS, TAMRA–NHS, and ROX–NHS), 5 mg/60 μl dimethyl sulfoxide (Applied Biosystems)

NAP-25 gel-filtration columns (Pharmacia-LKB) (homemade 10-ml Sephadex G-25 columns can be used as well)

Triethylammonium acetate (TEAAc) (0.1 M), pH 6.5–7.0 (made by dilution of 1 M stock solution[23])

Procedure. For each of the four dye-labeling reactions, combine amino oligonucleotide solution (50 μl; about 50 nmol of DNA) and 1 M sodium bicarbonate, pH 9 (25 μl). Mix and add the appropriate activated dye solution [6 μl (0.5 mg activated dye); about a 10-fold molar excess]. Vortex well and let sit in the dark for 2 hr. If the reaction cannot be worked up at this point, store it at 4° in the dark.

Gel filtration is used to remove excess dye (low molecular weight) from the desired product (high molecular weight). The separation should be performed under subdued lighting. Equilibrate a NAP-25 column with 0.1 M TEAAc. Apply the reaction mixture to the column, and elute with 0.1 M TEAAc. The colored material rapidly separates into two bands. The faster moving band contains the dye-labeled oligonucleotide along with unreacted nucleic acid material. This band is collected; it should elute in approximately 1.0 to 1.5 ml. Usually this colored band is sufficiently intense to observe its progress visually. If not, collection can be simplified by *occasionally* and *briefly* observing the effluent of the column under a long-wave UV (366 nm) lamp for the appearance of fluorescent material. The separation of the two bands should be distinct, although often the separation of TAMRA-labeled oligonucleotide from residual dye is poorly resolved. In this case, collect only the first 1.0–1.5 ml of fluorescent column effluent, which will contain the desired product contaminated by a little free dye. The crude dye-labeled oligonucleotide solutions are stored at 4° in the dark prior to further purification.

Purification of Fluorochrome-Labeled Oligonucleotides

Purification of the crude dye-labeled oligonucleotides is most easily accomplished using reversed-phase high-performance liquid chromatography. Standard analytical (4.6 × 250 mm) C_{18} or C_8 reversed-phase columns suitable for this application are available from a variety of sources. Alterna-

---

[23] The 1 M TEAAc stock buffer is prepared by combining distilled, deionized H_2O (800 ml) and reagent-grade triethylamine (1 mol, about 140 ml), and chilling the mixture in an ice–water bath. The mixture is stirred briskly during the slow, dropwise addition of glacial acetic acid (about 60 ml) to give a final pH of 6.5 to 7.0. The solution is filtered through a 0.45-μm nylon membrane and is stored at 4°.

tively, the products can be purified by preparative polyacrylamide gel electrophoresis.

Reagents

Gradient buffer A [98 : 2 (v/v) 0.1 M TEAAc : HPLC-grade acetonitrile]
Gradient buffer B [50 : 50 (v/v) 0.1 M TEAAc : acetonitrile]

Apparatus

HPLC system (injector with 1.0-ml or larger loop, pump, controller, gradient mixer, and detector)
Analytical C_8 or C_{18} column (Axxiom, Calabasas, CA)
Fraction collector (optional)

Procedure. Load the injector with about half of the crude product solution obtained in the previous step. Perform the chromatographic separation using a flow rate of 1.0 ml/min and a biphasic linear gradient of 90% buffer A : 10% buffer B to 70% A : 30% B over 30 min, then 70% A : 30% B to 30% A : 70% B over 15 min. Hold at 30% A : 70% B for 5 min, then return to 90% A : 10% B in 5 min. Collect the colored product peak, which elutes between 35 and 45 min using this gradient. Be conservative in collecting the product fraction, because any contaminating species can significantly degrade the performance of the labeled oligonucleotide in the sequencing reactions. Reequilibrate the column under these conditions for at least 20 min before the next injection. Figure 4 is a representative chromatogram for the purification of a FAM-labeled 25-mer.

The amount of product obtained from this procedure is sufficient for several hundred sequencing reactions. The purified dye–primers are best stored as lyophilized solids at $-20°$ in the dark. We generally store 80-pmol aliquots. To determine the concentration of the purified dye–oligonucleotide in the HPLC buffer, dilute an aliquot with 0.1 M sodium bicarbonate, pH 9, sufficiently so that the A_{260} of the resulting solution is in the linear range of the spectrophotometer. Measure the absorbance spectrum from about 240 to 640 nm. Two absorbance maxima should be observed, one for the nucleic acid, around 260 nm, and one for the fluorochrome, in the visible region. Table II gives the spectral characteristics of typical primer-length oligonucleotides labeled with each of the four dyes. Use the following formula, based on Beer's Law and the molar absorptivities of each of the bases, to calculate the concentration of the product solutions:

$$C \; (\text{pmol}/\mu\text{l}) = 100(A_{260})(D)/(1.54n_A + 0.75n_C + 1.17n_G + 0.92n_T)$$

where n_X is the number of bases of type X in the sequence, and D is the dilution factor. Remove sufficient volume to contain 80 pmol of each dye–primer, and dry completely in a vacuum centrifuge in the dark.

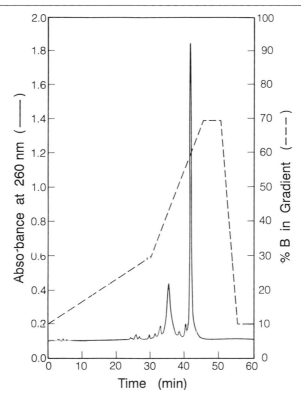

FIG. 4. Typical preparative scale reversed-phase high-performance liquid chromatogram of a dye-labeled oligonucleotide using a 4.6 × 250 mm ODS (C_{18}) column (Axxiom). The gradient conditions are described in the text. The flow rate is 1.0 ml/min. The chromatogram corresponds to the FAM-labeled oligonucleotide sequence 5'-CAGGAGCTGC-TCCATTTCCAAACTG-3'. The major peak at about 42 min is the desired product. The small peak at about 36 min is due to residual unlabeled nucleic acids (failure sequences and unreacted aminooligonucleotide).

Sequencing Reactions

The procedures for the use of custom labeled specific primers in DNA sequencing are exactly those given previously for use with the commercial universal primer. Representative data for a specific primer selected from the sequence data in Fig. 2 are given in Fig. 5. As can be seen in Fig. 2, the first 450 bases of the sequence from the universal primer is sufficiently reliable for use in the selection of the next primer sequence. The first candidate, 5'-ATAATTCCCATCTTTCCCAA-3', was chosen near the 3' end of this sequence at nucleotide positions 414 to 433 to minimize the overlap between the initial round and the subsequent round of sequencing,

TABLE II

SPECTRAL CHARACTERISTICS OF DYE-LABELED PRIMERS[a]

Label	λ_{max} (DNA)[b]	λ_{max} (dye)[c]	A_{DNA}/A_{dye}[d]
FAM	255–260	495–496	2.7–3.1
JOE	251–256	525–528	3.0–3.7
TAMRA	254–258	556–562	2.7–3.4
ROX	255–262	583–586	2.4–3.2

[a] Based on values obtained from nine sets of dye–primers, lengths 23 to 25 bases. Spectral data were obtained in 0.1 M NaHCO$_3$, pH 9.0.

[b] Wavelength (nm) of maximum absorbance due to DNA bases.

[c] Wavelength (nm) of maximum absorbance due to dye.

[d] Ratio of the absorbance at the λ_{max} (DNA) to the absorbance at the λ_{max} (dye). This number is characteristic of dye-labeled oligonucleotide. Values greater than this are characteristic of unlabeled oligonucleotide; values less than this are characteristic of free dye.

while maintaining a suitably high T_m (48°) and a reasonable G/C content (35%) in an A/T-rich region. However, a search revealed significant similarity between the 3' end of this primer (TTCCCAA) and two other sites in the sequence, positions 418–423 (TTCCCA) and positions 98–104 (TTC-CCAA). Thus, this sequence was discarded. The next alternative of sufficient T_m (52°) and appropriate G/C content (50%) was chosen at positions 320–339 (5'-GCTGCTCCATTTCCAAACTG-3'). This candidate showed no significant complementarity to any known sequence of the clone, and conforms very well to the rules presented previously. The dye–primer set corresponding to this sequence, appended with the sequence CAGGA at its 5' end for mobility correction, was synthesized and utilized in the next round, again employing the cycle sequencing protocol. This primer yielded over 400 bases of quality analyzed data. Comparison of the overlapping region of the two runs, bases 340–450 of the sequence obtained from the universal primer and bases 1–109 of the sequence obtained with the specific primer, gives excellent correlation.

Concluding Comments

This chapter has described in detail the current state of the synthesis and use of fluorochrome-labeled primers in automated DNA sequence analysis. The Applied Biosystems instrument has undergone significant

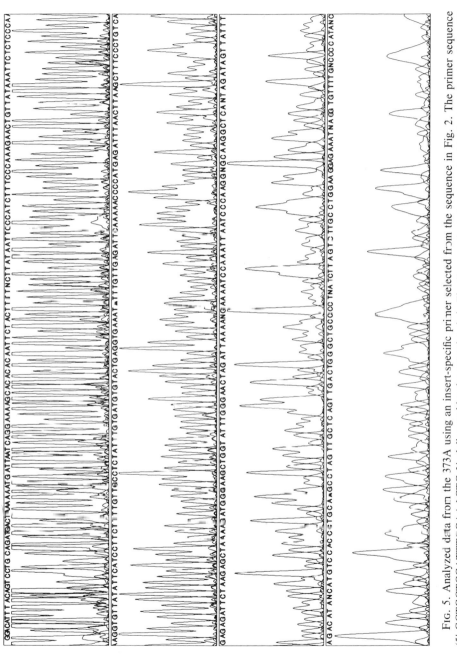

FIG. 5. Analyzed data from the 373A using an insert-specific primer selected from the sequence in Fig. 2. The primer sequence (5'-GCTGCTCCATTTCCAAACTG-3'; delineated by the arrowheads in Fig. 2) was selected as described in the text. Depicted are 436 bases (from the end of the primer) of automatically analyzed sequence obtained using the cycle sequencing protocol.

development since its inception, particularly in the areas of sequencing protocols and data analysis, and has become a reliable tool for the generation of large volumes of DNA sequence information. Accuracies in automated base-calling routinely exceed 95% for at least 450 bases from the priming site when the procedures described herein are followed carefully and completely.

The routine applicability of dye–primers in conjunction with four-color fluorescence detection is determined by the time, effort, and expense required to synthesize and purify the required labeled primer sets. As described previously, the minimum time for the synthesis and purification of a single set of dye–primers is about 48 hr. The majority of this time is spent in the synthesis of the oligonucleotide itself, and is thus a necessary expenditure in any specific primer-directed project. Once the crude amino oligonucleotide is obtained, the dye-labeling reactions are rapid and efficient. Purification of the labeled primers is currently a rate-limiting step, requiring a significantly more parallelized procedure than HPLC, or a less tedious and time-consuming process than gel electrophoresis, extraction, and desalting. Additionally, a tremendous excess of each dye–primer is produced using these protocols at a considerable cost. All of these considerations indicate that the use of labeled primers in conjunction with the 373A is currently best suited for the rapid analysis of multiple sequences using universal primers, and not for gap closure or gene walking, except in a limited way or under extraordinary circumstances. It simply takes too long and costs too much to make a set of four dye-labeled primers for a one-time use. The techniques described above can be productively applied to the synthesis and use of dye–primer sets as universal primers for vector systems other than M13. In this regard, Applied Biosystems now markets several universal primer sets for use either in M13 or common plasmid vectors. We have used these procedures successfully in our laboratory for generating universal dye–primer sets for various cosmid vectors.

A further constraint on the utilization of specific primers is the efficiency with which a candidate sequence functions as a primer in the enzymatic sequencing process. Once again, this constraint is much less restrictive in the case of universal primers, because one can afford to spend some time optimizing the primer sequence, time that will later be made up by the large number of sequences obtained for this single primer set. Optimization of each candidate primer in either a gap-filling or gene-walking strategy is once again too time consuming to be practical. Although following the selection rules presented above does not guarantee 100% success, they do serve as useful guidelines. Our experience with the use of specific primers and *Taq* DNA polymerase indicates that a majority of the primer sequences chosen according to these selection rules will afford significant (>400 bases) quality sequence from the automated sequencer.

Two developments in DNA sequence analysis using fluorescence detection may help overcome some of the limitations discussed above. As mentioned previously, following the lead of Du Pont, Inc.,[4] Applied Biosystems has introduced a new series of sequencing reagents and protocols in which the dideoxynucleotide analogs are themselves labeled with the four fluorochromes. This chemistry has several advantages over the dye–primer approach for SPDS. First, this "dye-terminator" chemistry uses standard, unlabeled primers. Second, it can simplify the processing of the DNA sequencing reactions because all four reactions can be performed in a single vessel. At present, the protocols for the use of dye-labeled dideoxynucleotides are still under intensive study, because the current versions typically yield runs of poorer length and quality than the dye–primer protocols. The second development is the marketing of an automated sequencer based on the work of Ansorge and co-workers at the European Molecular Biology Laboratory[5] by Pharmacia-LKB. The automatic laser fluorescent (ALF) automated DNA sequencer essentially replaces the single radioactive label used in conventional DNA sequencing with a single fluorescent label. The four reactions are performed separately with a single dye–primer, and each set is electrophoresed in four adjacent gel lanes, one reaction per lane. SPDS using the ALF instrument thus requires the synthesis and purification of a single dye–primer, instead of a set of four. This procedure is further simplified by the use of a fluorescent amidite derivative to introduce the dye label using standard procedures on the automated DNA synthesizer. This instrument is also undergoing continuous development, particularly relating to data analysis, in an attempt to overcome the problems related to lane-to-lane data comparison and registration inherent in the single-label, four-lane approach. Significant improvements in either of these two major technologies, or in the procedures for the fluorescence labeling of primers for use on the 373A, could improve the attractiveness of the gap-filling and gene-walking strategies for large-scale DNA sequencing projects

Acknowledgments

We thank Barbara Otto-Perry for expert technical assistance with the automated sequencer. We also thank Dr. Ben Koop for the generous donation of the M13-cloned DNA used for the purposes of illustration in this chapter, and for numerous helpful discussions on DNA sequence analysis. We greatly appreciate the valuable collaborative assistance and advice of the scientists at Applied Biosystems, Inc. This work has been supported by a grant from the National Science Foundation (#DIR8809710).

[12] High-Speed DNA Sequencing by Capillary Gel Electrophoresis

By John A. Luckey, Howard Drossman, Tony Kostichka, and Lloyd M. Smith

Introduction

The capabilities of current methods of novel and automated DNA sequence analysis are insufficient for the needs of large-scale sequencing projects such as the Human Genome Initiative. There is accordingly great interest in the development of faster, less expensive, and more highly automated methods for DNA sequencing. Approaches to this problem fall into two general categories: Gel-based approaches, which are evolutionary variations of existing methods for DNA sequencing; and non-gel-based approaches, in which radically new methods for DNA sequencing are sought. Although the latter approaches are appealing, particularly given the insatiable demands for sequence information evident in the research and medical community, they remain at this time largely speculative and unproven. Thus, in the near term, at least, some variation on standard gel-based sequencing technologies seems likely to remain the method of choice.

Current fluorescence-based automated sequencing instruments are, however, rather expensive. Much of the cost of the instruments is unavoidable, reflecting the use of expensive laser sources, computer and data acquisition systems, custom optical systems, and in some cases complex mechanical subassemblies. Inasmuch as it is difficult to pare these systems down substantially, efforts to decrease the cost of sequencing must turn rather to seeking ways of increasing their throughput.

The slow step in the operation of automated DNA sequencers is electrophoresis. Typically, the separation requires 12–14 or more hours to take place, during which the expensive instrument is fully occupied. If the speed of the separation could be substantially increased, the throughput of the instrument could be increased correspondingly, decreasing the cost of the process.

The speed of these separations can be increased either by (1) increasing the strength of the electric field in the gel, (2) shortening the gel, or (3) both. The latter approach decreases the resolution between the DNA fragments, causing fewer sequence data to be obtained in a given analysis, and is thus in most cases not preferred. The former approach is not feasible in conventional electrophoresis, as the increased field causes increased ohmic, or Joule heating, leading to gel breakdown and aberrant results. However, by performing the gel electrophoretic separations in extremely thin capillaries, in a technique known as capillary gel electrophoresis (CGE), a much greater field may be applied to the gels without deleterious effects, yielding correspondingly more rapid separations.

Separations up to 10–20 times more rapid may be performed by CGE. In addition, the ability to fabricate very long capillaries may eventually permit very long sequence reads to be obtained, and the extremely low absolute detection limits may permit sequence analysis to be effectively performed on much smaller amounts of template DNA than are commonly used. We describe here the apparatus, gel fabrication, and biochemical and data acquisition and analysis methods needed to perform fluorescence-based automated DNA sequence analysis[1,2] by capillary gel electrophoresis.

Capillary Gel Electrophoresis Instrumentation

Overview

The major components of a typical CGE system are identical to those used in open-tube capillary electrophoresis. These components include the high-voltage power supply, two buffer reservoirs, and a detection scheme that is suited to the sample under study. The major difference between the two techniques is in the use of a semisolid gel matrix that is cast inside the fused-silica capillary.

Figure 1 shows a block diagram of the CGE instrument used in fluorescence-based DNA sequence analysis. The instrument consists of two Plexiglas buffer chambers (equipped with platinum wire electrodes), which are each enclosed in a separate Plexiglas box as shown. The boxes are connected to a safety interlock that automatically shuts off the high voltage when either door is opened. The voltage is supplied from a +30-kV power supply (model RC10-30P; Gamma High Voltage Research, Mt. Vernon, NY) that is regulated by an internal 5-V DC power supply, connected to a potentiometer.

For the purposes of DNA sequencing, the separations are performed in capillaries that are filled with polyacrylamide gel. This allows the sample to be separated based on its molecular weight rather than on its free

[1] L. M. Smith, S. Fung, M. Hunkapillar, T. Hunkapillar, and L. E. Hood, *Nucleic Acids Res.* **13**, 2399 (1985).
[2] L. M. Smith, J. Z. Sanders, R. J. Kaiser, P. Hughes, C. Dodd, C. R. Connell, C. Heiner, S. B. Kent, and L. E. Hood, *Nature (London)* **321**, 674 (1986).

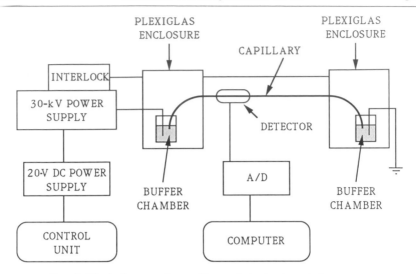

FIG. 1. Block diagram of a capillary gel electrophoresis instrument.

solution electrophoretic mobility, as is commonly done in open-tube capillary electrophoresis. Because the capillary is filled with a gel, the sample cannot be loaded hydrostatically,[3] either by creating a partial vacuum at the anodic end of the capillary or by applying a pressure to the sample solution at the cathodic end of the capillary. Instead, the sample must be injected electrokinetically.[4,5] This method of injecting sample relies on an electric field applied across the sample solution between the cathodic end of the capillary and the cathode to provide a driving force on the sample anions to enter the capillary. Although in principle this type of injection can bias the sample injection by loading more of the higher mobility constituents,[6] this effect is not noticeable for DNA samples because the free solution mobilities of DNA molecules do not vary appreciably with length.[7]

Sample Injection. The electrokinetic injection is performed by removing the cathodic end of the capillary from the buffer chamber and inserting it into the sample solution. The cathode is then placed in the same sample solution and an electric potential is applied across the capillary and the sample for a short period. During this time, the electric field in the sample drives the sample anions into the gel matrix. After the voltage has been applied, the capillary is removed from the sample and replaced in the

[3] S. Fujiwara and S. Honda, *Anal. Chem.* **58,** 1811 (1986).
[4] J. W. Jorgenson and K. D. Lukacs, *Anal. Chem.* **53,** 1298 (1981).
[5] K. D. Lukacs and J. W. Jorgenson, *J. High Resolut. Chromatogr.* **8,** 407 (1985).
[6] X. Huang, M. J. Gordon, and R. N. Zare, *Anal. Chem.* **60,** 375 (1988).
[7] B. M. Olivera, P. Baine, and N. Davidson, *Biopolymers* **2,** 245 (1964).

cathodic buffer chamber. The high voltage can then be turned back on, and electrophoresis begins.

Because the electrokinetic injection relies on the electric field in the sample, there are several factors that will affect the loading efficiency of the sample. These variables must be held constant if reproducible injections are to be made. Although a direct determination of the electric field in the sample is difficult, a great deal of insight can be obtained by considering the behavior of the drop in electric potential that occurs across the sample. If the buffer–capillary–sample system is treated as a set of resistors in series, one can obtain the following equations relating the voltage drop across the sample to the applied voltage, and the resistances of the sample, buffer and capillary.

$$V_{app} = I(R_B + R_C + R_S) \tag{1}$$
$$V_S = IR_S \tag{2}$$
$$V_S = V_{app}R_S/(R_B + R_C + R_S) \tag{3}$$

In Eqs. (1)–(3), I is the current in the circuit, R_B refers to the buffer resistance between the anode and the anodic end of the capillary, R_C refers to the resistance of the gel-filled capillary, R_S is the resistance of the sample solution between the cathodic end of the capillary and the cathode, V_{app} is the voltage applied to the electrodes, and V_S is the potential drop across the sample. From Eq. (3) it is apparent that the potential in the sample solution varies linearly with the applied voltage and the sample resistance, provided R_S is negligible compared to $R_B + R_C$. This latter stipulation is usually the case because the resistance of the capillary is on the order of 10^{10} Ω. Another factor that will influence the voltage in the sample is the buffer that is used for electrophoresis. A more conductive buffer will lower both R_B and R_C, which will consequently raise the voltage in the sample solution and result in a higher sample electric field.

These factors illustrate the need for careful control of the sample, gel, and buffer conductivities. Geometric factors such as the volume of the sample, container shape, and the relative orientation of the electrode and capillary during injection and electrophoresis are also important for reproducible results. When these variables are controlled, reproducible injections of samples are readily achieved.

Detection. The detector for our experiments is a laser-induced fluorescence detector, which is illustrated in detail in Fig. 2 and has been described previously.[8] A 15-mW beam from an air-cooled argon ion laser (Cat. No. 532; Omnichrome, Chino, CA) operated in multiline mode is brought to a height of approximately 20 cm above and parallel to an optical

[8] J. A. Luckey, H. Drossman, A. J. Kostichka, D. A. Mead, J. D'Cunha, T. B. Norris, and L. M. Smith, *Nucleic Acids Res.* **18,** 4417 (1990).

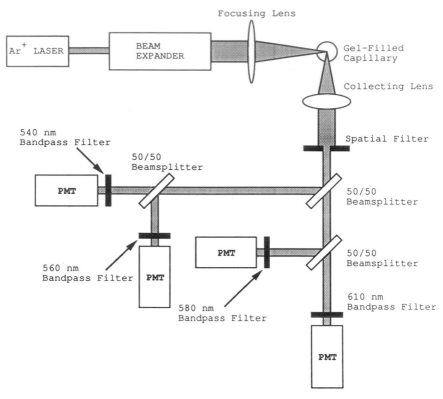

FIG. 2. Schematic diagram of the optical system employed for multiple-wavelength fluorescence detection in the capillary gel electrophoresis instrument. (Figure 2 reprinted with permission from Ref. 8, copyright 1990, *Nucleic Acids Res.*)

table (Cat. No. MST46 with Cat. No. NN4-28 legs; Newport Research Corporation, Fountain Valley, CA) using a beam director (Cat. No. 670; Newport Research Corporation). The beam is then expanded with a beam expander (Cat. No. 15280; Oriel Corporation, Stratford, CN) to a final beam width of about 40 mm and focused to a 50-μm spot size using a long focal length planoconvex lens (Cat. No. 01LPX277; Melles Griot, Irvine, CA).

Because the capillary gel is most easily mounted so that it runs parallel to the optical table and because the detection optics must be perpendicular to the excitation beam and the capillary, the focusing laser is directed in an upward direction to keep the detection optics parallel to the optical table. This is accomplished using a flat mirror between the focusing element and the capillary mount, as illustrated in

Fig. 3A. Briefly, the flat mirror is held fixed beneath the capillary mount and it directs the focusing beam through a 1/4-in. diameter hole on the mount. The focused beam thus strikes the capillary from below at the detection window, where the polyimide coating has been stripped away. The mount itself is machined from 1/4- and 1/2-in. thick aluminum. Teflon cylinders hold the capillary rigidly between the two aluminum posts. The entire apparatus rests on an XY translation stage (Cat. No. 400; Newport Research Corporation), which allows the position of the capillary to be adjusted.

The emitted fluorescence is collected perpendicular to the capillary and the excitation beam. Depending on the orientation of the capillary, the detection optics may be placed at any angle in the plane perpendicular to the incident laser light. For convenience, the capillary may be mounted at a 45° angle with respect to the path of the laser before striking the final mirror, as shown in Fig. 3B. This allows the detection optics to be easily aligned on the optical table. A 60× microscope objective (Cat. No. 79233 ELWD 60×; Nikon, Garden City, NY) is placed in a fine-focusing microscope body (Cat. No. 04TFF002; Melles Griot) and positioned with its focal point at the detection window of the capillary. The objective has an extra long working distance (4.90 nm), which allows even thick walled capillaries to be easily imaged at high magnification. The microscope body is firmly attached to a translation stage (Cat. No. 420; Newport Research Corporation), which allows movement parallel to the capillary and to an elevating platform with a fine positioner [Newport Research Corporation platform (Cat. No. 300) with positioner (Cat. No. 32A)], which allows the objective to be raised or lowered. Focusing adjustments are made possible with the coarse and fine focusing knobs on the microscope body. With the objective in proper position, the magnified image of the capillary is focused onto a horizontal slit that spatially filters the reflection of the excitation beam off the inner and outer walls of the capillary.

Finally, the spatially filtered emission is split into four equal parts using three 50/50 beamsplitters (Cat. No. 03BTF023; Melles Griot). Each of these parts is directed to a separate photomultiplier tube (Cat. No. R928; Hamamatsu Corporation, Bridgewater, NJ) that has one of four 10-nm bandpass filters (Cat. No. xxx-DF-10 series; Omega Optical, Inc., Brattleboro, VT; xxx denotes the peak transmission wavelength). The four bandpass emission filters have peak transmission wavelengths of 540, 560, 580, and 610 nm. These filters allow the discrimination of the four dyes used in fluorescent DNA sequencing.

In addition, it is crucial that the detection system be rigidly fixed so that it functions in a reproducible manner. Also, the system must be well

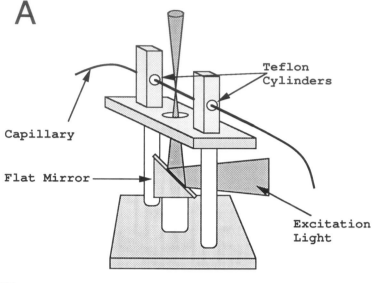

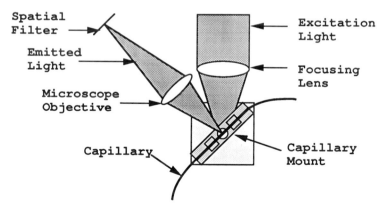

FIG. 3. (A) Side view of the capillary mount used in the CGE instrument. Excitation light enters from the right side and is reflected upward through the capillary. (B) Top view of the same mount, illustrating both the incident light as well as the emitted fluorescence that is collected and imaged onto the spatial filter.

baffled from stray light to minimize background. Simple baffling can be constructed from heavy duty poster board with the primary purpose of isolating the photomultiplier tubes as much as possible from the scattered excitation light from the capillary as well as the light rejected by the spatial filter. A more elegant baffle can be constructed from either aluminum or Plexiglas boxes.

Data Acquisition. The data collection hardware and software are designed around an eight-channel, 12-bit analog-to-digital converter (ADC) board (Cat. No. AD1000; Real Time Devices, State College, PA). This ADC is operated as an expansion board for an IBM PC/AT-compatible computer with enhanced graphics adapter (EGA) graphics.

The interface circuit that converts the electrical current from each photomultiplier tube into a voltage compatible with the ADC is diagrammed in Fig. 4. There are four identical interface circuits for each photomultiplier tube (PMT). The first step in the interface is accomplished by U1, which converts the current output into a voltage signal. The transformation is such that a 1-μA current from the PMT is converted to a 10-V output voltage. This signal is then sent to U2A, which is a third-order Bessel low-pass filter. The filter has a cutoff frequency of 1 Hz, which blocks any high frequency noise in the signal while passing the lower frequencies. This filter also prevents false signals due to beats between the signal frequency and the sampling frequency, which is set at 10 Hz. The filtered 0- to 10-V voltage is then offset to match the -5- to $+5$-V range of the ADC, using a 5-V reference from U3 and U2B, which performs the DC offset. Finally, the ADC converts the signal voltages into digital form.

The data collection program, written in Turbo Pascal, collects the signals from the four channels every 100 msec and converts them into integer form (2 bytes/integer). The program also displays a real-time plot of the signals from the four channels as they are collected. A second plot is also shown that illustrates the entire length of the run and the data collected from time zero to the current moment. The four channels are superimposed in distinct colors on the monitor screen. Data are stored on the hard disk of the computer during the run to prevent accidental losses. Once the run is completed, the four channels can be graphed with different time and intensity scales by selecting the first and last time points and setting the upper and lower bounds for each of the channels. The screen can also be split vertically into as many as eight windows, allowing a greater total time to be shown at once. Hard copies of the resultant data can be generated either on a laser printer or an HP 7475 pen plotter.

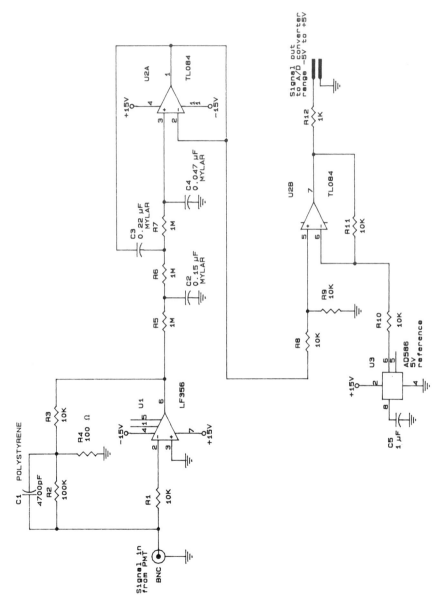

FIG. 4. Circuit diagram showing one of four identical circuits used for each photomultiplier tube interface.

Gel Fabrication

Overview

The procedures for the production of gel-filled capillaries are simple but can prove to be a problem if reagent quality and storage procedures are not carefully monitored.[9] Reproducible methods and freshly prepared reagents should be used to ensure consistent results. A major problem often encountered with capillary gels is the formation of bubbles within the capillary tube that lead to a loss of resolving power or a complete failure to elute samples. The formation of bubbles in the capillary during polymerization probably originates either from outgassing of the solvent or from a shrinkage of the polymer solution as it polymerizes.[10]

To avoid the formation of bubbles while pouring cross-linked capillary gels, procedures relying on pressure,[10] point source polymerization,[11] underivatized capillary tubes,[12,13] γ radiation-initiated polymerization,[14] and low-temperature polymerization have been employed.[13] Alternatively, the use of linear acrylamide (no cross-linking)[16] has been demonstrated for separating DNA fragments.[15,17] We have successfully used the pressure method and the low-temperature method of polymerization. The procedures for both of these methods are provided below.

Fused silica capillary tubing with a polyimide coating is preferred for capillary electrophoresis due to the tensile strength provided by the coating and the excellent optical properties of the fused silica. The tubing we employ has a 375-μm o.d. with inner diameters ranging from 50 to 150 μm [Cat. No. TSP(i.d.)375; Polymicro Technologies, Phoenix, AZ; i.d. indicates the desired inner diameter]. Thinner wall tubing allows improved heat transfer and is less expensive. The polyimide coating of the capillary is easily removed by burning a 1- to 10-mm section with a small flame. An alternative procedure employs a simple electrical heating device.[18] The coating can also be removed after the gel is formed either chemically, by a few hours of exposure to concentrated sulfuric acid, or mechanically,

[9] Bio-Rad Application Note, Bulletin No. 1156 (1984).
[10] P. F. Bente and J. Meyerson, inventors and H.-P. Co., Eur. Pat. Appl. EP272925, June 29, 1988; U.S. Pat. Appl. 946,568, Dec. 24, 1986.
[11] V. Dolnik, K. A. Cobb, and M. Novotny, *J. Microcolumn Sep.* **3**, 155 (1991).
[12] H. Yin, J. A. Lux, and G. Schomburg, *J. High Resolut. Chromatogr.* **13**, 624 (1990).
[13] Y. Baba, T. Matsuura, K. Wakamoto, and M. Tsuhako, *Chem. Lett.* p. 371 (1991).
[14] J. Lux, H. Yin, and G. Schomburg, *J. High Resolut. Chromatogr.* **13**, 436 (1990).
[15] D. White, H. Drossman, and H. Drayna, *Biotechniques* **13**, 232 (1992).
[16] D. Tietz, M. H. Gottlieb, J. S. Fawcett, and A. Chrambach, *Electrophoresis* **7**, 217 (1986).
[17] D. N. Heiger, A. S. Cohen, and B. L. Karger, *J. Chromatogr.* **516**, 33 (1990).
[18] J. A. Lux, U. Hausig, and G. Schomburg, *J. High Resolut. Chromatogr.* **13**, 373 (1990).

with the use of a minilathe.[19] Care must be taken when handling the capillary after the removal of the polyimide because the bare fused silica is quite fragile.

Derivatization of Capillary Wall

Reagents

γ-Methacryloxypropyltrimethoxysilane (Sigma, St. Louis, MO)

Procedure. Capillary tubes cut to lengths from 40 to 100 cm are normally employed in the sequencing experiments. Alumina wafers give a particularly clean cut of the capillary ends. After trimming the capillaries and burning the polyimide to form detection windows, the capillaries are rinsed either by mild suction (approximately 100 mmHg) or by low pressure (syringe) for 10 min with 0.1 M NaOH, 10 min with deionized water, and 10 min with acetone. The acetone is then flushed out with a stream of nitrogen or air. A 0.2% (v/v) solution of a bifunctional reagent, γ-methacryloxypropyltrimethoxysilane, in 50% (v/v) ethanol/water is then added to the capillary and allowed to sit for 1 hr. The capillary tube is then flushed with air to remove the residual reagent and dried under a heat lamp for 45 min. The derivatized capillaries may be stored indefinitely before the addition of acrylamide reagents.

Low-Temperature Polymerization

Reagents

Urea (enzyme grade; Bethesda Research Laboratories, Gaithersburg, MD)

Acrylamide (electrophoresis grade; Bethesda Research Laboratories)

N,N'-Methylenebisacrylamide (electrophoresis grade; Bethesda Research Laboratories)

Tris(hydroxymethyl)aminomethane (reagent grade; Sigma, St. Louis, MO)

Boric acid (Mallinckrodt, Paris, KY)

Ammonium persulfate (Aldrich, Milwaukee, WI)

Tetramethylethylenediamine (electrophoresis grade; Bethesda Research Laboratories)

Procedure. The preparation of the acrylamide solution follows standard procedures for denaturing slab gels,[9] with the exception that only small volumes are required for each capillary (a few microliters per capillary). For 10 ml of acrylamide solution 70 mmol of urea, 285 mg of acryl-

---

[19] R. M. McCormick and R. J. Zagursky, *Anal. Chem.* **63**, 750 (1991).

amide monomer, 15 mg of bisacrylamide cross-linker, 1 mmol Tris base, and 1 mmol boric acid are added and the volume brought to 10 ml with deionized water. The mixture is filtered through a 0.2-μm filter and then carefully degassed by vacuum and placed at 4° while remaining under vacuum. To the 10 ml of sparged solution, 50 μl of 10% (v/v) ammonium persulfate (APS) and 50 μl of 10% (w/v) tetramethylethylenediamine (TEMED) in water is added with careful mixing. The mixture is immediately syringed into the capillary tubes. After five or more column volumes have passed through each capillary, the ends are capped and allowed to polymerize at 4°. Total polymerization time for the bulk solution is approximately 30 min.

Polymerization under Pressure

Reagents

AG501 mixed-bed resin (Bio-Rad, Richmond, CA)

TBE buffer (10×): 0.89 M Tris base, 0.89 M boric acid, 22 mM disodium ethylenediaminetetraacetic acid (EDTA); approximately pH 8.3

Procedure. This procedure employs pressure instead of low-temperature polymerization to reduce bubble formation. Typically, we pour 4% (w/v) total acrylamide gels with 5% cross-linking. This is done by dissolving 380 mg of acrylamide, 20 mg of bisacrylamide, and 5.0 g of urea in deionized water. The mixture is then deionized on a bed of AG501 beads. The solution is filtered through a 0.2-μm filter, 1 ml of 10× TBE is added, and the volume is brought to 10 ml. The acrylamide mixture is then initiated with 50 μl of 10% APS and 4.5 μl TEMED. The polymerizing solution is then drawn up into the derivatized capillaries using a partial vacuum. Once the gel solution has filled all of the capillaries, the vacuum is released and the capillaries transferred to a stainless steel tube filled with water that can be pressurized either by a high-performance liquid chromatography (HPLC) pump or a hydraulic amplification pump.[10] The capillary is allowed to polymerize under a pressure of 100 atm for at least 1 hr before the pressure is slowly released.

Storage

After polymerization, the capillary tubes are inspected for bubbles under a microscope and then prerun at 150 V/cm for 1–2 hr to equilibrate the capillaries with the running buffer (1× TBE). The capillary tubes are typically stored at 4°, capped at both ends. The capillaries are stable for at least 1 month under these conditions.

TABLE I

COMPOSITION OF STOCK SOLUTIONS OF DEOXY- AND DIDEOXYNUCLEOTIDES FOR USE IN
FLUORESCENCE-BASED DNA SEQUENCING

	Sequencing reaction			
Nucleotide	A (mM)	C (mM)	G (mM)	T (mM)
dNTP[a] mix	dCTP (10)	dATP (10)	dATP (10)	dATP (10)
	dGTP (10)	dGTP (10)	dCTP (10)	dCTP (10)
	dTTP (10)	dTTP (10)	dTTP (10)	dGTP (10)
	dATP (1)	dCTP (1)	dGTP (1)	dTTP (1)
ddNTP[b] mix	ddATP (1)	ddCTP (1)	ddGTP (1)	ddTTP (1)

[a] dNTP, 2'-Deoxyribonucleoside 5'-triphosphate.

[b] ddNTP, 2',3'-Dideoxyribonucleoside 5'-triphosphate.

Sequencing Reactions

Preparation of Template DNA

Standard procedures for the preparation of template DNA can be found in Ref. 18. Commercial products from Pharmacia (Piscataway, NJ) and New England BioLabs (Beverly, MA) have shown no significant differences from template prepared in our laboratory.

Sequencing Reactions

Reagents

Deoxynucleotide triphosphates (dTTP, dATP, dCTP, and dGTP; Pharmacia)

Dideoxynucleotide triphosphates (ddTTP, ddATP, ddCTP, and ddGTP; Pharmacia)

M13 template solution (U.S. Biochemical, Cleveland, OH)

Fluorescent primer solutions: REG(A), RHO(C), TAM(G), ROX(T), (Applied Biosystems, Foster City, CA)

Bst buffer (10×): 100 mM Tris-HCl (pH 8.5), 100 mM MgCl$_2$

Bst polymerase (1 U/μl) Bio-Rad)

Procedure. The compositions of the various stock solutions of deoxynucleotide triphosphates (dNTPs) and dideoxynucleotide triphosphates (ddNTPs) for each sequencing reaction (A, C, G, and T) are given in Table I. Each solution is made up in deionized water and stored frozen at $-20°$ until needed.

The following describes the general procedure for preparing four-dye sequencing reactions from a 2-μg sample of template DNA. This procedure may be scaled up fourfold with no adverse effects.

The fluorescent primers are all used as 0.4 μM stock solutions in water and are kept shielded from light as much as possible during the procedure. Into each of four labeled Eppendorf tubes (A, C, G, and T), a 2-μl aliquot of the appropriate d/ddNTP mix is placed. An aliquot (0.5 μl) of the appropriate dye primer is then added to each tube and thoroughly mixed using the pipette. In a separate Eppendorf tube, the template DNA (4 μl of a 0.5-μg/μl solution of template), reaction buffer (1 μl of 10× Bst buffer), and 3.0 μl of deionized water are mixed together. Finally, a 0.5-μl aliquot of the Bst enzyme solution is added to the solution containing the template and buffer, resulting in a total volume of 8.5 μl. Into each of the four labeled tubes is placed a 2-μl aliquot from this solution. Immediately after addition of the aliquot, the tubes are capped and placed at 65° for 5 min.

After the 5-min incubation at 65°, the tubes are removed from the heating block one at a time, and 1 μl of a 100 mM (pH 8.0) solution of EDTA is added to each to terminate the sequencing reaction. The contents of the four tubes may now be combined into a single tube for the remainder of the procedure.

Once the reactions have been pooled, a 6-μl aliquot of 10 M ammonium acetate is added. This is followed by an 85-μl aliquot of cold (−20°) ethanol and the placement of the reaction tube at −20° for 15 min. The tube is then centrifuged in a Beckman (Fullerton, CA) microcentrifuge for 15 min. The supernatant is removed and replaced with 100 μl of cold ethanol. The tube is centrifuged for five more minutes, and the supernatant is carefully removed again. Finally, the tube is dried by evaporating the remaining ethanol. The resulting DNA pellet can then be resuspended in 5 μl of 50 mM EDTA (pH 8.0) and 25 μl of deionized formamide. The final reaction mixture yields five 6-μl samples that can either be electrophoresed immediately or stored indefinitely at −20°.

Electrophoresis

Capillary Placement

Before an experiment begins, the capillary must be equilibrated with the electrophoresis buffer so that the reaction products that derive from the ammonium persulfate and TEMED do not interfere with gel reproducibility. This preequilibration is accomplished by simply placing the capillary between the two buffer chambers and allowing a field of about 100 V/cm to be applied for 1 hr. Following the equilibration with buffer, the gel can be inspected under a microscope to determine whether the gel has maintained its integrity by not forming air bubbles. Any bubbles that are formed may be removed by trimming the capillary; however, this is usually not a problem.

The prepared capillary gel can now be threaded through the capillary mount (Fig. 3A) so that the detection window is in the path of the incident laser. Minor alignment adjustments are made on the translation stage that the mount is attached to so that the focused beam passes directly through the center of the capillary. The magnified image is then focused onto the spatial filter and the height of the objective is varied so that only the image of the gel passes through the slit. The remaining detection optics are not adjusted because they have already been optimized for the position of the slit, which remains fixed.

Sample Preparation and Loading

Sequencing reactions are withdrawn from the stored stock solutions, which are maintained at $-20°$. A 6-μl aliquot of the reaction mixture is dispensed in a 500-μl Eppendorf tube. The DNA sample is then denatured by heating at 90° for 4 min. The tube is then immediately placed underneath the cathodic end of the capillary and the cathode of the CGE instrument. An electrokinetic injection is then made, usually at an applied electric field between 100 and 300 V/cm for 10 to 20 sec. Following injection, the sample tube is discarded and the capillary replaced in the buffer solution. Electrophoresis commences by turning the voltage on and beginning the data collection program. For the duration of the experiment, the system is operated at constant voltage and at ambient temperature.

Data Collection

Sample data points are collected every 100 msec from each of the four PMTs. Experiments may last anywhere from 30 min to over 5 hr. The sampling period can be either raised or lowered to suit the estimated speed of the sample peaks through the detection window. Ideally, one should have about 20 data points per peak so that the sampling period does not adversely affect the resultant resolution. Once data collection is complete, the raw data may then be viewed and permanently stored on 5.25-in. floppy disks.

Data Analysis

Figure 5A shows representative raw data obtained from the CGE DNA sequencer. Due to the overlapping of the emission spectra of the four dyes

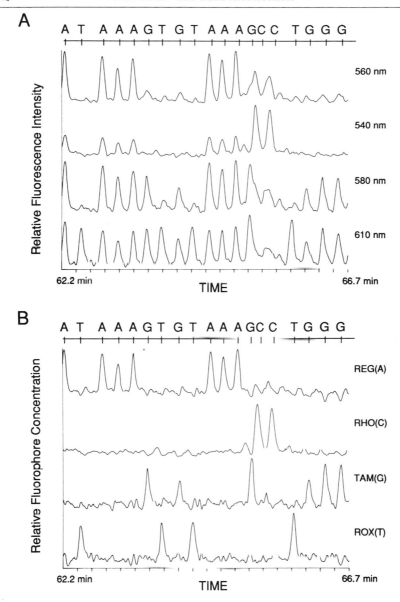

FIG. 5. Fluorescence data obtained from a mixture of four sequencing reactions. The mixture was applied to a gel-filled capillary and was detected using the optical configuration illustrated in Figs. 2 and 3, and as described in the text. The sequence of the illustrated region is printed above and corresponds to base 137 to base 155 past the primer. (A) Raw data; (B) analyzed data (see text for details of the data analysis).

employed, the resultant sequence data gives peaks that have components from more than one photomultiplier tube. To obtain a clear DNA sequence, one must therefore perform a multicomponent analysis to determine which dye primer is responsible for each peak. This first step involves performing the multicomponent analysis, which has been described elsewhere.[20] Briefly, a region of the raw data is isolated and single dye peaks are identified. The various components of each of the four PMTs in the peaks are then measured and written into a 4 × 4 matrix. The inverse of this matrix is then calculated, which is then multiplied by the raw data to convert the detector information to dye information. This determination of the matrix components is equivalent to a calibration of the detection system, and need not be repeated once performed unless changes are made in the dyes or optical system. A data-filtering step, employing a finite-impulse low-pass digital filter with a Hamming window,[21] is employed next to reduce high-frequency noise in the data. The cutoff frequency for the filter is set at 0.5 Hz, and only every fifth point is retained, permitting the data to be condensed.

The results of the matrix multiplication and the digital filtering steps are shown in Fig. 5B. The finished sequence data can be read from left to right and compared to the known sequence printed above. The complete sequence run is illustrated in Fig. 6. The four bases are displayed in the order A, C, G, T from top to bottom in each of the four windows to allow an easy identification of the sequence. The time shown is from 33 to 133 min and 490 bases of sequence were obtained in this time.

The illustrated separation was carried out in a 4% total acrylamide gel-filled capillary with 5% cross-linking, which was poured using the high-pressure method detailed above. The sample was a 6-μl aliquot from a 2-μg sequencing reaction, using the bacteriophage vector M13mp19. This sample was applied to the gel-filled capillary by an electrokinetic injection of 20 sec at 150 V/cm and then electrophoresed for 40 cm at a constant field of 150 V/cm.

Comments

Increased Speed

Currently, the principal advantage of this method is its speed. In this system, the speed of separations has been increased by as much as 25-fold

[20] L. M. Smith, R. J. Kaiser, and L. E. Hood, this series, Vol. 155, p. 260.
[21] R. A. Roberts and C. T. Mullis, "Digital Signal Processing." Addison-Wesley, Reading, Massachusetts, 1987.

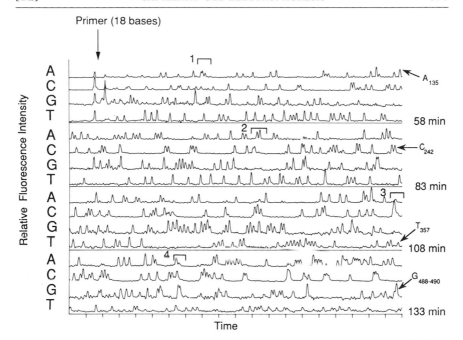

FIG. 6. DNA sequence data obtained in the capillary gel electrophoretic analysis of M13mp19. The conditions employed are those given in the text. The length of various DNA fragments are indicated at the right. The total time shown is 100 min with resolvable peaks out to base 480. Four GC compressions are marked on the plot in brackets and are discussed in the text.

over conventional electrophoresis. Whereas current automated sequencers require roughly 8 hr of electrophoresis to obtain 300 bases of information, data of similar quality can be obtained using capillary gel electrophoresis (CGE) in only 30 min.[8] This increased speed of separation is due to the high electrical resistance of these thin capillaries, which allows fields as high as 400 V/cm (compared to approximately 30 V/cm used in conventional DNA sequencing gels) to be applied. These large fields result in a corresponding increase in the migration velocity of the DNA fragments, with a concomitant reduction in the analysis time.

GC Compressions

For the sequencing run illustrated in Fig. 6, the operating temperature of the gel was approximately 25°. This temperature was insufficient to melt out all of the secondary structure present in the sequencing reaction. At low enough temperatures, regions in the DNA that are GC-rich have the tendency to form stable hairpin loops that alter the migration velocity of these fragments. These regions are termed "GC compressions" because their presence is usually associated with stretches of DNA that can form the more stable GC hydrogen bond pairs.

Four of these GC compressions are highlighted in Fig. 6. The first of these is at the sequence [57]CCTGCA<u>GGCATGC</u>AAG[72] (where the underlined sequence represents the observed compressed region), which is able to form several small hairpin loops. These include CCTGxCAGG, where "x" denotes where the DNA bends to form the hairpin; GCAxTGC; CATGxCAAG, which has one mismatch; and CCTGCAGGxCATG-CAAG, which has two mismatches. The result of these hairpins is the compression of the doublet G to a singlet and the jumbling of the remaining sequence.

The next three compressions all form simple, single hairpins in GC-rich regions. The compressions are located at the following sequences: [185]GTGAGCTxAACT<u>CAC</u>[198], [338]GGGCAACAGxCTGATTG<u>CCC</u>[356], and [394]CCA<u>C</u>xGCTGG[402]. Ideally, one would operate a gel at a high enough temperature that these secondary structure artifacts were completely eliminated. Unfortunately, gel stability and lifetime are decreased substantially when operated at higher temperatures. DNA sequencing by this method thus involves a tradeoff between gel stability and increased artifacts due to secondary structure.

An alternative approach to heating the gel is to employ modified nucleotides, such as 7-deazaguanosine and 2-deoxyinosine,[22,23] in the sequencing reactions. These nucleotides do not form the stable hydrogen bond pairs that are responsible for GC compressions. We have obtained good results with these analogs using a variety of polymerases in conventional electrophoresis, and recommend their use for addressing secondary structure problems in CGE.

---

[22] P. J. Barr, R. M. Thayer, P. Laybourn, R. C. Najarian, F. Seela, and D. R. Talan, *BioTechniques* **4**, 428 (1986).

[23] D. R. Mills and F. R. Kramer, *Proc. Natl. Acad. Sci. U.S.A.* **76**, 2232 (1979).

[13] Preparation and Fluorescent Sequencing of M13 Clones: Microtiter Methods

By Victoria Smith, Molly Craxton, Alan T. Bankier, Carol M. Brown, William D. Rawlinson, Mark S. Chee, and Barclay G. Barrell

Introduction

The shotgun sequencing strategy, in which randomly generated DNA subfragments are cloned into bacteriophage M13 vectors and sequenced using dideoxynucleotide chain terminators,[1] is a proven and reliable scheme. A major rate-limiting step is the generation of single-stranded M13 DNA templates. In the past, this involved growth and processing of 1.5-ml cultures by precipitation of phage with polyethylene glycol and phenol extraction to remove protein. We have modified existing methodology[2,3] to create a reliable and rapid protocol for the growth and preparation of M13 DNA templates (and phagemids containing an f1 origin of replication[4]) in 96-well microtiter plates, using either standard laboratory equipment or a robot pipettor.[5] We have compiled data from several thousand templates produced in this manner, which have been sequenced by a standard protocol using $[\alpha\text{-}^{35}S]dATP$[6] or on an automated DNA sequencer (model 373A; Applied Biosystems, Foster City, CA) using linear amplification sequencing[7] and fluorescently tagged primers.[8] Although microgram quantities of highly purified template were originally necessary to generate good results with the automated sequencer,[9] less than one-tenth of this amount is needed if linear amplification sequencing is used.[7] At this time, we find fluorescent sequencing the fastest and most efficient way to generate sequence data, when combined with microtiter protocols for template preparation and sequencing reactions.

[1] F. Sanger, S. Nicklen, and A. R. Coulson, *Proc. Natl. Acad. Sci. U.S.A.* **74**, 5463 (1977).
[2] I. C. Eperon, *Anal. Biochem.* **156**, 406 (1986).
[3] L. P. Eperon, I. R. Graham, A. D. Griffiths, and I. C. Eperon, *Cell* **54**, 393 (1988).
[4] W. D. Rawlinson, M. S. Chee, V. Smith, and B. G. Barrell, *Nucleic Acids Res.* **19**, 4779 (1991).
[5] V. Smith, C. M. Brown, A. T. Bankier, and B. G. Barrell, *DNA Sequence* **1**, 73 (1990).
[6] A. T. Bankier, K. M. Weston, and B. G. Barrell, this series, Vol. 155, p. 51.
[7] M. Craxton, *Methods: Companion Method Enzymol.* **3**, 20 (1991).
[8] L. M. Smith, J. Z. Sanders, R. J. Kaiser, P. Hughes, C. Dodd, C. R. Connell, C. Heiner, S. B. H. Kent, and L. E. Hood, *Nature (London)* **321**, 674 (1986).
[9] Applied Biosystems, 373A DNA Sequencing System, Users' Manual, 1990.

Generation of Random Library in Bacteriophage M13

The M13 library of the DNA to be sequenced is generated by sonication as previously described.[6] Sonication conditions are adjusted to give different size fractions as required; for example, libraries used for fluorescent sequencing contain inserts in the 500- to 2500-bp size range. It is strongly recommended that a sonication time course be carried out to determine the minimum conditions that produce DNA fragments in the desired size range. Self-ligation of DNA fragments prior to sonication, fragment end repair, size fractionation, ligation in M13, and transformation procedures have been described in detail.[6] It is important in the size-fractionation step not to use molecular weight markers generated by a blunt-end restriction digest (such as a HaeIII digest of ϕX DNA), as these will clone much more efficiently than the sonicated DNA.

Preparation of M13 DNA Templates in 96-Well Microtiter Plates

This method can be carried out by hand using a standard multichannel pipette (e.g., 8 or 12 channel) or it can be semiautomated by the use of a robot pipetting device.[5] The semiautomated protocol we describe here is for a Beckman (Fullerton, CA) Biomek 1000 automated workstation. It is most convenient to prepare templates in batches of two or four microtiter plates; subsequent to the growth phase, the processing of two microtiter plates of templates takes approximately 2.5–3 hr. This method uses sodium dodecyl sulfate (SDS) and heat to denature phage protein coats instead of phenol.[2] Filtration and precipitation steps have been removed from the previously described protocols for preparation of M13 in microtiter plates[2,3] to create a more simple and reliable protocol that is readily automated.

The 96-well microtiter plates used should be composed of rigid polystyrene, with round-bottomed wells of about 300-μl capacity. We find that either tissue culture-treated or untreated plates can be successfully used (e.g., No. 25850 or 25855; Corning, Corning, NY). The plates are sealed for mixing steps and storage with plastic plate sealers (such as Falcon pressure-sensitive film No. 3073; Becton Dickinson, Oxnard, CA). Other necessary equipment includes a centrifuge fitted with a microtiter plate rotor that is capable of speeds of up to 4000 rpm. Centrifugation steps can be carried out at 2000 rpm for double the length of time described, but better pellet formation is observed with a higher-speed rotor [e.g., a Centra-3C fitted with a rotor (No. 244; International Equipment Co., Needham Heights, MA)]. If the rotor trays are not lined with a rubber pad, it is necessary to support the microtiter plates during centrifugation with a 1-

cm slab of polystyrene foam cut to fit just inside the bottom of the plate. Mixing steps are most conveniently carried out using a multitube vortexer (VWR Scientific, South Plainfield, NJ) on a medium or high setting.

Protocol 1: Manual Preparation of M13 DNA Templates

Solutions

YT (2×): 1% (w/v) Bacto-tryptone (Difco, Detroit, MI), 1% (w/v) yeast extract, 0.5% (w/v) NaCl

PEG/NaCl: 20% (w/v) polyethylene glycol (PEG) 8000, 2.5 M NaCl

TE/SDS: 10 mM Tris-HCl (pH 8.0), 0.1 mM ethylenediaminetetraacetic acid (EDTA)(Na$_2$), 1% (w/v) SDS

Ethanol/sodium acetate: 95% (v/v) ethanol, 0.12 M sodium acetate (pH 5.0)

TE: 10 mM Tris-HCl (pH 8.0), 0.1 mM EDTA(Na$_2$)

Procedure

1. Toothpick each recombinant plaque, avoiding the transfer of agar, into 250 μl of a 1/100 dilution in 2× YT of an overnight TG1 bacterial culture,[10] in a rigid 96-well microtiter plate. Alternatively, plaques can be toothpicked into 15 μl of TE in a microtiter plate and stored at 4° for up to several weeks prior to processing. The plate should be well sealed during storage.

2. Affix the microtiter plate securely, without its lid, in a sealed container of dimensions 11.5 × 17.5 × 6 cm and incubate at 37°, 300 rpm for 5.5–6 hr. It is important that the temperature remains at 37° or just below, as at higher temperatures the yield of M13 phage is significantly reduced.

3. Centrifuge at 4000 rpm for 5 min to pellet bacterial cells. Carefully transfer 150 μl of phage-containing supernatant to a new microtiter plate; the same set of pipette tips can be used across the plate. During removal of supernatant, lower the pipette tips no further than two-thirds of the depth of the wells to avoid any contact with the cell pellet. This is sufficient to exclude bacterial contamination. For an inexperienced user, however, we recommend a second purification step in which the transferred supernatant is diluted to 250 μl with 2× YT, the plate centrifuged at 4000 rpm for 5 min, and 200 μl of supernatant transferred to a new microtiter plate.

4. Add PEG/NaCl to a final dilution of one-sixth in each well; that is, 30 μl of PEG/NaCl to 150 μl of supernatant or 40 μl of PEG/NaCl to 200

---

[10] T. J. Gibson, Ph.D. Thesis, Cambridge Univ., 1984.

μl of supernatant. Seal the plate, mix thoroughly, and incubate at room temperature for 15 min.

5. Centrifuge at 4000 rpm for 10 min. The phage pellets are usually visible as a white ring or diffuse smear in the bottom of the walls. Aspirate the supernatant by dipping a yellow tip attached to a vacuum line (e.g., as shown in Sambrook et al.[11]) briefly down the inside edge of each well. The trace of PEG/NaCl that remains in the wells does not cause any deterioration in sequence quality.

6. Add 50 μl of TE/SDS to each well,[12] cover the plate with a sealer and a rigid lid, and mix vigorously to ensure complete resuspension of the phage pellets. Place the plate in an oven at 80° for 20 min. The rigid lid is necessary to prevent the plate sealer from peeling off during heating.

7. Allow the plate to cool to room temperature, centrifuge briefly to collect condensation, and add 125 μl of ethanol/sodium acetate to each well. Seal the plate with a fresh sealer, mix, and incubate at room temperature for 15 min. This is sufficient to precipitate the M13 DNA; incubation at −20° increases the risk of precipitating other material that may reduce sequence quality.

8. Centrifuge at 4000 rpm for 10 min. Remove the ethanol by inverting the plate gently and draining it for a few seconds on a clean tissue. Wash the pellets by adding 200 μl of 95% ethanol to each well, then immediately removing it, and draining the plate as before.

9. Dry the DNA pellets under vacuum for 5 min and resuspend them in 20 μl of TE. Store the sealed plate at −20°.

Protocol 2: Semiautomated Preparation of DNA Templates

A Beckman Biomek 1000 automated laboratory workstation, shown in schematic layout in Fig. 1, can be programmed to perform template preparation.[5] All the necessary manipulations can be achieved using Beckman Biotest software. A robot arm attaches either an eight-channel pipetting or aspirating tool, and transfer, addition, or removal of solutions is carried out by movement of the arm in the vertical plane and front to back (Y and Z axes) and of the workstation tablet left to right (X axis). The program consists of linked routines for supernatant transfer, PEG/NaCl

---

[11] J. Sambrook, E. F. Fritsch, and T. Maniatis, "Molecular Cloning: A Laboratory Manual." Cold Spring Harbor Press, Cold Spring Harbor, New York, 1989.

[12] Alternatively, a solution of 4 M NaI, 10 mM Tris-HCl, pH 8.0, 1 mM Na$_2$ EDTA can be used to denature phage (mix thoroughly after addition). The heat incubation step is then unnecessary. DNA is precipitated by the addition of ethanol alone (i.e., do not add sodium acetate). [R. P. Alderton, L. M. Eccleston, R. P. Howe, C. A. Read, M. A. Reeve, and S. Beck, *Anal. Biochem.* **210**, 166 (1992).]

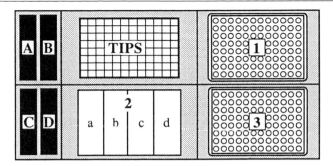

Fig. 1. The layout of the robot workstation. B, pipetting tool; D, aspirating tool; 1, microtiter plate; 2, quarter tray modules; 3, microtiter plate; A, vacant tool position; C, vacant tool position. [After V. Smith, C. M. Brown, A. T. Bankier and B. G. Barrell, *DNA Sequence* **1**, 73 (1990).]

addition, aspiration of PEG/NaCl, and addition of TE/SDS and ethanol reagents. Prompts are provided for the filling of trays with the appropriate solutions. At the end of each pipetting step, the pipette tips are touched to the sides of the microtiter wells for drainage. This minimizes cross-contamination and ensures that all wells receive equal volumes of reagents. The microtiter plate is transferred manually between the robot and centrifuge and oven. Although the robot can perform mixing steps, it is easier in practice to remove the plate and mix it on a multitube vortexer, as all mixing steps precede centrifugation or heating. Different numbers of plates can be processed at a time.

To compensate for the slightly jerky motion of the tablet, two centrifugation steps are used for the removal of phage supernatant. After the first centrifugation step, 200 μl of phage is transferred to a new microtiter plate, centrifuged, and returned to the robot. No dilution is made, and 150 μl of supernatant is transferred directly to a new plate for the addition of 30 μl of PEG/NaCl. The robot arm is programmed to remove the required volume of supernatant while maintaining the maximum possible distance from the bacterial pellet. The supernatant transfer is probably the most critical step in obtaining clean sequence, and the one most likely to suffer variability between users. A multiple pipettor in microtiter format (e.g., a 96-well cartridge device) with an adjustable height may be a useful investment for ease and consistency, if a programmable pipetting device is not available.

Sequencing Reactions

The sequence produced for analysis on an automated fluorescent sequencer with the available software, such as the Applied Biosystems 373A, must meet three main requirements. First, sufficient signal intensity must be generated for efficient real-time detection. Second, the template DNA must be of high-enough quality to produce a clean signal that is free of spurious peaks. Third, the signal or peak heights must be reasonably uniform across the sequence if reading lengths of up to 450 nucleotides are to be obtained consistently. We find that linear amplification sequencing of templates produced in microtiter plates, using *Taq* polymerase, routinely produces high-quality results. In the protocol described in this section,[7] 30 rounds of denaturation, annealing, and extension provide high and even signal intensities. As most of the fluorescently tagged primer becomes incorporated in the course of the reaction, the fluorescent background is reduced to a low level. The Applied Biosystems 373A requires a set of four primers of identical sequence but different 5' fluorophore tags to sequence one template. Each primer is used in conjunction with a specific dideoxynucleotide termination mix, and the four reactions are pooled and loaded in one lane of the sequencing gel. The fluorophores have different intensities, but it is not necessary to compensate for this by adding different quantities of primer. The linear amplification sequencing procedure can also be used with ^{32}P-labeled primers or automated systems using a single fluorescent tag [e.g., the Pharmacia (Piscataway, NJ) ALF[13]]; the reactions are not pooled but are precipitated and loaded in individual wells.[7]

We also provide a protocol for double-strand sequencing from M13 templates using a reverse primer,[14] a technique that is particularly useful near the end of a sequencing project for determining the sequence completely on both strands. This involves a simple modification of the cycle-sequencing reaction to include an extension step in the presence of an unlabeled forward primer and dNTPs. The termination mixes and fluorescently labeled reverse primer are then added with a repetitive dispenser and the reactions cycled as normal. As the molar concentration of dNTPs in the extension mix is similar to that in the termination mix, there is little danger of significantly altering the ddNTP/dNTP ratio in the sequencing mix. This method produces high-quality sequence data from M13 and phagemid clones containing inserts of up to 3 kb in size, although the best results are obtained with inserts of 300 to 800 bp.

Finally, we include an updated protocol for sequencing DNA templates produced in microtiter plates with Klenow enzyme and [α-^{35}S]dATP.

All of the sequencing reactions are carried out in flexible 96-well microtiter plates. For linear amplification sequencing, we use Techne (Princeton, NJ) Dri-Block cyclers (MW-1 and PHC-3) fitted with multiwell

[13] T. Hawkins, personal communication, 1991.
[14] V. Smith and M. S. Chee, *Nucleic Acids Res.* **19**, 6957 (1991).

heat blocks. Reaction mixes are prepared in 1.5-ml microfuge tubes on ice. The DNA polymerase should be removed from the freezer for the shortest time necessary and added last to the premixed reaction mix. Wherever possible, solutions are added to microtiter wells using repetitive dispensers. For example, a Hamilton (Reno, NV) PB600-1 fitted with a 1710LT syringe and an adapter to attach a yellow tip is used to dispense DNA templates and other reagents that require multiple dispensation in volumes of 2 or 4 μl. The solutions are added to the sides of the wells, close to the rim, and the plate is centrifuged briefly to pool the contents.

Protocol 3: Linear Amplification Sequencing Using Fluorescently Tagged Primers

  Solutions

  Nucleotide stock solutions: The preparation of dNTP and ddNTP stock solutions has been described previously.[6] Stock solutions of dNTPs and ddNTPs are also available commercially (e.g., from Pharmacia)
  dNTP mix (0.5 mM): To make a solution of 0.5 mM dTTP, 0.5 mM dCTP, 0.5 mM dGTP, and 0.5 mM dATP, add 10 μl of each 50 mM dNTP stock (in TE) to 960 μl TE
  dNTP termination mixes (5×):

	T	C	G	A
dNTP mix (0.5 mM)	25	25	25	25
ddTTP (10 mM)	62.5	—	—	—
ddCTP (10 mM)	—	40	—	—
ddGTP (10 mM)	—	—	8	—
ddATP (10 mM)	—	—	—	40
TE	412.5	435	467	435

All volumes are in microliters and each mix has a final volume of 500 μl. The 10 mM ddNTP stocks are in TE. All nucleotide stock solutions and mixes are stored at −20°.

  Taq buffer (10×): 100 mM Tris-HCl (pH 8.5), 500 mM KCl, 15 mM MgCl$_2$
  5′-Tagged fluorescent primers: Sets of four tagged primers for M13 universal and reverse priming sites, as well as T7 and SP6, can be purchased commercially. We use the primers supplied by Applied Biosystems, which are sold either predissolved at 0.4 pmol/μl or as a dried concentrate of 80 pmol/tube. Dried primers are redissolved in 100 μl of TE with vigorous vortexing to give a stock concentration of 0.8 pmol/μl. All primers are stored in the dark at −20°

Formamide loading solution: Deionized formamide (which has been filtered through a 0.2-μm pore filter) and 50 mM EDTA (pH 8.0), at a ratio of 5:1. Store in aliquots at $-20°$

Procedure

1. Assign 4 rows of a 96-well heat-resistant microtiter plate (e.g., FMW11; Techne) as T, C, G, and A. Dispense 2 μl of single-stranded template DNA into each of the four wells. In this manner, 24 clones can be sequenced in 1 microtiter plate.

2. Prepare four base-specific termination mixes (T, C, G, and A) on ice. For each template add (volumes for 24 clones are given in parentheses): 2 μl of 10× *Taq* buffer (48 μl), 4 μl of the appropriate 5× dNTP termination mix (96 μl), 0.25 pmol of the corresponding 5′-tagged fluorescent primer (15 μl of a 0.4-pmol/μl solution), 0.5 U *Taq* DNA polymerase (Perkin-Elmer Cetus, Norwalk, CT) (2.4 μl of a 5-U/μl solution), and deionized water to 19 μl (295 μl). Dispense 18 μl into each well.

3. Overlay each well with a drop of paraffin oil from a 1-ml pipettor fitted with a blue tip, or with two drops of oil from a multichannel pipettor. Cover the plate with a flexible lid (e.g., Falcon 3913) or a plate sealer and centrifuge briefly.

4. Place the plate in a preheated thermal cycler and cycle 30 times at 92° for 1 min, 55° for 2 min, 72° for 2.5 min. The reactions can be stored frozen on completion of the 30 cycles.

5. Pool the four base reactions for each clone by removing 15 μl from under the paraffin with a standard yellow tip into one well of a rigid, 96-well microtiter plate (with a well volume of about 300 μl). Avoid transferring an excessive amount of oil, but a small quantity does not present any problem. Precipitate the DNA by the addition of 7 μl of 3 M sodium acetate pH 5.0, and 200 μl of ethanol. Seal the plate, mix thoroughly, and incubate at $-20°$ for at least 1 hr. The reactions can also be stored at $-20°$ at this stage.

6. Centrifuge at 4000 rpm for 20 min and remove the ethanol by gently inverting the plate and draining it on a tissue. Wash the pellets by adding 200 μl of 95% ethanol, centrifuge at 4000 rpm for 2 min, invert the plate gently to remove the ethanol, and drain it briefly on a tissue. Turn the plate upright and allow the DNA pellets to dry in air for 10–15 min. The dry pellets can be stored in a sealed plate at room temperature or at $-20°$ for up to a week.

7. Add 6 μl of formamide loading solution to each sample. Resuspend the pellets by pipetting with a multichannel pipette. The formamide solution seems to distribute more easily around the wells of a tissue culture-treated microtiter plate rather than an untreated plate, although either can be used.

8. Incubate the plate uncovered in an oven at 80° for 5 min, then immediately load 5 μl of the samples (the remaining volume) into the rinsed wells of the sequencing gel. As alternate lanes of the sequencing gel are loaded in two stages on the automated fluorescent sequencing machine (protocol 6), it is convenient to precipitate the reactions in 2 sets of 12 in 2 microtiter plates (different rows of the plates can be used each day to avoid wastage). If all 24 reactions are precipitated in 1 microtiter plate, the plate should be stored on ice during loading time and the 5-min wait period before the next loading. Alternatively, the second set of samples can be returned to the oven for 2 min prior to loading.

Protocol 4: Reverse-Primer Sequencing Using Fluorescently Tagged Primers

Solutions

dNTP mix (50 μM): Make a solution of 50 μM dTTP, 50 μM dCTP, 50 μM dGTP, and 50 μM dATP by 1/10 dilution of 0.5 mM dNTP mix in TE

The other necessary solutions are described in protocol 3.

Procedure

1. Assign 4 rows of a 96-well heat-resistant microtiter plate (e.g., FMW11; Techne) as T, C, G, and A. Dispense 2 μl of single-stranded template DNA into each of the four wells.

2. Prepare one extension mix to be added to every well. For each template add (volumes for 24 clones are given in parentheses): 8 μl 10× *Taq* buffer (192 μl), 4 μl 50 μM dNTP mix (96 μl), 0.8 pmol unlabeled universal (forward) primer (19.2 μl of a 1-pmol/μl solution), 2 U *Taq* DNA polymerase (Perkin-Elmer Cetus) (9.6 μl of a 5-U/μl solution), and water to 55 μl (1 ml). Dispense 13 μl into each well.

3. Overlay each well with a drop of paraffin oil from a pipettor fitted with a blue tip, or with two drops of oil from a multichannel pipettor. Cover the plate with a flexible lid or a plate sealer and centrifuge briefly.

4. Place the plate in a preheated thermal cycler and cycle once at 92° for 1 min, 55° for 2 min, 72° for 3 min.

5. Remove the plate from the cycler and, using a repetitive dispenser, add 4 μl of each 5× termination mix to the side of the wells, above the paraffin oil. Also add 2 μl of each 5' fluorescently tagged reverse primer (diluted to a concentration of 0.125 pmol/μl with TE) to the appropriate wells. Centrifuge the plate briefly.

6. Return the plate to the heat block and cycle 30 times at 92° for 1 min, 55° for 2 min, 72° for 2.5 min. The reactions can be stored frozen on completion of the 30 cycles.

Continue with the pooling and precipitation of the samples as described in steps 5 to 8 of protocol 3.

Protocol 5: Radioactive Sequencing Using [α-^{35}S]dATP and Klenow Enzyme

Solutions

TM: 100 mM Tris-HCl (pH 8.5), 50 mM MgCl$_2$
dNTP termination mixes

	T	C	G	A
dTTP (0.5 mM)	25	500	500	500
dCTP (0.5 mM)	500	25	500	500
dGTP (0.5 mM)	500	500	25	500
ddTTP (10.0 mM)	50	—	—	—
ddCTP (10.0 mM)	—	8	—	—
ddGTP (10.0 mM)	—	—	16	—
ddATP (10.0 mM)	—	—	—	1
TE	1000	1000	1000	1000

All volumes are in microliters. The 0.5 mM dNTP stocks and 10 mM ddNTP stocks are in TE. All nucleotide stock solutions and mixes are stored at −20°.

   dNTP mix (0.5 mM): Make a solution of 0.5 mM dTTP, 0.5 mM dCTP, 0.5 mM dGTP, and 0.5 mM dATP by addition of 10 μl of each 50 mM dNTP stock (in TE) to 960 μl TE
   Formamide dye mix: 100 ml deionized formamide, 0.1 g xylene cyanol FF, 0.1 g bromphenol blue, 2 ml 0.5 M EDTA(Na$_2$) (pH 8.0)

Procedure

1. Assign 4 rows of a 96-well microtiter plate (e.g., Falcon 3911) into rows of T, C, G, and A. Dispense 2 μl of single-stranded template DNA into each of the four wells.

2. For each template to be sequenced, prepare a mix containing (volumes in parentheses are for 24 clones): 0.4 pmol of universal primer (9.6 μl of a 1-pmol/μl solution), 1 μl of TM (24 μl), and water to 9 μl (180 μl). Add 2 μl to each well.

3. Centrifuge the plate briefly, cover with plastic film (e.g., Saran wrap; Dow Chemical Co.) or a plate sealer and incubate at 55° for 30 to 60 min. The reactions can be stored at −20° at this stage.

4. Centrifuge the plate briefly to collect condensation before removing the Saran wrap. To each well, add 2 μl of the appropriate dNTP termination mix.

5. For each template to be sequenced, prepare a mix on ice containing (volumes in parentheses are for 24 clones): 4 μCi of [α-^{35}S]dATP (9.6 μl of a 10-μCi/μl solution), 1 μl 0.1 M dithiothreitol (DTT) (24 μl), 2 U Klenow fragment DNA polymerase I (Boehringer Mannheim, Indianapolis, IN) (9.6 μl of a 5-U/μl solution), and water to 9 μl (170 μl). Dispense 2 μl to each well, on the opposite side to the droplet of dNTP termination mix.

6. Centrifuge briefly and incubate at 37° for 15 min.

7. Add 2 μl of 0.5 mM dNTP mix to each well and centrifuge briefly. Incubate at 37° for 15 min. The reactions can be stored at this stage at −20° for up to a few days.

8. Add 2 μl of formamide dye mix to each well, centrifuge briefly, and incubate the plate uncovered at 80° for 15 min.

9. Load all of the remaining sample immediately into the rinsed wells of a sequencing gel.

The sequencing reactions can also be carried out using an automated robot workstation to dispense templates and reaction mixes in microtiter format.[15] The pouring and handling of gradient acrylamide gels has already been described in detail.[6] It is convenient to load the sequencing reactions for 12 clones (i.e., 48 wells) on one 20× 50 cm buffer gradient gel using either a shark's-tooth (e.g., 1045SH; Bethesda Research Laboratories, Gaithersburg, MD) or slot-former comb.

Gel Handling for Fluorescent Sequencing

It is important to prevent the contamination of gel mixes and electrophoresis solutions with fluorescent compounds. Contamination can be avoided by preparing solutions with purified water (such as Milli-Q-purified water) and by handling gel plates with a set of gloves (e.g., standard neoprene rubber gloves) that are reserved only for manipulations involving those plates. Avoid using disposable latex laboratory gloves, as some brands contain powder that fluoresces.[16] Gel mixes are prepared in 500-ml quantities from commercially available 40% (w/v) acrylamide stock

[15] A. T. Bankier and B. G. Barrell, in "Nucleic Acid Sequencing: A Practical Approach" (C. J. Howe and S. E. Ward, eds.), p. 37. IRL Press, Oxford, 1989.
[16] R. McClory, personal communication, 1990.

solutions and they are stable for many months when stored at 4°. Preparation of a bulk gel mix solution minimizes effort and ensures reproducibility from run to run. Glass plates can be cleaned with soaps such as Alconox or Decon and they should be rinsed well with deionized water. However, we also find that the gel plates can be kept clean without the use of any soaps by thorough rinsing with deionized water and wiping with a tissue after each run.

Protocol 6: Gel Electrophoresis

Solutions

TBE (10×): 108 g Tris base, 55 g boric acid, 8.3 g EDTA(Na$_2$). Adjust volume to 1 liter with Milli-Q water and filter through a 0.45-μm pore nitrocellulose membrane. Prepare the 1× TBE working solution with Milli-Q water. The working solution is 89 mM Tris–borate, 89 mM boric acid, 22 mM EDTA (pH 8.3)

TBE gel mix (1×): 75 ml 40% acrylamide (acrylamide : bisacrylamide, 19 : 1), 250 g urea. Add Milli-Q water until the volume is close to 450 ml (this is necessary for the urea to dissolve). When the urea has dissolved, adjust the volume to 450 ml, add 5 g mixed-bed resin, and stir gently for 20 min. Filter the solution through a 0.45-μm pore nitrocellulose membrane and add 50 ml 10× TBE. The gel mix is stable for many months at 4°. The final concentrations are 1× TBE, 6% acrylamide, 8.3 M urea

25% APS (25%): 25% (w/v) ammonium persulfate (APS) in Milli-Q water; stable for many months at 4°

Procedure

1. Assemble clean gel plates and spacers and tape the plates together with polyester tape. Clamp a bulldog clip over the spacers at each side of the plates, about a third of the distance up from the bottom, to ensure uniform gel thickness.

2. Pipette 50 ml of 1× TBE gel mix into a 150-ml beaker. Make sure that all of the urea is in solution by gentle warming of the gel mix if necessary. Add 125 μl of 25% APS and 50 μl of TEMED (N,N,N',N'-tetramethylethylenediamine), swirling gently to mix after each addition.

3. Lift the plates to an angle of between 45 to 60° with the eared plate topmost and pour the gel mix between them, lowering them gradually back to a horizontal position as the gel mix nears the top. Insert a slot former and clamp two bulldog clips, evenly spaced, over the top of the gel plates across the slot former. Lie the plates flat and allow the gel to set for at least 30 min.

4. Remove the polyester tape and rinse the plates thoroughly with deionized water to remove any traces of dried gel mix. Carefully remove the slot former under running deionized water to flush out any unpolymerized acrylamide. Remove extraneous polymerized acrylamide from the rim of the eared plate with a razor blade.

5. Dry the plates with ethanol or a rubber window wiper and attach the gel assembly in the electrophoresis apparatus. Check that the region scanned by the laser is clean (using the "plate check" option on the Applied Biosystems 373A automated DNA sequencer).

6. Add 1× TBE to the buffer chambers and immediately flush the wells thoroughly using a syringe. If a shark's-tooth comb is being used, flush the well before carefully inserting the comb.

7. Prerun the gel at 30 W for 30 min. This warms the gel and allows any low molecular weight fluorescent contaminants to run off.

8. Flush the wells thoroughly just prior to loading samples. Load samples in alternate lanes using a pipette fitted with a standard yellow tip. Electrophorese at 30 W for 5 min before flushing the wells again to load the second set of samples. This staggered loading assists the lane tracking of the automated sequencer, and commercial alignment braces, which define and number each lane, are a useful guide when loading.

9. Electrophorese at 30 W for 14 hr.

Data Compilation

We currently use the program xdap[17] (formerly SCREENV, SCREENR, DBAUTO, and DBUTIL[18]) for the compilation of sequence data on a Sun SPARCstation (this software is available from the authors[17]). New sequence data are screened for vector sequences and then assembled automatically within the program. xdap also includes a mouse-operated editor that permits rapid examination and editing of discrepancies in the aligned sequences by the concurrent display of the fluorescent traces for the gel readings in question. Conversion of the output from the automated sequencer to a suitable format for xdap is done using a script that incorporates a trace editor, tcd.[19] ted provides a visual display of the traces, which can be clipped at the 5' end to remove vector sequence close to the priming site and at the 3' end to remove sequence near the end of the run that is too inaccurate for inclusion in the sequence database. xdap provides a facility for viewing and extending the previously trimmed sequence in the

[17] S. Dear and R. Staden, *Nucleic Acids Res.* **19**, 3907 (1991).
[18] R. Staden, *Nucleic Acids Res.* **10**, 4731 (1982).
[19] T. Gleeson and L. Hillier, *Nucleic Acids Res.* **19**, 6481 (1991).

database if desired. Sequence data generated by standard radioactive methods[6] can also be assembled and edited in xdap, although autoradiographic images are not displayed.

We find that the current base-calling software on the Applied Biosystems 373A (version 1.0.2) tends to overcall the number of bases once lengths of about 500 nucleotides from the priming site are reached. After trimming of 5' vector sequence (approximately 30 nucleotides), we routinely obtain good-quality reading lengths of around 420 nucleotides (350 to 450) from templates prepared in microtiter plates and sequenced with *Taq* polymerase.

Discussion

The random strategy is a fast and efficient means of completing large-scale sequencing projects (see, e.g., Chee *et al.*[20]; Davison[21]). DNA fragments up to 134 kb in size[21] have been sequenced in a single shotgun, in which the DNA fragment is subcloned in M13 and assembled as one contiguous sequence. We routinely sequence λ clone inserts, of approximately 20 kb in size, in one M13 shotgun. Having established that the cloning efficiency is good, it is convenient to prepare several microtiter plates of DNA templates over the course of a few days. One day is spent in toothpicking recombinant plaques into TE in a microtiter plate, and over the next 2 or 3 days these phage are processed to purified templates in batches of two or four plates. Six microtiter plates are more than sufficient to sequence one 20-kb DNA fragment. The templates are then sequenced in batches of 24, which is the maximum number of clones that can be loaded on the Applied Biosystems 373A sequencer at this time with the available hardware and software (version 1.0.2). A comfortable routine involves cycle sequencing and pooling the completed reactions for an overnight precipitation at −20°. The next day, after the 1× TBE sequencing gel has been poured, the pooled DNAs are collected by centrifugation, dried, and resuspended in formamide loading buffer while the gel is setting and preelectrophoresing. After the gel is loaded, the next set of templates can be sequenced for the following day.

The 14-hr electrophoresis and automatic data analysis is carried out overnight. Analyzed traces are transferred to a Sun SPARCstation the next morning for processing and assembly into the sequence database,

[20] M. S. Chee, A. T. Bankier, S. Beck, R. Bohni, C. M. Brown, R. Cerny, T. Horsnell, C. A. Hutchison, III, T. Kouzarides, J. A. Martignetti, S. C. Satchwell, P. Tomlinson, K. M. Weston, and B. G. Barrell, *Curr. Top. Microbiol. Immunol.* **154**, 125 (1990).
[21] A. J. Davison, *DNA Sequence* **1**, 389 (1991).

which usually takes less than an hour. The database should be monitored for random distribution of sequences; accumulation of specific sequences, which indicates nonrandom sonication or some cloning problem, will become apparent early on. In practice, however, sonication seems to produce random libraries from DNAs varying widely in base composition and from a variety of organisms. Some minor editing may be necessary during data accumulation, but it is most efficient to collect the bulk of the sequence data prior to editing. The redundancy of data at this stage of the project makes it easier to resolve problem regions.

The development of a semiautomated microtiter protocol for template preparation overcomes a rate-limiting step in large sequencing projects. The current limitation in the procedure described, using the Applied Biosystems 373A, is gel throughput. However, this method offers significant advantages. In particular, the manual digitizing of autoradiographic images is no longer necessary and we consistently obtain greater than 400 nucleotides of accurate sequence per clone. The combination of this base-calling capability with high-throughput microtiter methods for template preparation and sequencing reactions is an effective implementation of the shotgun strategy.

Acknowledgments

We thank Simon Dear and Rodger Staden for advice and assistance with computer software and members of the C. elegans sequencing project for interesting discussions. We are also grateful to Applied Biosystems for the use of a 373A automated DNA sequencer. V.S. is supported by a scholarship from the Association of Commonwealth Universities, M.C. in part by grants from the NIH Human Genome Center and the MRC HGMP, W.D.R. by a Sir Robert Menzies Memorial Scholarship, and M.S.C. by Applied Biosystems.

[14] DNA Sequencing with Direct Transfer Electrophoresis and Nonradioactive Detection

By PETER RICHTERICH and GEORGE M. CHURCH

Introduction

In DNA sequencing by base-specific chemical degradation[1] or chain termination,[2] two steps that are critical to the overall efficiency and conve-

[1] F. Sanger, S. Nicklen, and A. R. Coulson, *Proc. Natl. Acad. Sci. U.S.A.* **74,** 5463 (1977).
[2] A. M. Maxam and W. Gilbert, *Proc. Natl. Acad. Sci. U.S.A.* **74,** 560 (1977).

nience of the procedure are the separation of the DNA fragments by polyacrylamide gel electrophoresis and the subsequent visualization of the sequencing patterns. Increasing the number of readable nucleotides per sequencing reaction reduces the number of template DNA preparations and sequencing reactions; furthermore, the sequence assembly from individual reactions can be significantly simplified. To maximize the number of readable bases, very thin gels, very long gels, low acrylamide concentrations, and multiple loadings have been used.[3-6]

One limiting factor in conventional gel electrophoresis is the unequal spacing of bands over the length of the gel. This results from the inverse relationship between DNA size and electrophoretic mobility. Lowering the electrical field toward the bottom of the gel by employing buffer or thickness gradient gels[7-9] reduces this effect, but does not eliminate it completely.

Constant spacing between bands can be obtained by a different observation mode: instead of looking at the whole gel at a fixed point in time, a fixed point (or line) in the gel is observed over a prolonged time period. Examples of separation methods that employ such "spatially fixed" detection are preparative electrophoresis methods and direct transfer electrophoresis.[10] In direct transfer electrophoresis, a membrane that is moved along the bottom of the gel immobilizes the DNA as it elutes out of the gel. This transforms the temporal order of elution into a spatial one on the membrane. Because the detection (or rather elution) time depends linearly on the size of the DNA fragments, a constant membrane transfer speed yields equal interband distances. The original description of direct transfer electrophoresis for DNA sequencing[10] showed the potential benefits of this method; however, the band patterns that were generated with 14-cm long, 0.35-mm thick gels were insufficiently resolved and made reliable sequence interpretation difficult. The subsequent use of thinner and longer gels allowed for the resolution of more than 350 nucleotides per lane set.[11] Here, we describe optimizations of apparatus design and gel parameters

[3] F. Sanger and A. R. Coulson, *FEBS Lett.* **87,** 107 (1978).
[4] W. Ansorge and L. deMayer, *J. Chromatogr.* **202,** 45 (1980).
[5] D. R. Smith and J. M. Calvo, *Nucleic Acids Res.* **10,** 2255 (1980).
[6] W. Ansorge and R. Barker, *J. Biochem. Biophys. Methods* **9,** 33 (1984).
[7] M. D. Biggin, T. J. Gibson, and G. F. Hong, *Proc. Natl. Acad. Sci. U.S.A.* **80,** 3963 (1983).
[8] W. Ansorge and S. Labeit, *J. Biochem. Biophys. Methods* **10,** 237 (1984).
[9] A. Olsson, T. Moks, M. Uhlén, and A. B. Gaal, *J. Biochem. Biophys. Methods* **10,** 83 (1984).
[10] S. Beck and F. M. Pohl, *EMBO J.* **3,** 2905 (1984).
[11] P. Richterich, C. Heller, H. Wurst, and F. M. Pohl, *BioTechniques* **7,** 52 (1989).

that now allow us to resolve fragments up to more than 800 nucleotides long.

While the membrane-bound sequence patterns produced by direct transfer electrophoresis can be detected by conventional radioactive labeling, they also offer a starting point for other detection methods that require a membrane-bound pattern, such as genomic sequencing,[12] multiplex sequencing,[13] or sequencing with enzyme-linked detection.[14] The use of nonradioactive labeling and detection methods offers a number of potential advantages; these include reduced costs, fewer health and regulatory risks, increased efficiency, and longer reagent shelf life.

DNA sequencing requires a high detection sensitivity because a typical band contains about 0.02–1 fmol of DNA (12 million to 600 million molecules). These requirements have been met by radioactive,[1,2] fluorescent,[15–18] and enzyme-linked detection systems.[14,19–21] Enzyme-linked detection offers advantages for both high throughput and occasional needs. For high throughput, automatic film readers[22,23] and development automatons[11] can be considerably faster and less costly than fluorescent devices. For occasional sequencing, the higher stability of reagents and sequencing reaction products of nonradioactive systems can be helpful; again, these advantages can be achieved with significantly lower equipment costs for enzyme-linked detection systems than with fluorescent devices.

For DNA sequencing with enzymatic detection, oligonucleotide primers end labeled with biotin[14] or digoxigenin[24] have been used. A good signal-to-noise ratio is critical for the performance of the enzymatic detection method; good results using the biotin-based system have been ob-

[12] G. M. Church and W. Gilbert, *Proc. Natl. Acad. Sci. U.S.A.* **81**, 1991 (1984).

[13] G. M. Church and S. Kieffer-Higgins, *Science* **240**, 185 (1988).

[14] S. Beck, *Anal. Biochem.* **164**, 514 (1987).

[15] L. M. Smith, J. Z. Sandes, R. J. Kaiser, P. Hughes, C. Dodd, C. R. Connell, C. Heiner, S. B. H. Kent, and L. E. Hood, *Nature (London)* **321**, 674 (1986).

[16] J. M. Prober, G. L. Trainor, R. J. Dam, F. W. Hobbs, C. W. Robertson, R. J. Zagursky, A. J. Cocuzza, M. A. Jensen, and K. Baumeister, *Science* **238**, 336 (1987).

[17] W. Ansorge, B. Sproat, J. Stegemann, C. Schwager, and M. Zenke, *Nucleic Acids Res.* **15**, 4593 (1987).

[18] J. A. Brumbaugh, L. R. Middendorf, D. L. Grone, and J. L. Ruth, *Proc. Natl. Acad. Sci. U.S.A.* **85**, 5610 (1988).

[19] P. Richterich, *Nucleic Acids Res.* **17**, 2181 (1989).

[20] S. Beck, T. O'Keeffe, J. M. Coull, and H. Köster, *Nucleic Acids Res.* **17**, 5115 (1989).

[21] R. Tizard, R. L. Cate, K. L. Ramachandran, M. Wysk, J. C. Voyta, O. J. Murphy, and I. Bronstein, *Proc. Natl. Acad. Sci. U.S.A.* **87**, 4514 (1990).

[22] J. K. Elder, D. K. Green, and E. M. Southern, *Nucleic Acids Res.* **14**, 417 (1986).

[23] M. J. Eby, *Bio/Technology* **8**, 1046 (1990).

[24] F. M. Pohl, unpublished results, 1990.

tained by using streptavidin and biotinylated phosphatase. Nonspecific binding of streptavidin is efficiently minimized by high concentrations of sodium dodecyl sulfate (SDS) in the incubation buffer,[11] and nonspecific binding of biotinylated alkaline phosphatase is easily blocked by several detergents or proteins. In the digoxigenin system, milk protein fractions work effectively as blocking agents.[25]

One significant advantage of nonradioactive labels is the possibility of combining several different labels in one experiment, an approach that might be called "label multiplexing." The use of different primer oligonucleotides labeled with different fluorophores in fluorescent sequencing machines[15,16] can be seen as one example. Another example is the use of two hybridization probes labeled with different enzymes, alkaline phosphatase and peroxidase, for nucleic acid hybridizations.[26] We describe the simultaneous use of two different labels, biotin[27] and digoxigenin,[25] for enzymatically detected DNA sequencing. This procedure doubles the amount of sequence information that can be obtained from a gel.

We also compare different commercially available chemiluminescent enzyme–substrate systems: peroxidase-based chemiluminescence with luminol and enhancers,[28–30] and two chemiluminescent dioxetane substrates for alkaline phosphatase.[31–33] A protocol is described for combining peroxidase-based enhanced chemiluminescence detection of digoxigenin-labeled primers and phosphatase-based chemiluminescence detection of biotinylated primers to generate two sequence patterns from one gel in a total time of less than 3 hr.

Future extensions of this approach toward the use of more labels, thus going from "label duplexing" to "label multiplexing," may require the repeated use of one enzyme with different labeling systems. Such a proto-

[25] C. Kessler, H.-J. Höltke, R. Seibl, J. Burg, and K. Mühlegger, *Biol. Chem. Hoppe-Seyler* **371**, 917 (1990).

[26] M. Renz and C. Kurz, *Nucleic Acids Res.* **12**, 3435 (1984).

[27] J. J. Leary, D. J. Brigati, and D. C. Ward, *Proc. Natl. Acad. Sci. U.S.A.* **80**, 4045 (1983).

[28] T. P. Whitehead, G. H. G. Thorpe, T. J. N. Carter, C. Groucutt, and L. J. Kricka, *Nature* (*London*) **305**, 158 (1983).

[29] G. H. G. Thorpe, L. J. Kricka, E. Gillespie, S. Moselfy, R. Amess, N. Baggett, and T. P. Whitehead, *Anal. Biochem.* **145**, 96 (1985).

[30] I. Durrant, L. C. A. Benge, C. Sturrock, A. T. Devenish, R. Howe, S. Roe, M. Moore, G. Scozzafava, L. M. F. Proudfoot, T. C. Richardson, and K. G. McFarthing, *BioTechniques* **8**, 564 (1990).

[31] A. P. Schaap, M. D. Sandison, and R. S. Handley, *Tetrahedron Lett.* **28**, 1159 (1987).

[32] I. Bronstein, J. C. Voyta, and B. Edwards, *Anal. Biochem.* **180**, 95 (1989).

[33] I. Bronstein, R. R. Juo, Voyta, and B. Edwards, *in* "Bioluminescence and Chemiluminescence: Current Status" (P. E. Stanley and L. J. Kricka, eds.), p. 73. Wiley, Chichester, England, 1991.

col would necessitate the removal or inactivation of the enzyme between sequential detection steps. We show the feasibility of sequential detections using differently labeled primers with phosphatase-based chemiluminescence.

Principle of Direct Transfer Electrophoresis

In direct transfer electrophoresis (DTE), a membrane is moved along the lower surface of the gel; molecules that elute from the gel and move toward the lower electrode are immobilized on the membrane. The membrane is usually an unmodified or positively charged nylon membrane, but other materials, such as nitrocellulose, have also been used.[10,34] Most apparatus designs use a conveyor belt made of nylon or polyester mesh to transport the membrane. The conveyor belt is moved by a stepper motor that is connected to axles with timing pulleys and a timing belt. The stepper motor is controlled via electronic timing circuits or, more conveniently, by a small computer. The conveyor belt is usually moved with constant speed (see Introduction), but different speed gradients or different speeds can also be used.[10]

Because the spacing of bands on the membrane can be controlled by the transport speed in direct transfer electrophoresis, gels as short as 14–20 cm can be used for DNA sequencing; they offer a convenient alternative to normal long sequencing gels. When extended reads and higher resolution, especially for longer fragments, are the objectives, longer DTE gels can give superior results (see Discussion).

Materials

Enzymes, Plasticware, Machines

Anti-digoxigenin–phosphatase conjugates (Fab fragments) and anti-digoxigenin–peroxidase conjugates (Fab fragments) (Boehringer Mannheim, Indianapolis, IN)

Biotinylated alkaline phosphatase: Purchase from Molecular Probes (Eugene, DR) and Millipore/New England BioLabs (Beverly, MA) or prepare as described;[11] note that after biotinylation, two subsequent purifications are essential to obtain a maximal signal intensity, and that lyophilization or freezing of the biotinylated phosphatase leads to increased background and largely reduced signal intensities

[34] F. M. Pohl and S. Beck, this series, Vol. 155, p. 250.

Modified T7 DNA polymerase (Sequenase): Purchase from United States Biochemicals (Cleveland, OH)

Polyester plastic material of various thicknesses: Purchase from AIN Plastics and LKB (Rockville, MD)

Thin shark's-tooth combs (0.125 mm) with 2.25-, 3.0-, and 4.5-mm wide wells (Owl Scientific)

Twelve-channel syringe pipettors (Cat. No. 0155200; Hamilton, Reno, NV)

Direct transfer electrophoresis machines and large Plexiglas trays for nonradioactive developments: Ours were made in the Harvard Medical School machine shop.

Portable computers (Tandy 102): Used as pulse generators for the stepper motor

Power supply (6000 V) (Fisher Scientific, Pittsburgh, PA)

Programmable, tilt-adjustable roller for four bottles (14-cm i.d., 38 cm long for wide membranes): From BioComp (Fredericton, New Brunswick, Canada)

Chemicals

The sources we have used are given below; other chemicals from other suppliers may be adequate.

3-(Trimethoxysilyl)propyl methacrylate: Aldrich (Milwaukee, WI)

BCIP (5-bromo-4-chloro-3-indolyl phosphate, p-toluidine salt) and NBT (Nitro Blue Tetrazolium): Sigma (St. Louis, MO) and GIBCO/Bethesda Research Laboratories (Gaithersburg, MD)

Biotin-XX-NHS ester: Clontech (Palo Alto, CA)

Chemiluminescent dioxetane substrates for alkaline phosphatase {disodium 3-[4-methoxyspiro(1,2-dioxetane-3,2'-tricyclo[3.3.1.1.^{3,7}]decan-4-yl)phenyl phosphate]}: Called AMPPD (Tropix, Bedford, MA) or LumigenPPD (Millipore/New England BioLabs)

AMPPD-Cl (5-chloro-substituted adamantyl dioxetane): Tropix

Chemiluminescent substrates for horseradish peroxidase: ECL gene detection system (Amersham, Arlington Heights, IL)

Diethanolamine: Sigma (ACS reagent) and Tropix

NHS-digoxigenin (digoxigenin-3-O-methylcarbonyl-ε-aminocaproic acid-N-hydroxysuccinimide ester), blocking reagent on milk protein basis: Boehringer Mannheim

Nylon membrane: Biodyne A, 1.2-$μ$m pore size, BNNG3R: Pall (Glen Cove, NY)

Sephadex G-25 columns (NAP-5 and NPA-10): Pharmacia (Piscataway, NJ)

Streptavidin: Sigma, Millipore/New England BioLabs, and others
Strains-all, Tris (Trizma base), Tris-HCl (Trizma hydrochloride), and
Triton X-100: Sigma
Triethylamine acetate, pH 7.0 (2 M solution): Applied Biosystems (Foster City, CA)

Solutions and Buffers

For oligonucleotide labeling

NaCO$_3$ (1 M), pH 9.0
Stains-all (20×): 1 mg/ml Stains-all in formamide; store in a dark bottle
(Stains-all is 1-ethyl-2-[3-(1-ethylnaphthol[1,2-d]thiazolin-2-ylidene)-2-methylpropenylnaphthol[1,2-d]thiazolium bromide)
Stains-all (1×): Dilute 20× Stains-all in 50% formamide shortly before
use

For dideoxy sequencing

Dithiothreitol (DDT) (0.1 M)
Annealing buffer (6.5×), 1 ml: 27.4 μl 1 M MnCl$_2$, 65.5 μl 1 M trisodium
isocitrate, 274 μl 1 M Tris-HCl, pH 7.5 (at 25°), 40 μl 5 M NaCl, 590
μl H$_2$O; store in screw cap vial at −20°, replace when dark brown
TE: 10 mM Tris-HCl (pH 8.0), 0.1 mM ethylenediaminetetraacetic acid
(EDTA)
Nucleotide mixes: dATP, dCTP, dGTP, and dTTP (250 μM each) and
the respective ddNTP (1 μM) in water; store at −20°
Formamide–dye mix: 95% (w/v) deionized formamide, 5 mM
Na$_2$EDTA, pH 8, 0.04% (w/v) bromphenol blue, 0.04% (w/v) xylene
cyanol FF; store at −20°

For gel electrophoresis[34]

Bind silane dilution: 0.5% (v/v) 3-(Trimethoxysilyl)propyl methacrylate,
0.3% (v/v) acetic acid, 3% (v/v) water, 96.2% (v/v) ethanol; store dark
at 4°
TBE (10×): 154.5 g Tris base, 26.2 g boric acid, 9 g Na$_2$EDTA, 810 ml
H$_2$O
Urea buffer: 486.5 g H$_2$O g urea, 93.5 g 10× TBE; store at room temperature for up to 2 months
Ammonium persulfate (10%, w/w): Prepare fresh before use or store
aliquots at −20°
Acrylamide solution [3.5% (6%), w/v]: 3.4 g (5.7 g) acrylamide, 0.1 g
(0.3 g) bisacrylamide, 97 (95) ml urea buffer; filter, use within 24 hr:
degas before adding N,N,N',N'-tetramethylethylenediamine
(TEMED) and ammonium persulfate

For nonradioactive detection

MgCl$_2$ (1 *M*), sterilized

SDS, 20%

Triton buffer (10×): 44.4 g Tris-HCl (Sigma), 26.5 g Tris base, 73 g NaCl, 3.7 g Na$_2$EDTA, 818 ml water; dissolve, add 100 ml Triton X-100; store at 4° for up to 4 weeks

Triton buffer (1×): Dilute 10× Triton buffer to 1× with distilled water, and filter; use within 24 hr

SDS buffer: 100 ml 10× Triton buffer, 250 ml 20% (w/v) SDS, 650 ml H$_2$O; filter, use within 24 hr

Block buffer (for digoxigenin-based detections): Dissolve 10 g of blocking reagent (purified casein fractions; Boehringer Mannheim) in 1 liter 0.1 *M* Tris-HCl (pH 7.6), 150 m*M* NaCl by stirring at 55–65° for 1 hr

DEA buffer (10×): 885 ml water, 32.2 g Tris-HCl, 105.2g diethanolamine (store the diethanolamine at 37°); store at 4°

DEA buffer (1×): Dilute 10× DEA buffer to 1× with distilled water, add 1/1000 vol of 1 *M* MgCl$_2$, and filter; use within 24 hr

Dioxetane substrate solution: 0.1 mg/ml AMPPD (or AMPPD-Cl) in 1× DEA buffer with 50 m*M* MgCl$_2$; prepare immediately before use

Tris-HCl (10×), pH 9.5: 4.33 g Tris-HCl, 87.5 g Tris base, 43.9 g NaCl, 9.68 g MgCl$_2$ · 6H$_2$O; use within 24 hr

BCIP stock solution: 50 mg/ml in dimethylformamide (DMF); store at −20° (only the *p*-toluidine salt of BCIP will dissolve in DMF)

NBT stock solution: 75 mg/ml in 70% (v/v) dimethylformamide; store at −20°

BCIP/NBT substrate solution: Add 33 μl NBT stock solution and 33 μl BCIP stock solution per 10 ml 1× Tris-HCl, pH 9.5; mix after each addition; prepare shortly before use

Stripping buffer: 1% (w/v) SDS, 1 m*M* EDTA, 20 m*M* Tris-HCl (pH 8.0), 125 m*M* NaCl

Direct Transfer Electrophoresis Protocols

Apparatus Design

To obtain the best transfer results, a constantly good contact between the lower gel surface and the membrane is essential. Several concepts have been used previously to achieve this goal, for example, brakes on the rear axle,[34] tensioning motors on front and rear axle, and metal weight cylinders on the conveyor belt.[11] A novel method is used in the machine

FIG. 1. An apparatus for direct transfer electrophoresis. The conveyor belt is moved from left to right by a stepper motor that is connected to two axles via a timing belt and timing pulleys. A string that is tensioned by a spring motor (front right) via a pulley connects front and rear axle. The outer dimensions of the machine are 55 × 70 × 13 cm.

shown in Fig. 1: a string pulls on both the front and the rear axle of the machine; the string is tensioned by a spring motor via a pulley. By pulling from both ends, any slack of the conveyor belt that might build up during the run is effectively removed. The spring motor exerts similar levels of force, regardless of its extension. Therefore, the system can take up a considerable amount of lengthening (or shortening) of the conveyor belt, providing superior performance and reliability over long periods of use.

Glass Plate and Thermostatting Design

One point that needs special attention in direct transfer electrophoresis is the lower surface of the gel itself: the gel polymerization must be uniform at the bottom, the glass plates must be flat over the whole width of the gel, and they need to be aligned carefully. Even polymerization can be achieved by sealing the gel bottom with Parafilm and pressing it against a vertical surface[11] as described below. The lower edges of glass plates usually show slight curvature after being cut, often an inverted U shape; when such plates are used, contact in the middle of the gel can be insuffi-

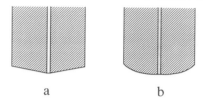

a   b

Fig. 2. Schematic drawing of the polished bottom edge of glass plates for DTE. (a) Side view of machine-polished plates; (b) manual polishing gives a rounded shape that makes it easier to determine if the plates are perfectly aligned.

cient. Machine-polished glass plates with an angle of 5–15° at the bottom as shown in Fig. 2a offer better performance and can be obtained by precision glass suppliers. Alternatively, plates can be manually polished to the shape shown in Fig. 2b. Because the rounded shape makes it easier to feel if the plates are correctly aligned, we usually subject even machine-polished plates to a short manual fine-polishing.

Manual Polishing of Glass Plates. Use freshly cut glass plates with unseamed bottom surfaces. Clean and align a set of plates; especially with 40-cm wide glass plates, check that they have either no curvature or identical curvature over their width. If the curvature differs slightly, the thickness of the gel will differ from the side toward the middle. When such plates are used, it can be difficult to insert the spacer, the shark's-tooth combs may leak, or bands in the middle of the gel may be wider than on the sides.

Clamp the plates together and polish them on wet, waterproof sanding paper, fixed to a flat bench with adhesive tape; use long stretches of sanding paper, starting with coarse (grit 120–220) paper. Try to obtain a shape as shown in Fig. 2b. Switch to finer sanding paper (320, finally 400) when they show an evenly rounded, well-aligned surface.

Glass Plate Setup for Thermostatted Gels. Although aluminum plates on one side of the gel assembly can be used to eliminate smiling, they can produce electrical short circuits, especially when voltages above 2000 V are used. Upper buffer chambers that extend almost over the whole length of the gel offer a convenient alternative. They can be easily made with glass spacers and silicone rubber adhesive.

To maintain a constant temperature independent from the applied voltage and the room temperature, we use a simple thermostatting design. Plastic tubing (Tygon, o.d. 6 mm, wall thickness 0.8 mm; VWR Scientific, South Plainfield,NJ) is mounted onto a plastic sheet with cable ties and connected to a pumping water bath (Model 1130, VWR Scientific) through quick coupling connectors (see Fig. 1). The glass plate, spacer, and buffer

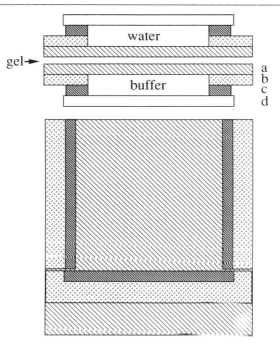

gel→

a
b
c
d

FIG. 3. Glass plate design for thermostatted DTE gels. A cross-section through the gel plate assembly is shown on the top. It consists of gel plates (a), wide spacers (b), narrow spacers (c), and buffer plate (d). Two sets of 4-mm thick glass spacers are used to give enough space for insertion of the thermostatting tubing (see text). The wide spacers extend to the side of the gel plate to minimize temperature gradients; the second set of narrow spacers allows the use of normal clamps. A view from the top during glueing of the plate assembly is shown below; the buffer plate is omitted.

chamber dimensions setup is shown in Fig. 3, and the dimensions are given in Table I.

To glue spacers and buffer plate to the gel plate, the inner side of the gel plate is covered with a piece of bench coat to shield it, and spacers and buffer plate are glued to it with generous amounts of clear silicone rubber adhesive. The plate assemblies can be used after 2 days.

Casting of Gels for Direct Transfer Electrophoresis

General Comments. Several casting methods have been successfully used, including pouring from the top, sliding, and clapping. Especially for 0.1-mm thin long gels, the sideways clapping method described below works best for us.

TABLE I

DIMENSIONS OF GLASS PLATES, SPACERS, AND BUFFER PLATES FOR
THERMOSTATTED DIRECT TRANSFER ELECTROPHORESIS GELS

Apparatus	For 20 × 60 cm plate (cm)	Formula
Gel plates[a]	22 × 62.5	$w \times h$
Wide spacers		
Side	2.5 × 59	$2.5 \times (h - 3.5)$ cm
Bottom	2.5 × 22	$2.5 \times (w)$ cm
Narrow spacers		
Side	1.2 × 59	$1.2 \times (h - 3.5)$ cm
Bottom	1.2 × 19.5	$1.2 \times (w - 2.5)$ cm
Buffer plate	20 × 59.5	$(w - 2) \times (h - 3)$ cm

[a] All glass plates are 4.7 mm (3/16 in.) thick; ears are 3–4 cm wide and 2.5 cm high.

Whatever method is used, it is essential to seal the lower surface from air to allow homogeneous polymerization; this is usually done with Parafilm and pressing of the gel assembly against a vertical surface. The second essential point is the careful alignment of glass plates at the bottom; badly aligned glass plates will give blurred band patterns.

While untreated glass plates can give good results, most consistent results were obtained when both glass plates were treated with bind silane. On reuse of bind silane-treated plates, sometimes significantly more blurry band patterns were observed; treatment of the glass plates with 0.5 M NaOH for 15–60 min after cleaning reverses these effects.

The usual precautions while handling acrylamide and bisacrylamide should be followed (wear gloves!).

Preparation of Glass Plates. Clean plates thoroughly with distilled water, ethanol, and Kimwipes. Lay plates horizontally and spread 0.5–1.0 ml of bind silane dilution onto each plate with Kimwipes. After 5–15 min at room temperature, wash off excess bind silane with ethanol and Kimwipes.

Remove any dust particles with flint-free paper tissue, slightly wet with ethanol.

Now attach the spacers to the glass plate without ears. When using clear, water-resistant adhesive tape (Tesafilm 4204, Baiersdorf, Germany gives best results; Scotch tape 3750-G, 3M, St. Paul, MN, also works), press one or two strips onto the glass without introducing air bubbles between the tape and the glass. Especially for gels longer than 20 cm, the minimum width of the spacer should be 3 cm.

Alternatively, conventional plastic spacers can be used. Grease a 2-cm wide stripe on the sides of the glass plates without ears and one side of the spacers; use a minimal amount of silicone vacuum grease and spread it evenly (wear gloves!). Put the spacer with the greased side down onto the glass plate and press it on over the whole length of the gel. The greasing prevents slippage of the spacers and reduces the introduction of air bubbles from the side when the sideways clapping technique is used.

Cover the glass plate–spacer assembly with the other (eared) glass plate. Align the plates and clamp them, using two clamps in the lower third. Insert the precomb (a flat piece of spacer material) about 3 mm deep into the top of the assembly, and fix it to the bottom plate with adhesive tape.

Now the bottom seal: turn the plate assembly around so that the eared plate is laying down. Fix a piece of Parafilm to the bottom of plate assembly with adhesive tape, leaving 2 cm at the sides (spacers) free. Bend the Parafilm around the lower edge of the gel, flip the plate assembly around, and lay it flat onto the casting rack or store it on a dust-free place for up to several days (a simple casting rack consists of a 3- to 5-cm high, 18 × 18 cm piece of Plexiglas with a second 18-cm wide piece of Plexiglas vertically screwed onto it[11]; dimensions are for 22 × 22 cm glass plates).

Casting by Sideways Clapping. Prepare fresh acrylamide solution by dissolving acrylamide and bisacrylamide in urea buffer, and filter the solution. Before use, degas with stirring for 5–10 min; gels with at least 6% (w/v) acrylamide do not have to be degassed.

Check that you have the lower glass plate leveled, enough clamps (e.g., 15 for 60 × 20 cm gels), a rubber band to press the plate assembly against the vertical surface, and a stop watch.

Add TEMED (e.g., 40 μl/50 ml) and 10% ammonium persulfate (e.g., 140 μl/50 ml), mix, start the stop watch, and take 1 ml into an Eppendorf tube to control the polymerization time.

Lift the top glass plate (with ears) into a vertical position on the near side of the lower plate. Pour the acrylamide solution onto the lower plate next to the near spacer, starting in the middle of the plate and extending toward the top and the bottom. Slowly lower the upper glass plate, avoiding introduction of air bubbles.

Align the bottom of the plates carefully, checking the alignment with your fingertips. Set the first four clamps near the bottom of the gel. Check the alignment after each clamp and adjust it, if necessary. Now set the other clamps.

Press the gel assembly against the vertical piece of the casting rack and attach the rubber band.

Finally, set a weak clamp on the top of the gel. It should exert pressure onto the precomb; if this clamp is too strong, it may be difficult to insert the shark's-tooth comb later. For gels wider than 35 cm, we prefer a different method to ensure a tight fit of the precomb. Using a 22-cm wide glass plate and a 3-kg weight, pressure is exerted over a wider region onto the notched glass plate directly on top of the precomb.

Check if the gel solution in the Eppendorf tube has polymerized yet. The casting should take 5–7 min, and the solution should start to polymerize in 12–18 min. If the polymerization times are significantly shorter or longer, adjust both TEMED and ammonium persulfate concentrations. When the polymerization starts while the gel is still being clamped, poor resolution will result due to a deformed gel matrix; it is probably better to discard these gels.

Let the gel polymerize for 1–24 hr (if the polymerization did not start within 15 min, let it sit for at least 2 hr). If the gel is stored overnight, squirt some water into the buffer chamber after the gel has polymerized, and cover the top with Saran wrap to prevent drying out of the upper gel surface; also recheck that the bottom is pressed tightly against a vertical surface.

Setup and Electrophoresis. Prewarm the electrophoresis buffer to 35–45°. Degas the buffer for the lower buffer chamber for at least 5 min with stirring in a vacccum sidearm flask. Clean the gel assembly and remove any excess acrylamide from the lower edges of the glass plates, but do not touch the lower surface of the gel.

Pour the degassed buffer into the lower buffer chamber and mount the gel in the electrophoresis unit; let it touch the buffer with one corner first to avoid trapping of air bubbles between gel and conveyor belt. Insert the thermostatting tubing, and fill prewarmed buffer into the upper electrophoresis chamber and prewarmed water into the thermostatting chamber, if thermostatting is used.

Remove the precomb. Carefully remove all excess acrylamide from the top of the gel, using a syringe and, if necessary, a piece of comb material. Flush the upper gel surface with buffer.

The tips of the shark's-tooth comb may be greased lightly with silicone grease. Ungreased combs can give more problems with horizontal leakage, but too much grease interferes with sample loading and electrophoresis.

Insert the shark's-tooth comb; it should just touch the gel or go into it for about 0.5 mm. If there are problems inserting it, removing the uppermost clamps and supporting the comb with stiffer plastic may help. Practice loading or checking the wells with 50% stop buffer may be done now.

Preelectrophorese for 30–60 min at 70 V/cm for 0.1-mm gels and 60 V/cm for 0.2-mm gels. If you do not use thermostatting, set an ampere (or watt) limit at about 25% more than the initial current (or wattage) to prevent thermal runaway.

TABLE II
TYPICAL RUNNING CONDITIONS AND RESOLVED BASES IN DIRECT
TRANSFER ELECTROPHORESIS

Length (cm)	Thickness (mm)	Gel concentration (%)	Voltage (V)	Membrane speed (cm/hr)	Resolved fragments[a]
20[b]	0.2	6	1200	14	20–320
30[b]	0.1	6	1800	9	20–420
30[c]	0.1	4	2100	20	50–550
45[c]	0.1	4	3150	13	50–600
60[c]	0.1	3.5	4200	11	60–750

[a] Typical results are given; optimal results can be better, and the interpretable range can also be larger. The acrylamide : bisacrylamide ratio is 19 : 1, except for 3.5% gels, for which a 34 : 1 ratio is used.
[b] Gels run with one aluminum plate for temperature distribution; gel temperature is approximately 37°.
[c] Gels run with thermostatting from both sides at 50°.

Denature the samples immediately before loading for 2–3 min in an 80° waterbath, and quick-chill them on ice water. Flush the wells to remove urea, and load the samples with a duckbill pipette tip, a Hamilton syringe, or with a Hamilton 12-channel pipettor; the 12-channel pipettor is by far the easiest and most reliable of these. Start electrophoresis (60–70 V/cm; see above) immediately (if the loading takes longer than 5–10 min, start electrophoresis for 1–2 min between samples).

After the samples enter the gel and before the bromphenol blue reaches the gel bottom, shut off the power supply and move the nylon membrane under the gel. First mark the membrane by cutting an edge or writing on it with a suitable pen. Depending on the machine used, a dry membrane can be taped to the conveyor belt with waterproof tape, or a membrane wetted with distilled water is put underneath a strap on the conveyor belt.[34] Move the membrane below the gel and further under the first glass axle; this eliminates floating of the membrane in the buffer chamber.

Start the transport of the membrane shortly before the bromphenol blue reaches the membrane. Table II gives an overview of typical membrane speeds for several setups. Make sure that the membrane is transported out of the buffer at the end of the run.

Membrane Treatment after Electrophoresis. When radioactively labeled sequencing reactions are loaded, the membrane can be wrapped in thin plastic foil and exposed to X-ray film at room temperature; after the exposure, the nylon membrane can be stripped and reused as described.[34]

For nonradioactive DNA sequencing and genomic or multiplex sequencing, put the membrane on clean Whatman (Clifton, NJ) 3MM paper and dry it at 70–90° for 5–15 min. Covalently bind the DNA to the membrane by ultraviolet (UV) cross-linking.[12] Normal exposure times are 20 sec for multiplex and genomic sequencing and 30–120 sec for sequencing with end-label primers at 1.3 mW/cm^2 (UV boxes as described[12] and many UV transilluminators have light intensities of 0.6–1.5 mW/cm^2). Store the membrane between clean Whatman 3MM paper sheets.

Especially when using nonradioactive deletion methods, always handle the membrane with clean gloves and avoid unnecessary mechanical stress and exposure to dirt; dust, fingerprints, bacterial contamination, forceps marks, and so on are likely to result in increased nonspecific background.

Apparatus Maintenance. After the run, take the gel out of the apparatus and replace the buffer in the lower buffer chamber with distilled water. Move the conveyor belt through the water four to six times to remove buffer residuals from the conveyor belt; change the water once between the washes.

Disassembly of Gel Sandwich. Insert a gel knife [#2117-161 (Pharmacia), the small blade of a Swiss army knife, or a razor blade] between the glass plates, about 2 cm below the ears. Insert slowly until the plates are far enough apart to allow the insertion of plastic wedges (#SE 1514; Hoefer, San Francisco, CA). Working with three or four plastic wedges from one side of the sandwich, pry the glass plates totally apart; take care that the bottoms of the plates are not damaged.

Use a piece of thin plastic (e.g., 0.35-mm thin polyester) to scrape the acrylamide from the glass plates; washing with warm water helps. Remove any remaining acrylamide with water and paper towels. Clean plates thoroughly with ethanol, water, and Kimwipes. Detergents are not necessary.

It may be necessary to immerse the glass plates in 0.5 M NaOH for 15–60 min to remove excess bind silane before reuse, followed by washing with water and ethanol.

Protocols for Nonradioactive DNA Sequencing

Oligonucleotide Labeling with Biotin and Digoxigenin

For biotin labeling of oligonucleotides, biotinylated phosphoramidites that can be used on an automatic synthesizer are most convenient. For labeling with digoxigenin and as a low-cost alternative to biotin phosphoramidites, oligonucleotides can be 5'-aminated during synthesis (e.g., with N-TFA-C$_6$-amino modifier; Clontech), followed by reaction with NHS-biotin or NHS-digoxigenin.

Both biotin and digoxigenin are hydrophobic, enabling purification by reversed-phase high-performance liquid chromatography (HPLC).[35,36] Because both labels also reduce the electrophoretic mobility of oligonucleotides by an apparent size of 1–2 nucleotides, polyacrylamide gel electrophoresis can also be used for analysis of the reaction and for purification.[37] However, a purification is normally not necessary.

For maximum detection sensitivity, a sufficiently long spacer must separate the label from the oligonucleotide.[27,36] With N-TFA-C$_6$-amino modifier and NHS-XX-biotin (also called NHS-LC-biotin) or NHS-digoxigenin, this requirement is fulfilled.

Labeling of Aminated Oligonucleotides with Biotin. Dissolve the deprotected and lyophilized oligonucleotide in water (180 μl for 0.2-μmol synthesis scale, 900 μl for 1.0-μmol synthesis). Mix 90 μl (0.1 μmol) with 10 μl 1 M sodium carbonate, pH 9.0, and add 100 μl of biotin-XX-NHS ester (20 mg/ml in dimethylformamide, freshly prepared). Incubate at 42° for 2 hr; the formation of a precipitate is normal and can be ignored. Add 300 μl 0.1 M Tris-HCl, pH 8.0, and incubate at room temperature for 1 hr.

Remove excess biotin by gel filtration through Sephadex G-25 (NAP-5 columns; Pharmacia) with 0.1 M triethylamine acetate, pH 7.0, as buffer, lyophilize the eluate, and resuspend in TE. The oligonucleotide can be analyzed by gel electrophoresis as described below; we usually obtain biotinylation efficiencies of 50–90%. The oligonucleotides can be used directly or purified by HPLC or gel electrophoresis.

Labeling of Aminated Oligonucleotides with Digoxigenin. The labeling is done essentially as described for the biotinylation, except that 200 μl of NHS-digoxigenin (6.25 mg/ml, freshly dissolved in dimethylformamide) is added to 90 μl of oligonucleotide (0.1 μmol) and 10 μl of 1 M sodium carbonate, pH 9.0; again, a precipitate may form and can be ignored. After reaction overnight at room temperature, add 250 μl of 0.2 M Tris-HCl, pH 8.0, incubate for 1 hr at room temperature, and purify through NAP-5 columns as above. The reaction yield is about 50%.

Analysis of Labeling Reaction by Gel Electrophoresis and Staining with "Stains-All." Mix 0.5–1 nmol of labeled oligonucleotide with an equal volume of deionized formamide containing 0.04% bromphenol blue, and load onto a 15% denaturing acrylamide gel (e.g., 20 cm × 20 cm × 0.8 mm). Electrophorese at 40 V/cm until the bromphenol blue reaches the lower quarter of the gel (about 1 hr). Disassemble the glass plates, and

[35] A. J. Cocuzza, *Tetrahedron Lett.* **30**, 6287 (1989).
[36] K. Mühlegger, E. Huber, H. von der Eltz, R. Rüger, and C. Kessler, *Biol. Chem. Hoppe-Seyler* **371**, 953 (1990).
[37] A. Chollet and E. H. Kawashima, *Nucleic Acids Res.* **13**, 1529 (1985).

stain the gel in 1× Stains-all until bands are clearly visible (about 5–20 min). Destain the gel in distilled water for 15–60 min; prolonged exposure to daylight reduces signal intensities. Document the results by photography, or by photocopying the gel sandwiched between clear plastic sheets.

Sequencing Reactions

The sequencing reactions can be done with a one-step protocol because no labeling reaction is necessary. Best results are obtained with single-stranded template DNA from recombinant phagemid clones (in pTZ18R; Pharmacia) or M13, using modified T7 DNA polymerase[38] and manganese isocitrate buffers.[39] The uniform band intensities with this system simplify interpretation of the sequence patterns and often allow reliable reading even in regions where multiple bands are badly resolved. Double-stranded template DNA, and different DNA polymerases and protocols can also be used.[11,14]

The ratio of deoxynucleotides to dideoxynucleotides must be adjusted; with manganese isocitrate buffers and T7 DNA polymerase, a ratio of 250:1 gives good sequence patterns up to more than 1000 bases. If the main interest is the first 300 bases, a ratio of 125:1 can be used.

Primer Annealing

1. Combine 0.5 pmol of single-stranded template DNA, 0.5 pmol of end-labeled primer and 2 μl of annealing buffer in a total volume of 13 μl.
2. Place in a 65–75° water bath for 2–3 min.
3. Cool slowly to room temprature over 15–45 min; use directly or store at −20°.

Sequencing Reactions

1. Add 1 μl of 0.1 M DTT to the annealing mixture and mix carefully; the color changes to dark brown and then to clear.
2. Distribute 2.5 μl of the nucleotide mixes into reaction vials (Eppendorf tubes, Terazaki plates, polypropylene plates, or polystyrene microtiter plates) and prewarm in a 37–40° water bath for 1–2 min.
3. Dilute modified T7 DNA polymerase to 1.5 U/μl with ice-cold TE.
4. Add 2 μl of enzyme to the annealing mixture, mix carefully, and distribute 3.5-μl aliquots to the prewarmed nucleotide mixes; incubate at 37–40° for 2–15 min.
5. Stop the reactions by adding 6 μl formamide–dye mix; the reactions can be stored at −20° for several months and loaded repeatedly.

[38] S. Tabor and C. C. Richardson, *Proc. Natl. Acad. Sci. U.S.A.* **84,** 4767 (1987).
[39] S. Tabor and C. C. Richardson, *Proc. Natl. Acad. Sci. U.S.A.* **86,** 4076 (1989).

6. Immediately before loading, denature the sequencing reactions for 2 min in an 80° waterbath, chill on ice–water, and load directly.

Nonradioactive Detection of DNA Sequencing Patterns

General Comments. All buffers used should be made with the highest purity water available and kept free of contamination with external enzymes; fingerprints and bacteria, for example, are good sources of alkaline phosphatase. Most of the buffers used, including SDS buffer, support growth of bacteria. Addition of 1 mM EDTA to Triton and SDS buffers shows sufficient bacteriostatic effects, but it can also lead to twofold reduced signal; the most foolproof way is to prepare all 1× buffers immediately before use. Additional background problems can arise from lint and dirt and from mechanical damage of membranes. Membranes should be handled carefully with clean gloves and kept between Whatman 3MM paper (or in hybridization bags) before the detection.

The detection protocols can be carried out in hybridization bags, large plastic trays, or large cylinders.[11,40] All methods avoid overlapping of membrane parts that can lead to reduced signal and increased background in nonradioactive protocols.

Development in hybridization bags with a small "chimney" for exchange of solutions offers the advantage that chemiluminescent exposures can be done by putting an X-ray film on top of the bag; however, the thickness of the bags leads to more blurry sequence patterns. Further problems can arise from uneven distribution of enzyme or substrate solutions and trapped air bubbles.

Large, flat plastic trays on gyratory shakers are convenient for the development of small numbers of membranes at a time. They can give the most efficient washing from both sides of the membrane; however, somewhat higher volumes of all solutions are needed.

Large cylinders on horizontal rollers are ideal to minimize the volume of valuable incubation solutions. Simultaneous development of several membranes is easy, and automation of the development process is straightforward.[11] A variety of simple home-made rollers, cell culture rollers, or hybridization ovens can be used for this purpose. A more sophisticated machine that uses up to four polycarbonate bottles with magnetic bottoms and is programmable for tilt angle, revolution speed, and step time for up to 484 steps (Western Roller-4; BioComp) is routinely used in our laboratory.

[40] N. Thomas, C. N. Jones, and P. L. Thomas, *Anal. Biochem.* **170,** 393 (1988).

For All Detection Protocols. Unless otherwise indicated, all washing steps are done for 2–5 (up to 20) min with 250–350 ml of buffer (approximately 500 ml in trays) for 32 × 40 cm membranes. The numbers of washing steps given below are minimum numbers; if background is a problem, up to four washing steps at each point and careful removal of solutions between steps can be helpful.

Developments in cylinders are done at 5–10 rpm. When trays are used, the shaker speed is adjusted so that the membrane is completely covered with solution (25 to 75 rpm); in incubations with streptavidin, enzymes, and substrates, it is advantageous to distribute the solutions by manual shaking every 2 to 3 min.

1. After drying and UV cross-linking, fit the membrane in a plastic or siliconized glass cylinder with the DNA facing inward.

2. Rehydrate the membrane with distilled water; try to avoid or remove air bubbles between the membrane and the cylinder.

Protocol 1: Detection of Biotin-Labeled Sequence Patterns

3. Incubate with SDS buffer for 5–30 min.

4. Incubate with streptavidin in SDS buffer (1 μg/ml, 50 ml) for 10 min.

5. Wash two times with SDS buffer.

6. Wash once with Triton buffer.

7. Incubate with biotinylated alkaline phosphatase in Triton buffer (1 μg/ml, 50 ml) for 10 min.

8. Wash two times with Triton buffer.

Protocol 1a: Chemiluminescent Detection

9. Wash two times with 1× DEA buffer.

10. Incubate with 50 ml dioxetane substrate solution for 5 min.

11. Discard the substrate solution, take the membrane out of the cylinder, and let it drip dry for 30–60 sec. Wrap it in clear plastic foil (Saran wrap), avoiding wrinkles and air bubbles between membrane and foil.

12. Expose to X-ray film; exposure times depend on the amount of DNA loaded; 5–30 min is typical. Please note that the signal per time unit increases for a couple of hours with dioxetane substrates[33]; a 15-min exposure after 60 min will give about four times more intense bands than an immediate 15-min exposure.

Protocol 1b: Colorimetric Detection

This protocol can also be used after chemiluminescent detection; two additional washes with Triton buffer can then be added before step 9.

9. Wash the membrane twice with 1× Tris-HCl, pH 9.5.

10. Incubate with BCIP/NBT solution until bands are sufficiently dark (20 min to overnight).

11. Wash the membrane twice with Triton buffer for at least 15 min (up to overnight).

12. Wash the membrane with distilled water.

13. Dry the membrane on Whatman 3MM paper at 70–90° for 5–15 min. Store the membrane in a hybridization bag in a dark place (e.g., between black cartons).

Protocol 2: Detection of Digoxigenin-Labeled Sequence Patterns with Phosphatase–Antibody Conjugates

Start with step 3.

3. Incubate the membrane with 100 ml block buffer for 30–60 min (shorter times may be sufficient).

4. Incubate with 50 ml of anti-digoxigenin–alkaline phosphatase conjugates (150 mU/ml) in block buffer for 30 min.

5. Wash twice with block buffer.

6. Proceed with colorimetric or chemiluminescent development as described in protocols 1a and 1b, respectively.

Protocol 3: Detection of Digoxigenin-Labeled Sequence Patterns with Peroxidase–Antibody Conjugates

Start with step 3.

3. Incubate the membrane with 100 ml block buffer for 30–60 min (shorter times may be sufficient).

4. Incubate with 50 ml of anti-digoxigenin–peroxidase conjugates (150 mU/ml) in block buffer for 30 min.

5. Wash twice with block buffer.

6. Wash twice with 0.1 M Tris-HCl (pH 8.0), 150 mM NaCl.

7. Decant, add 50 ml ECL detection solution 2, and mix; add 50 ml ECL detection solution 1, and mix for 1 min.

8. Wrap membrane in clear plastic foil (Saran wrap) and expose to X-ray film. An initial exposure of 1 min can give an indication of the exposure time needed (usually 10–45 min). Please note that the signal intensity of peroxidase-based enhanced chemiluminescence is maximal in the first minutes after substrate addition and then decays over 1–2 hr.[30]

Protocol 4: Combined Protocols for Successive Detection of Biotin- and Digoxigenin-Labeled Sequence Patterns

Before denaturing and loading onto a polyacrylamide gel, combine equal amounts of two sequencing reactions, one with a biotinylated primer

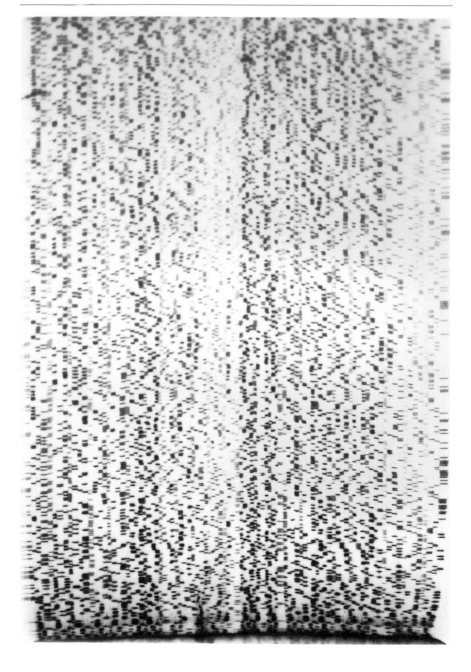

and one with a digoxigenin-labeled primer. Alternatively, differently labeled "universal" and "reverse" primer can be used for sequencing of double-stranded templates. Dry, cross-link, and rehydrate the membrane as described above. Start with step 3.

3. Incubate with SDS buffer for 5–30 min.
4. Incubate with streptavidin in SDS buffer (1 μg/ml, 50 ml) for 10 min.
5. Wash two times with SDS buffer.
6. Wash once with Triton buffer.
7. Incubate with biotinylated alkaline phosphatase (0.5 μg/ml, 50 ml) in block buffer for 20 min.
8. Incubate with anti-digoxigenin–peroxidase conjugates (150 mU/ml, 50 ml) in block buffer for 30 min.
9. Wash two times with block buffer.
10. Wash twice with 0.1 M Tris-HCl (pH 8.0), 150 mM NaCl.
11. Decant, add 50 ml ECL detection solution 2, and mix; add 50 ml ECL detection solution 1 and mix for 1 min.
12. Wrap membrane in clear plastic foil (Saran wrap) and expose to X-ray film (15–45 min).
13. Wash the membrane two times with Triton buffer.
14. Wash twice with 1× DEA buffer.
15. Incubate with 50 ml dioxetane substrate solution for 5 min.
16. Wrap membrane in clear plastic foil and expose to X-ray film.

Protocol 5: Membrane "Stripping" after Phosphatase-Based Detections

To denature (or remove) alkaline phosphatase from a previous development, the membrane is treated two times with stripping buffer, preheated to 65–75°, for 15 min. Most conveniently, this step is done either in a sealed hybridization bag in a water bath or hybridization oven, or in a tray that has been prewarmed with hot distilled water and covered with a plastic sheet to reduce temperature drops due to evaporation. The effectiveness of the stripping procedure can be checked by washing the membrane with 1× DEA buffer, incubation with dioxetane substrate, and exposure to X-ray film.

FIG. 4. Chemiluminescence-detected sequencing patterns from direct transfer electrophoresis with 45 × 45 cm gels. Twenty-four sequencing reactions with biotinylated primers were loaded onto a 4% acrylamide gel and detected as described in protocol 1a, using a plastic tray. The original size of the membrane was 32 × 40 cm.

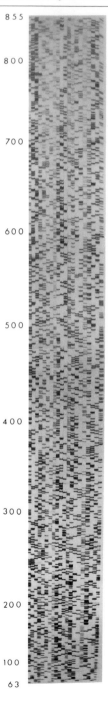

Results

Figures 4 and 5 show sequencing patterns obtained with direct transfer electrophoresis and chemiluminescent detection. In Fig. 4, 24 reactions were loaded onto a 45 × 45 cm 4% acrylamide gel; about 400 nucleotides per lane set can be read, giving almost 10,000 bases of sequence information from one film. The pattern shown in Fig. 5 was generated with a 60-cm long, 20-cm wide 3.5% acrylamide gel. A constant interband spacing for fragments between 200 and 850 nucleotides in length was obtained with a constant membrane speed. For shorter fragments, the bands are more closely spaced; because these bands also tend to be sharper, the sequence can be read accurately. With increasing fragment length, the bands become broader until multiple bands are poorly resolved; an enlarged section from the top of Fig. 5 is shown in Fig. 6 to illustrate this point. Due to the constant and sufficient spacing between bands and the even band intensities obtained with modified T7 polymerase and manganese buffers,[39] the correct sequence can be obtained even in this top region, giving about 790 nucleotides of sequence information per lane set. For different gel parameters, exemplary numbers of resolved bases are given in Table II.

Figures 4 and 5 also show that virtually background-free sequence patterns can be obtained with chemiluminescent detection. For the data shown in Fig. 5, three sequencing reactions with differently labeled primers and 1 pmol of template DNA per reaction were combined, giving a total volume of 36 μl; about 1 μl, that is 1000 fmol/(4 lanes × 36 μl total) = 7 fmol was loaded per lane. With a 250 : 1 ratio of deoxy- to dideoxynucleotides and no discrimination between deoxy- and dideoxynucleotides,[39] the amount of DNA in the first bands is about 0.03 fmol (20 million molecules).

The patterns shown in Figs. 5 and 6 were the second patterns produced from this membrane; the membrane has been developed with the combined protocol for the successive detection of digoxigenin- and biotin-labeled

---

FIG. 5. Chemiluminescence-detected sequencing patterns from a 60 × 20 cm, 3.5% acrylamide gel. The length of the original membrane was 82 cm. Five lane sets are shown; the fragment length is indicated on the left. Three different sequencing reactions with biotin-, fluorescein-, and digoxigenin-labeled primers were mixed before loading. After electrophoresis at 4200 V with a constant membrane speed of 11 cm/hr, the sequencing patterns were detected as described in protocol 4, except that 1% casein in 1× Triton buffer was used instead of block buffer ("essentially vitamin-free" casein from Sigma was dissolved in 0.1 M NaOH and neutralized by addition of 100 ml 1 M Trizma-HCl before addition of 10× Triton buffer and filtration). The pattern shown is from the streptavidin–biotinylated alkaline phosphatase detection with AMPPD; the exposure time was 20 min, starting 33 min after substrate addition. Enlarged sections of this film are shown in Fig. 6 (lanes 5–8 from the left, nucleotides 735–855) and Fig. 7b (nucleotides 336–417).

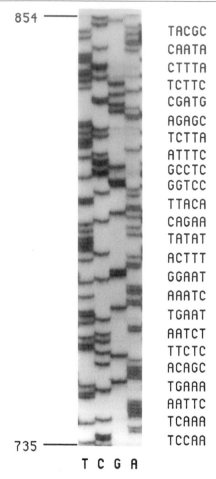

854 ——

	TACGC
	CAATA
	CTTTA
	TCTTC
	CGATG
	AGAGC
	TCTTA
	ATTTC
	GCCTC
	GGTCC
	TTACA
	CAGAA
	TATAT
	ACTTT
	GGAAT
	AAATC
	TGAAT
	AATCT
	TTCTC
	ACAGC
	TGAAA
	AATTC
	TCAAA

735 —— TCCAA

T C G A

FIG. 6. An enlarged section from the top of Fig. 5, showing the region where poor band separation begins to limit reliable reading. The total length of the fragments is indicated on the left, the correct sequence on the right.

primers (protocol 4). Figures 7a and b show the results of these detections for a section of the membrane; two independent sequencing patterns can be obtained within less than 4 hr after electrophoresis.

After these first two detections, the membrane has been "stripped" to remove phosphatase activity and anti-digoxigenin–peroxidase conjugates from the membrane (protocol 5) and then developed with anti-digoxigenin–phosphatase conjugates and AMPPD-Cl (protocol 2). The results in Fig. 7c show that the phosphatase activity from the biotin-based detec-

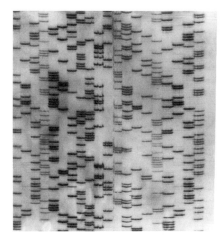 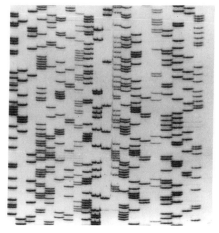

a. anti-DIG-HRP
   ECL

b. streptavidin
   biotin-CIAP
   AMPPD

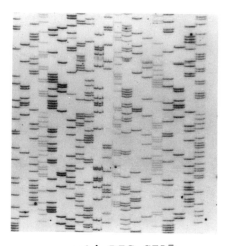

c. anti-DIG-CIAP
   AMPPD-Cl

FIG. 7. Sequential detection of DNA sequencing patterns with digoxigenin-labeled primer, anti-digoxigenin antibody–peroxidase conjugates and enhanced chemiluminescence (a), biotin-labeled primer, streptavidin–biotinylated alkaline phosphatase and AMPPD (b), and anti-digoxigenin antibody–alkaline phosphatase conjugates and AMPPD-Cl (c). Exposure times were (a) 45 min, (b) 20 min (starting 33 min after substrate addition), and (c) 30 min (starting 30 min after substrate addition). The detection was done with the combined protocol (protocol 4 with modifications as noted in the caption to Fig. 5) for (a) and (b), followed by stripping as described in protocol 5 and detection as described in protocol 2. The total length of the DNA fragments in the section shown ranges from 336 nucleotides at the bottom to 417 nucleotides at the top.

tion can be removed well enough to allow two subsequent phosphatase-based detections.

Discussion

Comparison of Direct Transfer Electrophoresis and Other Electrophoresis Methods for DNA Sequencing

In DNA sequencing, the main priorities can vary significantly. When teaching or learning the technique or when sequencing is done only occasionally, simple and reliable systems can be a priority. For high-throughput projects, highly optimized systems that emphasize long reads for the resulting higher overall efficiency are useful. Under both circumstances, nonradioactive detection methods may be preferred to radioactive labeling because of reduced health hazards, fewer regulatory problems, more stable reagents and reactions, and faster detection.

Direct transfer electrophoresis offers considerable advantages for all these needs. For learning and occasional sequencing, very short gels that are easy to cast and handle can be useful; 20-cm long gels can resolve fragments up to 400 nucleotides in length, comparable to 40- to 55-cm long gels in conventional electrophoresis. When higher performance is desired, one 60-cm long gel can resolve fragments up to 800 nucleotides in length; to obtain similar results with conventional electrophoresis, multiple loadings on two or three 80- to 100-cm long gels are necessary.[5]

The membrane-bound sequencing patterns generated by direct transfer electrophoresis are also an ideal starting point for enzyme-linked detection methods. Colorimetric detections are cost competitive, and chemiluminescent detection can give results within less than 2 hr after electrophoresis. If high efficiency is of importance, the combination of two different label-detection systems as shown offers a simple way to double the information per sequencing gel, independent of other parameters like slot width or resolved bases; for highest demands, direct transfer electrophoresis can be combined with multiplex sequencing.[13]

Another step toward higher sequencing efficiency can be the elimination of the manual sequence reading; this can be achieved by using film scanners[23] or fluorescence-based machines. Because machines for direct transfer electrophoresis are technically much simpler than those for fluorescent sequencing, several DTE machines and a film scanner can be obtained for the price of one fluorescence-based automate. Because most film scanners can analyze at least 10 films/day,[23] the DTE-plus-scanner setup could give a significantly higher performance.

Practical Aspects: Gel Length, Gel Thickness, and Thermostatting

Direct transfer electrophoresis is highly flexible; to learn the method, short, convenient gels can be used, possibly in combination with standard radioactive sequencing. Later, the system can be easily changed and expanded according to the preferences of the experimenter, for example, to nonradioactive duplex sequencing with "high performance" 60-cm gels.

A critical point in DTE is the lower gel surface; it may take a few trials to produce reproducibly good gels. Our success rate on casting 0.1-cm thin, 20 × 60 cm or 45 × 45 cm gels is above 90%.

For learning and some routine applications, 0.2-mm thick gels can be more convenient than 0.1-mm gels, for which sample loading can be somewhat tedious. The use of 12-fold Hamilton syringes simplifies loading considerably, especially on 0.1-mm gels; 96 lanes can be loaded in less than 10 min. If longer reads or sharper band patterns are desired, we recommend using 0.1-mm thin gels.

Thermostatting of direct transfer electrophoresis gels is not mandatory. Good results can be achieved by using an aluminum plate or a buffer chamber that extends toward the bottom of the gel for heat distribution. However, we prefer to use thermostatted gels because of several advantages.

1. Gels can be run at higher temperatures without increased problems with "thermal runaway" (with voltage regulation, the power generated heats the gel, the electrical conductance increases, higher power heats the gel even more . . . until the buffer boils or glass plates crack). The normal countermeasure against thermal runaways is to use power instead of voltage regulation; however, this often leads to reduced voltages during the run, resulting in increasing interband distances in DTE. Higher temperature reduces the number of compression artifacts and also leads to faster runs; the mobility of DNA increases by about 2%/°C.[41]

2. Gel temperature and voltage can be regulated independently. This helps to obtain reproducible results when experiments are run at different room temperatures, and it is especially useful for systematic studies.

3. Thermostatting from both sides reduces the temperature gradient that can form over the width of the gel; such temperature gradients can lead to broader band patterns.[41]

The thermostatting method as described here is simple to implement, electrically insulated and efficient. A simple, noncooling, pumping water bath is sufficient and can be used for two DTE machines simultaneously.

[41] R. J. Wieme, *in* "Chromatography" (E. Heftman, ed.), 2nd Ed., p. 210. Reinhold, New York, 1967.

Theoretical Aspects of Gel Parameters in Direct Transfer Electrophoresis

The following discussion is simplified, intended to answer some common questions about direct transfer electrophoresis; more explicit discussions of some principles can be found elsewhere.[41–45]

What determines the range of constant interband spacing in direct transfer electrophoresis? The linear range in DTE is mainly determined by the gel concentration and the electric field. Lowering the acrylamide concentration increases the linear range toward longer fragments, whereas increased field strength reduces it (see below). Because lower acrylamide concentrations also reduce the resolution between fragments, an optimal concentration for practical purposes can be found; with 60-cm long gels, 3.5% acrylamide gives best results, whereas 3.0% gels are slightly inferior. For shorter gels, higher concentrations can be optimal. Similar optimal gel parameters were found in capillary electrophoresis.[46]

Does the transfer process itself lead to the blurry bands? Why do thinner gels give better results. With carefully polished and aligned glass plates and DTE machines that give good contract between lower the gel surface and the membrane, the broadening of bands during the transfer process due to diffusion is negligible. This has been shown with machines that move the membrane in steps slightly larger than the gel thickness,[11] resulting in "striped" bands due to membrane movements during the transfer of one band. Typical step sizes in our DTE machines are around 30 μm.

However, bands on the membrane cannot be thinner than the thickness of the gel. With 0.35-mm thick gels and normal membrane speeds, the data quality is significantly reduced; 0.2-mm gels give much sharper bands, and a slight further enhancement of data quality can be found with 0.1-mm thin gels.

Thinner gels also reduce the temperature gradient that can form over the width of the gel, thereby further improving the data quality.

How much (and how) can longer gels improve the readable range? With high-quality sequencing reactions and electrophoresis systems such as DTE, the length of the readable patterns is not restricted by the gel size

[42] J. C. Giddings, *Sep. Sci.* **4**, 181 (1969).

[43] J. J. Kirkland, W. W. Yau, H. J. Stoklosa, and C. H. Dilks, Jr., *J. Chromatogr. Sci.* **15**, 303 (1977).

[44] J. W. Jorgenson and K. D. Lukacs, *Anal. Chem.* **53**, 1298 (1981).

[45] H. Swerdlow and R. Gesteland, *Nucleic Acids Res.* **18**, 1415 (1990).

[46] J. A. Luckey, H. Drossman, A. J. Kostichka, D. A. Mead, J. D'Cunha, T. B. Norris, and L. M. Smith, *Nucleic Acids Res.* **18**, 4417 (1990).

itself; instead, the reliable interpretation of sequencing pattern becomes tedious or impossible because of badly overlapping, unresolved bands. Important factors that contribute to the broadness of bands are (1) random distortions, for example diffusion, (2) loading of samples, (3) the transfer process, (4) the detection process, and (5) possibly other factors.

To examine the expected dependence of the gel length on the single factors contributing, we do a *Gedankenexperiment* (thought experiment) with direct transfer electrophoresis, keeping all conditions such as field strength, and so on, constant, but varying the gel length and the membrane speed; the membrane speed is adjusted so that the same distance between bands is achieved regardless of the gel length.

The elution time t_e for fragments is proportional to the gel length g and inversely proportional to the electrophoretic mobility m; in the linear range, t_e can be described as

$$t_e = g(k_1 n + k_2) = g/m \tag{1}$$

Random distortion: Diffusion (and similar random effects during electrophoresis) induce variance according to Einstein's law, $\sigma_{dif}^2 = 2Dt$; D is the effective diffusion coefficient under electrophoresis conditions.[42] Starting with a band of negligible width, a gaussian peak with width

$$4\sigma_{dif} = (32Dt)^{1/2} \tag{2}$$

will result at time t. Increasing the gel length g increases the elution time proportionally; also, the distance and elution time difference between two adjacent bands increases in proportion to g.

The resolution between two bands is defined as the ratio of their distance to their average width[42]; an equivalent definition can be made on the time scale:

$$R = gk_1/t_w \tag{3}$$

where gk_1 is the elution time difference between two bands and t_w is the "time width," $4\lambda/m$. Thus,

$$R = mgk_1/4\sigma \tag{4}$$

or, with Eq. (2),

$$R = mgk_1(32Dt_3)^{-1/2} \tag{5}$$

We define $k_3 = mk_1(32D/m)^{-1/2}$; because $t_e = g/m$, this gives

$$R = k_3 g^{1/2} \tag{6}$$

This gives a square root relation between resolution and gel length at a given fragment length n. Increasing the gel length fourfold would result in twofold better resolution at a given point; for gaussian bands of equal intensity, previously unresolved peaks (with $R = 0.5$) would be clearly resolved ($R = 1.0$, the intensity between two peaks would drop to 27% of the maximum).

Loading effects: samples are loaded at a given height, usually 1–3 mm; while entering the gel, this zone is compressed, but of course does not become one-dimensional. This compression is independent of the gel length; however, the membrane speed in our *Gedankenexperiment* is indirectly proportional to the gel length. If loading effects were the only broadening effects, a fourfold longer gel with fourfold lower membrane speed would compress the band width to 25%. The loading effects are inversely proportional to the gel length.

Transfer process: As discussed before, broadening due to the transfer can be easily minimized by using thin, well-aligned gels. The theoretical treatment is just as simple: transfer broadening is independent of the gel length. Its effects on the resolution are, however, dependent on the membrane speed under given conditions and gel length (a variation of the above *Gedankenexperiment*): faster transport would increase the distance between bands, simultaneously increasing the resolution.

Detection process: The detection process can contribute to the band broadening by giving wider bands; think, for example, of overexposures with ^{32}P or chemiluminescent substrates. The dependences on gel length and membrane speed are the same as for transfer broadening: independent of the gel length, but increasing interband distances by higher membrane speeds would reduce these effects.

Combined effects of different factors: The variance introduced by different mechanisms such as diffusion (σ_{dif}^2), loading (σ_{ld}^2), transfer (σ_{tr}^2), detection (σ_{det}^2), and other possible sources (σ_{oth}^2) is additive:[43]

$$\sigma_{tot}^2 = \sigma_{dif}^2 + \sigma_{ld}^2 + \sigma_{tr}^2 + \sigma_{det}^2 + \sigma_{oth}^2 \qquad (7)$$

The width of a band, assuming gaussian shape, is four times the square root of the variance, $4\sigma_{tot}$. If we assume that random distortions such as diffusion contribute mainly to the broadness of bands, we would expect a square root dependency of the resolution on the gel length; this would obviously give sharper band patterns for all fragment lengths and thus increase the readable range. Another way to reduce diffusion effects would be to use higher field strengths to reduce elution time; again, a square root relation between resolution and field strength would be expected. However, Fig. 8 shows that increasing field strengths also lead to reduced distances between bands at long fragments, as is well known for agarose

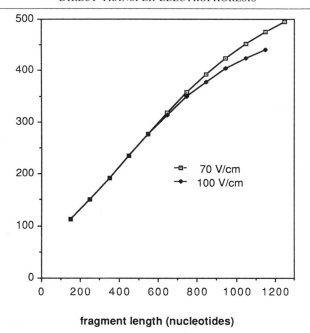

fragment length (nucleotides)

FIG. 8. Elution time as a function of fragment length in direct transfer electrophoresis at 70 V/cm (□) and 100 V/cm (◆); 0.1-mm thin, 60-cm long 3% acrylamide gels (3% total/5% bisacrylamide) were used. To allow a comparison of the linearity of the elution times, the times at 100 V/cm were normalized to the 70 V/cm times by multiplication with 1.551.

gels.[47,48] This can practically limit the beneficial effects of higher field strength and may (at least partly) explain why capillary electrophoresis can give sharp bands for short fragments, but badly overlapping bands at longer fragments.[45]

Another implication of the curvature for long fragments in Fig. 8 is that it becomes increasingly difficult to obtain well-separated bands with increasing fragment length. However, the optimal gel parameters given in this chapter do not represent a physical limit; further improvements are likely.

How is the transfer speed determined? Can the readable range be increased by spreading the bands with higher membrane speeds? Higher membrane speeds increase the distances between bands, but they also spread the bands out farther; the resolution as defined by the distance

[47] E. M. Southern, *Anal. Biochem.* **100**, 319 (1979).
[48] N. C. Stellwagen, *Biochemistry* **22**, 6180 (1983).

between bands divided through their width is not improved—if the inter-band distances are large enough to make broadening due to the transfer and the detection process negligible. Increasing the distances between bands also increases exposure times and reduces the data density, both unwelcome effects. As a compromise, the transfer speed is usually chosen to give interband distances between 0.8 and 1.4 mm. Thin gels (0.1 mm) and colorimetric detection (or ^{35}S) allow distances closer to 0.8 mm, whereas thicker gels and chemiluminescent detection (or ^{32}P) can benefit from slightly larger distances.

Why are bands from short fragments straight, but become more and more distorted with increasing fragment length? The most typical reasons for band shape distortions are small irregularities in the upper (and some-times lower) surface of the gel; when, for example, shark's-tooth combs are used, the slots and bands often show a U shape. Such shapes are imprinted into the band shape. Within the gel, they have the same form for all bands regardless of fragment size. During transfer at the bottom of the gel, these distortions become enlarged according to the speed of the membrane in relation to the speed of a band; with constant membrane speed, the distortions become increasingly enlarged with decreasing mobil-ity for longer fragments.

Increasing the gel length and keeping interband distances constant allows reduction of the membrane speed; therefore, images of distortions on the gel edges are reduced, and the sequencing patterns from longer gels are easier to interpret, especially for longer fragments.

Enzyme-Linked Detection Methods in DNA Sequencing: Common Questions

How much DNA is needed for enzymatically detected DNA sequenc-ing? We routinely use standard amounts, 0.5 pmol of single-stranded template DNA, per reaction; less than one-tenth of a reaction is loaded on 0.1-mm thin gels, and the nucleotide mixes are optimized for long reads—the amount per band in typical radioactive or fluorescent se-quencing gels is often severalfold higher. An indication of the possible sensitivity of the methods was obtained when we determined the biotinylation rate of a primer that gave a weak, but clearly readable, sequence: only about 1.2% of the primer was biotinylated; the amount of biotinylated DNA per band was calculated to be below 0.001 fmol (0.6 million molecules).

How reproducible, efficient, and cost effective is nonradioactive sequencing? Following the protocols and general guidelines given in this chapter, we routinely obtain easily interpretable sequencing patterns, often completely free of background; however, we usually use slightly more washes than given as minimum numbers in the protocol. Most problems, for example, in undergraduate courses, could be explained by use of old, bacterially contaminated buffers or inadequate membrane treatment.

When done routinely, the development process takes about 1 hr before the exposure or the start of the color reaction; efficiency can be easily increased by developing more membranes in parallel; four membranes can be developed in about 90 min. The combined detection protocol for differently labeled primers offers even more efficiency by reducing the time per protocol and the number of gels.

Even with lower data density, the costs for chemiluminescent detections[49] are unlikely to contribute significantly to the overall sequencing costs. Because the larger part of the expenses is for chemiluminescent substrates, costs could be reduced by using lower concentrations of substrate, or with colorimetric detection.

How do the biotin and digoxigenin systems and peroxidase vs phosphatase compare? Both labels have considerable advantages: biotinylated phosphoramidites are commercially available and make primer walking strategies easy, and we find it easier to obtain totally background-free patterns. The separate use of streptavidin and biotinylated alkaline phosphatase necessitates the use of an additional incubation and a few washing steps compared to digoxigenin; however, the use of streptavidin–phosphatase conjugates has been reported.[50]

Digoxigenin offers detection protocols with fewer steps, and anti-digoxigenin–antibody conjugates with peroxidase or phosphatase can be used. Peroxidase-based enhanced chemiluminescence is sufficiently sensitive for normal sequencing; however, dioxetane substrates are more stable in solution and can give higher sensitivities at longer exposure times.

A convenient alternative to biotin- or digoxigenin-labeled primers is the use of oligonucleotides that are complementary to the primer and covalently cross-linked to alkaline phosphatase[51,52]; this approach can also be used for multiplex sequencing.[21,53]

[49] The cost of one chemiluminescent detection is about $50 for a 30 × 40 cm membrane, with 24 clones and 400 readable bases per clone, the cost per base pair of raw data is well below one cent.

[50] C. Martin, L. Bresnick, R.-R. Juo, J. C. Voyta, and I. Bronstein, *BioTechniques* **11,** 110–113 (1991).

[51] E. Jablonski, E. W. Moomaw, R. H. Tullis, and J. L. Ruth, *Nucleic Acids Res.* **14,** 6115 (1986).

[52] A. Murakami, J. Tada, K. Yamagata, and J. Takano, *Nucleic Acids Res.* **17,** 5587 (1989).

[53] P. Richterich and G. M. Church, unpublished results, 1990.

Acknowledgments

We thank Drs. I. Bronstein, C. Martin, and J. Voyta for help in initial experiments with AMPPD and AMPPD-Cl, Dr. F. M. Pohl and Dr. D. Sullivan for helpful discussions, Millipore for supplying LumigenPPD, and Amersham for samples of ECL substrate. The work was supported by the European Community and DOE Grant DE-FG02-87ER60565.

[15] Adenine-Specific DNA Chemical Sequencing Reaction

By BRENT L. IVERSON and PETER B. DERVAN

DNA sequence determination according to the method of Maxam and Gilbert utilizes base-specific chemical modification reactions followed by a workup that causes cleavage of the sugar–phosphate backbone at the site of the modified base.[1] Reactions have been reported that are capable of selective cleavage of DNA at G, G + A, A > G, A > C, C, C + T, and T residues.[1-3] We describe here a simple protocol specific for adenine (A) that may be a useful addition to current chemical sequencing reactions. Reaction of DNA with K_2PdCl_4 at pH 2.0 followed by heating in the presence of piperidine produces an A-specific DNA cleavage reaction.[4] The K_2PdCl_4 reaction involves selective depurination at adenine, affording an excision reaction analogous to the other chemical DNA sequencing reactions.

Materials and Methods

Cleavage of DNA at Adenine with K_2PdCl_4

Potassium tetrachloropalladate(II), K_2PdCl_4 (Aldrich, Milwaukee, WI) is dissolved to a final concentration of 10 mM in a 100 mM HCl solution previously adjusted to pH 2.0 with NaOH. This solution (40 μl) is added to 160 μl of a solution containing the ^{32}P-labeled DNA fragment[5] and 1 μg of sonicated calf thymus DNA in distilled H_2O. The reaction is mixed and incubated at room temperature for 20 min and stopped by adding 50 μl of a solution containing 1.5 M sodium acetate, 1.0 M 2-mercaptoethanol and

[1] A. Maxam and W. Gilbert, *Proc. Natl. Acad. Sci. U.S.A.* **74**, 560 (1977).
[2] A. Maxam and W. Gilbert, this series, Vol. 65, p. 499.
[3] T. Friedmann and D. M. Brown, *Nucleic Acids Res.* **5**, 615 (1978).
[4] B. L. Iverson and P. B. Dervan, *Nucleic Acids Res.* **15**, 7823 (1987).
[5] M. W. Van Dyke, R. P. Hertzberg, and P. B. Dervan, *Proc. Natl. Acad. Sci. U.S.A.* **79**, 5470 (1982).

20 μg/ml of sonicated calf thymus DNA. (The stop solution and workup is the same as the Maxam–Gilbert G-specific sequencing reaction with dimethyl sulfate.[2]) Ethanol (750 μl) is added and the solution is chilled in dry ice for 10 min, then spun at 12,000 rpm for 6–10 min. [A large yellow pellet (likely a palladium salt) appears, which does not dissolve in the 70% (v/v) wash. This may seem disconcerting, but does not affect the sequencing result.] The supernatant is removed, the DNA pellet washed with 70% (v/v) ethanol, and dried briefly *in vacuo*. The pellet is redissolved in 100 μl 10% (v/v) aqueous piperidine, (background cleavage can be minimized by decreasing 10% aqueous piperidine to 1% aqueous piperidine), heated at 90° for 30 min, frozen in dry ice, and lyophilized. (Care must be taken to remove all the piperidine. This can be accomplished by repeated resuspension and lyophilization or lyophilization for 12 hr.) The lyophilized DNA is then dissolved in formamide buffer and loaded onto an 8% denaturing polyacrylamide [50% (w/v) urea] sequencing gel.

Results

A restriction fragment of double-stranded DNA uniquely labeled with ^{32}P at the 5′ or 3′ end was allowed to react with 2 mM K$_2$PdCl$_4$ in 20 mM HCl/NaCl, pH 2.0 buffer at 25° for 20 min followed by a piperidine workup. This produced an A-specific sequencing lane (Fig. 1). The cleavage efficiency at each A is sufficiently uniform to enable unambiguous assignment of sequence. The electrophoretic mobilities of the fragments on a 20% polyacrylamide gel produced by the K$_2$PdCl$_4$ reaction are identical to the A cleavage fragments from the formic acid-catalyzed G + A depurination reaction,[2] which is known to produce 3′- and 5′-phosphate termini. The K$_2$PdCl$_4$ cleavage is therefore an excision reaction that can be used together with other base-specific reactions for complete DNA sequence determination. The K$_2$PdCl$_4$ reaction is stopped with thiol to coordinate the Pd(II), which might otherwise remain bound to the DNA and interfere with electrophoretic mobility.

High-performance liquid chromatography (HPLC) analysis of the products of the K$_2$PdCl$_4$ reaction with calf thymus DNA revealed that the only apparent product released is adenine.[4] Thus at low pH K$_2$PdCl$_4$ causes selective depurination of adenine residues.

Discussion

Spectroscopic studies with purine bases, nucleosides, and nucleotides have indicated that Pd(II) binds the N-7 and/or N-1 positions.[6] For adeno-

[6] R. B. Martin, *Acc. Chem. Res.* **18**, 32 (1985), and references cited therein.

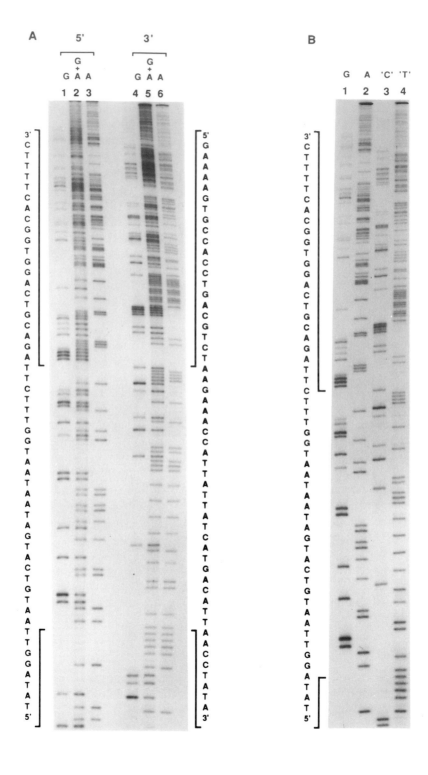

sine (and guanosine) 5'-monophosphate, Martin and co-workers have reported that the stability constants for (dien)Pd^{2+} binding to N-1 and N-7 are similar.[7] The stability constants for Pd(II) binding to guanine are higher than those for adenine.[7] The pK_a values for N-1 and N-7 in adenosine are 3.6 and -1.6, respectively.[6,8] Therefore, at low pH, Pd(II) binding to N-7 should predominate, because metal binding must compete with protonation at the N-1 position. The pK_a values for N-1 and N-7 of guanosine are 9.2 and 2.2, respectively, so at low pH Pd(II) would be expected to bind N-7 almost exclusively.[6]

The specific depurination of adenine by the reaction of DNA with K_2PdCl_4 at low pH could be produced by suppressing G reactivity, by enhancing specific depurination at A, or a combination of both. Because transition metals such as Pd(II) have been shown to bind N-7 of guanine,[6] we cannot rule out that the Pd(II) is binding to N-7 of guanine and inhibiting to some extent the acid-catalyzed depurination reaction at G.[2] However, we do find that the K_2PdCl_4 reaction at pH 2.0 causes cleavage at A with higher efficiency than occurs under the same pH conditions in the absence of K_2PdCl_4, suggesting that enhanced adenine depurination is occurring. Consistent with this, adenine is the major product released from the reaction of K_2PdCl_4 with calf thymus DNA at low pH. Presumably, the Pd(II) is binding to the adenine base in a manner that assists the hydrolysis of the glycosidic bond at low pH.

In a formal sense, the Pd(II) could bind to any of the nitrogen lone pairs on adenine; N-1, N-3, N-7, or the exocyclic amine N-6. Cleavage of DNA at sites containing N^6-methyladenine was found to decrease by 35% when compared to unmethylated adenine in an identical sequence.[4] The observed decrease in cleavage efficiency of N^6-methyladenine might be interpreted as resulting from the methyl group at N-6, creating steric inhibition to binding by Pd(II) at N-6, N-1, or N-7. Direct Pd(II) coordination to N-6 is unlikely for Pd(II)[9] and N-1 is expected to be protonated at

[7] K. H. Scheller, V. Scheller-Krattiger, and R. B. Martin, *J. Am. Chem. Soc.* **103**, 6833 (1981).

[8] S. H. Kim and R. B. Martin, *Inorg. Chim. Acta* **91**, 19 (1984).

[9] R. B. Martin and Y. H. Mariam, *Metal Ions Biol. Syst.* **8**, 57 (1979).

FIG. 1. (A) Comparison of the G, G + A, and A reactions on a *Eco*RI/*Rsa*I 517-bp restriction fragment of DNA from pBR322. Lanes 1–3 contain reactions on the 517-bp fragment labeled with ^{32}P at the 5' end. Lanes 4–6 contain reactions on the 517-bp fragment labeled with ^{32}P at the 3' end. Lanes 1 and 4, Maxam–Gilbert G reaction. Lanes 2 and 5, Maxam–Gilbert G + A reaction. Lanes 3 and 6, K_2PdCl_4 A reaction. (B) Comparison of G and A sequencing reaction on opposite strands. Each base position is represented by a band in only one of the four lanes, illustrating a convenient sequencing strategy.

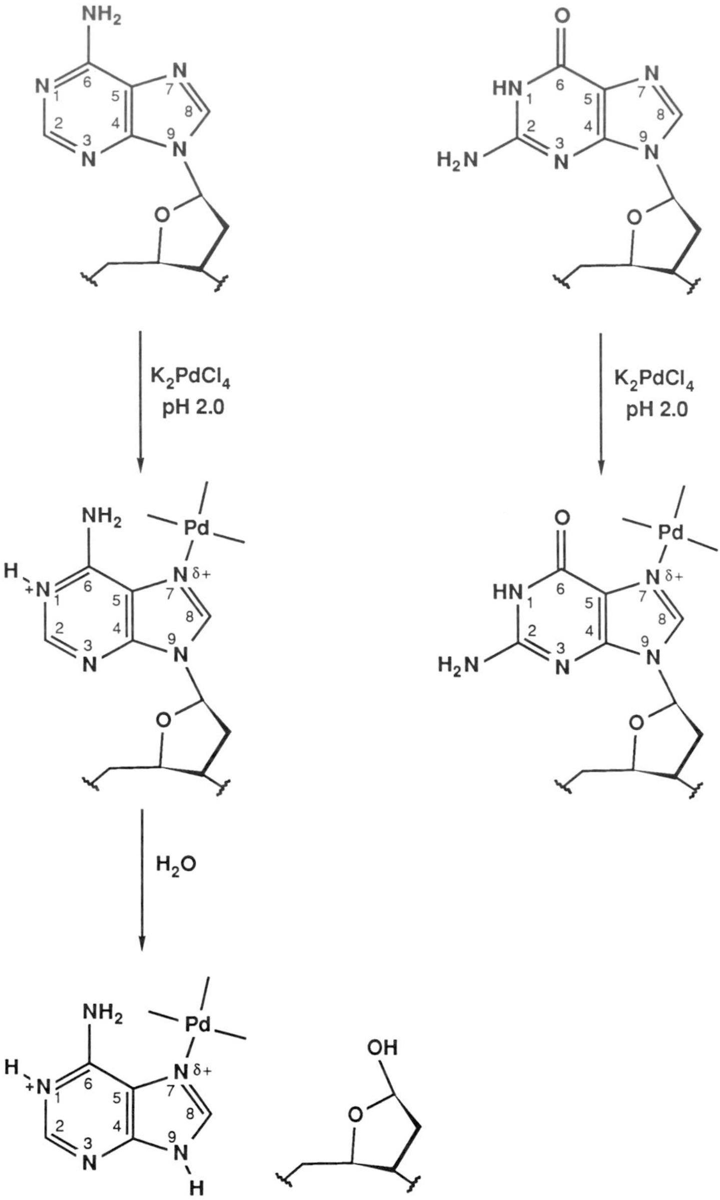

FIG. 2. Proposed mechanistic scheme for the enhanced specific depurination of A ≫ G by K_2PdCl_4 at pH 2.0.

pH 2.0. Therefore N-7 emerges as the likely site of binding during the depurination reaction. The absence of significant neighboring base dependence on the observed A cleavage indicates that the depurination reaction caused by Pd(II) probably does not involved bridging bonds between N-7 of adenine and adjacent bases.

An adenine bound by Pd(II) at N-7 and protonated at N-1 at pH 2.0 would contain a significant amount of positive charge on the adenine ring, affording a labile glycosidic bond and allowing release of the adenine–Pd(II) complex (Fig. 2). Even though Pd(II) is probably binding to N-7 of G, a similar depurination mechanism is not likely, because the protonated N-1 position of G does not create a corresponding positive charge on the guanine ring (Fig. 2). This mechanistic scheme could be analogous to the Maxam–Gilbert reaction, which involves dimethyl sulfate alkylation of guanine (at N-7) and adenine (at N-3) followed by acid treatment (protonation at N-1 of A), which results in selective depurination of A > G.[1]

In summary, the reaction of K_2PdCl_4 with DNA at low pH is a convenient and reliable method for the production of an adenine-specific chemical sequencing lane. The reaction of K_2PdCl_4 with DNA at pH 2.0 probably involves binding of Pd(II) to N-7 and protonation at N-1 of adenine, which results in specific depurination of adenine.

Acknowledgments

We are grateful to the American Cancer Society for support of this research and for a National Research Service Award to B.L.I. from the National Institute of General Medical Sciences.

[16] Direct DNA Sequencing of Polymerase Chain Reaction-Amplified Genomic DNA by Maxam–Gilbert Method

By JAN P. KRAUS and TAKAHIRO TAHARA

Introduction

Polymerase chain reaction (PCR) amplification has greatly simplified and accelerated the analysis of mutations in eukaryotic genes. Various methods to detect mutations in PCR-amplified products have been described. These include allele-specific oligonucleotide hybridization,[1] restriction enzyme digestion,[2] sequencing of subcloned PCR products,[3] transcript sequencing of amplified DNA,[4] and direct DNA sequencing using single-stranded[5] or double-stranded[6] PCR-amplified products employing the chain termination method of Sanger *et al.*[7] The amplified products are not always easily subcloned, however. Furthermore, direct sequencing of single- or double-stranded DNA by the method of Sanger *et al.*[7] often yields unreadable sequenes, particularly with shorter templates.

In the course of our study of a mutation in patients with propionic acidemia, we circumvented these problems by carrying out direct sequencing of PCR amplified genomic DNA by the chemical method of Maxam and Gilbert.[8] This procedure yields clearly readable DNA sequences, 100–400 base pairs (bp) in length, derived from human genomic DNA, in 4 days.

Principle

The amplified DNA was radiolabeled in the polymerase chain reaction itself by labeling one of the two primers at its 5' end with $[\gamma\text{-}^{32}P]ATP$ prior to the amplification. Following the removal of excess primers, the purified PCR product was directly sequenced.

Materials and Reagents

Dimethyl sulfate (DMS), hydrazine, and piperidine are purchased from Aldrich (Milwaukee, WI); Centricon-10 and Centricon-30 microconcentrators are from Amicon (Danvers, MA); *Taq* DNA polymerase and DNA thermal cycler are from Perkin-Elmer Cetus (Norwalk, CT); T4 polynucleotide kinase is from Boehringer Mannheim (Indianapolis, IN); calf thymus DNA and yeast tRNA are from Sigma (St. Louis, MO); and $[\gamma\text{-}^{32}P]ATP$ is from Amersham (Arlington Heights, IL). All other chemicals are of the highest purity available.

[1] R. K. Saiki, T. L. Bugawan, G. T. Horn, K. B. Mullis, and H. A. Erlich, *Nature (London)* **324**, 163 (1986).
[2] R. K. Saiki, S. Scharf, F. Falloona, K. B. Mullis, G. T. Horn, H. A. Erlich, and N. Arnheim, *Science* **230**, 1350 (1985).
[3] S. J. Scharf, G. T. Horn, and H. A. Erlich, *Science* **233**, 1076 (1986).
[4] E. S. Stoflet, D. D. Koeberl, G. Sarkar, and S. S. Sommer, *Science* **239**, 491 (1988).
[5] U. B. Gyllensten and H. A. Erlich, *Proc. Natl. Acad. Sci. U.S.A.* **85**, 7652 (1988).
[6] L. J. Huber and G. McMahan, *Comments (U.S. Biochem. Newsl.)* **15**, 3 (1988).
[7] F. Sanger, S. Nicklen, and A. R. Coulson, *Proc. Natl. Acad. Sci. U.S.A.* **74**, 5463 (1977).
[8] A. M. Maxam and W. Gilbert, this series, Vol. 65, p. 499.

Solutions

Buffer A: 10 mM Tris-HCl, pH 8.0, 2 mM ethylenediaminetetraacetic acid (EDTA), 15% (w/v) sucrose

Buffer B: 10 mM Tris-HCl, pH 8.0, 2 mM EDTA

Buffer C: 10 mM Tris-HCl, pH 8.0

DMS buffer: 50 mM sodium cacodylate, pH 8.0, 10 mM MgCl$_2$, 1 mM EDTA

DMS stop buffer: 1.5 M sodium acetate, pH 7.0, 1 M 2-mercaptoethanol, yeast tRNA (100 μg/ml)

Hydrazine stop buffer: 0.3 M sodium acetate, 0.1 M EDTA, yeast tRNA (25 μg/ml)

Phenol: Phenol from Bethesda Research Laboratories (Gaithersburg, MD) is melted at 68° and saturated with STE buffer containing 50 mM Tris-HCl, pH 8.0, 100 mM NaCl, 1 mM EDTA. The STE-saturated phenol is kept at 4°

Chloroform: Chloroform and isoamyl alcohol (Baker, Phillipsburg, NJ) are mixed at a ratio of 24 : 1 (v/v)

TBE buffer: 89 mM Tris–borate, pH 8.0, 2 mM EDTA

Acrylamide gel: 150 μl of 30% (w/v) ammonium persulfate and 15 μl of N,N,N',N' tetramethylethylenediamine (TEMED) are added to 25 ml of a mixture consisting of 5% (w/v) acrylamide, 0.25% (w/v) bis-acrylamide, and 25% (v/v) glycerol in TBE buffer. The gel is polymerized for at least 30 min prior to use

Gel loading buffer: 95% (v/v) formamide, 20 mM EDTA, 0.05% (w/v) bromphenol blue, 0.05% (w/v) xylene cyanol FF

Apparatus

SpeedVac Concentrator (Savant Instruments, Farmingdale, NY)

DNA thermal cycler (Perkin-Elmer Cetus)

Methods

End Labeling of Primer

One of the two primers for PCR is labeled at the 5′ end by incubation at 37° for 1 hr in a 50-μl mixture consisting of 100 mM Tris-HCl, pH 8.0, 20 mM 2-mercaptoethanol, 10 mM MgCl$_2$, 100 μCi of [γ-^{32}P]ATP, 200 ng of the primer, and 30 units of T4 polynucleotide kinase. Following the reaction, the sample is diluted to 800 μl with buffer A. The sample is applied to a Centricon-10 microconcentrator and overlaid with 500 μl of buffer B. The sample is concentrated for 90 min at 3000 g (4°), then 750

μl of buffer C is added to the concentrated sample, followed by another 30-min centrifugation. The volume of the dialyzed sample is reduced to 50 μl in a SpeedVac concentrator.

Polymerase Chain Reaction Amplification

The PCR is carried out in 100 μl employing the standard reaction mixture (10 mM Tris-HCl, pH 8.4, 50 mM KCl, 1.5 mM MgCl$_2$, 100 μg/ml gelatin, 0.2 mM deoxynucleotide triphosphates), 800 ng of genomic DNA (extracted from cultured skin fibroblasts), and 200 ng of the 5'-end labeled primer ($\sim$3.7 $\times$ 10^7 cpm). Another 400 ng of the same primer (unlabeled) and 600 ng of the other primer are also added to the PCR mixture. After the reactions have been incubated at 65° for 5 min, 2.5 units of *Taq* DNA polymerase is added. The reaction mixture is overlaid with 100 μl of mineral oil, and temperature cycling is carried out in a DNA thermal cycler as follows: 1 min at 94°, 2 min at 55°, 1 min at 72° for 40 cycles.

Purification of Polymerase Chain Reaction-Amplified Products

The amplified DNA samples are purified by sequential extractions with 1 vol of a mixture of phenol and chloroform (1 : 1, v/v), then with 1 vol of chloroform alone, followed by spin dialysis in a Centricon-30 microconcentrator using the same conditions as described above. A 1/10 vol of 3 M sodium acetate and 2.5 vol of ethanol are added to the dialyzed samples and the DNA is precipitated on dry ice. After precipitation, samples are dried in a SpeedVac and resuspended in 20 μl of water. To remove minor contaminating bands and the genomic DNA, the PCR products are further purified by electrophoresis on an acrylamide–glycerol gel, and the excised band containing the PCR product is electroeluted for 3 hr at 100 V in 0.5 $\times$ TBE buffer. After addition of 6 μg of calf thymus DNA, the samples are ethanol precipitated as described above, and resuspended in 26 μl of water.

Chemical Cleavage

The chemical cleavage was carried out as described[8] with some modifications. The flow chart of the sequencing reactions is shown in Table I. Following the steps shown in Table I, the G, C, and C + T reactions are centrifuged for 10 min at 12,000 g at 4°. The supernatants are carefully removed with drawn-out Pasteur pipettes and discarded. This is monitored with a Geiger counter to make sure that the radioactive pellets are not touched. The pellets are redissolved in 250 μl of 0.3 M sodium acetate, pH 5.0, and the DNA is precipitated with 750 μl of ethanol ($-20°$) on dry

TABLE I
DNA Sequencing Reactions

Reaction G	Reaction A > C	Reaction C	Reaction C + T
Mix 5 μl ^{32}P-labeled DNA, 195 μl DMS buffer	Mix 5 μl ^{32}P-labeled DNA, 100 μl 1 M NaOH/1 mM EDTA	Mix 5 μl ^{32}P-labeled DNA, 15 μl 5 M NaCl	Mix 10 μl ^{32}P-labeled DNA, 15 μl H$_2$O
Chill on ice	Incubate 8 min submerged at 90°	Chill on ice	Chill on ice
Add 1 μl DMS and mix	Chill on ice	Add 30 μl hydrazine	Add 30 μl hydrazine
Incubate 20 min on ice	Add 150 μl 1 M acetic acid, 1 μl yeast tRNA (10 mg/ml), and mix	Incubate 10 min on ice	Incubate 12 min on ice
Add 50 μl DMS stop buffer and 750 μl of ethanol (− 20°)		Add 3 μl calf thymus DNA (1 mg/ml), 200 μl hydrazine stop buffer and mix	Add 3 μl calf thymus DNA (1 mg/ml), 200 μl hydrazine stop buffer and mix
Incubate 10 min or more on dry ice	Add 750 μl of ethanol (− 20°)	Add 750 μl of ethanol (− 20°)	Add 750 μl of ethanol (− 20°)
	Incubate 10 min or more on dry ice	Incubate 10 min or more on dry ice	Incubate 10 min or more on dry ice

ice for 10 min. The G, A > C, C, and C + T reactions are centrifuged for 10 min at 12,000 g at 4° and the supernatants are discarded as above. One milliliter of ethanol (− 20°) is added to the pellets and the mixtures are vortexed. After centrifugation for 5 min at 12,000 g at 4° the supernatants are carefully removed and the pellets lyophilized in a SpeedVac. Piperidine (100 μl of a 1 M solution) is added and the tubes submerged at 90° for 30 min. The mixtures are then frozen on dry ice and lyophilized in a Speed-Vac overnight. The dry pellets are dissolved in 15 μl of water, frozen on dry ice, and relyophilized to dryness. The samples are dissolved in a few microliters of the loading buffer to achieve a concentration of approximately 5000 cpm/μl. After boiling for 2 min, aliquots of the four mixtures, containing 10–15 × 10³ cpm each, are electrophoresed in a 6 or 18.5% (w/v) polyacrylamide, 8.3 M sequencing gel. Following electrophoresis, the gel is mounted on a sheet of used film and exposed to X-ray film at − 70° for ~6 hr, using an intensifier screen.

Results and Discussion

Figure 1 shows DNA sequences derived from genomic DNAs of a control and a patient (572) with propionic acidemia. As can be seen, the resolution of each nucleotide is clear; both the normal and the extensively altered patient DNA sequences are easily readable. The sequences derived

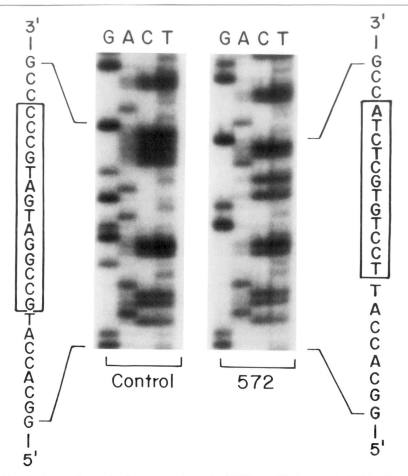

FIG. 1. Autoradiograph of a sequencing gel of PCR-amplified genomic DNAs directly sequenced by the Maxam–Gilbert method. The abbreviations G, A, C, and T specify the four chemical reactions: G, A > C, C, C + T, respectively.[8] The boxed sequences show the differences between the control and patient 572.

from these experiments have been subsequently confirmed,[9] after subcloning of the PCR products, by the chain-termination method.

As a further modification, instead of end labeling the primer prior to PCR, we labeled the PCR product after the amplification and gel purification at a unique restriction site using the Klenow "filling in" reaction.[10]

[9] T. Tahara, J. P. Kraus, and L. E. Rosenberg, *Proc. Natl. Acad. Sci. U.S.A.* **87,** 1372 (1990).
[10] T. Maniatis, E. F. Fritsch, and J. Sambrook, "Molecular Cloning: A Laboratory Manual." Cold Spring Harbor Press, Cold Spring Harbor, New York, 1982.

Following removal of the unincorporated label in the Centricon-30 micro-concentrator, the Maxam–Gilbert chemical cleavage reactions were performed. This method gave us equally readable sequences. Thus, for larger PCR products, the DNA sequencing can be initiated from within the sequence rather than from either end.

There are several advantages of the Maxam–Gilbert sequencing method for analysis of PCR products. First, it has worked consistently in our hands, yielding clear DNA sequences of fragments 100–400 bp long, whereas multiple attempts at direct sequencing of these PCR products, either single or double stranded by the chain termination method, often yielded unreadable sequences. This was probably either due to rapid reassociation of the short linear DNA strands after denaturation or due to the presence of small amounts of remaining PCR primers. Second, the procedure is relatively fast, allowing us to obtain the sequence of a 400-bp fragment amplified from genomic DNA within 4 days. This is a clear improvement over any procedure requiring subcloning. It may even be better than direct chain-termination sequencing, particularly in those cases in which second amplification steps with internal primers or asymmetric primer concentrations are used. Third, it allows examination of all sequences present in the PCR-amplified products. The sequencing gels obtained are often of sufficient quality that the presence of alleles differing at a single base pair can be observed as a position on the sequence ladder where a band is present in two lanes. Patient 572, shown in Fig. 1, was homozygous for the mutation. Finally, because the sequence starts with the first (5′) nucleotide of the primer, there is no loss of sequence information immediately adjacent to the primer and no modifications to the sequencing protocol are required to obtain these data.

[17] Direct Sequencing of λgt11 Clones

By DAVID L. STEFFENS and RICHARD W. GROSS

Introduction

Since the initial description of the bacteriophage expression vector λgt11,[1] it has become the most widely utilized vector for generating cDNA libraries. Typically, cDNA is cloned into the EcoRI site in λgt11 situated within the β-galactosidase coding region, which allows discrimination be-

[1] R. A. Young and R. W. Davis, Proc. Natl. Acad. Sci. U.S.A. 80, 1194 (1983).

tween recombinant and wild-type phage. Subsequent induction of the *lacZ* gene results in expression of β-galactosidase fusion proteins corresponding to the N-terminal portion of β-galactosidase and the cDNA insert. Due to the strength of the promoter of the *lacZ* gene in *Escherichia coli* and the high copy number of λgt11 during lytic infection, substantial amounts of fusion proteins (usually 0.1–4% of total cellular protein) are produced during induction, thus facilitating expression screening of λgt11 libraries.[2] Typically, expression screening of λgt11 libraries is accomplished through utilization of an antibody directed against the polypeptide of interest.[3] Alternatively, library screening can also be accomplished by hybridization to a labeled nucleic acid probe.

Due to the large genome size of the phage and the difficulties commonly encountered in isolation and manipulation of λgt11 DNA, sequence analysis has generally been preceded by subcloning the insert into a much smaller vector such as M13, a plasmid, or a phagemid. Both M13 and phagemid vectors allow production of single-stranded sequencing templates, whereas plasmids must be sequenced in the double-stranded form. A major trend in DNA sequencing technology is toward direct sequencing of double-stranded DNA from isolated clones, utilizing recently developed methods[4,5] to obviate the need for subcloning into a single-stranded generating vector.

This strategy has been exploited to facilitate the direct sequencing of λgt11 template DNA by development of methods that can generate template of sufficient purity to allow direct sequence analysis. Through purification of plate phage lysates by polyethylene glycol (PEG) precipitation, phenol extraction, and ethanol precipitation, the isolation of λgt11 DNA sufficiently pure for direct sequence analysis is feasible. The DNA template is subsequently denatured, annealed to an oligonucleotide primer, and directly sequenced utilizing modified T7 polymerase methodology.[6]

Materials and Methods

Bacterial Strain

Escherichia coli Y1090:Δ(*lac*)U169 Δ(*lon*) *araD139 strA supF mcrA* [trpC22:Tn*10*] (pMC9)

[2] M. Snyder, S. Elledge, D. Sweetster, R. A. Young, and R. W. Davis, this series, Vol. 154, p. 107.
[3] R. A. Young and R. W. Davis, *Science* **222**, 778 (1983).
[4] R. J. Zagursky, K. Baumeister, N. Lomax, and M. L. Berman, *Gene Anal. Tech.* **2**, 89 (1985).
[5] E. Y. Chen and P. H. Seeburg, *DNA* **4**, 165 (1985).
[6] S. Tabor and C. C. Richardson, *Proc. Natl. Acad. Sci. U.S.A.* **84**, 4767 (1987).

Media

Y1090 plates: 1% (w/v) Bacto-tryptone and 0.5% (w/v) yeast extract (Difco Laboratories, Detroit, MI), 1% (w/v) NaCl, 10 mM MgCl$_2$, 50 μg/ml ampicillin, and 1.5% (w/v) ultrapure agarose (GIBCO-Bethesda Research Laboratories, Grand Island, NY)
Y1090 top agar: As above, except containing 0.7% ultrapure agarose and lacking ampicillin

Solutions

Plaque storage buffer (PSB): 10 mM Tris-HCl, pH 7.4, 100 mM NaCl, 10 mM MgCl$_2$, 0.05% (w/v) gelatin (Difco)
TE: 10 mM Tris-HCl, pH 8.0, 1 mM ethylenediaminetetraacetic acid (EDTA)
Enzymes: DNase (Sigma Chemical Co., St. Louis, MO); RNase (Sigma); proteinase K (International Biotechnologies, Inc., New Haven, CT)

Most other reagents were obtained from Sigma.

Methods

Phage Growth and Isolation

Sufficiently pure template is the key to obtaining quality DNA sequence data from any procedure. λgt11 template DNA purity is primarily affected by the method of growth and isolation of the recombinant phage. In our experience and that of others,[7] plate phage lysates routinely yield template of higher quality than those obtained from liquid phage lysates. Phage growth in liquid culture presents difficulties in optimizing the multiplicity of infection and the resulting phage titer. Additionally, template purity may be compromised due to bacterial lysis and release of cellular contaminants directly into the medium, resulting in their subsequent copurification with template DNA and/or its covalent modification. In contrast, plate growth is easily visualized and optimized at the point of confluent lysis. Contaminants may also be trapped within the agarose matrix of the growth medium, allowing template preparations of higher homogeneity and facilitating superior sequencing results. The solid medium should contain ultrapure agarose because some preparations of reagent-grade agar contain impurities that elute and contaminate the phage DNA.

Recombinant λgt11 phages are grown and isolated as follows: *E. coli*

[7] B. A. White and S. Rosenzweig, *BioTechniques* **7**, 694 (1989).

strain Y1090 bacteria are infected with the λgt11 recombinant phage and plated out on 150-mm Y1090 media petri plates at a density of 30,000 to 50,000 plaque-forming units (pfu)/plate. Plates are incubated at 37° for 6–8 hr until they are confluently lysed. Ten milliliters of plaque storage buffer (PSB) is pipetted onto each plate and phage are eluted from the plates overnight at 4° prior to agitation on an orbital shaker at room temperature for 3 hr the following day. The lysate is carefully pipetted off, an additional 5 ml of PSB is added, and the mixture is agitated on an orbital shaker for an additional 30 min. The supernatant is pipetted off and combined with the previous lysate. Alternatively, a 1- to 2-hr room-temperature elution on an orbital shaker immediately following phage growth should also produce an acceptable yield of phage.

Phage DNA Template Isolation

Template DNA for sequencing is isolated from the phage as follows: 25 ml of plate lysate is placed into a 50-ml Oak Ridge tube and centrifuged at 8000 g for 10 min at 4° to remove debris. The supernatant is transferred to a clean Oak Ridge tube, 50 μl each of 10 mg/ml DNase I and 10 mg/ml RNase A are added, and the tube is incubated at 37° for 30 min. The phages are precipitated by adding one-half volume (12.5 ml) of a solution of polyethylene glycol 8000 (15%, w/v) containing 1.5 M NaCl and this mixture is subsequently incubated for 4–16 hr at 4°. Phages are harvested by centrifugation at 11,000 g for 30 min (4°) after carefully discarding the supernatant to remove as much of the polyethylene glycol solution as possible. The phage pellet is resuspended in 0.75 ml of PSB and transferred to a 1.5-ml Eppendorf tube. To the resuspended phage pellet, 7.5 μl of 0.5 M EDTA (pH 8.0), 2.0 μl of 20 mg/ml proteinase K solution, and 7.5 μl of a 10% (w/v) sodium dodecyl sulfate solution are added. The resultant solution is subsequently incubated at 68° for 15 min prior to centrifugation in a microcentrifuge for 2 min. The supernatant is aliquotted into two 1.5-ml Eppendorf tubes prior to addition of 375 μl of a 7.5 M solution of guanidine hydrochloride and 3.75 μl of a 1 M solution of dithiothreitol. DNA is obtained by three extractions with 1 vol of a 1 : 1 phenol : chloroform solution (v/v) (previously saturated with TE buffer) followed by a single extraction with 1 vol of chloroform. Next, one-half volume of 7.5 M ammonium acetate is added to each tube, the contents are thoroughly mixed, and the solution is incubated for 10 min on ice. The phage DNA is precipitated by addition of 2 vol of ethanol and subsequent incubation at $-70°$ for 5 min. A visible, well-formed precipitate appears on mixing the ethanol with the DNA solution. Phage DNA is subsequently recovered by pelleting the precipitate for 5 min (4°) in a microcentrifuge. The pellet is briefly dried under vacuum and phage DNA is subsequently resuspended in 0.5 ml of 50 mM Tris-HCl, pH 8.0, 100 mM sodium acetate, 1 mM EDTA. If the pellet does not readily dissolve, resuspension may be promoted by overnight incubation at 4° (as necessary). Next, the DNA is reprecipitated by addition of 2 vol of ethanol and subsequent incubation at −70° for 5 min. The DNA is repelleted by centrifugation at 4° for 5 min in a microcentrifuge and the pellet is washed with 0.5 ml of ice-cold 70% ethanol. The final DNA pellet is dried briefly under vacuum and dissolved in 100 μl of TE buffer.

The final DNA solution is fairly viscous and the yield from 25 ml of lysate obtained from two plates is approximately 40 μg. The ratio of absorbance at 260 vs 280 nm is approximately 1.7, indicating that only diminutive amounts of contamination by phenol or RNA are present. The DNA may be visualized by agarose gel electrophoresis at this juncture if desired. We have found that this procedure routinely yields template DNA suitable for direct dideoxy sequencing. Other isolation procedures utilizing DEAE anion exchange,[7–9] ultracentrifugation,[10,11] Cetyltrimethylammonium bromide precipitation,[9] hydroxylapatite chromatography,[12,13] and cesium chloride gradients[14] should increase purity, but we have not found that these measures are necessary (in most cases) for obtaining quality sequence data. λgt11 DNA has also been isolated utilizing immunoadsorption methods. However, most immunoadsorption methods, as well as the present procedure applied to liquid lysates, result in ultraviolet (UV)-absorbing contaminants that coprecipitate with the phage DNA, resulting in unsatisfactory sequencing results in our hands.

Sequencing λgt11 Templates

The isolated λgt11 DNA is alkali denatured, neutralized, and ethanol precipitated prior to direct sequencing utilizing a Sequenase kit (United States Biochemical, Cleveland, OH). First, 5.0 μg of template DNA is brought up to a volume of 18 μl with distilled water and placed in a 0.5-ml Eppendorf tube. The template is denatured by the addition of 2.0 μl of freshly prepared 2 N NaOH, 2 mM EDTA and subsequently incubated at room temperature for 5 min. The solution is neutralized by addition of 3.0

[8] K. J. Reddy, T. Kuwabara, and L. A. Sherman, *Anal. Biochem.* **168,** 324 (1988).
[9] G. Manfioletti and C. Schneider, *Nucleic Acids Res.* **16,** 2873 (1988).
[10] D. Grossberger, *Nucleic Acids Res.* **15,** 6737 (1987).
[11] J. P. Dumansky, M. Ruttledge, and S. Datta, *Nucleic Acids Res.* **16,** 9044 (1988).
[12] T. R. Johnson and J. Ilan, *Anal. Biochem.* **132,** 20 (1983).
[13] I. Ivanov and L. Gigova, *Anal. Biochem.* **146,** 389 (1985).
[14] J. Sambrook, E. F. Fritsch, and T. Maniatis, "Molecular Cloning: A Laboratory Manual," 2nd Ed. Cold Spring Harbor Press, Cold Spring Harbor, New York, 1989.

μl of 3 M sodium acetate (pH 5.0) and 7.0 μl of distilled water. Denatured DNA is precipitated by addition of 75 μl of ethanol and subsequent incubation for 15 min at $-70°$. The DNA is pelleted by centrifugation for 15 min at 4°. Although the pellet may be difficult to see at this point, it is carefully washed with 200 μl of ice-cold 70% ethanol prior to recentrifugation for 5 min at 4°. The supernatant is removed and the pellet is dried briefly under vacuum.

To the denatured DNA template, 2 μl of 5 × Sequenase reaction buffer and 10 ng of the desired sequencing primer (dissolved in distilled H_2O) are added prior to bringing the final volume of the solution to 10 μl with distilled H_2O, followed by redissolving the pellet. The primer is annealed to template during subsequent incubation at 37° for 15 min. The λgt11 DNA is suitable for direct sequencing, which is performed utilizing a Sequenase kit (United States Biochemical) according to instructions provided by the manufacturer. Briefly, the labeling mixture is diluted 1 : 5 with water and the enzyme is diluted 1 : 8 with ice-cold enzyme dilution buffer. To the annealed mixture of phage and primer DNA, 1 μl of 0.1 M dithiothreitol, 1 μl of fresh [^{32}P]dATP (10 μCi/μl, 3000 Ci/mmol; Amersham Corporation, Arlington Heights, IL), 2 μl of diluted labeling mix, and 2 μl of diluted Sequenase are added. Termination tubes are prewarmed at 37° for 1–2 min while the labeling reaction is incubated at room temperature for 5 min. Sequencing reactions are initiated by addition of 3.5 μl of the labeling mixture into each of the four termination tubes containing 2.5 μl of termination mix and subsequent incubation for 10–15 min at 37°. Reactions are terminated by addition of 4 μl of stop solution to each tube. Next, the samples are denatured in a 70° water bath for 3 min and placed immediately on ice prior to loading 1–2 μl of each sample into individual lanes. Typically, we utilize two loadings (separated by 2–3 hr) of each reaction set. Standard 6% polyacrylamide sequencing gels with 7 M urea and 1 × Tris–borate buffer, prewarmed to 45°, are routinely utilized. After electrophoresis the gel is directly transferred to Whatman (Clifton, NJ) 3MM Chr paper, covered with Saran wrap, dried, and exposed for 15–20 hr to Kodak (Rochester, NY) SB-5 diagnostic film. The results of this procedure are illustrated in Fig. 1.

Discussion

Approximately 300–350 bases of sequence information can be obtained directly from λgt11 inserts employing the procedure described herein.[15] This result is equivalent to that obtained with other double-stranded tem-

[15] D. L. Steffens and R. W. Gross, *BioTechniques* **7**, 674 (1989).

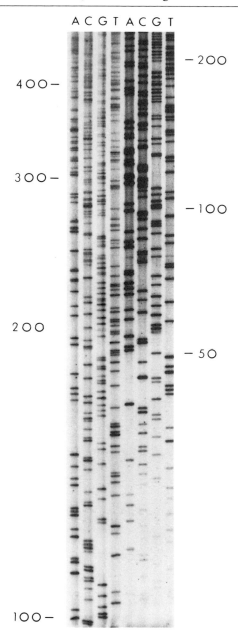

FIG. 1. Autoradiogram of a sequence obtained from a λgt11 recombinant clone. Two loadings of the sequencing reactions can be visualized on the gel, the leftmost four lanes containing the first load.

plates, but is somewhat less than that that can be obtained from single-stranded templates. Nevertheless, direct sequencing of a λgt11 clone, either completely or in part, is quite feasible and presents a viable alternative to subcloning.

The relatively large size of the λgt11 vector (44 kb) probably causes minor difficulties encountered in the sequencing procedure, such as increased background in all gel lanes. This could be due to host nucleic acid contamination, damaged template DNA,[16,17] or to the increased opportunity for spurious priming on the large vector. We typically utilize primers that are 17–18 bases in length with satisfactory results. Primers should routinely be checked for homology to vector sequences before use, especially at the 3′ end of the oligonucleotide where polymerization is initiated. Our experience includes a 17-mer that serendipitously matched 14 out of 17 bases in the vector and primed almost exclusively at this alternate site, thus demonstrating the need for careful primer selection.

Several potential improvements of this method are currently under evaluation. For example, utilization of alternate sources of polymerase for sequence analyses (including the thermostable *Taq* and *Bst* polymerases) may improve the efficacy of direct λgt11 sequencing. High-temperature sequencing reactions have resulted in improvement with other templates and could potentially also be utilized with λgt11. Thermostable enzymes allow application of linear polymerase chain reaction (PCR) methodology (''cycle sequencing''),[18] having the benefit of requiring much smaller quantities of template. Automated DNA sequencers have the potential to produce more sequence data (up to 500 bases or more per run) than manual methods, which may also improve the efficiency of this method. Any or all of these methodologies could further enhance sequence analysis from this method, thereby further increasing the value of direct phage sequencing.

Acknowledgments

    This research was supported by National Institutes of Health Grants HL35864 and HL41250. R.W.G. is the recipient of an Established Investigator award from the American Heart Association.

[16] G. F. Hong, *Biosci. Rep.* **2,** 907 (1982).
[17] B. S. Kim and C. Jue, *BioTechniques* **8,** 156 (1990).
[18] V. Murray, *Nucleic Acids Res.* **17,** 8889 (1989).

[18] DNA Fingerprinting by Sampled Sequencing

By TERRI L. MCGUIGAN, KENNETH J. LIVAK, and SYDNEY BRENNER

Introduction

One approach to the physical mapping of DNA is to construct a contiguous sequence (contig) map consisting of overlapping clones spanning the DNA region of interest. Contig maps, at various degrees of completion, have been constructed for the entire genomes of *Escherichia coli,*[1,2] yeast,[3] *Caenorhabditis elegans,*[4,5] *Arabidopsis,*[6] and *Drosophila melanogaster.*[7] Many groups are constructing contig maps for various regions of the human genome. These maps eliminate the need for the tedious steps of chromosome walking, reducing greatly the time and effort required to isolate any gene of interest. The set of contigs can also be the basis for complete integration of physical and genetic maps. Finally, a contig map is a good starting point for large-scale sequencing projects.

The key factor in assembling a contig map is deciding if two clones overlap or not. For the genomic maps referenced above, restriction enzymes were used to produce various types of fingerprints for the clones involved. After comparing the fingerprints, if two clones had a statistically significant number of restriction fragments in common, they were deemed overlapping. The simplest type of fingerprint is that used by Olson *et al.*[3] in the yeast project. For each clone, they performed a double digest with *Eco*RI and *Hin*dIII, analyzed the resulting fragments on an agarose gel, and recorded fragment lengths in the range of 400–7500 bp. One drawback of this type of fingerprint is that the ability to discriminate and accurately size closely spaced fragments on an agarose gel is somewhat limited.

[1] Y. Kohara, K. Akiyama, and K. Isono, *Cell* **50,** 495 (1987).
[2] D. L. Daniels and F. R. Blattner, *Nature (London)* **325,** 831 (1987).
[3] M. V. Olson, J. E. Dutchik, M. Y. Graham, G. M. Brodeur, C. Helms, M. Frank, M. MacCollin, R. Scheinman, and T. Frank, *Proc. Natl. Acad. Sci. U.S.A.* **83,** 7826 (1986).
[4] A. Coulson, J. Sulston, S. Brenner, and J. Karn, *Proc. Natl. Acad. Sci. U.S.A.* **83,** 7821 (1986).
[5] A. Coulson and J. Sulston, *in* "Genome Analysis: A Practical Approach" (K. Davies, ed.), p. 19. IRL Press, Oxford, 1988.
[6] B. M. Hauge, J. Giraudat, S. Hanley, I. Hwang, T. Kohchi, and H. M. Goodman, *J. Cell. Biochem., Suppl.* No. 14E, 259 (1990).
[7] I. Sidén-Kiamos, R. D. C. Saunders, L. Spanos, T. Majerus, J. Treanear, C. Savakis, C. Louis, D. M. Glover, M. Ashburner, and F. C. Kafatos, *Nucleic Acids Res.* **18,** 6261 (1990).

METHODS IN ENZYMOLOGY, VOL. 218

To address the resolution problem, Coulson *et al.*[4,5] developed a two-step restriction digest fingerprint that produces fragments in the size range 50–2000 bp. These fragments are small enough to be analyzed on a denaturing polyacrylamide (sequencing) gel, which has much better resolution than an agarose gel. The Coulson *et al.* fingerprint consists of primary cleavage with a 6-base cutter such as *Hin*dIII, use of a fill-in reaction to label the primary cleavage ends with ^{32}P, and then secondary cleavage with a 4-base cutter such as *Sau*3A. Although digestion with a 4-base cutter generates a large number of fragments, only those fragments that contain a labeled primary cleavage end will be detected by autoradiography after electrophoresis in a sequencing gel. In other words, the primary cleavage site determines which of the restriction fragments are sampled from the clone to be part of the fingerprint.

Principle of Sampled Sequence Fingerprinting

In both the Olson *et al.*[3] and Coulson *et al.*[4,5] fingerprinting procedures, each clone is characterized by a list of restriction fragment sizes. The amount of information about each clone is limited by the number of fragments in the fragment-size list and the resolution of the separation process used. If the fragments could be differentiated in some other way besides just size, more information would be available for making comparisons. Increasing the information content of each fragment in the fragment-size list provides better discrimination in deciding which overlaps between clones are significant and greatly increases the rate at which the contig map is assembled.[8]

Sampled sequence fingerprinting increases information content by adding a small amount of sequence information to the size information of each restriction fragment in the fingerprint. The sequence information is obtained by taking advantage of two reagents. First, there are the 5'-ambiguous end restriction enzymes, listed in Fig. 1, that cleave DNA and generate 5' overhangs with some degree of degeneracy. In a sampled sequence fingerprint, one end of each labeled fragment is created with one of the 5'-ambiguous end restriction enzymes. It is by determining the nucleotide sequence of these degenerate 5' overhangs that sequence information is added to the fingerprint. These 5'-ambiguous ends can be sequenced by taking advantage of the second reagent—the four fluorescent dideoxynucleotides (SF–ddNTPs) developed for automated DNA sequencing.[9] Each dideoxynucleotide terminator is labeled with a slightly

[8] E. S. Lander and M. S. Waterman, *Genomics* **2**, 231 (1988).
[9] J. M. Prober, G. L. Trainor, R. J. Dam, F. W. Hobbs, C. W. Robertson, R. J. Zagursky, A. J. Cocuzza, M. A. Jensen, and K. Baumeister, *Science* **238**, 336 (1987).

Enzyme	Sequence	Enzyme	Sequence	Enzyme	Sequence
AccI	GT^MK AC / CA KM^TG	DraII (EcoO109I)	RG^GNC CY / YC CNG^GR	PleI	GAGTCNNNN^ / CTCAGNNNNN^
AflIII	A^CRYG T / T GYRC^A	DsaI	C^CRYG G / G GYRC^C	PpuMI	RG^GWC CY / YC CWG^GR
AsuI (Sau96I)	G^GNC C / C CMG^G	Eco31I (BsaI)	GGTCTCN^ / CCAGAGNNNNN^	RsrII	CG^GWC CG / GC CWG^GC
AvaI	C^YCGR G / G RGCY^C	EcoNI	CCTNN^N NNAGG / GGANN N^NNTCC	SauI (Bsu36I)	CC^TNA GG / GG ANT^CC
AvaII	G^GWC C / C CWG^G	BstNI (MvaI)	CC^W GG / GG W^CC	ScrFI	CC^N GG / GG N^CC
BbvI	GCAGCNNNNNNN^ / CGTCGNNNNNNNNNNNN^	EspI (CelII)	GC^TNA GC / CG ANT^CG	SsoII (DsaV)	^CCNGG / GGNCC^
BbvII (BbsI)	GAAGACNN^ / CTTCTGNNNNNN^	Fnu4HI	GC^N GC / CG N^CG	SecI (BsaJI)	C^CNNG G / G GNNC^C
BinI (AlwI)	GGATCNNNN^ / CCTAGNNNN^	FokI	GGATGNNNNNNNNN^ / CCTACNNNNNNNNNNNNN^	SfaNI	GCATCNNNNN^ / CGTAGNNNNNNNNN^
BsmAI	GTCTCN^ / CAGAGNNNNN^	HgaI	GACGCNNNNN^ / CTGCGNNNNNNNNNN^	StyI	C^CWWG G / G GWWC^C
BspMI	ACCTGCNNNN^ / TGGACGNNNNNNNN^	HgiCI (BanI)	G^GYRC C / C CRYG^G	TfiI	G^AWT C / C TWA^G
BstEII	G^GTNAC C / C CANTG^G	HinfI	G^ANT C / C TNA^G	Tth111I	GACN^N NGTC / CTGN N^NCAG
CauII (NciI)	CC^S GG / GG S^CC	Ksp632I (EarI)	CTCTTCN^ / GAGAGNNNN^		
DdeI	C^TNA G / G ANT^C	MaeIII	^GTNAC / CANTG^		

FIG. 1. Commercially available 5'-ambiguous end restriction enzymes.

different succinylfluorescein dye. By filling in the 5'-ambiguous overhangs with a mixture of SF–ddNTPs and unlabeled dNTPs, a small sequencing ladder is generated for each overhang. When these fluorescence-labeled fragments are separated on a sequencing gel, the Du Pont (Wilmington, DE) GENESIS 2000 DNA analysis system can distinguish the four succinylfluorescein dyes and thus identify which terminator is at the 3' end of each labeled fragment. In this way, the sequence of each ambiguous overhang is determined.

Again, the key component of sampled sequence fingerprinting is that, for each restriction fragment analyzed, terminal sequence information is added to the size information being compared in other fingerprinting methods. It is useful to consider how much information is provided by size designation as compared to terminal sequence specification. On a sequencing gel, people routinely sequence 50–300 nucleotides. Thus, it is certainly possible to designate 256 ($= 2^8$) different sizes on a sequencing gel. This means that specifying the size of a fragment distinguishes it from 255 other possible sizes. By using a 5'-ambiguous end restriction enzyme such as FokI, 4-base degenerate overhangs are created. There are $4^4 = 256$ possible 4-base sequences, which means that determining the 4-base terminal sequence of a fragment distinguishes it from 255 other possible sequences. Thus, designating the 4-base terminal sequence of a fragment provides about the same amount of information as designating the size of the fragment. Now consider the consequences of trying to increase information content. With due care, it is perhaps possible to designate 512 ($= 2^9$) sizes on a sequencing gel, thus doubling the amount of information provided by size. Digestion with the restriction enzyme HgaI generates 5-base degenerate overhangs. Because there are $4^5 = 1024$ possible 5-base sequences, designating just 1 additional base of terminal sequence increases the amount of information by a factor of 4. Clearly, it is much more productive to devise means for increasing the sequence information in a fingerprint than it is to strive for marginal improvements in size resolution.

The nine boxed enzymes in Fig. 1 consist of one enzyme that produces 3-base degenerate overhangs, seven that produce 4-base degenerate overhangs, and one that produces 5-base degenerate overhangs. Because these are the most degenerate overhangs, these 5'-ambiguous end restriction enzymes generate the greatest amount of terminal sequence information. Thus, these nine enzymes are the ones that are most useful in sampled sequence fingerprinting.

The first sampled sequence fingerprints were produced using the direct fingerprinting procedure described by Brenner and Livak.[10] This proce-

---

[10] S. Brenner and K. J. Livak, *Proc. Natl. Acad. Sci. U.S.A.* **86**, 8902 (1989).

dure is a straightforward extension of Coulson *et al.*[4,5] fingerprinting, involving the same basic steps of primary cleavage, labeling, and secondary cleavage. In direct sampled sequence fingerprinting, primary cleavage is with one of the 5′-ambiguous end restriction enzymes and labeling is with the fluorescent dideoxynucleotides. This direct method, however, has a number of disadvantages. First, primary cleavage must be with a 5′-ambiguous end restriction enzyme and thus, for maximum information content, is restricted to the nine boxed enzymes in Fig. 1. As with Coulson *et al.*[4,5] fingerprinting, it is the primary cleavage that determines how often the clone is sampled and thus how many labeled fragments are part of the fingerprint. This means it is highly desirable to have as much flexibility as possible in the primary cleavage step so that the experimenter can adjust approximately how often each clone is cleaved depending on average insert size, GC content, and other particulars of the clone library being analyzed. This flexibility is especially important with DNAs that are cosmid size or larger. Second, after fragments are labeled with the fluorescent SF–ddNTPs, it is necessary to remove most of the unincorporated SF–ddNTPs before the sample is electrophoresed past the laser detection system. In the direct method, this requires an additional cleanup step. In the capture procedure outlined below, labeled fragments are immobilized on a solid support so that SF–ddNTPs and other contaminants that adversely affect electrophoresis can be efficiently washed away. The washing also eliminates the bulk of the DNA so that there is no problem with overloading the gel.

Because of the limitations of the direct procedure, almost all our work has been done using the capture fingerprinting procedure outlined in Fig. 2, and this will be the only procedure described in detail below. In the capture procedure, primary cleavage is flexible because it can be performed with any restriction enzyme or combination of enzymes. These primary cleavage ends are then derivatized by adding the affinity reagent biotin. The biotin can be attached either by using a DNA polymerase to fill in 5′ overhangs with biotinylated nucleotides or by using terminal transferase to add biotinylated nucleotides to blunt or 3′-overhang ends. We have exclusively used restriction enzymes that generate 5′ overhangs in the primary cleavage step because it is possible to combine primary cleavage and biotin attachment in a single reaction. It is the secondary cleavage that is then performed using one of the 5′-ambiguous end restriction enzymes. Next, the DNA fragments are incubated with streptavidin immobilized on a solid support. DNA fragments with a biotinylated primary cleavage end bind to the immobilized streptavidin and all other fragments are washed away. In Coulson *et al.*[4,5] and direct sampled sequence fingerprinting, primary cleavage end fragments are sampled because only the primary cleavage ends are labeled with a radioactive or

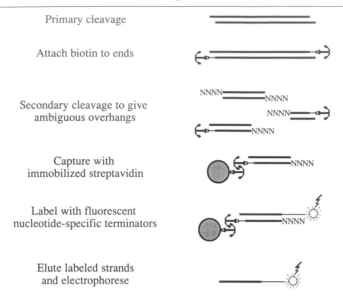

Primary cleavage

Attach biotin to ends

Secondary cleavage to give
ambiguous overhangs

Capture with
immobilized streptavidin

Label with fluorescent
nucleotide-specific terminators

Elute labeled strands
and electrophorese

FIG. 2. Capture procedure for sampled sequence fingerprinting.

fluorescent reporter. In capture sampled sequence fingerprinting, sampling
is accomplished by using the biotin–streptavidin affinity pair to physically
separate primary cleavage end fragments from all other fragments. The
next step is to label the secondary cleavage ends of the bound fragments
with the SF-ddNTPs. The labeled fragments are then eluted from the solid
support and analyzed on the GENESIS 2000 system. The size of labeled
fragments can be determined by comparison to size standards electrophoresed in a parallel lane. Terminal sequence is obtained by using the capabilities of the GENESIS 2000 to determine the identity of the SF–ddNTP
attached to each labeled fragment. In this way, capture sampled sequence
fingerprinting provides both size and terminal sequence information that
can be used to compare clones for overlaps.

Materials and Reagents

Labeled Dideoxynucleotides

All of our work has been done using the four fluorescence-labeled
dideoxynucleotides SF-505–ddGTP, SF-512–ddATP, SF-519–ddCTP,
and SF-526–ddTTP, which were obtained from New England Nuclear
(Boston, MA) as 125 μM stocks. These fluorescent dideoxynucleotides

are detected and discriminated using the Du Pont GENESIS 2000 DNA Analysis System. For the basic operation of the GENESIS 2000, the reader is referred to the manufacturer instructions and application updates. Applied Biosystems (Foster City, CA) has developed a set of four fluorescence-labeled dideoxynucleotides that are called DyeDeoxy terminators. Using these terminators, it should be possible to produce sampled sequence fingerprints that can be detected on the Applied Biosystems model 373A automated DNA sequencing system.

Enzymes

5'-Ambiguous End Restriction Enzymes. As discussed above, the most useful 5'-ambiguous end restriction enzymes are the nine boxed enzymes in Fig. 1. All nine of these enzymes are available from New England BioLabs (Beverly, MA). *Bbv*I, *Fok*I, and *Ksp*632I are available from other vendors as well. Because of its 5-base degenerate overhang, *Hga*I provides the most sequence information to a sampled sequence fingerprint, but this enzyme is very expensive. We have used *Fok*I in our studies because it is the least expensive of the nine preferred enzymes.

DNA Polymerases. The SF–ddNTPs are accepted by two DNA polymerases—reverse transcriptase and modified T7 DNA polymerase (Sequenase; U.S. Biochemical, Cleveland, OH). In our initial studies, only reverse transcriptase would produce usable fingerprints. So, sampled sequence fingerprinting was developed using reverse transcriptase as described by Brenner and Livak.[10] Then Tabor and Richardson discovered that substituting Mn^{2+} for Mg^{2+} reduces the discrimination against dideoxynucleotides for T7 DNA polymerase.[11] This prompted a reinvestigation of the use of Sequenase. Once optimized, Sequenase produced fingerprints that are slightly stronger and more even than the fingerprints obtained with reverse transcriptase. Thus, we switched to using Sequenase in Mn^{2+} buffer for the preparation of sampled sequence fingerprints and this is the procedure described below. We have used both Sequenase 1.0 and 2.0 (U.S. Biochemical).

Streptavidin Magnetic Particles

The capture fingerprinting described in Brenner and Livak[10] was performed using magnetic streptavidin–CrO_2 particles obtained from the Diagnostic Systems group of the Du Pont Medical Products Department. Magnetic particles with streptavidin attached are now commercially avail-

---

[11] S. Tabor and C. C. Richardson, *Proc. Natl. Acad. Sci. U.S.A.* **86,** 4076 (1989).

able, so we have switched to using Dynabeads M-280 streptavidin (Cat. No. 112.05 or 112.06; Dynal, Great Neck, NY). These are superparamagnetic, polystyrene beads activated by p-toluenesulfonyl chloride and coated with streptavidin. Dynal also sells a magnetic rack (MPC-E, Cat. No. 120.04) that holds six 1.5-ml microcentrifuge tubes. It should also be possible to prepare capture-sampled sequence fingerprints using other forms of immobilized streptavidin or avidin. The magnetic particles are especially convenient for rapid washing.

Methods

Determination of Dideoxy/Deoxy Ratio

The experiments used to find an appropriate dideoxy/deoxy ratio for preparing sampled sequence fingerprints with Sequenase will be briefly described. This information is included to serve as a guide for adapting the sampled sequence fingerprinting procedure to use different DNA polymerases or to use labeled ddNTPs other than the SF-ddNTPs we have used.

In sampled sequence fingerprinting, a very short sequencing run is performed at the end of every restriction fragment of the fingerprint. This requires a much greater ratio of dideoxynucleotides to deoxynucleotides than is used in conventional sequencing. For Sequenase in Mn^{2+} buffer and SF-ddNTPs, conventional sequencing is performed using the following nucleotide concentrations: 0.02 μM SF-505–ddGTP, 0.027 μM SF-512–ddATP, 0.027 μM SF-519–ddCTP, 0.27 μM SF-526–ddTTP, and 7.5 μM concentrations of each dNTP. For the first attempt at sampled sequence fingerprinting with Sequenase, the dideoxy/deoxy ratio was increased 50×, 100×, or 200×. This was accomplished by increasing the concentration of each SF–ddNTP 10-fold, and decreasing the dNTP concentration 5-, 10-, or 20-fold. Using these ratios, sampled sequence fingerprints were prepared following the basic protocol described in detail below. Each fingerprint was prepared using 1 μg λ DNA, primary cleavage with BstEII and BamHI, and secondary cleavage with FokI. (The labeled fragments expected for this BstEII–BamHI/FokI fingerprint of λ DNA are shown in Brenner and Livak.[10]) The 50×, 100×, and 200× fingerprints were compared to see which gave the best combination of strong signal and evenness of labeling across the 4-base pattern of each FokI end. If the dideoxy/deoxy ratio is too low, then very low signal is observed because most of the FokI overhangs are filled in completely with unlabeled dNTPs. If the dideoxy/deoxy ratio is too high, very strong signal is observed, but the first peak in any 4-base pattern is much stronger than any of the subsequent peaks. In this case, the 200× fingerprint had the strongest signal and still showed even labeling across all the peaks in each 4-base pattern. Thus, even higher dideoxy/deoxy ratios were tried.

For the next batch of fingerprints, SF–ddNTP concentrations were increased by another factor of 3 and dNTP concentrations of 1.5, 0.75, 0.5, or 0.25 μM were used. This produced dideoxy/deoxy ratios that are 150×, 300×, 450×, or 900× greater than that used in conventional sequencing. Again, BstEII–BamHI/FokI fingerprints of λ DNA were prepared and compared. The 450× and 900× fingerprints showed skewed patterns with strong first peaks, indicating an excessive dideoxy/deoxy ratio. The 300× fingerprint showed the best compromise of strong signal and even distribution. In all the fingerprints, however, the G peaks were stronger than any of the other three bases. Therefore, the 300× conditions were modified by decreasing the SF-505–ddGTP concentration threefold. This gave the final conditions we have used for sampled sequence fingerprinting with Sequenase, which are 0.2 μM SF-505–ddGTP, 0.8 μM SF-512–ddATP, 0.8 μM SF-519–ddCTP, 8 μM SF-526–ddTTP, and 0.75 μM concentrations of each dNTP.

Preparation of Cosmid DNA

The preparation of DNA is an important factor in the success of the sampled sequence fingerprinting procedure. We found that the following modification of the Birnboim and Doly alkaline lysis method[12] gave consistently good fingerprints. Generally, cosmids were prepared in batches of 12.

Recipes

TB ("terrific broth")[13]: Prepare a sterile solution containing 12 g Bacto-tryptone (Difco, Detroit, MI), 24 g Bacto-yeast extract, 4.0 ml glycerol, and H_2O to a final volume of 900 ml. Just prior to use, supplement with 10% (v/v) sterile 0.17 M KH_2PO_4/0.72 M K_2HPO_4

GTE: 50 mM glucose, 25 mM Tris-HCl (pH 8.0), 10 mM ethylenediaminetetraacetic acid (EDTA)

Potassium acetate, pH 4.8 (3 M with respect to K^+; 5 M with respect to acetate): To 60 ml 5 M potassium acetate, add 11.5 ml glacial acetic acid and 28.5 ml H_2O

TE-8.0: 10 mM Tris-HCl (pH 8.0), 1 mM EDTA

50 μg/ml RNase A: A concentrated stock free of DNase is prepared as

[12] H. C. Birnboim and J. Doly, *Nucleic Acids Res.* **7**, 1513 (1979).
[13] K. D. Tartof and C. A. Hobbs, *Gene* **67**, 169 (1988).

described by Maniatis *et al.*[14] Dissolve pancreatic RNase (RNase A) at a concentration of 10 mg/ml in 10 mM Tris-HCl, pH 7.5/15 mM NaCl. Heat to 100° for 15 min and allow to cool slowly to room temperature. Dispense into aliquots and store at −20°. This stock is diluted 1 : 200 in TE-8.0 to prepare the 50-μg/ml working solution that can be stored at 4°

1. Streak colonies from frozen cultures on 25-μg/ml kanamycin plates. Incubate at 37° overnight.

2. Using the smallest healthy colony from each plate, streak again on a 25-μg/ml kanamycin plate. Incubate at 37° overnight. (Picking small colonies helps to minimize deletions. Streaking the clones twice seems to be essential for a high DNA yield.)

3. Again using small colonies, inoculate 2-ml cultures of TB containing 50 μg/ml kanamycin. (Our cultures were grown in 18-mm glass tubes.) Place culture tubes at an angle in a 37° air shaker set at 220 rpm and incubate for 15 to 18 hr.

4. Chill the cells for 15 to 20 min, then transfer 1.5 ml of culture to microcentrifuge tubes. (It is important to keep samples on ice as much as possible throughout this procedure until pellets are at the air-drying step.)

5. Centrifuge the cells at 3500 rpm for 5 min. Return tubes to ice and remove supernatants with a Pasteur pipette.

6. Resuspend the pellets in 100 μl of a cold solution of GTE plus 4 mg/ml lysozyme (lysozyme should be added to cold GTE just prior to use). Incubate on ice for 5 min.

7. Add 200 μl of a freshly made solution of 0.2 M NaOH/1% (w/v) sodium dodecyl sulfate (SDS). Mix contents of tubes by inverting three or four times. Incubate on ice for 5 min. (Make up 3.5 ml of this solution with 2.45 ml H$_2$O, 0.7 ml 1 M NaOH, 0.35 ml 10% SDS; keep at room temperature to prevent the SDS from precipitating out of solution.)

8. Add 150 μl of a cold potassium acetate solution. Mix by inverting the tubes. Incubate on ice for 15 min. (This incubation can be extended up to 60 min.)

9. Centrifuge at 12,000 rpm for 10 min at 4°.

10. Transfer the supernatants to fresh microcentrifuge tubes containing 400 μl phenol : chloroform : isoamyl alcohol (50 : 49 : 1, by volume). Briefly vortex and centrifuge at 12,000 rpm for 2 min.

11. Leaving all interface behind, transfer the aqueous layers to fresh microcentrifuge tubes containing 460 μl 2-propanol. Invert the tubes and incubate on ice for 5 min.

---

[14] T. Maniatis, E. F. Fritsch, and J. Sambrook, "Molecular Cloning: A Laboratory Manual," p. 451. Cold Spring Harbor Press, Cold Spring Harbor, New York, 1982.

12. Centrifuge the samples at 12,000 rpm for 10 min. Remove as much supernatant as possible and allow pellets to air dry. (This step should yield a large, white pellet.)

13. Dissolve the pellets in 50 μl TE-8.0. Add 10 μl 50 μg/ml RNase A and incubate at 37° for 30 min.

14. Add 60 μl phenol : chloroform : isoamyl alcohol (50 : 49 : 1, by volume) and extract as described in step 10. Carefully load the aqueous layers onto the centers of prespun Sephadex G-50 columns (Select-D columns from 5 Prime → 3 Prime, Boulder, CO; columns are mixed, drained, and prespun at 4000 rpm for 2 min). Collect DNA by centrifugation at 4000 rpm for 4 min into fresh collection tubes.

15. Transfer collected DNA into sterile, screw-cap microcentrifuge tubes. Label and store at 4°.

Capture Fingerprinting Using Sequenase

Recipes

One-Phor-All PLUS buffer (Pharmacia, Piscataway, NJ): 100 mM Tris–acetate (pH 7.9), 100 mM magnesium acetate, 500 mM potassium acetate

Biotin–dNTP mix: 5 μM biotin-11-dUTP (ENZO, New York, NY), 5 μM biotin-11-dCTP (ENZO), 50 μM dATP, 50 μM dGTP

10× FokI buffer: 200 mM KCl, 100 mM Tris-HCl (pH 7.5), 100 mM MgCl$_2$, 100 mM 2-mercaptoethanol; 2 mg/ml bovine serum albumin

5× Mn^{2+} buffer: 250 mM NaCl, 100 mM morpholinepropanesulfonic acid (MOPS) (pH 7.5), 50 mM sodium isocitrate, 25 mM MnCl$_2$

2-8-8-80 SF–ddNTP mix: 2 μM SF-505–ddGTP, 8 μM SF-512–ddATP, 8 μM SF-519–ddCTP, 80 μM SF-526–ddTTP

TE-7.4: 10 mM Tris-HCl (pH 7.4), 1 mM EDTA

1× SSC/0.1% SDS: 150 mM NaCl, 15 mM sodium citrate (pH 8.0), 0.1% (w/v) sodium dodecyl sulfate

Triton wash: 0.17% (w/v) Triton X-100, 100 mM NaCl, 10 mM Tris-HCl (pH 7.5), 1 mM EDTA

10× MTB: 162 g Tris base, 27.5 g boric acid, 9.3 g Na$_2$EDTA · 2H$_2$O dissolved in H$_2$O to a total volume of 1 liter

Primary Cleavage Plus Biotinylation

Primary cleavage mix (for 12 samples; BglII and HindIII used as examples): 102 μl H$_2$O, 52 μl One-Phor-All PLUS buffer, 52 μl biotin–dNTP mix, 5 μl 100 mM dithiothreitol (DTT), 25 μl BglII (~400 U), 20 μl HindIII (~400 U), 4 μl avian myeloblastosis virus (AMV) reverse transcriptase (20–40 U)

1. To tubes containing 20 μl of cosmid DNA, add 20 μl of primary cleavage mix. Incubate in a 37°C for 3 hr.

2. Precipitate DNA by adding 40 μl 5 *M* ammonium acetate and 200 μl cold ethanol. Mix by inverting tubes. Incubate on dry ice for 20 to 30 min.

3. Centrifuge the samples at 12,000 rpm for 10 to 15 min. Pour off the supernatants and allow the pellets to air dry. (Pellets should be visible.)

4. Dissolve the pellets in 16 μl of TE-7.4. Heat at 65° for 10 min. Cool to room temperature.

Secondary Cleavage

Secondary cleavage mix (for 12 samples; *Fok*I used as an example): 26 μl 10 × *Fok*I buffer, 26 μl *Fok*I (200 U)

5. After a brief centrifugation of the 16-μl samples, add 4 μl secondary cleavage mix, cover with a layer of mineral oil, and incubate at 37° overnight.

6. Add the aqueous layer of each sample to a tube containing 50 μl 1× SSC/0.1% SDS. Load these samples carefully onto prespun Sephadex G-50 columns (Select-B columns for biotinylated DNA from 5 Prime → 3 Prime). Collect by centrifugation at 4000 rpm into 1.5-ml siliconized microcentrifuge tubes. Samples can be stored at this point for at least 1 week.

Attachment to Solid Support

7. To the collected samples, add 75 μl Triton wash and 15 μl 10 mg/ml freshly washed Dynabeads M-280 streptavidin. Incubate at 37° for 30 min. Tap the tubes to resuspend beads every 5 min for the first 15 min of incubation. (Prior to use, beads are washed three times with 2 vol of Triton wash. Washing is achieved by resuspending the beads in the wash solution, followed by placing the tube in a magnetic rack and allowing the beads to collect on the side of the tube. While keeping the beads collected against the side of the tube, remove the supernatant with a pipettor. Care should be taken to avoid generating bubbles or allowing the beads to dry between washes. If bubbles are generated, pipette the bubbles off the top of the solution to prevent them from dislodging the beads while the supernatant is being removed. Resuspend the beads in Triton Wash to their original volume.)

8. After incubation at 37°, place the tubes in a magnetic rack and discard liquid. Wash the beads three times with 100 μl Triton wash.

9. Resuspend beads in 12 μl TE-8.0.

Labeling with Fluorescent Terminators

Labeling mix (for 12 samples): 56 μl 5× Mn^{2+} buffer, 28 μl 2-8-8-80 SF–ddNTP mix, 14 μl 15 μM dNTPs, 14 μl 0.1 M DTT, 4 μl Sequenase (50 U)

10. To the resuspended samples, add 8 μl labeling mix. Incubate at 37° for 20 min. Tap the tubes to resuspend beads every minute for the first 5 min of incubation and occasionally thereafter.

11. Wash the beads three times with 100 μl Triton wash.

12. Suspend the samples in 8 μl GENESIS loading solution [NEK 531, New England Nuclear; contains 95% (v/v) formamide, 25 mM EDTA, 0.004% (w/v) Crystal Violet, SF-505-labeled 20-mer tracking primer]. Incubate in a boiling water bath for 5 min. Quickly spin and place the tubes in a magnetic rack. Transfer the liquid into fresh 1.5-ml siliconized microcentrifuge tubes. Store at −70° until ready for use. (Samples lose most of their signal after about 3 weeks of storage.)

13. Run prepared fingerprint samples on a GENESIS 2000 DNA analysis system. We run 2–3 μl of each sample on 6% 19:1 acrylamide : bisacrylamide, 8 M urea gels, and electrophorese at 25–27 W for 7 hr using MTB buffer. To ensure favorable gel quality, several steps are taken:

a. The MTB buffer is left out of the stock acrylamide : bisacrylamide : urea solution.

b. The stock acrylamide : bisacrylamide : urea solution is routinely deionized by stirring with Bio-Rad (Richmond, CA) mixed-bed resin AG 501-X8 (D) and then filtered. Sufficient solution to pour a gel was adjusted to 1× MTB just before use.

c. A blanket of nitrogen is blown over the working area until the gels are polymerized. (This procedure is especially important if humidity levels are high.)

Comparison of Rhodobacter Cosmids

Relying predominantly on the radioactive fingerprinting method of Coulson *et al.*,[4,5] Williams constructed a physical genome map of the photosynthetic bacterium *Rhodobacter capsulatus* (Williams, unpublished results, 1989). Over 1000 random cosmid clones were fingerprinted and analyzed for overlaps. The final map consisted of a canonical set of 200 cosmids selected to provide complete coverage of the genome. To test the validity of this map, we prepared sampled sequence fingerprints of the 200 cosmids in the canonical set.

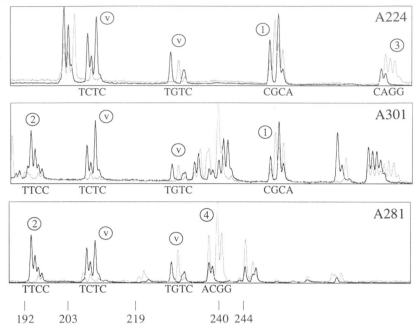

FIG. 3. Sampled sequence fingerprints comparing three cosmids with inserts of *Rhodobacter capsulatus* DNA. Fluorescence signal intensities are plotted versus electrophoresis time. Each window shows approximately 100 min of electrophoresis time. At the bottom are shown the position and size (in nucleotides) of pBR322 *Msp*I marker fragments electrophoresed in a parallel lane. The nucleotide sequence indicated by selected patterns is shown below each pattern. Patterns marked with a "v" are derived from the vector, Lorist2, used to construct these cosmids. These vector patterns are used as internal guides to line up the fingerprints. Patterns showing possible matches among the cosmids are marked 1, 2, 3, and 4. These are additional matches in other parts of the fingerprints. The raw data for these fingerprints were processed using the GENESIS program Base Caller 4.0 to obtain the orthogonal displays shown here. Although Base Caller 4.0 processes the fingerprint data, it sometimes does not make base calls throughout the entire run. This is because Base Caller 4.0 is designed to interpret continuous sequence data, so the program does not know how to handle the many blank spots in fingerprint data. This is not a problem because, once the data are processed into orthogonal form, it is easy to make base calls by visual inspection.

After the 200 fingerprints were prepared and processed, they were visually compared to find overlaps between cosmids.[15] We were able to use visual comparison because we were dealing with a relatively small number of cosmids and we had an existing map to guide us. Figure 3 shows portions of three of the sampled sequence fingerprints aligned to show

[15] This analysis was done in collaboration with Anne Kubelik and Barry Marrs of Du Pont.

overlaps. Overlaps were found by looking for cosmids that had fragments that were both the same size and had the same terminal sequence pattern. To have a size standard, the fingerprint samples were electrophoresed along with a lane containing SF-505-labeled pBR322 *Msp*I fragments. Also, sequence patterns derived from the vector (in this case Lorist2[16]) were used as internal controls to compare samples from gel to gel.

The analysis of the sampled sequence fingerprints revealed a number of errors in the original *Rhodobacter* map. Rather than a single circular contig, the data indicated 13 unconnected contigs, ranging in size from 5 to 32 cosmids. Because of the increased information content of the sampled sequence fingerprints, it was often possible to determine what types of errors had been made. A number of mistakes in the original map were due to double clones, that is, cloning artifacts in which two unrelated segments were packaged in the same cosmid. An example of this is shown in Fig. 3. Based on comparison with other cosmids, it is clear that cosmids A224 and A281 are in separate contigs. A301 is probably a double clone because it contains some DNA fragments found in A224 and others found in A281. The CGCA pattern (marked 1 in Fig. 3) of A224 and the TTCC pattern (2) of A281 are both clearly in A301. It is likely that the CAGG pattern (3) of A224 and perhaps the ACGG pattern (4) of A281 are also in A301. In these latter two cases, the presence of additional fragments in A301 precludes the assignment of definitive matches, but it is still possible to discern some relatedness. This type of mapping error was readily detected because, in a genome the size of *R. capsulatus*, the finding of just one matching pattern is significant.

Concluding Remarks

Choice of Enzymes

As discussed above, it is the primary cleavage that determines how many fragments will be part of the fingerprint. Optimally, a sampled sequence fingerprint should have about 15 patterns in the size range 50 to 350 nucleotides. This means that primary cleavage should cut each clone 10–20 times (assuming *Fok*I is the secondary cleavage enzyme). For cosmids, this generally means using a combination of two 6-base cutters. GC content and information about nonrandom distribution of nucleotides will help determine which enzymes should be used to achieve the appropriate number of cleavages. We have also found it useful to have two or four patterns derived from vector DNA to serve

[16] T. J. Gibson, A. R. Coulson, J. E. Sulston, and P. F. R. Little, *Gene* **53**, 275 (1987).

as internal size controls, so this should be taken into account in picking primary cleavage enzymes.

Analysis of Larger DNAs

One of the chief attributes of the capture fingerprinting method is that it can be used to analyze very large DNAs. This was demonstrated by Brenner and Livak,[10] who showed fingerprints of herpesvirus DNA (150 kb) and E. coli DNA (4500 kb). We have now also prepared sampled sequence fingerprints of YACs (yeast artificial chromosomes) containing large human DNA inserts (100–500 kb). First, the technique of Alu PCR[17] was used to specifically amplify a limited number of human segments in a preparation of total yeast DNA. The Alu PCR was performed with an Alu-specific primer carrying biotin at its 5' end. In essence, performing Alu PCR with a biotinylated primer accomplishes both primary "cleavage" of the YAC and biotinylation of the primary cleavage ends. Secondary cleavage with FokI, capture, and labeling with SF–ddNTPs were then performed as in the cosmid fingerprints. Each human YAC was characterized by a distinctive pattern of fragment sizes and terminal sequences. The inclusion of PCR in sampled sequence fingerprinting both takes the place of primary cleavage with a restriction enzyme and eliminates any problems with detection sensitivity. It has been found that short primers of arbitrary sequence will reproducibly amplify a limited number of segments from the genomic DNA of a wide variety of species.[18] These amplified segments have been called RAPD markers (for random amplified polymorphic DNA). Using RAPD amplification as the primary cleavage step, it should be possible to prepare sampled sequence fingerprints for genomes of any size.

Future of Fingerprinting Methods in Genome Mapping

A number of projects have been completed or are in progress that use the fingerprinting of random clones to build physical genome maps. For a number of reasons, it is not likely that any new, large-scale random fingerprinting projects will be initiated. One reason is the redundancy involved in random fingerprinting. In general, the number of clones that must be fingerprinted to build the map is at least five times the number of

---

[17] D. L. Nelson, S. A. Ledbetter, L. Corbo, M. F. Victoria, R. Ramírez-Solis, T. D. Webster, D. H. Ledbetter, and C. T. Caskey, Proc. Natl. Acad. Sci. U.S.A. **86,** 6686 (1989).
[18] J. G. K. Williams, A. R. Kubelik, K. J. Livak, J. A. Rafalski, and S. V. Tingey, Nucleic Acids Res. **18,** 6531 (1990).

clones that corresponds to one genome equivalent. One way of looking at this problem is to consider that each sampled sequence fingerprint requires one sequencing lane. Building the map of a complex genome means that thousands of sequencing lanes are going to be used. Perhaps it would be more efficient to use each sequencing lane to obtain 300 nucleotides of sequence for a cDNA clone selected from a normalized library. Accumulating thousands of cDNA sequences would be a good start at enumerating the total number of genes in the genome. The sequenced cDNAs would then serve as jumping-off points for completing a physical map, using some sort of gap-filling method. One of the main reasons for building a genome map is to know where the genes are, so perhaps it makes sense to identify a large number of genes first, rather than search for genes within an already established map.

Another reason that large-scale fingerprinting is becoming less attractive is the growing acceptance of using sequence-tagged sites (STSs) as the basis for defining physical maps.[19] Sequence-tagged sites are small sequence landmarks (200–500 nucleotides) that can be amplified using defined primers and PCR conditions. For the human genome, the goal is to have an STS map with average spacing of 100 kb by 1995. The power of the STS concept is that it provides a common language so that mapping data generated by diverse methods can be integrated into a single physical map. Thus, to have a consistent map, it is no longer necessary to map the entire genome or chromosome in a single, large mapping program. Nor is it necessary to have a single, permanent clone archive that is shared by all workers. With an STS map, it makes more sense to identify important landmarks first and then integrate these landmarks into an evolving map using physical mapping techniques on relatively small regions.

Although random fingerprinting will probably not be used to map entire genomes, fingerprinting will still be used in conjunction with other physical mapping techniques. Fingerprinting could be used to order clones that cover relatively small segments, such as ordering subclones derived from a YAC insert. As shown with the *Rhodobacter* cosmids, sampled sequence fingerprinting provides a sensitive method for validating a map generated by another method because it can detect rearrangements and other cloning artifacts that might have been missed. Fingerprinting can be a rapid means for using an already existing map. After defining a canonical set of clones that cover a region, these clones can be fingerprinted by the sampled sequence method to generate a reference set of patterns. Then sampled sequence fingerprinting of new clones or amplified products provides a

[19] M. Olson, L. Hood, C. Cantor, and D. Botstein, *Science* **245**, 1434 (1989).

rapid means for locating DNA segments on the map. Finally, by using PCR as the primary "cleavage" step, sampled sequence fingerprinting can be used not only to order clones, but also in other areas where DNA fingerprinting is important.

[19] Transposon-Based and Polymerase Chain Reaction-Based Sequencing of DNAs Cloned in λ Phage

By B. Rajendra Krishnan, Dangeruta Kersulyte, Igor Brikun, Henry V. Huang, Claire M. Berg, and Douglas E. Berg

Introduction

More efficient methods for DNA sequencing are needed for studies of the structure of individual genes and gene clusters and for analyses of entire genomes. In this chapter we describe a combination of bacterial transposon[1] and polymerase chain reaction (PCR)[2,3] technologies that should increase the speed and efficiency of sequencing DNA segments cloned in λ phage vectors.

Transposable elements, the "jumping genes" of prokaryotes and eukaryotes, are segments of DNA that can move from one position in a genome to other, nonhomologous sites. They have been of great biological interest for nearly half a century, and have also proved to be superb tools in molecular genetics. Bacterial transposons, in particular, can insert at many positions in target DNAs when they hop between donor and target DNA molecules. They also generate nested deletions by hopping within the same DNA molecule. Although many transposons are more than 5 kb long, their ability to move between and within DNA molecules depends on short terminal sequences that are only 19–40 bp long. All other segments can be removed [in many cases, however, the element-specific transposase protein is cis acting, and must be provided from a gene that is in the donor DNA molecule (if it is located outside of the transposon)]. Transposons appropriate for sequencing contain unique sequences near each end that can serve as binding sites for PCR and sequencing primers. The transposition of these elements figuratively divides a large target DNA segment into many smaller segments with-

[1] D. E. Berg and M. M. Howe, eds., "Mobile DNA." Am. Soc. Microbiol., Washington, D.C., 1989.
[2] K. B. Mullis and F. A. Faloona, this series, Vol. 155, p. 335.
[3] H. A. Erlich, D. Gelfand, and J. Sninsky, *Science* **252**, 1643 (1991).

out actually disrupting physical linkage, each segment delimited by the transposon ends that provide universal primer-binding sites. This is the special feature that makes transposon-based approaches valuable as alternatives to extensive shotgun subcloning or primer walking, in both highly automated, large-scale genome sequencing efforts, and in "cottage-industry" style analysis of individual genes and gene clusters.

Principle of Method

Figure 1 schematically depicts applications of PCR technology for both mapping sites of transposon insertion and generating templates for DNA sequencing. Any transposon insertion close enough to a junction between the vector and the cloned DNA can be used to amplify the region of cloned DNA between that junction and the site of transposon insertion using appropriate vector- and transposon-specific PCR primers (Fig. 1B). The size of the PCR product reflects the site of transposon insertion. The PCR product itself can be used as a template for DNA sequencing, using the same primers that had been used for amplification. The efficiency of sequencing in this strategy depends in large measure on the ease of obtaining transposon insertions, the randomness of their distribution in target DNA, and also on the reach of PCR, which determines what fraction of insertions in cloned DNA can be identified in one attempt. In the case of λ phage clones, this use of PCR avoids the need to prepare high-titer phage stocks and extract their DNAs for restriction mapping or DNA sequencing, an important feature, because such steps are time consuming and difficult to automate.

Insertions too far from the junction to be mapped by direct PCR can be mapped relative to previously located insertions by cross-over (biparental, jumping) PCR (Fig. 1C), a type of amplification in which hybrid products arise by priming at sites on different molecules, and template switching during the multiple cycles (Fig. 2).

Transposon-based strategies have been developed for λ phage (this chapter), and for plasmids.[4] In the case of λ, most phage used as vectors kill infected cells; hence, genetic markers are needed that can be selected in λ plaques without requiring host cell viability. In addition, the size of the transposon must not exceed the difference between the size of the target phage DNA (λ arms plus cloned segment) and the capacity of the λ

---

[4] C. M. Berg, G. Wang, L. D. Strausbaugh, and D. E. Berg, this volume [20].

A. Transposon Insertions

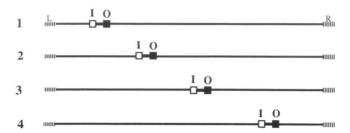

B. Products of Direct PCR

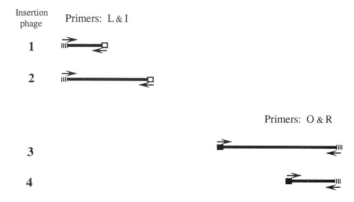

C. Products of Crossover PCR

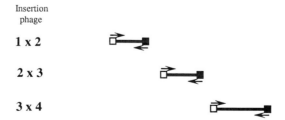

FIG. 1. Direct and cross-over PCR amplification from transposon-containing phage. (A) Schematic presentation of transposon-containing derivatives of a λ phage clone. (□) and (■), binding sites for different PCR primers that partially overlap the divergent I end and O end sequences that constitute essential sites for transposition; solid line, cloned target DNA;

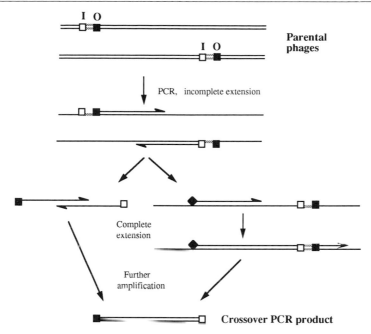

FIG. 2. Probable mechanism of cross-over PCR amplification: incomplete extension of primer strands (half-arrows) during one cycle of amplification; denaturation and chance annealing with a strand derived from phage with an insertion at another site; completion of these extensions to generate hybrid molecules; and further amplification of this hybrid product, as in direct PCR.

head (~52 kb).[5] Transposon-mediated nested deletion strategies that are useful for plasmids[6,7] are not appropriate for λ phage because a minimum DNA size is needed for λ packaging (~38 kb), and plaque formation requires specific genes in each λ arm. The method of analyzing λ phage clones described here is based on intermolecular transposition of a minide-

[5] M. Feiss and D. A. Siegele, *Virology* **92**, 190 (1979).
[6] A. Ahmed, this series, Vol. 155, p. 177.
[7] C. M. Berg, D. E. Berg, and E. Groisman, *In* "Mobile DNA" (D. E. Berg and M. M. Howe, eds.), p. 879. Am. Soc. Microbiol., Washington, D.C., 1989.

hatched line, vector sequences to the left and right of the cloned DNA segment. (B) Products of direct PCR using primers specific for transposon ends and vector sequences adjacent to cloning sites. Each transposon insertion gives access to a different portion of the cloned target DNA. (C) Products of cross-over (biparental) PCR using Tn5supF-specific primers. For proposed mechanism, see Fig. 2.

rivative of transposon Tn5 called Tn5supF[8] (Fig. 3) and methods for long-distance PCR amplification.

Tn5 and its many derivatives, including Tn5supF, have been widely used for mutational and DNA sequence analysis.[7-10] These elements transpose with relatively little specificity. They generate simple insertions that are free of vector sequences by an apparently cut-and-paste, conservative transposition mechanism. Nine base pairs of target sequence are duplicated at the site of insertion, probably reflecting staggered cuts made in the target, but the element itself is not replicated during transposition. The ability of Tn5 to transpose without making a cointegrate intermediate is an advantage for use with λ targets (in contrast to plasmids[4]).

Tn5 wild type is not convenient for sequencing for several reasons: it is 5.8 kb long, which is too large for many λ clones; its antibiotic resistance genes, although good for selecting plasmid and chromosomal insertions, are not suitable for selecting insertions in typical phage vectors, which kill infected cells; and Tn5 contains terminal 1.5-kb inverted repeats (the insertion sequence IS50), and thus does not provide unique binding sites for PCR and sequencing primers. A Tn5 derivative with just 19 bp from each Tn5 or IS50 end will transpose, however, if the Tn5 transposase protein is provided in cis. This is exemplified by Tn5supF, which is only 264 bp long, and consists of just a supF (suppressor tRNA) gene, flanked by BamHI restriction sites and the 19-bp segments from the ends of IS50.[8]

Tn5supF insertion in λ cloning vectors with nonsense (amber) mutations in essential phage genes, for example, the $A_{amber} B_{amber}$ phage (Charon4A,[11] is selected by plating on bacterial strains that do not carry nonsense suppressors (Sup°): only supF-containing derivatives of amber mutant phage form plaques on Sup° lawns.[8] Selection for Tn5supF insertion in λ vectors that do not contain nonsense mutations, for example,

[8] S. H. Phadnis, H. V. Huang, and D. E. Berg, *Proc. Natl. Acad. Sci. U.S.A.* **85,** 6468 (1989).

[9] C. M. Berg and D. E. Berg, in "*Escherichia coli* and *Salmonella typhimurium*—Cellular and Molecular Biology" (F. C. Neidhardt, J. L. Ingraham, K. B. Low, B. Magasanik, M. Schaechter, and H. E. Umbarger, eds.), p. 1071. Am. Soc. Microbiol., Washington, D.C., 1987.

[10] D. E. Berg, in "Mobile DNA" (D. E. Berg and M. M. Howe, eds.), p. 185. Am. Soc. Microbiol., Washington, D.C., 1989.

[11] F. R. Blattner, B. G. Williams, A. E. Blechl, K. Dennison-Thompson, H. E. Faber, L.-A. Furlong, D. J. Grunwald, D. O. Kiefer, D. D. Moore, J. W. Schumm, E. L. Sheldon, and O. Smithies, *Science* **196,** 161 (1977).

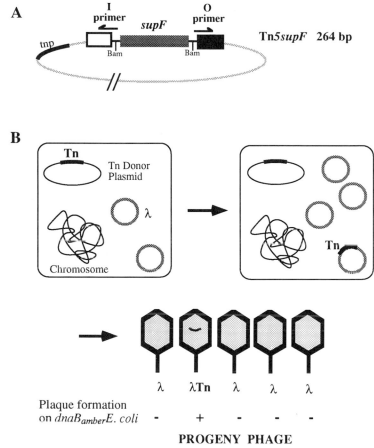

FIG. 3. Transposition of the 264-bp Tn5supF element. (A) the Tn5supF donor plasmid. *tnp*, Transposase gene. (B) Selection for Tn5supF insertion into λ by plaque formation on the *dnaB_amber* E. coli strain DK21.

EMBL4,[12] is based on the need for the host DnaB protein for λ DNA replication, and the availability of a bacterial strain with an amber mutant allele of *dnaB*: only *supF*-containing derivatives form plaques on this strain.[8,13] In the case of Tn5supF insertions in λ containing cloned *Escherichia coli* DNAs, the resulting insertion mutant alleles can be recombined

[12] A.-M. Frischauf, H. Lehrach, A. Poustka, and N. Murray, *J. Mol. Biol.* **170**, 827 (1983).
[13] D. M. Kurnit and B. Seed, *Proc. Natl. Acad. Sci. U.S.A.* **87**, 3166 (1990).

easily into the *E. coli* chromosome (alleleic replacement).[14,15] This allows efficient functional tests, hand in hand with DNA sequencing, a feature that is important in genome analysis projects.

Materials

Escherichia coli Strains

DB4496[8]: Carries pBRG1310, the Tn5*supF*-donor plasmid, and p3, a compatible Kan[r] plasmid that also carries amber mutant alleles of *amp* and *tet*.[16] Plasmid p3 is used to select for maintenance of the Tn5*supF* donor plasmid (by suppression of *amp* and *tet* alleles of p3, and thus host cell resistance to ampicillin and tetracycline, respectively). DB4496 also carries a *dam* mutation; Tn5*supF* transposition is more frequent in Dam[−] than in Dam[+] cells

MC1061[17]: MC1061 is Sup°, and is used to select Tn5*supF* insertions in amber mutant phages, as well as for routine growth of nonamber λ phage[8]

DK21[13]: DK21 is *sup° lacZ*$_{amber}$*dnaB266*$_{amber}$ and carries a gene called *ban* (*dnaB* analog from phage P1) in a λ*imm*[21] prophage. The *ban* gene allows bacterial growth, but not phage growth. DK21 is used to select Tn5*supF* insertions in nonamber λ phage[8]

Growth Media

LN broth: 10 g NZ-amine (Sheffield Products, Norwich, NY), 5 g Difco (Detroit, MI) yeast extract, 10 g NaCl, 990 ml distilled water; adjust pH to 7.2 (1.2 ml of 4 *M* NaOH); sterilize by autoclaving

LN agar: LN broth, with 15 g/liter Difco Bacto-agar. Antibiotics are

---

[14] S. H. Phadnis, S. Kulakauskas, B. R. Krishnan, J. Hiemstra, and D. E. Berg, *J. Bacteriol.* **173**, 896 (1991).

[15] Tn5*supF* insertions obtained in λ clones containing *E. coli* chromosomal DNA can be moved efficiently to the *E. coli* chromosome by homologous recombination (allelic replacement).[14] Operationally, this entails simple infection of a Sup° strain carrying the p3 (*amp*$_{amber}$ *tet*$_{amber}$) plasmid with λ::Tn5*supF* phage, selection for Sup[+] transductants by plating on ampicillin- or tetracycline-containing medium, and then purification of transductants by restreaking. When nonlysogens are used as recipients all transductants are haploid, and potentially mutant. This is because partial diploids, which are also formed, retain a copy of the λ vector phage and are eventually killed by expression of its lytic functions.[14] These properties make Tn5*supF*-insertion derivatives of λ phage clones useful for functional analyses and the identification of essential genes.

[16] B. Seed, *Nucleic Acids Res.* **11**, 2427 (1983).

[17] M. J. Casadaban and S. N. Cohen, *J. Mol. Biol.* **138**, 179 (1980).

used when needed at the following concentrations: ampicillin, 150 μg/ml; kanamycin, 50 μg/ml; and tetracycline, 10 μg/ml

N agar: 10 g NZ-amine, 10 g NaCl, 990 ml distilled water; adjust pH to 7.2; sterilize by autoclaving

N soft agar: 10 g NZ-amine, 10 g NaCl, 6 g Difco Bacto-agar, 990 ml distilled water; adjust pH to 7.2; sterilize by autoclaving

Other Reagents

Maltose (20%, w/v)

$MgSO_4$ (1 *M*)

Isopropyl-β-ᴅ-thiogalactopyranoside (IPTG) (100 m*M*)

5-Bromo-4-chloro-3-indolyl-β-ᴅ-galactoside (X-Gal), 2% (w/v) in for-mamide (store at $-20°$)

SM buffer: 5 g NaCl, 2 g $MgSO_4$, 50 ml 1 *M* Tris-HCl (pH 7.5), 5 ml 2% (w/v) gelatin, distilled water to 1 liter; sterilize by autoclaving

Chloroform

PCR buffer II (Perkin-Elmer Cetus, Norwalk, CT)

AmpliTaq DNA polymerase (Perkin-Elmer Cetus)

Hot Tub DNA polymerase (Amersham Corporation, Arlington Heights, IL)

Vent DNA polymerase (New England BioLabs, Beverly, MA)

Pfu DNA polymerase (Stratagene, La Jolla, CA)

Geneclean kit (Bio 101, Inc., La Jolla, CA)

$MgCl_2$ (25 m*M*)

Bovine serum albumin (100 mg/ml)

Deoxynucleoside triphosphates

Dideoxynucleoside triphosphates

Kinase buffer: 10 m*M* Tris-HCl (pH 8.0), 1 m*M* ethylenediaminetetra-acetic acid (EDTA), 10 m*M* $MgCl_2$

Cycle sequencing buffer: 10 m*M* Tris (pH 8.3), 50 m*M* KCl, 1.5 m*M* $MgCl_2$

STOP solution: 95% formamide, 20 m*M* EDTA, 0.05% (v/v) bromphenol blue, 0.05% (v/v) xylene cyanol

Double-stranded DNA cycle sequencing kit (Life Technologies, GIBCO/Bethesda Research Laboratories, Gaithersburg, MD)

[γ-^{32}P]ATP (~6000 Ci/mmol; Amersham)

Sequenase kit (U.S. Biochemical Corporation, Cleveland, OH)

Oligonucleotide Primers

λ left arm cloning site: EMBL4, λ2001, and related vectors contain the *b*189 deletion, which removes the segment ~19,700 to 27,730 nucleotides, which is to the left of the cloning site at λ nucleotide position 27,950.[18-20] Many oligonucleotides that match sequences between b189 and the cloning site, or just to the left of b189, would be suitable primers for PCR. In most experiments we used the oligonucleotide 5'-GGCCATAGAGTCTTGCAGACAAACTGC (underlined, λ nucleotides 27,928–27,951)

λ right arm cloning site: The right arm of EMBL4, λ2001, and related vectors begins at position 34,505[19,20] and many oligonucleotides that match sequences to the right of this position would be suitable for PCR. In most experiments we used the oligonucleotide 5'-GCCTAACGATCATATACATGGTTCTCTCC (underlined, λ nucleotides 34,569–34,544)

Tn5supF I end: 5'-TAGGATCCCGAGATCTGATC[8]

Tn5supF O end: 5'-TAGGATCCCCTACTTGTGTA[8]

Thermal Cycler

TC1 or TC480 (Perkin-Elmer Cetus)

Methods

The descriptions below detail (1) methods for obtaining Tn5supF insertions in λ, (2) PCR amplification strategies to map transposon insertions and generate sequencing templates, and (3) a method for sequencing PCR products. Following these descriptions, we present representative results obtained using these methods.

Isolating and Mapping Transposon Tn5supF Insertions in λ Clones

To Transpose Tn5supF to λ Phage

1. Inoculate a fresh colony of the Tn5supF-donor strain DB4496 into 5 ml of LN broth containing 0.2% maltose and 10 mM MgSO₄. Incubate overnight with shaking at 37°.
2. Prepare stocks of the λ clone of interest on strain DB4496 as follows: Add ~10⁶ λ phage (a suspended single plaque or aliquot of a larger stock) to 0.2 ml of the DB4496 overnight culture. Leave the mixture at room temperature for 5 min for adsorption of phage to cells, then

[18] D. L. Daniels, J. L. Shroeder, W. Szybalski, F. Sanger, A. R. Coulson, G. F. Hong, D. F. Hill, G. B. Petersen, and F. R. Blattner, *in* "Lambda II" (R. W. Hendrix, J. W. Roberts, F. W. Stahl, and R. A. Weisberg, eds.), p. 519. Cold Spring Harbor Press, Cold Spring Harbor, New York, 1983.

[19] J. Sambrook, E. F. Fritsch, and T. Maniatis, "Molecular Cloning: A Laboratory Manual." Cold Spring Harbor Press, Cold Spring Harbor, New York, 1989.

[20] K. Isono, personal communication, 1991.

add 2.5 ml of N soft agar, and pour quickly on an N agar plate. To obtain independent insertions, set up a number of separate infections.

3. Incubate plates (noninverted) at 37° for 6–8 hr, or in the case of slow-growing phage until there is confluent lysis of the bacterial lawn.

4. Once lysis is confluent, harvest phage stocks in one of two ways:
   a. Add 5 ml of SM buffer to the plates, and scrape off the mixture of soft agar overlay and buffer using the base of a bent glass rod. Collect this mixture in a Corex glass or polypropylene tube. Add 100 μl of chloroform, and vortex the suspension for 1 min. Chill on ice, centrifuge at ~10,000 g in a Sorvall (Norwalk, CT) refrigerated centrifuge for 5 min at 4°, and transfer the supernatant to a fresh test tube.
   b. Alternatively, add 5 ml of SM buffer over the soft agar overlay, and leave at 4° overnight. The following morning, collect the buffer (into which phage have diffused) with a sterile Pasteur pipette and transfer it to a tube. Add chloroform and remove cell debris and agar by centrifugation, as above.

To Select Tn5supF Insertions in Phage That Do Not Have Amber Mutations, Such as EMBL4

1. Prepare an overnight culture of MC1061 or other Sup° host (on which the target phage does not form plaques) in LN broth with 0.2% maltose and 10 mM MgSO$_4$.

2. Add ~0.1 ml (~10^8–10^9 pfu) of phage from the lysate prepared on the Tn5supF-donor strain (above) to 0.1 ml (~10^8 cfu) of the overnight culture of Sup° cells. After 5 min for adsorption, add 2.5 ml of N soft agar, and pour quickly on an N agar plate. Incubate overnight at 37°.

To Select Tn5supF Insertions in Phage That Do Not Have Amber Mutations, Such as EMBL4

1. While the λ stocks are growing on DB4496, inoculate E. coli DK21 (the dnaB$_{amber}$ strain for selecting supF-containing phage) from a frozen working stock[21] into 5 ml of LN broth containing 0.2% maltose and 10 mM MgSO$_4$. Incubate with shaking at 37° until late log phase.

---

[21] The E. coli DK21 cultures used to select Tn5supF insertions in λ should be freshly prepared by inoculation from a frozen stock. This is because DK21 grows rather slowly at 37°, and grows poorly at 30° and lower temperatures. Faster growing variants that do not allow Tn5supF-containing phage to plate, or that do not discriminate well between Tn5supF-containing and Tn5supF-free phage, tend to accumulate in cell populations maintained by subculturing.[8]

2. Add ~0.1 ml (~10^8–10^9 pfu) of phage from the lysate prepared on the Tn5supF-donor strain (above) to 0.2 ml of late log-phase DK21. After 5 min for adsorption, add 2.5 ml of N soft agar (containing X-Gal and IPTG, when using a blue plaque phenotype to help identify Tn5supF-insertion phage[22]), and pour quickly on an N agar plate. Incubate overnight at 37°.

Polymerase Chain Reaction Mapping of Tn5supF Insertions in λ

Most current PCR protocols have been optimized for amplification of at most a few kilobases; the few attempts to amplify longer distances have used purified DNAs as substrates. We have been optimizing conditions for amplifying longer DNA segments using material from phage plaques, in order to efficiently map Tn5supF insertions in cloned DNA. The following protocol resulted in amplification of ~6 kb routinely and more than 10 kb occasionally (see Results and Discussion, below).

Preparation of λ Phage for Polymerase Chain Reaction

Using a sterile Pasteur pipette, pick fresh plaques from lawns of DK21 and suspend them in 100 μl sterile distilled water. Use 5 μl of the suspension for PCR, and store the remainder at 4° for further PCR or phage propagation.

Direct Polymerase Chain Reaction Amplification

Set up reactions in 25-μl volumes with (1) 5 μl phage suspension, (2) 10 pmol left (L) or right (R) arm vector primer, (3) 10 pmol Tn5supF I end or O end primer, (4) 2.5 μl 10 × PCR buffer (supplied by the manufacturer), with additional Mg^{2+} if needed,[23] (5) 200 μM concentrations of each dNTP, (6) 1 U of a thermostable polymerase (AmpliTaq, Vent, Hot Tub, or Pfu)

---

[22] Because λ::Tn5supF phage form very small plaques on lawns of DK21, some workers prefer to detect these phage using a blue plaque phenotype, which is based on the lacZ$_{amber}$ allele in DK21. Growth of λ::Tn5supF in DK21 results in suppression of this allele prior to cell lysis, and thus a blue plaque on a white background if the lac operon inducer, IPTG, and the chromogenic β-galactosidase substrate, X-Gal, are present in the medium. To detect Tn5supF-containing phage by a blue plaque phenotype, add 10 μl of 100 mM IPTG and 50 μl of 2% X-Gal to the mixture of phage, cells, and soft agar just before pouring on an N agar plate. We do not recommend use of this blue plaque screen when the efficiency of Tn5supF transposition is very low, however, because control experiments showed that addition of IPTG lowered the efficiency of plating of Tn5supF-containing phage on DK21 approximately 10-fold when IPTG was added.

[23] In experiments using AmpliTaq and its Mg^{2+}-free PCR buffer, we usually added 1.5 mM MgCl$_2$. With other thermostable polymerases, we tested for optimal Mg^{2+} concentrations as recommended by the manufacturers, and obtained the best amplifications using their buffers without additional Mg^{2+}.

(add 1 mg/ml bovine serum albumin when using Vent DNA polymerase) and (7) sterile distilled water. Overlay the solution with a drop of mineral oil, and transfer to a thermal cycler programmed as follows: 1 cycle (94°, 1 min; 53°, 30 sec; 72°, 5–7 min), then 34 cycles (94°, 20 sec; 53°, 30 sec; 72°, 5–7 min), and finally 1 cycle (94°, 20 sec; 53°, 30 sec, and 72°, 10 min) (times and temperatures optimized for Perkin-Elmer TC1 and TC480 instruments). Electrophorese 5- to 10-μl aliquots from each reaction in a 1% (w/v) agarose gel. Infer the sites of Tn5supF insertion from the sizes of PCR products.

Cross-Over (Biparental) Polymerase Chain Reaction Amplification

For cross-over PCR (Fig. 2), set up 25-μl reactions, as above, but with 5 μl of each of two phage plaque suspensions and with 5 pmol of each Tn5supF primer (O end and I end). Program the thermal cycler as follows: 2–5 cycles (94°, 1 min; 53°, 30 sec; 72°, 30 sec), then 34 cycles (94°, 20 sec; 53°, 30 sec; 72°, 5–7 min), and 1 cycle (94°, 20 sec; 53°, 30 sec; 72°, 10 min).

λ DNA Purification for Restriction Mapping and Sequencing from Tn5supF Insertions

To purify λ::Tn5supF DNAs for restriction mapping and sequencing, we use the procedure of Helms et al.,[24] which entails ion exchange on a DE-52 minicolumn to purify λ phage, proteinase K, and sodium dodecyl sulfate treatment to remove the protein phage coat, and acetate–2-propanol precipitation and phenol–chloroform extraction to complete the DNA purification. λ plate stocks are routinely diluted with an equal volume of sterile distilled water before pouring on the DE-52 column. We have not found a coprecipitant to be necessary for efficient λ DNA recovery. The resulting DNA is suitable for restriction mapping and sequencing sites of Tn5supF insertion. The protocol for restriction mapping depends on the target DNA and the resolution needed, but can involve BamHI, which cleaves near each end of Tn5supF, either alone or in combination with other enzymes that do not cleave Tn5supF.[8]

For double-stranded chain termination sequencing using modified T7 DNA polymerase[25] (Sequenase; U.S. Biochemical Corp.), add 2–5 pmol of the appropriate Tn5supF-specific (or other) primer to 0.3–0.4 μg of λ DNA in a 10-μl volume containing sequencing buffer (from the Sequenase kit). Heat denature the mix in boiling water for 5 min, quench on ice for

[24] C. Helms, J. E. Dutchik, and M. V. Olson, this series, Vol. 153, p. 69.
[25] S. Tabor and C. C. Richardson, Proc. Natl. Acad. Sci. U.S.A. **84**, 4767 (1987).

10 min, and then carry out sequencing reactions as recommended by the supplier.

Large numbers of DNA samples could also be prepared for restriction analysis using a minilysate method, in which all steps (growth of individual stocks, phage harvests, and DNA extractions) are carried out in wells of microtiter plates.[26] The resulting DNA should also be a good template for linear amplification (cycle) sequencing (below).

Linear Amplification (Cycle) Sequencing for Analysis of Uncloned Polymerase Chain Reaction Fragments

This sequencing method[27-29] entails multiple cycles of DNA synthesis from one end-labeled primer in the presence of chain-terminating nucleotides, using a thermostable DNA polymerase such as AmpliTaq. The reaction is simple and easy to carry out on many samples. Very little template is needed, because of the manyfold (albeit linear) amplification (even aliquots of unpurified phage plaques can be used[30]). The reaction is quite insensitive to potential secondary structures, because most are melted at the high temperatures used for primer extension (72°). It is also insensitive to the "strong stops" seen when *Taq* polymerase is used for just one cycle, because free 3' ends that result from chance dissociations of Taq can prime further synthesis in later cycles. The use of cycle sequencing to analyze PCR products allows high-quality data to be obtained without the need for cloning, and thus without the risk of a mutant sequence ladder, or the need to isolate a specific single strand after amplification.

Purification of Amplified DNA

Polymerase chain reaction-amplified DNA is purified using the Geneclean kit: (1) mix an aliquot (10–20 μl) of PCR products with a threefold larger volume of the NaI solution; (2) leave this mixture at room tempera-

---

[26] Y. Kohara, K. Akiyama, and K. Isono, *Cell* **50**, 495 (1987).

[27] V. Murray, *Nucleic Acids Res.* **17**, 8889 (1989).

[28] M. Craxton, *Methods: Companion Methods Enzymol.* **3**, 20 (1991).

[29] B. R. Krishnan, R. W. Blakesley, and D. E. Berg, *Nucleic Acids Res.* **19**, 1153 (1991).

[30] The linear amplification protocol[28] also allows a few hundred base pairs of sequence to be read from primer-binding sites using DNA from phage plaques without any purification steps.[29] For direct sequencing from λ plaques, suspend the plaque in 10 μl of sterile distilled water, vortex vigorously, and incubate at 37° for 15 min. Centrifuge the suspension at 13,000 g for 1 min in a microfuge, transfer the supernatant to a fresh tube, and use 2 μl for linear amplification sequencing, as above. Partially purified λ DNAs[24,26] can also be used for linear amplification sequencing when ladders longer than those obtained from plaques are needed.

ture for 5 min; (3) add 5 μl of "glassmilk" slurry; (4) wash the slurry-bound DNA twice with 2 vol of the "NEW" solution (NaCl, ethanol, water); (5) suspend the slurry in 20 μl of distilled water and incubate at 55° for 5 min to elute bound DNA; (6) centrifuge the slurry at 13,000 g for 1 min, and recover the supernatant DNA solution.

Alternatively, if multiple PCR products are present, separate the products by electrophoresis in an 0.8% agarose gel. Stain DNA with ethidium bromide and visualize it with long-wavelength ultraviolet (UV) light. Cut the DNA band of interest from the gel using a razor blade, dissolve the slice of agarose in the NaI solution at a ratio of 1 mg agarose per 300 μl of NaI solution, and purify the DNA in it with the Geneclean glass slurry, as above.

Protocol for Cycle Sequencing of Polymerase Chain Reaction Products

To carry out sequencing with one of the primers used for PCR amplification, 5' end label 2 pmol of this oligonucleotide with a threefold molar excess of $[\gamma\text{-}^{32}P]ATP$ using T4 polynucleotide kinase in a 10-μl volume[19] and use it without purification. Set up 10-μl sequencing reactions containing (1) ~5–50 ng of purified PCR fragment (depending on fragment size), (2) 5 μM concentration of each of the four dNTPs, (3) 0.5 mM ddATP, ddCTP, or ddTTP (for the A, C, and T reactions respectively) or 0.1 mM ddGTP (for the G reaction), (4) PCR-sequencing buffer, and (5) 1 U of AmpliTaq DNA polymerase. Alternatively, a double-stranded DNA cycle sequencing kit (Life Technologies, Inc.), containing components 2–5, above, can be used. Overlay the reaction mixes with mineral oil and transfer to a thermal cycler programmed as follows: 1 cycle (95°, 2 min), then 30 cycles [95°, 1 min; 53° (or other annealing temperature appropriate for the T_m of the primer), 1 min; 72°, 2 min]. Add 2 μl of STOP solution, heat to 95° for 1 min, load 4 μl from each reaction on a sequencing gel, and electrophorese. Expose X-ray film to gel without drying.

Results and Discussion

Frequency of Tn5supF Insertion

The frequencies of Tn5supF insertions recovered in λ phage clones grown on Tn5supF-donor strain DB4496, and selected by plaque formation on the $dnaB_{amber}$ strain DK21, varied 5- to 10-fold, depending on the phage clone used as target. For example, the frequencies were ~2 × 10^{-7} and ~10^{-8} in λ656 cysQ::kan and λ138, respectively (phages that carry different segments of the E. coli chromosome[14,26]). λ656 plaques are larger

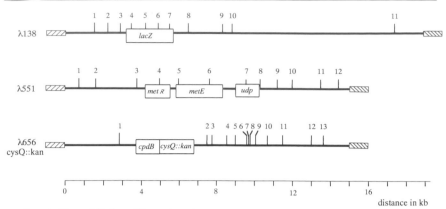

FIG. 4. Sites of Tn5supF insertions in λ phage used in present experiments. [Reproduced from B. R. Krishnan, D. Kersulyte, I. Brikun, C. M. Berg, D. E. Berg, *Nucleic Acids Res.* **19,** 6177 (1991).] The λ phages used contain cloned *E. coli* chromosomal DNAs: λ138 (10A6), λ551(1C10); and a derivative of λ656 (5B5) that carries a *kan* insertion at a central location [Y. Kohara, K. Akiyama, and K. Isono, *Cell* **50,** 495 (1987); Y. Kohara, *in* "The Bacterial Chromosome" (K. Drlica and M. Riley, eds.), p. 29. Am. Soc. Microbiol., Washington, D.C., 1990; S. Kulakauskas, P. M. Wikström, and D. E. Berg, *J. Bacteriol.* **173,** 2633 (1991)]. Sites of insertion were determined by direct and cross-over PCR methods as described in the text. Left (▨) and right (▧) arms, respectively, of the λ vector.

than λ138 plaques on standard host strains; because all λ::Tn5supF plaques are small on DK21, differences in vigor of phage growth may explain differences in the efficiency of recovery of insertions. λ::Tn5supF phage are easier to detect on medium containing X-Gal, where they form blue plaques.[22]

Mapping Tn5supF Insertions by Direct Polymerase Chain Reaction

Tn5supF insertions can be mapped by traditional restriction analyses. Because PCR-based mapping methods should generally be more efficient (Fig. 1), we have been developing methods for long-distance PCR amplification from phage plaques to expedite transposon-based sequencing. The transposon insertions shown in Fig. 4 were mapped during these experiments, and they were used to further improve the PCR mapping method.

If amplification for at least ~10 kb (half the length of the largest segments cloned in typical λ vectors[19]) was routine, all useful Tn5supF insertions in a λ clone could be mapped in one attempt, using primers that match Tn5supF ends and λ vector sequences adjacent to the cloning site. The appearance and size of an amplification product would indicate the position and orientation of the transposon (Fig. 1A and B). Because most

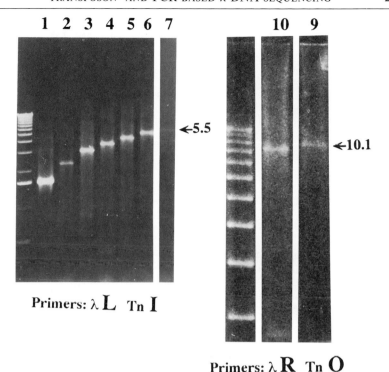

Primers: λ L Tn I

Primers: λ R Tn O

FIG. 5. Products of representative direct PCR amplifications by AmpliTaq DNA polymerase. [Reproduced in part from B. R. Krishnan, D. Kersulyte, I. Brikun, C. M. Berg, and D. E. Berg, *Nucleic Acids Res.* **19,** 6177 (1991).] Numbers above each lane indicate sites of Tn*5supF* insertion in λ138 (see Fig. 4). The 1-kb molecular weight standard (Life Technologies, Inc., Gaithersburg, MD) was used as size standards in electrophoresis (fragment sizes: 12.2, 11.2, 10.2, 9.15, 8.15, 7.13, 6.1, 5.1, 4.1, 3.1, 2.0, 1.64, 1.02, 0.52, 0.40, 0.35, 0.30, 0.20, 0.13, and 0.08 kb).

current PCR protocols were developed for amplification of, at most, a few kilobases, or have used purified DNA,[3] we sought to optimize conditions for amplification over long distances from single phage plaques. Using the methods described above and the transposon insertions diagrammed in Fig. 4, up to 6 kb was usually amplified with AmpliTaq, Hot Tub, Vent, and *Pfu* DNA polymerases (more than 80% of attempts); on occasion, amplification proceeded for up to 9 kb with Hot Tub, 10 kb with AmpliTaq, and 12 kb with Vent polymerase. Representative results are shown in Figs. 5 and 6.

These best-case amplifications were not obtained reproducibly. Enzyme aliquots from different batches, and also from different vials of the

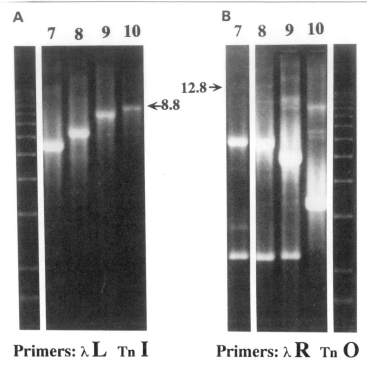

Primers: λ L Tn I   **Primers: λ R Tn O**

FIG. 6. True and spurious products formed during long-distance direct PCR amplification with Vent DNA polymerase. [Reproduced from B. R. Krishnan, D. Kersulyte, I. Brikun, C. M. Berg, and D. E. Berg, *Nucleic Acids Res.* **19,** 6177 (1991).] Numbers above lanes indicate sites of Tn5*supF* insertion in λ138 (Fig. 4). The largest amplification products from the left and right ends of the cloned DNA (8.8 and 12.8 kb, respectively) are indicated. Other true products of amplification from left end of segment in λ138 were 6.4, 8.2, and 8.8 kb, and products of amplification from right end were 11.9, 10.1, and 9.5 kb. The predominant bands at ~5.9, 5.8, 5.0, and 3.2 kb in lanes 7 through 10, respectively (right), may correspond to internal deletion products; the bands at 2.2 kb in lanes 7, 8, and 9 (right) probably resulted from false priming. The size standards were the same as used in Fig. 5.

same batch of AmpliTaq, differed in ability to amplify DNA segments in the 6- to 10-kb range, despite equivalent amplification of 1- to 2-kb segments. Variability was also seen in less extensive tests with Hot Tub and Vent polymerases. Attempts to amplify more than 8–10 kb sometimes resulted in spurious shorter DNA fragments (fast-migrating bands) (Fig. 6). These bands were more abundant with Vent than with AmpliTaq or Hot Tub polymerases. Further tests showed that the spurious bands generated by AmpliTaq and Vent polymerases on the same template dif-

A
B

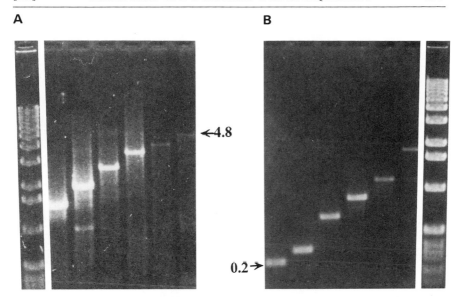

FIG. 7. Cross-over PCR amplification. [Reproduced from B. R. Krishnan, D. Kersulyte, I, Brikun, C. M. Berg, and D. E. Berg, *Nucleic Acids Res.* **19**, 6177 (1991).] (A) Long-distance cross-over PCR amplification (here, up to 4.8 kb), using mixtures of Tn5*supF* insertion derivatives of λ138; (B) Short-distance cross-over PCR amplification (down to 0.2 kb). The insertions in λ138 used, and product sizes in (A) are (from left to right): 1 and 3 (1.5 kb), 1 and 4 (2.0 kb), 1 and 5 (2.8 kb), 1 and 6 (3.5 kb), 1 and 7 (4.1 kb), and 1 and 8 (4.8 kb). Similarly, the insertions in λ656 *cysQ::kan* and product sizes in (B) are 7 and 5 (0.2 kb), 7 and 8 (0.3 kb), 7 and 4 (0.6 kb), 7 and 9 (0.8 kb), 7 and 3 (1.1 kb), and 7 and 10 (1.8 kb). The size standards were the same as used in Fig. 5.

fered in size, which probably reflects the different specificities of these polymerases. High-titer phage stocks or purified phage DNA gave results equivalent to those obtained with plaques; they did not increase the length of DNA that was amplified, decrease the variability in success of the longest amplifications, or affect the accumulation of spurious fragments. Spurious bands probably arise by false priming, and by internal deletion due to slippage of nascent and template strands during amplification from the intended primer-binding sites. In either case, the abundance of these spurious fragments must also reflect a selection for shorter DNAs that will operate during attempts to amplify very long DNA segments. Possible mismapping due to spurious long-distance PCR products can be avoided by cross-over PCR mapping against another insertion (Fig. 1C; below), or by use of another thermostable polymerase (which gives different patterns of spurious fragments).

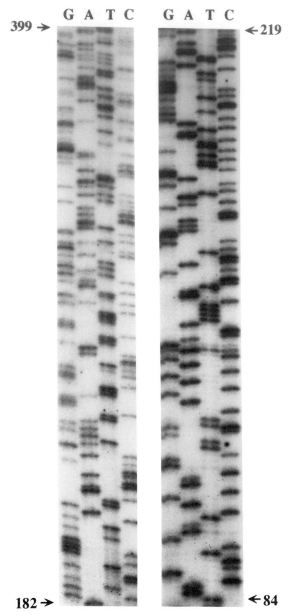

FIG. 8. Autoradiograph illustrating linear amplification DNA sequencing of cross-over PCR products. [Reproduced from B. R. Krishnan, D. Kersulyte, I. Brikun, C. M. Berg, and D. E. Berg, *Nucleic Acids Res.* **19,** 6177 (1991).] The template was generated by cross-over PCR using aliquots of plaques of Tn*5supF*-containing derivatives 4 and 5 of phage λ138 (Fig. 4). Sequencing was carried out with the Tn*5supF* I primer. Numbers indicate the distance in base pairs from the 3′ end of the primer, and correspond to *E. coli lacZ* positions 1064 (84), 935 (219), 964 (182), and 747 (399) [A. Kalnins, K. Otto, U. Rüther, and B. Müller-Hill, *EMBO J.* **2,** 593 (1983)].

Cross-Over Polymerase Chain Reaction for Mapping and Template Generation

Although amplification usually proceeds from a single DNA species, hybrid products also arise due to priming at sites on different molecules (cross-over PCR; Fig. 2). This reflects the relatively low processivity of commonly used thermostable DNA polymerases (e.g., *Taq* DNA polymerase incorporates ~60–100 nucleotides per single extension, but reinitiates synthesis many times during single amplification cycles[31]). The melting and reannealing steps allow strands partially extended on one template to prime further synthesis on other template strands, thereby generating a hybrid product. Although cross-over PCR has sometimes been regarded as a nuisance, as in the amplification of single genes from heterozygous diploid individuals, it has also been exploited to construct specific recombinant genes *in vitro*.[32,33] We have exploited cross-over PCR to (1) map insertions near the center of cloned DNA segments, beyond the reach of direct PCR using λ vector primers, (2) verify and refine map locations inferred from direct PCR, and (3) generate DNA sequencing templates.

Cross-over PCR amplification proceeded for up to ~3 kb routinely, and up to 6 kb on occasion. At the other extreme, cross-over PCR products between sites as close as 200 bp were also obtained, and each of the 21 intervals of <1 kb tested gave cross-over PCR products (Fig. 7). This indicates that no special rare sites are needed to trigger the jumping between templates that results in cross-over PCR amplification.

DNA Sequencing Using Direct and Cross-Over Polymerase Chain Reaction Products as Templates

The products of direct PCR obtained with an array of Tn*5supF* insertions and transposon- and vector-specific primers can be used to sequence different portions of one target DNA strand systematically (Fig. 1B). It had seemed that cross-over PCR products should be more useful: those products formed with insertions in the same orientation have different left and right ends, corresponding to the two Tn*5supF* ends (Fig. 1C), and could be sequenced from each end using the two Tn*5supF* primers.

---

[31] R. D. Abramson, P. D. Holland, R. Watson, and D. H. Gelfand, *FASEB Meet. Abstr.* No. 386 (1991).

[32] R. M. Horton, H. D. Hunt, S. N. Ho, J. K. Pullen, and L. R. Pease, *Gene* **77**, 61 (1989).

[33] A. A. Yolov and Z. A. Shabarova, *Nucleic Acids Res.* **18**, 3983 (1990).

To test this application of cross-over PCR, we sequenced uncloned pools of cross-over PCR products using the Tn5supF-specific primers (after quick DNA purification by adsorption to the Geneclean glass slurry). Trials with DNA fragments amplified using Tn5supF insertions in *lacZ* gave about 400 bp of readable sequence from each primer site (e.g., see Fig. 8); the 2 kb of sequence read in these tests matched the *E. coli lacZ* sequence perfectly. This result indicates that the jumping between, and copying of, parental templates during cross-over PCR occurs with fidelity sufficient for sequencing pooled (uncloned) PCR products. Cross-over and direct PCR products were equal to or better than highly purified λ DNA as templates for DNA sequencing. Their use gave easy access to each DNA strand.

We have found that more than 90% of Tn5supF insertions in cloned DNA segments were in the orientation that results in transcription from the *supF* and λ P_L vector promoters in the same direction. This implies that cross-over PCR products generated with most pairs of insertions should be useful as sequencing templates.

Concluding Remarks and Perspectives

Transposon- and PCR-based methods for sequencing DNAs cloned in λ phage vectors hold great promise for many DNA sequencing projects. Three sets of steps are involved, each of which is automatable: (1) transposon insertion, (2) direct and cross-over PCR from transposon-containing phage to map insertions and generate sequencing templates, and (3) linear amplification sequencing of appropriate PCR fragments. This strategy yields accurate high-resolution sequence ladders and is cost- and labor-efficient. More development is needed, however, to realize the full potential of this transposon- and PCR-based approach. The two most serious current limitations are (1) an undersupply of Tn5supF insertions in certain target regions, and (2) an inability to reproducibly PCR amplify over long enough distances to map all insertions of interest in λ clones in one attempt. Each of these limitations will probably be overcome in the near future. A solution to the specificity problem may emerge from the combined use of Tn5supF and an analogous γδ-*supF* element that is under development. Tn5 and γδ insert with different low specificities, and hence their joint distribution in λ clones should be much closer to random than the pattern obtained using either element alone. The nonrandomness will become less important as the length of readable sequence ladders improves. With respect to PCR, because amplification for >10 kb from phage plaques is now seen on occasion, we can expect that methodological improvements will soon make such long-distance amplification routine, and thereby allow all insertions of interest to be mapped in one attempt. The improvements may come variously from the development of better conditions for purification, storage, or use of existing thermostable polymerases, discoveries of new polymerases and cofactors, or their creation by protein engineering. Sets of transposon insertions, such as described here, should be useful in developing these conditions: they allow the amplification of templates of diverse sequence and length from a small number of universal primers, using conditions that are appropriate for high-throughput sequence analyses. In the interim, the best general sequencing approach entails transposon insertion, PCR mapping, sequencing from the ends of cross-over PCR fragments generated using appropriate Tn5supF insertions, and then sequencing interior portions of these fragments, perhaps using small preformed libraries of oligonucleotide primers[34] or PCR products generated with arbitrary primers.[35,36] Similar combinations of transposition, PCR amplification, and primer walking should become increasingly valuable for analyses of the larger DNAs cloned in cosmid and other plasmid vectors.

Acknowledgments

We thank Drs. R. W. Blakesley, D. H. Gelfand, K. Isono, S. H. Phadnis, E. A. Rose, and J. A. Sorge for stimulating discussions, and Perkin-Elmer Cetus Instruments for the loan of a model TC480 thermal cycler. This work has been supported by Grants DEFGO2-89ER60862 and DEFGO2-90ER610 from the U.S. Department of Energy, and GM-37138 and HG-00563 from the U.S. Public Health Service.

[34] F. W. Studier, *Proc. Natl. Acad. Sci. U.S.A.* **86**, 6917 (1989).
[35] J. Welsh and M. McClelland, *Nucleic Acids Res.* **19**, 861 (1991).
[36] I. Brikun and D. E. Berg, unpublished observations, 1991.

[20] Transposon-Facilitated Sequencing of DNAs Cloned in Plasmids

By CLAIRE M. BERG, GAN WANG, LINDA D. STRAUSBAUGH, and DOUGLAS E. BERG

Introduction

Most strategies for sequencing genomic DNA involve an initial *in vitro* cloning into large-capacity vectors in yeast or *Escherichia coli,* and one or more rounds of shotgun subcloning in *E. coli* before the sequences of short segments are determined from "universal" primer binding sites in the vector. These short segments are then aligned with the help of computer programs that identify sequence overlaps. Primer walking, using oligonucleotides that match the 3' end of the newly determined sequences, can reduce the amount of subcloning required or fill gaps. Difficulties inherent in both shotgun subcloning and primer walking methods stimulated us and others to develop bacterial transposons as tools to insert universal primer binding sites throughout cloned target DNA.[1-5] Transposon insertion does not fragment or rearrange target DNA; thus the λ, cosmid, or plasmid targets that contain a transposon inserted at one site retain the linkage information that is lost during subcloning. Transposon insertion sites can be readily mapped, so redundant sequencing can be avoided, and gap closure, if necessary, is simplified. In addition, problems associated with sequencing repetitive DNA are minimized.

Ideally, transposon insertion sites should be evenly spaced with a minimum of hot spots or blank regions. We have chosen to concentrate on two unrelated transposons, γδ and Tn5, because they insert more randomly into plasmids in *E. coli* than do other well-studied transposons.[6,7] γδ and Tn5 transpose by different mechanisms and with different specificities. γδ has been most widely used for plasmid mutagenesis, in part because it inserts preferentially into plasmids and can be easily delivered by simple bacterial matings. Tn5 has been widely used to mutagenize chromosomal DNA and DNA cloned in bacteriophage λ, but is also useful for plasmid mutagenesis.[6,7]

We describe here some genetically engineered derivatives of γδ and Tn5, and present protocols for transposon mutagenesis of plasmids and for mapping insertions to be used for DNA sequencing. A derivative of Tn5 specifically designed for mutagenesis of λ is described in an accompanying

[1] A. Ahmed, this series, Vol. 155, p. 177.
[2] T. Adachi, M. Mizuuchi, E. A. Robinson, E. Appella, M. H. O'Dea, M. Gellert, and K. Mizuuchi, *Nucleic Acids Res.* **15,** 771 (1987).
[3] L. Liu, W. Whalen, A. Das, and C. M. Berg, *Nucleic Acids Res.* **15,** 9461 (1987).
[4] D. K. Nag, H. V. Huang, and D. E. Berg, *Gene* **64,** 135 (1988).
[5] S. H. Phadnis, H. V. Huang, and D. E. Berg, *Proc. Natl. Acad. Sci. U.S.A.* **86,** 5908 (1989).
[6] C. M. Berg and D. E. Berg, *in* "*Escherichia coli* and *Salmonella typhimurium*—Cellular and Molecular Biology" (F. C. Neidhardt, J. L. Ingraham, K. B. Low, B. Magasanik, M. Schaechter, and H. E. Umbarger, eds.), p. 1071. Am. Soc. Microbiol., Washington, D.C., 1987.
[7] C. M. Berg, D. E. Berg, and E. Groisman, *in* "Mobile DNA" (D. E. Berg and M. M. Howe, eds.), p. 879. Am. Soc. Microbiol., Washington, D.C., 1989.

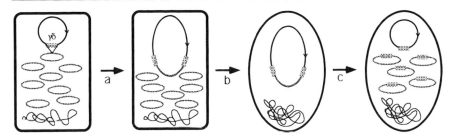

FIG. 1. Mutagenesis by γδ and mini-γδ. (a) Transposition from the donor F factor (for wild-type γδ) or pOX38::mini-γδ (for mini-γδ) to a nonconjugative plasmid, such as pBR322, forming an F::plasmid or pOX38::plasmid cointegrate. (b) Conjugation with a plasmid-free F⁻ recipient strain, selecting for a plasmid-borne marker (usually Ampʳ). (c) Resolution of the cointegrate in the recipient to yield a plasmid containing one copy of γδ or mini-γδ (e.g., pBR322:γδ or pBR322::mini-γδ), and reforming the donor molecule (F or pOX38::mini-γδ), followed by growth of the transconjugants.

chapter.[8] Other transposons that have been used for sequencing, as well as aspects of γδ and Tn5 biology not covered here, are reviewed elsewhere.[6-12]

Principles of Method

The final product of transposition to a target plasmid by both γδ and Tn5 is a simple insertion of the element, bracketed by short direct repeats of target DNA (5 bp for γδ and 9 bp for Tn5). However, γδ transposes by a two-step replicative process (involving a cointegrate intermediate) (Fig. 1), whereas Tn5 transposes by a one-step conservative (break and join) process (Fig. 2). Conjugation (mating, involving transfer of the γδ-target cointegrate from the donor to the recipient bacterial cell) can be used to easily select plasmids that have a γδ insertion. Plasmids that have a Tn5 insertion are usually selected by infecting plasmid-containing cells with a λ::Tn5 suicide vector.

[8] B. R. Krishnan, D. Kersulyte, I. Brikun, H. V. Huang, C. M. Berg, and D. E. Berg, this volume [19].

[9] M. S. Guyer, this series, Vol. 101, p. 362.

[10] N. D. F. Grindley and R. R. Reed, Annu. Rev. Biochem. 54, 863 (1985).

[11] D. E. Berg, in "Mobile DNA" (D. E. Berg and M. M. Howe, eds.), p. 185. Am. Soc. Microbiol., Washington, D.C., 1989.

[12] D. Sherratt, in "Mobile DNA" (D. E. Berg and M. M. Howe, eds.), p. 163. Am. Soc. Microbiol., Washington, D.C., 1989.

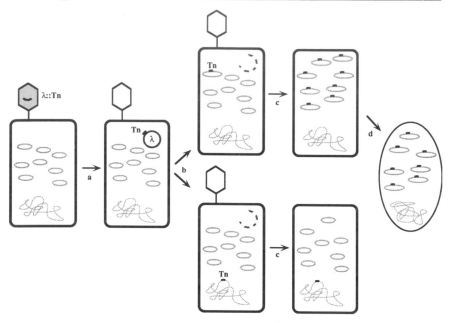

FIG. 2. Mutagenesis by Tn5 or an engineered derivative. (a) Infection of a plasmid-containing strain by λ::Tn5. (b) Transposition of Tn5 from λ to a plasmid (top) or to the chromosome (bottom), with degradation of the donor λ molecule. (c) Growth of pooled Kan^r cells generated by Tn5seq1 or Tn5tac1 transposition. (d) Transformation of a Kan^s strain using plasmid DNA from pooled Kan^r survivors of λ::Tn5 infection and selection for Kan^r transformants, followed by growth of the transformants.

γδ Mutagenesis

The easiest way to generate γδ insertions in plasmids exploits the presence of γδ in the *E. coli* K12 F (fertility) factor. Cells containing both F and a nonconjugative cloning vector (such as pBR322) transfer the plasmid to recipient cells during mating as part of the F-plasmid cointegrate that is produced by γδ transposition. The cointegrate is resolved in the recipient cell to yield the plasmid containing a single copy of γδ.[9,13,14] Selection for insertion-containing plasmids is technically simple and straightforward, and has been widely used for γδ insertion mutagenesis (Fig. 1). It should be noted that this procedure involves selection for a marker on the target plasmid, not on γδ (in fact, wild-type γδ does not contain a selectable marker).

[13] M. S. Guyer, *J. Mol. Biol.* **126**, 347 (1978).
[14] L. Liu and C. M. Berg, *J. Bacteriol.* **172**, 2814 (1990).

TABLE I
PRIMERS AND PROBES FOR TRANSPOSON-FACILITATED SEQUENCING

Transposon	Site, position	Primer/probe sequence	Ref.
Wild-type γδ	γ end, bp 53–37	5'-TCAATAAGTTATACCAT	(3)
(primer)	δ end, bp 48–32	5'-GAATTATCTCCTTAAACG	
mγδ-1	res end, bp L75–L59	5'-GTAGGGAGCCTGATATG	(18)
(primer)	kan end, bp R87–R71	5'-GCTATCCGCGCATCCAT	
mγδPLEX[a]	Primer A, bp 375–358	5' CACAGGACGGGTGTGGTC	(19)
(primer)	Primer B, bp 381–396	5'-TCTACGCCGGACGCATG	
mγδPLEX-1	Tag 1E (E10)	5'-TATATATAGGGTATTAGGTG	(30–32)
(probe[b])	Tag 1P (P10)	5'-TGAGTATATTGATGATTAGG	
mγδPLEX-3	Tag 3E (E19)	5'-AGAAGTTAATGTAGGGTTGG	(30–32)
(probe)	Tag 3P (P19)	5'-GTGATAAGTAGAGTTGGTTG	
mγδPLEX-4	Tag 4E (E04)	5'-AGTTTAGTGGATGGGTGTTG	(30–32)
(probe)	Tag 4P (P04)	5'-AGTGTGAGGTTTAAATATTG	
mγδPLEX-5	Tag 5E (E05)	5'-TTTGAGATTTGAGTTATGTG	(30–32)
(probe)	Tag 5P (P05)	5'-AGGGTTTAGGTTATATTATG	
Tnsseq1	T7	5'-TAATACGACTCACTATAGGGG	(4)
(primer)	SP6	5'-CATACGATTTAGGTGACACTATAG	
	O end	5'-GATCCTACTTGTGTATAAGAGTCAGGGTAC	
Tn5supF	I end	5'-TAGGATCCCGAGATCTGATC	(5)
(primer)	O end	5'-TAGGATCCCCTACTTGTGTA	
Tn5tac1	I end	5'-CAGAATTCCCGGGGATCCCC	(37)
(primer)	O end	5'-GATAAGCTGTCAAAC	

[a] The "universal" primer sites for these plasmids come from the indicated pBR322 sequences in tet, except that the 3' G of primer B is the first base of the adjacent NotI site.

[b] The tag (probe) sequences are a subset of those developed previously.[30] The tag numbers are those assigned by a commercial supplier of chemiluminescent tags,[32] and the parenthetical numbers are those assigned by Church and Kieffer-Higgins.[30] The sequences given are those used to probe the products of dideoxy sequencing, which are complementary to the published tag sequences.[30]

The inverted repeats at the ends of wild-type γδ are only 35 bp in length; internal to these ends are unique sequences that can be used as primer-binding sites for probe (blot) hybridization, polymerase chain reaction (PCR) amplification, and DNA sequencing (Table I). Our initial demonstration of transposon-facilitated DNA sequencing used wild-type γδ to provide primer-binding sites,[3] as have implementations of the method,[15–17] but γδ is now being supplanted by genetically engineered mini-γδ derivatives[18–20] (see Materials, below).

[15] L. D. Strausbaugh, M. T. Bourke, M. T. Sommer, M. E. Coon, and C. M. Berg, Proc. Natl. Acad. Sci. U.S.A. 87, 6213 (1990). Probe mapping is essentially a blot analysis to map insertion sites in plasmid DNA.

[16] S. M. Thomas, H. M. Crowne, S. C. Pidsley, and S. G. Sedgwick, J. Bacteriol. 172, 4979 (1990).

[17] M. Strathmann, B. A. Hamilton, C. A. Mayeda, M. I. Simon, E. M. Meyerowitz, and M. J. Palazzolo, Proc. Natl. Acad. Sci. U.S.A. 88, 1247 (1991).

[18] C. M. Berg, N. V. Vartak, G. Wang, X. Xu, L. Liu, D. J. MacNeil, K. M. Gewain, L. A. Wiater, and D. E. Berg, Gene 113, 9 (1992).

The wild-type F factor cannot be used as a delivery vehicle for genetically engineered mini-γδ derivatives because it already contains wild-type γδ. Rather, a deletion derivative of F that lacks γδ but remains conjugation proficient (such as pOX38[21]) is converted into an "F" factor analog" by introducing the mini-γδ element into it (by transposition). The resultant pOX38::mini-γδ plasmid is then used, like the wild-type F factor, to deliver the transposon to target plasmids by mating (Fig. 1).

Existing mini-γδ derivatives do not contain genes for transposition or cointegrate resolution. Therefore, transposase must be supplied in the donor, and resolvase in the recipient. Transposase is supplied in the donor from a gene cloned in a multicopy plasmid that is compatible with F and pBR322; resolvase is supplied either from a gene cloned in another F- and pBR322-compatible plasmid in the recipient or from wild-type γδ elements resident in the chromosome of some *E. coli* strains. Although most mini-γδ elements encode antibiotic resistance, this resistance cannot be used after mating because the donor plasmid (pOX38::mγδ) also carries the resistance gene, and it is transferred to the recipient as part of the donor–target plasmid cointegrate. Cointegrate transfer is selected simply, however, using the resistance (usually *amp*) of the target plasmid, as is done with wild-type γδ (Fig. 1). The resistance genes in mini-γδ derivatives become useful later for monitoring the presence of the transposon in the plasmid, for subcloning, and for *in vivo* replacement of chromosomal DNA by the insertion-tagged fragment.[18]

Tn5 Mutagenesis

The easiest way to obtain Tn5 insertions in plasmids is to use a Tn5-containing λ phage vector that does not integrate efficiently or replicate in the strain harboring the target plasmid (Fig. 2). Tn5 and all derivatives used for insertion into plasmids contain one or more antibiotic resistance genes. Antibiotic-resistant survivors of λ::Tn5 infection that are λ free can be readily obtained on selective medium. Because Tn5 transposes to both the chromosome and to plasmids, most antibiotic-resistant survivors will have chromosomal insertions. Cells containing Tn5 on a multicopy plasmid can be selected on the basis of their resistance to higher levels of kanamy-

[19] G. Wang, X. Xu, J.-M. Chen, C. Chin, D. E. Berg, L. D. Strausbaugh, and C. M. Berg, in preparation.

[20] R. Weiss and R. Gesteland, personal communication, 1991.

[21] M. S. Guyer, R. R. Reed, J. A. Steitz, and K. B. Low, *Cold Spring Harbor Symp. Quant. Biol.* **45,** 135 (1980).

cin (or neomycin) than those with chromosomal Tn5,[22,23] but it is more efficient to isolate plasmid DNA from cells that had been infected by λ::Tn5 and transform an antibiotic-sensitive strain.

The inverted repeats of wild-type Tn5 are 1.5 kb long; thus wild-type Tn5 does not contain unique sites close enough to its ends for it to be useful for double-stranded sequencing without subcloning. However, only about 19 bp at each end of Tn5 and the Tn5-encoded transposase are required for transposition. The transposase gene must be in cis for transposase to work efficiently, but it can be cloned outside of Tn5.[5,8] Many Tn5 derivatives with unique subterminal sequences have been constructed[7] and some, with special features useful for genetic manipulation and analysis, are described under Materials (below).

Randomness of Transposition

The relative randomness of γδ and Tn5 transposition was the major consideration in choosing to develop these elements as sequencing tools. Wild-type γδ transposes quite randomly into most plasmids.[7,9] Tn5 and its derivatives also insert into many sites, but a few sites of preferential insertion (hot spots) are found superimposed on this background.[11] Curiously, γδ transposes poorly into the *tet* gene in several pBR322-related plasmids,[14,24,25] but its mγδ-1 derivative does not discriminate against *tet* in the same plasmids.[18,26] The factors responsible for this regional specificity of wild-type γδ, but not of mγδ-1, are not known, but this observation suggests that mini-γδ elements may be better than wild-type γδ for sequencing plasmid clones. No differences have been detected between the insertion pattern of wild-type Tn5 and engineered Tn5 derivatives.

The 5-bp target duplication at the γδ insertion site is usually AT rich,[3,15,27] but no consensus sequence is evident.[28] This preference for AT richness raised concerns that γδ might insert poorly into GC-rich DNA, which is abundant in eukaryotic and many prokaryotic genomes. However, γδ inserted quite uniformly throughout a cloned *Drosophila* fragment containing distinct GC-rich and AT-rich regions.[15] In addition, mγδ-1 inserted readily into a cloned *Streptomyces* fragment[18] (*Streptomyces* DNA

[22] D. E. Berg, M. Schmandt, and J. B. Lowe, *Genetics* **105**, 813 (1983).

[23] D. E. Berg and C. M. Berg, *Bio/Technology* **1**, 417 (1983).

[24] L. A. Wiater and N. D. F. Grindley, *J. Bacteriol.* **172**, 4959 (1990).

[25] A. Schwacha, J. A. Cohen, K. B. Gehring, and R. A. Bender, *J. Bacteriol.* **172**, 5991 (1990).

[26] X. Xu, G. Wang, and C. M. Berg, in preparation.

[27] R. R. Reed, R. A. Young, J. A. Steitz, N. D. F. Grindley, and M. S. Guyer, *Proc. Natl. Acad. Sci. U.S.A.* **76**, 4882 (1979).

[28] J. Zhang, L. D. Strausbaugh, and C. M. Berg, in preparation.

FIG. 3. Map of mγδ-1 insertions in pVE1322.[18] The thin line between the left and middle BamHI (B) sites depicts cloned *Streptomyces* DNA, the heavy line between the EcoRI (E) site and the bar (pBR322 DNA) depicts a selectable *Streptomyces* gene present in the vector. In the pBR322 DNA the stippled region is the origin and the filled-in region the *amp* gene. The thin vertical lines indicate single insertions and the heavy lines two or more insertions (if more than two, the number is given). Insertions above the line are in one orientation and those below are in the opposite orientation. Four insertions were mapped to either of two sites indicated by the heavy dashed line. An additional restriction enzyme digestion would have been required to resolve their positions.

is about 70% G + C) (Fig. 3). This seeming paradox was resolved by the finding that γδ inserts preferentially into local AT-rich sites, or "AT valleys" (that need be only a few base pairs long)[15,28] (Fig. 4), which are abundant, even in GC-rich DNAs.

Most of the 9-bp target duplications at Tn5 insertion sites have GC pairs at one or both ends. Beyond that, no consensus sequence or special feature is evident, although the use of one major hot spot in pBR322 depends on high negative supercoiling.[11,29]

Mapping Strategies

Several strategies can be used to locate and orient transposon insertions to identify the subset to be used for DNA sequencing. Restriction mapping was used initially[3,4] and is still the most widely used method, but it suffers from the need to construct a restriction map. This is time consuming, especially for large cloned fragments, and becomes less effective as fragment size increases. We have therefore been developing probe (blot) mapping[15] and PCR mapping[8] of insertion sites. Both probe and PCR mapping have the advantage that little prior knowledge of the cloned fragment is required to map insertions.

The number of random transposon insertions needed for a particular probability of complete coverage depends on the size of the plasmid and the amount of sequence read from each primer. A statistical analysis of theoretical mapping requirements is presented elsewhere,[15] but a good rule of thumb for plasmids in the range of 5–15 kb, assuming sequence runs of 250–350 bp, is to begin with a number of purified colonies equal to at least

[29] J. K. Lodge and D. E. Berg, *J. Bacteriol.* **172**, 5956 (1990).

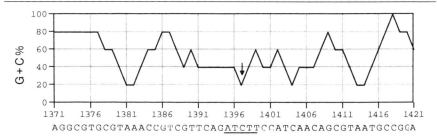

AGGCGTGCGTAAACCGTCGTTCAG<u>ATCTT</u>CCATCAACAGCGTAATGCCGCA

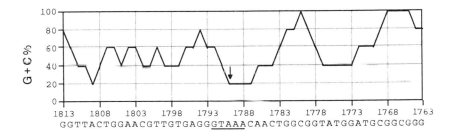

GGTTACTGGAACGTTGTGAGG<u>GTAAA</u>CAACTGGCGGTATGGATGCGGCGGG

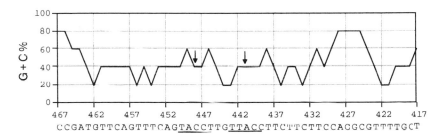

CCGATGTTCAGTTT<u>C</u>AGTACCTTG<u>TTAC</u>CTTCTTCTTCCACGCGTTTTGCT

FIG. 4. Representative γδ insertion sites in cloned DNA.[78] The G + C composition using a 5-bp window is depicted. The arrows point to the center of the 5-bp γδ target site duplication (underlined). *Top*: Insertion in the cloned *E. coli avtA* gene[3]. *Middle*: Insertion in pBR322. *Bottom*: Insertion in the cloned *E. coli alaA* gene [M.-D. Wang, L. Buckley, and C. M. Berg, *J. Bacteriol.* **169**, 5610 (1987); C. M. Berg, L. Liu, and G. Wang, unpublished observations, 1990.]

eight times the size of the plasmid (in kilobases). A fraction of these will contain insertions in the vector (dependent on how much nonessential DNA is in the vector). Of the subset of plasmids mapped to the cloned fragment, a yet smaller portion will be used to obtain overlapping sequences.

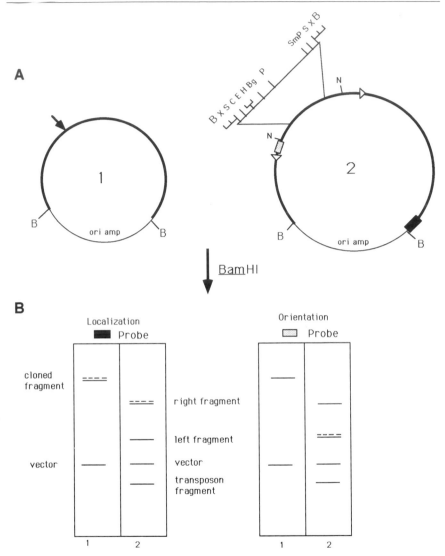

FIG. 5. Example of probe mapping of a myδPLEX insertion in a standard cloning vector (such as a pBR322 or pUC derivative) containing a fragment cloned at the *Bam*HI site (not drawn to scale). (A) Plasmid 1 (heavy line, cloned fragment; light line, cloning vector) with the site of the myδPLEX insertion in plasmid 2 designated by the arrow. Plasmid 2 is a detailed depiction of a myδPLEX insertion plasmid, illustrating relevant recognition sites in the transposon and the locations of the probe for the "right" end of the cloned fragment (filled box), which is used to map the insertion, and the probe for one end of the transposon (stippled box), which is used to orient the insertion. B, *Bam*HI; Bg, *Bgl*II; C, *Cla*I; E, *Eco*RI, H, *Hin*dIII; P, *Pst*I; R, *Rsa*I; Rs, *Rsa*II; S, *Sal*I; Sm, *Sma*I; Sp, *Sph*I; X, *Xba*I; Xh, *Xho*I.

The choice of mapping approach will depend on local expertise and equipment available, and on the size and degree of prior knowledge of the cloned fragment. For small fragments whose restriction map is already known, restriction mapping of insertions is probably most efficient. For other fragments up to the size of the maximum PCR product routinely obtainable (currently 6 kb from λ in our hands), PCR mapping may be the method of choice because PCR products are excellent sequencing substrates. In this case, two PCR reactions are set up for each insertion; both reactions contain primers specific for the transposon ends and each reaction has a unique vector-specific (or fragment-specific) primer. For insertions in small cloned fragments, both PCR reactions will yield products whose lengths will reflect the position of the transposon in the fragment. Although the orientation of the insertion is not known, each PCR product (unlike the plasmid it came from) contains only a single transposon end, which provides a unique primer-binding site. In this way cloned DNA is sequenced in each direction from one insertion using the two PCR products. Insertions in larger fragments can be mapped by a combination of probe mapping (below) and direct and "cross-over" ("biparental") PCR (whose products also serve as sequencing templates), as discussed elsewhere.[8,29a]

In essence, probe mapping is a blot analysis that uses appropriate combinations of short DNA (or RNA) hybridization probes and restriction enzyme digestion products to map transposon insertion sites against the

[29a] B. R. Krishnan, D. Kersulyte, I. Brikun, C. M. Berg, and D. E. Berg, *Nucleic Acids Res.* **19**, 6177 (1991).

(B) Results of probe mapping using *Bam*HI digestion. Each set of lanes depicts the bands identified by ethidium bromide staining (solid lines) and by hybridization (dashed lines). *Left*: Probe mapping for localization of an insertion. Lane 1 illustrates results for the parental plasmid DNA. Digestion of γδ-containing plasmid DNA with *Bam*HI produces four fragments (lane 2): two of constant size for every plasmid with an insertion in the cloned fragment (representing the intact cloning vector and an internal transposon fragment), and two variable-sized fragments. Hybridization with the probe identifies one of the variable-sized fragments as the "right" portion of the cloned DNA; its length represents the distance of the insertion point from the right cloning juncture. By definition, the other variable-sized fragment is the left portion and its length represents the distance of the insertion point from the left cloning juncture. *Right*: Probe mapping to determine orientation of the insertion. A multiplex tag probe (specific for one end of the transposon) will identify that end on one of the two variable-sized fragments; by definition, the other variable-sized fragment contains the opposite end of the transposon. Using this same logic, probe mapping is accomplished easily for fragments cloned into the *Xho*I or *Sal*I sites, and with minor modifications (digestion with an additional enzyme or hybridization with an additional probe), for those cloned into the *Cla*I, *Eco*RI, *Hin*dIII, *Sma*I, *Bgl*II, or *Pst*I sites.

cloning site (illustrated in Fig. 5). Probe mapping is especially valuable for mapping fragments in the 6- to 15-kb range, but can also be useful for mapping insertions in plasmids with larger or smaller fragments. In the case of large plasmids, efficient probe mapping generally requires use of transposons and vectors with rare restriction sites to minimize the chances of cleavage at sites in the cloned fragment (illustrated in Fig. 6).

For insertions in cloned fragments larger than 15 kb, it will be most efficient to map by a combination of probe and PCR mapping. First, low-resolution probe mapping can locate insertions approximately; then, cross-over PCR can map insertions with high resolution and the product can be used as a sequencing template. For probe mapping, which is dependent on gel electrophoresis, the resolution decreases as the fragment size increases. This technical limitation is, however, not a practical argument against mapping insertion sites in very large fragments (>20 kb), because in such fragments the insertions will most often be used to provide anchor points, and for limited sequencing. In addition, the ability to sample discrete mapped regions by sequencing from transposon insertions makes it possible to identify dispersed coding tracts in a large cloned fragment, without sequencing the entire fragment. Obviously, the number of insertions that must be mapped for such "sampling" applications will be quite small.

Materials

The strains, plasmids, phage, and media used are shown in Table II and the primers that have been used for transposon-based sequencing and PCR amplification are shown in Table I.

Small Transposon Derivatives Useful for Sequencing

$\gamma\delta$. Wild-type $\gamma\delta$, which contains unique subterminal sequences, can be used directly for sequencing.[3] Some small derivatives, which lack the genes for transposase and resolvase, are described below.

1. m$\gamma\delta$-1[18] is a 1.8-kb element that contains only the $\gamma\delta$ resolution site (*res*) and the *kan* gene from Tn5, bracketed by 40-bp repeats of the δ end of $\gamma\delta$ (Fig. 7). m$\gamma\delta$-1 appears to transpose even more randomly than wild-type $\gamma\delta$ (see above).

2. m$\gamma\delta$PLEX-1, -3, -4, and -5[19] are a set of 1.9-kb m$\gamma\delta$-1 derivatives (Fig. 8) that combine the advantages of transposon- and multiplex-based sequencing. The multiplex method was originally developed for sequencing small, randomly subcloned DNA fragments in specialized

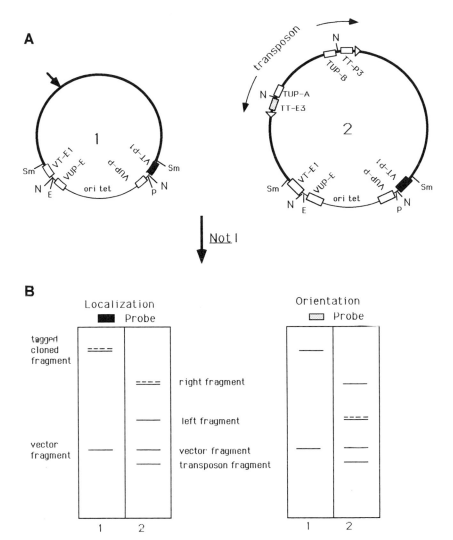

FIG. 6. An example of probe mapping a myδPLEX insertion in a compatible Plex plasmid (not to scale). As in Fig. 5, except that a Plex plasmid vector is used; the variable-sized fragments are excised with vector (VT-P1 or VT-E1) and transposon (TT-P3 or TT-E3) tag sites when the plasmid is digested by NotI. Two different "universal" primer binding sites for DNA sequencing occur in the transposon (TUP-A and TUP-B) and two in the vector (VUP-P and VUP-E). (A) Plex plasmid 1 with the site of a myδPLEX-3 insertion in plasmid in next panel designated by the arrow. (B) Results of probe mapping using NotI digestion. Left: Hybridization with a probe specific for the vector tag, VT-P1 (filled in), which identifies the "right" portion of the cloned fragment and maps the insertion. Right: Hybridization with a probe specific for the "left" transposon tag, TT-E3 (stippled), to orient the transposon. The logic used to convert hybridization results to transposon locations and orientations is as described in the caption to Fig. 5.

TABLE II
Escherichia coli K12 Strains, Plasmids, Phage, and Media

Component	Description
Strains	
CBK884	F⁻, γδ free, Δ(*srl-recA*)*306*::Tn*10*Δ(*tet*)*277* derivative of RR1; Rec⁻; sensitive to Amp, Cam, Kan, Tet, Nal; used as donor strain for mγδ mutagenesis (Strʳ)[18]
CBK904	F⁻, Spontaneous Nalʳ mutant of χ697 (Strʳ)[18]
CBK928	F⁻, derivative of SURE (Strategene, La Jolla, CA)[18]
DH2	*RecA*[42] (Strˢ, Nalˢ)
LE392	F⁻ *hsdR⁻ hsdM⁺ supE supF*[a]
LW49	MG1047, Nalʳ [18,24]
LW68	LW49, Strʳ [18,24]
MC1061	F⁻, *araD139* Δ(*ara leu*)*7697* Δ*lacY74 galU galK hsr⁻ hsm⁺ rpsl*[45] (Strʳ)
MG1047	F⁻, *recA56*, chr::γδ; contains three copies of γδ in the chromosome[21]
MG1063	F⁺, *recA56 thi;* sensitive to Amp, Cam, Kan, Tet, Str; used as donor strain for γδ mutagenesis[13]
RR1	F⁻, γδ free, *lacY1 rpsL20*[42] (Strʳ)
WM1100	MC1061 Δ*recA⁻* [b] (Strʳ)
Plasmids	
pIF200	pOX38::mγδ-1; used for mγδ-1 mutagenesis[18]
pIF341-pIF345	pOX38::mγδPLEX-1 to mγδPLEX-5; used for mγδPLEX mutagenesis[19]
pNG54	pACYC184-*tnpR;* Camʳ; γδ resolvase gene cloned with its own promoter; pBR322 compatible[43]
pOX38	"F⁺"; deletion derivative of F factor that is conjugation proficient, but deleted for γδ; used as donor plasmid for mγδ derivatives[21]
pVE1322	pUC19 derivative containing *Streptomyces* DNA[18]
pXRD4043	pACYC184-*tnpA*; Camʳ; γδ transposase gene cloned downstream of a *lac* promoter, making it IPTG inducible; pBR322 compatible[41]
Phage	
λ::Tn*5seq1*	*b*221 *c*I857 *P*am80[4]
λ::Tn*5tac1*	*b*221 *c*I857 *O*am29 *P*am80[37]
Media	
For γδ mutagenesis	
L broth	10 g Difco (Detroit, MI) tryptone, 5 g Difco yeast extract, 5 g NaCl, 1 g glucose, 990 ml distilled water; adjust pH to 7.0[c]
L agar	L broth solidified with 1.5% (w/v) agar
Medium E	For 1.5 liters of 20× stock: 6 g MgSO₄ · 7H₂O, 60 g citric acid · H₂O, 300 g K₂HPO₄ (anhydrous), 105 g NaNH₄ HPO₄ · 4H₂O, 1.33 liters distilled H₂O, pH 7.0[d]

TABLE II (*continued*)

Component	Description
For Tn5 mutagenesis[e]	
LN broth	10 g NZ amine (Humko-Sheffield Products, Norwich, NY), 5 g Difco yeast extract, 10 g NaCl, 990 ml distilled water; adjust pH to 7.2 (1.2 ml of 4 *M* NaOH)
LN agar	LN broth, with 15 g/liter Difco Bacto-agar
N agar	10 g NZ-amine, 10 g NaCl, 10 g Difco Bacto-agar, 990 ml distilled water; adjust pH to 7.2
N soft agar	10 g NZ-amine, 10 g NaCl, 6 g Difco Bacto-agar, 990 ml distilled water; adjust pH to 7.2
Antibiotics and other additives[f]	
Amp	Ampicillin (100–150 μg/ml)
Cam	Chloramphenicol (30 μg/ml); freeze stock in 95% (v/v) ethanol
Kan	Kanamycin (25–50 μg/ml)
Meth	Methicillin (100 μg/ml)
Nal	Nalidixic acid (20 μg/ml), in 1 *M* NaOH
Tet	Tetracycline (10–25 μg/ml); freeze stock in 95% ethanol
Str	Streptomycin (200 μg/ml)
Suc	Sucrose (5%)
IPTG	Isopropylthio-β-galactoside (1 m*M*)

[a] N. Murray, W. J. Brammar, and K. Murray, *Mol. Gen. Genet.* **150**, 53 (1977).
[b] H. Huang, personal communication (1991). This strain is Tet[r].
[c] E. S. Lennox, *Virology* **1**, 190 (1955).
[d] H. J. Vogel and D. M. Bonner, *J. Biol. Chem.* **218**, 97 (1956).
[e] D. E. Berg, A. Weiss, and L. Crossland, *J. Bacteriol.* **142**, 439 (1980).
[f] Final concentration. Antibiotics prepared in sterile distilled water at 250 times the final concentration and refrigerated, unless noted otherwise.

plex plasmids.[30,31] Tags from four of the original plex plasmids have been cloned in m$\gamma\delta$-1 and adapted for dideoxy sequencing by inserting "universal" primer sites (derived from the pBR322 *tet* gene) upstream of the tag sites. These transposons are amenable to multiplex probe mapping and

[30] G. M. Church and S. Kieffer-Higgins, *Science* **240**, 185 (1988).
[31] The Plex plasmids are a series of cloning vectors with "universal" primer sites and different DNA "tag" sites bracketing the cloning site.[30] After a library is made in each vector, one clone from each of a number of differently tagged vectors is pooled and processed for sequencing. After Maxam–Gilbert[30] or Sanger (G. Church, personal communication, 1990) sequencing, the mixture of sequence ladders is electrophoresed on a sequencing gel, transferred to a nylon membrane, and each sequence ladder in a set of lanes is developed sequentially by hybridization with a different multiplex probe.

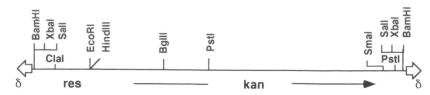

FIG. 7. Detailed restriction map of mγδ-1.[18] A computer-generated list of restriction enzymes that do not cleave mγδ-1 includes the following: *Aat*II, *Bgl*I, *Eco*RV, *Kpn*I, *Mlu*I, *Not*I, *Sac*I, *Sfi*I, and *Xho*I. Symbols: Arrowhead, 38-bp δ end inverted repeat.

multiplex sequencing. They contain the same tag sites as the Plex plasmid subset for which chemiluminescent probes are commercially available.[32,33]

3. A 260-bp set of mγδ transposons[20] has been constructed in which each transposon contains the *lac* universal and reverse primer sites and a set of tag sites (unrelated to the original[30] tag sites). These transposons do not carry a selectable marker, and use the bacteriophage P1 *loxP* site, rather than γδ *res*, for resolution of cointegrates.

Tn5.[11] Although the long inverted repeats of wild-type Tn5 preclude its use for sequencing without subcloning, any derivative with unique subterminal sequences can be used to provide primer binding sites. Some Tn5 derivatives with the requisite unique sequences near each end are described below.

1. Tn5*supF*[5] is a 264-bp element that contains the 19-bp Tn5 inverted repeats and a *supF* suppressor tRNA gene; it is particularly useful for analysis of λ phage clones.[8,8a] Transposase is supplied from *tnp* cloned outside of the transposon in the donor plasmid.

2. Tn5*seq1*[4] contains outward facing T7 and SP6 promoters that allow high-level transcription of adjacent genes *in vitro* and *in vivo*, and are useful for generating RNA probes for DNA flanking any insertion.

3. Tn5*lac*[34] contains a promoterless *lacZ* reporter gene close to one end and has been used to detect promoters and to monitor regulation in various species.

4. Tn5ORF*lac*[35] and Tn*lacZ*[36] contain a promoterless *lacZ* gene close to one end and have been used to generate protein fusions.

[32] Millipore/New England BioLabs Plex Luminescent kits product catalog.
[33] A. Creasey, L. D'Angio, Jr., T. S. Dunne, C. Kissinger, T. O'Keefe, H. Perry-O'Keefe, L. S. Moran, M. Roskey, I. Schildkraut, L. E. Sears, and B. Slatko, *BioTechniques* **11,** 102 (1991).
[34] L. Kroos and D. Kaiser, *Proc. Natl. Acad. Sci. U.S.A.* **81,** 5816 (1984).
[35] M. P. Krebs and W. S. Reznikoff, *Gene* **63,** 277 (1988).
[36] C. Manoil, *J. Bacteriol.* **172,** 1035 (1990).

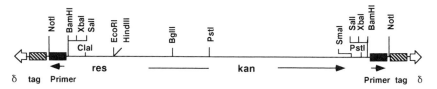

FIG. 8. Detailed restriction map of mγδPLEX-1 to mγδPLEX-5.[19]

5. Tn5*tacl*[37,38] contains a regulated *lac* promoter that allows insertion mutations with novel isopropylthio-β-galactoside (IPTG)-regulated phenotypes to be isolated.

6. Tn*phoA*[39] contains a promoterless *phoA* gene close to one end and has been used to obtain protein fusions that detect membrane or exported proteins.

7. Mini-Tn5(pfm1)[40] contains the 19-bp Tn5 inverted repeats, *cam, kan,* and several rare restriction enzyme recognition sites. Transposase is supplied from *tnp* cloned outside of the transposon in the donor plasmid.

Escherichia coli Strains and Donor Plasmids for γδ and Mini-γδ Mutagenesis

The standard donor strain used to obtain insertions of wild-type γδ is the F⁺ *recA* strain MG1063.[9] Almost any F⁺ or F′ strain that is *recA,* readily transformable, and easily killed after mating (by sensitivity to an antibiotic, or by auxotrophy) can be used. Any F⁻ strain can be used as recipient, as long as it contains a resistance determinant or prototrophy not shared with the donor.

The donor strain used to obtain insertions of mini-γδ elements should also be *recA,* readily transformable, and easily killed after mating. In addition, the strain should contain a source of transposase (pXRD4043, a pBR322-compatible plasmid containing the cloned transposase gene[41]) and a conjugative plasmid that is free of γδ and other mobile elements. The deletion derivative of F called pOX38[21] meets these criteria. As donors we have used CBK884, CBK928, and DH2 (Table II). Although many donor strains have been used for γδ and mγδ mutagenesis, when we

[37] W.-Y. Chow and D. E. Berg, *Proc. Natl. Acad. Sci. U.S.A.* **85,** 6468 (1988).
[38] B. T. Cookson, T. Tomcsanyi, D. E. Berg, and W. E. Goldman, *Proc. Int. Symp. Pertussis, 6th* (C. R. Manclark, ed.), p. 243. Dep. Health Hum. Serv., Bethesda, Maryland, 1990.
[39] C. Manoil and J. Beckwith, *Science* **233,** 1403 (1986).
[40] K. K. Wong and M. McClelland, *J. Bacteriol.* **174,** 3807 (1992).
[41] M.-M. Tsai, R. Y.-P. Wong, A. T. Hoang, and R. C. Deonier, *J. Bacteriol.* **169,** 5556 (1987).

attempted to use strain DH5α,[42] the strain rapidly segregated infertile derivatives.[18] The basis of this instability has not been examined.

The recipient strain should contain a source of resolvase so that the incoming cointegrate can resolve quickly into separate conjugative and target plasmids. Resolvase can be provided by pNG54[43] (a pBR322-compatible plasmid containing the $\gamma\delta$ *tnpR* gene), or by chromosomal copies of $\gamma\delta$[13,21] in a strain such as LW68.[18,24]

It is important that the target plasmid in the donor strain be a monomer. If the plasmid is a dimer, or present as a mixture of monomers and dimers, as is common in Rec[+] strains,[44] recovery of insertion-containing plasmids will be inefficient because a transposon-containing dimer will have one $\gamma\delta$-free component that can segregate from its $\gamma\delta$-containing sibling after recombination and may be selected during bacterial growth to yield a colony that does not contain $\gamma\delta$ in the target plasmid.[14]

Escherichia coli Strains and λ Phage for Tn5 Mutagenesis

The λ phage vectors for Tn5*seq1* and Tn5*tac1* are integration defective (due to a *b*221 deletion that removes the phage attachment site) and replication defective in Sup° hosts such as MC4100[45] (due to amber mutations in the phage replication gene *P* in λ::Tn5seq1, or in the replication genes *O* and *P* in λ::Tn5tac1). These phage also contain the *cI*857 temperature-sensitive allele of the *cI* repressor gene, and can establish immunity (repression) at 30°, even though they do not integrate into the chromosome. As a result, most cells infected at a multiplicity of 3–5 phage/cell at 30° survive infection. After several hours (needed for full expression of Kan[r]), approximately 10^{-4} of infected cells form Kan[r] colonies. About 1% of the plasmid DNA extracted from pools of these transductants carry a transposon insertion, and can be selected by their ability to transform recipient cells to Kan[r] (Fig. 2).[29]

Methods

$\gamma\delta$ Mutagenesis Conditions

Transposition of m$\gamma\delta$-1 from pIF200 (pOX38::m$\gamma\delta$-1) to a nonconjugative plasmid is obtained by selecting for conjugal transfer of that plasmid to

[42] D. Hanahan, *J. Mol. Biol.* **166,** 557 (1983).
[43] B. J. Newman and N. D. F. Grindley, *Cell* **38,** 463 (1984).
[44] C. M. Berg, L. Liu, M. Coon, L. D. Strausbaugh, P. Gray, N. B. Vartak, M. Brown, D. Talbot, and D. E. Berg, *Plasmid* **21,** 138 (1989).
[45] M. J. Casadaban and S. N. Cohen, *J. Mol. Biol.* **138,** 179 (1980).

an F^- recipient (Fig. 1). The target plasmid must have a unique selectable marker (Amp^r in most vectors). Generally, streptomycin or nalidixic acid is used to kill donor cells after conjugal transfer of the cointegrate and ampicillin is used to kill unmated recipient cells. Selection for the kanamycin resistance determinant of m$\gamma\delta$-1 is not useful at this step because both donor and recipient cells will be Kan^r.

The target plasmid can be mutagenized by selecting for its conjugal transfer during liquid or plate matings. Liquid matings are easier than plate matings, but a number of transconjugants from the same mating may have identical insertions because the plasmids they contain are all descended from the same cointegrate.[46] Plate matings can virtually ensure that each insertion is independent. Suggested protocols using the donor strains CBK884 (Str^r Nal^s) in a liquid mating, or DH2 (Str^s Nal^s) in a plate mating (both strains containing pIF200, pXRD4043, and an Amp^r pBR322-based target plasmid), and the recipient strain CBK904 (Str^r Nal^r) (containing pNG54) are given below.

Liquid Mating. Cultures should be grown aerobically by incubating the donor and the mating mixture in shallow cultures in large test tubes or small flasks without shaking to avoid breaking the F pili needed for conjugation.

1. Transform a monomer form of the target plasmid into CBK884 (pIF200, pXRD4043) and select transformants on L agar plus Amp, Kan, and Cam, to ensure the presence of these three plasmids.

2. Streak this donor strain for single colonies on L agar plus Amp, Kan, and Cam and incubate overnight at 37°.

3. Pick several independent single colonies of the donor strain and inoculate each into 1 ml of L broth plus Cam (but not Amp or Kan, which would kill the recipients in step 4) and IPTG (to induce transposase synthesis), and incubate at 37° for about 3 hr without shaking (to about 5×10^7 cells/ml). Concurrently, grow a single culture of the recipient in L broth containing Cam (to select for pNG54) to early stationary phase (to about 2×10^8/ml).

4. For each donor culture, mix 0.5 ml of donor cells and 0.2 ml of recipient cells in an 18-mm diameter test tube or a small flask. Incubate for 30 min in a 37° water bath without shaking. Add 5 ml of prewarmed L broth containing Cam plus IPTG and incubate for 3 hr without shaking.

[46] Although the transposase gene is cloned downstream of a *lac* promoter in pXRD4043,[41] we have found that some transposition occurs in uninduced cells. A cointegrate, once formed, will replicate with the result that sibling cointegrates can be transferred during mating. This is less of a problem with wild-type $\gamma\delta$ because resolvase, which breaks down cointegrates, is present in donor cells. The problem of sibling cointegrates is significantly reduced in plate matings, because most cointegrates will arise during growth of the donor colonies and only one transconjugant is chosen per donor colony.

5. Plate 0.1 ml of mating mixture on L agar containing Amp[47] plus Nal (to kill the unmated recipient and donor, respectively).

6. There should be 100 to 1000 transconjugants per plate. Purify a few transconjugants from each mating on the same medium, extract plasmid DNA from individual colonies by an alkaline lysis method,[48] and map insertions. Those insertions that are at the same site and that came from the same mating mixture (step 4) can be assumed to be siblings and all but one should be discarded.

Plate Mating. This protocol can be used with Str but not Nal to kill the donor, because Nal inhibits conjugal transfer.

1. About 16 hr before the mating, streak the donor strain on an L agar plate containing Cam, Kan, Amp, and IPTG at a cell density that yields both isolated single colonies and a region of confluent growth (to allow for fluctuations in mating efficiency). Incubate overnight at 37°.

2. The next morning spread a selection plate (Cam, Amp, Str, and IPTG) with 0.1 ml of recipient cells from an early stationary-phase overnight broth culture (not washed) to form a lawn.

---

[47] The satellite colonies that often appear on Amp selection plates can be avoided by the addition of 100–200 μg/ml of methicillin to Amp plates (D. Gelfand, personal communication, 1991).

[48] This procedure for preparing plasmid DNA is modified from the rapid alkaline lysis method [R. Kraft, J. Tardiff, K. S. Krauter, and L. A. Leinwand, *BioTechniques* **6**, 544 (1988)]: Inoculate a single purified colony into 5 ml of L broth plus Amp; incubate overnight with shaking. Transfer 1.5 ml of the culture to a 1.5-ml microfuge tube and pellet the cells in a microfuge at top speed for 2 min. Following removal of the supernatant, add an additional 1.5 ml of culture to the pellet and again spin the microfuge to pellet the cells (the remaining 2 ml from the overnight culture is used to establish −70° freezer stocks by adding glycerol to the tube). Following removal of the supernatant, resuspend the cells by brief vortexing in 100 μl of ice-cold TGE solution (25 mM Tris-HCl, pH 8.0, 50 mM glucose, 10 mM EDTA) and incubate the mixture at room temperature for 5 min. Add 200 μl of freshly made 0.2 N NaOH, 1% SDS, and mix gently by inversion; incubate this highly viscous mixture on ice for 5 min. Add 150 μl of ice-cold 3 M potassium acetate, pH 4.8, and after gentle mixing by inversion incubate the suspension on ice for 15 min. Pellet cellular debris and precipitated proteins for 10 min at top speed in a microfuge at 4° and carefully transfer the supernatant to a clean 1.5-ml microfuge tube (if a clear supernatant is not obtained, a second spin should be performed). Digest RNA with RNase A (added to a final concentration of 50 μg/ml) at 37° for 20 min. Extract the solution with an equal volume of 1 : 1 (v/v) phenol : chloroform by shaking and separate the phases by centrifugation for 5 min at top speed in a microfuge. Transfer the aqueous phase to a clean 1.5-ml microfuge tube and precipitate the DNA by the addition of 2.5 vol (1 ml) of ice-cold 100% ethanol and place at −70° for 10 min. Pellet the DNA at 4° for 10–20 min at top speed in a microfuge; rinse the DNA once with 1 ml of ice-cold 70% ethanol and repellet in the microfuge. Dry the pellet under vacuum (preferably in a vacuum centrifuge) for 20 min (or air dry for 2 hr). Dissolve the pellet, which should be visible, in 50 μl of sterile water. Aliquots of 5 μl are used for restriction enzyme digestion for probe mapping.

3. Replica plate the young, growing colonies from the donor plate onto the prespread selection plate, from step 2, and incubate at 37°. Transconjugants appear as patches corresponding to the positions of about 5% of the colonies. If there is background growth of unmated recipient cells after overnight incubation (because of β-lactamase excreted by transconjugants), methicillin (an Amp analog) should be added to the selection plates to reduce the background growth.[47] If more than 25% of the donor colonies yield transconjugants, it is possible that many of the plasmids in the transconjugants are siblings that are descended from the same cointegrate; so only one or a few colonies should be picked from that plate and insertions that are found to be at the same site should be assumed to be siblings.

4. Purify transconjugants on the same medium without IPTG, extract plasmid DNA,[48] and map insertions.

Tn5 Mutagenesis Conditions

Transposition of Tn5seq1 or Tn5tac1 from λ to a plasmid is obtained by infecting the plasmid-containing strain with λ::Tn5seq1 or λ::Tn5tac1 and selecting Kan^r colonies under conditions in which λ will be lost (Fig. 2).

1. Transform a monomer form of the target plasmid (which must not carry kan) into a recA sup° strain such as WM1100.

2. Inoculate this strain into 2 ml of LN broth containing 0.2% (w/v) maltose, 10 mM MgSO$_4$, and the antibiotic used to select the plasmid. Incubate overnight with shaking at 37° in the presence of antibiotic.

3. Add 0.1 ml of the overnight culture and 0.1 ml of a stock of the donor phage (λ::Tn5seq1 or λ::Tn5tac1) to 0.5 ml N broth, and let stand for 15 min at room temperature for adsorption.

4. Dilute the infected cells with 2 ml N broth, and aerate at 30° for 3 hr.

5. Pellet the cells by centrifugation and spread on LN kanamycin agar plate.

6. Incubate transductant plates at 30° for 40 hr. Expect a Kan^r frequency of about 10^{-4}/cell plated.

7. Extract plasmid DNA from large pools of Kan^r transductants (to reduce the probability of recovering siblings), and use this DNA to transform Kan^S recipient cells. Select for the Kan^r trait of the transposon and the Amp^r or other selectable marker of the target plasmid. Expect about 1% of these plasmids to contain the Tn5 element (the Kan^r cells that yielded only Kan^S plasmids were due to chromosomal Tn5 insertion).

8. Extract plasmid DNA from individual Kan^r transformant colonies.

Mapping Insertions

Restriction Mapping. The choice of which restriction enzymes to use depends on restriction sites available in the transposon, the vector, and the cloned fragment. If the fragment was generated by digestion with one enzyme, and cloned into a site for the same enzyme, then insertions in the vector or cloned fragment can be distinguished by digestion with that enzyme and identification of the disrupted fragment. The sizes of new fusion fragments provide the first mapping information; insertion is inferred to be at one of two possible sites if the transposon is cut symmetrically, and at one of four possible sites if the transposon is cut asymmetrically. One additional enzyme that cuts the transposon and the target fragment asymmetrically can be used to map and orient most insertions.

Restriction mapping is illustrated by the results of mapping mγδ-1 insertions in a plasmid carrying a 5.2-kb *Bam*HI fragment (from *Streptomycesa vermitilis*), with a unique *Eco*RI site near one end of the vector[18] (Fig. 3). mγδ-1 contains a pair of *Bam*HI sites symmetrically located at the boundaries of its inverted repeats and a unique, asymmetrically located *Eco*RI site (Fig. 7). First, the plasmids containing mγδ-1 were cut with *Bam*HI and the DNA fragments were separated by electrophoresis to determine whether the vector (4.1 kb) or cloned fragment (5.2 kb) was interrupted by γδ insertion. Because mγδ-1 also contains two *Bam*HI sites, insertion results in four fragments: two of constant size (from within the vector and the transposon) and two fragments of variable sizes. These data indicated how far the insertion was from a fragment end, but not which end. Comparison of the results of a further single digestion using *Eco*RI with those of the *Bam*HI digests permitted unambiguous mapping and orientation of all insertions except those within about 200 bp of the neighboring *Bam*HI and *Eco*RI sites.

Probe Mapping. Probe mapping entails the hybridization of one or more DNA (or RNA) probes with electrophoresed products of restriction enzyme digestion (blot analysis). It is a powerful and cost-effective method for localizing transposons in cloned DNA fragments of all sizes. With appropriate restriction sites in the transposon and vector, only a single hybridization step is required before sequencing. This determines the location, but not the orientation of the insertion. The sequence will usually be determined in both directions, and the orientation will be revealed by opposite strand overlap with sequences read from neighboring insertions. Probe mapping is straightforward in plasmids, such as illustrated here, where the cloned fragment does not contain internal recognition sites for the enzyme(s) used in the analysis. However, probe-mapping strategies can be easily designed to accommodate fragments with one or more internal sites by the use of additional enzymes and probes.

mγδ-1 (Fig. 7) and its PLEX (Fig. 8) derivatives contain a number of symmetric and asymmetric restriction sites that can be used to map, and also to orient, insertions. For preexisting libraries, insertions into fragments cloned into sites that are also present as central or symmetrical sites in the transposons (*Bam*HI, *Xba*I, and *Sal*I for both mγδ-1 and mγδPLEX; also *Not*I for mγδPLEX transposons) can be mapped using one enzyme in a single hybridization (Fig. 5). The insertions can be oriented, if needed, in a second hybridization. Mapping of insertions into fragments cloned into sites cleaved by enzymes that also cleave the transposon asymmetrically (*Cla*I, *Eco*RI, *Hind*III, *Bgl*II, *Pst*I, and *Sma*I) requires digestion by one enzyme and two hybridization steps (this also orients the insertion). Regardless of the enzyme used in cloning, insertions into Plex plasmid[30–32] clones can be mapped using *Not*I (because there are *Not*I sites between the vector tag and primer sites) (Fig. 8).

A sample protocol for probe-mapping insertions in a pBR322- or pUC-based plasmid clone that contains a fragment cloned at the *Bam*HI site (with no internal *Bam*HI sites) is illustrated in Fig. 5 and described below.

1. Determine the number of insertions to be mapped, as described above.[15] For mapping purposes, 40–60 digested DNA samples may be run on a single 20- to 25-cm agarose gel. The DNA preparations used in mapping can also be used for sequencing.

2. Following transposon mutagenesis (above), prepare plasmid DNA by a modification of the rapid alkaline lysis method.[48]

3. Restriction enzyme analysis: Digest plasmid DNA (5 µl) in a total volume of 10 µl with *Bam*HI and electrophorese the digestion products in agarose gels [0.6–1.2% (w/v) agarose; Tris-borate buffer, 0.45 M Tris-HCl, 0.45 M boric acid, 0.01 M ethylenediaminetetraacetic acid (EDTA), pH 8.0]. Include standard size markers such as 1 kb ladder (GIBCO/Bethesda Research Laboratories, Gaithersburg, MD) and *Eco*RI or *Hind*III digests of λ in order to determine fragment sizes. After electrophoresis, stain gels with a dilute solution of ethidium bromide for 1 min, then destain with double-distilled water for 20–30 min prior to photographing under ultraviolet (UV) illumination.

Plasmids with transposons in the cloned fragment are identified by the presence of an intact fragment the size of the cloning vector: only these molecules are mapped and used for sequencing. The insertion of mγδ-PLEX into the cloned fragment introduces *Bam*HI recognition sites into it and four fragments result from digestion: two variable-sized fragments from the interrupted cloned fragment, one constant vector fragment, and one constant internal transposon fragment.

4. Hybridization analysis: Photograph the gel, and transfer the DNA to filters for hybridization. The precise protocols for transfer and fixation of the DNA will depend on the membrane filter chosen; detailed protocols are supplied by the manufacturers. Hybridize the membrane-bound fragments with a probe representing one end of the cloned fragment (see discussion below for selection of probes and labeling methods). The size of the band visualized with this probe reflects the distance of the insertion from that end of the cloned fragment (the additional length contributed by the transposon fragment is negligible). The other variable-sized fragment reflects the distance of the insertion from the other end of the cloned fragment. When required, the orientation of the transposon-borne primer-binding sites can be determined by a second round of hybridization with a probe from one end of the transposon (e.g., a transposon multiplex tag).

When the enzyme used for analysis has also been used for cloning (as in the example in Fig. 5), a specific probe is required for each different cloned fragment. The best probes for mapping purposes are oligonucleotides or small restriction fragments of a few hundred base pairs (larger fragments will span a number of insertion sites, complicating the analysis). If a restriction map for the cloned fragment exists, a small, gel-isolated fragment from the end of the cloned DNA can be labeled by nick translation or random hexanucleotide priming, using any of a number of commercially available kits and either radioactive or nonradioactive labels. If there is no prior knowledge of the fragment, an oligonucleotide probe (15–30 bp) should be synthesized using sequence information obtained with a universal vector primer. Oligonucleotide probes are end labeled using commercially available kits using either radioactive or nonradioactive methods. For radiolabeled probes, standard procedures are used for hybridization, rinsing, and autoradiography.[49] Procedures for hybridization and visualization with nonradioactive probes are documented in the various commercially available kits.

5. Select insertions appropriate for sequencing. Sequence by a double-strand modification of dideoxy sequencing, or cycle sequencing.[8,50]

The preceding example illustrates the probe mapping strategy for insertions in fragments cloned in standard vectors. Probe mapping should be

[49] J. Sambrook, E. F. Fritsch, and T. Maniatis, "Molecular Cloning: A Laboratory Manual." Cold Spring Harbor Press, Cold Spring Harbor, New York, 1989.

[50] M. Craxton, *Methods: Companion Methods Enzymol.* **3,** 20 (1991).

most effective when multiplex transposons are coupled with multiplex vectors. This combination offers three important advantages over the example shown in Fig. 5: (1) it is not necessary to make decisions about the appropriate enzyme to use because *Not*I (with its rare recognition sites in most genomic DNAs) is used in all digestions; (2) the different multiplex tags in plasmid and transposon enable the use of different "universal" probes for mapping (new fragment-specific probes need not be made for each fragment analyzed); (3) independent insertions can be pooled for mapping as well as for sequencing.[30] For example, as illustrated in Fig. 6, the insertion of an mγδPLEX transposon into a fragment cloned into a Plex plasmid with a different tag introduces two more *Not*I sites and results in four *Not*I fragments: two variable fragments that reflect the insertion site, one constant internal vector fragment, and one constant internal transposon fragment. With multiplex transposons inserted into multiplex vectors with different tags, both vectors and transposons contain appropriate internal restriction sites (*Not*I), and vector specific probes are used to map insertions. To also orient the insertions, transposon-specific probes would be used in a second hybridization. If a fragment with an internal *Not*I site has been cloned, both vector tags would be used to map the insertion. This strategy can be readily adapted to multiplex DNA preparations, multiplex mapping, and multiplex sequencing. Nonradioactive probes to the multiplex tags in both transposons and vectors are commercially available.[32,33]

Overview of Methods

In designing transposon-based sequencing protocols for plasmid clones, a number of considerations should be taken into account.

1. Which transposon should be used? In our experience, γδ and its derivatives insert most randomly in plasmid DNA, and are easiest to deliver; mγδ-1 appears to transpose even more randomly than γδ. Our second choice for plasmids (and first choice for λ phage clones and chromosomal DNA) is one of the Tn5 derivatives because they transpose readily to any DNA, and more randomly into plasmids than most transposons (except γδ).

2. How many insertions are enough? We recommend aiming for levels of coverage that are adequate for direct sequencing if the plasmid is relatively small (<15 kb). If the plasmid is larger, transposons are most useful as sources of anchoring or sequence tagged sites (STS) for PCR amplification, primer walking, directed subcloning, or the identification of coding regions.

3. How should insertions be mapped? The best mapping protocol will depend on the size of the target DNA, and also on whether its restriction map is already known. Restriction mapping is efficient for small targets, or for targets where some restriction sites are already known. PCR mapping is efficient for small to moderately sized (up to 6 kb[8,8a]) fragments. Probe mapping can be used for any sized target, and is the method of choice for targets too large for easy mapping by restriction or PCR analysis.

4. When should a combination of mapping methods be used? For large insertions, the most effective strategy may well involve probe mapping for approximate placement and orientation of insertions, followed by direct and cross-over PCR for high-resolution mapping and template preparation.

5. How should gaps be filled? If the gap is due to nonrandom insertion by either γδ or Tn5, the gap could be closed using the other transposon because they have different insertion specificities.[7,11,28] Small gaps in the sequence could best be filled by limited primer walking on the PCR products, perhaps using primers from preformed libraries of short oligonucleotides. Larger gaps could be filled by isolating additional insertions, by primer walking, or by subcloning.

Concluding Remarks and Discussion

Bacterial transposons show great promise as tools for both small- and large-scale sequencing efforts, by virtue of their ability to move and carry useful specific primer-binding sites to many locations in cloned DNA targets. Several naturally occurring transposons and specially engineered derivatives have been used in this way.[1–7,15–20,37,38,51] These elements facilitate genetic and physical analyses because they tag any target with selectable markers, known restriction sites, and probe- or primer-binding sites. Some transposon derivatives also contain regulatable promoters, reporter genes, and/or replication origins.[6,7] DNAs cloned in *E. coli* that are tagged by γδ or Tn5 can be returned to the original prokaryotic or eukaryotic host for complementation and gene replacement studies.[7,18,52–55]

Many new methods should interface well with transposon-based template-generating methods. Among them are (1) direct PCR for amplifying DNA segments, using a transposon to provide one primer-binding site[8,8a];

[51] Z.-G. Peng and R. Wu, this series, Vol. 155, p. 214.

[52] M. F. Hoekstra, H. S. Seifert, J. Nickoloff, and F. Heffron, this series, Vol. 194, p. 329.

[53] J. D. Noti, M. N. Jagadish, and A. Szalay, this series, Vol. 154, p. 197.

[54] S. H. Phadnis, S. Kulakauskas, B. R. Krishnan, J. Hiemstra, and D. E. Berg, *J. Bacteriol.* **173,** 896 (1991).

[55] S. Kulakauskas, P. M. Wikstrom, and D. E. Berg, *J. Bacteriol.* **173,** 2633 (1991).

(2) cross-over PCR to amplify DNA segments delimited by independent transposon insertions in different molecules[8,8a]; (3) multiplex sequencing[30] using transposon-carried multiplex tags; (4) linear amplification (cycle) sequencing, a one-ended "PCR" reaction, using a single transposon primer in the presence of chain-terminating nucleotides[8,8a,50,56]; (5) primer walking from transposon anchor sites using preformed oligomer libraries[57–59]; and (6) exometh sequencing using a transposon to provide appropriate restriction sites.[60]

The methods described here deal with the isolation and analysis of insertions obtained following hopping into a target plasmid (intermolecular transposition). An alternative, promising strategy involves the use of nested deletions, generated by intramolecular transposition in special "deletion factory" cloning vectors, in which a transposon end is brought next to a new site. Because transposition is a rare event, it is necessary to select *against* a marker that is located between the transposon end and the cloned fragment and, hence, is lost in deletion derivatives. This approach was developed and first used by Ahmed, who constructed a set of cloning vectors containing the transposon Tn9, or its component IS*1* element.[1,51] However, Tn9 (IS*1*) transposes too nonrandomly to be generally useful.[61] We have tested Tn5- and γδ-based nested deletion formation, and found both elements to transpose quite randomly to sites within the same molecule.[62–64] Deletions formed by Tn5 or γδ intramolecular transposition provide efficient means of bringing all portions of DNA in plasmids or cosmids close to binding sites for sequencing primers.

Large-scale sequencing of complex genomes will require improvement on many fronts, including DNA preparation, sequencing and data analysis, and the automation of many steps. Transposons can play a major role in these efforts both directly, by providing mobile primer-binding sites, and indirectly, by reducing the need for redundant sequencing or for sophisticated computer programs and extensive analysis for sequence alignment. It should even be possible to use transposons to access the next generation

[56] V. Murray, *Nucleic Acids Res.* **17**, 8889 (1989).
[57] F. W. Studier, *Proc. Natl. Acad. Sci. U.S.A.* **86**, 6917 (1989).
[58] W. Szybalski, *Gene* **90**, 177 (1990).
[59] Deleted in proof.
[60] J. A. Sorge and L. A. Blinderman, *Proc. Natl. Acad. Sci. U.S.A.* **86**, 9208 (1989).
[61] D. Zerbib, P. Gamas, M. Chandler, P. Prentki, S. Bass, and D. Galas, *J. Mol. Biol.* **185**, 517 (1985).
[62] T. Tomcsanyi, C. M. Berg, S. H. Phadnis, and D. E. Berg, *J. Bacteriol.* **172**, 6348 (1990).
[63] G. Wang, J. Chen, D. E. Berg, and C. M. Berg, in preparation.
[64] G. Wang, D. E. Berg, R. W. Blakesley, and C. M. Berg, in preparation.

of large-capacity (>90 kb) vectors, such as P1[65,66] and F,[67,68] and use a combination of probe mapping, cross-over PCR, and primer walking to sequence target DNA in these vectors without any *in vitro* subcloning.

As presently practiced, transposon-facilitated sequencing is very much "hands on," but mutagenesis, DNA preparation, PCR mapping, probe mapping, and sequencing are all amenable to the scaleup and automation that will increasingly dominate genome sequencing efforts in the future.

Acknowledgments

We thank our past and present colleagues, especially J. Chen, P. Gray, C. Green, L. Liu, D. MacNeil, N. Vartak, X. Xu, and J. Zhang, for stimulating discussions and for permission to cite unpublished work, and Perkin-Elmer Cetus Instruments (Norwalk, CT) for the loan of model TC480 thermal cyclers. This work has been supported by Grants DEFGO2-89ER60862 and DEFGO2-90ER610 from the U.S. Department of Energy, HG-000563 from the U.S. Public Health Service, and DMB8802310 and BSR-9009938 from the National Science Foundation, and grants from the University of Connecticut Research Foundation.

[65] N. Sternberg, *Proc. Natl. Acad. Sci. U.S.A.* **87,** 103 (1990).
[66] J. C. Pierce and N. L. Sternberg, this series, Volume 216 [47].
[67] F. Hosoda, S. Nishimura, H. Uchida, and M. Ohki, *Nucleic Acids Res.* **18,** 3863 (1990).
[68] E. D. Leonardo and J. M. Sedivy, *Bio/Technology* **8,** 841 (1990).

Section II

Polymerase Chain Reaction for Amplifying and Manipulating DNA

[21] Use of Polymerase Chain Reaction to Amplify Segments Outside Boundaries of Known Sequences

By HOWARD OCHMAN, FRANCISCO JOSÉ AYALA, and DANIEL L. HARTL

Introduction

The polymerase chain reaction (PCR) is an effective method for selectively amplifying specific DNA segments without resorting to conventional cloning procedures. Oligonucleotide primers are designed to anneal to complementary strands and oriented such that polymerization by one primer creates a template for DNA synthesis by the other primer. Repeated cycles of synthesis result in a geometric increase in the number of copies of the region bounded by the primers. The PCR, as originally described,[1,2] does not allow the amplification of segments that lie outside the primers because an oligonucleotide that primes synthesis into a flanking region has no primer in the reverse direction and therefore produces only a linear increase in the number of copies.

Several techniques incorporating the PCR have been developed to amplify flanking or anonymous DNA fragments when primers specific to the region are not available.[3] Unlike standard PCR, the utility and efficiency of each of these methods have largely depended on the particular application. In this chapter we review the procedures adopted by several investigators to obtain fragments outside a region of known sequence and provide a protocol for one technique—inverse PCR—used by our laboratory.

Prior to the implementation of the PCR, the acquisition of specific DNA fragments usually entailed the construction and screening of DNA libraries, and the traditional approach for "walking" from regions of known sequence into flanking DNA involved the successive probing of libraries with clones obtained from prior screenings.[4] Although the manipulation of DNA libraries is a relatively time-consuming procedure, it still offers the best approach for obtaining an ordered array of DNA fragments

[1] R. K. Saiki, S. J. Scharf, F. Faloona, K. B. Mullis, G. T. Horn, and H. A. Erlich, *Science* **230,** 1230 (1985).

[2] R. K. Saiki, D. H. Gelfand, S. Stoffel, S. J. Scharf, R. G. Huguchi, G. T. Horn, K. B. Mullis, and H. A. Erlich, *Science* **239,** 23 (1988).

[3] T. J. White, N. Arnheim, and H. A. Erlich, *Trends Genet.* **5,** 185 (1989).

[4] W. Bender, P. Spierer, and D. S. Hogness, *J. Mol. Biol.* **168,** 17 (1983).

encompassing a very large region. Nevertheless, several applications, including the recovery of the flanking sequence and the generation of end-specific hybridization probes for chromosome walking, can be greatly facilitated by the PCR.

Methodology

Many techniques for amplifying flanking regions of DNA are based on the creation of new primer-binding sites onto potential PCR templates either by enzymatic synthesis ("tailing") or by ligating oligonucleotides of known sequences ("cassettes") to the ends of DNA fragments. For the amplification of cDNAs, it is usually sufficient to use a unique primer—one based on a region of previously determined sequence—along with a primer complementary to the enzymatically synthesized tail to generate a specific product. However, this is not generally the case with protocols employing cassettes at the ends of restriction fragments. Using one specific primer and one that pairs with the ligated primer binding site will normally yield many nonspecific products that result from fragments with cassettes ligated onto both ends. Therefore, the DNA must be modified, or purified, prior to the PCR, to promote the amplification of a specific region.

The following techniques have been used to amplify segments of DNA adjacent to a core region of known sequence. The specificity of the reaction is, in most cases, supplied by the unique primer designed to the core region. No single method can be recommended for every application because the source (DNA, RNA, or inferred from protein sequence) and size of the core region, as well as the requirements for flanking information, vary with each experiment.

Anchored Polymerase Chain Reaction

The anchored PCR was originally designed to amplify the variable ends of T cell antigen receptors and can be extended to obtain the 5′ ends of any transcript.[5] RNA is reverse transcribed with either poly(dT) or a unique primer complementary to a known sequence. The cDNA is then tailed with poly(dG) so that an "anchor" primer that includes a poly(C) stretch can be used along with the specific primer for amplifying a specific region by the PCR.

Rapid Amplification of cDNA Ends

Details of the rapid amplification of cDNA ends (RACE), used to generate copies of a specific cDNA between a point of known sequence

[5] E. Y. Loh, J. F. Elliot, S. Cwirla, L. L. Lanier, and M. M. Davis, *Science* 243, 217 (1989).

and either end, can be found in a series of papers by Frohman and collaborators.[6-9] To generate the 3' end of the transcript, the mRNA is reverse transcribed using an "adapter" primer containing a 3' stretch of poly(dT) that anneals to the poly(A) of the message. Amplification is performed using oligonucleotides to the unique [nonpoly(dT)] portion of the adapter primer and one specific to the gene of interest. For the 5' end, the mRNA is reverse transcribed with the gene-specific primer and the cDNA products are poly(dA) tailed by terminal transferase. Complementary copies of the tailed products are synthesized by the adapter primer and amplification proceeds as described for the 3' end.

One-Sided Polymerase Chain Reaction

As in other procedures designed to amplify cDNAs, the one-sided PCR technique uses unique primers in conjunction with nonspecific primers directed to either the poly(A) region of the message or to be enzymatically synthesized poly(dA) tail.[10] Depending on the combination of primers used for the PCR, it should be possible to amplify regions either upstream or downstream of the point of known sequence.

Inverse (Inverted or Inside-Out) Polymerase Chain Reaction

The inverse (inverted or inside-out) PCR for amplifying anonymous flanking regions was developed independently by three groups.[11-13] The technique involves the digestion of source DNA, circularization of restriction fragments, and amplification using oligonucleotides that prime DNA synthesis directed away from the core region of known sequence (Fig. 1). For inverse PCR, all primers are designed to anneal to the region of known sequence and are oriented in directions opposite to those normally employed in the PCR. Although the conditions for ligations that favor the formation of monomeric circles are well established,[14] the major difficulty with this procedure seems to involve the circularization of restriction fragments.

[6] M. A. Frohman, M. K. Dush, and G. R. Martin, Proc. Natl. Acad. Sci. U.S.A. **85**, 8992 (1988).

[7] M. A. Frohman and G. R. Martin, Techniques **1**, 165 (1989).

[8] M. A. Frohman, Amplifications **5**, 11 (1990).

[9] M. A. Frohman, in "PCR Protocols: Methods and Applications" (M. A. Innis, D. H. Gelfand, J. J. Sninsky, and T. J. White, eds.), p. 28. Academic Press, San Diego, 1989.

[10] O. Ohara, R. L. Dorit, and W. Gilbert, Proc. Natl. Acad. Sci. U.S.A. **86**, 5673 (1989).

[11] H. Ochman, A. S. Gerber, and D. L. Hartl, Genetics **120**, 621 (1988).

[12] T. Triglia, M. G. Peterson, and D. J. Kemp, Nucleic Acids Res. **16**, 8186 (1988).

[13] J. Silver and V. Keerikatte, J. Virol. **63**, 1924 (1989).

[14] F. S. Collins and S. M. Weissman, Proc. Natl. Acad. Sci. U.S.A. **81**, 6812 (1984).

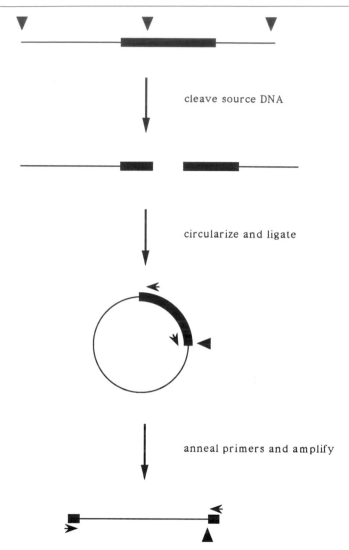

FIG. 1. Amplification of flanking DNA by the inverse PCR. In the application of the procedure described in this chapter, source DNA is treated with a restriction enzyme that cleaves within the region of known sequence. (Filled triangles denote restriction enzyme recognition sites.) Resulting restriction fragments are ligated under conditions that favor the formation of monomeric circles. Primers are oriented such that DNA synthesis proceeds into regions of unknown sequence.

Inverse PCR has been used to evaluate viral and transposon integration sites,[11–13] generate end-specific probes for chromosome walking,[15–18] and directly clone unknown cDNA sequences from total RNA.[19] Variations of these procedures have been reported. For example, Rich and Willis[20] obtained the regions flanking a Tn5 insertion using a single primer complementary to the inverted terminal repeats of the transposable element. However, their method does not allow the direct sequencing of the amplified fragment because the end-specific primer binds at two locations.

Ligation-Mediated Polymerase Chain Reaction

Ligation-mediated PCR provides a method for directly sequencing or obtaining fragments beyond the boundaries of a known sequence.[21] The procedure is based on the ligation of an oligonucleotide cassette to specific cleavage fragments. The double-stranded cassette consists of a 24-mer and an 11-mer that is complementary to the 3' end of the longer oligonucleotide. The flush side of the cassette is attached onto specific DNA fragments whose ends were made blunt by primer extension using a unique oligonucleotide constructed from a known sequence. This technique has enabled several researchers to walk from targeted regions into flanking DNA in organisms with complex genomes and experimental details of the procedure are described by Mueller and Wold,[21] Pfeifer et al.,[22] and Fors et al.[23]

Vectorette (Bubble) Polymerase Chain Reaction

The vectorette (bubble) PCR was developed to isolate the terminal sequences of inserts cloned as yeast artificial chromosomes (YACs).[24] Yeast cells harboring artificial chromosomes are treated with restriction

[15] D. Garza, J. W. Ajioka, J. P. Carulli, R. W. Jones, D. H. Johnson, and D. L. Hartl, *Nature* (*London*) **340**, 577 (1989).

[16] H. Ochman, J. W. Ajioka, D. Garza, and D. L. Hartl, in "PCR Technology: Principles and Applications for DNA Amplification" (H. A. Erlich, ed.), p. 105. Stockton Press, New York, 1990.

[17] G. A. Silverman, R. D. Ye, K. M. Pollack, J. E. Sadler, and S. J. Korsmeyer, *Proc. Natl. Acad. Sci. U.S.A.* **86**, 7485 (1989).

[18] G. A. Silverman, J. I. Jockel, P. H. Domer, R. M. Mohr, P. Taillon-Miller, and S. J. Korsmeyer, *Genomics* **9**, 219 (1991).

[19] S. Huang, Y. Yu, C. Wu, and J. Holcenberg, *Nucleic Acids Res.* **18**, 1922 (1990).

[20] J. J. Rich and D. K. Willis, *Nucleic Acids Res.* **18**, 6673 (1990).

[21] P. R. Mueller and B. Wold, *Science* **246**, 780 (1989).

[22] G. P. Pfeifer, S. D. Steigerwald, P. R. Mueller, B. Wold, and A. D. Riggs, *Science* **246**, 810 (1989).

[23] L. Fors, R. A. Saavedra and L. Hood, *Nucleic Acids Res.* **18**, 2793 (1990).

[24] J. Riley, R. Butler, D. Ogilvie, R. Finniear, S. Powell, R. Anand, J. C. Smith, and A. F. Markham, *Nucleic Acids Res.* **18**, 2887 (1990).

enzymes known to have sites conveniently located in the YAC vector. Double-stranded oligonucleotide cassettes ("vectorettes") containing a tract of noncomplementary base pairs are ligated onto the ends of cleavage products. Amplification proceeds with one primer specific to the YAC vector and the other homologous to the noncomplementary ("bubble") region of the ligated vectorette. The extension products from the vector-specific primer includes a segment that perfectly complements the bubble region of the vectorette, which helps prevent the recovery of nonspecific products.

Targeted Gene Walking

Targeted gene walking amplifies anonymous flanking regions by using nonspecific "walking" primers, 18 to 28 base pairs (bp) in length, along with primers unique to the targeted core region of known sequence.[25] Parallel amplification reactions are performed with several different walking primers and assayed on agarose gels after primer–extension with a radiolabeled oligonucleotide internal to the original amplification primer. Polymerase chain reaction products verified as walks from the targeted sequence are purified and excised from the gel and subjected to secondary amplification to generate a specific product.

Alu Polymerase Chain Reaction

Segments of human DNA cloned into cosmids, or as yeast artificial chromosomes,[26] frequently contain copies of *Alu,* a short interspersed repetitive element present in approximately 10^6 copies in the genomes of primates.[27] Primers homologous to the region upstream of the vector cloning site and to *Alu* elements allow the enzymatic amplification of end-specific fragments for sequencing, screening libraries, and chromosome walking. Because the primer that anneals to the repeated sequence may amplify regions between *Alu* elements, it is necessary to set up control reactions with and without the vector-specific primer to identify those products corresponding to inter-*Alu* regions. Descriptions of the technique, as well as protocols and primer sequences, can be found in Nelson *et al.*,[28] Breukel *et al.*,[29] and Ledbetter *et al.*[30]

[25] J. D. Parker, P. S. Rabinovitch, and G. C. Burmer, *Nucleic Acids Res.* **19,** 3055 (1991).

[26] D. T. Burke, G. F. Carle, and M. V. Olson, *Science* **236,** 806 (1987).

[27] M. A. Batzer and P. L. Deininger, *Genomics* **9,** 481 (1991).

[28] D. L. Nelson, S. A. Ledbetter, L. Corbo, M. F. Victoria, R. Ramirez-Solis, T. D. Webster, D. H. Ledbetter, and C. T. Caskey, *Proc. Natl. Acad. Sci. U.S.A.* **86,** 6686 (1989).

[29] C. Bruekel, J. Wijnen, C. Tops, H. van der Klift, H. Dauwerse, and P. Meera Khan, *Nucleic Acids Res.* **18,** 3097 (1990).

[30] S. A. Ledbetter, D. L. Nelson, S. T. Warren, and D. H. Ledbetter, *Genomics* **6,** 475 (1990).

Transposon Walking

Analogous to the procedures described for *Alu* PCR, it is possible to obtain flanking regions by the PCR using primers directed to the ends of a transposable element and a region of known sequence.[31,32] Transposon insertions are screened by PCR: transposable elements are usually so widely spaced in the genome that no interelement regions are amplified.

Oligonucleotide Cassette-Mediated Polymerase Chain Reaction

Oligonucleotide cassette-mediated PCR has been used to walk outside regions of known sequence in complex samples, including total human, nematode, and yeast genomic DNA.[33] DNA samples are treated with a restriction enzyme producing sticky ends to which 28-bp double-stranded cassettes are ligated. The specificity of the method is furnished by 50 rounds of primer extension using a biotinylated primer complementary to the core region, and the biotinylated products of this linear amplification are isolated with streptavadin-coated magnetic beads. The isolated product is then subjected to conventional PCR using a unique internal, but nonbiotinylated, primer paired with one complementary to the oligonucleotide cassette.

Application of Inverse Polymerase Chain Reaction

The distribution and abundance of insertion sequences in populations of *Escherichia coli* have been well characterized[34] but little is known about the integration sites of these small translocatable elements in natural isolates. As an example of the procedures currently employed to amplify DNA fragments outside the boundaries of known sequences, we provide a detailed protocol for inverse PCR as used to obtain the regions flanking an insertion sequence, IS*30*, from isolates of *E. coli*. The regions upstream and downstream of the IS element can be manipulated to resolve their map position on the *E. coli* chromosome and the nucleotide sequence of the insertion site. To facilitate our analysis, we selected four strains of *E. coli* known to contain a single copy of IS*30*.

One-microgram samples of genomic DNA from these strains (ECOR reference collection numbers 24, 51, 56, and 69) are treated with *Taq*I or *Cla*I according to the specifications of the supplier and electrophoresed through a 0.9% (w/v) agarose gel in 0.5× TBE (45 m*M* Tris-borate, pH

[31] D. G. Ballinger and S. Benzer, *Proc. Natl. Acad. Sci. U.S.A.* **86,** 9402 (1989).
[32] T. M. Barnes, *Nucleic Acids Res.* **18,** 6741 (1990).
[33] A. Rosenthal and D. S. Jones, *Nucleic Acids Res.* **18,** 3095 (1990).
[34] S. A. Sawyer, D. E. Dykhuizen, R. F. DuBose, L. Green, T. Mutangadura-Mhlanga, D. F. Wolczyk, and D. L. Hartl, *Genetics* **115,** 51 (1987).

8.0, 5 mM EDTA). IS30 has a single recognition site for ClaI and two TaqI sites, and we perform Southern blots using either end of IS30 as a hybridization probe to determine which restriction enzyme produces fragments of appropriate size for circularization and amplification.

To obtain hybridization probes corresponding to only one side of IS30, we amplify this insertion element from genomic DNA using primers specific to the terminal sequences of IS30, digest the PCR product with ClaI, and isolate the two resulting fragments, designated the right-hand and left-hand fragments, on DEAE membranes (NA-45; Schleicher & Schuell, Keene, NH). The nylon filters containing ClaI- or TaqI-digested genomic DNAs from the ECOR strains are probed with radiolabeled fragments[35] and the resulting autoradiogram using the left-hand probe indicates that ClaI fragments in strains 56 and 69, and TaqI fragments in strains 24 and 51, are of suitable size for the PCR.

Genomic DNAs from each of the strains of E. coli are treated for 1 hr with the proper restriction enzyme (as ascertained above by the diagnostic Southern blots), followed by phenol/chloroform/isoamyl alcohol (25 : 24 : 1, v/v/v) extraction and ethanol precipitation to remove proteins and salts. [All ethanol precipitations are performed by the addition of one-tenth volume of cold 2.5 M ammonium acetate, 1 μl of a linear polyacryl-amide carrier,[36] and 2 vol of cold ethanol. Nucleic acids are precipitated at $-20°$ and collected by centrifugation. DNA pellets are washed twice with 70% ethanol and dried under vacuum.]

Ligations are set up in a buffer containing 50 mM Tris-HCl (pH 7.4), 10 mM MgCl$_2$, 10 mM dithiothreitol (DTT), and 1 mM ATP[37] at three DNA concentrations (500, 100 or 20 ng/ml) in 100-μl volumes in order to increase the likelihood of finding dilutions favorable to the formation of monomeric circles. Reactions are initiated by the addition of 1 Weiss unit of T4 DNA ligase per microliter, incubated at 14° for 16 hr, followed by heat inactivation at 65° for 15 min, phenol/chloroform extraction, and ethanol precipitation.

We carry out inverse PCR in 30 μl of reaction buffer [50 mM KCl, 10 mM Tris-HCl (pH 8.4), 2.5 mM MgCl$_2$, 0.01% (w/v) gelatin] containing 0.2 mM concentrates of each dNTP, 0.5 units of Taq polymerase, and 50 pmol of each primer. To obtain left-hand flanking regions, amplification primers are designed to anneal to the start, and to the region immediately upstream of the internal ClaI/TaqI site, of IS30 and oriented such that

[35] A. P. Feinberg and B. Vogelstein, *Anal. Biochem.* **132,** 6 (1983).
[36] C. Gaillard and F. Strauss, *Nucleic Acids Res.* **18,** 378 (1990).
[37] T. Maniatis, E. F. Fritsch, and J. Sambrook, "Molecular Cloning: A Laboratory Manual." Cold Spring Harbor Press, Cold Spring Harbor, 1989.

DNA synthesis progresses away from the IS element. Similarly, primers to the 3' portion of IS30 are constructed to amplify right-hand flanking sequences. Synthetic oligonucleotides are 20 bp in length and approximately 50% G + C, yielding a theoretical average T_d of 60° (see the next section for calculation of T_d).

Amplification reaction mixes containing primers, template, dNTPs, and buffer are heated to 95° for 10 min prior to the addition of Taq polymerase to introduce nicks, which facilitate amplification of circular templates. We typically employ 30 rounds of amplification by denaturing DNA at 95° for 40 sec, annealing primers at 56° for 30 sec, and extending primers at 72° for 100 sec. When long amplification products (>3 kb) are anticipated, we lengthen the denaturation and extension times by 50%. Occasionally, the PCR yields more than one product but the fragment of correct size, as determined by the initial Southern blots, can often be recovered by electrophoresis onto DEAE membranes.[38]

The inverse PCR product generated by the above procedures is examined to determine the integration site of IS30 in a natural isolate of E. coli. A partial sequence of the 2-kb fragment from ECOR strain 69 shows the 5' end of IS30 and upstream flanking DNA (Fig. 2). We determine the chromosomal location of this insertion sequence by hybridization to an ordered array of λ clones.[39,40] These λ clones, comprising the Kohara library, span the entire E. coli W3110 chromosome and the complete set of nearly 500 clones has been spotted onto nitrocellulose membranes. DNA flanking the IS element from strain 69 hybridizes with two clones from the Kohara library corresponding to 60 min on the E. coli linkage map (Fig. 3). The strain used to construct the original phage library (W3110) harbors six copies of IS30, the position of one of which is identical to that of the IS30 in the natural isolate.

Considerations in Inverse Polymerase Chain Reaction Methodology

In some applications, the standard procedures for inverse PCR, as described in the previous section, are not adequate for amplifying the proper DNA fragment. Unlike conventional PCR, only about half of the initial attempts to apply inverse PCR are successful for obtaining DNA outside the known region of interest. Several aspects of the technique can be varied and we review some of the modifications.

---

[38] J. G. Lawrence, H. Ochman, and D. L. Hartl, J. Gen. Microbiol. **137**, 1911 (1991).
[39] Y. Kohara, K. Akiyama, and K. Isono, Cell **50**, 495 (1987).
[40] A. Noda, J. B. Courtwright, P. F. Denor, G. Webb, Y. Kohara, and A. Ishihama, BioTechniques **4**, 474 (1991).

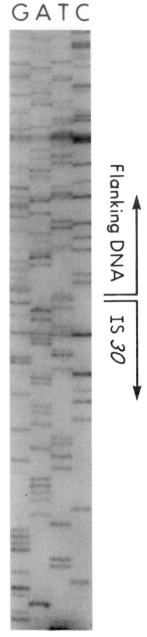

FIG. 2. Nucleotide sequence of a DNA fragment flanking IS*30*. The arrow shows the integration site of the insertion sequence. Protocols used to sequence this inverse PCR product are described in Chapter [3] of this volume.

FIG. 3. Mapping the IS30 from ECOR strain 69 by probing a filter containing an ordered array of λ clones spanning the *E. coli* chromosome with the fragment flanking the insertion sequence. Positive signals represent hybridization with two overlapping Kohara clones.

1. Successful application of inverse PCR depends on the complexity of the starting material; smaller genomes tend to yield more reliable results. Suitable inverse PCR products have been obtained from organisms with genome sizes up to about 200 Mb (*Drosophila* and *Caenorhabditis*), but there are few published reports of inverse PCR products recovered from genomes containing over 10^9 bp. It is possible to enrich for a fragments of a particular size by fractionating genomic DNA prior to circularization. Restriction fragments of defined size, as determined by the initial Southern blots, can be extracted from agarose gels using glass powder, electroelution, or DEAE membranes. Whether total genomic or size-fractionated DNAs is used, we try to recover a total of about 1 μg of cleaved DNA in order to conduct circularization and PCR at several DNA concentrations. Certain applications of inverse PCR, such as the identification of the integration site of a specific transposable element from a population of similar elements, require the initial selection of fragments of a restricted size class.

2. Some researchers have found that the efficiency of inverse PCR is enhanced by the amplification of linear rather than circular molecules. Prior to the PCR, the circularized fragments are treated with a restriction

enzyme known to cleave in the region between the 5' ends of both primers. This reopens the monomeric circles and often results in superior amplifications. The disadvantage of this procedure is identifying a unique restriction site within the core region, using an endonuclease that does not cleave within the anonymous flanking region. We prefer to introduce nicks into the circular molecules by heating DNA samples to 95° for 10 min prior to the PCR. In addition, this "hot start" initiation of the PCR eliminates many nonspecific amplification products.

3. Several factors, such as the restriction map of the region and the size of the resulting DNA fragments, influence the choice of restriction endonucleases for inverse PCR. We usually try to identify fragments containing less than 3 kb of flanking DNA, a limitation imposed by the size of a region that can be efficiently amplified by the PCR. Both upstream and downstream flanking regions can be obtained in a single reaction by selecting enzymes that have no recognition sites within the region of known sequence. Use of enzymes that cleave within the core region allows the recovery of either 3' or 5' flanking regions in separate reactions, as demonstrated in the study described above. To obtain a convenient fragment, it is sometimes necessary to use two restriction enzymes that produce incompatible ends. In these cases, we make the ends of these restriction fragments blunt with Klenow or T4 polymerase prior to ligations.

4. Several procedures can aid the specificity of the PCR. When multiple or nonspecific products are generated by the PCR, and the annealing temperature has already been elevated to the computed T_d of the amplification primers $[T_d = 2(A + T) + 4(G + C)]$, we synthesize an internal oligonucleotide to the 3' side of one of the original amplification primers. Reactions with the nested primers follow either of two procedures: (1) The PCR products are analyzed on an agarose gel and the fragment of correct size is extracted from the gel by poking with the narrow end of a Pasteur pipette. The agarose plug (5 to 10 μl) is dispensed into 100 μl of distilled H_2O and heated to 95° for 5 min to melt the agarose and disperse the DNA. One microliter of this diluted sample is used as the template in second-stage reactions with the internal nested primer and the original primer designed to anneal to the complementary strand. (2) Alternatively, 1 μl of the reaction mixture can be used as the template for a second reaction containing the nested and the original complementary primers without intermediate purification by gel electrophoresis. (The amount of original primers carried over into the secondary reactions is not sufficient to interfere with the subsequent amplification of a specific fragment.) The primary reaction is typically carried out for only 10 to 15 cycles in a 10-μl reaction volume.

5. Once reliable results are obtained, the protocol for inverse PCR can be optimized to eliminate DNA purification procedures.[41] Following restriction endonuclease digestions, samples are heated to inactivate the enzyme and diluted to the proper concentration for ligations without the intervening phenol extractions and ethanol precipitations. Templates used for the PCR from the ligation reaction can be aliquotted directly into the reaction mixes because the reagents in the ligation buffer do not interfere with the activity of *Taq* polymerase.

[41] H. Ochman, M. M. Medora, D. Garza, and D. L. Hartl, *in* "PCR Protocols: Methods and Applications" (M. A. Innis, D. H. Gelfand, J. J. Sninsky, and T. J. White, eds.), p. 219. Academic Press, San Diego, 1989.

[22] Amplification of Complementary DNA from mRNA with Unknown 5' Ends by One-Way Polymerase Chain Reaction

By MARY JANE GEIGER, MICHAEL BULL, DAVID D. ECKELS, and JACK GORSKI

Introduction

The polymerase chain reaction (PCR)[1] is a powerful technique for *in vitro* amplification of specific DNA or RNA sequences. Conventional PCR requires that the sequences flanking the target of interest be known. Specific oligonucleotide primers corresponding to these flanking regions are used to amplify the target sequence through repetitive cycles of denaturation, annealing, and extension. This requirement, however, limits conventional PCR to only target sequences that are known *a priori*. We report here a detailed method for amplifying sequences with uncharacterized 5' termini but known 3' termini, such as immunoglobulin and T cell receptor (TCR) sequences whose 5' termini are encoded by variable gene segments. Poly(A)+ RNA is reverse transcribed and a poly(dG) tail is added to extend the cDNA. This tail effectively substitutes for the unknown 5' sequence information. Using an oligonucleotide complementary to the tail in conjunction with a 3'-specific primer, the target of interest can be amplified. The amplified products are readily ligated into a sequencing vector as a result of restriction enzyme sites engineered into the primers. We have

[1] R. D. Saiki, D. H. Gelfand, S. Stoffel, S. J. Scharf, R. Higuchi, G. T. Horn, D. B. Mullis, and H. A. Erlich, *Science* **239**, 487 (1988).

METHODS IN ENZYMOLOGY, VOL. 218

employed this technique, referred to as one-way PCR, to amplify the variable encoded domains of TCR α and β chain cDNAs from both alloreactive T lymphocyte clones (TLCs) and peripheral blood mononuclear cells (PBMCs). This method has widespread application to the analysis of any target for which some 3' sequence is known.

Material

T lymphocyte clones have previously been studied as part of the Tenth International Histocompatibility Workshop.[2] Oligonucleotide primers and probes are synthesized using an automated DNA synthesizer (Gene Assembler; Pharmacia, Piscataway, NJ). Oligonucleotides are purified using Poly-Pac cartridges according to manufacturer protocols (Glen Research, Sterling, VA). The DNA vector PGEM 7 is from Promega (Madison, WI), and the DH5α-competent cells are from Bethesda Research Laboratories (Gaithersburg, MD).

Moloney murine leukemia virus (MOMLV) reverse transcriptase is purchased from Bethesda Research Laboratories. RNAguard is from Pharmacia. Terminal deoxynucleotidyltransferase and tailing buffer are from Bethesda Research Laboratories. Restriction enzyme SacI is purchased from New England BioLabs (Beverly, MA); XbaI is from Molecular Biology Resources (Milwaukee, WI); and BamHI is from Promega. AmpliTaq is purchased from Perkin-Elmer Cetus (Norwalk, CT).

DNA amplification is performed using a DNA thermal cycler from Perkin-Elmer Cetus. Centricon-100 microconcentrators are purchased from Amicon (Danvers, MA). The DNA sequencing kit is from Pharmacia. LE and NuSieve agarose are from FMC BioProducts (Rockland, ME). [α-^{32}P]DGTP (3000 Ci/mmol) is purchased from New England Nuclear Research Products, Du Pont (Boston, MA). Oligo(dT)- and oligo(dC)-cellulose are from Pharmacia.

Solutions include the following:

Loading buffer (LB): 0.5 M LiCl, 10 mM Tris-HCl (pH 7.5), 1 mM ethylenediaminetetraacetic acid (EDTA), 0.1% (w/v) sodium dodecyl sulfate (SDS)

Middle wash buffer (MWB): 0.15 M LiCl, 10 mM Tris-HCl (pH 7.5), 1 mM EDTA, 0.1% (w/v) SDS

elution buffer (EB): 2 mM EDTA, 0.1% (w/v) SDS

TBE buffer (stock): 1.0 M Tris, 1.0 M boric acid, 20 mM EDTA

[2] D. D. Eckels, M. J. Geiger, T. W. Sell, and J. A. Gorski, Hum. Immunol. **27,** 240 (1990).

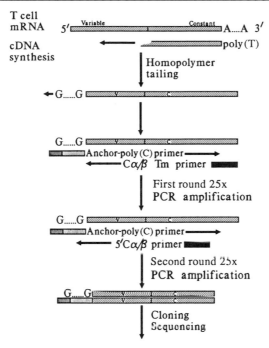

FIG. 1. Diagram of the one-way PCR technique. Poly(A)$^+$ RNA is isolated, reverse transcribed using an oligo(dT) primer, and a poly(dG) tail is added to the 3' terminus of the cDNA. The tailed cDNA is purified by passage through an oligo(dC) column and amplified by PCR using primers complementary to the poly(dC) tail and the constant α or β gene. One-third of the first PCR reaction mixture is added to fresh buffer and further amplified using the poly(dC) oligonucleotide and a constant α or β primer located upstream of those employed in the first amplification. The amplified products are subcloned and sequenced.

Methods

Described below are the cloning and sequencing of TCR genes from alloreactive TLCs as an example of one-way PCR; however, this procedure can be applied to the amplification of any gene whose 3' terminus is known (Fig. 1).

Cells

T cell clones can be used as either frozen pellets or harvested from culture. Peripheral blood mononuclear cells are isolated from buffy coat preparations by centrifuging whole blood over Ficoll-Paque (Pharmacia).

Poly(A)⁺ RNA Isolation

Isolate total RNA from 10^7 cells (TLCs or PBMCs) by homogenization in 4 M guanidinium isothiocyanate buffer and ultracentrifugation through a discontinuous cesium chloride gradient.[3] Recover RNA from the pellet by resuspension in 300 μl of distilled H_2O. Ethanol precipitate, centrifuge in an Eppendorf tabletop microcentrifuge at 14,000 rpm for 30 min at 4°, and dry the pellet under vacuum. Resuspend precipitated RNA in 95 μl of distilled H_2O, heat to 68° for 10 min, and cool to room temperature. Add 5 μl of 10 M LiCl to the RNA sample (final concentration, 0.5 M LiCl) and apply to an oligo(dT) column.[4] Rinse the tube with 100 μl of loading buffer (LB) and apply to the column, followed by an additional 800 μl of LB. Collect the flow-through into a siliconized, diethylpyrocarbonate (DEPC)-treated microcentrifuge tube and recycle it through the column two more times to ensure maximal binding of the poly(A)⁺ RNA. Apply fresh LB to the column until the OD_{260} returns to background. Rinse the column with 2.7 ml of middle wash buffer (MWB). To elute RNA off the column, pass 300-μl aliquots of elution buffer (EB) through the column until the OD_{260} returns to background. Collect all elutions into siliconized, DEPC-treated microcentrifuge tubes. Precipitate poly(A)⁺ RNA from the appropriate fractions by adding sodium acetate until the solution is 0.3 M sodium acetate and then adding 2.5 vol of 100% ethanol. To regenerate the column, pass several column volumes of 0.1 N NaOH through the column followed by several column volumes of distilled H_2O, then LB until the pH is 7.0–8.0.

cDNA Synthesis

Centrifuge and dry precipitated poly(A)⁺ RNA. Resuspend the RNA in distilled H_2O, bring to 68° for 2 min, and cool on ice. Perform a 50-μl cDNA synthesis reaction in 50 mM Tris-HCl (pH 8.3), 75 mM KCl, 3 mM MgCl$_2$, 10 mM dithiothreitol (DTT), 500 units MoMLV reverse transcriptase, 8.5 units RNAguard, 80 pM oligo(dT) primer, and dATP, dCTP, dGTP, and dTTP (500 μM each). Incubate for 45–60 min at 42°. To ensure the mRNA did not degrade and the cDNA synthesis worked efficiently, we use conventional PCR to amplify the constant portion of the TCR cDNA from 2 μl of the cDNA synthesis reaction. The primers used are directed against the TCR constant α (TCR-Cα1' and TCR-Cα3) or β (TCR-

[3] J. M. Chirgwin, A. E. Przybyla, R. J. MacDonald, and W. J. Rutter, *Biochemistry* **18**, 5294 (1979).
[4] F. M. Ausebel, R. Brent, R. E. Kingston, D. D. Moore, J. G. Seidman, J. A. Smith, and K. Struhl, eds., "Current Protocols in Molecular Biology." Wiley, New York, 1987.

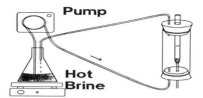

FIG. 2. Heated oligo(dC) column. Schematic diagram of apparatus used for heating the oligo(dC) column. Brine (NaCl-saturated water) was used as the circulating liquid to ensure against heat loss in the tubing since the boiling point is raised.

Cβ1' and TCR-Cβ3) (Table I) genes. The PCR conditions are 10 mM Tris-HCl (pH 8.3), 50 mM KCl, 1.5 mM MgCl$_2$, 0.01% (w/v) gelatin, dNTPs (200 mM each), primers (1 μM each), and 2.5 units of AmpliTaq. Twenty-five to 50 cycles of amplification are performed using a DNA thermal cycler with 1.0-min denaturations at 95°, 1.0-min annealings at 50°, and 1.0-min extensions at 73°. The amplified products are separated by electro-phoresis on a 1% (w/v) LE–2% (w/v) NuSieve gel in a 1 : 10 dilution of TBE buffer (results not shown).

Spermine Precipitation of cDNA

Add spermine (100 mM stock) to the cDNA reaction mixture for a final concentration of 10 mM (total volume, 100 μl), vortex, and incubate on wet ice for 1.5 hr. Centrifuge the sample in an Eppendorf tabletop centrifuge at 14,000 rpm for 25–30 min at 4°. Carefully pipette off the supernatant, resuspend the pellet in 25 μl of extraction buffer (75% ethanol, 0.3 M sodium acetate, 10 mM magnesium acetate) and incubate on ice 1 hr, vortexing every 20 min. Centrifuge and dry the sample as above.

Homopolymer Tailing of cDNA

Dissolve the pellet in 5× tailing buffer supplied by Bethesda Research Laboratories [100 mM potassium cacodylate (pH 7.2), 2 mM CoCl$_2$, 200 μM DTT]. Place in a 37° water bath. Add 1 μl of 500 mM dGTP, 2.5 μl of [α-^{32}P]dGTP, (3000 Ci/mmol), and 28 units of terminal deoxynucleotidyl-transferase (final volume, 20 μl). Incubate at 37° for 30 min. Freeze at −80° to inactivate the enzyme. Do not heat inactivate.

Passage of Homopolymer Tailed cDNA through Oligo(dC) Column

Assemble an oligo(dC) column according to the protocol for an oli-go(dT) column, but a heated water jacket is used as shown in Fig. 2.[4] Check the binding characteristics of the column with poly(dG). Use the same solutions to run this column as were used to run the oligo(dT) column. Rinse the column with LB until the pH is 7.0–8.0. Adjust the sample to 0.5 M LiCl using 10 M LiCl (final volume, 100 μl). Apply the sample to the column, followed by an additional 1.2 ml of LB. Collect the flow-through into a siliconized, DEPC-treated tube and pass it through the column two more times. Wash the column with an additional 1.2 ml of LB followed by 1.2 ml of MWB. Seal off the column and circulate hot water (95–100°) around it. Heat EB to 100° and apply 300 μl to the column. After 5 min remove the column plug and collect the eluate into a siliconized, DEPC-treated tube. Pass additional 300-μl aliquots of prewarmed EB through the column until the tailed cDNA is eluted, as determined by counting each fraction until background levels are achieved. Precipitate fractions containing the tailed cDNA by adjusting each to 0.1 M KCl with 2 M KCl, adding 2 vol of 100% ethanol and placing at $-20°$ overnight. Centrifuge the samples in an Eppendorf tabletop centrifuge at 14,000 rpm at 4° for 30 min. Rinse the pellets with ethanol, dry under vacuum, resuspend the pellets in distilled H_2O, and pool so that the final volume is 20 μl.

To determine if precipitation of the tailed cDNA was successful we employ conventional PCR to amplify a 1-μl aliquot, using primers directed against either the TCR Cα or Cβ region. Fifty cycles of amplification are performed with 1.0-min denaturations at 95°, 1.0-min annealings at 53°, and 1.0-min extensions at 73°. The amplified products are analyzed by electrophoresis on a 1% (w/v) LE–2% (w/v) NuSieve gel (in a 1 : 10 dilution of TBE buffer stock).

One-Sided Amplification of Homopolymer Tailed cDNA

Stage I. Amplify 5 μl of poly(dG) tailed cDNA in 1× PCR buffer [1× PCR buffer: 10 mM Tris-HCl (pH 8.3), 50 mM KCl, 1.5 mM MgCl$_2$, 0.01% (w/v) gelatin, dNTPs (200 mM each), primers (1 μM each)] in a final volume of 100 μl. The 3' primer is complementary to either the TCR Cα (TCR-Cα3) or Cβ (TCR-Cβ3) transmembrane region and the 5' primer consists of three restriction enzyme sites linked to a series of poly(dC) residues [poly(C-An1)] (Table I). Heat the reaction mixture to 95° for 2 min to denature the tailed cDNA–mRNA complex, cool on ice, and add 2.5 units of AmpliTaq. Perform 25 rounds of amplification with 1.5-min denaturations at 95°, 1.5-min annealings at 53°, and 2.25-min extensions at 73°.

Stage II. Combine one-third of the reaction mixture from stage I with fresh reagents to make a 1× PCR buffer (final volume, 100 μl) containing 1 μM concentrations of the poly(C-An1) primer and either a Cα (TCR-Cα4) or Cβ (TCR-Cβ1) nested primer located at the 5' terminus of the

TABLE I
OLIGONUCLEOTIDE PRIMER SEQUENCES[a]

Primer	Sequence
TCR-Cα1	5'-GGAGCTCAGCTGGTACACGGCAGG-3'
TCR-Cα1'	5'-CCTGCCGTGTACCAGCTGAGA-3'
TCR-Cα3	5'-GGGATCCTGGACCACAGCCGCAGCGTCAT-3'
TCR-Cα4	5'-CTGAGCTCACTGGATTTAGAGTC-3'
TCR-Cβ1	5'-GGGATCCAGATCTCTGCTTCTGATGGCTC-3'
TCR-Cβ1'	5'-AGAATTCGAGCCATCAGAAGCAGAGATCT-3'
TCR-Cβ2	5'-GGGATCCGACCTCGGGTGGGAACA-3'
TCR-Cβ3	5'-GGGATCCTTTCTCTTGACCATGGCCAT-3'
Poly(C-An1)	5'-AAGAGCTCTAGAGCGGCCGC(C)₁₃-3'
Reverse	5'-GGAAACAGCTATGACCATGATTACGCC-3'
Universal	5'-CACGACGTTGTAAAACGACGGCCAGTG-3'

[a] Restriction enzyme sites are underlined: BamHI (Cα3, Cβ1, Cβ2, Cβ3); EcoRI (Cβ1'); NcoI (Cβ3); NotI (An1); PvuII (Cα1); SacI (Cα1, Cα4, An1); and XbaI (Cβ1, An1).

respective constant genes (Table I). Perform 25 rounds of amplification with 2.5 units of AmpliTaq, using the same conditions employed for stage I PCR. Analyze the amplified products by electrophoresis through a 1% (w/v) LE–2% (w/v) NuSieve gel in a 1 : 10 dilution of TBE buffer. If desired, the DNA can be blotted onto Gene Screen or comparable nylon membrane and hybridized with TCR Cα or Cβ chain-specific probes.

Centricon-100 Purification of Amplified Tailed cDNA

Take one-half of the stage II PCR reaction mixture and bring it to a final volume of 100 μl with distilled H_2O. Extract with an equal volume of phenol : chloroform : isoamyl alcohol (50 : 50 : 1) and then with chloroform. Back extract with 100 μl of distilled H_2O. Rinse a Centricon-100 microconcentrator with 2 ml of 10 mM Tris-HCl (pH 7.4), 1 mM EDTA (TE) in a Clay Adams SERO-FUGE II centrifuge (Becton-Dickenson, San Jose, CA) at 3500 rpm for 2 min. Add both the sample and the back extraction to the prerinsed Centricon-100. Bring to a final volume of 2 ml with TE. Centrifuge as above for approximately 3 min. Discard the flow-through. Bring the volume to 2 ml again, using TE, and spin as above. Repeat this process a third time until the retentate volume is minimized. Cap the tube, invert, and centrifuge for 2 min at 3500 rpm to collect retentate. To precipitate the amplified products from the retentate, make the sample 0.3 M in sodium acetate and add 2 vol of 100% ethanol. Store at $-20°$ overnight. Centrifuge the sample in an Eppendorf tabletop centrifuge at 14,000 rpm at 4° for 30 min. Rinse the pellets with 80% ethanol, dry under vacuum, and resuspend in 15 μl of distilled H_2O.

Subcloning and Screening of Amplified Tailed cDNA

Restriction sites designed into the PCR primers facilitated the subcloning procedure (Table I). Sequentially digest TCR α chain amplified cDNAs with *Sac*I, then *Xba*I. Simultaneously digest TCR β chain amplified products with *Bam*HI and *Xba*I. Ligate digested DNA into PGEM 7[5] and transform into DH5α-competent cells according to the Bethesda Research Laboratories product insert. Screen positive (white) colonies for inserts using [γ-^{32}P]ATP-labeled probes specific for the 5' region of either the Cα (TRC-Cα1) or Cβ (TCR-Cβ2) regions (Table I).[5] These primers are internal to those used for the amplification.

Additional Screening by Amplifying Directly from Bacterial Colonies

Because the purification procedure eliminates only amplified products 100 bp or smaller, colonies testing positive with radiolabeled oligonucleotide probes are further screened for inserts >400 bp in length. These would correspond to cDNAs containing the entire variable region. Using a toothpick, swab a small number of bacteria off the master plate and place them in 40 μl of 1× PCR buffer containing a primer specific for the insert [TCR-Cα1, TCR-Cβ1, or Poly(C-An1)] and one targeted against the vector (universal or reverse) (Table I). Boil the bacteria for 2 min to lyse them. Add 2 units of AmpliTaq and perform 30 cycles of amplification with 1.0-min denaturations at 95°, 1.0-min annealings at 50° and 1.0-min extensions at 73°. Electrophorese through a 1% (w/v) LE–2% (w/v) NuSieve gel in a 1 : 10 dilution of TBE buffer to analyze the amplified products.

Large-Scale Plasmid Preparations and DNA Sequencing

Those colonies with inserts >400 bp in length are grown overnight at 37° in 50-ml cultures of terrific broth.[5] Plasmids are isolated using large-scale plasmid preparation techniques.[5]

Purified DNA is sequenced using the dideoxy chain termination method and a commerical DNA sequencing kit.[6] Both templates were sequenced

[5] J. Sambrook, E. F. Fritsch, and T. Maniatis, "Molecular Cloning: A Laboratory Manual," 2nd ed. Cold Spring Harbor Press, Cold Spring Harbor, New York, 1989.
[6] F. Sanger, S. Nicklen, and A. R. Coulson, *Proc. Natl. Acad. Sci. U.S.A.* **74,** 5463 (1977).

using primers specific to the vector as well as the $C\alpha$ or $C\beta$ region. Sequences were compared with known TCR gene sequences to identify the gene segments.

Isolation of Full-Length cDNA

It is difficult to obtain full-length cDNA clones (i.e., complete V genes) when starting with T cells isolated from PBMCs. This becomes problematic when a new V gene has been identified and only a partial sequence has been generated. To overcome this, the following procedure has been of use.

Synthesize an oligonucleotide primer corresponding to a unique portion of the $3'$ terminus of the V gene. Care should be taken that this does not include possible N region sequence. Determine the optimal PCR conditions for the primer by amplifying a stretch of DNA from the clone from which the new V gene was identified, using a primer corresponding to the vector as the second primer. Amplify the full-length cDNA from the unused stage I material (2–4 μl) employing the V-specific primer and the poly(C-An1) primer for 25 cycles. Analyze a portion of the amplification on a gel. If the V gene-specific amplification results in DNA of sufficient length, it can be subcloned or directly sequenced. If the V gene-specific amplification results in heterogeneously sized DNA, perform a second electrophoresis using a large amount of amplified material (>4 μg) and visualize with ethidium bromide. Excise the highest molecular weight DNA and trim away excess nonfluorescent agarose. Amplify a portion of the gel slice with a partially overlapping nested primer (50% overlap) for another 25 cycles. The amplified material should consist of only full-length cDNA that can be used for subcloning and sequencing.

Results

Using one-way PCR we have analyzed both the TCR α- and β-chain cDNAs expressed by six alloreactive TLCs (Fig. 3). The size of the amplified products ranged from approximately 350 to 650 bp, which corresponds to cDNAs extending from the $5'$ terminus of the C gene to the V region and, in some instances, to part of the $5'$-untranslated sequence. Following transformation, the colonies were screened with either a $C\alpha$- or $C\beta$-specific probe. More than 75% of the colonies tested positive. However, further screening by amplification from the bacterial colonies showed that only 1 or 2 out of 10 colonies selected for amplification actually contained an insert of 350 bp or greater. These in turn were sequenced and compared

1 2 3 4 5 6 7 8 9 10 11 12 13 14

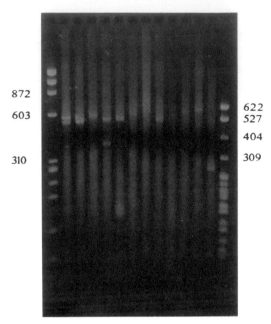

872

603

310

622
527

404

309

FIG. 3. TCR α- and β-chain PCR products derived from amplification of poly(dG) cDNA. The products range from 350 to 650 bp. Lane 1, φX174/HaeIII marker. Lanes 2–7 represent TCR β chains from the following TLCs: Lane 2, HA1.70; lane 3, AL62.119; lane 4, AL62.173; lane 5, AL62.178; lane 6, AL63.7; and lane 7, AL63.124. Lanes 8–13 represent TCR α chains from the following TLCs: Lane 8, HA1.70; lane 9, AL62.119; lane 10, AL62.173; lane 11, AL62.178; lane 12, AL63.7; and lane 13, AL63.124. Lane 14, PBR322/Msp. Molecular weight markers are indicated.

with known TCR families to identify the gene sequences.[7] Antigen-specific T cell clones have also been used with this method (M. J. Geiger, J. Gorski, and D. D. Eckels, unpublished observations, 1991).

An example of cloning TCR α- and β-chain cDNA from PBMCs derived from 30 ml of blood is also shown (Fig. 4). Peripheral blood mononuclear cells were obtained from buffy coat preparations. Poly(A)$^+$ RNA was isolated from the PBMCs and reverse transcribed using an oligo(dT) primer. The products were tailed and amplified as described. After two-stage amplification we were able to visualize a smear corresponding to cDNAs of appropriate size on an ethidium bromide-stained gel (Fig. 4A). The amplified products were spotted and hybridized with specific TCR Cα

[7] M. J. Geiger, J. Gorski, and D. D. Eckels, *J. Immunol.* **147,** 2082 (1991).

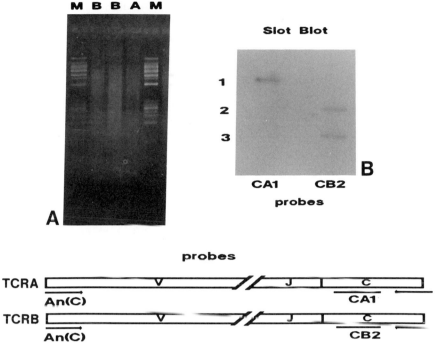

FIG. 4. Total RNA was prepared from PBMCs from 30 ml of blood. Poly(A)$^+$ RNA was reverse transcribed, tailed with dGTP residues, and amplified using the one-way PCR method. (A) An aliquot of the amplified material was electrophoresed and stained with ethidium bromide (Lanes: M, pBR322 digested with *Bst*NI; B, TCR β-chain; A, TCR α-chain). (B) Slot blot of the amplified material hybridized with either a TCR Cα- or Cβ-specific probe.

and Cβ oligonucleotides (Fig. 4B). This method has been employed to generate over 100 TCR α and β clones from 2 individuals.[8]

Discussion

One-way PCR broadens the applicability of conventional PCR in that RNA or DNA can be amplified in instances where only some of the sequence is known, namely the 3' terminus. This is a more efficient method than generating a cDNA library and requires considerably less starting material. In addition, when sequencing genes that encode proteins with variable regions, such as TCRs and immunoglobulins, one-way PCR is less costly than synthesizing primers against all known sequences and

[8] M. Yassai, M. Bull, and J. Gorski, *Hum. Immunol.,* in press.

trying various primer combinations to amplify the unknown cDNA. The multiple or degenerate primer approach also risks missing TCR genes whose sequences differ from the known sequences used for primer generation. Of course, certain steps in this protocol may need to be modified. For example, annealing temperatures will vary according to the oligonucleotides selected and extension times will depend on the length of target DNA to be amplified.

One-way PCR is not without technical difficulties. Initially, we began with a cDNA pool generated from total RNA, but the tailing reactions were inefficient, so we switched to poly(A)$^+$ RNA as the starting material. Oligo(dT) was selected over a TCR constant gene-specific primer for first-strand synthesis because, with a TCR-specific oligonucleotide, too few cDNAs were generated to permit terminal deoxynucleotidyltransferase to catalyze the tailing reaction effectively. Incomplete cDNA synthesis also was a problem. Precipitation with spermine proved to be the best method for purifying the cDNAs away from unincorporated nucleotides and enzyme. Alkaline treatment and passage through a Centricon-30 microconcentrator were tested but little cDNA was recovered, presumably due to adherence of the single-stranded cDNA to the dialysis membrane. Elimination of unused nucleotides must be efficient or else they will be incorporated by the terminal transferase in the tailing step.

The homopolymer tailing reaction introduced additional dilemmas. Selection of the deoxyribonucleotide triphosphate proved critical. Initially dCTP was used for polymerization of the tails, as in the Okayama and Berg tailing procedure,[9] then dATP. We found that the tails were quite long, approximately 50 bp, regardless of the concentration of dNTPs used. It was difficult to amplify products that would be visible as a discrete band on a gel, probably because the poly(C-An1) oligonucleotide was binding at various positions along the tail during second-strand synthesis. We then tailed with dGTP, because it has been suggested that the homopolymer tail terminates spontaneously after approximately 15–20 nucleotides.[10] Heating is generally used for inactivating terminal deoxynucleotidyltransferase; however, when we did this, our amplifications were unsuccessful. We therefore inactivated the enzyme by freezing at $-80°$. To simplify the one-way PCR protocol, we originally tried to amplify an aliquot of the tailing reaction but the cobalt present in the tailing buffer precipitated out of solution when an aliquot was added to the PCR buffer. To eliminate the unincorporated dGTP residues and tail buffer and, more specifically, to enrich for the population of tailed cDNAs, the tailing reaction mixture was passed through an oligo(dC) column. Although [α-^{32}P]dGTP was utilized

[9] H. Okayama and P. Berg, *Mol. Cell. Biol.* **2,** 161 (1982).
[10] W. H. Eschenfeldt, R. S. Puskas, and S. L. Berger, this series, Vol. 152, p. 337.

to monitor the efficiency of the tailing reaction and to act as a tracer during passage through the oligo(dC) column, it can be eliminated, and the tailing reaction works just as effectively with the same final concentration of cold dGTP (25 mM) alone.

The two-stage amplification procedure with half-nested primers increased the ability to amplify TCR-specific cDNAs preferentially. We tried amplifying the same reaction mixture using the same set of primers for 50 cycles; however, we were able to visualize amplified products only by Southern blotting, not on an ethidium bromide-stained gel. Similar to Loh et al.,[11] a primer was designed composed of only the three restriction enzyme sequences found in the poly(C-An1) primer. Varying ratios of this oligonucleotide and the poly(C-An1) were used to enhance the amplification of TCR-specific tailed cDNAs during the first 25 cycles of amplification. During the second 25 rounds, only the restriction enzyme-specific primer was employed. We found no enhancement of the amplification using this procedure in comparison to amplifying with the poly(C-An1) primer alone in both PCR stages. Also like Loh et al.,[11] we electrophoresed an aliquot of the first PCR reaction mixture, excised a gel slice corresponding to the expected size of TCR-amplified products, and amplified the DNA contained in about one-third of the slice. We had limited success with cDNA from TLCs but moderate success with cDNA from PBMCs.

It would be ideal to purify only the amplified products corresponding to the expected molecular weight size. We attempted this by excising the appropriately sized band from the gel and purifying the DNA away from the agarose using a GeneClean kit (Bio 101, La Jolla, CA). However, the DNA permanently bound to the silica matrix used in the kit, perhaps due to the high hydrophobicity inherent to the tail. Electroelution was not tried but is an alternative. The best purification we could attain was with the Centricon-100 microconcentrator, which eliminated unincorporated nucleotides, excess primers, primer dimers, and any DNA less than about 100 bp. The Ultrafree-MC filter unit (100,000 NMWL polysulfone PTHK membrane; Millipore, Bedford, MA), a dialysis system similar to the Centricon-100 microconcentrator, was tested but the amplified products bound irreversibly to the dialysis membrane. An alternative to agarose gel purification that could be used, if the apparatus is available, is high-performance liquid chromatography (HPLC) purification using an anion-exchange column.

Because purification on the basis of size was not possible, a double-screening process was employed. Hybridization with radiolabeled, constant region-specific oligonucleotides selected those plasmids with TCR-specific inserts. Unfortunately such primers hybridized to all TCR inserts

[11] E. Y. Loh, J. F. Elliott, S. Cwirla, L. L. Lanier, and M. M. Davis, Science 243, 217 (1989).

regardless of size. Thus, bacterial colony screening via PCR was employed to identify plasmids with inserts greater than 400 bp, because these corresponded to TCR cDNA composed of *V, D, J,* and *C* gene segments.

To facilitate the ligation procedure, restriction sites were incorporated into the PCR primers. The recognition sites were selected based on the observation that most of the known TCR genes do not contain these sequences. Nonetheless, this was not always true. Several TCR β-chain sequences were found to contain either a *Bam*HI or an *Xba*I restriction site. These would continuously give small inserts after digestion and ligation. Thus, selection of appropriate restriction enzyme sites is imperative. Occasionally very large inserts were isolated that resulted from the ligation of several small DNA fragments.

When amplifying cDNA from either T or B cells, it must be remembered that the PCR methodology is powerful enough to amplify partial or aberrant transcripts that originate from unspliced RNA, from the nonproductively rearranged chromosome, or from alternately spliced RNAs. Because this protocol will amplify such species, examples of these should be expected during sequence analysis.

Sequence analysis of the cDNAs provided additional insight into the DNA amplification process and cDNA synthesis. Although infrequent, a few deoxynucleotides not encoded by the variable gene were identified at the 5' terminus of the cDNA, preceding the poly(dG) tail. These may have been incorporated during the tailing reaction or during the cDNA synthesis. Therefore, the last several bases of the 5' terminus were not considered part of the sequence. *Taq* polymerase has been reported to have a relatively high error rate[1,12,13] with T–C transitions occurring more often than A–G substitutions.[14] Comparison of TCR variable sequences with those previously published showed occasional nucleotide changes. Some were silent; others would result in amino acid substitutions. To determine whether these changes represented *Taq* replication errors, sequences were derived from two independent amplifications or from a sample amplified by conventional PCR, after *V* gene usage had been determined. The majority of these base pair substitutions were T–C transitions.

Several similar protocols have been previously described to amplify cDNAs with uncharacterized 5' termini.[11,15,16] Ours most closely resembles

[12] A. M. Dunning, P. Talmud, and S. E. Humphries, *Nucleic Acids Res.* **16,** 10393 (1988).
[13] J. W. Larrick, L. Danielsson, C. A. Brenner, E. F. Wallace, M. Abrahamson, K. E. Fry, and C. A. K. Borrebaeck, *Bio/Technology* **7,** 934 (1989).
[14] M. Krawczak, J. Reiss, J. Schmidtke, and U. Rosler, *Nucleic Acids Res.* **17,** 2197 (1989).
[15] M. A. Frohman, M. K. Dush, and G. R. Martin, *Proc. Natl. Acad. Sci. U.S.A.* **85,** 8998 (1988).
[16] O. Ohara, R. L. Dorit, and W. Gilbert, *Proc. Natl. Acad. Sci. U.S.A.* **86,** 5673 (1989).

that developed by Loh *et al.*[11] Frohman *et al.*[15] coined the "rapid amplifi-cation of cDNA ends" (RACE) procedure, and Ohara *et al.*[16] the "one-sided PCR." Several of the subtle differences surround selection of the starting material [total RNA vs poly(A)$^+$ RNA], the primer used for first-strand synthesis [oligo(dT) vs a specific primer], the procedures used to purify the cDNA prior to the tailing reaction, and the dNTP selected for polymerization of the tail (dATP vs dGTP). The major difference is the manner in which the tailed cDNA is repetitively amplified. Most continue to amplify the cDNA from agarose gel slices but, as noted earlier, we have tried this and met with limited success. Several other differences have been previously mentioned. Nonetheless, the use of nested primers is essential to amplify the target sequence specifically, as evidenced by all protocols.

The advantages of the method detailed here are the ability to start with a small amount of material and yet obtain visible proof of amplification in a straightforward manner. When starting with blood, smaller amounts of RNA are obtained and tailing of a specifically reverse transcribed RNA is inefficient. For this reason, all poly(A)$^+$ RNA is converted into cDNA. Full-length cDNA is favored in the tailing reaction by leaving the cDNA–RNA hybrid in place. Finally, the selection of cDNA whose dGTP tails are long enough to permit binding to an oligo(dC) column overcomes inefficiencies in the tailing reaction and probably contributes to the ease with which this technique can be implemented.

Since the submission of this protocol in 1991, a number of publications have appeared using the "anchor-PCR" approach. These have included analysis of T cell clones responding to antigen,[17] analysis of T cells isolated from the site of an autoimmune response,[18] and general analysis of periph-eral T cell repertoires.[19–22]

Acknowledgments

    This work was supported by Grants AI22832 (D.D.E.), AI26085 (J.G.), P01HI44612, and by the Blood Center Research Foundation. D.D.E. is the recipient of Research Career Development Award AI00799. We would like to thank Ms. Marcia Iverson for editorial assistance.

[17] P. A. H. Moss, R. J. Moots, W. M. C. Rosenberg, S. J. Rowland-Jones, H. C. Bodmer, A. J. McMichael, and J. I. Bell, *Proc. Natl. Acad. Sci. U.S.A.* **88,** 8987 (1991).
[18] G. Pluschke, G. Richen, H. Taube, S. Kroninger, I. Melchers, H. H. Peter, K. Eichmann, and U. Krawinkel, *Eur. J. Immunol.* **21,** 2749 (1991).
[19] M. A. Robinson, *J. Immunol.* **146,** 4392 (1991).
[20] S. Roman-Roman, L. Ferradini, J. Azocar, H. Michalaki, C. Genevee, T. Hercend, and F. Triebel, *Eur. J. Immunol.* **21,** 927 (1991).
[21] W. M. C. Rosenberg, P. A. H. Moss, and J. I. Bell, *Eur. J. Immunol.* **22,** 541 (1992).
[22] C. Lunardi, C. Marguerie, and A. K. So, *Immunogenetics* **36,** 314 (1992).

[23] Amplification of Bacteriophage Library Inserts Using Polymerase Chain Reaction

By DAVID M. DORFMAN

Introduction

DNA inserts cloned in λgt bacteriophage vectors can be amplified directly from bacteriophage plaques using the polymerase chain reaction (PCR)[1] with oligonucleotide primers corresponding to vector sequences flanking the cloning site.[1,2] The amplified insert can be subcloned into a plasmid vector following digestion with restriction enzyme that cleaves at the cloning site, or it can be used for other purposes, for example, to generate radioactively labeled probe, or as sequencing template. This procedure eliminates the need to grow bacteriophage, to perform bacteriophage DNA preparations following routine library screening, or even to elute bacteriophage from agarose plugs.

Principle of Method

Bacteriophage plaques contain unpackaged DNA in sufficient quantities that PCR can be performed directly on small quantities of plaque material. Polymerase chain reaction oligonucleotide primers that correspond to bacteriophage DNA sequences flanking the foreign DNA cloning site are used. This allows PCR amplification irrespective of the foreign DNA sequence, and also results in the generation of PCR products containing the cloning site restriction enzyme recognition sequence, which can be used to generate sticky ends for subcloning the insert into plasmid vector.

Materials and Reagents

Polymerase chain reaction is performed using *Taq* polymerase, available from Perkin-Elmer Cetus (Norwalk, CT) or other suppliers. Forward and reverse λgt10 [5'-d(AGCAAGTTCAGCCTGGTTAAG)-3' and 5'-d(CTTATGAGTATTTCTTCCAGGGTA)-3', respectively] or λgt11 5'-

---

[1] R. K. Saiki, D. H. Gelfand, S. Stoffel, S. J. Scharf, R. Higuchi, G. T. Horn, K. B. Mullis, and H. A. Erlich, *Science* **239,** 487 (1988).

[2] D. M. Dorfman, L. I. Zon, and S. H. Orkin. *BioTechniques* **7,** 568 (1989).

METHODS IN ENZYMOLOGY, VOL. 218

d(GGTGGCGACGACTCCTGGAGCCCG)-3' and 5'-d(TTGACACCA-GACCAACTGGTAATG)-3', respectively] oligonucleotide sequencing primers are used for PCR, and are generated using a 380B DNA synthesizer (Applied Biosystems, Inc., Foster City, CA) or similar apparatus, or are available from New England BioLabs (Beverly, MA). A 10× PCR buffer stock, which is stable at room temperature, consists of 500 mM KCl, 100 mM Tris-HCl, pH 8.3, 15 mM MgCl$_2$, 0.1% (w/v) gelatin.[1] A mixed stock of deoxynucleotide triphosphates (dATP, dCTP, dGTP, and TTP), 2.5 mM for each, is prepared and stored at −20°.

Methods

Phage plaques to be amplified are touched with a sterile toothpick, which is then agitated in a standard 50-μl PCR reaction mix consisting of 1× PCR buffer, 200 μM dATP, dCTP, dGTP, and TTP, 1.0 mM concentrations for each oligonucleotide primer, and 2.5 U of Taq polymerase.[1,2] The mixture is overlaid with mineral oil, and PCR is allowed to proceed for 30 cycles in a Perkin-Elmer Cetus thermal cycler or similar apparatus (programmed for 15 sec at 94°, 15 sec at 55°, and 30 sec at 72° per cycle, followed by a final step of 7 min at 72°).[2] A 5–10% aliquot of each reaction can then be electrophoresed on an ethidium bromide-stained agarose gel to visualize the PCR product. The PCR reaction mix can be extracted with phenol, then chloroform, followed by ethanol precipitation,[3] after which it is suitable for restriction endonuclease digestion, for subcloning.

Polymerase chain reaction can also be performed directly on bacteriophage lysates prepared from relevant plaques,[2] in which case 1–5 μl of lysate in a 50-μl PCR reaction mix is usually sufficient. The fast PCR program described above has worked for λgt bacteriophage inserts as large as 6.5 kb.[2] Alternatively, it may be necessary to use slower PCR programs and/or lower annealing temperatures (e.g., 1 min at 94°, 1 min at 37–55°, and 3 min at 72°), or to optimize cycle times and temperatures for specific inserts and vectors. The use of Perfectmatch (0.5 U/50-μl PCR reaction; Stratagene, La Jolla, CA) may improve yields. For DNA inserts with a high GC content, the substitution of 50 μM 7-deaza-dGTP for 50 μM dGTP improves the yield considerably.[4]

[3] J. Sambrook, E. F. Fritsch, and T. Maniatis, "Molecular Cloning: A Laboratory Manual." Cold Spring Harbor Press, Cold Spring Harbor, New York, 1989.
[4] L. McConlogue, M. A. D. Brow, and M. A. Innis, $Nucleic Acids Res.$ **16**, 9869 (1988).

Example

A λgt11 human endothelial cell library[5] was screened for human prepro-endothelin[6] cDNA inserts, using an oligonucleotide probe corresponding to the endothelin peptide-coding sequence. Tertiary screen-positive plaques were used to prepare bacteriophage lysates by standard methods, or, alternatively, were amplified by PCR, as described above. Ten microliters of each reaction was electrophoresed on an ethidium bromide-stained 1.0% (w/v) agarose gel, which was then Southern blotted and probed with preproendothelin cDNA, using standard methods.[3] Figure 1A shows the PCR products generated from the two tertiary screen-positive plaques, using either plaque material or phage lysate as PCR template. Polymerase chain reaction was also performed on a primary screen-positive plaque, which did not result in a visualizable PCR product, although one was apparent by Southern blotting (see Fig. 1B). This suggests that PCR performed on a primary screen-positive plaque produces sufficient quantities of a desired phage library insert for detection and sizing by Southern blotting as well as for other uses, such as plasmid subcloning (if one adds an additional screening step for the desired insert). Others have reported that it is possible, using insert-specific probes, to generate specific PCR products out of libraries without any screening steps.[7] Figure 1B shows a Southern blot demonstrating that PCR products generated from either plaque material or phage lysates contain sequences that hybridize to the appropriate cDNA probe. Figure 1A also demonstrates that Blue-Gal (Bethesda Research Laboratories, Gaithersburg, MD)/isopropyl-β-D-thio-galactopyranoside (IPTG) selection can be used to select insert-containing plaques for PCR. Polymerase chain reaction performed on random clear, β-galactosidase-negative insert-containing plaques generates PCR products of various sizes, whereas no PCR products are generated from random blue, β-galactosidase-positive plaques lacking cDNA inserts.

Conclusion

A procedure is described to expedite bacteriophage library screening using PCR. The method enables PCR of bacteriophage library inserts to be performed directly from phage plaque material, the product of which can be used for a number of purposes as noted above. It should be possible

[5] D. Ginsburg, R. I. Handin, D. T. Bonthron, T. A. Donlon, G. A. P. Bruns, S. A. Latt, and S. H. Orkin, *Science* **228**, 1401 (1985).

[6] Y. Itoh, M. Yanagisawa, S. Ohkubo, C. Kimura, T. Kosaka, A. Inoue, N. Ishida, Y. Mitsui, H. Onda, M. Fujino, and T. Masaki, *FEBS Lett.* **231**, 440 (1988).

[7] K. D. Friedman, N. L. Rosen, P. J. Newman, and R. R. Montgomery, *Nucleic Acids Res.* **16**, 8718 (1988).

A

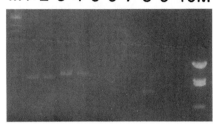

B

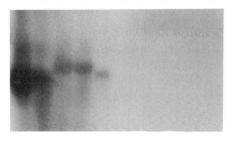

FIG. 1. (A) PCR products electrophoresed in an ethidium bromide-stained 1.0% agarose gel. Lanes 1 and 3, PCR of two different tertiary screen-positive plaques (also independently grown, subcloned, and sequenced by standard methods[3]); lanes 2 and 4, PCR of corresponding tertiary screen phage lysates (5 µl); lane 5, PCR of a primary screen-positive plaque (corresponding to lanes 1 and 3); lanes 6–8, PCR of random clear plaques from the above library plated on a Blue-Gal/IPTG plate; lanes 9 and 10, PCR of random blue plaques from the same plate. Marker positions: M, *Hin*dIII-cut λ bacteriophage DNA (23.1, 9.4, 6.6, 4.4, 2.3, 2.0, and 0.6 kb), M', *Bst*NI-cut pBR322 plasmid DNA (1.9, 1.1, 0.9, and 0.4 kb). (B) Corresponding Southern blot probed with radioactively labeled preproendothelin cDNA. [Reproduced, with modifications, from D. M. Dorfman, L. I. Zon, and S. H. Orkin, *BioTechniques* **7**, 568 (1989).]

to extend the procedure to DNA inserts cloned in other bacteriophage vectors, the only limitation being the size of the DNA insert to be amplified. Additional applications are possible using one or two insert-specific primers for PCR, for example, to determine the orientation of the insert without sequencing or restriction endonuclease mapping, or to map the distance between exons in bacteriophage library genomic inserts. It should also be possible to perform other PCR-based procedures[8] directly from bacteriophage plaques.

[8] M. A. Innis, D. H. Gelfand, J. J. Sninsky, and T. J. White, eds., "PCR Protocols: A Guide to Methods and Applications." Academic Press, San Diego, 1990.

[24] Rapid Amplification of Complementary DNA Ends for Generation of Full-Length Complementary DNAs: Thermal RACE

By MICHAEL A. FROHMAN

Introduction

The generation of a full-length cDNA can represent the most challenging part of a cloning project, particularly with respect to obtaining the 5' end of the cDNA. The initial step—reverse transcription of mRNA—is also the most critical one, and many variations on it have been described to maximize the likelihood of obtaining fully extended cDNAs. The different protocols can profoundly affect the chances of obtaining full-length cDNA clones, but it is generally impractical to establish optimal conditions for reverse transcription of a given mRNA and construction of a library, because weeks to months are required to prepare and screen a single cDNA library and to isolate and analyze candidate cDNAs.

Thermal RACE (rapid amplification of cDNA ends)[1,2] differs markedly in that cDNAs derived from the reverse transcription of mRNA can be analyzed within 1–2 days of the start of the experiment. Consequently, the reverse transcription step can be modified and repeated until it is clear that the desired cDNA has been obtained and is fully extended. Furthermore, such analyses can be used to determine whether alternate splicing or alternate promoter use is occurring and, if so, the corresponding cDNAs produced from different transcripts can

[1] M. A. Frohman, M. K. Dush, and G. R. Martin, *Proc. Natl. Acad. Sci. U.S.A.* **85,** 8998 (1988).
[2] M. A. Frohman and G. R. Martin, *Techniques* **1,** 165 (1989).

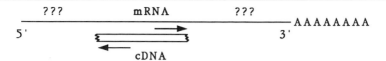

Fig. 1. Schematic representation of the circumstances under which thermal RACE plays a role in cDNA cloning strategies. Depicted is an mRNA for which a cDNA representing only a small and internal portion of the transcript has been cloned. Such circumstances arise frequently, for example, (1) when closely related family members are cloned using PCR and degenerate primers encoding sequences homologous to amino acids found in all known members of the family, and (2) when incomplete cDNAs are obtained after screening conventional or PCR libraries.

be separated prior to cloning steps. Finally, RACE products can be characterized directly on a population level or efficiently cloned into standard high-copy plasmid vectors, allowing many independent isolates to be obtained rapidly.

Thermal RACE [also known as single-sided or "anchored" polymerase chain reaction (PCR)[3]] is a PCR technique through which previously unobtained 3' and 5' ends of a cDNA can be amplified starting with the knowledge of a small stretch of sequence from within an internal region of the cDNA (Fig. 1). This technique provides an alternative to constructing and screening conventional libraries in order to obtain the remainder of the sequence for a partially cloned cDNA.

Principles

Partial cDNAs are generated using PCR to amplify copies of the region between a single point in a mRNA transcript and its 3' or 5' end. To use the RACE protocol, a short, internal stretch of sequence must already be known from the mRNA of interest. From this site, primers are chosen oriented in the 3' or 5' direction. These primers provide specificity during the amplification steps. Extension of cDNAs from the ends of the message back to the region of known sequence is accomplished using primers that anneal to the natural (3' end) or a synthetic (5' end) poly(A) tail. By use of this procedure, enrichments of the desired cDNA in the range of 10^6- to 10^7-fold can be obtained. As a result, relatively pure partial cDNA "ends" are generated, which can be cloned easily or characterized rapidly using conventional techniques.

[3] E. L. Loh, J. F. Elliott, S. Cwirla, L. L. Lanier, and M. M. Davis, *Science* **243,** 217 (1989).

Figure 2 illustrates this strategy in more detail. To generate 3' end partial cDNA clones, mRNA is reverse transcribed using a "hybrid" primer (Q_T) that consists of 17 nucleotides (nt) of oligo(dT) followed by a unique 35-base oligonucleotide sequence (Q_I–Q_O; Fig. 2a and c). Amplification is subsequently carried out using a primer containing part of this sequence (Q_O), which now binds to each cDNA at its 3' end, and using a primer specific to the gene of interest (GSP1). A second round of amplification cycles is then carried out using "nested" primers (Q_I and GSP2) to achieve greater specificity.[2] To generate 5' end partial cDNA clones, reverse transcription (primer extension) is carried out using a gene-specific primer (GSP-RT; Fig. 2b). Then a poly(A) tail is appended using terminal deoxynucleotidyltransferase (TdT) and dATP to tail the first-strand reaction products. Amplification is subsequently carried out using the hybrid primer Q_T described above to form the second strand of cDNA, the Q_O primer, and a gene-specific primer upstream of and distinct from the one used for reverse transcription. Finally, a second round of PCR cycles is carried out using nested primers (Q_I and GSP2) to increase specificity.

Experimental Procedures

Materials

The enzymes required for this procedure can be purchased, along with the appropriate 5× or 10× reaction buffers, from most major suppliers. Moloney Murine leukemia virus (MMLV) reverse transcriptase is obtained from Bethesda Research Labs (Gaithersburg, MD), heat-stable reverse transcriptase and *Taq* polymerase from Perkin-Elmer Cetus (Norwalk, CT), TdT from either BRL or Boehringer Mannheim (Indianapolis, IN), and RNasin from Promega Biotech (Madison, WI). Enzymes are used as directed by the suppliers, except for *Taq* polymerase: Instead of using the recommended reaction mixture, a 10× buffer consisting of 670 mM Tris-HCl (pH 9.0), 67 mM $MgCl_2$, 1700 μg/ml bovine serum albumin (BSA), and 166 mM $(NH_4)_2SO_4$ is substituted, and reaction conditions are altered as further described below. Oligonucleotide primer sequences are listed in the legend to Fig. 2. Primers can be used "crude" except for Q_T, which should be purified to ensure that all of it is full length. dNTPs are purchased as 100 mM solutions from P-L Biochemicals/Pharmacia (Milwuakee, WI).

3' End cDNA Amplification

Step 1: Reverse Transcription to Generate cDNA Templates

Procedure. Assemble reverse transcription components on ice: 4 μl of 5× reverse transcription buffer, 1.3 μl of dNTPs (stock concentration is

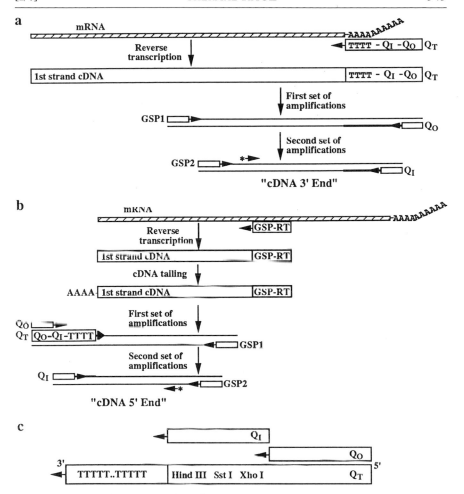

FIG. 2. Schematic representation of thermal RACE. Explanations are given in text. At each step, the diagram is simplified to illustrate only how the new product formed during the previous step is utilized. GSP1, Gene-specific primer 1; GSP2, gene-specific primer 2; GSP-RT, gene-specific primer used for reverse transcription; *→, GSP-Hyb/Seq or gene-specific primer for use in hybridization and sequencing reactions. (a) Amplification of 3' partial cDNA ends. (b) Amplification of 5' partial cDNA ends. (c) Schematic representation of the primers used in thermal RACE. The 52-nt Q_T primer (5'-Q_O-Q_I-TTTT-3') contains a 17-nt oligo(dT) sequence at the 3' end followed by a 35-nt sequence encoding HindIII, SstI, and XhoI recognition sites. The Q_I and Q_O primers overlap by 1 nt; the Q_I primer contains all three of the recognition sites. Primers:

Q_T: 5'-CCAGTGAGCAGAGTGACGAGGACTCGAGCTCAAGCTTTTTTTTTTTTTT-TTTT-3'

Q_O: 5'-CCAGTGAGCAGAGTGACG-3'

Q_I: 5'-GAGGACTCGAGCTCAAGC-3'

[Figure used with permission from Amplifications Newsletter (September 1990). Copyright Perkin Elmer.]

15 mM of each dNTP), 0.25 μl (10 units) of RNasin, and 0.5 μl of Q$_T$ primer (100 ng/μl). Heat 1 μg of poly(A)$^+$ RNA or 5 μg of total RNA in 14 μl of water at 80° for 3 min and cool rapidly on ice. Add to reverse transcription components. Add 1 μl (200 units) of MMLV reverse transcriptase, and incubate for 2 hr at 37°. Dilute the reaction mixture to 1 ml with TE [10 mM Tris-HCl (pH 7.5), 1 mM ethylenediaminetetraacetic acid (EDTA)] and store at 4° ("3' end cDNA pool").

Comments. The use of clean, intact mRNA is obviously important for the successful production of cDNAs. RNA can be isolated using any one of several recent protocols.[4] Poly(A)$^+$ RNA is preferentially used for reverse transcription to decrease background, but it is unnecessary to prepare it if only total RNA is available.

The cDNAs generated will be extended to varying degrees, because complete reverse transcription of mRNA is never entirely efficient. As depicted in Fig. 3, any cDNAs that extend past the location of the GSP1 primer will be truncated to a uniform length in the subsequent amplification reaction, whereas any cDNAs that do not make it to that point will fail to participate in the amplification reaction, because they lack a binding site for GSP1. Thus, the efficiency of extension during the reverse transcription step is relatively unimportant with respect to generating 3' end partial cDNAs, as long as at least a small number of cDNAs are extended past the location of GSP1.

A more important factor in the generation of full-length 3' end partial cDNAs concerns the stringency of the reverse transcription reaction. Such reactions are typically carried out at relatively low temperatures (37–42°) using a vast excess of primer (about one-half the mass of the mRNA, which represents an ~30 : 1 molar ratio). Under these low-stringency conditions, a stretch of A residues as short as 6–8 nt will suffice as a binding site for an oligo(dT)-tailed primer. This may result, as depicted in Fig. 4, in cDNA synthesis being initiated at sites upstream of the poly(A) tail, leading to truncation of the desired amplification product. One should be suspicious that this has occurred if a canonical polyadenylation signal sequence[5] is not found near the 3' end of the cDNAs generated. The phenomenon can be minimized by controlling two parameters: primer concentration and reaction temperature. The primer concentration can be reduced dramatically without decreasing the amount of cDNA synthesized significantly[6] and at some point it will begin to bind preferentially to

[4] P. Chomczynski and N. Sacchi, *Anal. Biochem.* **162,** 156 (1987).
[5] M. Wickens and P. Stephenson, *Science* **226,** 1045 (1984).
[6] C. Coleclough, this series, Vol. 154, p. 64.

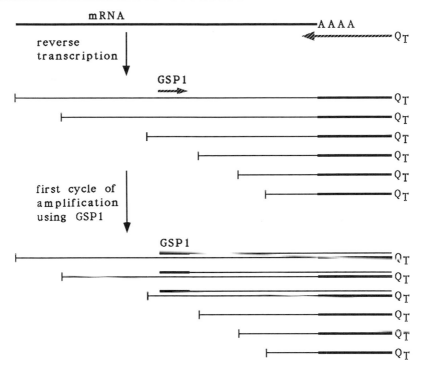

FIG. 3. Amplification of 3' partial cDNA ends. The diagram depicts the fate of first strands of cDNA of differing length. Strands that extend past the location of the GSP1 primer serve as templates for subsequent amplification but are truncated by the GSP1 primer. Strands that do not extend as far as GSP1 do not participate in the subsequent amplification reaction. Additional explanations are given in text. Primers to be used for amplification or reverse transcription reactions are shown as shaded arrows. Primers already incorporated into cDNA strands are shown as thick lines. mRNA is also shown as a thick line.

the longest A-rich stretches present [i.e., the poly(A) tail]. The quantity recommended above represents a good starting point; it can be reduced five fold further if significant truncation is observed.

Until recently, reaction temperatures could not be increased because reverse transcriptase became inactivated by elevated temperatures. However, heat-stable reverse transcriptases are now available from several suppliers [Perkin-Elmer Cetus, Amersham (Arlington Heights, IL) and others]. As in PCR reactions, the stringency of reverse transcription can thus be controlled by adjusting the temperature at which the primer is annealed to the mRNA. The optimal temperature depends on the specific reaction buffer and reverse transcriptase used and should be determined

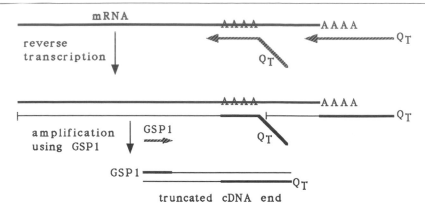

Fig. 4. Truncation of cDNA ends corresponding to the 3′ end of the mRNA. The diagram depicts the fate of first strands of cDNA that are initiated both at the poly(A) tail and at an A-rich sequence upstream of the tail. Explanations are given in text.

empirically, but it will usually be found to be in the range of 48–56° for a primer terminated by a 17-nt oligo(dT) tail.

Step 2: Amplification

Procedure. First round: Add an aliquot of the "3′ end cDNA pool" (1 μl) and primers (25 pmol each of GSP1 and Q_O) to 50 μl of PCR cocktail [1× *Taq* polymerase buffer (described above), each dNTP at 1.5 mM, and 10% dimethyl sulfoxide (DMSO)] in a 0.6-ml microfuge tube and heat in a DNA thermal cycler for 5 min at 97° to denature the first-strand products. Cool to 75°. Add 2.5 U of *Taq* polymerase, overlay the mixture with 30 μl of mineral oil [400-5 (Sigma, St. Louis, MO); preheat it in the thermal cycler to 75°] and incubate at the appropriate annealing temperature (52–60°) for 2 min. Extend the cDNAs at 72° for 40 min. Carry out 30 cycles of amplification using a step program (94°, 1 min; 52–60°, 1 min; 72°, 3 min), followed by a 15-min final extension at 72°. Cool to room temperature.

Second round: Dilute amplification products from the first round 1 : 20 in TE. Amplify 1 μl of the diluted material with primers GSP2 and Q_I using the procedure described above, but eliminate the initial 2-min annealing step and the 72°, 40-min extension step.

Comments. The following points concerning the PCR reaction should be noted. First, it is important to add the *Taq* polymerase *after* heating the mixture to a temperature above the T_m of the primers ("hot start" PCR). Addition of the enzyme prior to this point allows one "cycle" to take place at room temperature, promoting the synthesis of nonspecific background products dependent on low-stringency interactions.

Second, the high-dNTP, DMSO, and $(NH_4)_2SO_4$-containing PCR cocktail described above has been found empirically to work more frequently (without titrating Mg^{2+} concentrations) and to result in greater yields of product than the buffer suggested by the supplier (Perkin-Elmer Cetus). It should be noted, however, that this buffer results in a much higher polymerase error rate during amplification. Thus, although it is recommended for the initial isolation and characterization of partial end cDNAs, other conditions should be employed in subsequent amplifications to generate full-length cDNA clones (below), for which a low error rate is essential. It should also be noted that primer T_m values are 5–6° lower in this buffer as compared to the one suggested by Perkin-Elmer Cetus.

Third, an annealing temperature close to the effective T_m of the primers should be used. The Q_I and Q_O primers work well at 60° under the PCR conditions recommended here, although the actual optimal temperature may depend on the PCR machine used. Gene-specific primers of similar length and GC content should be chosen. Computer programs to assist in the selection of primers are widely available and should be used.

Fourth, an extension time of 1 min/kb expected product should be allowed during the amplification cycles. If the expected length of product is unknown, try 3–4 min initially.

Fifth, *very* little substrate is required for the PCR reaction. One microgram of poly(A)$^+$ RNA typically contains ~5×10^7 copies of *each* low-abundance transcript. The PCR reaction described here works optimally when 10^3–10^5 templates are present in the starting mixture; thus as little as 0.002% of the reverse transcription mixture suffices for the PCR reaction! Addition of too much starting material to the amplification reaction will lead to production of large amounts of nonspecific product and should be avoided. The RACE technique is particularly sensitive to this problem, because every cDNA in the mixture, desired and undesired, contains a binding site for the Q_O and Q_I primers.

Sixth, it was found empirically that allowing extra extension time during the first amplification round (when the second strand of cDNA is created) sometimes resulted in increased yields of the specific product relative to background amplification and, in particular, increased the yields of long cDNAs versus short cDNAs when specific cDNA ends of multiple lengths were present.[1] Prior treatment of cDNA templates with RNA hydrolysis or a combination of RNase H and RNase A seldomly improves the efficiency of amplification of specific cDNAs, but is a reasonable step to carry out if the specific product is not initially detected.

5' End cDNA Amplification

Step 1: Reverse Transcription to Generate cDNA Templates

Procedure. Assemble reverse transcription components on ice: 4 μl of 5× reverse transcription buffer, 1.3 μl of dNTPs (stock concentration is 15 mM of each dNTP), 0.25 μl (10 units) of RNasin, and 0.5 μl of GSP-RT primer (100 ng/μl). Heat 1 μg of poly(A)$^+$ RNA or 5 μg of total RNA in 14 μl of water at 80° for 3 min and cool rapidly on ice. Add to reverse transcription components. Add 1 μl (200 units) of MMLV reverse transcriptase, and incubate for 2 hr at 37°. Dilute the reaction mixture to 1 ml with TE and store at 4° ("5' end cDNA pool-1").

Comments. Many of the remarks made above in the section on reverse transcribing 3' end partial cDNAs are also relevant here and should be noted. There is, however, one major difference. The efficiency of cDNA extension is now critically important, because each specific cDNA, no matter how short, is subsequently tailed and becomes a suitable template for amplification (Fig. 5). Thus, the PCR products eventually generated directly reflect the quality of the reverse transcription reaction. Extension can be maximized by using clean, intact RNA, by selecting the primer for reverse transcription to be near the 5' end of region of known sequence, and by using heat-stable reverse transcriptase at elevated temperatures or a combination of MMLV and heat-stable reverse transcriptase at multiple temperatures. Synthesis of cDNAs at elevated temperatures diminishes the amount of secondary structure encountered in GC-rich regions of the mRNA.

Step 2: Appending Poly(A) Tail to First-Strand cDNA Products

Procedure. Remove excess primer using Centricon-100 spin filters (Amicon Corp., Danvers, MA). (*Note:* Neither Centricon-30 spin filters nor Sephadex G-50 columns separate primers from cDNAs effectively.) Dilute reverse transcription mixture to 2 ml with TE and centrifuge at 1000 g for 20 min. Repeat using 2 ml of 0.2× TE and collect retained liquid. Concentrate to 10 μl using SpeedVac (Savant, Hicksville, NY) centrifugation.

Add 4 μl of 5× tailing buffer [125 mM Tris-HCl (pH 6.6), 1 M potassium cacodylate, 1.25 mg/ml BSA], 1.2 μl of 25 mM CoCl$_2$, 4 μl of 1 mM dATP, and 10 U of TdT.[7] Incubate for 5 min at 37° and 5 min at 65°. Dilute to 500 μl with TE ("5' end cDNA pool-2").

Comments. To attach a known sequence to the 5' end of the first-strand cDNA, a homopolymeric tail is appended using TdT. We prefer appending

---

[7] R. Roychoudhury, E. Jay, and R. Wu, *Nucleic Acids Res.* **3**, 863 (1976).

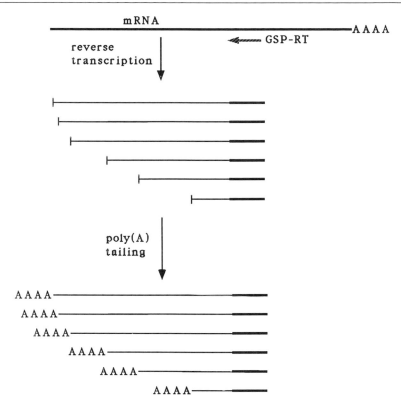

FIG. 5. A schematic diagram illustrating the importance of achieving uniform full-length extension of future 5′ partial cDNA ends. All primer-extended gene-specific cDNAs, regardless of their length, are tailed and become legitimate templates for subsequent amplifications. Additional explanations are given in text.

poly(A) tails rather than poly(G) tails for several reasons. First, the 3′ end strategy is based on the naturally occurring poly(A) tail; thus the same adapter primers can be used for both ends, decreasing variability in the protocol and cost. Second, because A:T binding is weaker than G:C binding, longer stretches of A residues (~2×) are required before the oligo(dT)-tailed Q_T primer will bind to an internal site and truncate the amplification product. Third, vertebrate coding sequences and 5′-untranslated regions tend to be biased toward G/C residues; thus, use of a poly(A) tail further decreases the likelihood of inappropriate truncation.

Unlike many other situations in which homopolymeric tails are appended, the actual length of the tail added here is unimportant, as long as it exceeds 17 nt. This is because although the oligo(dT)-tailed primers subsequently bind all along the length of the appended poly(A) tail, only the innermost one becomes incorporated into the amplification product, and consequently the remainder of the poly(A) tail is lost. The conditions described in the procedure above result in the addition of 30–400 nt.

Step 3: Amplification

Procedure. First round: Add an aliquot of the "5' end cDNA pool-2" (1 μl) and primers [25 pmol each of GSP1 and Q_O (shown in Fig. 2b), and 2 pmol of Q_T] to 50 μl of PCR cocktail [1× *Taq* polymerase buffer (described above), each dNTP at 1.5 mM, and 10% DMSO] in a 0.6-ml microfuge tube and heat in a DNA thermal cycler for 5 min at 97° to denature the first-strand products. Cool to 75°. Add 2.5 U of *Taq* polymerase, overlay the mixture with 30 μl of mineral oil [400-5 (Sigma); preheat it in the thermal cycler to 75°] and incubate at the appropriate annealing temperature (48–52°) for 2 min. Extend the cDNAs at 72° for 40 min. Carry out 30 cycles of amplification using a step program (94°, 1 min; 52–60°, 1 min; 72°, 3 min), followed by a 15-min final extension at 72°. Cool to room temperature.

Second round: Dilute amplification products from the first round 1 : 20 in TE. Amplify 1 μl of the diluted material with primers GSP2 and Q_I (Fig. 2b) using the procedure described above, but eliminate the initial 2-min annealing step and the 72°, 40-min extension step.

Comments. Many of the remarks made above in the section on amplifying 3' end partial cDNAs are also relevant here and should be noted. There is, however, one major difference. The annealing temperature in the first step (48–52°) is lower than that used in successive cycles (52–60°). This is because cDNA synthesis during the first round depends on the interaction of the appended poly(A) tail and the oligo(dT)-tailed Q_T primer, whereas in all subsequent rounds amplification can proceed using the Q_O primer, which is composed of ~60% GC and which can anneal at a much higher temperature to its complementary target.

Analysis of Amplification Products during Execution of RACE Protocol

The production of specific partial cDNAs by the RACE protocol is assessed using Southern blot hybridization analysis. After the first set of amplification cycles, the reaction products are electrophoresed in a 1% (w/v) agarose gel, stained with ethidium bromide, denatured, and transferred to nylon membrane. After hybridization with a labeled oligomer or gene fragment derived from a region necessarily contained within the amplified fragment (e.g., primer GSP2 or GSP-Hyb/Seq in Fig. 2a and b), gene-specific partial cDNA ends should be detected easily. At this stage, yields of the desired product relative to nonspecific amplified cDNA vary from <1% of the amplified material to nearly 100%, depending largely on the stringency of the amplification reaction, the amplification efficiency of the specific cDNA end, and the relative abundance of the specific transcript within the mRNA source. If specific hybridization is not observed, then troubleshooting steps should be initiated.

After the second set of amplification cycles, ~100% of the cDNA detected by ethidium bromide staining should represent specific product. Note that the GSP2 primer cannot be used as a hybridization probe at this point, because it can be incorporated into nonspecific cDNA fragments during the second amplification reaction.

Information gained from this analysis can be used to optimize the RACE procedure. If low yields of specific product are observed because nonspecific products are being amplified efficiently, then annealing temperatures can be raised gradually (~2° at a time) and sequentially in each stage of the procedure until nonspecific products, and almost until specific products, are no longer observed. Optimizing the annealing temperature is also indicated if multiple species of specific products are observed, which could indicate that truncation of specific products is occurring. If multiple species of specific products are observed after the reverse transcription and amplification reactions have been fully optimized, then the possibility should be entertained that alternate splicing or promoter use is occurring. If a nearly continuous smear of specific products is observed up to a specific size limit after 5' end amplification, this suggests that polymerase pausing occurred during the reverse transcription step. To obtain nearly full-length cDNA ends, the amplification mixture should be electrophoresed and the longest products recovered by gel isolation. An aliquot of this material can then be reamplified for a limited number of cycles.

Further Analysis and Use of RACE Products

Cloning. RACE products can be cloned like any other PCR products. To assist in this step, the Q_I primer that encodes *Hind*III, *Sst*I, and *Xho*I restriction enzyme sites can be used. Products can be efficiently cloned into vectors that have been double-cut with one of these enzymes and with a blunt-cutting enzyme such as *Sma*I. If clones are not obtained, determine whether the restriction enzyme chosen is cutting the amplified gene fragment a second time, at some internal location in the new and unknown sequence. There is a large literature that concerns the cloning of PCR products and this should be consulted if difficulties persist.

Sequencing. RACE products can be sequenced using a variety of protocols. The products can be sequenced directly on a population level from the end at which the gene-specific primers are located. [Note that the products cannot be sequenced on a population level from the unknown end, because individual cDNAs contain different numbers of A residues in their poly(A) tails and, as a consequence, the sequencing ladder falls out of register after reading through the tail.] The following protocol works well and involves a modification of the asymmetric PCR protocol first described by Gyllensten and Erlich.[8]

Separate the amplification products from residual primers twice by diluting them each time to 2 ml with 0.2× TE and passing them over a Centricon-100 spin column. Quantitate the specific product by electrophoresing it in a 1% (w/v) agarose gel and comparing it to DNA standards of known concentration. Prepare a microfuge tube containing 100 μl of PCR cocktail as described above and add 50 pmol of primer Q_I and the amplification products to a final concentration of 1 ng/μl. Amplify as described above for 20 cycles, using 5 U of *Taq* polymerase. Separate the amplification products from residual Q_I primer and free nucleotides three times by diluting them each time to 2 ml with 0.2× TE and passing them over a Centricon-100 spin column. Compare the asymmetric DNA to the double-stranded product by electrophoresing aliquots of each in a 1% agarose gel. An additional diffuse band migrating at a slightly lower apparent molecular weight should be apparent in the asymmetric sample. Sequence 7 μl of the asymmetric sample using primer GSP2 or primer GSP-Hyb/Seq with a single-stranded DNA sequencing protocol.

Individual cDNA ends, once cloned into a plasmid vector, can be sequenced from either end using gene-specific or vector primers. The unknown end can also be sequenced from the internal end of the poly(A) tail, using the following set of primers simultaneously[9]: TTTTTTTTTTT-TTTTTA, TTTTTTTTTTTTTTTTG, TTTTTTTTTTTTTTTTC. The non-T nucleotide at the 3' end of the primer forces the appropriate primer to bind to the inner end of the poly(A) tail. The other two primers do not participate in the sequencing reaction.

Hybridization Probes. RACE products are generally pure enough that they can be used as probes for RNA and DNA blot analyses. It should be kept in mind that small amounts of contaminating nonspecific cDNAs will always be present. It is also possible to include a T7 RNA polymerase promoter in one or both primer sequences and to use the RACE products in *in vitro* transcription reactions to produce RNA probes.[2] Primers encod-

---

[8] U. B. Gyllensten and H. A. Erlich, *Proc. Natl. Acad. Sci. U.S.A.* **85,** 7652 (1988).
[9] R. Thweatt, S. Goldstein, and R. J. S. Reis, *Anal. Biochem.* **190,** 314 (1990).

ing the T7 RNA polymerase promoter sequence do not appear to function as amplification primers as efficiently as the ones listed in the legend to Fig. 2 (Frohman, unpublished observations, 1989). Thus, the T7 RNA polymerase promoter sequence should not be incorporated into RACE primers as a general rule.

Construction of Full-Length cDNAs. It is possible to use the RACE protocol to create overlapping 5' and 3' cDNA ends that can later, through judicious choice of restriction enzyme sites, be joined together through subcloning to form a full-length cDNA. It is also possible to use the sequence information gained from acquisition of the 5' and 3' cDNA ends to make new primers representing the extreme 5' and 3' ends of the cDNA, and to employ them to amplify a *de novo* copy of a full-length cDNA directly from the "3' end cDNA pool." Despite the added expense of making two more primers, there are several reasons why the second approach is preferred.

First, a relatively high error rate is associated with the PCR conditions for which efficient RACE amplification takes place, and numerous clones may have to be sequenced to identify one without mutations. In contrast, two specific primers from the extreme ends of the cDNA can be used under inefficient but low error rate conditions[10] for a minimum of cycles to amplify a new cDNA that is likely to be free of mutations. Second, convenient restriction sites are often not available, thus making the subcloning project difficult. Third, by using the second approach, the synthetic poly(A) tail can be removed from the 5' end of the cDNA. Homopolymer tails appended to the 5' ends of cDNAs have in some cases been reported to inhibit translation.[6] Finally, if alternate promoters, splicing, and polyadenylation signal sequences are being used and result in multiple 5' and 3' ends, it is possible that one might join two cDNA halves that are never actually found together *in vivo*. Employing primers from the extreme ends of the cDNA as described confirms that the resulting amplified cDNA represents an mRNA actually present in the starting population.[11]

Comments

The RACE protocol offers several advantages over conventional library screening to obtain additional sequences for cDNAs already partially cloned. The procedure is simpler, cheaper, and much faster. It requires very small amounts of primary material, and provides rapid feedback on

[10] K. A. Eckert and T. A. Kunkel, *Nucleic Acids Res.* **18**, 3739 (1990).
[11] C. W. Ragsdale, M. Petkovich, P. B. Gates, P. Chambon, and J. P. Brockes, *Nature (London)* **341**, 654 (1989).

the generation of the desired product. Information regarding alternate promoters, splicing, and polyadenylation signal sequences can be obtained and a judicious choice of primers (e.g., within an alternately spliced exon) can be used to amplify a subpopulation of cDNAs from a gene for which the transcription pattern is complex. Furthermore, differentially spliced or initiated transcripts can be separated by electrophoresis and cloned separately,[1] and essentially unlimited numbers of independent clones can be generated to examine rare events.[3] Finally, for 5' end amplification, because a primer extension library is created instead of a general purpose one, the ability of reverse transcriptase to extend cDNAs all the way to the 5' end of a message is greatly increased.

A number of modifications of the RACE protocol have been developed. It has been suggested that ligation of oligonucleotides to first-strand cDNA[12] or linkers to double-stranded cDNA[13,14] provides an alternative to appending a poly(A) tail to cDNA ends produced in the 5' amplification procedure. The use of a 3' end reverse transcription primer terminating in random nucleotides instead of oligo(dT) has also been suggested to be of use when the poly(A) tail is too far away from the region of sequence already known to be amplified in a single RACE step.[15] It is probable that these and future modifications will continue to increase the utility of the RACE protocol for cDNA cloning.

Troubleshooting and Controls

Problems with Reverse Transcription

Damaged RNA: Electrophorese RNA in a 1% (w/v) formaldehyde minigel and examine the integrity of the 18S and 28S ribosomal bands. Discard the RNA preparation if the ribosomal RNA bands are not sharp.

Contaminants: Ensure that the RNA preparation is free of agents that inhibit reverse transcription, for example, LiCl and sodium dodecyl sulfate (SDS) (see Ref. 16 regarding the optimization of reverse transcription reactions).

Bad reagents: To monitor reverse transcription of the RNA, add 20 μCi of [^{32}P]dCTP to the reaction, separate newly created cDNAs using gel electrophoresis, wrap the gel in Saran wrap, and expose it to X-ray film. Accurate estimates of cDNA size can best be determined using alkaline

---

[12] J. B. Dumas, M. Edwards, J. Delort, and J. Mallet, *Nucleic Acids Res.* **19,** 5227 (1991).
[13] P. R. Mueller and B. Wold, *Science* **246,** 780 (1989).
[14] M. S. H. Ko, *Nucleic Acids Res.* **18,** 5706 (1990).
[15] J. D. Fritz, M. L. Greaser, and J. A. Wolff, *Nucleic Acids Res.* **19,** 3747 (1991).

agarose gels,[16] but a simple 1% agarose minigel will suffice to confirm that reverse transcription took place and that cDNAs of reasonable length were generated. Note that adding [^{32}P]dCTP to the reverse transcription reaction results in the detection of cDNAs synthesized both through the specific priming of mRNA and through RNA self-priming. When a gene-specific primer is used to prime transcription (5' end RACE) or when total RNA is used as a template, the majority of the labeled cDNA will actually have been generated from RNA self-priming. To monitor extension of the primer used for reverse transcription, label the primer using T4 DNA kinase and [γ-^{32}P]ATP prior to reverse transcription.[16] Much longer exposure times will be required to detect the labeled primer–extension products than when [^{32}P]dCTP is added to the reaction.

To monitor reverse transcription of the gene of interest, one may attempt to amplify an internal fragment of the gene containing a region derived from two or more exons, if sufficient information is available.

Problems with Tailing

Bad reagents: Tail 100 ng of a DNA fragment (approximately 100–300 bp in length) for 30 min. In addition, mock tail the same fragment (add everything but the TdT). Run both samples in a 1% agarose minigel. The mock-tailed fragment should run as a tight band. The tailed fragment should have increased in size by 20–200 bp and should appear to run as a diffuse band that trails off into higher molecular weight products. If this is not observed, replace the reagents.

To monitor tailing of experimental cDNAs, add 20 μCi of [^{32}P]dCTP to the cDNA tailing reaction and examine the distribution of an aliquot of the labeled cDNAs by gel electrophoresis.

Verification: Alternatively, mock tail 25% of the cDNA pool (add everything but the TdT). Dilute to the same final concentration as the tailed cDNA pool. This serves two purposes. First, although amplification products will be observed using both tailed and untailed cDNA templates, the actual pattern of bands observed should be different. In general, discrete bands are observed using untailed templates, and a broad smear of amplified cDNA accompanied by some individual bands is typically observed using tailed templates. If the two samples appear different, this confirms that tailing took place and that the oligo(dT)-tailed Q_T primer is annealing effectively to the tailed cDNA during PCR. Second, observing specific products in the tailed amplification mixture that are not present in the untailed amplification mixture indicates that these products are being

---

[16] J. Sambrook, E. F. Fritsch, and T. Maniatis, "Molecular Cloning: A Laboratory Manual," 2nd Ed. Cold Spring Harbor Press, Cold Spring Harbor, New York, 1989.

synthesized off the end of an A-tailed cDNA template, rather than by annealing of the dT-tailed primer to an A-rich sequence in or near the gene of interest.

Problems with Amplification

No product: If no products are observed for the first set of amplifications after 30 cycles, add additional *Taq* polymerase and carry out an additional 15 rounds of amplification (extra enzyme is not necessary if the entire set of 45 cycles is carried out without interruption at cycle 30). Product is always observed after a total of 45 cycles if efficient amplification is taking place. If no product is observed, carry out a PCR reaction using control templates and primers to ensure the integrity of the reagents.

Smeared product from the bottom of the gel to the loading well: This is most likely due to too many cycles, or too much starting material.

Nonspecific amplification, but no specific amplification: Check the sequence of cDNA and primers. If all are correct, examine the primers (using a computer program) for secondary structure and self-annealing problems. Consider ordering new primers. Determine whether too much template is being added, or if the choice of annealing temperatures could be improved.

Alternatively, secondary structure in the template may be blocking amplification. Consider adding formamide[17] or 7-deaza-GTP (in a 1 : 3 ratio with dGTP) to the reaction to assist polymerization. 7-deaza-GTP can also be added to the reverse transcription reaction.

The last few base pairs of the 5' end sequence do not match the corresponding genomic sequence: Be aware that reverse transcriptase can add on a few extra template-independent nucleotides.

Inappropriate templates: To determine whether the amplification products observed are being generated from cDNA or whether they derive from residual genomic DNA or contaminating plasmids, pretreat an aliquot of the RNA with RNase A.

[17] G. Sarker, S. Kapelner, and S. S. Sommer, *Nucleic Acids Res.* **18,** 7465 (1990).

[25] Cloning of Polymerase Chain Reaction-Generated DNA Containing Terminal Restriction Endonuclease Recognition Sites

By VINCENT JUNG, STEVEN B. PESTKA, and SIDNEY PESTKA

Using the polymerase chain reaction (PCR)[1] to generate DNA containing terminal restriction endonuclease recognition sites to permit cloning usually relies on the use of unphosphorylated primers incorporating a restriction endonuclease recognition site of choice plus three or four extra 5' bases flanking that site.[2,3] Various sites (e.g., NotI, XhoI, and XbaI) incorporated into the termini of PCR products have proved difficult to cut with their respective restriction endonucleases.[4-6] There are several possible explanations for this difficulty; first, Taq polymerase might be inefficient for certain terminal sequences, producing frayed ends that cannot be cleaved by the restriction endonuclease. Alternatively, the "breathing" of terminal sequences might prevent the stable association of restriction endonucleases with terminal sites. Also, Taq polymerase or other contaminants might bind to the ends of the PCR products, blocking restriction endonuclease activity.

The above three explanations are unlikely because various combinations of steps, as described in Table I, designed to deblock (proteinase K), to stabilize (spermidine), or to repair (Klenow or T4 polymerase) the ends, failed to increase the cleavage efficiency and hence the cloning efficiency. Also, prolonged or overnight digestion had no significant effect.

The extra 3–4 bases at the terminal restriction endonuclease recognition site may be insufficient to allow stable association with and cutting by certain restriction endonucleases.[5,6] In proving that this is indeed the case, we describe a method by which terminal sites can be efficiently cleaved for cloning into appropriate vectors. The method involves the simple conversion of terminal restriction endonuclease recognition sites into internal sites by concatamerization of the PCR product with T4 DNA

[1] K. B. Mullis and F. Faloona, this series, Vol. 155, p. 335.
[2] H. A. Erlich, ed., "PCR Technology: Principles and Applications for DNA Amplification" Stockton Press, New York, 1989.
[3] S. J. Scharf, G. T. Horn, and H. A. Erlich, Science 233, 1076 (1986).
[4] Cetus Corp., personal communications.
[5] V. Jung, S. B. Pestka, and S. Pestka, Nucleic Acids. Res. 18, 6156 (1990).
[6] D. L. Kaufman and G. A. Evans, BioTechniques 9, 304 (1990).

METHODS IN ENZYMOLOGY, VOL. 218

TABLE I
FACTORS AFFECTING CLONING EFFICIENCY[a]

Experimental conditions	Cloning efficiency (%)
Control	≤0.05
Proteinase K	≤0.05
Klenow	≤0.05
T4 DNA polymerase	≤0.05
Spermidine	≤0.05
Prolonged digestion with restriction endonuclease XhoI	
4 hr	≤0.05
Overnight	≤0.05
Concatamerization	50

[a] Enzymes were used according to company specifications. The primers contained a XhoI site with three extra bases (G or C) and, after digestion with XhoI, the product was used for cloning portions of a human IFN-γ receptor cDNA into a vector derived from pVJ3.[8] The cloning efficiency is the ratio of the number of positive colonies determined by colony hybridization to the total number of colonies obtained. Combinations of proteinase K treatment of the PCR product, followed by phenol extraction and filling the ends with either the Klenow fragment of *E. coli* DNA polymerase I (Klenow) or with T4 DNA polymerase, were also tried, with results identical to the use of each alone. Phosphorylated primers used instead of unphosphorylated primers to prepare the PCR product yielded the same results as the control. The data reflect results with the use of unphosphorylated primers, except for the concatamerization reactions. (From Jung *et al.*[5])

ligase. These internal sites can then be readily cleaved by the restriction endonuclease of choice. We also describe a variation of this strategy, in which primers are used that contain restriction enzyme half-sites. Such sites, if they are palindromic, can be reconstituted by the concatamerization reaction.

Materials and Methods

The following oligonucleotides are synthesized in a DNA synthesizer (model 380B; Applied Biosystems, Foster City, CA) by phosphoramidite chemistry and purified by a reversed-phase column:

1. 5'-CGA **CTC GAG** GGT AGC AGC **ATG** GCT CTC CTC-3'
2. 5'-GCT **CTC GAG** **TCA** ACA GAT GAA TAC CAG GCT-3'
3. 5'   **GAG** GGT AGC AGC **ATG** GCT CTC CTC-3'
4. 5'   **GAG** **TCA** ACA GAT GAA TAC CAG GCT-3'

These primers are designed according to the primary sequence of the human interferon γ (IFN-γ) receptor[7] to generate a truncated cDNA that can be used for expression of the first 267 amino acids of the receptor. **CTCGAG** represents the restriction endonuclease *Xho*I recognition site. **ATG** is the start codon. **TCA** represents the complement to the stop codon TGA in the opposite strand. Oligonucleotides 1 and 2 are a primer pair, each containing three extra nucleotides appended to the 5' side of the entire restriction endonuclease site. The second primer pair, oligonucleotides 3 and 4, is identical to the first primer pair except that the oligonucleotides contain only half the *Xho*I restriction endonuclease site: **GAG**.

Plasmid pVJ3 is a modification of the vector pVJ2.[8] This plasmid is a shuttle vector that allows eukaryotic expression from the SV*lac0* promoter[9] and contains an antibiotic G418 selectable marker. A *Xho*I site just downstream of the promoter permits the insertion of the PCR product.

Efficient Cloning from Terminal Restriction Enzyme Sites

Primers are phosphorylated in a reaction containing 0.1 nmol of each primer pair in a 20-μl reaction with T4 DNA kinase (Amersham, Arlington Heights, IL) at 37° for 30–60 min [10× kinase buffer: 0.5 mM Tris-HCl (pH 7.6), 0.1 M MgCl$_2$, 10 mM ATP, 50 mM dithiothreitol (DTT), 1 mM spermidine, 1 mM ethylenediaminetetraacetic acid (EDTA)]. The kinase reaction is inactivated by heating at 65° for 10 min. To the kinase reaction is added 10 μl of 10× PCR buffer [10× PCR buffer: 100 mM Tris-HCl (pH 8.3), 500 mM KCl, 30 mM MgCl$_2$, 0.1% (w/v) gelatin], 8 μl of 2.5 mM dNTP, 5 μl of a template DNA solution (5 ng/μl), 58.5 μl H$_2$O, and 0.5 μl of *Taq* polymerase (Perkin-Elmer, Norwalk, CT). The PCR reaction is carried through 30 cycles: denaturation at 94° for 1 min 30 sec, annealing at 50° for 2 min, and polymerization at 72° for 3 min. The PCR reaction is extracted with phenol/CHCl$_3$ and precipitated by adjusting to 2 M ammonium acetate and adding 2 vol of ethanol. The PCR product is concatemerized in a 10-μl reaction containing 1–2 Weiss units of T4 DNA ligase, 2 μl

[7] M. Aguet, Z. Dembic, and G. Merlin, *Cell* **55**, 273 (1989).
[8] V. Jung, C. Jones, C. S. Kumar, S. Stefanos, S. O'Connell, and S. Pestka, *J. Biol. Chem.* **265**, 1827 (1990).
[9] M. Brown, J. Figge, U. Hansen, C. Wright, K.-T. Jeang, G. Khoury, D. M. Livingston, and T. M. Roberts, *Cell* **49**, 603 (1987).

of 5× ligation buffer [5× ligation buffer: 0.25 M Tris-HCl (pH 7.6), 50 mM MgCl$_2$, 5 mM ATP, 5 mM DTT, 25% (w/v) polyethylene glycol (PEG)] and incubated at 23° overnight. The ligation reaction mixture is heated at 65° for 10 min to inactivate ligase, adjusted to 20 μl by the addition of 2 μl of 10× restriction enyzme buffer [0.5 M Tris-HCl (pH 7.5), 0.1 M MgCl$_2$, 1 M NaCl, 10 mM dithioerythritol] and 8 μl H$_2$O, then digested with 10 units of XhoI restriction endonuclease. The digested DNA is then extracted with phenol/CHCl$_3$, precipitated with ethanol, and ligated into dephosphorylated plasmid pVJ3 that was previously linearized by XhoI restriction endonuclease.

As shown in Table I, the 1000-fold increase in cloning efficiency of the PCR product by this method indicates that certain terminal restriction endonuclease sites are less susceptible to cleavage and that this inefficiency can be eliminated by converting the terminal sites into internal sites through concatamerization of the DNA.

Cloning with Restriction Endonuclease Half-Sites

For palindromic restriction enzyme sites, it was predicted that concatamerization of PCR products containing restriction enzyme half-sites would reconstitute the site, allowing for cleavage by the respective restriction endonuclease. Primers used to test this idea, numbers 3 and 4, are identical to primers 1 and 2 described above except that they contain half-sites for XhoI restriction endonuclease. To test this hypothesis, the PCR product is generated with the use of 5'-phosphorylated primers as described above (Fig. 1, lanes 1 and 3). The PCR product is then treated with T4 DNA ligase as described to generate concatamers of the monomeric PCR product (lane 2). The predominant products are dimers and trimers. This inefficiency in ligation of PCR-generated DNA can apparently be relieved by pretreatment with T4 DNA polymerase.[6] The reason for this and its implications for the procedure are discussed later. That the concatamerized PCR product has reconstituted XhoI sites is evidenced by its susceptibility to XhoI restriction enzyme cleavage (lane 4). The cleavage products can then be conveniently ligated to vectors of choice.

Concluding Remarks

It has been shown that, in the case of the restriction endonuclease XhoI, the incorporation of a 10-bp tail was still insufficient for the cloning of PCR-generated DNA and that 20-bp tails were required.[10] This restric-

---

[10] S. N. Ho, J. K. Pullen, R. M. Horton, H. D. Hunt, and L. R. Pease, *DNA Protein Eng. Tech.* **2,** 50 (1990).

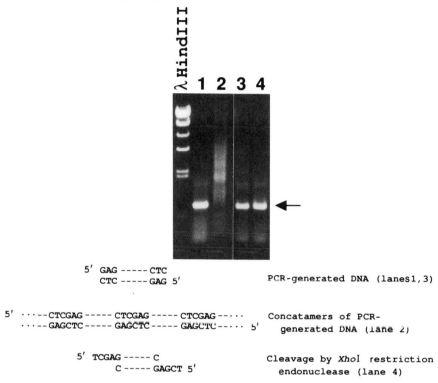

FIG. 1. Reconstitution of *Xho*I restriction endonuclease sites by concatamerization of half-sites. Primers 3 and 4, containing only half of the *Xho*I restriction endonuclease site, were used for the PCR reaction. Lanes 1 and 3 represent agarose gel electrophoresis of the PCR products that contain half of the *Xho*I sites at each end. Lane 2 represents the analysis of the ligation of the PCR product shown in lane 1. Lane 4 represents the analysis of the ligated PCR product of lane 2 after digestion with *Xho*I restriction endonuclease. The marker is λ phage DNA digested with *Hin*dIII restriction endonuclease.

tion endonuclease site is therefore ideal for the development of strategies to overcome this predicament. We have shown here that difficulty in cleaving the terminal restriction endonuclease sites can be eliminated by the internalization of such sites by concatamerization. This indicates a relationship between the ability of certain restriction enzymes to associate stably with their consensus sequences and the span of DNA flanking the site.

As described earlier, the limitation of the concatamerization procedure is the relative inefficiency of the ligation event when dimers and trimers predominate. This relatively inefficient concatamerization, however, still permits a significant increase in cloning efficiency, as shown in Table I.

Kaufman and Evans have described a similar procedure,[6] and have further observed that for certain PCR products, greater self-ligation occurred if the PCR product was first treated with T4 DNA polymerase in the presence of 200 μM concentrations of the four dNTPs. This would seem to indicate either incomplete polymerization or modification of the ends of the PCR product by TaqI polymerase. The latter explanation is most likely because TaqI polymerase has been found to have a "terminal transferase"-like activity involving the preferential addition of dATP to the 3' ends of blunt-ended DNA, with no template required.[11] In fact, this preferential addition of dATPs has led to the design of a strategy to clone PCR products directly into linearized vectors containing a 5'-deoxythymidylate overhang.[12] Pretreatment of the PCR product with T4 polymerase would therefore remove the 3'-deoxyadenylate overhang, allowing more efficient concatamerization. Because the generality and rules governing this "terminal transferase"-like activity of TaqI polymerase have not been well established, it would be prudent to treat PCR products with T4 DNA polymerase prior to concatamerization by T4 ligase.

The procedure described does not require additional nucleotides beyond the restriction endonuclease recognition site. Moreover, for palindromic restriction endonuclease recognition sites, only half the recognition site need be incorporated at the end of each primer, because concatamerization would reconstitute the site. Therefore, the oligodeoxynucleotide primers can contain fewer extraneous nucleotides that do not hybridize to the target DNA sequence. In the primer pairs described above, 12 fewer bases were used in the construction of half-site primers (3 and 4) as compared to conventional primers (1 and 2). This technique is also applicable to primers with different restriction sites. In that case, only one-quarter of the reconstituted sites will be cleavable. The procedure, therefore, increases the efficiency, specificity, and economy of the PCR used for cloning with flanking restriction endonuclease recognition sites.

Acknowledgment

This research was supported in part by National Institutes of Health Grants AI-25914 and CA46465 awarded to Sidney Pestka.

[11] J. M. Clark, *Nucleic Acids Res.* **16,** 9677 (1988).
[12] D. Mead, N. K. Pey, C. Herrnstadt, R. A. Marcil, and L. Smith, *BioTechnology* **9,** 657 (1991).

[26] Characterization of Recombinant DNA Vectors by Polymerase Chain Reaction Analysis of Whole Cells

By Gurpreet S. Sandhu, James W. Precup, and Bruce C. Kline

Introduction

The polymerase chain reaction[1] (PCR) for amplifying specific targeted sequences of DNA is a versatile tool. Functional studies, transcription of RNA, or the expression of protein from PCR-amplified DNA fragments requires cloning into plasmid, phage, or viral vectors. Conventional analysis of recombinant vectors, be they in bacterial colonies, viral plaques, or infected eukaryotic cells, has traditionally involved a subculture followed by extraction and purification of recombinant DNA. Subsequently, the vectors are analyzed for the presence and the orientation (if required) of the inserted sequences by restriction enzyme digestion and gel electrophoresis. Short fragments of DNA without convenient restriction sites may require DNA sequencing to obtain this information.

The use of PCR on whole cells has enabled us to shorten the detection and characterization time for recombinants from days to within 6 hr of the appearance of a bacterial colony, viral plaque, or an infected cell on a culture plate.[2] Whole bacterial cells from a colony, packaged phage from a plaque, or eukaryotic cells containing putative recombinant vectors are added to a PCR mix containing primers designed to specifically amplify the insert DNA. The entire reaction mix is boiled for 10 min to lyse cells, denature DNA, inactivate DNases, and destroy proteases. *Taq* polymerase is added and 25–30 cycles of PCR are carried out. The reaction mix is run on an agarose gel. The presence of an appropriately sized band indicates the presence of the insert.

Principle of Method

Oligonucleotide primers for sequencing DNA cloned into most common cloning vectors (pBR322, pUC, M13, λ, etc.) are readily available.

[1] R. K. Saiki, S. Scharf, F. Faloona, K. B. Mullis, G. T. Horn, H. A. Erlich, and N. Arnheim, *Science* **230**, 1350 (1985).
[2] G. S. Sandhu, J. W. Precup, and B. C. Kline, *BioTechniques* **7**, 689 (1989).

METHODS IN ENZYMOLOGY, VOL. 218

These primers flank the common cloning sites, and by using these sequencing primers for a PCR the vectors with insert DNA can be readily identified. A vector without an insert will give an amplified product equal in length to the distance between the annealing sites for the two sequencing primers. A recombinant vector, on the other hand, yields an amplified product having an additional length equivalent to the expected size of the insert (Fig. 1a and b).

Characterizing the orientation of an insert (Fig. 1c and d) requires one primer that will anneal to the insert DNA (primer B) along with another that anneals to the flanking vector sequence (sequencing primer A). A PCR under these conditions is sufficient to show both the presence as well as the orientation of the insert. Amplification occurs only if the insert is present in the correct orientation, because primers A and B are on opposite strands. If the orientation of the insert is reversed, both the primers are on the same strand and extend in the same direction; therefore, amplification is not possible.

Examples of these reactions using recombinant bacterial clones as well as recombinant plaques of λgt11 and baculovirus are shown in Fig. 2, which is a photograph of products analyzed by agarose gel electrophoresis.

Materials and Reagents

Polymerase Chain Reaction Reagents

Reagents	Amount for 100-μl reaction
PCR buffer ($10\times$): 500 mM KCl, 100 mM Tris-HCl (pH 8.3), 15 mM MgCl$_2$	10 μl
dNTP mix: 1.25 mM concentrations of each dNTP in water	16 μl
Primer 1	100 pmol
Primer 2	100 pmol
H$_2$O, use to adjust final volume to:	100 μl
AmpliTaq DNA polymerase (5 U/μl) (Perkin-Elmer Cetus, Norwalk, CT)	2.5 units

Other Reagents and Equipment

PCR thermal cycler: Perkin-Elmer Cetus (Norwalk, CT) DNA thermal cycler N801-0150

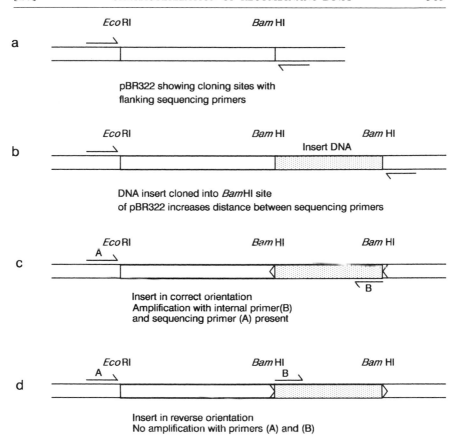

Fig. 1. Location of primers for the detection of insert DNA. (a) Segment of pBR322 showing *Eco*RI and *Bam*HI cloning sites along with the *Eco*RI proximal clockwise sequencing primer and the *Bam*HI proximal counterclockwise sequencing primer. (b) Recombinant pBR322 with an insert in the *Bam*HI cloning site will cause an increase in size of the amplified DNA band if the above primers are used for the PCR reaction. (c and d) Location of primers for determining the presence and orientation of insert DNA. (c) The insert is in the correct orientation and internal primer B, when used with sequencing primer A, will give rise to an appropriately sized band on the gel. (d) The insert is in the reverse orientation and, because both primers A and B extend in the same direction, no amplification takes place.

PCR primers: Detritylated oligonucleotide primers are synthesized on Applied Biosystems (Foster City, CA) 380A, 391, or 394 DNA/RNA synthesizers[3,4]

[3] J. W. Efcavitch, "Macromolecular Sequencing and Synthesis. Select Methods and Applications," p. 221. Alan R. Liss, New York, 1990.
[4] "ABI—Model 392 and 394 DNA/RNA Synthesizer. User's Manual," Part No. 901237, Revision B, July 1990.

size (kb)

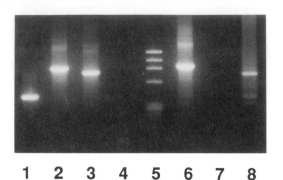

−1.35
−1.07
−0.87
−0.60

1   2   3   4   5   6   7   8

FIG. 2. Analysis of direct PCR-amplified DNA on a 1% agarose gel. Lanes 1 and 2: Direct PCR on bacterial colony with flanking sequencing primers as in Fig. 1a and b. 1, transformed bacterial colony with no insert; lane 2, colony with insert DNA. Lanes 3 and 4: PCR with one internal and one sequencing primer as in Fig. 1c and d. Lane 3, bacterial colony with insert in correct orientation; lane 4, insert in reverse orientation. Lane 5, ϕX174 DNA size markers (*Hae*III digest); lane 6, direct PCR on λgt11 viral plaque, showing size of insert DNA (900-bp band) using flanking primers. Lanes 7 and 8: PCR on baculovirus-infected *Spodoptera* cells with one internal and one flanking primer. Lane 7, wild-type baculovirus with no insert; lane 8, recombinant baculovirus with insert DNA. The amplicon is the 750-bp band.

Primer purification: NAP-25 columns (Pharmacia, Piscataway, NJ) are rinsed with 20 ml of sterile double-distilled water and 2.5 ml (~450 μg/ml) of deprotected primers in ammonium hydroxide is added to the column. Water (3.5 ml) is used to elute the primer off the column and the concentration determined by measuring optical density at A_{260}

Agarose gel, buffer, and electrophoresis conditions: 1% (w/v) SeaKem GTG agarose gel (FMC BioProducts, Rockland, ME) in TAE buffer containing 500 μg/liter of ethidium bromide electrophoresed at 110 V for 60 min

TAE buffer (10×): 0.4 M Tris base, 0.2 M sodium acetate (trihydrate), 0.01 M Na$_2$EDTA (dihydrate). Adjust pH to 8.0 with glacial acetic acid

ϕX174 *Hae*III size markers: 1–2 μg of ϕX174 *Hae*III digest (1.35, 1.07, 0.87, 0.60, 0.31, 0.28, 0.27, 0.23, 0.19, 0.11, and 0.07 kb) is used as a DNA size marker

Loading dye (5×): 20% (w/v) Ficoll, 0.1% (w/v) bromphenol blue, 0.1% (w/v) xylene cyanol

Methods

A master mix of water, buffer, dNTPs, and primers is made to ensure constancy of reaction conditions and is aliquotted out at 20 μl/0.5-ml PCR tube. Depending on the number of templates to be analyzed, up to 48 tubes (total capacity of the machine) can be prepared.

Each bacterial colony or viral plaque is gently touched in the middle with a sterile 200-μl pipette tip and rinsed once in the 20 μl of PCR solution. The tube is capped and heated to 100° for 10 min to lyse cells, destroy proteases, and release denatured vector DNA. The tubes are cooled to room temperature and spun down for 5 sec. For each tube, 0.5 U of *Taq* polymerase (diluted to 1 U/μl in 20 mM Tris, pH 8.0, and 0.1 mM EDTA) is added followed by a thin overlay of mineral oil (one drop from a 19-gauge hypodermic needle).

Cycling Protocol

Twenty-five to 30 cycles of PCR are carried out: denaturation at 94° for 1 min, annealing at 50° for 2 min, and synthesis at 72° for 3 min. Following the PCR, the 20-μl mix is transferred to a fresh tube containing 5 μl of loading dye. The entire reaction mix is loaded onto the 1% (w/v) agarose gel and electrophoresed for 1 hr at 110 V. The gel is photographed under ultraviolet (UV) light to visualize the DNA in the ethidium bromide-stained gel.

Concluding Remarks and Discussion of Problems

The advantage in the use of PCR to analyze recombinant DNA molecules is that it is a rapid and reliable technique. It is a technique that can obviate the need for purification, restriction mapping, or sequencing. In terms of sensitivity, the ability to amplify DNA from single eukaryotic cells is well established.[5] A serial dilution analysis by us has shown that approximately 100 bacterial cells in a PCR mix are sufficient to yield consistent and accurate results.[2] The larger number of bacterial cells is required because presumably only a small fraction of cells is lysed by heat alone.

Other advantages are a considerable saving of time, labor, and chemicals, as up to 48 samples can be analyzed at one time, using a solitary procedure that decreases the risk of experimental error. Finally, in the case of recombinant λ plaques, the amplified DNA provides an appropriate template for subcloning as well as direct sequencing.

As in the case with all PCR reactions, there are several sources of error an investigator must control. To prevent false positives due to contamination of PCR reagents, it is essential to change pipette tips each time any reagent is withdrawn from a stock solution. A cell (DNA)-free tube must be included in each set of PCR reactions as a check against contamination.

[5] N. Arnheim, H. Li, and X. Cui, *Genomics* **8,** 415 (1990).

Exceptionally long primers (>45 bp) and those with a high GC content (>60%) may give anomalous bands on the gel due to false priming and amplification. One way to overcome this problem is to raise the annealing temperature to 60° or more.

Failure to boil the cells for a full 10 min can lead to a failure of amplification due to either incomplete cell lysis and release of vector DNA, residual DNA nucleases, or the destruction of *Taq* polymerase by residual cellular proteases. The addition of an exceptionally large number of cells (and their accompanying culture media) to the PCR reaction mix is another source of difficulty as these can seriously alter the buffer conditions in the 20-μl reaction mix and inactivate *Taq* polymerase. A few dozen cells is all that is needed for analysis. If a cell pellet is visible following the 10-min lysis and 5-sec centrifugation, too much sample has been used.

Direct PCR analysis also saves time in identifying recombinant baculovirus.[6] Following the cotransfection of *Spodoptera* cells with recombinant vector and wild-type baculovirus, recombination occurs within 0.1 to 5% of cells creating recombinant baculovirus. Distinguishing recombinant from wild-type plaques is a time-consuming process that requires careful visual analysis under a light microscope. Cells infected with wild-type baculovirus have refractile occlusion bodies whereas recombinants appear gray. Confirmation of suspected positive plaques on the basis of morphology alone requires over 1 week, followed by further plaque purification and screening for protein synthesis. By using PCR analysis with one recombinant specific internal primer and another baculovirus-specific flanking primer we were able to confirm the identity of a suspected plaque without waiting days or weeks to be absolutely certain of our visual identification (Fig. 2).

[6] D. W. Miller, P. Safer, and L. K. Miller, *Genet. Eng.* **8,** 277 (1986).

[27] Genetic Analysis Using Polymerase Chain Reaction-Amplified DNA and Immobilized Oligonucleotide Probes: Reverse Dot-Blot Typing

By ERNEST KAWASAKI, RANDALL SAIKI, and HENRY ERLICH

Introduction

Polymerase Chain Reaction

A variety of different methods have been used to analyze nucleotide sequence variation in genomic DNA. The application of these various methods of genetic analysis has been greatly facilitated by the introduction of the polymerase chain reaction (PCR), an *in vitro* method for the enzymatic amplification of a specific DNA segment.[1-3] The reaction is based on the annealing and extension of two oligonucleotide primers that flank the target region in duplex DNA; after denaturation of the DNA, each primer hybridizes to one of the two separated strands such that extension from each 3'-OH end is directed toward the other. The annealed primers are then extended on the template strand by using a DNA polymerase. These three steps (denaturation, primer binding, and DNA synthesis) represent a single PCR cycle. Although each step can be carried out at a discrete temperature (e.g., 94–98, 37–65, and 72°, respectively), a reaction cycled between the denaturation and the primer-binding temperatures generally allows sufficient time for polymerase activity to amplify short PCR products. If the newly synthesized strand extends to or beyond the region complementary to the other primer, it can serve as a primer-binding site and template for subsequent primer extension reactions. Consequently, repeated cycles of denaturation, primer annealing, and primer extension result in the exponential accumulation of a discrete fragment whose termini are defined by the 5' ends of the primers. This exponential amplification results because under appropriate conditions the primer extension products synthesized in cycle "*n*" function as templates for the other primer in cycle "*n* + 1." The PCR can amplify double- or single-stranded DNA and, with the reverse transcription of RNA into a cDNA

[1] K. B. Mullis and F. A. Faloona, this series, Vol. 155, p. 335.
[2] R. K. Saiki, S. Scharf, F. Faloona, K. Mullis, G. Horn, H. A. Erlich, and N. Arnheim, *Science* **230**, 1350 (1985).
[3] R. K. Saiki, D. H. Gelfand, S. Stoffel, S. Scharf, R. Higuchi, G. T. Horn, K. B. Mullis, and H. A. Erlich, *Science* **239**, 487 (1988).

copy, RNA can also serve as a target. Because the primers become incorporated into the PCR product and mismatches between the primer and the original genomic template can be tolerated, new sequence information (specific mutations, restriction sites, regulatory elements) and labels can be introduced via the primers into the amplified DNA fragment.

In this chapter, we discuss the analysis of PCR-amplified DNA with sequence-specific oligonucleotide hybridization probes. It is the ability of PCR to amplify the target DNA sequences to high concentration that has allowed us to "reverse" the conventional dot-blot approach (immobilized target DNA and labeled oligonucleotide probes in solution at high concentration) and use a labeled target DNA sequence (present after amplification at high concentration) in solution to hybridize with a panel of immobilized oligonucleotide probes.

Analysis by Oligonucleotide Probe Hybridization

Analysis of nucleic acid sequences by hybridization with complementary probes is probably the most widely used technique in general molecular biological research. There are many variations on this theme and some commonly used methods include the Southern blot[4] for DNA fragments and the Northern blot[5,6] for RNA molecules. Although indispensable for structural studies of nucleic acids, these methods are somewhat cumbersome and time consuming when large numbers of samples need to be analyzed. If the analysis of DNA fragment *size* is not critical and the detection and/or quantitation of a specific sequence is required, then the extensive DNA purification, restriction enzyme digestions, and gel electrophoresis steps of the Southern blot can be omitted. Simplification of the detection schemes was accomplished with the advent of "dot-blot" protocols.[7,8] In this case, the DNA or RNA samples (purified or unpurified) are simply dotted onto membranes, fixed, and then hybridized with specific probes. Further simplification came about as short, sequence-specific oligonucleotides were found to be useful as probes.[9-11] Prior to this, probes

---

[4] E. M. Southern, *J. Mol. Biol.* **98,** 503 (1975).

[5] P. S. Thomas, *Proc. Natl. Acad. Sci. U.S.A.* **77,** 5201 (1979).

[6] P. S. Thomas, this series, Vol. 100, p. 255.

[7] F. C. Kafatos, C. W. Jones, and A. Estratiadis, *Nucleic Acids Res.* **7,** 1541 (1979).

[8] B. A. White and F. C. Bancroft, *J. Biol. Chem.* **257,** 8569 (1982).

[9] R. B. Wallace, J. Shaffer, R. F. Murphy, J. Bonner, T. Hirose, and K. Itakura, *Nucleic Acids Res.* **6,** 3453 (1979).

[10] S. V. Suggs, R. B. Wallace, T. Hirose, E. H. Kawashima, and K. Itakura, *Proc. Natl. Acad. Sci. U.S.A.* **78,** 6613 (1981).

[11] R. B. Wallace, M. J. Johnson, T. Hirose, T. Miyake, E. H. Kawashima, and K. Itakura, *Nucleic Acids Res.* **9,** 879 (1981).

were usually obtained from cDNA and genomic clones labeled by nick translation. As DNA synthesis capabilities improved, the oligonucleotide probes were not only easier to obtain and use but, more importantly, they could be used, under appropriate conditions, to discriminate nucleic acid sequences that differed by only one nucleotide.[12–15] One disadvantage of the short oligonucleotide probes, however, is that it is often difficult to analyze DNAs where the nucleotide complexity is high, as in the mammalian genomes. The analysis of complex genomes using allele-specific hybridization requires more technical expertise than other hybridization protocols in that the separation of genomic restriction fragments is required prior to hybridization. Probes labeled to high specific activity were also required for the analysis of single-copy genes in genomic DNA.

The difficulty of analyzing single-copy genes in genomic DNA has been overcome by use of the PCR to first amplify the target sequence to high abundance, followed by hybridization with the allele-specific oligonucleotide probe.[16] Thus the dot-blot method in conjunction with PCR amplification and oligonucleotide probes has greatly simplified the analysis of any DNA or RNA sequence, including those involved in genetic diseases, HLA polymorphisms, cancer, and so forth. If the sample size is large, the PCR/dot-blot method is a convenient format, especially when the number of probes required is small. However, as the number of probes required for genetic typing increases, this method becomes cumbersome because the PCR product must be immobilized on a number of membranes, each of which is hybridized to a different labeled oligonucleotide probe. Examples where the number of alleles and, hence, of oligonucleotide probes can complicate analysis are the mutants of the RAS oncogene system and cystic fibrosis mutations; in both cases more than 60 oligonucleotides are required to detect the known variants. Analysis of HLA class II polymorphisms is another area where many probes are required and complete genetic typing by the dot-blot method is complicated and time consuming.[17] To alleviate some of these difficulties, a method known as the

[12] M. Pirastu, Y. W. Kan, A. Cao, B. J. Conner, R. L. Teplitz, and R. B. Wallace, N. Engl. J. Med. 309, 284 (1983).

[13] J. L. Bos, M. Verlaan-de Bries, A. M. Jansen, G. H Veeneman, J. H. van Boom, and A. J. van der Eb, Nucleic Acids Res. 12, 9155 (1984).

[14] H. H. Kazazian, Jr., S. H. Orkin, A. F. Markham, C. R. Chapman, H. Youssoufian, and P. G. Waber, Nature (London) 310, 152 (1984).

[15] V. J. Kidd, M. S. Golbus, R. B. Wallace, K. Itakura, and S. L. Woo, N. Engl. J. Med. 310, 639 (1984).

[16] R. K. Saiki, T. L. Bugawan, G. T. Horn, K. B. Mullis, and H. A. Erlich, Nature (London) 324, 163 (1986).

[17] H. Erlich, T. Bugawan, A. B. Begovich, S. Scharf, R. Griffith, R. Saiki, R. Higuchi, and P. S. Walsh, Eur. J. Immunogenet. 18, 33 (1991).

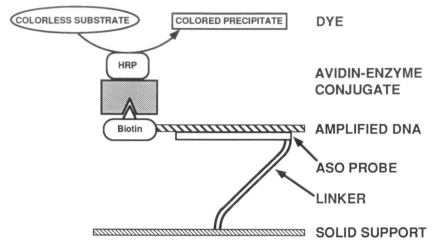

FIG. 1. Schematic of reverse dot blot. A biotinylated PCR product is hybridized to an allele-specific oligonucleotide (ASO) probe. The probe is bound to a nylon membrane by poly(dT) and UV cross-linking or by covalent attachment with amino linkers as described in text. The hybrids may be detected by reaction of a colorless substate with HRP (horseradish peroxidase) to form a colored precipitate.

"reverse dot blot" was developed.[18] In this technique, illustrated in Fig. 1, the allele- or gene-specific oligonucleotide probes are bound to the filter instead of the PCR products, and the amplified DNA labeled during the PCR is used to hybridize to the immobilized array of probes, hence the name, reverse dot blot. In the next section we describe two ways in which to attach probe oligonucleotides to filters for use in the reverse dot-blot method.

Reverse Dot-Blot Methodologies

Synthesis of 5' Biotinylated Polymerase Chain Reaction Primers

For purposes of nonradioactive detection, a single biotin molecule is incorporated into the 5' end of each of the PCR primer pairs. After amplification by PCR, each of the product DNA strands will have a 5' end labeled biotin that can be detected by the streptavidin–horseradish peroxidase methods outlined below. Having the biotin on both strands allows the use of oligonucleotide probes complementary to either strand.

[18] R. S. Saiki, P. S. Walsh, C. H. Levenson, and H. A. Erlich, *Proc. Natl. Acad. Sci. U.S.A.* **86,** 6230 (1989).

Moreover, the strand of the PCR product hybridized to the immobilized oligonucleotide probe also appears to hybridize to the primer, thus each bound strand of PCR product contains *two* biotin molecules, one incorporated and one in the annealed primer. The exact method for attachment of biotin to the oligonucleotides has been previously described in detail.[19]

Probe Attachment through Poly(dT) Tails

Enzymatic dT Tailing of Oligonucleotides. Oligonucleotides are synthesized on a Biosearch 8700 (San Rafael, CA) DNA synthesizer using phosphoramidite nucleosides (American Bionetics, Hayward, CA). Synthesis and purification of the oligonucleotides are done by using protocols provided by the manufacturer. Oligonucleotides (200 pmol) are tailed at the 3' end in 100 μl of buffer containing 100 mM potassium cacodylate, 25 mM Tris-HCl, 1 mM CoCl$_2$, 0.2 mM dithiothreitol (DTT), pH 7.6,[20] with 80 nmol dTTP (Pharmacia, Piscataway, NJ), and 60 units (50 pmol) of terminal deoxyribonucleotidyltransferase (Ratliff Biochemicals, Los Alamos, NM). Reactions are done at 37° for 60 min and stopped by the addition of 100 μl of 10 mM ethylenediaminetetraacetic acid (EDTA). The lengths of the homopolymer tails are controlled by limiting the dTTP; a nominal tail length of 400 dT residues is obtained by using 80 nmol of dTTP in the above reaction.

Machine Synthesis of dT-Tailed Oligonucleotides. The sequence-specific portion of the oligonucleotides are synthesized as above, but 100–200 T's are added to the 5' end after the probe part of the oligonucleotide has been completed. When the T's are added, the capping steps are omitted to allow higher efficiency of poly(dT) synthesis.

Attachment of Oligonucleotides to Filters. The tailed oligonucleotides are diluted into 100 μl of TE buffer (10 mM Tris-HCl, 1 mM EDTA, pH 8.0) and applied to nylon membranes [Genetrans-45 (Plasco, Woburn, MA) or Biodyne B (Pall Biosupport, East Hills, NY)] by a dot-blot apparatus (Bio-Dot; Bio-Rad, Richmond, CA). The damp filters are then placed in a metered ultraviolet (UV) light box (Stratalinker 1800; Stratagene, San Diego, CA) on TE-soaked paper pads and irradiated at 254 nm. Maximum fixation of the oligonucleotides occurs at an energy dosage of about 240 mJ/cm^2. The irradiated membranes are then washed in 200 ml of 5 × SSPE (1 × SSPE is 180 mM NaCl, 10 mM NaH$_2$PO$_4$, 1 mM EDTA, pH 7.2) with 0.5% (w/v) sodium dodecyl sulfate (SDS) for 30 min at 55° to remove

---

[19] C. H. Levenson and C. A. Chang, *in* "PCR Protocols: A Guide to Methods and Applications" (M. Innis, D. H. Gelfand, J. J. Sninsky, and T. J. White, eds.), p. 99. Academic Press, San Diego, 1990.

[20] R. Roychoudhury and R. Wu, this series, Vol. 65, p. 43.

unbound oligonucleotides. If not used immediately, the filters can be stored at room temperature after rinsing in water and air drying. We have subsequently found that UV fixation is not absolutely necessary for binding of the probes to filters. Simply drying the filters at room temperature appears to allow sufficient attachment of oligonucleotides for most hybridization/detection requirements.

Probe Attachment by 5'-Reactive Amino Groups[21]

Oligonucleotide Synthesis with Addition of 5'-Reactive Amino Groups. Oligonucleotides were synthesized on a MilliGen/Biosearch (Novato, CA) model 8750 DNA synthesizer using reagents and protocols obtained from the manufacturer. A primary, reactive amine group is added to their 5' ends.[19] Briefly, the terminal primary amino groups (amino linkers) are introduced during the final coupling step on the DNA synthesizer using N-trifluoroacetyl-6-aminohexyl-2-cyanoethyl-N',N',diisopropyl-phosphoramides,[22,23] which can be purchased from MilliGen/Biosearch. To add distance between the amine and the 5' base, "spacer" groups are added just before attachment of the amino linker. The spacer group reagent is a hexaethylene glycol-based cyanoethyl diisopropylphosphoramidite,[24] and is added immediately prior to addition of the 5' amino group. The combined length of a single spacer and the amino linker is approximately 28 Å if the chain is fully extended. Several spacer groups can be added to increase the distance between the aminolinker and the 5' base of the oligonucleotide. The crude amino-labeled oligonucleotides are converted to their lithium salts by precipitation from 4 M lithium chloride using 5 vol of cold ethanol : acetone (1 : 1). The precipitates are pelleted by centrifugation at 4000 g for 15 min at 4°, and the pelleted nucleic acid reconstituted in water and stored at −20°. The oligonucleoides may be purified by polyacrylamide gel electrophoresis followed by elution and quantitation by determining the absorbance at 260 nm. The spacer and the amino linker do not contribute significantly to the absorbance at this wavelength.

Covalent Attachment of Oligonucleotides to Filters.[21] Biodyne C membranes (Pall Biosupport) are rinsed briefly in 0.1 N HCl as a preactivation step (see Fig. 2). The filters are then soaked in freshly prepared 20% (v/v) EDC [1-ethyl-3-(dimethylaminopropyl)carbodiimide hydrochloride (Aldrich, Milwaukee, WI)] in water and rinsed once with water to remove

[21] Y. Zhang, M. Y. Coyne, S. G. Will, C. H. Levenson, and E. S. Kawasaki, *Nucleic Acids Res.* **19**, 3929 (1991).

[22] S. L. Beaucage and M. H. Caruthers, *Tetrahedron Lett.* **22**, 1859 (1981).

[23] N. D. Sinha, J. Biernat, J. McManus, and H. Koster, *Nucleic Acids Res.* **12**, 4539 (1984).

[24] C. H. Levenson, C. A. Chang, and F. Oakes, U.S. Pat. No. 4,914,210 (1990).

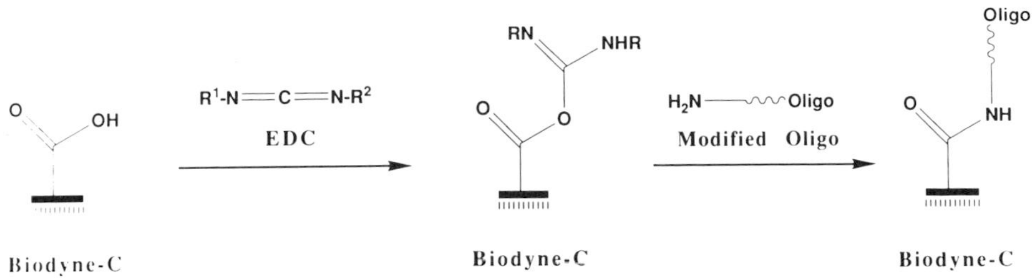

Fig. 2. Covalent immobilization of oligonucleotides on Biodyne-C. Carboxyl groups on Biodyne-C are activated with EDC to form O-acylureas. These groups are reacted with the amine groups on the amino-modified oligonucleotides to form stable amide bonds.

excess EDC. At this stage the active groups on the membrane are unstable and the membrane must be processed in less than half an hour. The preactivated membrane is placed immediately in a Bio-Dot apparatus and 1–2 pmol of amino-modified oligonucleotides is applied in 20 μl of 0.5 M sodium bicarbonate buffer. The probes are allowed to react with the filter for 15 min before applying vacuum and washing with TBS, 0.1% (v/v) Tween 20 (TBS: 150 mM NaCl, 50 mM Tris-HCl, pH 7.5). To destroy remaining active groups, the filters are rinsed thoroughly with 0.1 N NaOH for 10 min. The filters are then neutralized with rinses in TBS and water. The filters may then be air dried for storage or used immediately for hybridization purposes.

Hybridization to Covalently Attached Oligonucleotides

Filters are hybridized with denatured PCR products in 5 × SSPE, 0.5% (w/v) SDS for 30 min at varying temperatures depending on the T_m values of the immobilized probes. In general, temperatures in the range of 45 to 55° provide sufficiently stringent conditions for probes in the range of 15 to 20 bases in length with an average GC content. The filters can be washed with TMACl (tetramethylammonium chloride)[25] to reduce the influence of base composition variation, or with 2× SSPE, 0.1% (w/v) SDS when cross-hybridization background is not a problem. Again, washing temperatures will vary, but 45 to 55° is generally sufficient for most probes. For detection, filters with biotinylated products are incubated with 1–2 ml of 2× SSPE, 0.1% (w/v) SDS containing 10–20 μl of streptavidin-HRP (5 mg/ml; Perkin-Elmer Cetus, Norwalk, CT) for 15 min, and then washed in the same buffer of 10 min. Equal volumes of ECL (enhanced chemiluminescence) gene detection reagents A + B (Amersham, Arlington Heights,

[25] A. G. DiLella and S. L. C. Woo, this series, Vol. 152, p. 447.

IL) are applied to the filters for 1 min. The filters are then removed, covered with Saran wrap, and exposed to Hyperfilm-ECL (Amersham) or Kodak (Rochester, NY) XRP film for a few seconds or minutes to detect photon emission. This approach is illustrated in the detection of various *RAS* mutations in three tumor cell lines (Fig. 3).

Detection (Color Development). The membranes are incubated at 50° for 10 min in 5 ml of the buffer containing 10–20 μl of streptavidin–horseradish peroxidase conjugate (5 mg/ml; Perkin-Elmer Cetus). The conjugate solution is removed and the filters washed twice for 5 min at 50° with the same buffer. The filter is then rinsed twice with color development buffer (100 mM sodium citrate, pH 5.0); if necessary the filters can be stored at 4° in buffer until ready to proceed. The buffer is remove and 5 ml of color development solution freshly prepared (no more than 10 min before use) is added. This solution is 0.1 mg/ml TMB, 0.003% H_2O_2 made up with the color development buffer [TMB is 3,3′,5,5′-tetramethylbenzidine at 2 mg/ml in 100% ethanol (Fluka, Ronkonkoma, NY)]. The filters are incubated with shaking for 10–30 min at room temperature. The solution is removed and the color development stopped by rinsing filters three times for 5 min each with distilled water. The membranes are photographed for permanent records over a black background to improve contrast (see Fig. 4).

Chromogenic detection using tetramethylbenzidine (TMB) can also be carried out as described above.

Discussion and Summary

Choice of Probe Sequence and Length

In general, designing immobilized sequence-specific probes is similar to the process used to create nonimmobilized, solution-based probes. Although still largely an empirical exercise (much like devising PCR primers), there are several guidelines that should be followed to help ensure success.

1. Length of probe: Because they are typically used to detect the presence of single base pair mismatches, most probes are between 15 and 23 bases in length. Oligonucleotides at the lower end of this range often provide better discrimination because of the relatively greater destabilization effect of a mismatch.

2. Nature of mismatch: Some mismatches are less destabilizing than others. In particular, G : T mismatches appear to have little effect on

K-ras

12	WT	SER	CYS	ARG	ASP	ALA	VAL	
13		SER	CYS	ARG	ASP	ALA	VAL	
61	WT	GLU	LYS	ARG	LEU	PRO	HIS1	HIS2

N-ras

12	WT	SER	CYS	ARG	ASP	ALA	VAL	
13		SER	CYS	ARG	ASP	ALA	VAL	
61	WT	GLU	LYS	ARG	LEU	PRO	HIS1	HIS2

H-ras

12	WT	SER	CYS	ARG	ASP	ALA	VAL	
13		SER	CYS	ARG	ASP	ALA	VAL	
61	WT	GLU	LYS	ARG	LEU	PRO	HIS1	HIS2

FIG. 3. Analysis of *Ras* mutations in three tumor cell lines. DNA sequences around codons 12, 13, and 61 of the K-, N-, and H-*ras* oncogenes were amplified. The biotinylated products were hybridized to filters with covalently bound mutation-specific oligonucleotides. Hybrids were detected by the ECL system described in text. K-*ras* wild-type (WT) sequence for codon 12 is 5'-GGAGCTGGTGGCGTA-3'; the Cys mutant is 5'-GGAGCTTGTGGCGTAG-3'. N-*ras* wild-type sequence for codon 12 is 5'-GAGCAGGTGGTGTTGG-3'; the Asp mutant is 5'-GAGCAGATGGTGTTGG-3'. H-*ras* WT sequence for codon 12 is 5'-GGCGCCGGCGGTG-3'; the Val mutant is 5'-GGCGCCGTCGGTGT-3'. Mia-PaCa is a pancreatic carcinoma cell line, PA-1 is a teratocarcinoma, and T-24 is a bladder cancer cell line.

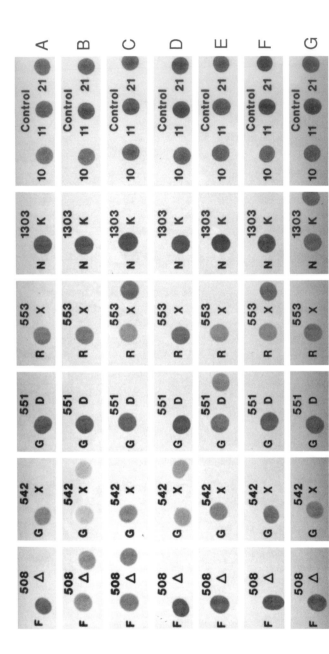

Fig. 4. Detection of cystic fibrosis (CF) mutations with the immobilized probe method. Biotinylated primers were used to simultaneously amplify exons 10, 11, and 21 of the cystic fibrosis transmembrane conductance regulator (CFTR) gene in seven human genomic DNA samples. These three exons contain the sites of five relatively common CF mutations: delF508, G542X, G551D, R553X, and N1303K. Probes specific for the normal and mutant sequences at each position were synthetically tailed and spotted onto nylon strips in pairs—normal probes on the left, mutant probes on the right. Also included were three positive control probes that always hybridize to CFTR exons 10, 11, and 21. Amplified DNA was denatured and individually hybridized to the strips. Following hybridization, the strips were incubated with streptavidin–HRP, then color developed with TMB and hydrogen peroxide. Genotypes of the samples can be readily determined by observing the pattern of colored dots. Sample A, homozygous normal (two copies of the normal, wild-type gene); sample B, delF508/G542X compound heterozygote (one copy each of the delF508 and G542X mutations); sample C, delF508/R553X compound heterozygote (one copy each delF508 and R553X); sample D, G542X carrier (one copy each normal and G542X); sample E, G551D carrier (one copy each normal and G551D); sample F, R553X carrier (one copy each normal and R553X); sample G, N1303K carrier (one copy each normal and N1303K).

duplex stability.[26] These mismatches can be avoided by simply switching the probe to the opposite strand, creating a more potent C : A mismatch.

3. Position of mismatch: The maximum destabilizing effect is obtained by placing the site of the mismatch in the central portion of the probe, at least 3 bases from either end.

Estimation of T_d. The classic formula for predicting the dissociation temperature T_d of an oligonucleotide, 2° for each A/T and 4° for each G/C,[27] is simple but not necessarily accurate. In recent years several algorithms have been developed that appear to give much better estimates and can assist in the design of probes for particular hybridization conditions.[28]

In addition to these guidelines, two other considerations specific to the reverse dot-blot format should be kept in mind.

1. Adjacent sequence: For probes to be tailed with poly(dT), the sequence immediately adjacent to the target site on the PCR product must be examined for complementarity to the tail. If one or more successive adenines are present, the effective length of the probe will be increased by that amount. This problem is most conveniently overcome by adding noncomplementary bases, such as dC, directly to the end of the probe during synthesis, creating a compound tail of a few cytosines followed by the long thymidine tract.

2. Amplicon secondary structure: Because they are not bound to the membrane, the single-stranded PCR products are free to assume secondary structures. On occasion, particular amplification products will form stable hairpin structures that occlude the target site and prevent hybridization to the immobilized probe. DNA folding programs, such as Squiggles,[29] can be used to identify these situations and design alternate PCR primers that avoid the complementary region. Another option is to increase the length of the probe so that it can efficiently compete with the secondary structure in the amplicon.

Because all of the probes are exposed to the PCR product simultaneously, it is clearly necessary that each of the probes hybridize specifically under the identical reaction conditions. Although the T_d estimation algorithms mentioned previously are helpful in designing a series of disparate probes with equivalent annealing characteristics, it is often necessary to

[26] S. Ikuta, K. Takagi, R. B. Wallace, and K. Itakura, *Nucleic Acids Res.* **15**, 797 (1987).
[27] S. V. Suggs, T. Hirose, E. H. Miyake, M. J. Kawashima, K. I. Johnson, and R. B. Wallace, ICN–UCLA Symp. Dev. Biol. **23**, 683 (1981).
[28] W. Rychlik and R. E. Rhoads, *Nucleic Acids Res.* **17**, 8543 (1989).
[29] Univ. of Wisconsin Genet. Comput. Group, Madison.

adjust the sequences to balance the signal intensities. In most cases, it is simply a matter of adding or subtracting a base or two.

Influence of Spacer Length. As described above, the poly(T) tails serve both as a spacer and a means of immobilizing the oligonucleotides. Hexaethylene glycol can also serve as a spacer for the covalent attachment approach (see above). A single hexaethylene glycol spacer plus the amino linker is about 28 Å in length. We previously tested the influence of the spacer on the sensitivity of hybridization (see Ref. 21). In this experiment we found that a single spacer gave an approximate fourfold increase in sensitivity of hybridization when hybridizing PCR products against two-fold dilutions of the probe oligonucleotides. As little as 0.06 pmol of oligonucleotide was sufficient to give a strong signal with one spacer, whereas 0.25 pmol was required without the spacer. Addition of more spacers did not increase the sensitivity to any significant degree. It is possible that inclusion of the spacer allows a more efficient hybridization by lowering steric hindrance imparted by the filter.

Pooling of Oligonucleotides in One Dot. The sensitivity of the covalent attachment approach makes attractive the idea of pooling several different allele specific oligonucleotides in one spot. One may wish to do this when screening genetic diseases such as cystic fibrosis, or the activated *ras* oncogene implicated in many cancers; in both cases more than 60 different mutations have been identified.[30,31] In large-scale screening a yes/no answer is usually sufficient for the first analysis and pooling can be advantageous here by reducing the complexity of the filters. We have found that up to a 100-fold excess of another probe can be bound in the same spot without losing discrimination.[21] Even *ras* oligonucleotides with single base pair differences can be combined and still give acceptable signal-to-noise ratios.

Summary

The reverse dot-blot method is a simple and rapid diagnostic procedure that allows screening of sample for a variety of mutations/polymorphisms in a single hybridization reaction. Several methods of immobilizing the oligonucleotide probes are discussed. The reverse dot-blot method has several unique properties that are valuable in a diagnostic setting: (1) the typing results from a single sample can be located on a single strip. This facilitates scanning and interpretation of the probe reactivity patterns and minimizes the potential for user error. (2) The test can utilize premade

[30] L. Roberts, *Science* **250,** 1076 (1990).
[31] J. L. Bos, *Science* **250,** 1076 (1990).

typing strips. This minimizes user labor as well as error potential and allows the use of standardized reagents. (3) Unlike dot-blot/oligonucleotide typing, only the PCR product is labeled, eliminating the potential problem of probes labeled to different specific activities. This method has already been used in the areas of forensic genetic typing (the HLA-DQα Amplitype test),[32] tissue typing for transplantation (the HLA-DRβ) test,[33] cystic fibrosis screening,[21,34] as well as in a variety of research applications.

[32] E. Blake, J. Mihalovich, R. Higuchi, P. S. Walsh, and H. A. Erlich, *J. Forensic Sci.* **37**, 700 (1992).
[33] H. A. Erlich, unpublished observations, 1992.
[34] H. A. Erlich, D. Gelfand, and J. J. Sninsky, *Science* **252**, 1643 (1991).

[28] Removal of DNA Contamination in Polymerase Chain Reaction Reagents by Ultraviolet Irradiation

By GOBINDA SARKAR and STEVE S. SOMMER

Introduction

Contamination of reagents for various biochemical or biological reactions is especially problematic in experiments that involve the exponential growth of microorganisms such as bacteria or yeast. DNA contamination during the polymerase chain reaction (PCR) represents a parallel situation because the template DNA also amplifies exponentially.[1] Contamination of reagents with previously amplified material poses the most serious problem in the routine use of PCR. Therefore, caution is necessary in handling PCR reagents, especially in situations in which the input DNA is from only a few cells.

We have been exploring methods to eliminate contamination of DNA from PCR reagents without compromising their reactivity.[2-4] Several alternative methods for decontaminating samples include the following: digestion with DNase or restriction endonucleases,[5,6] γ irradiation,[7] psoralen

[1] K. B. Mullis and F. A. Faloona, this series, Vol. 155, p. 335.
[2] G. Sarkar and S. Sommer, *Nature (London)* **343**, 27 (1990).
[3] G. Sarkar and S. Sommer, *Nature (London)* **347**, 341 (1990).
[4] G. Sarkar and S. Sommer, *BioTechniques* **10**, 590 (1991).
[5] B. Furrer, U. Candrian, P. Weiland, and J. Luthy, *Nature (London)* **346**, 324 (1990).
[6] F. M. DeFilippes, *BioTechniques* **10**, 26 (1991).
[7] J.-M. Deragon, D. Sinnett, G. Mitchell, M. Potier, and D. Labuda, *Nucleic Acids Res.* **18**, 26 (1990).

treatment followed by ultraviolet (UV) irradiation,[8] and cleavage by uracil N-glycosylase (commercially available from Perkin-Elmer Cetus, Norwalk, CT).

As UV irradiation has long been known to damage DNA, we have investigated the effectiveness of UV exposure to remove DNA contamination in PCR reagents. Ultraviolet damages DNA in various ways, including the formation of cyclobutyl pyrimidine dimers, the formation of pyrimidine–pyrimidine photoadducts, the oxidization of bases, and the induction of single- and double-strand breaks (reviewed in Cadet et al.[9]). These UV effects often are more pronounced on double-stranded DNA than on single-stranded DNA.

The multiple effects of UV irradiation on DNA suggest that these reactions may be dependent on sequence and size, with the larger molecules more likely to be highly UV sensitive. The single-stranded short oligonucleotides used in PCR may not be as susceptible to the damaging effect of UV irradiation as would be the double-stranded template DNA. Thus, PCR reagents, including the primers, but excluding the actual template DNA, potentially could be decontaminated with UV irradiation. The experiments described below demonstrate the efficacy of UV irradiation as a decontamination agent. The method involves mixing the reagents, irradiating with UV for the desired time, adding template DNA, and amplifying with PCR. Although the experiments focus on PCR, UV irradiation can inactivate undesired nucleic acids in other protocols such as in vitro transcription and certain alternative amplification systems.

Materials and Reagents

AmpliTaq (the Taq polymerase used in all experiments) is purchased from Perkin-Elmer Cetus (Emeryville, CA). Ethidium bromide is purchased from Sigma Chemical Company (St. Louis, MO). Oligonucleotides are synthesized in an Applied Biosystems (Foster City, CA) automated DNA synthesizer. The DNA thermal cycler used for PCR is from Perkin-Elmer Cetus. The Fotodyne (New Berlin, WI) 1000 transilluminator is used to irradiate the samples at wavelengths of 254 and 300 nm.

Oligonucleotide primers for the human factor IX gene are used to amplify either template DNA (human genomic DNA) or "target" ("contaminant") DNA. This target DNA consists of a segment of the complete human factor IX cDNA cloned into pSP65 and linearized with an appro-

[8] Y. Jinno, K. Yoshiura, and N. Niikawa, Nucleic Acids Res. 18, 6739 (1990).
[9] J. Cadet, M. Berger, C. Decarroz, J. R. Wagner, J. E. Van Lier, Y. M. Ginot, and P. Vigny, Biochimie 68, 813 (1986).

priate enzyme. The primers are listed below with informative names (see
Refs. 3 and 10); one-letter designations are used in text for ease of reading.
The numbering system is from Ref. 11.

A: F9(Hs)-(*Bam*-T7/TI-37)E5(20365)-51D
   5'-GGATCCTAATACGACTCACTATAGGGAGACCACCATGCC
   ATTTCCATGTGG-3'
B: F9(Hs)E8(31215)-16U
   5'-GGGTCCCCCACTATCT-3'
C: F9(Hs)E5(20537)-14U
   5'-TCTTCTCCACCAAC-3'
D: F9(Hs)E8(31022)-17U
   5'-TGAGGAAGATGTTCGTG-3'
E: F9(Hs)(*Kpn*/SP6-29)E8(31346)-47U
   5'-GGTACCAATTAGGTGACACTATAGAATAGTAATCCAGTT
   GACATACC-3'
F: (T7-23/T-20)-43(D/U)
   5'-TAATACGACTCACTATAGGGAGATTTTTTTTTTTTTTTTTT
   TT-3'

Methods

Polymerase Chain Reaction Conditions. A standard PCR protocol[3] is
used that includes in a final volume of 20 μl: 10 mM Tris-HCl (pH 8.3), 50
mM KCl, 1.5 mM MgCl$_2$, 200 μM concentrations of each dNTP, 0.1%
(w/v) gelatin, and 0.1 μM concentrations of each of two primers. The
complete PCR mix with or without UV irradiation (minus *Taq* polymerase)
is heated at 94° for 10 min. One-half unit of the AmpliTaq [diluted in 10
mM Tris-HCl (pH 8.0), 1 mM ethylenediaminetetraacetic acid (EDTA)] is
added and mixed followed by an overlay of mineral oil, and then the tubes
were placed into the thermal cycler. The PCR cycles are set at 94° for 1 min
(denaturation), 50° for 2 min (annealing), and 72° for 3 min (elongation),
followed by a final 10-min elongation at 72° after 30 cycles. The tubes then
are kept at 4° or frozen at −20°.

Irradiation of Samples and Components. The PCR mix without the
actual template DNA and (except where noted) *Taq* polymerase, but
with "target" or "contaminating DNA," is added to colorless 0.5-ml
polypropylene microcentrifuge tubes. The tubes are placed in direct contact with the platform of the Fotodyne 1000 transilluminator. Styrofoam

[10] G. Sarkar and S. Sommer, *Science* **244**, 331 (1989).
[11] S. Yoshitake, B. G. Scatch, D. C. Foster, E. W. Davie, and K. Kurachi, *Biochemistry* **24**,
3736 (1985).

S 1 2 3 4 5 6 7 8 S 9 10 11 12 13 14 15 16 S

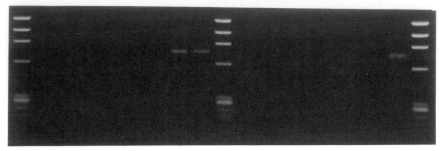

FIG. 1. Effect of UV irradiation in inactivating PCR contamination. A standard protocol[3] (see Methods) was used to prepare a 20-μl PCR mix with primers A and B to amplify a 750-bp segment of human factor IX complementary DNA. Various amounts of a linearized plasmid (6 kb in length) that contain the full-length human factor IX cDNA were added, so that the amount of target DNA was as indicated below. The samples then were irradiated for 5 min (lanes 2–8) or 20 min (lanes 10–16). After irradiation, the samples were denatured for 10 min, *Taq* polymerase (0.5 U/tube) was added, and 30 cycles of PCR were performed. Samples (5 μl from each tube) were then electrophoresed on a 1.5% agarose gel in standard TAE (Tris–acetate–EDTA) buffer (T. Maniatis, E. F. Fritsch, and J. Sambrook, "Molecular Cloning: A Laboratory Manual" Cold Spring Harbor Press, Cold Spring Harbor, New York, 1982) for approximately 1 hr. The electrophoresis buffer contained 0.1 μg/ml of ethidium bromide. Lanes: S, DNA standards obtained by digesting ϕX174 DNA with *Hae*III; 1 and 9, no UV irradiation and no target; 2 and 10, UV irradiation but no target; 3 and 11, 3 pg target DNA; 4 and 12, 30 pg target DNA; 5 and 13, 300 pg target DNA; 6 and 14, 3 ng target DNA; 7 and 15, 30 ng target DNA; 8 and 16, 3 pg nonirradiated target DNA added in irradiated reagents (positive control). (Reprinted with permission from Sarkar and Sommer.[2] Copyright © 1990 Macmillan Magazines, Ltd.)

holders are used to maintain the tubes in a vertical position on the surface of the transilluminator. To this end, holes are punched in a 1.5-cm thick piece of Styrofoam such that the top of a 0.6-ml Eppendorf tube cannot pass through. The tubes are uncapped, inserted into the holes as far as possible, and placed upright on the transilluminator platform. Ultraviolet irradiation is carried out for a desired time with the 254- and 300-nm wavelength bulbs separately or in combination in the standard instrument configuration (for routine use the two types of bulbs are used simultaneously).

Amount of Contaminating DNA. Figure 1 illustrates an experiment in which 3 to 30,000 pg of target DNA in 20 μl of PCR mix was irradiated as above with a combination of 254- and 300-nm wavelength bulbs for 5 min (lanes 1–8) or 20 min (lanes 9–16) and then amplified for 30 rounds with primers A and B. These primers amplify a 750-bp segment from the factor IX cDNA. (Lanes 1 and 9 in Fig. 1 are negative controls with no target DNA and no UV irradiation.) The positive controls (lanes 8 and 16, Fig.

1) received 3 pg of nonirradiated target DNA that was amplified using the irradiated PCR mix. A band of normal intensity can be seen in these lanes, indicating that irradiation of the PCR mix does not noticeably compromise the ability of the reagents to generate a strong signal after 30 cycles of PCR. No amplified product is detectable in lanes 2–6 and 10–15 (Fig. 1), indicating that a 5-min UV exposure can inactivate 3000 pg of target DNA (lane 6) and a 20-min UV exposure can inactivate up to 30,000 pg of target DNA (no band in lane 15). In further experiments, amplification of as little as 0.03 pg of target DNA could be detected routinely using the irradiated PCR mix. Thus, irradiation with UV eliminates 10^5-fold (in 5 min) or 10^6-fold (in 20 min) more target than an amount of target that can be detected routinely in a 30-cycle PCR. This level of inactivation is usually more than adequate for the level of contamination generated in routine practice by aerosoled DNA or trace carryover of previously amplified material.

Decontamination of Individual Polymerase Chain Reaction Reagents. In additional experiments to assess the effects of UV on individual components of PCR, 1.5 pg/μl of the factor IX cDNA plasmid was introduced into concentrated stock solutions of PCR buffer (10×), primers (10×), or deoxynucleoside triphosphates (7×) and irradiated with UV for 5 min with 254 and 300-nm light. In each case the contaminant failed to amplify, whereas DNA added (0.015 pg/μl) after irradiation amplified efficiently.

Irradiation of *Taq* polymerase (AmpliTaq formulation) was carried out at two dilutions (1 and 5 U/μl) with the 300- and 254-nm bulbs separately and in combination. After 5 min of irridiation, the enzyme amplified a product of normal intensity as visualized with ethidium bromide staining. However, after 20 min of irradiation, the enzyme was inactivated. Because 5 min of irradiation can remove 10^5-fold excess of target DNA (of 750 bp), routine contamination of the *Taq* enzyme should be inactivated with this exposure time. Contamination of PCR reagents also was efficiently removed by a 5-min irradiation when the contaminating DNA was 200 ng of human genomic DNA. In these experiments, a 542-bp segment was amplified because the 750-bp segment cannot be amplified from genomic DNA because of the length of the introns. The expected segment was not detected in the irradiated sample, but a control PCR tube that was not irradiated did show the 542-bp amplified product.

Size of Contaminating DNA. The effect of UV irradiation was evaluated for contaminant DNA segments of different length.[4] Four 5′ coterminal DNA segments of increasing lengths (209, 551, 904, and 2277 bp) were amplified with PCR (with primers A + C, A + D, A + E, and A + F, respectively) using the factor IX cDNA template described earlier. Equimolar aliquots of the four amplified segments were serially diluted and used as a source of known contaminants in subsequent experiments

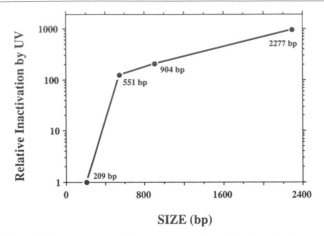

SIZE (bp)

FIG. 2. Effect of UV exposure on DNA target length in PCR. Irradiation was performed for 5 min with a combination of 254- and 300-nm wavelength bulbs. Four DNA targets of 209, 551, 904, and 2277 bp were generated with PCR. For each target, the amplified product of the first PCR was quantitated, serially diluted, irradiated with UV, and reamplified. The efficiency of amplification for each segment was verified by amplification of each of these nonirradiated segments at 10[6]-fold dilution. The relative susceptibility of these segments to UV inactivation is plotted against the logarithm of size. The data are the average of three experiments. (Reprinted with permission from Sarkar and Sommer.[4])

that employed irradiation with UV. The irradiation was carried out for 5 min with 254- and 300-nm wavelength bulbs in combination. The relative equimolar susceptibility to UV irradiation of these segments was then calculated from the highest dilution of each of these segments in which no amplification was detected by ethidium staining. The relative susceptibilities of these segments were then plotted against size on a semilog graph (Fig. 2). A fourfold decrease in size of the larger segments (2277 to 551 bp) resulted in an eightfold decrease in UV susceptibility, and an additional 2.6-fold decrease in size (to 209 bp) reduced the sensitivity to UV irradiation by 125-fold.

The length of the flanking sequence around a target was not found to have an effect on the efficiency of UV inactivation of the target. For example, UV inactivation was virtually equivalent when either a 600-bp segment or a 6-kb plasmid containing this segment was irradiated, followed by PCR amplification of the 600-bp target (data not shown). This suggests that long-range interstrand or intrastrand cross-links are not the major source of UV inactivation.

Sequence of Contaminating DNA. Sequence as well as size should affect susceptibility to UV inactivation. In total, at least 11 segments from

TABLE I
PYRIMIDINE DINUCLEOTIDES IN SEGMENTS WITH MARKEDLY DIFFERENT
ULTRAVIOLET SENSITIVITIES[a]

| | | | Pyrimidine dimers[c] | | | | | | | |
| | | | Sense strand | | | | Antisense strand | | | |
Segment[b]	Size (bp)	G + C content (%)	TT	TC	CT	CC	TT	TC	CT	CC
A	151	43	14	15	13	7	10	9	12	8
B	202	44	17	19	16	8	12	15	19	12
C	431	56	15	34	38	51	22	20	27	32

[a] Reprinted with permission from Ref. 4.
[b] Segment A is at least 25-fold more sensitive to UV than segment B or C. Segments A and B are derived from the human factor IX gene and segment C is derived from the dopamine D_2 receptor gene.
[c] The number of dimers per strand was calculated from frequencies generated by the DNASTAR computer program, Distribution.

regions of high and low G + C content in the human genome have been examined.[4] The susceptibility varied markedly, with some of the small segments being essentially resistent to UV. Marked variability in suscepti- bility to UV inactivation also was reported by Cimino et al.,[12] who found that a 121-bp segment was much less susceptible to UV than an unrelated 500-bp segment.

In multiple experiments, a 151-bp segment was found to be at least 25- fold more susceptible to UV irradiation than two unrelated segments of 202 and 431 bp. Because the lesions produced by UV light are most commonly at pyrimidine dimers,[9] the number of pyrimidine dinucleotides were determined in the sense and antisense strand as a function of sensitiv- ity to UV (Table I). No correlation could be found. In fact, the more UV- sensitive segment had *fewer* pyrimidine dimers than the other two UV- resistant segments in both strands.

State of Contaminating DNA. To investigate whether exposure to UV would inactivate a dried DNA, DNA from the 551-bp segment was lyophilized, irradiated with UV, and then rehydrated and amplified with PCR.[4] When irradiated dry, 0.01 pg of this target amplified well. However, 46 pg of the same target failed to amplify when irradiated in solution (data not shown). This observation suggests that DNA contaminants on laboratory equipment may be refractory to inactivation with UV light. Therefore, effective potentiators of UV inactivation that work in atmo-

[12] G. D. Cimino, K. Metchette, S. T. Isaacs, and Y. S. Zhu, *Nature (London)* 345, 773 (1990).

spheric as well as in aqueous environments would be useful to rapidly decontaminate PCR reagents or laboratory equipment.

Conclusions

Two major practical conclusions emerge from these studies. First, it is advantageous not to amplify small segments of DNA whenever feasible because these are likely to be poorly inactivated by UV if these segments become contaminants. We found five out of five DNA segments <700 bp to be highly susceptible ($>10^3$ inactivation) to UV inactivation in contrast to only one of four segments <250 bp. Of the remaining segments < 250 bp, one was moderately sensitive (more than 25-fold) and two were resistant to UV. If it is necessary to amplify small segments, uracil N-glycosylase, UV with psoralen treatment, DNase digestion, restriction endonuclease digestion, or γ radiation may be preferable.[5–8] Second, UV inactivation is much less effective in eliminating dried DNA, suggesting that decontamination of laboratory equipment may require the aid of a photosensitizer or even a completely different approach.

Acknowledgments

Plasmid pSP6-9A was the generous gift of Charles Shoemaker of the Genetics Institute. We thank Mary Johnson for typing the manuscript. The work was aided by March of Dimes Grant 6-581.

[29] Polymerase Chain Reaction Amplification of Specific Alleles: A General Method of Detection of Mutations, Polymorphisms, and Haplotypes

By Cynthia D. K. Bottema, Gobinda Sarkar, Joslyn D. Cassady, Setsuko Ii, Charyl M. Dutton, and Steve S. Sommer

Introduction and Principle of Method

The polymerase chain reaction (PCR) is a method that utilizes two oligonucleotide primers to amplify a segment of DNA more than 1 million-fold. The PCR can be adapted for the rapid detection of known single-base changes in DNA by using specially designed oligonucleotides in a method we call PCR amplification of specific alleles (PASA).[1,2] The method is also known as allele-specific amplification (ASA), allele-specific PCR (ASP), and amplification refractory mutation system (ARMS).[3-6] The principle is to design an oligonucleotide primer that preferentially will amplify one allele over another. Specificity is obtained if the oligonucleotide matches the desired allele, but mismatches the other allele(s) at or near the 3' end of the allele-specific primer.[1,2] The desired allele is readily amplified, whereas the other allele(s) is poorly amplified if at all. The poor amplification is a result of the mismatch between the DNA template and the oligonucleotide, which prevents efficient 3' elongation by *Taq* polymerase (Fig. 1).[7,8]

PASA has been used to perform population screening, haplotype analysis, patient screening, and carrier testing for over 62 different single-base changes in our laboratory.[1,2,9-14] This method has also been adapted to obtain haplotypes in the absence of samples from family members.[14,15] This "double PASA" utilizes pairs of allele-specific PCR primers to amplify each haplotype differentially (Fig. 2). Four amplifications can distinguish the haplotypes produced by a pair of biallelic polymorphisms.

[1] S. S. Sommer, J. D. Cassady, J. L. Sobell, and C. D. K. Bottema, *Mayo Clinic Proc.* **64**, 1361 (1989).

[2] G. Sarkar, J. Cassady, C. D. K. Bottema, and S. S. Sommer, *Anal. Biochem.* **186**, 64 (1990).

[3] C. R. Newton, A. Graham, I. E. Heptinstall, S. J. Powell, C. Summers, and N. Kalsheker, *Nucleic Acids Res.* **17**, 2503 (1989).

[4] D. Y. Wu, L. Ugozzoli, B. K. Pal, and R. B. Wallace, *Proc. Natl. Acad. Sci. U.S.A.* **86**, 2757 (1989).

[5] W. C. Nichols, J. J. Liepnieks, V. A. McKusick, and M. D. Benson, *Genomics* **5**, 535 (1989).

[6] H. Okayama, D. T. Curiel, M. L. Brantly, M. D. Holmes, and R. G. Crystal, *J. Lab. Clin. Med.* **114**, 105 (1989).

[7] S. Gustafson, J. A. Proper, E. J. W. Bowie, and S. S. Sommer, *Anal. Biochem.* **165**, 294 (1987).

[8] R. A. Gibbs, P.-N. Nguyen, and C. T. Caskey, *Nucleic Acids Res.* **17**, 2437 (1989).

[9] C. D. K. Bottema, D. D. Koeberl, R. P. Ketterling, E. J. W. Bowie, S. A. Taylor, P. J. Bridge, D. Lillicrap, A. Shapiro, G. Gilchrist, and S. S. Sommer, *Br. J. Haematol.* **75**, 212 (1990).

[10] S. Ii, S. Minnerath, K. Ii, P. J. Dyck, and S. S. Sommer, *Neurology* **41**, 893 (1991).

[11] R. P. Ketterling, C. D. K. Bottema, D. D. Koeberl, S. Ii, and S. S. Sommer, *Hum. Genet.* **87**, 333 (1991).

[12] R. P. Ketterling, C. D. K. Bottema, J. P. Phillips, III, and S. S. Sommer, *Genomics* **10**, 1093 (1991).

[13] G. Sarkar, S. Kapelner, D. K. Grandy, O. Civelli, J. Sobell, L. Heston, and S. S. Sommer, *Genomics* **11**, 8 (1991).

[14] S. Ii, J. Sobell, and S. S. Sommer, *Am. J. Hum. Genet.* **50**, 29 (1992).

[15] G. Sarkar and S. S. Sommer, *BioTechniques* **10**, 436 (1991).

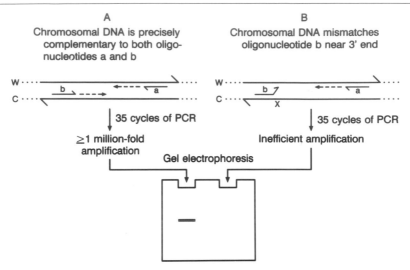

FIG. 1. Diagram of polymerase chain reaction (PCR) amplification of specific alleles (PASA). The two antiparallel strands of chromosomal DNA are indicated by W, the Watson strand, and C, the Crick strand. The 5' to 3' directions are indicated by the half-arrows. Strands have been melted apart by high temperature, and PCR oligonucleotide primers, a and b, have been annealed. In this diagram, elongation with DNA polymerase, the third step of PCR, is underway. This process is represented by dashed lines that originate from the 3' end of the oligonucleotides. If the oligonucleotides are precisely complementary to the chromosomal DNA (A), elongation initiates efficiently and results in four strands in the region where two were initially. If there is a mismatch in chromosomal DNA [X in (B)], elongation cannot be initiated efficiently for oligonucleotide b. After 35 cycles, a 1- to 10 million-fold amplification occurs in (A), whereas much less amplification occurs in (B). When an aliquot of the material is electrophoresed, an abundance of amplified sequence of appropriate size can be detected in (A) by staining with ethidium bromide. In contrast, no segment is detected in (B). Thus, a mutation or polymorphism can be detected by using an appropriately designed oligonucleotide that promotes efficient elongation from a mutant chromosome and inefficient elongation from a normal chromosome. (Reprinted with permission from Sommer et al.[1])

PASA has been further modified to include *two* allele-specific primers in the same PCR reaction.[16] By varying the lengths of the allele-specific oligonucleotides, the amplified products can be distinguished from each other by gel electrophoresis. This method, called PCR amplification of multiple specific alleles (PAMSA), allows the rapid detection of more than one allele in a single PCR reaction.

[16] C. Dutton and S. S. Sommer, *BioTechniques* **11,** 700 (1991).

A

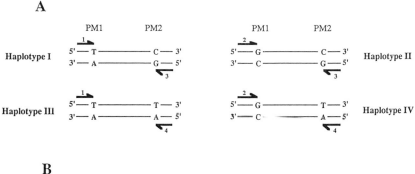

B

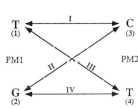

FIG. 2. Schematic of double PASA. (A) In this example, there are two biallelic polymorphic sites. T and G are present on the sense strand at polymorphic site 1 (PM1), and C and T are present at polymorphism 2 (PM2). PCR primers 1, 2, 3, and 4 are synthesized. The half-arrowhead signifies the 3′ end of the oligonucleotides. Primers 1 and 2 are identical except for T and G at their 3′ ends, respectively. Likewise, primers 3 and 4 are identical except for G and A at their 3′ ends, respectively. (B) The possible haplotypes (I–IV) can be differentially amplified by four PCRs with the indicated primers. (Reprinted with permission from Sarkar and Sommer.[15])

Methods

Adjunct Methods

DNA Extraction. DNA is extracted as previously described from blood drawn in acid citrate dextrose (ACD) solution B, an anticoagulant that enhances the stability of DNA in blood.[7] The extraction protocol and the anticoagulant are important for obtaining reproducible PCR amplification for each sample.

Polymerase Chain Reaction. Genomic DNA (250 ng) was added to a 25-μl total volume of 50 mM potassium chloride, 10 mM Tris-HCl (pH 8.3), 1.5 to 2.5 mM magnesium chloride, 200 μM of each deoxyribonucleotide, and 0.25–0.50 units of AmpliTaq polymerase (Perkin-Elmer Cetus, Norwalk, CT). Thirty to 40 cycles of PCR were performed with little observable difference. Consequently, 35 cycles of PCR were used routinely. The cycles (denaturation, 1 min at 94°; annealing, 2 min at 50°;

and elongation, 3 min at 72°) were performed in a Perkin-Elmer Cetus automated thermal cycler with one final 10-min elongation at 72°. The PCR amplification products were electrophoresed through a 2–3% (w/v) agarose gel and visualized by staining with ethidium bromide. Optimization and troubleshooting are discussed in the final section.

Oligonucleotides. Initially, the standard concentration of each primer was 1 μM. However, it was observed that decreasing the total oligonucleotide concentration to 0.05–0.25 μM could increase specificity. Subsequently, the oligonucleotide concentration was determined experimentally for each primer pair. Generally, 0.10 μM of each primer is sufficient to obtain detectable and yet specific amplification. For each pair of PCR primers, the allele-specific PCR oligonucleotide is generally designed to have an estimated melting temperature (T_m) in the range of 44–48° under standard conditions (1 M sodium chloride). The T_m is calculated as 4° $(G + C) + 2°(A + T)$.[17] The other primer that does not anneal at the polymorphic site is designed to have a melting temperature of 48°.

PASA

Allele-Specific Oligonucleotides.[17a] It has been hypothesized that a mismatch at or near the 3' end of the oligonucleotide would be likely to hinder the 3' elongation of *Taq* polymerase. A transition in the X-linked factor IX gene at nucleotide 31,311 substitutes threonine for isoleucine at amino acid 397 and causes hemophilia B in males with this mutation. Two PCR primers have been synthesized to be identical to the antisense strand for the normal and mutant alleles by differing at the 3' base (Table I). Primer 18 (A^n) is specific for the normal I^{397} allele in the factor IX gene, whereas primer 17 (G^n) is specific for the mutant T^{397} allele. If primer 18 (A^n) is used, specific amplification occurs only in normal males (Fig. 3). If primer 17 (G^n) is used, specific amplification occurs only in male patients with the mutation. In females carrying this mutation, amplification occurs with both primers because they are heterozygous at this site (Fig. 3).

To examine whether single-base mismatches can reproducibly and dramatically interfere with DNA amplification by PCR, additional oligonucleotides are synthesized with mismatches to the target DNA located at different positions (Table I).[2] When PCR is performed on genomic DNA with these oligonucleotides, specific amplification is independent of magnesium chloride concentration (Table II). However, when the mismatch is 3 or 4 bases from the 3' end of the primer, only a narrow window of magnesium concentrations produces specificity. If specificity is desired

---

[17] C. G. Miyada and R. B. Wallace, this series, Vol. 154, p. 94.
[17a] This section was adapted from Sarkar *et al.*[2]

TABLE I
OLIGONUCLEOTIDES SPECIFIC FOR TRANSITION AT AMINO
ACID 397 OF HUMAN FACTOR IX GENE

Oligonucleotide	Sequence[a]
17 (G^n)	GGA TAC CTT GGT ATA TG
17 (G^{n-1})	GA TAC CTT GGT ATA T$\overline{G}$T
17 (G^{n-2})	A TAC CTT GGT ATA T$\overline{G}$T T
17 (G^{n-3})	TAC CTT GGT ATA T$\overline{G}$T TC
18 (A^n)	G GGA TAC CTT GGT A$\overline{TA}$ TA

[a] The G that is underlined matches the T^{397} mutant factor IX allele and mismatches the normal I^{397} allele. The A that is underlined in oligonucleotide 18 (A^n) matches the normal allele and mismatches the mutant allele. These sequences are in the antisense direction. (Reprinted with permission from Sarkar et al.[2])

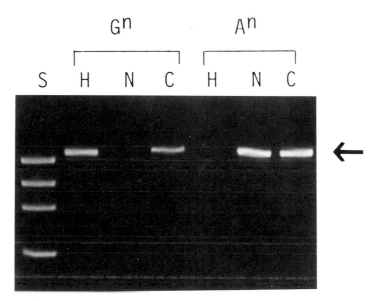

FIG. 3. Carrier testing by PASA in the family of a hemophiliac (HB16) with the $I^{397} \rightarrow T$ mutation in the factor IX gene. The first set of PCRs (bracketed by G^n) were performed with oligonucleotides 1 and 17 (G^n). This pair of oligonucleotides will amplify only the T^{397} mutant allele. The second set of reactions (bracketed by A^n) were performed with oligonucleotides 1 and 18 (A^n). This pair of oligonucleotides will amplify only the normal I^{397} allele. S, Standard ϕX174 HaeIII restriction fragments. Lane H, hemophiliac HB16; lane N, granddaughter 1 (noncarrier); lane C, granddaughter 2 (carrier). The arrow indicates the expected size of the amplified DNA segments. (Reprinted with permission from Bottema et al.[9])

TABLE II

Specificity of Mismatches Within 2 Bases of 3' Terminus of Oligonucleotide

| | [Mg^{2+}] (mM) | | | | | | | | | | | |
| | T^{397} DNA | | | | | | I^{397} DNA | | | | | |
Oligonucleotide[a]	1.5	2	2.5	3	5	8.5	1.5	2	2.5	3	5	8.5
17 (G^n)	−[b]	−	+[b]	+	+	ND[c]	−	−	−	−	−	−
17 (G^{n-1})	−	+	+	+	+	ND	−	−	−	−	−	−
17 (G^{n-2})	−	−	+	+	+	+	−	−	−	−	−	−
17 (G^{n-3})	−	+	+	+	+	ND	−	−	−	−	+	ND
18 (A^n)	−	−	−	−	−	ND	−	−	−	+	+	ND

[a] Concentrations of the oligonucleotides were 0.1 μM. Reprinted with permission from Sarkar et al.[2]

[b] (+), Specific amplification; (−), no specific amplification.

[c] ND, Not determined.

for mismatches three or more bases from the 3' end, allele-specific oligonucleotides can be differentially labeled, mixed, and subsequently used for amplification by competitive oligonucleotide priming.[8]

The following conclusions emerge from these studies[1,2]: (1) when the 3' or the 3' penultimate base of the oligonucleotide mismatches a target allele, no amplification product can be detected; (2) when the mismatches are 3 or 4 bases from the 3' end of the primer, differential amplification is still observed, but only at certain concentrations of magnesium chloride; and (3) a primer as short as 13 nucleotides (T_m 36°) is effective but a T_m of 48° is routinely used in the laboratory.

Screening of Populations. To determine if genomic samples from multiple individuals can be simultaneously analyzed for screening populations, 250 ng of normal genomic DNA is mixed with decreasing concentrations of genomic DNA with a rare polymorphism (Fig. 4). With as little as 6.25 ng of genomic DNA with the rare polymorphism, a specific product can be seen. Thus, the matched allele can be detected in the presence of a 40-fold excess of the mismatched allele. It is possible to screen over 400 chromosomes in 50 tubes by amplifying 4 individuals per PCR reaction.[2] Recently 800 transthyretin alleles were screened by 1 person in 1 day for mutations associated with familial amyloidotic polyneuropathy.[14]

To ascertain whether high concentrations of target DNA can overcome the specificity of PASA, a normal cDNA clone of factor IX is serially diluted by factors of 10. An amplified product of appropriate size can be detected with the mutant oligonucleotide 17 (G^n) when the concentration of target DNA is 10,000-fold or higher than the minimum concentration of

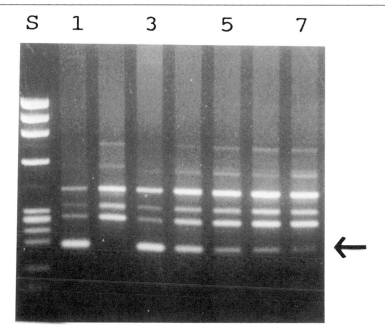

Fig. 4. Effects of allele concentration on PASA specificity. PASA was performed using allele-specific oligonucleotides for a rare variant (E91M) with decreasing concentrations of DNA in the presence of 250 ng of normal DNA (E100M). S, Standards; 250 ng ϕX174 *Hae*III restriction fragments. Lane 1, E91M (rare variant) (250 ng); lane 2, E100M (normal) (250 ng); lane 3, E91M/E100M (250 ng/250 ng); lane 4, E91M/E100M (83 ng/250 ng); lane 5, E91M/E100M (25 ng/250 ng); lane 6, E91M/E100M (12.5 ng/250 ng); lane 7, E91M/E100M (6.25 ng/250 ng). The arrow indicates the size of the expected amplified segment. (Reprinted with permission from Sarkar *et al.*[2])

DNA required for amplification with the normal matched primer, 18 (A^n).[2] At other sites, lower concentrations of DNA are able to overcome specificity (see below).

Generality of PASA. We have successfully used PASA at 41 sites (see Refs. 1, 2, 9–16, and unpublished observations, 1992) to perform (1) haplotyping in the transthyretin gene (10 sites), the factor IX gene (2 sites), the dopamine D_2 receptor gene (3 sites), and the factor VIII gene (1 site); (2) population screening for rare variants in the factor IX gene (4 sites), and (3) carrier testing and population screening for mutations in the phenylalanine hydroxylase gene (6 sites), the transthyretin gene (6 sites), the factor IX gene (7 sites), and the factor VIII gene (2 sites). One or two alleles were assayed at each site for a total of 69 allele-specific assays

TABLE III
Generality of PASA[a]

PASA primer[c]	Mismatched template allele[b]				Number of transitions[d]	Number of transversions[d]
	T	A	C	G		
A	—	1	15	4	15	5
T	1	—	4	11	11	5
G	9	4	—	4	9	8
C	2	9	5	—	9	7
					44	25

[a] Reprinted with permission from Sommer et al.[21]
[b] Sequence of the mismatched DNA template.
[c] Sequence of the allele-specific oligonucleotide.
[d] Relationship of two template alleles being distinguished. For example, distinguishing C and T alleles in genomic DNA would involve obtaining specificity for A:C and G:T primer:template mismatches.

(Table III). Reproducible discrimination between single-base changes has been obtained using PASA from a total of 25 transversion alleles and 44 transition alleles when the 3' or the 3' penultimate base of the oligonucleotide primer matched the desired allele. These results differ from that of Kwok et al.[18] who made all the mismatched combinations at one site of a human immunodeficiency virus type 1 (HIV-1) sequence. They found that differential amplification of the perfect match occurred only when the mismatch was A:G, G:A, C:C, or A:A. The reasons for the discrepancy are unclear. Optimization of DNA concentrations and dNTP concentrations were reported by Kwok et al.,[18] but the oligonucleotide concentration was high (0.5 μM). It is possible that lowering the oligonucleotide concentration and other optimization measures (see below) are critical to obtaining specificity at this site.

Double PASA[18a]

Haplotypes are useful in population genetics and medicine. However, determining linkage of haplotypes in the absence of DNA samples from appropriate family members can be difficult and laborious. PASA can be adapted to provide a rapid and reproducible method for haplotyping an

[18] S. Kwok, D. E. Kellog, N. McKinney, D. Spasic, L. Goda, C. Levenson, and J. J. Sninsky, Nucleic Acids Res. **18,** 999 (1990).
[18a] This section was adapted from Sarkar and Sommer.[15]

individual in the absence of relatives.[15,19] This method, termed "double PASA," uses four pairs of allele-specific PCR primers to differentially amplify each of the four possible haplotypes from two biallelic polymorphisms (Fig. 2).

Haplotypes are selectively amplified with PCR primers specific for the relevant alleles. For example, in Fig. 2, haplotype I can be detected with primers 1 and 3, haplotype II can be detected with primers 2 and 3, and so on. We have demonstrated double PASA with two polymorphisms (PM1 and PM2) in the human dopamine D_2 receptor.[15] DNA from six unrelated individuals was amplified with each of the four sets of primers (one set per gel quadrant in Fig. 5). The technical success of each PCR was internally controlled by coamplifying an unrelated segment from the factor IX gene. An amplified segment at the position of the arrow indicated the presence of the relevant haplotype. Amplified segments are detected for individuals 4, 5, and 6, in the upper left quadrant of the gel (Fig. 5), indicating that these individuals have haplotype I. Likewise, amplified segments in the upper right quadrant of the gel are seen for individuals 1, 2, and 4, indicating that they have haplotype II.

Double PASA is an important tool for haplotyping doubly heterozygous individuals (e.g., individual 6) because the physical linkage of alleles on a strand of DNA is necessary to determine the haplotype. In the example given above, physical linkage is required to distinguish doubly heterozygous individuals with haplotypes I and IV from those with haplotypes II and III. Double PASA should be generally applicable for haplotyping, provided that the segment between the polymorphisms can be amplified at least to a modest extent. If the polymorphic sites are separated by too great a distance to allow PCR amplification, double ARMS inverse PCR can be employed.[19] By circularization, the genomic targets can be placed close enough together to allow inverse PCR.

PAMSA[19a]

PASA is a general method that can be optimized to detect all possible single base changes. However, PASA has certain disadvantages:

1. If the specific allele is absent and spurious bands of other sizes are not produced, another set of compatible primers must be added to generate a constant band that serves as an internal control for the technical success of the PCR.

[19] Y.-M. D. Lo, P. Patel, C. R. Newton, A. F. Markham, K. A. Fleming, and J. S. Wainscoat, *Nucleic Acids Res.* **19,** 3561 (1991).
[19a] This section was adapted from Dutton and Sommer.[16]

S N 1 2 3 4 5 6 S N 7 8 9 10 11 12 S

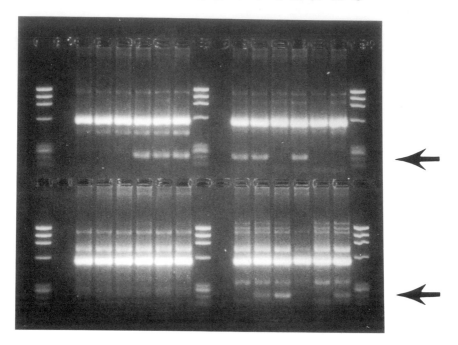

S N 13 14 15 16 17 18 S N 19 20 21 22 23 24 S

FIG. 5. Haplotyping six unrelated individuals with double PASA. The gel is divided into four quadrants. In each quadrant, a different set of PCR primers detects one of the four possible haplotypes (see Fig. 2). The arrows indicate the size of the specifically amplified segment. The intense amplified segment below the 610-bp size standard (fourth largest of the size markers) is from an additional pair of oligonucleotides that serve as an internal control for assessing the technical success of the PCR. PCR was performed with the following primers: lanes 1–6, primers 1 and 3; lanes 7–12, primers 2 and 3; lanes 13–18, primers 1 and 4; and lanes 19–24, primers 2 and 4. For each set of primers, the individuals are in order 1 through 6. Lanes S, DNA standards obtained by HaeIII digestion of φX174 DNA; lanes N, no DNA template added (control for DNA contamination in reagents). (Reprinted with permission from Sarkar and Sommer.[15])

2. Two PCR reactions are required to determine if a patient is heterozygous or homozygous for one of the alleles.
3. The specificity of PASA can be overwhelmed by high concentrations of template DNA. If template DNA concentration varies substantially, this can be a major problem.

PCR amplification of multiple specific alleles (PAMSA), a modification

S 1 2 3 S

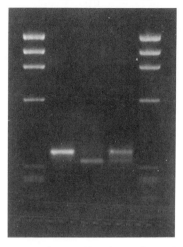

FIG. 6. PAMSA of the *Alu* 4a polymorphism in the factor IX gene. An amplified region that includes the *Alu* 4a polymorphism was diluted 10,000-fold and reamplified. The 3' end of allele-specific primers 1 and 2 are G and A, respectively, thereby specifically annealing to the C and T alleles, respectively. In addition, primer 2 is an extra 31 bases longer to increase the length of the T allele amplified product. Lane S, *Hae*III digestion of φX174; lane 1, male with the T allele; lane 2, male with the C allele; lane 3, heterozygous female. (Reprinted with permission from Dutton and Sommer.[16])

of PASA, can eliminate the above three problems.[16] For PAMSA, 1 allele-specific primer is designed to be longer than the other allele-specific primer by adding 30 or more noncomplementary bases. Because the additional bases will not hybridize to the DNA template, the T_m of the allele-specific primer will not be affected. The difference in size of the two resulting amplification products can be detected by 4% agarose gel electrophoresis, allowing both allele-specific primers to be used in one reaction tube. Thus, both heterozygotes and homozygotes can be detected in one PCR amplification reaction (Fig. 6). Moreover, because an amplified product is always produced, no internal control is necessary.

The competition of the two allele-specific oligonucleotides can also prevent high concentrations of template DNA from overwhelming specificity. Enhancement of specificity by competitive PCR with mismatches in the middle of the PCR primer has also been described.[20] Specificity was retained with PAMSA for a set of primers even with a greater than 1000-fold increase in template DNA concentration. However, the specificity

[20] R. A. Gibbs, P.-N. Nguyen, and C. T. Caskey, *Nucleic Acids Res.* **17**, 2437 (1989).

could be overwhelmed by a 5- to 10-fold increase in template DNA concentration for this set of primers when used separately with PASA.[16]

We have successfully used PAMSA for specific amplification at three different sites.[16] PAMSA is useful for detecting mutations and polymorphisms in situations in which two alleles are commonly observed. However, PASA is better suited for simultaneously screening multiple samples for a rare allele (see above). In heterozygotes, the heteroduplexes formed between the two amplified products migrated at or near the homoduplexes with our conditions of electrophoresis. However, it is possible that up to four distinct bands can be observed. Moreover, in one case the first allele-specific primer appeared to amplify more efficiently than the second allele-specific primer. It was, therefore, necessary to decrease the relative molar ratio of the 2 allele-specific primers 40-fold to obtain equal amplification.[16] This may reflect the ability of an amplified product to serve as a primer for further cycles of PCR.[15] Such "megapriming" can artifactually convert the short amplified segment to the larger size.

Optimization and Troubleshooting[21]

In the majority of cases, a standard magnesium titration (1.5, 2.5, 3.5, and 4.5 mM) and oligonucleotide titration (1, 0.25, 0.1, and 0.05 μM) are sufficient to provide conditions for both a robust and specific amplification. Ideally, magnesium and oligonucleotide concentrations are adjusted to also produce spurious amplified segments that do not interfere with allele detection but provide an internal control for the technical success of the amplification when the specific band is absent.

Occasionally, further optimization is required for specificity. In such cases, a variety of parameters can be optimized to increase specificity, including:

1. Magnesium concentration: Specificity can sometimes be achieved by lowering magnesium concentration below 1.5 mM. Adding ethylenediaminetetraacetic acid (EDTA) is a simple way of decreasing the "effective" magnesium concentration without making a different PCR buffer/salt stock solution. Likewise, increasing magnesium concentrations above 4.5 mM can produce noninterfering spurious amplification products that are useful internal controls, as discussed above.

2. Oligonucleotide concentration: Decreasing the oligonucleotide concentration to 0.05 μM may increase specificity. However, below 0.025 μM the amplification signal generally becomes weak.

---

[21] This section was adapted from S. S. Sommer, A. Groszbach, and C. D. K. Bottema, *BioTechniques* **12,** 82 (1992).

3. DNA concentration: A 10-fold dilution of the standard genomic DNA concentration (i.e., 1 ng/μl) can increase specificity and still provide an adequate amplification signal. Diluting the template can also avoid problems caused by contamination of the DNA with any PCR inhibitors.

4. Nonspecific oligonucleotide: Occasionally a given pair of primers will not be specific. Surprisingly, replacement of the nonspecific primer with another oligonucleotide at a new location often will provide specificity. Generally, product sizes between 300 and 600 bases are chosen. This, of course, is not always possible as in the case of double PASA. We have specifically amplified segments from 200 to 2700 bases in length.

5. Allele-specific oligonucleotide: Designing the allele-specific primer is critical. Lowering the T_m to 42–44° and placing the mismatch at the 3' base will increase specificity. Occasionally, a segment does not amplify well. Designing new primers using the *other* strand for the allele-specific mismatch may provide better amplification.

6. Competitive oligonucleotide priming: See PAMSA (above).

7. Deoxynucleotide concentration: Decreasing the concentrations of dNTPs to 25–50 μM can prevent spurious amplification.

8. Formamide: Inclusion of formamide (typically 2–5%) can increase the signal strength and eliminate undesired spurious bands, especially at high G + C content.[22] In addition, allele specificity may be enhanced.

9. Additional pair of primers: Inclusion of a second pair of compatible nonspecific primers can be added to generate a constant band. This serves both as an internal control for the technical success of the PCR *and* to increase specificity by providing another substrate for *Taq* polymerase.

10. *Taq* polymerase: Decreasing the amount of enzyme in each reaction (0.2–0.3 U/25 μl) can increase specificity and reduce reagent costs, whereas adding more enzyme can create spurious bands to serve as internal controls.

11. Source of DNA template: Using a PCR product as the source of DNA template for a *nested* PASA can provide specificity, particularly if the region to be amplified is highly repetitive. Note, however, that the concentration of DNA is critical. Prior to the nested amplification, a 10^6-fold or greater dilution may be required of the original PCR product.

---

[22] G. Sarkar, S. Kapelner, and S. S. Sommer, *Nucleic Acids Res.* **17**, 7465 (1990).

12. Number of amplification cycles: Decreasing the number of PCR cycles may reduce detection of any minor amplification of the mismatched allele. However, usually the number of cycles makes little difference in specificity.

13. Annealing temperature: Raising the PCR annealing temperature can increase specificity, but this is undesirable because these amplification reactions will be incompatible with reactions optimized at different cycle parameters. In our experience with more than 100 different segments, it has never been necessary to deviate from our standard cycle times except to increase elongation times for segments greater than 1.5 kb.

PASA is a generally applicable technique for detection of point mutations or polymorphisms. The method may also be used to detect the presence of small deletions or insertions. PASA has the advantages of being rapid, reproducible, nonisotopic, and amenable to automation.

[30] Chromosome Assignment by Polymerase Chain Reaction Techniques

By Craig A. Dionne and Michael Jaye

Introduction

The assignment of newly identified genes to specific human chromosomes is an important process in the elucidation of the human genetic map and in analyzing the potential association of specific genes with inherited human diseases. Chromosome assignment often involves analysis of the genomic DNAs from a panel of somatic cell hybrids that preferentially segregate the chromosomes of only one of the parent species.[1-4] In our experiments we have used a panel of human/rodent somatic cell hybrids,[5] in which each hybrid contains the normal complement of rodent chromosomes and one or more human chromosomes.

[1] V. A. McKusick and F. H. Ruddle, *Science* **196,** 390 (1977).
[2] S. J. O'Brien and W. G. Nash, *Science* **216,** 257 (1982).
[3] F. H. Ruddle and R. P. Cregan, *Annu. Rev. Genet.* **9,** 407 (1975).
[4] S. J. O'Brien, J. M. Simonson, and M. Eichelberger, *in* "Techniques in Somatic Cell Genetics" (J. W. Shay, ed.), p. 513. Plenum, New York, 1982.
[5] S. J. O'Brien, W. G. Nash, J. L. Goodwin, D. R. Lowy, and E. Chang, *Nature (London)* **302,** 839 (1983).

Conventional methods of chromosome assignment usually entail restriction digestion of all the genomic DNAs of the panel followed by Southern analysis with radiolabeled probes of interest. The presence or absence of the target gene in each one of the hybrid cell lines is compared to the previously determined chromosome content for each hybrid and an assignment is made. Although the Southern technique obviously works quite well, and the Southern blots can be reused several times, the procedure is relatively labor intensive. In addition, because there is variability among the hybrid cells for retention of a particular chromosome,[4] the Southern hybridization signals can vary, even to the point of being below detection limits. Because the greatest part of the effort in the analysis is expended in the characterization of the hybrids, and relatively little DNA is obtained per passage, we have chosen to use the polymerase chain reaction (PCR) approach to chromosome assignment. The technique is rapid, convenient, and sensitive, requiring <0.1 μg DNA per analysis.[6]

We describe the application of the technique for the chromosome assignment of the human *BEK* gene, which codes for a tyrosine kinase-linked receptor for fibroblast growth factors.[7-11] We previously mapped human *BEK* to chromosome 10q25.3–10q26 by a combination of PCR techniques and *in situ* hybridizations.[12]

Strategies

Instead of Southern analysis of somatic cell hybrid DNAs, we chose to PCR amplify the human gene of interest in each of the DNA samples. Like most analytical PCR techniques, the objective of this application is the detection of minute amounts of specific nucleic acid sequences in different samples and therefore requires high specificity and sensitivity. However, the requirement for high specificity is somewhat more stringent in this application because the target sequence is contained within a back-

[6] C. A. Dionne, R. Kaplan, J. Seuánez, S. J. O'Brien, and M. Jaye, *BioTechniques* 6, 190 (1990).

[7] S. Kornbluth, K. E. Paulson, and H. Hanafusa, *Mol. Cell. Biol.* 8, 5541 (1988).

[8] C. A. Dionne, G. Crumley, F. Bellot, J. M. Kaplow, G. Seaross, M. Ruta, W. H. Burgess, M. Jaye, and J. Schlessinger, *EMBO J.* 9, 2685 (1990).

[9] Y. Hattori, H. Odagiri, H. Nakatani, K. Miyagawa, K. Naito, H. Sakamoto, O. Katoh, T. Yoshida, T. Sugimura, and M. Terada, *Proc. Natl. Acad. Sci. U.S.A.* 87, 5983 (1990).

[10] E. Houssaint, P. R. Blanquet, P. Champion-Arnaud, M. C. Gesnel, A. Tarriglia, Y. Courtois, and R. Breathnach, *Proc. Natl. Acad. Sci. U.S.A.* 87, 8180 (1990).

[11] T. Miki, T. P. Fleming, D. P. Bottaro, J. S. Rubin, D. Ron, and S. A. Aaronson, *Science* 251, 72 (1991).

[12] C. A. Dionne, W. S. Modi, G. Crumley, H. Seuánez, S. J. O'Brien, J. Schlessinger, and M. Jaye, *Cytogenet. Cell Genet.* 60, 34 (1992).

ground of similar sequences from another species. Obviously, the more one knows about the sequence and genomic arrangement of the gene in the target and background species, the easier it is to succeed in placing an oligomer pair that gives an interpretable answer. Because, in general, the exact cDNA or DNA sequence from only one species is known, several different approaches can be used to facilitate the detection of specific sequences with a minimum of effort and cost in mind.

Amplification of 3'-Untranslated Sequences

This approach has been described earlier[6] and is in some ways the simplest way to begin an evaluation because it is the most direct. In many cases, the 3'-untranslated region is less conserved than the coding region and contains fewer, if any, intervening sequences. Consequently, there is a greater probability of amplifying only the human sequence and having a predicted size corresponding to the cDNA sequence. We have used this approach quite readily for the *aFGF* and *FGF-5* genes[6] and the *PLC-γ* gene (unpublished observations, 1991) and have used a modification of this approach for the *BEK* gene (see Experiment 2, below).

Amplification Across an Intron

In general, one has little knowledge of genomic structure when cloning a cDNA for a new gene product. However, if one has sufficient resources or access to knowledge of intron placement then amplification across an intron is a good approach if the intron is relatively small. The advantage of this approach is that absolute species specificity for annealing of primers need not be achieved because intron length alone may be sufficient to determine species origin. The disadvantage is that PCR amplification becomes somewhat inefficient above several kilobases in length and introns can be relatively large. A second major disadvantage is that intron placement is not predictable from the cDNA sequence. Despite these potential drawbacks, PCR amplification across an intron has been used successfully in the chromosome assignment of the nuclear antigen *p68*[13] and of human *BEK*[12] as described below.

Amplification Followed by Restriction Digestion

It is possible that the first several choices for oligomer pairs do not distinguish between the DNAs from different species. In this case, it is desirable to compare restriction sites between the amplified DNA from

[13] R. Iggo, A. Gough, W. Xu, D. P. Lane, and N. K. Spurr, *Proc. Natl. Acad. Sci. U.S.A.* **86,** 6211 (1989).

the two species in order to generate a distinctive diagnostic pattern. The advantages of this approach are that it conserves oligomer costs and is easy to perform. Most restriction digests can be performed directly in the PCR reaction solution after amplification with only a small adjustment in salt. In most cases the sequence of one PCR product is known and can guide the choice of enzymes. It is desirable to find a site that is present only in the product of the target species yielding a product unique to the target; in this way, incomplete digestion will still yield interpretable results. Even if a unique site cannot be found, the approach can work well, as described below for the human *BEK* gene.

Materials

DNA thermal cycler: PCR reactions are performed in a Perkin-Elmer Cetus (Norwalk, CT) DNA thermal cycler according to the instructions of the manufacturer

Somatic cell hybrids: The panel of somatic cell hybrid DNAs[5] was obtained as a kind gift from Dr. Steven O'Brien at the National Cancer Institute (Frederick, MD)

Taq polymerase: Purchased from Perkin-Elmer Cetus

Restriction enzymes and T4 polynucleotide kinase: Purchased from Boehringer Mannheim (Indianapolis, IN)

Deoxyribonucleotides: Purchased from Pharmacia-LKB Biotechnology (Piscataway, NJ)

Ultrapure agarose: Purchased from Bethesda Research Laboratories (Gaithersburg, MD)

Gelatin: Purchased from Difco Laboratories (Detroit, MI)

Paraffin oil: Purchased from Fisher Scientific (Fair Lawn, NJ)

Microcentrifuge tubes (0.65-ml Clickseal): Purchased from National Scientific (San Rafael, CA)

GeneScreen Plus hybridization membrane: Purchased from Du Pont/ New England Nuclear (Boston, MA)

X-ray Film (Kodak X-Omat AR): Purchased from Eastman Kodak (Rochester, NY)

PCR buffer (10×): 500 mM KCl, 15 mM MgCl$_2$, 0.1% (w/v) gelatin, 100 mM Tris-HCl (pH 8.3)

dNTPs (2 mM): 2 mM dATP, 2 mM dCTP, 2 mM dGTP, and 2 mM TTP in 5 mM Tris-HCl (pH 7.5)

SSPE, Blotto, SSC, and 50× TAE solutions are prepared as described by Sambrook *et al.*[14]

[14] J. Sambrook, E. F. Fritsch, and T. Maniatis, "Molecular Cloning: A Laboratory Manual," 2nd Ed. Cold Spring Harbor Press, Cold Spring Harbor, New York, 1989.

Oligonucleotides: All oligonucleotides are synthesized on an Applied Biosystems (Foster City, CA) 380A DNA synthesizer using cyanoethyl phosphoramidate chemistry according to the directions provided by the manufacturer. They are dissolved in 10 mM Tris-HCl, 0.1 mM ethylenediaminetetraacetic acid (EDTA) (pH 7.5) at a concentration of 50 pmol/μl. The oligonucleotide primers have the following sequences:

Bek16A	5'-GCCGCCGGTGTTAACACCACGGAC-3'
BekR6	5'-GCTATCTCCAGGTAGTCT-3'
Bek12B	5'-AACCTCAATCTCTTTGTCCGTGGT-3'
Bekterm	5'-ATAAACGGCAGTGTTAAAACATGA-3'
Bek3'881non	5'-GCTGCCTGCATAGAAATGCCAC-3'

Method I: Polymerase Chain Reactions

Each PCR reaction contains, in a final volume of 50 μl:

H_2O	36.5 μl
Buffer (10×)	5 μl
dNTPs (2 mM)	5 μl
Oligomer 1 (50 pmol/μl)	1 μl
Oligomer 2 (50 pmol/μl)	1 μl
Genomic DNA (100 ng/μl)	1 μl
Taq DNA polymerase (5 U/μl)	0.5 μl

Reagents are added in the order indicated to 0.65-ml microcentrifuge tubes on ice. After gentle mixing, 25 μl of paraffin oil is added to the top, using a fresh pipette for each sample. The samples are centrifuged briefly to force the oil onto the sample and the reactions are placed into the thermal cycler apparatus. Amplification is obtained by 30 cycles of heating to 94° for 1.5 min, 60° for 1.5 min, and 72° for 4 min. After amplification is completed, the samples are stored at 4° until time for analysis. When a large number of PCR reactions utilizing the same oligomers but different DNAs are being assembled, it is easier to make a "master mix" of all components and then add the DNA last.

Method II: Analysis of Polymerase Chain Reactions

After amplification, the PCR reactions are analyzed by gel electrophoresis under standard conditions.[14] Three microliters of 50% (v/v) glycerol is added to 0.015 ml of the reaction product and the mixture is loaded onto horizontal submarine gels containing 1% (w/v) agarose, 1× TAE, and 0.5 μg/ml ethidium bromide. The electrophoresis buffer is 1× TAE containing 0.5 μg/ml ethidium bromide. After electrophoresis the gel is examined with long-range ultraviolet (UV) light and it is recommended that a photograph of the transilluminated gel be taken for documentation. Usually a visual scoring of the results is sufficient for analysis.

When restriction analysis of the PCR reaction products is required, additional salt and restriction enzymes are added as necessary. For the example described in Fig. 2, digestion was achieved by the addition of 1.5 μl 0.1 M MgCl$_2$ and 0.5 μl DraI (4 U/μl) followed by incubation at 37° for 2 hr. The samples are then processed as described above without any phenol extractions.

When blotting of the gel and radioactive probing is required after electrophoresis, we use the alkaline dry blot protocol. The gel is soaked in 0.4 N NaOH, 0.6 N NaCl for 20 min, then placed on a sheet of Whatman (Clifton, NJ) 3MM paper soaked in the same solution with no buffer reservoir and blotted to GeneScreen Plus membrane. Because PCR reactions yield so much product, a 2-hr transfer is sufficient although overnight blotting is usually done for convenience. The blot is marked for loading slots and orientation, then neutralized in two 15-min washes of 1 M Tris-HCl (pH 7.8), 1.5 M NaCl. The blot is then UV cross-linked to the transferred DNA by illumination in a UV Stratalinker 1200 (Stratagene, La Jolla, CA) according to the instructions of the manufacturer. Alternatively the blot is heated in vacuo at 80° for 2 hr. The blot is then probed under standard conditions appropriate to the type of labeled probe that is used. For the experiment described in Fig. 1, 25 pmol of oligomer was end labeled with 1 U of T4 polynucleotide kinase and 80 μCi [γ-^{32}P]ATP (specific activity, 3000 Ci/mmol) in a final volume of 25 μl containing 50 mM Tris-HCl (pH 7.6), 5 mM dithiothreitol, 10 mM MgCl$_2$, 0.2 mM EDTA, and 0.1 mM spermidine, for 30 min at 37°. The blot was prehybridized in 20 ml of a solution containing 5× SSPE, 2× Blotto, and 0.5% (w/v) SDS for 1 hr at 50°. The probe was added to 20 ml of the prehybridization solution and hybridization was allowed to occur at 50° for 16 hr. The blot was washed twice in 500 ml of 2× SSC, 0.5% (w/v) SDS for 15 min each wash, followed by two 15-min washes in 500 ml 0.5× SSC, 0.5% (w/v) SDS. The blot was wrapped in plastic film and exposed to Kodak X-Omat AR film at −80° for 30 min.

Practical Considerations

Oligomer Size. We have found that oligonucleotide primers 22 to 24 bases in length usually work well in amplification of specific genomic sequences, although primers 17 to 19 bases long sometimes work as well. It is usually worth the effort to try the shorter oligomers that have been

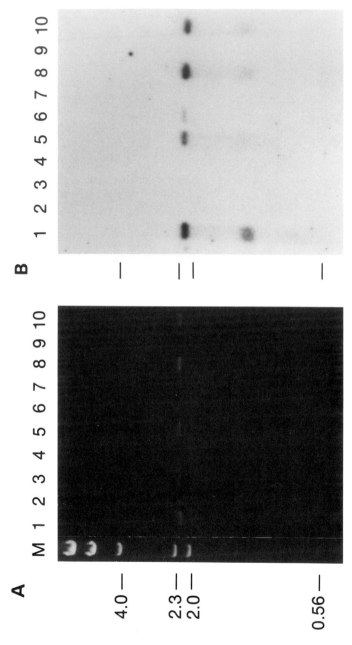

Fig. 1. Amplification of Human *BEK*: PCR across an intron. Somatic cell hybrid DNAs were subjected to PCR amplification with the primers Bek16A and BekR6 then analyzed by agarose gel electrophoresis as described in the text. (A) Ethidium bromide-stained gel. Note the specific products at ~2.1 kb. (B) Autoradiogram of blot of the gel in (A), which was probed with the bek-specific oligonucleotide Bek12B. Note specific hybridization at ~2.1 kb. The DNAs used in the analysis were derived from the following cell lines: 1, human lymphocyte; 2, hamster E36; 3, mouse RAG; 4, 70M4; 6, 70M11; 7, 80H1; 8, 80H7; 9, 81P8; 10, 81P16. The molecular weight markers in lane M were derived from *Hind*III-digested λ DNA.

synthesized as sequencing primers before synthesizing longer ones. We try, if possible, to match oligomers in length and GC content.

Annealing Temperatures and Reaction Times. For the first pilot reactions we usually use an annealing temperature of 60° for 24-mers and 50° for 17-mers. Although the proper choice can be calculated from the length and base composition of the primers, we have found these starting temperatures to be useful guidelines. Depending on the size of the anticipated product we use extension times of 1 to 4 min. Empirically, we have obtained better results with extension times of at least 2.5 min for this PCR application.

Enzymes. We have tried *Thermus aquaticus* DNA polymerase enzyme from several sources and have found the *Taq* polymerase from Perkin-Elmer Cetus and Boehringer Mannheim to be consistently high in quality and reliable.

Pilot Reactions. It is helpful to work out the PCR reaction conditions on genomic DNA from the parent cells used to generate the somatic cell hybrids in the test panel, although genomic DNAs from the species in question are usually adequate. The test reactions provide a quick assessment of the proper combination of primary oligomers and reaction conditions before proceeding to the reactions on the valuable DNAs of the test panel.

Cleanliness and Contamination. The problems of contamination for PCR analysis are well documented and are particularly acute for this application because the researcher usually has extensive experience with the gene in question and consequently has worked with many fragments of the DNA. In addition to the precautions described elsewhere,[14] we find it advisable to rigorously clean pipettes before and after use and to use plugged pipette tips to prevent aerosol carryover from sample to sample. These tips are now available from numerous suppliers. The use of gloves is highly recommended.

Experiment 1. Assignment of Human *BEK*: PCR across an Intron

The isolation and sequence of human BEK cDNA clones have been described by several groups[8][11] and sequence differences among the various isolates indicates that some have arisen by alternative splicing events. The location of putative genomic intron/exon boundaries can be inferred from the point of divergence among the different clones. We have chosen to amplify a portion of the human *BEK* gene that crosses an intron/exon junction corresponding to amino acid Pro-361 in the human bek sequence.[8] The oligonucleotide primers Bek16A and BekR6 correspond to sequences that are 196 bp apart in the BEK cDNA clone and yield a 196-bp PCR product when the BEK cDNA is used as template (data not shown). However, when the same primers are used with a human genomic DNA template we observe a 2150-bp product that is not observed when genomic DNA from mouse or hamster cells is used as template (Fig. 1A, lanes 1–3). These results suggest that the 2150-bp product arises from amplification of a portion of the human *BEK* gene. Similar products are observed in some, but not all, of the PCR reactions that used somatic cell hybrid DNAs for templates (Fig. 1A, lanes 4–10).

To verify that the visual products were actually an amplified portion of the *BEK* gene, the gel was blotted to GeneScreen Plus and probed with the oligonucleotide Bek12B, which lies between the PCR primer sequences in the BEK cDNA. The autoradiogram of the resulting blot confirms the visual results and indicates more clearly the specificity of the reaction between human vs. rodent DNA templates.

This particular experiment indicates the advantage of following the visual results with hybridization results. Besides confirming the identity of the products, the hybridization is much more sensitive. Note that the visual product is barely detectable in lane 6 (Fig. 1) but is clearly seen in the autoradiogram. In addition, the hybridization can clarify inconclusive visual results. For instance, an identical experiment, performed with an enzyme obtained from a different manufacturer, gave 10–12 bands per lane (data not shown). Although the visual results were uninterpretable, the hybridization results were as clear as the ones presented in Fig. 1.

Experiment 2. Amplification Followed by Restriction Digestion

The DNA sequence corresponding to the bek 3′-untranslated region (EMBL data library #52832, nucleotides 2647 to 3362) was amplified with the oligonuleotide primers Bekterm and Bek3′881non. Because of high homology between the rodent and human genes in this region, a single band of 715 bp was found to be amplified in DNAs from human, hamster, or mouse (Fig. 2, lanes 1–3). Detection of a human-specific PCR product was ascertained by subjecting the PCR products to complete digestion with the restriction endonuclease *Dra*I, which cleaves the human DNA into 485- and 232-bp fragments but cleaves the rodent PCR products into approximately 483-, 190-, and 40-bp products (Fig. 2, lanes 4–6). Consequently, detection of the 232-bp product in complete *Dra*I digests of the PCR relations is indicative of the presence of the human chromosome bearing the *BEK* gene. Application of this approach to the panel of 45 somatic cell hybrids gave results identical to those obtained with the method described in experiment 1 above.

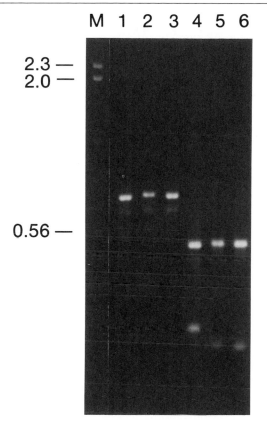

Fɪɢ. 2. PCR amplification followed by restriction digests. The portion of the *BEK* gene corresponding to the 3'-untranslated region was amplified with the primers Bek5′nn and Bek3′881non. A portion of each reaction was digested with *Dra*I as described in text. Undigested reaction products (lanes 1–3) and *Dra*I-digested products (lanes 4–6) were subjected to electrophoresis though a 2% agarose gel. A picture of the ethidium bromide-stained gel is shown. The DNAs used for PCR templates were derived from the following: lanes 1 and 4, human lymphocytes; lanes 2 and 5, hamster E36 cells; lanes 3 and 6, mouse RAG cells. The molecular weight markers in lane M were derived from *Hin*dIII-digested λ DNA. Note the presence of the human specific 232-bp product arising from complete *Dra*I digestion in lane 4.

Analysis and Discussion

After the PCR reactions have been scored positive or negative for the presence of the target gene, the results are compared to the previously characterized chromosome content of each cell hybrid. In its simplest form, the analysis is as follows: the presence of the BEK PCR product in

TABLE I
PRESENCE OF BEK POLYMERASE CHAIN
REACTION PRODUCT AND HUMAN CHROMOSOME
10 IN HYBRID CELL LINES

Hybrid	BEK PCR product	Chromosome 10
70M3	−	−
70M4	+	+
70M11	+	−
80H1	−	−
80H7	+	−
81P8	−	+
81P16	+	P

each cell line is compared to the presence of each human chromosome in the same cell line, and discordances for each chromosome are calculated as a percentage of the total informative hybrids. For instance, when the presence of the BEK PCR product is compared to the presence of human chromosome 10 in the eight cell lines used in experiment 1, we find two cell lines that have both *BEK* and chromosome 10, two that have neither, and three discordant hybrids that have either *BEK* or chromosome 10, but not both (Table I). This selection of hybrid cell lines was chosen to emphasize the discordant hybrids. When the analysis is extended to the 45 cell lines of our panel, we find that the presence of *BEK* shows least discordance with human chromosome 10 and much greater discordance with all other chromosomes, and so is assigned to human chromosome 10.[12]

It is important to note that we simplify matters by assuming that all cell hybrids designated M (for uncertain negative) and P (for positive but at low frequency) are really positive for that particular chromosome but at a level too low for confident detection by karyotype or enzyme analysis. Another approach is to consider these as noninformative hybrids and to perform the analysis only with informative hybrid cells. We obtain the same assignment of *BEK* to human chromosome 10 in either case.

Ideally, one should obtain 0% discordancy with one chromosome and high discordances with all the others. This is almost never the case, even with conventional Southern blotting procedures. Discordancies using the PCR method are generally higher than those obtained by the Southern blot method[6] due to its extreme sensitivity in detecting a target chromosome that is retained at less than one copy per cell. Most of the discordancies arise when the PCR technique detects the chromosome in a cell line that was scored negative. For *BEK*, we found such discordancies with the hybrids 70M11 and 80H7. We also found one example, 81P8, of the opposite type of discordance, for which we fail to detect a chromosome that is scored positive in the cell line. These types of discordancies may arise when the target sequence, like *BEK*, is located at the very end of a chromosome, where small truncations are less noticeable. It is likely that when the more sensitive PCR techniques are used in characterization of the panels, both type of discordancies will diminish in number.

It is always useful, if possible, to verify the results of PCR-generated assignment by comparison to the results obtained for another gene mapped to the same chromosome. Theoretically, two genes on the same chromosome should cosegregate. Similarly, the product should be available from the appropriate chromosome specific genomic DNA library.

Even with the extra controls and verifications, chromosome assignment by PCR technique is much faster than by conventional methods and is much more conservative of the valuable somatic cell hybrid DNA.

Acknowledgments

The authors thank Dr. Stephen O'Brien and Dr. William Modi for the gift of somatic cell hybrid DNAs. We thank Robin McCormick and Patricia Gallagher for excellent preparation of the manuscript.

[31] Reverse Transcription of mRNA by *Thermus aquaticus* DNA Polymerase followed by Polymerase Chain Reaction Amplification

By MICHAEL D. JONES

Introduction

The structural analysis of messenger RNA (mRNA) plays an important role in the examination of gene structure and expression. The key step is the reverse transcription of mRNA into complementary DNA (cDNA), which subsequently can be cloned and analyzed. Central to this process are the reverse transcriptases (RNA-dependent DNA polymerases, EC 2.7.7.49).

One difficulty that can prevent the efficient copying of RNA sequences into cDNA is the ability of single-stranded RNA to adopt stable intramolecular stem–loop structures. This can cause pausing of the reverse transcriptase and thus result in premature termination of the reverse transcript.

Such secondary structure may account for the relatively poor yields of full-length cDNA clones obtained in the construction of recombinant cDNA libraries. One solution to this problem would be to perform the reverse transcription reaction at an elevated temperature at which such structures would be disrupted, allowing the enzyme to copy through to the 5' end of the RNA molecule.

Loeb et al.[1] reported that Escherichia coli DNA polymerase has the potential to use RNA as a template for the synthesis of DNA in vitro, and Chien et al.[2] have alluded to the possibility that Thermus aquaticus DNA polymerase may possess reverse transcriptase activity. The advent of the polymerase chain reaction (PCR) and the subsequent commercial availability of thermostable DNA polymerases[3] permitted an investigation into the utilization of T. aquaticus DNA polymerase to copy RNA into cDNA.[4]

Materials

Enzymes

Thermus aquaticus DNA polymerase and AmpliTaq (cloned T. aquaticus DNA polymerase) are from Perkin-Elmer Cetus (Norwalk, CT), avian myeloblastosis virus (AMV) reverse transcriptase is from Anglian Limited (Essex, UK) and IBI, Limited (New Haven, CT) RNase-free DNase is purchased from Stratagene (La Jolla, CA), RNasin is from Promega (Madison, WI), and RNase A is from Sigma (St. Louis, MO).

Reagents

dNTPs are from Pharmacia-LKB (Piscataway, NJ) and are dissolved in 10 mM Tris-HCl (pH 8.0), 0.1 mM ethylenediaminetetraacetic acid (EDTA), to 10 mM and stored at $-20°$. The oligodeoxynucleotide primers are synthesized on an Applied Biosystems (Foster City, CA) 380B DNA synthesizer, deprotected, precipitated with ethanol, and finally dissolved in water at ~1–2 $\mu g/\mu l$. They are used without further purification. The sequences of the primers are as follow: primer G, 5'-GGGAAGGAGGGT-GGCCGTG-3'; primer F, 5'-CTTCAACCCCGAGGAGT-3'[4,5]; primer 6SP1, 5'-CGCGTTCGTTTAACATATGG-3'; and primer 6SP2, 5'-GAAT-

[1] L. A. Loeb, K. D. Tartof, and E. C. Travaglini, Nature (London) New Biol. 242, 66 (1973).
[2] A. Chien, D. B. Edgar, and J. M. Trela, J. Bacteriol. 127, 1550 (1976).
[3] R. K. Saiki, D. H. Gelfand, S. Stoffel, S. Scharf, R. Huguchi, G. T. Horn, K. B. Mullis, and H. A. Erlich, Science 239, 487 (1988).
[4] M. D. Jones and N. S. Foulkes, Nucleic Acids Res. 17, 8387 (1989).
[5] M. G. Persico, G. Viglietto, G. Martini, D. Toniolo, G. Paonessa, C. Moscatelli, R. Dono, T. Vulliamy, L. Luzzatto, and M. D'Urso, Nucleic Acids Res. 14, 2511, 7822 (1986).

CAGCATCGGTTACGTG-3'. SeaKem ME agarose is purchased from FMC Corporation (Rockland, ME). Size markers for agarose gel electrophoresis are prepared from either pBS or pBluescript KS(−), purchased from Stratagene, by digestion with *Taq*I and *Sau*AI, separately, and then mixed together in equimolar amounts.[4]

Solutions

RNA lysis buffer: 140 mM NaCl, 1.5 mM MgCl$_2$, 10 mM Tris-HCl (pH 8.0), 0.5% (v/v) Nonidet P-40 (NP-40), 0.15% (w/v) Macaloid (Steetly Minerals Ltd.)

RT buffer (10×): 500 mM KCl, 80 mM MgCl$_2$, 500 mM Tris-HCl (pH 8.3)[6]

PCR buffer (10×): 500 mM KCl, 20 mM MgCl$_2$, 100 mM Tris-HCl (pH 8.3)

TBE buffer (10×): 108 g of Tris base, 55 g of boric acid, 9.3 g of Na$_2$EDTA per liter of water

dNTP mix: 2.5 mM concentrations of each dNTP

mRNA

Total cytoplasmic RNA from HeLa cells and HHV-6 infected HSB-2 cells are prepared by the method of Favaloro *et al.*[7] Essentially, washed cells are resuspended in lysis buffer, and after 5 min on ice the nuclei are pelleted and the supernatant removed. The supernatant is extracted three times with phenol–chloroform. Cytoplasmic RNA is precipitated with ethanol and dissolved in 10 mM Tris-HCl (pH 7.0) at approximately 1–5 mg/ml and stored at −70°.

Methods

1. RNA, 1–5 μg, in 5 μl diethyl pyrocarbonate (DEPC)-treated water is heated to 90° for 2–5 min and then rapidly chilled on ice. Two microliters of 10× RT buffer or 10× PCR buffer is added, together with 8 μl of 2.5 mM dNTP mix (final concentration for reverse transcription is 1 mM for each dNTP), 40 units of RNasin, 50 ng of the reverse oligodeoxynucleotide primer, and either 40 units of AMV reverse transcriptase or 2.5 units of AmpliTaq (*T. aquaticus* DNA polymerase). Then DEPC-treated water is added to a final volume of 20 μl. For AMV reverse transcriptase, the reaction mixture is incubated at 42° for 60 min and for *T. aquaticus* DNA polymerase at 68–72° for 30–60 min. The reaction mixture is then heated to 90° for 5 min and rapidly chilled on ice.

---

[6] C. J. Watson and J. F. Jackson, *in* "DNA Cloning: A Practical Approach" (D. M. Glover, ed.), Vol. 1, p. 79. IRL Press, Oxford, 1985.

[7] J. Favaloro, R. Treisman, and R. Kamen, this series, Vol. 65, p. 718.

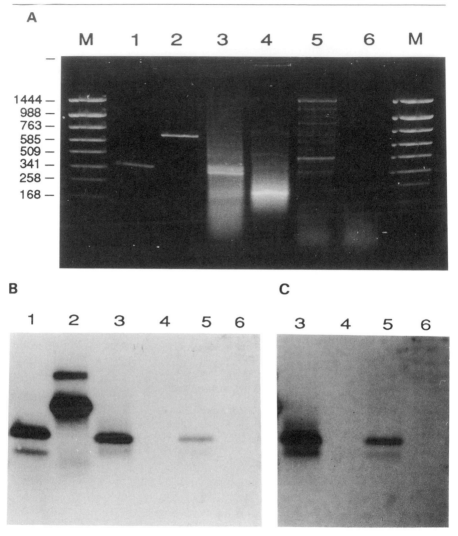

FIG. 1. G6PD reverse transcription by AMV reverse transcriptase and *T. aquaticus* DNA polymerase. Aliquots of HeLa cell total cytoplasmic RNA (~5 μg) were treated as follows: lane 3, with 20 units of AMV reverse transcriptase and 1 μg of primer G in 20 μl RT buffer at 42° for 45 min, and subsequent PCR amplification, for 30 cycles, with 1 μg each of primers F and G, and 2.5 units of *T. aquaticus* DNA polymerase; lane 4, as for lane 3 but with prior incubation with 10 μg of RNase A; lane 5, reverse transcription with 2.5 units of *T. aquaticus* DNA polymerase in 20 μl of PCR buffer with 1 μg of primer G, for 30 min at 68°. The sample was then treated to PCR amplification as for lane 3. Lane 6, direct PCR amplification with 2.5 units of *T. aquaticus* DNA polymerase, 1 μg each of primers G and F in 50 μl PCR buffer. Lanes 1 and 2 were plasmid cDNA and genomic DNA clones (~50 ng each), respectively, amplified directly. Lanes M are size markers of an equimolar mixture of *Taq*I- and *Sau*3AI-

2. Ten microliters of the cDNA mixture from step 1 above is amplified by PCR. The buffer is adjusted to 50 mM KCl, 2 mM MgCl$_2$, 10 mM Tris-HCl (pH 8.3) (1× PCR buffer), 200 μM for each dNTP. One microgram of both forward and reverse oligodeoxynucleotide primers and 2.5 units of AmpliTaq are added, and the final volume adjusted to 50 μl with water. Mineral oil (50 μl) is overlaid on top of the reaction mix to prevent evaporation during the PCR cycles.

3. The reaction mixture is subjected to amplification for 30–35 cycles. The actual parameters will vary depending on the length of the target sequence and the GC content of the primers. The values used in the examples shown here were as follow: 94°, 1 min; 55°, 1 min; 72°, 4 min, for 30 cycles using a Perkin-Elmer Cetus DNA thermal cycler (Fig. 1) and 94°, 30 sec; 55°, 30 sec; 72°, 2 min, for 30–35 cycles using an M J Research (Watertown, MA) programmable thermal controller (Fig. 2). The last cycle includes an incubation at 72° for an extra 5 min.

4. After completion of the PCR cycles, the mineral oil is carefully extracted twice with 500 μl of chloroform and the aqueous phase stored at −20°. Eight microliters of the reaction mix is analyzed by agarose gel electrophoresis. Fifty-milliliter agarose gels (1.4–1.6% agarose) in 1× TBE buffer, containing 50 μg/ml ethidium bromide, are run submarine fashion at 60-mA constant current. The PCR-amplified DNA fragments can be analyzed by standard Southern blot analysis[8] using cloned plasmid sequences containing the PCR target sequences.

Discussion

The experiments shown were designed to reverse transcribe mRNA into cDNA with primers specific for the human glucose-6-phosphate dehydrogenase gene (G6PD)[4,5] and a conserved viral gene from human herpesvirus 6 (HHV-6).[9] The primers were chosen in each case to cross an intron such that amplification from mRNA could be distinguished from genomic DNA. Reverse transcription of G6PD-specific RNA with either AMV

[8] T. Maniatis, E. F. Fritsch, and J. Sambrook, "Molecular Cloning: A Laboratory Manual." Cold Spring Harbor Press, Cold Spring Harbor, New York, 1982.
[9] G. L. Lawrence, M. Chee, M. A. Craxton, U. A. Gompels, R. W. Honess, and B. G. Barrell, *J. Virol.* **64,** 287 (1990).

digested pBS. (A) Ethidium bromide-stained agarose gel; (B) and (C) are two different exposures of the Southern blot probed with the plasmid cDNA clone. Primers G and F amplify a genomic DNA fragment of 657 bp and an mRNA fragment of 358 bp. (Reproduced from Jones and Foulkes[4] with permission of Oxford University Press.)

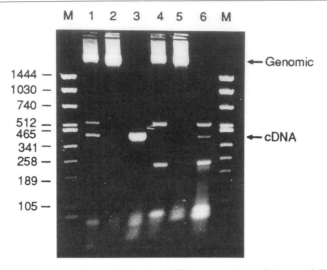

FIG. 2. HHV-6 reverse transcription by AMV reverse transcriptase and *T. aquaticus* DNA polymerase. Aliquots of total cytoplasmic RNA from HHV-6-infected HSB-2 cells, ~2 μg, were reverse transcribed in 1× PCR buffer (final volume, 20 μl) with 50 ng primer 6SP1, 40 units AMV reverse transcriptase at 42° for 60 min (lanes 1–3) or 2.5 units of AmpliTaq at 72° for 60 min (lanes 4–6). Samples (10 μl) were then submitted to amplification in 1× PCR buffer (50-μl final volume) with 0.5 μg each of primers 6SP1 and 6SP2, and 2.5 units of AmpliTaq. Samples were amplified through 30 cycles (lanes 1–3) or 35 cycles (lanes 4–6). In lanes 2 and 5 the RNA was incubated with 10 μg of RNase A for 60 min at 37° prior to reverse transcription, and lanes 3 and 6 were pretreated with 10 units of RNase-free DNase for 60 min at 37°. Lanes M are size markers of an equimolar mixture of *Taq*I- and *Sau*3AI-digested pBluescript KS(−). The arrowed genomic PCR fragment is 3623 bp in size and the cDNA fragment is 426 bp.

reverse transcriptase or *T. aquaticus* DNA polymerase resulted after PCR amplification in a smear of DNA fragments (Fig. 1A). Specific amplification of the G6PD cDNA sequences was confirmed by Southern blotting (Fig. 1B and C). No contaminating nuclear genomic DNA was detected, as shown by the absence of a DNA band of 657 bp. RNase digestion prior to PCR completely destroyed any G6PD cDNA amplification (Fig. 1, lane 4). Reverse transcription by *T. aquaticus* DNA polymerase is capable of giving rise to large fragments, as some of the artifactual bands generated are greater than 1000 bp in size (Fig. 1A, lane 5).

Human herpesvirus 6 infection of cells leads to viral particles in the cytoplasm, and preparation of cytoplasmic RNA concomitantly isolates viral genomic DNA. The intron present in the gene is over 3000 bp in size.[9] This "contaminating" DNA was conclusively seen in the RNA sample and was successfully eliminated by pretreatment of the RNA with DNase

(Fig. 2). This in effect released primers sequestered by genomic DNA, which could be utilized to amplify cDNA (Fig. 2, compare lanes 1 and 3, and 4 and 6). The correct cDNA band was confirmed by Southern blot analysis with an HHV-6 recombinant plasmid clone (data not shown). Thus, if contaminating DNA is present in an RNA sample, DNase digestion will improve the yield of PCR-amplified cDNA.

The efficiency of reverse transcription by *T. aquaticus* DNA polymerase is low compared with AMV reverse transcriptase, approximately ≤1%. This is not surprising, as *T. aquaticus* DNA polymerase has evolved to utilize a DNA template rather than a RNA one. A simple way to overcome the inefficient reverse transcription step is to perform the PCR amplification for 40 or more cycles. The gene for *T. aquaticus* DNA polymerase has been cloned and expressed in *E. coli*,[10] and with site-directed mutagenesis it should be possible to engineer an enzyme with an improved reverse transcriptase activity. Analysis of the specificities of other thermostable DNA polymerases, for example, the enzyme isolated from *Thermus thermophilus*, may reveal polymerases with better reverse transcriptase activity.[11] A requirement for 2–3 mM $MgCl_2$ was found for the reverse transcription of the G6PD sequences by *T. aquaticus* DNA polymerase.[4] Thus, as with PCR, care must be exercised in the magnesium ion concentration in the reaction buffer. Investigation into the replacement of magnesium with manganese may improve the reverse transcriptase activity of *T. aquaticus* DNA polymerase.[11]

Tse and Forget[12] have also shown the ability of *T. aquaticus* DNA polymerase to carry out reverse transcriptions. They obtained reverse transcription by directly amplifying the RNA under standard PCR conditions, without a transcription preincubation step. We have not been able to amplify RNA directly by PCR (Fig. 1, lane 6),[4] but this may be a reflection of the actual RNA sequences involved.

In conclusion, *T. aquaticus* DNA polymerase has been shown to possess reverse transcriptase activity and this enzyme (or engineered variants and other thermostable polymerases) should prove extremely useful for the analysis of mRNA structure.

Acknowledgments

The work described was supported by grants from the Wellcome Trust, Cancer Research Campaign, the Nuffield Foundation, and the Society for General Microbiology.

[10] F. C. Lawyer, S. Stoffel, R. K. Saiki, K. Myambo, R. Drummond, and D. H. Gelfand, *J. Biol. Chem.* **264,** 6427 (1989).
[11] Perkin-Elmer Cetus is marketing *T. thermophilus* DNA polymerase as a single enzyme for both reverse transcription and PCR amplification.
[12] W. T. Tse and B. G. Forget, *Gene* **88,** 293 (1990).

[32] Absolute Levels of mRNA by Polymerase Chain Reaction-Aided Transcript Titration Assay

By MICHAEL BECKER-ANDRÉ

Introduction

An impressive number of methods designed specifically to detect minor amounts of mRNA targets have been reported.[1] They all are based on molecular technologies that have been introduced and refined to achieve the greatest sensitivity while retaining fidelity. Conventional techniques based on hybridization-mediated detection of nucleic acid targets have been used with much success. These techniques rely on the principles of either mixed-phase hybridization or solution hybridization. However, the detection limit of these methods is only about 10^6 target molecules. To improve its sensitivity, the solution hybridization technique has been modified to reduce the background and/or to amplify the amount of specifically hybridized probe. From these efforts assays using sandwich hybridization with capture probes evolved [e.g., reversible target capture (RTC), and RTC in combination with an amplifiable reporter RNA (MDV-1 RNA and its cognate $Q\beta$-replicase)]. In the best case the detection limit was reported to be around 10,000 target molecules.[2] Another general approach is the specific amplification of the target itself via polymerase chain reaction (PCR) to amounts that allow direct analysis for quantification.[3-5] Because of its extraordinarily high sensitivity, the PCR technology is now widely used to detect low-abundance nucleic acid targets. The application of PCR technology for mRNA analysis first requires the conversion of target mRNA into cDNA by reverse transcription (RT-PCR).[6,7] However, it has been difficult to use PCR on its own for quantitative evaluations of specific mRNA targets. Only under well-defined conditions ("quantitative PCR"),

[1] M. Becker-André, *Methods Mol. Cell. Biol.* **2,** 189 (1991).

[2] H. Lomeli, S. Tyagi, C. G. Pritchard, P. M. Lizardi, and F. R. Kramer, *Clin. Chem.* **35,** 822 (1989).

[3] R. K. Saiki, S. Scharf, F. Faloona, K. B. Mullis, G. T. Horn, H. A. Erlich, and N. Arnheim, *Science* **230,** 1350 (1985).

[4] R. K. Saiki, D. H. Gelfand, S. Stoffel, S. J. Scharf, R. Higuchi, G. T. Horn, K. B. Mullis, and H. A. Erlich, *Science* **239,** 487 (1988).

[5] K. B. Mullis and F. A. Faloona, this series, Vol. 155, p. 335.

[6] D. A. Rappolee, D. Mark, M. J. Banda, and Z. Werb, *Science* **241,** 708 (1988).

[7] P. J. Doherty, M. Huesca-Contreras, H. M. Dosch, and S. Pan, *Anal. Biochem.* **177,** 7 (1989).

or with the aid of internal standards ("competitive PCR") can PCR give reliable information about the amount of the original target present in the sample.

Quantitative Polymerase Chain Reaction

Theoretically, within the exponential phase of PCR the extent of amplification (Y) is proportional to the input amount as described by the equation:

$$Y = A(1 + R)^n$$

where A is the initial amount of target, R is the efficiency (between 0 and 1), and n is the number of cycles performed. The original protocol[6,8] was performed under conditions that keep the exponential amplification efficiency R as constant as possible by limiting the amount of input target and the number of cycles performed. Consequently, it could be shown that the logarithm of the number of input target molecules and the logarithm of the mass of amplified product follow essentially a linear relationship—although only within a range of two to three orders of magnitude.[6,8–15] The efficiency R (hereafter called the R value) has a theoretical maximum of 1, but varies due to several factors: the nature of sequence, the primers, the cycle number, and sample effects (e.g., impurities or position in the thermocycler[16]). Slight differences or changes of the R value can dramatically influence the amount of product being amplified near the end of the reaction, rendering any extrapolation of the initial amount of starting targets from the amount of amplified product highly unreliable.[17] Quantitative PCR is dependent on high reproducibility of the overall reaction from test tube to test tube. A way to circumvent the requirements for high reproducibility has been described through simultaneous, yet independent, coreverse transcription and coamplification of endogenous but unrelated

[8] D. A. Rappolee, A. Wang, D. Mark, and Z. Werb, *J. Cell. Biochem.* **39**, 1 (1989).
[9] M. A. Abbott, B. J. Poiesz, B. C. Byrne, S. Kwok, J. J. Sninsky, and G. D. Ehrlich, *J. Infect. Dis.* **158**, 1158 (1988).
[10] S. J. Arrigo, S. Weitsman, J. D. Rosenblatt, and I. S. Y. Chen, *J. Virol.* **63**, 4875 (1989).
[11] B. C. Delidow, J. J. Peluso, and B. A. White, *Gene Anal. Technol.* **6**, 120 (1989).
[12] S. Oka, K. Urayama, Y. Hirabayashi, K. Ohnishi, H. Goto, K. Mitamura, S. Kimura, and K. Shimada, *Biochem. Biophys. Res. Commun.* **167**, 1 (1990).
[13] J. Singer-Sam, M. O. Robinson, A. R. Bellvé, M. I. Simon, and A. D. Riggs, *Nucleic Acids Res.* **18**, 1255 (1990).
[14] M. O. Robinson and M. I. Simon, *Nucleic Acids Res.* **19**, 1557 (1991).
[15] M. F. Gaudette and W. C. Crain, *Nucleic Acids Res.* **19**, 1879 (1991).
[16] U. Linz, *BioTechniques* **9**, 286 (1990).
[17] G. Gilliland, S. Perrin, K. Blanchard, and H. F. Bunn, *Proc. Natl. Acad. Sci. U.S.A.* **87**, 2725 (1990).

"reporter" mRNAs serving as internal standard.[18–20] As a prerequisite for valid results, however, the R values of both the target and the standard must be established. If these R values turn out to be different, then the amplification must be analyzed kinetically.[18–20] The "reporter" mRNAs serve two functions: (1) they help to compensate for sample effects; and (2) they provide a means for relative or absolute quantitation. To quantify target molecules in an absolute manner the amount of "reporter" mRNA must be known (e.g., by quantitative Northern blotting). For this reason, the addition of known amounts of exogenous reference RNA to the sample is superior. This was first realized by adding an exogenous artificial reference RNA consisting of a set of primer annealing sites identical to those of a set of target mRNA to be investigated.[21,22] Under these conditions the detection limit is around 10,000 target mRNA molecules. This artificial reference RNA is preferable to the completely unrelated endogenous "reporter" mRNAs, because identical primers can be used for both the reference RNA and the target mRNA. However, the unrelated sequence between the primers still constitutes a reaction parameter not shared by the two types of templates.

Competitive Polymerase Chain Reaction

In 1989 we described a method that circumvents the disadvantages of quantitative PCR and allows quantitative evaluation of very low-abundance mRNA levels.[23] This method (with the acronym PATTY, standing for PCR-aided transcript titration assay) uses a titration system to measure specific mRNA concentration. To overcome sample effects, we used coreverse transcription and coamplification of an *in vitro*-synthesized reference RNA differing from the authentic mRNA only by a single base exchange that creates a new restriction endonuclease site. Because the reference RNA and the target mRNA behave identically, the ratio of target- and reference-specific DNA should remain constant during the whole process of amplification. A changing R value will affect the amplification of target-

[18] J. Chelly, J.-C. Kaplan, P. Maire, S. Gautron, and A. Kahn, *Nature (London)* **333,** 858 (1988).

[19] J. Chelly, J.-P. Concordet, J.-C. Kaplan, and A. Kahn, *Proc. Natl. Acad. Sci. U.S.A.* **86,** 2617 (1989).

[20] J. Chelly, D. Montarras, C. Pinset, Y. Berwald-Netter, J.-C. Kaplan, and A. Kahn, *Eur. J. Biochem.* **187,** 691 (1990).

[21] A. M. Wang, M. V. Doyle, and D. F. Mark, *Proc. Natl. Acad. Sci. U.S.A.* **86,** 9717 (1989).

[22] A. M. Wang and D. F. Mark, *in* "PCR Protocols: A Guide to Methods and Applications" (M. A. Innis, D. H. Gelfand, J. J. Sninsky, and T. J. White, eds.), p. 70. Academic Press, San Diego, 1990.

[23] M. Becker-André and K. Hahlbrock, *Nucleic Acids Res.* **17,** 9437 (1989).

and reference-specific DNA to an equivalent extent. Thus, the reference RNA is a truly equivalent template serving as a probe for the amount of target mRNA present in the original sample. Kinetic analysis of the PCR is unnecessary. PATTY is completely cycle independent and can be run well into the plateau phase to accumulate enough material for easy detection (e.g., by ethidium bromide-stained gel analysis). A similar approach to the quantitative evaluation of mRNA has been described by Gilliland et al.[17]; instead of using a mutated reference RNA, however, they use a mutated cDNA as a competitive template, not correcting for variations in the efficiency of reverse transcription.

The following section describes the principle of PATTY in more detail. Here are emphasized improvements that facilitate the generation of the reference RNA and its discrimination from the authentic mRNA.

Principle of PATTY

Figure 1 illustrates the technical principle of PATTY. Its concept is based on a titration system: identical aliquots of an RNA master sample are "spiked" with different known amounts of a site-specifically mutated RNA. This reference RNA serves as a probe for the amount of specific target mRNA present in the RNA sample. It is an in vitro-synthesized transcript representing a piece of the target mRNA to be quantified. In the original version of PATTY,[23] the reference RNA differs from the target mRNA by a single base exchange that creates a novel restriction endonuclease site. To distinguish it from the more recent protocol described below the original protocol will be referred to as drPATTY (dr standing for differential restriction). The different mixtures of target mRNA and reference RNA are subjected to cDNA synthesis and PCR using specific primers. The amplification step allows easy detection and analysis of the PCR products. With drPATTY, the amplified DNA fragments are digested with the appropriate restriction endonuclease to discriminate by virtue of differential susceptibility to restriction endonuclease digestion between DNA derived from endogenous target mRNA and DNA derived from exogenous reference RNA. After digestion, the two different DNA species are separated by gel electrophoresis. In a dilution series containing added reference RNA, the resultant PCR products will show progressively more "endogenous" DNA fragments and fewer "exogenous" DNA fragments. Only one aliquot will contain equal or nearly equal amounts of both types of DNA, reflecting equal or nearly equal starting amounts of the corresponding RNA species. The known amount of in vitro transcript added to this sample can be taken as equivalent to the amount of specific target mRNA present.

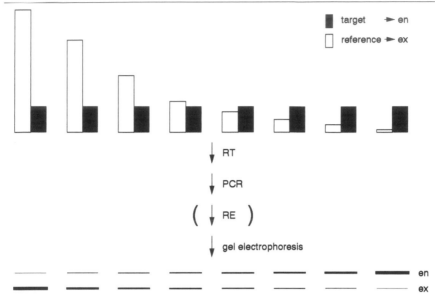

FIG. 1. Scheme illustrating the technical principle of PATTY. An RNA master sample is subdivided in a series of identical aliquots containing unknown, yet identical, amounts of the target mRNA to be quantified. "Titration" of these target molecules with decreasing amounts of internal reference RNA, reverse transcription (RT), amplification to approximately equivalent amounts of DNA (PCR), and analysis by gel electrophoresis result in a series of double bands. The upper band represents DNA fragments derived from the target mRNA (en, endogenous), the lower band represents DNA derived from the reference RNA (ex, exogenous). For drPATTY (see text and Fig. 2 for explanations), discrimination between the different types of DNA fragments requires restriction endonuclease digestion (RE) prior to gel electrophoresis. For dsPATTY, the amplified DNA fragments can be directly separated, because the reference RNA sequence contains a small deletion rendering the resulting DNA fragments shorter than those originating from the target mRNA.

Since the first description of drPATTY, significant improvements have been introduced to overcome some of the inherent disadvantages that are discussed below. The most important change affects the ease of analysis: a reference RNA is created that contains a deletion of about 15 nucleotides close to one end. This reference RNA simplifies the protocol, because the deletion allows discrimination of the reference RNA-specific PCR-amplified DNA products from those originating from the endogenous target mRNA by virtue of size difference, making restriction endonuclease treatment superfluous. The PCR-generated DNA fragments can be analyzed directly by gel electrophoresis. However, for this analytical format (differential-size PATTY, dsPATTY) native agarose gel electrophoresis is inappropriate. The size difference constitutes only about 5% of the overall length (with drPATTY it is 50%). Resolution of the double band requires polyacrylamide gel electrophoresis. Incorporation of radiolabeled nucleotides during the polymerase reaction allows detection of the bands by autoradiography. Alternatively, both drPATTY and dsPATTY can be performed with a final hybridization step. Hybridization with a radiolabeled probe can dramatically increase the sensitivity and specificity of the assay. The different possible analytical approaches are summarized in Fig. 2.

Materials and Reagents

Oligonucleotides

Oligonucleotides are prepared on an Applied Biosystems (Foster City, CA) 380 B automated synthesizer. Following synthesis, they are deprotected by incubating at 55° for 6 hr in ammonium hydroxide, cooled on ice, and desalted on a Sephadex G-50 (medium grade) column equilibrated with 10 mM Tris-HCl, pH 8.0/1 mM ethylenediaminetetraacetic acid (EDTA) (TE buffer), diluted to 25 pmol/μl in TE buffer, and stored in aliquots at 4 or $-20°$. Alternatively, deprotected oligonucleotides can be vacuum dried (most commercial oligonucleotides are shipped like this) and redissolved in TE buffer without a desalting step. Contaminating salt seems not to interfere with PCR or sequencing reactions. Primers are designed using Primer Designer software [Scientifical & Educational Software, State Line, PA, 1990; a Macintosh-compatible program (Gene Jockey) is available from Biosoft, Milltown, NJ]. Special care is taken to ensure that there is no 3' complementarity between primer pairs.

Enzymes

Polymerase chain reaction is performed using *Taq* DNA polymerase purchased from Cetus (Norwalk, CT), Serva (Westbury, NY), and Stratagene (La Jolla, CA) with equal success. However, it should be noted that *Taq* DNA polymerase from Serva did not work in the PCR buffers provided by the other two suppliers. Apparently, Triton X-100 at a concentration of 0.1% (v/v) is strictly required by *Taq* DNA polymerase from Serva. Restriction enzymes like *Bam*HI and *Eco*RI are from Stratagene or Boehringer Mannheim (Indianapolis, IN). *In vitro* transcription is performed via T3 and T7 RNA polymerase (Promega, Madison, WI) off linearized recombinant pBS$^{(+)}$ or pBluescript vector (Stratagene). RNase inhibitor and RQ1 RNase-free DNase are from Promega. Reverse transcription is performed using Moloney murine leukemia virus (M-MuLV) reverse

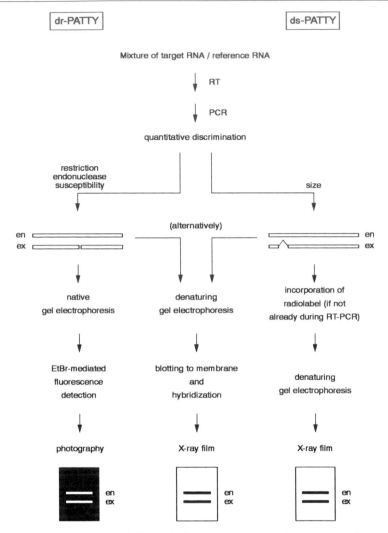

FIG. 2. Outline of the analytical principle of PATTY. A given amount of total RNA containing an unknown amount of a specific target mRNA is mixed with a known amount of reference RNA. This mixture is subjected to reverse transcription (RT) and PCR to give a population of DNA fragments defined in size and sequence by the specific oligonucleotide primers used. Quantitative discrimination between DNA fragments originating from target mRNA (en, endogenous) or reference RNA (ex, exogenous), respectively, can be performed in two ways. (1) drPATTY (dr standing for differential restriction): The reference RNA is an *in vitro*-synthesized mutated transcript that contains a central novel restriction endonuclease site. Digestion with the appropriate restriction endonuclease exclusively cuts DNA derived from the reference RNA. After native gel electrophoresis ethidium bromide-mediated fluorescence detection allows relative quantitation of digested versus undigested DNA fragments.

transcriptase from Life Technologies (Gaithersburg, MD). Synthesis of single-stranded DNA probe is carried out using DNA polymerase (Klenow fragment, labeling grade; Boehringer Mannheim).

Buffers, Solutions, and Media

For RNA and DNA Manipulations

Quartz-distilled water or water treated with diethylpyrocarbonate (DEPC)

Transcription buffer (5×) (Life Technologies): 200 mM Tris-HCl (pH 8.0), 40 mM MgCl$_2$, 10 mM Spermidine, 250 mM NaCl

PCR buffer (10×) (according to supplier's specification)

Universal buffer (10×) (Stratagene): 1000 mM potassium acetate, 250 mM Tris–acetate (pH 7.6), 100 mM magnesium diacetate, 5 mM 2-Mercaptoethanol, 100 μg/ml bovine serum albumin (BSA)

Ammonium acetate (4 M), pH 5.4

TE buffer: 10 mM Tris-HCl (pH 8.0), 1 mM EDTA

TAE buffer (50×): 2 M Tris–acetate (pH 8.0), 100 mM EDTA

TBE buffer (10×): 1 M Tris base, 830 mM Boric acid, 10 mM EDTA

Formamide loading buffer: 98% (w/v) Deionized formamide, 10 mM EDTA (pH 7.5), 0.2% (w/v) bromphenol blue, 0.2% (w/v) xylene cyanol

The following nucleotide solutions in water: 2 mM dNTP; 2 mM dATP/dGTP/dTTP; 0.2 mM dCTP; 10 mM NTP

Plasmids and Bacterial Strains

pBS$^{(+)}$ or pBluescript (Statagene) with promoters for T3 and T7 RNA polymerase flanking the multiple cloning site

Escherichia coli strains MC1061 or DH5α (Life Technologies) or SURE (Stratagene)

This ratio reflects the situation in the original mixture. The size of the cut "ex" DNA fragments is half the size of the "en" DNA fragments. (2) dsPATTY (ds standing for differential size): the reference RNA is an *in vitro*-synthesized mutated transcript that contains a small deletion of about 15 nucleotides close to one end. Incorporation of radiolabeled deoxynucleotides during PCR or during a second PCR run with nested primers (which increases sensitivity of the assay) allows direct detection of the DNA fragments after denaturing gel electrophoresis. The "ex" DNA fragments are approximately 5% shorter than the "en" DNA fragments. Alternatively, a hybridization step using a 5' end-specific single-stranded DNA probe can be added to either the ds or dr protocol as a means of increasing sensitivity of the assay.

For Identification and Analysis of Recombinant Plasmid DNA

LB plates supplemented with 100 μg/ml ampicillin

TB medium[24]: Dissolve 12 g Bacto-tryptone (Difco, Detroit, MI) 24 g yeast extract, and 4 ml glycerol in 900 ml water; autoclave and add 100 ml of autoclaved 0.17 M KH_2PO_4/0.72 M K_2HPO_4

Ampicillin stock in water (100 mg/ml)

Minibuffer I: 50 mM Tris-HCl, 20 mM EDTA

Minibuffer II: 0.2 N NaOH, 1% (v/v) sodium dodecyl sulfate (SDS)

Minibuffer III: 3 M potassium acetate, 1.8 M formic acid

Minibuffer IV: 4 M LiCl, 20 mM sodium acetate, pH 5.2

For Hybridization[25,26]:

(Pre)hybridization buffer: 0.5 M sodium phosphate (pH 7.2), 1 mM EDTA, 1% (w/v) BSA, 7% (w/v) SDS; filter through 0.45-μm membrane

Washing buffer: 40 mM sodium phosphate (pH 7.2), 1% (w/v) SDS, 100 mM NaCl per 5% GC content of the target template below 50%

[α-^{32}P]dCTP (>3000 Ci/mmol; Amersham, Arlington Heights, IL)

Gel Electrophoresis

Analysis of PCR fragments and *in vitro*-synthesized runoff transcripts is performed on native 2% (w/v) agarose gels in 1× TAE buffer containing 0.1–0.2 μg/ml ethidium bromide. Separation of radiolabeled PCR fragments is done in denaturing 6% (w/v) polyacrylamide (PAA) gels in 1× TBE buffer. For the preparation of the polyacrylamide gels, follow the instructions given in the "Guidelines for Quick and Simple Plasmid Sequencing" (edited by Boehringer Mannheim).

Acrylamide stock (40%, w/v): 95 g acrylamide (Fluka, Ronkonkoma, NY) and 5 g N,N'-methylenebisacrylamide (Fluka) dissolved in a final volume of 250 ml H_2O and filtered through a 0.45-μm membrane filter; store in the dark at 4°

PAA premix: 420 g urea (ultrapure; Fluka) and 145 ml 40% (w/v) acrylamide stock dissolved in a final volume of 900 ml H_2O and stirred with

---

[24] K. D. Tartof and C. A. Hobbs, *Focus* (*Bethesda Res. Lab.*) **9**(2), 12 (1987).

[25] G. Church and W. Gilbert, *Proc. Natl. Acad. Sci. U.S.A.* **81**, 1991 (1984).

[26] H. Saluz and J. P. Jost, *Gene* **42**, 151 (1986).

5 g of mixed-bed ion-exchange resin for 30 min at room temperature; store in the dark at 4°

Ammonium persulfate: A 10% (w/v) solution in H_2O (Sigma)

N,N,N',N'-Tetramethylethylenediamine (TEMED) (Sigma, St. Louis, MO)

To prepare a gel mix ready for casting, 54 ml of premix and 6 ml of 10× TBE buffer are combined and filtered through a 0.45-μm membrane filter before 230 μl of 10% ammonium persulfate solution and 75 μl of TEMED are added. The dimensions of the cast PAA gels are 400 × 300 × 0.4 mm. Treatment of the glass plates is according to the instructions given in the Boehringer Mannheim publication, "Guidelines for Quick and Simple Plasmid Sequencing." Briefly, the larger glass plate is coated with "cross-linker": a mixture of 15 μl γ-methacryloxy-propyltrimethylsilane (Sigma), 15 μl glacial acetic acid, and 5 ml ethanol (industrial grade) is spread evenly over the glass plate, allowed to dry for a few minutes, and washed away extensively with water. The glass plate is wiped dry. If the gel is to be used for electrotransfer, the larger glass plate must not be coated with "cross-linker" but should be left untreated. The smaller or notched glass plate is treated with "repellent": a 5% (w/v) dimethyldichlorosilane (Sigma) solution in $CHCl_3$ is spread evenly over the glass plate, allowed to dry for a few seconds, and washed away extensively with water. The glass plate is wiped dry. New glass plates or those not used for several weeks should be cleaned with chromosulfuric acid before coating.

Equipment

For PCR, the Perkin-Elmer/Cetus as well as the Techne PHC 2 (Princeton, NJ) thermocycler are used with equal success. For denaturing gel electrophoresis the vertical gel electrophoresis system sold by Bethesda Research Laboratories/Life Technologies (No. 1070, model V161) is used, with plate dimensions of 20 × 20 cm. For hybridizations the size-separated DNA fragments are electroblotted to nylon membranes (Gene-screen; New England Nuclear, Boston, MA) using an inhouse-built Western blot apparatus. Cross-linking of the DNA to the nylon membrane is achieved through ultraviolet (UV) illumination in a Stratalinker (Stratagene). Hybridization is carried out in a hybridization oven using glass cylinders (Bachofer, Reutlingen, FRG, or Hybaid, Teddington, Middlesex, UK).

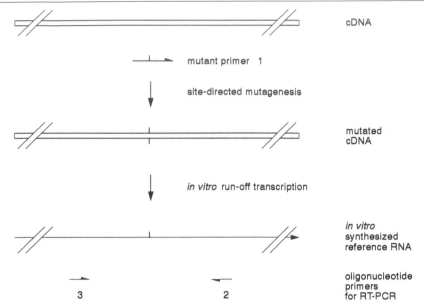

FIG. 3. Outline of the procedure to create mutated reference RNA for drPATTY. A cDNA representing the target mRNA (preferentially in full length) is site-directed mutated, using the mutant oligonucleotide primer 1. The resulting single-base exchange creates a novel restriction endonuclease site not present in the cDNA of the target mRNA. *In vitro* transcription off an appropriate plasmid DNA containing this mutated cDNA generates the reference RNA. The oligonucleotide primers 2 and 3 (18 to 20-mers) used for RT-PCR are shown paralleling their annealing sites within the cDNA.

Methods

Creation of Mutant cDNA Template by Site-Directed Mutagenesis

Differential Restriction PATTY

The original version of PATTY requires a site-directed mutated cDNA (derived from the mRNA to be analyzed) that carries a novel restriction endonuclease site. The mutated cDNA does not need to be a full-length version of the authentic cDNA but it must contain the region that will be amplified by RT-PCR. The mutation should be situated in the middle of the final PCR product. The overall size of the final PCR product should not be smaller than 0.3 kbp and not larger than 1 kbp. Smaller fragments will be difficult to analyze, whereas larger fragments will be amplified with reduced efficiency. Figure 3 shows schematically the generation of mutant cDNA templates suitable for runoff transcription.

At present, there are many different protocols for site-specific mutagenesis available, making it difficult to select the most convenient one.[27–30] I have had good experiences with the gapped-duplex approach,[31] but a purely *in vitro*-performed protocol might work as well.[32] *In vitro* mutagenesis kits can be purchased from Boehringer Mannheim, Stratagene, Promega, and Amersham. The mutation should be verified by sequencing of the entire fragment used for mutagenesis and, of course, by restriction endonuclease digestion.

Differential Size PATTY

In dsPATTY the mutant cDNA template represents a relatively short piece of the target mRNA that is to be quantified. In contrast to drPATTY this mutated cDNA template contains a small deletion of about 15 bp for later discrimination of the PCR-generated products. The deletion is introduced via PCR. Thus, the mutagenesis protocol is both easy and efficient. The overall size of the final PCR end product should be between 200 and 300 bp. Smaller sizes will increase the difference in length between mutant and authentic templates (15-bp deletion in a 200-bp fragment account for a 7.5% difference in length; greater differences may cause considerable differences in amplification efficiency). Larger fragment sizes will make separation of the amplification products by virtue of their gel mobility differences difficult. Figure 4 shows schematically the generation of mutant templates suitable for runoff transcription. Preparation of the relevant DNA piece from the target cDNA ready for subcloning and introduction of the deletion are simultaneously carried out by PCR. The upstream primer (1, Fig. 4) contains a cloning site (10 nucleotides, e.g., for *Bam*HI) as well as 18 nucleotides and 15 nucleotides specific for the target bridging a gap of 15 nucleotides. The downstream primer (2, Fig. 4) comprises the second cloning site (e.g., for *Eco*RI) and 18 nucleotides of target-specific sequence 200–300 bp downstream of the sequence position of the upstream primer in the cDNA. The PCR generates a cDNA template that can be directionally cloned into a transcription vector downstream of a promoter for T3/T7/SP6 RNA polymerases. Linearization of the resulting recombinant plasmid with an appropriate restriction endonuclease precedes *in vitro* runoff transcription. The same procedure can be followed to generate unmutated ("wild-type") cDNA template using instead of the mutant

[27] M. J. Zoller and M. Smith, this series, Vol. 154, p. 329.

[28] W. Kramer and H. J. Fritz, this series, Vol. 154, p. 350.

[29] T. A. Kunkel, J. D. Roberts, and R. A. Zakour, this series, Vol. 154, p. 367.

[30] P. Carter, this series, Vol. 154, p. 382.

[31] P. Stanssens, C. Opsomer, Y. M. McKeown, W. Kramer, M. Zabeau, and H. J. Fritz, *Nucleic Acids Res.* **17**, 4441 (1989).

[32] K. Nakamaye and F. Eckstein, *Nucleic Acids Res.* **14**, 9679 (1986).

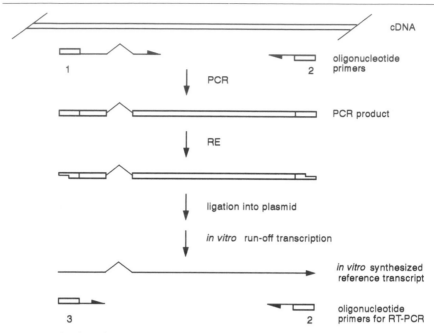

FIG. 4. Outline of the procedure to create mutated reference RNA for dsPATTY. A cDNA representing the target mRNA (not necessarily full length) is subjected to PCR using a set of special oligonucleotide primers: the upstream, mutant, primer 1 contains adjacent to a restriction enzyme recognition site (a "cloning box" of 10 nucleotides) 18- and 15-nucleotide target-specific sequences flanking a gap of 15 nucleotides. On annealing to the cDNA, this primer will introduce a deletion of 15 bp into the PCR-generated DNA fragment; the downstream primer 2 contains next to a "cloning box" of 10 nucleotides (specific for a different restriction endonuclease) 18 nucleotides of target-specific sequence 200–300 bases 3′ of the annealing site of the mutant primer 1. The PCR product is treated with the appropriate restriction endonucleases (RE) to generate stick ends for oriented subcloning into a transcription plasmid. *In vitro* transcription off this plasmid generates the reference RNA. The primers 2 and 3 used for RT-PCR are shown paralleling their annealing sites within the cDNA. This scheme is not drawn to scale. Primer 3 also serves to generate unmutated RNA that can be used to set up test experiments of dsPATTY.

primer 1 (Fig. 4) a primer that comprises only the cloning site and the first 18 nucleotides of the target-specific sequence (3, Fig. 4; used later also for PCR). It is optional to use PCR primers that introduce a cloning site. Primers representing only target-specific sequences will be shorter and cheaper. However, subcloning of PCR products into plasmid DNA would have to be performed via less efficient blunt-end ligation.

Procedure. Five nanograms of uncut plasmid containing the cDNA of the mRNA target in 5 μl TE buffer is combined with

PCR buffer (10×)	5 μl
Mutant primer 1 (or primer 3 for "wild-type" template)	1 μl
Primer 2	1 μl
dNTP (2 mM)	5 μl
H$_2$O	33 μl
Taq DNA polymerase	2.5 units

Twenty PCR cycles are performed with the following profile: 94° (1 min), 45° (1 min), 72° (20 sec). Ten microliters of the product is analyzed on a 2% agarose gel. After phenol extraction of the other 40 μl of the PCR reaction mixture, the DNA is precipitated with 20 μg glycogen (Boehringer Mannheim), 1 vol 4 M ammonium acetate, pH 5.4, and 4 vol ethanol, pelleted for 15 min at room temperature, washed with 80% ethanol, and redissolved in 50 μl TE buffer. Half of this DNA material is digested with restriction endonucleases cutting the flanking cloning sites (e.g., *Bam*HI and *Eco*RI), purified by phenol–chloroform extraction and ethanol precipitation, and subcloned into an appropriate transcription vector like pBluescribe (Stratagene). Identification of recombinant plasmids is performed by plasmid minipreparations: 10 individual bacterial colonies are picked and grown overnight in TB medium. One milliliter of each culture is transferred into an Eppendorf tube, and spun for 1 min (all centrifugations are at 12,000 rpm). Supernatant is removed and bacteria are resuspended in 200 μl of minibuffer I; 400 μl of freshly prepared minibuffer II is added and mixed by vigorous shaking for 3 sec (*no* vortexing), then 300 μl of minibuffer III is added and mixed again. The mixtures are centrifuged for 2–4 min at room temperature; 800 μl of each supernatant is transferred into a fresh tube, mixed with 400 μl 2-propanol, and spun for 5 min at room temperature. The pellets are washed with 0.5 ml 80% (v/v) ethanol, air dried, and redissolved in 100 μl TE buffer. Two to 5 μl will be sufficient for restriction digestion (including 10 μg RNase, DNase free) and gel analysis. One volume minibuffer IV is added and the mixtures are incubated for at least 1 hr on ice. Most of the contaminating RNA is removed by centrifugation for 10 min at room temperature. The supernatants are transferred to fresh tubes and mixed with 0.5 vol 2-propanol. The plasmid DNA is pelleted by centrifugation for 10 min at room temperature and redissolved in 20 μl TE buffer. This plasmid DNA preparation (usually at 0.5–1 μg/μl) is ready for sequencing and linearization.

Creation of Mutant Reference RNA by in Vitro Runoff Transcription

Linearization of the recombinant plasmid by digestion with a restriction endonuclease downstream of the insert must be complete. This should be verified by gel electrophoresis of the digestion products. Residual RNA is removed by digestion with RNase A (1 μg/50-μl reaction volume for 10 min at 37°). The cut plasmid DNA must be purified by phenol–chloroform extraction and ethanol precipitation.

   Procedure (Adapted from Stratagene). To 1 μg linearized plasmid DNA in 10 μl TE buffer add

Transcription buffer (5×)	5 μl
NTP (10 mM)	1 μl
Dithiothreitol (DTT) (0.75 M)	1 μl
H$_2$O (quartz distilled or DEPC treated)	7 μl
RNase inhibitor	20 units
RNA polymerase	10 units

Incubate at 37° for 30 min and add

RQ1 buffer (1×; Promega) containing 1 μl 0.75 M DTT	25 μl
RNase inhibitor	20 units
RQ1 DNase, RNase free (Promega)	2 units

Incubate at 37° for 10 min, purify through phenol–chloroform extraction, precipitate with ethanol, and redissolve in 100 μl DEPC-treated or quartz-distilled water. The same procedure can be followed to synthesize unmutated, "wild-type" *in vitro* runoff transcripts. After quantitation by OD$_{260}$ measurement, 0.2–0.5 μg of the *in vitro* runoff transcripts is heat treated in formamide loading buffer, chilled on ice, and immediately loaded on a native 2% agarose gel. The heat denaturation prior to loading prevents formation of multiple bands in the gel. Electrophoresis conditions are similar to those used for PCR fragments: 10 V/cm for 20–30 min. RNA fragments up to about 300 nucleotides in size will comigrate with the respective double-stranded DNA marker (e.g., 1-kb ladder; Life Technologies). However, with increasing size, RNA molecules will run progressively faster than the corresponding DNA molecules.[33] The gel analysis allows confirmation of molecular integrity and reestimation of relative amounts of different preparations (e.g., mutated versus unmutated transcript templates).

   At this point, *in vitro* transcripts of both mutant and "wild-type" sequence are ready for use. The first PATTY experiments should be run with these two RNA templates before doing experiments with rare RNA samples. This will allow testing of the system for sensitivity and accuracy. To do so, the two types of *in vitro* transcripts are mixed in different ratios and different starting amounts before they are subjected to RT-PCR.[23] Examples are as follow:

[33] *Promega Notes* No. 20 (Aug. 1989).

1. a dilution series of a 1 : 1 mixture of both transcript types, starting with 1 ng and going down in 10-fold dilution steps to 1 ag (1 ng equals about 10^{10} molecules, 1 ag equals about 10 molecules, assuming the transcripts are 200 nucleotides in size)

2. a series of different ratios between the two transcript types, such as 10 : 1, 5 : 1, 2 : 1, 1 : 1, 0 : 1, 1 : 0, 1 : 2, 1 : 5, and 1 : 10 (1 stands for 1 pg)

Reverse Transcription and Polymerase Chain Reaction

Using the downstream primer 2, the transcripts are converted into first-strand cDNA via reverse transcriptase. For the sake of convenience, the reaction is carried out in the PCR buffer in the presence of *Taq* DNA polymerase.[14,15] After the reverse transcription reaction the samples proceed directly to the PCR part without any further manipulation. To start quantitation of real RNA samples, 10–100 ng of total RNA is taken for each RT-PCR tube and spiked with 10-fold diluted amounts of reference RNA. The 10-fold dilution series will give an estimation of the abundance of the target mRNA within 10–100 ng of total RNA. Consecutive PATTY experiments can then be done with a more fine-tuned titration for a more precise estimation.

Procedure. Pipette into 0.5-ml test tubes 19 μl of the master mix with the following composition:

H₂O (quartz distilled)	13.5 μl
PCR buffer (10×) (e.g., Stratagene)	2 μl
dNTP (2 m*M*)	2 μl
Upstream primer 3	0.4 μl
Downstream primer 2	0.4 μl
DTT (0.1 *M*)	0.1 μl
Total RNA (10–100 ng)	1 μl

Then add to the individual samples

Reference RNA (beginning with 1 ng down to 100 ag)   1 μl

The mixtures are overlaid with 30 μl light mineral oil (Sigma), incubated in a water bath at 95° for 20 sec, and quick-chilled on ice. Three microliters 1× PCR buffer containing 100 units M-MuLV reverse transcriptase, 10 units RNase inhibitor, and 0.5 units *Taq* DNA polymerase is added to each sample underneath the oil phase to give a final volume of 23 μl. Reaction mixtures are incubated at 42° for 30 min. Reverse transcription is stopped by shifting the temperature to 95° for 1 min and the amplification is carried out through 30 PCR cycles with the profile 94° (30 sec), 50° (1 min), 72° (20 sec). This profile has been optimized using a Techne PHC2 apparatus that controls the actual sample temperature through a dummy tube probe. Another thermocycler might require a different profile.

Second Polymerase Chain Reaction

The second PCR serves two functions: (1) use of a nested primer enhances the sensitivity and the specificity of PATTY; and (2) inclusion of radiolabeled dCTP allows a direct analysis via autoradiography following gel electrophoresis. After RT-PCR 1 μl of each PCR sample is withdrawn and diluted into 200 μl TE buffer. From the resulting dilutions 2 μl is pipetted into 18 μl of fresh PCR mixture.

Procedure. Pipette into 0.5-ml tubes:

Quartz-distilled H_2O	12 μl
PCR buffer (10×)	2 μl
dATP, dGTP, dTTP (2 mM)	2 μl
dCTP (0.2 mM)	1 μl
[α-^{32}P]dCTP	0.2 μl (= 2 μCi)
Upstream primer 3	0.4 μl
Nested primer 4	0.4 μl
Taq DNA polymerase	0.5 units

Again, this solution can be prepared as a master mix and aliquotted into the tubes before adding 2 μl of the diluted RT-PCR products. The nested primer 4 is an oligonucleotide (18 to 20-mer) whose sequence is derived from the amplified DNA target. Its position is 5' of the downstream primer starting DNA synthesis in the same direction. The distance to the upstream primer 3 should not be smaller than 180 bp. Thirty more PCR cycles are performed with the profile 94° (30 sec), 50° (1 min), 72° (2 min). Alternatively, if more than 10^4 target copies are expected to be present in the RT-PCR mixture, then in the dsPATTY format the radiolabeled dCTP can be included earlier in the RT-PCR step and the RT-PCR products can be directly analyzed by gel electrophoresis.

Analysis of Amplified DNA Fragments for Quantitative Discrimination

Differential Restriction PATTY: Restriction Endonuclease Digestion and Native Agarose Gel Electrophoresis

After amplification, 5 μl of each sample is directly subjected to gel electrophoresis in 2% agarose (Fig. 5A). This step serves as a control for the efficiency of amplification and is used to adjust the samples to roughly identical amounts of amplified DNA prior to further manipulations. Another 10 μl of each sample is boiled for 10 min and cooled to room temperature. This step serves to standardize the samples with respect to heteroduplexes that may have formed during the last PCR cycles. Like DNA fragments with native sequence in both strands, these heteroduplexes are not cleaved with the restriction endonuclease. This leads to consistent overrepresentation of uncleaved PCR product. Heat denaturation maximizes the formation of heteroduplexes and simultaneously destroys *Taq* DNA polymerase activity.[17] Due to heteroduplex formation an actual 1 : 1 molar ratio between unmutated and mutated DNA fragments would be turned into an apparent 3 : 1 ratio. Thus, any uncertainty about the extent of heteroduplex formation is thereby avoided. In the original publication,[23] 2 μl of each sample was diluted 1 : 200 in fresh PCR reaction mixture and subjected to one additional PCR cycle. This served to remove the heteroduplexes. However, this also means additional manipulation of the samples. Finally, the PCR products are digested with 10 units of the appropriate restriction endonuclease and size separated by gel electrophoresis in 2% agarose (Fig. 5B).

Differential Restriction PATTY: Denaturing Gel Electrophoresis and Hybridization

Alternatively, for the sake of higher sensitivity and specificity, the restriction endonuclease-digested DNA fragments are size separated in a denaturing 6% polyacrylamide gel, electroblotted to nylon membrane, UV cross-linked, and hybridized to a single-stranded DNA probe.

Procedure. Ten microliters of digested sample is added to 40 μl of formamide loading buffer, heated to 95° for 1 min, and quick-chilled on ice. Two microliters is loaded on a 1-mm thick denaturing 6% polyacrylamide gel. Electrophoresis is carried out at 300–500 V until the xylene cyanol dye reaches the bottom of the gel (15 cm). An aluminum plate will help to level out temperature gradients across the glass plates, thus avoiding "smiling effects." The small (or notched) glass plate is removed and the gel is transferred to a dry sheet of Whatman (Clifton, NJ) paper. The paper support is carefully soaked in 0.5× TBE buffer, and a piece of nylon membrane (Genescreen) cut to size and equilibrated in 0.5× TBE is laid on top of the gel, avoiding air bubbles trapped in between. Electrotransfer is performed at 4 V/cm for 1 hr in a Western blot chamber containing 0.5× TBE buffer. The nylon membrane is removed and the DNA is cross-linked by UV illumination (Stratalinker). The membrane is prehybridized at 60° for 30 min in hybridization buffer. Specific detection of one end of the DNA fragments is achieved by hybridization to a single-stranded DNA probe that is synthesized according to the following protocol: 1 μg of single-stranded DNA template containing a subfragment of the target cDNA and 100 ng of specific synthesis primer are annealed for 5 min at

dr PATTY   ds PATTY

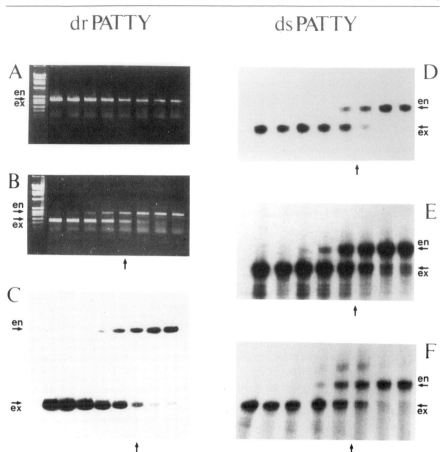

FIG. 5. Determination of absolute levels of specific mRNA in total RNA following either the protocol of drPATTY (A–C) or the protocol of dsPATTY (D–F). (A) Samples containing 2 μg total RNA isolated from cultured potato cells and various amounts of site-directed mutated 4-coumarate : CoA ligase (4CL)-specific reference RNA were subjected to drPATTY to "titrate" endogenous 4CL-specific mRNA. After 30 PCR cycles, a portion of each sample was analyzed on a 2% agarose gel. The amplified unmutated ("en") and mutated ("ex") 375-bp DNA fragments comigrated in one band. (B) Another portion of each sample was digested with *Eco*RI and analyzed the same way. The two subfragments of the reference-specific DNA fragment generated by *Eco*RI digestion (184 and 191 bp) migrated as a single band ("ex"). The vertical arrow indicates the transition point of the titration reflecting the apparent ratio between target mRNA and reference RNA. (C) The samples from (A) were diluted and subjected to one additional PCR cycle, digested with *Eco*RI, separated on a 6% polyacrylamide gel, blotted onto nylon membrane, and hybridized to a single-stranded DNA probe. Dilution and the final PCR cycle removed heteroduplexes; hence, the transition point reflects the real ratio between target mRNA and reference RNA. Titration was done with a threefold dilution series starting with 1000 pg and ending with 0.3 pg of reference RNA. The equivalence point is around 3 pg. (D) Samples containing 0.1 μg total RNA isolated from cultured human

50° in 5 μl 1× universal buffer (Stratagene) containing 1 mM concentrations each of dATP, dGTP, dTTP. Five microliters (= 50 μCi) [α-^{32}P]dCTP is added and the reaction is started with 2 units of DNA polymerase (Klenow fragment, labeling grade; Boehringer Mannheim). After 10 min at room temperature the synthesis reaction is stopped by adding 100 μl minibuffer I. Unincorporated deoxynucleotides are removed through a Sephadex G-50 medium spin column. The synthesis primer anneals close to the discriminative restriction endonuclease site and primes the DNA synthesis toward one of the master fragments ends.[23] After prehybridization, the heat-denatured probe is added without changing the buffer, and hybridization is carried out overnight. The nylon membrane is washed once at room temperature in 2× SSC/0.1% SDS (SSC: 0.15 M NaCl, 15 mM sodium citrate, pH 7.0) and twice at 60° in wash buffer. Exposure is for 30–60 min with an intensifying screen at −70°. An example is shown in Fig. 5C.

Differential Size PATTY: Denaturing Gel Electrophoresis and Autoradiography

Either after RT-PCR (performed in the presence of radiolabeled dCTP) or after the second PCR, 2 μl of each sample is mixed with 50 μl formamide loading buffer, heated to 90° for 2 min, and quick-chilled on ice. One microliter is loaded on a 0.4-mm thick, denaturing 6% polyacrylamide gel. After electrophoresis (1500 V; until the bromophenol blue dye has migrated approximately 30 cm) the small glass plate is removed, the gel—together with the glass plate to which it is covalently bound—is submersed in 10% acetic acid for 15 min, rinsed 2 min with water to remove urea, and dried in an oven at 80–100° for 1 hr. Exposure is for 1 hr or longer with an intensifier screen at −70°. Figure 5D–F shows examples.

Comments

Presented in this chapter are two versions of PATTY. Table I gives an overview on their respective advantages and disadvantages. The ds ver-

---

umbilical vein cells and different amounts of deletion-mutated human endothelial leukocyte adhesion molecule 1 (hELAM-1)-specific reference RNA were subjected to dsPATTY to "titrate" endogenous hELAM-1 mRNA. After the second PCR, a portion of each sample was analyzed by electrophoresis in a denaturing 6% polyacrylamide gel and by subsequent autoradiography (2 hr). (E) A 10 times longer exposed autoradiography of the same gel as in (D). (F) The same samples as in (D); however, 10 times more of each sample was loaded on the gel, causing a third, artifactual band in the area of the transition point. Titration was with a 10-fold dilution series starting with 200 pg and ending with 20 ag of reference RNA. The transition point is around 6 fg (equivalent to about 100 fg ELAM-1 mRNA, because this RNA is approximately 15 times larger than the reference RNA used).

TABLE I
Advantages and Disadvantages of Two Versions of PATTY[a]

Characteristic	drPATTY	dsPATTY
Simplicity of mutant generation	Moderate[b]	High
Similarity between reference RNA and target mRNA	High	Moderate[c]
Performance of analysis (manipulation of samples during course of procedure)	Moderate[d]	Easy
Nonisotopic performance	Possible	Not possible
Sensitivity	High	High
Reproducibility	High	High

[a] Based on different principles of discrimination between reference-specific and target-specific DNA fragments: differential restriction endonuclease susceptibility (dr) PATTY and differential size (ds) PATTY.
[b] Depending on the mutagenesis kit used, the procedure can be considerably time consuming.
[c] Mutant cDNA template is not designed to the size of the target mRNA.
[d] Restriction endonuclease digestion requires additional manipulation.

sion of PATTY is the first choice, because of its simple procedure for creating the mutant cDNA template and its direct quantitative discrimination scheme. However, if nonisotopic analysis is preferred, drPATTY will be the favored procedure. Both versions display equal sensitivity. Using a hybridization step or a second PCR with nested primers, the detection limit of these procedures is well below 100 target molecules. Provided that precautions against contaminations and pipetting errors are taken,[34,35] the intrinsic limit of sensitivity is dictated only by statistical principles: at low concentrations, the actual number of target molecules in a given portion of the sample volume varies with a Gaussian distribution. However, it might be argued that mRNA at or below this level of quantitative detection is of no biological significance anymore.[19]

An inherent problem especially with dsPATTY can arise from the fact that reference RNA and target mRNA are not identical with respect to their overall size. This discrepancy could evoke a differential tendency to form secondary structures that might interfere with the process of reverse transcription. For example, the authentic target mRNA could be less efficiently converted into cDNA as compared to the reference RNA. Manipulations that generally break secondary structures—such as briefly heating the RNA sample to 95° before adding the reverse transcriptase—should level out such differences. Favored by their high concentra-

[34] Y. M. Lo, W. Z. Mehal, and K. A. Fleming, *Lancet* **ii,** 699 (1988).
[35] S. Kwok and R. Higuchi, *Nature (London)* **339,** 237 (1989).

tion, the specific oligonucleotide primers can quickly occupy their complementary sequence segments within the target RNA templates on cooling of the samples, thereby avoiding obstruction of these sites through possible intramolecular base pairing. Alternatively, a thermostable reverse transcriptase such as, for example, *Tth* Pol,[36] might be used to perform the reaction at elevated temperature. This would not only destabilize potential secondary structures present in the RNA template but also increase the specificity of primer extension. For RNA templates forming extraordinarily stable secondary structures, however, the reverse transcription might be carried out in the presence of the strong denaturant methylmercury hydroxide (CH_3HgOH; Invitrogen, San Diego, CA).

The position of the 200- to 300-bp sequence from the target mRNA sequence is not critical, unless the target mRNA is subjected to a rapid turnover *in vivo*. Quantitative results obtained by PCR seem to be relatively unaffected by random target degradation, as long as short enough fragments of the target mRNA are amplified.[37] This is in one way an advantage over hybridization-dependent assays (in particular sandwich hybridization) when postmortem or very old samples are analyzed.[38] Alternatively, problems might arise in interpreting the results. Degradation of mRNA goes from the 3' to the 5' end via exonucleolytic activities.[39,40] Thus, biologically nonfunctional fragments of mRNA could be detected with a PATTY system based on mutant templates derived from a sequence region close to the 5' end. Therefore, a region close to the translation stop codon is recommended. Furthermore, it is advantageous to select a region containing a genomic intron. This allows detection of DNA contaminations.

For dsPATTY, the creation of the mutant RNA template can be further simplified. Using a deletion-mutant downstream primer and an upstream primer containing both target-specific sequence (e.g., 18 nucleotides) and an RNA polymerase-specific promoter sequence (17 nucleotides in length), the requirement for subcloning the PCR-generated mutant cDNA fragment is eliminated. This fragment can be used directly as a template for *in vitro* runoff transcription to yield reference RNA template [reminiscent of nucleic acid sequence-based amplification (NASBA)[41,42]]. This might offer an attractive alternative to researchers lacking facilities for microbiological

[36] T. M. Myers and D. H. Gelfand, *Amplifications (Perkin-Elmer Cetus)* No. 7, 5 (1991).
[37] T. E. Golde, S. Estus, M. Uslak, L. H. Younkin, and S. G. Younkin, *Neuron* **4**, 253 (1990).
[38] J. Tenhunen, *Mol. Cell. Probes* **3**, 391 (1989).
[39] G. Brawerman, *Cell* **48**, 5 (1987).
[40] G. Brawerman, *Cell* **57**, 9 (1989).
[41] C. Tuerk and L. Gold, *Science* **249**, 505 (1990).
[42] J. Van Brunt, *Bio/Technology* **8**, 291 (1990).

techniques such as bacterial transformation and growth of bacterial cultures.

The titration system itself can be simplified by using more than just one mutated reference RNA template, that is, part of the titration series is incorporated into each coamplification sample. For example, by performing dsPATTY with a mixture of three different reference RNA templates that can be size discriminated from each other as well as from the target RNA, the actual number of titration points necessary to cover a certain range can be cut by two to three. In the relevant lanes (close to the transition point) the signal intensities of four bands are compared instead of two. This is schematically illustrated in Fig. 6. Two representative examples are shown in Fig. 7.

For purification of RNA from biological samples I prefer the rapid guanidinium thiocyanate–phenol–chloroform extraction method,[43] which provides a pure preparation of undegraded RNA in high yield. For quantitative recovery of minute amounts of RNA I include glycogen (Boehringer Mannheim) as a precipitation carrier. Because the quantitation procedure is preferably carried out with purified samples, recovery of target mRNA during the purification process must be controlled independently. As a normalizing factor the amount of total RNA isolated can be used. For that purpose, a small aliquot of each RNA sample (1–10 ng) is analyzed by Northern blot hybridization using, for example, a cDNA of the 18S rRNA as specific probe. This RNA constitutes about 40% of the mass of the total RNA. Therefore, hybridization and exposure to an X-ray film can be carried out within 8 hr.

The evaluation of levels of target nucleic acid molecules by PATTY is based on the estimation of relative amounts (i.e., relative signal intensities) of coamplified radiolabeled DNA species in a given titration sample. Thus, the accuracy of PATTY depends on the quality of signal intensity measurement. Routinely, autoradiography is used. However, X-ray films do not respond linearly to irradiation. Several autoradiographs with different exposure times might be required to provide true imaging of signal intensities and their reliable measurement by densitometry. Therefore, where available, radioanalytic imaging systems (e.g., AMBIS, San Diego, CA) should be used that directly quantify radioactive samples. Radiolabeling of PCR products with $[\alpha\text{-}^{32}P]dNTPs$ allows their easy and rapid detection by autoradiography. However, for safety reasons a more preferable isotope is $[\alpha\text{-}^{33}P]dATP$ (Amersham BF1001).[44] The ^{33}P isotope emits a mild β particle having a maximum energy of 0.249 MeV (for comparison: ^{32}P,

[43] P. Chomczynski and N. Sacchi, *Anal. Biochem.* **162**, 156 (1987).
[44] M. R. Evans and C. A. Read, *Nature (London)* **358**, 520 (1992).

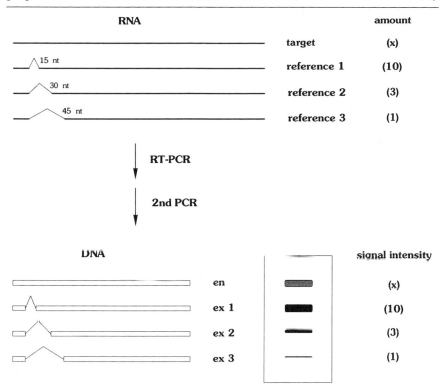

FIG. 6. A way to reduce the number of titration points for dsPATTY. Three different reference RNA templates with individual overall sizes are prepared via deletion of 15, 30, and 45 nucleotides (reference 1, reference 2, and reference 3, respectively) from the authentic sequence. These three templates are mixed in a ratio of 10:3:1. This way, a titration between, for example, 100 pg and 100 fg using threefold dilution steps can be carried out with only three instead of seven individual reaction samples. The first sample would contain 100/30/10 pg, the second sample 10/3/1 pg, and the third sample 1000/300/100 fg of the three respective reference templates.

1.71 MeV; ^{35}S, 0.167 MeV) and a half-life of 25.4 days. Working with this isotope substantially reduces exposure of the experimenter to hazardous radiation. However, a significantly reduced sensitivity, and therefore detection of ^{33}P labeled products, would be expected. Exposure times of samples at −70° to X-ray films with an intensifier screen normally requiring 1 to 2 hours would need 10 to 20 hours, when labeled with [33]P. Moreover, an intensifier screen is of little use, since most of the β particle energy is absorbed by the film itself.

I would also like to point out a general limitation of PATTY. The competitive PCR approach, as it is presented here with PATTY, might

A

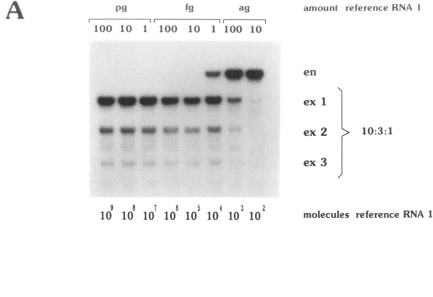

B

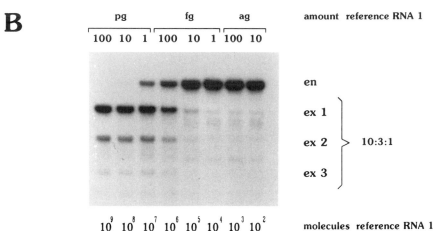

Fig. 7. Samples containing 0.1 μg total RNA isolated from the heart of an untreated mouse (A), or a mouse systematically treated with interleukin 1 for 4 hr (B) and different amounts of a 10:3:1 mixture of three deletion-mutated murine endothelial leukocyte adhesion molecule 1 (mELAM-1)-specific reference RNAs (0.38 kb in size) were subjected to dsPATTY to titrate endogenous mELAM-1 mRNA (3.8 kb in size). Both the amount and the number of molecules of reference RNA 1 added to each titration sample before RT-PCR are indicated. In (A) (untreated mouse), the transition point is around 0.4 fg—equivalent to 4 fg or 4×10^3 molecules of mELAM-1 mRNA. In (B) (interleukin 1-treated mouse), the transition point is around 200 fg—equivalent to 2 pg or 2×10^6 molecules of mELAM-1 mRNA.

not be suitable for the quantitation of the transcripts of a gene family. If these transcripts are too similar to be selectively amplified or too divergent to be uniformly amplified by RT-PCR, spurious results could be found.

Finally, besides the mRNA quantitation there are also other applications imaginable: PATTY can be used to estimate the efficiency of cDNA synthesis reactions. An RNA sample with a known amount of target mRNA is subjected to reverse transcription; titration is then performed with a reference cDNA template. I have found values between 25 and 50% of target mRNA converted to cDNA.[45] PATTY is by no means restricted to the quantitation of mRNA targets. Analysis of levels of DNA targets can be equally well performed with PATTY (for this purpose the acronym should be modified to PCR-aided template titration assay).

In addition to applications in research, quantitation of target nucleic acids is gaining increasing importance in clinical diagnosis as well.[46,47] Besides the detection of nucleic acids in a diagnostic sense, monitoring the changing amounts of target nucleic acids (e.g., either DNA of infectious agents or cellular mRNA of affected tissue) will allow previously unobtainable information on molecular events during the course of a disease. For this purpose, PATTY could make a valuable contribution: because absolute levels of target nucleic acids are estimated, comparisons between different evaluations for the same target as well as for different targets can be undertaken.

Acknowledgments

I am grateful to Dr. J. F. DeLamarter for critically reading the manuscript, and to Ms. M. Leemann for preparing the illustrations.

---

[45] M. Becker-André, unpublished results, 1989.
[46] D. H. Persing and M. L. Landry, *Yale J. Biol. Med.* **62,** 159 (1989).
[47] R. I. Fox, I. Dotan, T. Compton, H. M. Fei, M. Hamer, and I. Saito, *J. Clin. Lab. Anal.* **3,** 378 (1989).

[33] Polymerase Chain Reaction-Based mRNA Quantification Using An Internal Standard: Analysis of Oncogene Expression

By RICHARD H. SCHEUERMANN and STEVEN R. BAUER

Protooncogene expression is often regulated with respect to cellular growth rates, cell cycle progression, tissue specificity, and lineage differentiation. Regulated expression is extremely important for normal cell function because perturbation of oncogene expression frequently results in tumor formation. Although the details concerning the influence of oncogene expression on normal and abnormal cell function are beyond the scope of this chapter, it is clear that the accurate quantification of steady state mRNA levels could greatly enhance our understanding of the relationship between oncogene expression and normal or abnormal cell growth and differentiation. Indeed, quantification of mRNA levels has important application for the analysis of many biological phenomena.

Some of the earliest experiments designed to quantify mRNA levels involve the measurement of heteroduplex formation using solution hybridization (R_0t analysis). Although this technique can be accurate it is cumbersome, and would require great effort to measure multiple RNA species in multiple samples. In 1977, Alwine et al.[1] described a method for the detection of mRNA species by filter hybridization following gel electrophoresis, the "Northern blot." This technique is now widely used for estimating the abundance of specific mRNA species. However, this technique suffers from three main drawbacks: (1) a relatively large amount of RNA is needed (~ 10 μg total RNA $\approx 10^6$ cells/analysis); (2) if one wishes to estimate mRNA levels for several genes the technique can become labor intensive; and (3) because mRNA levels are measured in relative terms, it is difficult to compare the results from different experiments, let alone the results from different research groups. Although techniques like S1 protection,[2] RNase protection,[3] and primer extension[4] can be an order of magnitude more sensitive, they also suffer from producing measurements in relative terms.

[1] J. C. Alwine, D. J. Kemp, and G. R. Stark, *Proc. Natl. Acad. Sci. U.S.A.* **74,** 5350 (1977).
[2] A. J. Berk and P. A. Sharp, *Cell* **12,** 721 (1977).
[3] K. Zinn, D. DiMaio, and T. Maniatis, *Cell* **34,** 865 (1983).
[4] J. Sambrook, E. F. Fritsch, and T. Maniatis, "Molecular Cloning: A Laboratory Manual," 2nd Ed., Vol. 1, p. 7.79. Cold Spring Harbor Press, Cold Spring Harbor, New York, 1989.

The technique for the amplification of specific sequences utilizing temperature cycling and a thermostable polymerase,[5] the polymerase chain reaction (PCR), has revolutionized the field of molecular biology. The exquisite sensitivity and relative simplicity of the technique have stimulated the search for an application of this technique to quantify mRNA levels. However, it was clear very early that because the technique involves many repeated cycles, even small differences in reaction efficiencies between samples could result in large differences in the final measurements. The coamplification of an mRNA species such as β-actin, with roughly equal abundance from one RNA source to the next, in the same reaction mix as the test mRNA,[6,7] can control for these efficiency differences to some degree. However, because RNA levels are still quantified in relative terms, it is difficult to compare results obtained from separate experiments. Wang et al.[8] described a technique that can overcome some of these problems by including a synthetic RNA molecule as an internal standard for quantification. With this technique the same PCR primers are used to quantify the standard and mRNA molecules in the same reaction mix. The use of this internal, synthetic RNA standard controls for tube-to-tube efficiency differences. The main limitation to this modification is that the efficiency of amplification of the endogenous and standard molecules must be the same.

Here we describe a similar method for the quantification of mRNA levels for a set of protooncogenes thought to be involved in lymphocyte growth and transformation. However, in this chapter we also show how to detect efficiency differences between the standard and test RNA molecules, and suggest approaches to alleviate these problems. This method can easily be modified for the quantification of mRNA levels for any set of genes for which the sequences are known. In addition, quantification using PCR provides the high degree of sensitivity needed for the analysis of RNA expression from small numbers of cells or small amounts of RNA. Finally, due to the use of an internal standard, results are obtained in absolute terms, that is, number of mRNA molecules per microgram of total RNA or number of molecules per cell. This allows easy comparison between experiments or between laboratories.

[5] R. K. Saiki, D. H. Gelfand, S. Stoffel, S. J. Scharf, R. Higuchi, G. T. Horn, K. B. Mullis, and H. A. Erlich, *Science* **239**, 487 (1988).

[6] J. Chelly, J.-C. Kaplan, P. Maire, S. Gautron, and A. Kahn, *Nature (London)* **333**, 858 (1988).

[7] M. Kashani-Sabet, J. J. Rossi, Y. Lu, J. X. Ma, J. Chen, H. Miyachi, and K. J. Scanlon, *Cancer Res.* **48**, 5775 (1988).

[8] A. M. Wang, M. V. Doyle, and D. F. Mark, *Proc. Natl. Acad. Sci. U.S.A.* **86**, 9717 (1989).

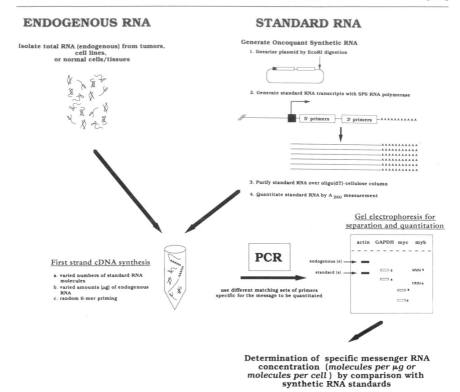

FIG. 1. Schematic representation of PCR-based mRNA quantification procedure using an internal standard. Details of this procedure are described in text.

Methods

Principles of Procedure

The main feature of this method is the addition of a synthetic RNA to serve as an internal standard in each PCR reaction mixture. This standard RNA contains PCR primer sites identical to those present in the specific mRNA to be analyzed. Using radioactive primers, the amount of PCR product generated from the endogenous mRNA can be quantified and compared to the amount of product generated from the standard RNA. A standard quantification curve derived from known amounts of synthetic RNA molecules can be used to determine the absolute number of mRNA molecules for the gene of interest.

The essential steps for this technique are depicted in Fig. 1. RNA can be isolated from any source, including cell lines, tissue samples, or tumor biopsies. Because the technique is quite sensitive, very little starting material is needed, and poly(A) selection is usually not necessary. The specific mRNA for the gene to be analyzed from these sources is referred to as the endogenous RNA, and the PCR product derived from this mRNA as the endogenous product.

In parallel, synthetic RNA is made and purified to act as an internal standard. Using SP6 RNA polymerase, we synthesize the standard RNA from linearized Oncoquant I plasmid DNA. The structure of Oncoquant I is described in detail in Fig. 2 and Table I.[9-31] The synthetic, standard RNA contains sequences homologous to PCR primer pairs, the 5' and 3'

[9] T. Kawakami, C. Y. Pennington, and K. C. Robbins, *Mol. Cell. Biol.* **6**, 4195 (1986).

[10] J. Y. Tso, X.-H. Sun, T.-H. Kao, K. S. Reece, and R. Wu, *Nucleic Acids Res.* **13**, 2485 (1985).

[11] K. Tokunaga, H. Taniguchi, K. Yoda, M. Shimizu, and S. Sakiyama, *Nucleic Acids Res.* **14**, 2829 (1986).

[12] Q. Chen, J.-T. Cheng, L.-H. Tsai, N. Schneider, G. Buchanan, A. Carroll, W. Crist, B. Ozanne, M. J. Siciliano, and R. Baer, *EMBO J.* **9**, 415 (1990).

[13] A. Johnsson, C.-H. Helden, A. Wasteson, B. Westermark, T. F. Deuel, J. S. Huang, P. H. Seeburg, A. Gray, A. Ullrich, G. Scrace, P. Stroobant, and M. D. Waterfield, *EMBO J.* **3**, 921 (1984).

[14] C. Oppi, S. K. Shore, and E. P. Reddy, *Proc. Natl. Acad. Sci. U.S.A.* **84**, 8200 (1987).

[15] T. I. Bonner, H. Oppermann, P. Seeburg, S. B. Kerby, M. A. Gunnell, A. C. Young, and U. R. Rapp, *Nucleic Acids Res.* **14**, 1009 (1986).

[16] T. P. Bender and W. M. Kuehl, *Proc. Natl. Acad. Sci. U.S.A.* **83**, 3204 (1986).

[17] R. Watson, M. Oskarsson, and G. F. VandeWoude, *Proc. Natl. Acad. Sci. U.S.A.* **79**, 4078 (1982).

[18] M. Negrini, E. Silini, C. Kozak, Y. Tsujimoto, and C. M. Croce, *Cell* **49**, 455 (1987).

[19] O. Bernard, S. Cory, S. Gerondakis, E. Webb, and J. M. Adams, *EMBO J.* **2**, 2375 (1983).

[20] R. Zakut-Houri, S. Hazum, D. Givol, and A. Telerman, *Gene* **54**, 105 (1987).

[21] D. K. Watson, M. J. McWilliams, P. Lapis, J. A. Lautenberger, C. W. Schweinfest, and T. S. Papas, *Proc. Natl. Acad. Sci. U.S.A.* **85**, 7862 (1988).

[22] T. Yamamoto, S. Ikawa, T. Akiyama, K. Semba, N. Nomura, N. Miyajima, T. Saito, and K. Toyoshima, *Nature (London)* **319**, 230 (1986).

[23] A. Tanaka, C. P. Gibbs, R. R. Arthur, S. K. Anderson, H.-J. Kung, and D. J. Fujita, *Mol. Cell. Biol.* **7**, 1978 (1987).

[24] M. P. Kamps, C. Murre, X.-H. Sun, and D. Baltimore, *Cell* **60**, 547 (1990).

[25] S. Katamine, V. Notario, C. D. Rao, T. Miki, M. S. C. Cheah, S. R. Tronick, and K. C. Robbins, *Mol. Cell. Biol.* **8**, 259 (1988).

[26] C. Van Beveren, F. van Straaten, T. Curran, R. Muller, and I. M. Verma, *Cell* **32**, 1241 (1983).

[27] J. R. Jenkins, K. Rudge, S. Redmond, and A. Wade-Evans, *Nucleic Acids Res.* **12**, 5609 (1984).

[28] L. Coussens, C. Van Beveren, D. Smith, E. Chen, R. L. Mitchell, C. M. Isacke, I. M. Verma, and A. Ullrich, *Nature (London)* **320**, 277 (1986).

[29] F. Sanger, A. R. Coulson, G. F. Hong, D. F. Hill, and G. B. Petersen, *J. Mol. Biol.* **162**, 729 (1982).

[30] E. Rouer, T. V. Huynh, S. L. de Souza, M.-C. Lang, S. Fischer, and R. Benarous, *Gene* **84**, 105 (1989).

[31] R. H. Scheuermann and S. R. Bauer, *Curr. Top. Microbiol. Immunol.* **166**, 221 (1990).

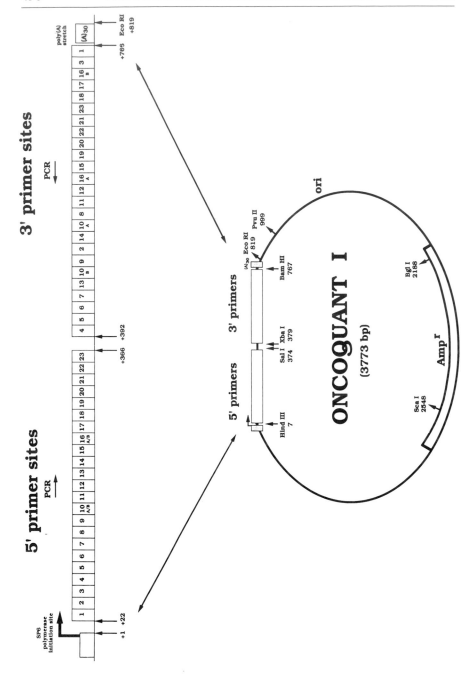

primers. These primer pairs will also hybridize to the mRNAs of interest (specific oncogenes). The primer pairs are separated in the standard RNA by a spacer that is of different length from the distance between these sites in the endogenous mRNA, so that the products arising from the standard and endogenous RNAs can be distinguished by their size. In addition, the standard RNA contains a poly(A) tail that allows purification on oligo(dT)-cellulose, and also ensures that the purified synthetic RNA is full length. The purified standard RNA can be quantified by ultraviolet absorbance and the exact concentration of the synthetic RNA preparation determined from its molecular weight.

The synthetic, standard RNA is mixed with the endogenous RNA in a standard cDNA reaction mix. cDNA synthesis is initiated using random 6-mer oligonucleotide primers. Random priming is preferred over oligo(dT) priming to avoid efficiency differences due to the distance of the PCR priming sites from the poly(A) tail. This cDNA mix can then be used to analyze expression of any gene whose sequences are also present on the standard RNA. A portion of the cDNA mix is added to a standard PCR reaction that contains a primer pair specific for the gene of interest. Following PCR amplification the products are separated by gel electrophoresis and the amounts quantified. Bands corresponding to the endogenous and standard RNA are easily distinguishable because the sizes of products derived from the endogenous mRNA differ from products derived from the standard RNA. A rough quantification can be obtained by comparing ethidium bromide staining intensities by eye; however, for more sensitive and accurate quantification we use ^{32}P end-labeled oligonucleotides as PCR primers. Using primer pairs specific for different genes, mRNA levels for a variety of genes can be quantified from the same cDNA mix. The amount of mRNA molecules present in the endogenous RNA preparation is simply determined by comparison with the amount of product generated from a known amount of standard RNA added to the cDNA reaction mix. This procedure gives absolute quantification in terms of the number of mRNA molecules per microgram of isolated RNA, or if the source of RNA is from counted numbers of cells, the number of molecules per cell.

FIG. 2. Structure of the plasmid Oncoquant I used to generate the standard RNA. Oncoquant I is a derivative of pSP64 poly(A) containing a 760-bp synthetic DNA segment inserted between the SP6 RNA polymerase initiation site and the poly(A) stretch. The insert is composed of 15-bp blocks corresponding to PCR primer pairs (numbered 1–23) specific for 22 different oncogenes and 2 control genes. Details of the genes and sequences used are described in Table I. The synthetic insert was constructed, as described in Fig. 3, with *Hind*III and *Bam*HI restriction sites at opposite ends for cloning into pSP64 poly(A). Distances, in base pairs, are measured from the start of SP6 polymerase transcription.

TABLE I
Oncoquant I Primers

Gene	Homologous gene[a]		GC content[b]	Nucleotide number[c]	Sequence[d]	Comments[e]	Endo[f]	Stand[f]	Ref.[g]
1. *fyn*	h	5'	0.53	1182	CTGCAGTTGATCAAG	Not homologous	695	744	9
		3'	0.53	1876	CCAGTGGATCATGAG	to *yes, src, fgr*[h]			
2. GADPH	h/r	5'	0.60	303	CACCATGGAGAAGGC	ex5; ex8	382	474	10
		3'	0.60	685	TGCCAGTGAGCTTCC				
3. *β*-actin	h/m	5'	0.47	346	CCTTCTACAATGAGC	ex2; ex4	594	699	11
		3'	0.47	940	ACGTCACACTTCATG				
4. *tal-1*	h	5'	0.47	33	CAATCGAGTGAAGAG	ex''1''/ex''2''[i]	251	339	12
		3'	0.53	515	CTGGTCATTGAGCAG				
5. *c-sis*	h	5'	0.60	58[j]	ACCGAGGTGTTCGAG	ex3; ex4	236	339	13
		3'	0.60	296[j]	CACTTGCATGCCAGG				
6. *c-abl*	h/m	5'	0.53	588	CTCTGATGGCAAGCT	ex3/4; —	285	339	14
		3'	0.47	873	CTTCAAGGTCTTCAC				
7. *c-raf*	h/v	5'	0.40	1240	GATGTTGCAGTAAAG	ex11; ex12/13	271	339	15
		3'	0.47	1511	GCATGCAAATAGTCC				
8. *c-myb*	h/m	5'	0.60	485	CAGATGTGCAGTGCC	near *v-myb* start;	331	429	16
		3'	0.47	816	GTGGTTCTTGATAGC	end region I			
9. *c-mos*	h/m	5'	0.60	542	GGAGTTTCTGGGCTG		306	354	17
		3'	0.53	848	TTCAGGTCCAAGTGC				
10A. *bcl-2α*	h/m	5'	0.67	2163	CAGCTGCACCTGACG	ex1; ex2	296	384	18
		3'	0.60	3380	AGAGACAGCCAGGAG				
10B. *bcl-2β*	h/m	5'	0.67	2163	CAGCTGCACCTGACG	ex1; ex1	241	324	18
		3'	0.60	2404	ATGCACCTACCCAGC				
11. *c-myc*	h/m	5'	0.67	758 (ex2)[k]	CCAGCAGCGACTCTG	ex2; ex3	345	399	19
		3'	0.53	330 (ex3)[k]	CCAAGACGTTGTGTG				
12. *pim-1*	h/m	5'	0.60	178	GACAACTTGCCGGTG	ex2/3; ex4	332	399	20
		3'	0.60	524	GTCCTTGATGTGCGCG				
13. *ets-2*	h/m	5'	0.60	355	ACACTCAAGCGGCAG		356	264	21
		3'	0.60	711	ACACAGCATCTGGCC				
14. *c-erb-B2*	h/r	5'	0.47	3131	TTTGTGGTCATCCAG	Cytoplasmic	380	309	22
		3'	0.53	3511	TGTACCGCTGTAGAG	domain			
15. *src*	h/m	5'	0.53	13	AAGAGCAAGCCCAAG	ex2; ex4/5	442	384	23
		3'	0.60	455	TACCACTCCTCAGCC				

		5'/3'	Fraction	Nucleotide	Sequence	Region			Ref
16A. E2A	h	5'	0.60	1269	GACATGCACACGCTG	5 and 3' of break point	462	354	24
		3'	0.53	1731	CTCGTTGATGTCACG				
16B. E2A.prl	h	5'	0.60	1269	GACATGCACACGCTG	5' and 3' of break point	556	489	24
		3'	0.60	1834[l]	TGCGGTGGATGATGC				
17. c-fgr	h/v	5'	0.60	1093	GACAAGCTGGTGCAG	ex8; ex11	365	459	25
		3'	0.53	1458	GATGGTGAATCTGCC				
18. c-fos	h/m/v	5'	0.60	1372	CTACGAGGCGTCATC	ex1; ex2/3	482	429	26
		3'	0.60	1854	TCTGTCTCCGCTTGG				
19. p53	h/m	5'	0.60	460	GGACAGCCAAGTCTG	ex3- ex6	429	339	27
		3'	0.53	889	GAGTCTTCCAGTGTG				
20. c-fms	h/v	5'	0.47	2464	ACCTATGTGGAGATG	Cytoplasmic region	414	339	28
		3'	0.53	2878	CTTGCTGTTCACCAG				
21. ets-1	h,c	5'	0.60	565	CATCCCATCAGCTCG	—; ex3/4	403	354	21
		3'	0.60	968	GCAGGAATGACAGGC				
22. phage λ	NA	5'	0.53	32753	AATCCATCCGACAC		383[m]	324	29
		3'	0.47	33136	CACAGCTATTTCAGG				
23. lck	h/m	5'	0.53	839	CTGCAAGACAACCTG	ex2/3; ex6	392	339	30
		3'	0.60	2023	GAAGCCACCGTTGTC				

[a] Sequences of the PCR primers were often identical between the human gene (h) and the homologous gene in mouse (m), rat (r), chicken (c), or the transforming retrovirus (v) from which the gene was originally identified.

[b] The fraction of nucleotides in the upstream (5') or downstream (3') PCR primer that is either guanidine or cytidine.

[c] The number of the first nucleotide for each PCR primer as described in the reference given.

[d] The sequence of the two 15-mer oligonucleotides used as PCR primers (5' → 3').

[e] Sequences for PCR primer pairs were usually picked in separate exons (ex) or at the boundary between two exons (e.g., ex3/4) in order to discriminate between mRNA and genomic DNA following the PCR reactions.

[f] The length of the PCR products derived from the cellular mRNA (endo) and the in vitro-generated synthetic RNA (stanc) is indicated in base pairs.

[g] Complete references are given in the text footnotes.

[h] While the fyn gene shows striking homology with other tyrosine kinases, PCR primer sequences were selected from regions where little homology exists.

[i] ex"1" and ex"2" correspond to the first two exons described in the reference, which may not correspond to the first two exons of the actual gene.

[j] Nucleotide numbers starting from the first nucleotide of the ATG translation initiation codon.

[k] Nucleotide numbers starting from the first nucleotide of the indicated exon.

[l] Nucleotide numbers starting from the first nucleotide of the prl gene.

[m] PCR product from a messenger RNA present in certain transgenic mouse constructs.[31]

Description of Oncoquant Polymerase Chain Reaction Standard and Construction of Oncoquant I

We designed a template for *in vitro* synthesis of standard RNA that would be useful for the analysis of a series of oncogenes in both mouse and human cells. The structure of Oncoquant I is depicted in Fig. 2. The essential feature is the presence of specific sequences homologous to PCR primer pairs for 22 oncogenes and 2 "housekeeping" genes, β-actin and glyceraldehyde-phosphate dehydrogenase (GAPDH), located between an SP6 promoter and a poly(A) stretch. Using SP6 polymerase and oligo(dT) chromatography the standard RNA can be synthesized and purified. Oncogenes were chosen for Oncoquant I that would be interesting for the analysis of expression in normal and transformed lymphoid cells, and are listed in Table I. In every case, 15-nucleotide sequences derived from the human gene were chosen. The sequences were chosen by the following criteria: (1) whenever possible sequences were used that were identical to the homologous gene of another species (usually mouse); (2) sequences were chosen that contain moderate G/C content (40–67%) with no obvious secondary structure, and a G or C at the 3′ end; (3) primer pairs were chosen to avoid complementary sequences, especially at the 3′ ends; (4) when possible, primers reside in separate exons, to allow distinction between products derived from authentic cDNA rather than contaminating genomic DNA; and (5) in cases in which the gene is a member of a multigene family, primers were chosen from regions with greatest sequence difference between family members. The relative positions of the primer pairs in Oncoquant I were designed to produce PCR products that differed from the endogenous products by 50–100 bp.

Oncoquant I was constructed by sequential PCR reactions with overlapping primers, as described in Fig. 3. Before synthesis, the nucleotide sequence was analyzed for the presence of internal repeats and restriction sites. It was anticipated that internal repeats could cause problems during the PCR synthesis of the Oncoquant insert and during subsequent quantification experiments with the standard RNA.

After synthesis and cloning, sequence analysis[32] of the 760-bp Oncoquant insert from eight independent clones revealed several point mutations in each. The best insert contained five errors. This clone was corrected by site-directed mutagenesis and sequenced again to verify mutation reversion. These point mutations presumably arose by insertional errors generated during *Taq* polymerase synthesis. This is not sur-

[32] F. Sanger, S. Nicklen, and A. R. Coulson, *Proc. Natl. Acad. Sci. U.S.A.* **74,** 5463 (1977).

prising because, under the conditions used here, the 5' end of the insert had undergone over 200 PCR cycles, whereas the 3' end had undergone only 30 cycles. In support of the idea that *Taq* polymerase was responsible, most of the mutations were in the 5' end of the insert.

To minimize the number of point mutations generated during synthesis, we have modified this procedure to reduce the number of PCR cycles required. Longer overlapping oligonucleotides can be used to reduce the number of steps for the synthesis of the insert. While the first step should still be done for 10 cycles, subsequent steps can be done for only 5 cycles. We have already begun the construction of Oncoquant II, which will contain a 1300-bp insert, by linking, in 15 steps, oligonucleotides 100 residues long and overlapping by 20 nucleotides. The combination of larger oligonucleotides (i.e., fewer steps) and fewer cycles will allow the synthesis with 80 PCR cycles for this 1300-bp insert, rather than the 200 cycles used for Oncoquant I. This modification should reduce the number of insertional errors introduced during the construction.

Synthesis, Purification, and Quantification of Oncoquant Standard RNA

To generate a template for *in vitro* RNA synthesis, Oncoquant I plasmid DNA is linearized by digestion with *Eco*RI, then purified by electroelution from a 1% (w/v) agarose gel. The electroeluted DNA is extracted with phenol, then phenol–chloroform, and then chloroform followed by precipitation with ethanol in the presence of 0.1 M sodium acetate, pH 4. After centrifugation to recover the linearized DNA, the pellet is rinsed with 70% (v/v) ethanol, air dried, and resuspended in RNase-free deionized, distilled H$_2$O. The RNA synthesis reaction includes 5 μg of linearized Oncoquant I plasmid DNA, 10 mM dithiothreitol (DTT), 6.25 μg of acetylated bovine serum albumin (BSA; GIBCO/Bethesda Research Laboratories, Gaithersburg, MD), 250 μM of ATP, CTP, GTP, and UTP, 40 mM Tris-HCl (pH 7.5 at 37°), 6 mM MgCl$_2$, 2 mM spermidine hydrochloride, 5 mM NaCl, 125 units (U) of human placental RNase inhibitor (Boehringer Mannheim, Indianapolis, IN), and 30 units of SP6 RNA polymerase (Boehringer Mannheim) in a volume of 50 μl. This mixture is incubated for 1 hr at 40° and another aliquot of 30 U of SP6 polymerase is added, followed by another 1-hr incubation at 40°. The mixture is then extracted with 50 μl of phenol (equilibrated to pH 4 with 0.1 M sodium acetate, pH 4) and twice with CHCl$_3$. The aqueous phase is finally precipitated by adding 1/10 vol of 3 M sodium acetate, pH 4, and 2.5 vol of ethanol, and storing at $-20°$ for 1 hr. RNA is pelleted by centrifugation for 15 min at high speed at 4° in an

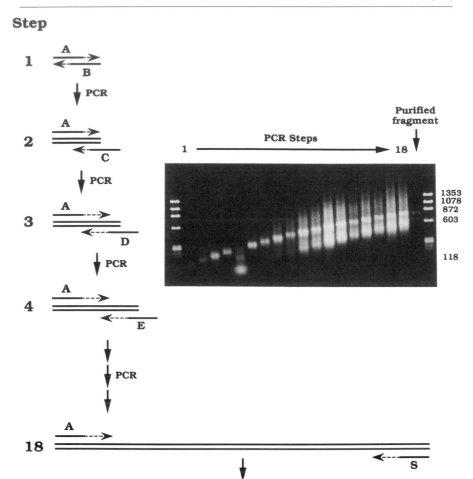

FIG. 3. Construction of the Oncoquant DNA insert. The synthetic insert was constructed using overlapping oligonucleotides under our standard PCR conditions except that the annealing step was done at 55°. In the first step, two 60-mer oligonucleotides (A and B) containing 20 complementary nucleotides at their 3′ ends are annealed, extended, and amplified with 10 PCR cycles. This generates a 100-bp product with sequences from both primers. In the second step, 1/50 of this material was used as template for 10 PCR cycles with oligonucleotide A and a new oligonucleotide, C, as PCR primers. The 20 nucleotides at the 3′ end of oligonucleotide C are identical to the 20 nucleotides at the 5′ end of oligonucleotide B. The use of oligonucleotides A and C as PCR primers in step 2 results in the extension of the step 1 product by 40 bp. Likewise, in step 3, 1/50 of the product from step 2 was used as template in combination with oligonucleotide A and a new oligonucleotide, D, to extend an additional 40 bp. Following 18 steps a 780-bp synthetic DNA was generated from nineteen 60-mer oligonucleotides. The product from each step was analyzed by electrophoresis on a 1.8%

Eppendorf centrifuge. The pellet is rinsed with 80% (v/v) ethanol, air dried, then resuspended in 20 μl of RNase-free H_2O. Agarose gel analysis showed that the Oncoquant standard RNA was a homogeneous band migrating at the correct size (820 nucleotides).

To ensure that the *in vitro*-synthesized standard RNA is full length and free from plasmid DNA contamination, the synthetic Oncoquant RNA is next purified by oligo(dT)-cellulose column chromatography. The RNA is incubated with 0.2 ml of oligo(dT)-cellulose beads (Pharmacia, Piscataway, NJ) in a high-salt buffer (HSB) containing 400 mM LiCl, 10 mM Tris-HCl (pH 8), and 1 mM ethylenediaminetetraacetic acid (EDTA) for 30 min on ice. This slurry is then loaded into a 1-ml syringe-column and washed with 10 ml of HSB. The poly(A)-containing RNA is then eluted into a 1.5-ml Eppendorf tube by loading four 100-μl aliquots of low-salt buffer [LSB: 10 mM Tris-HCl (pH 8), 1 mM EDTA] over the column. The purified RNA is precipitated by adding 40 μl of 3 M sodium acetate, pH 4, and 1 ml of ethanol and storing at $-20°$ for 8 hr. The RNA is recovered by centrifugation at high speed in an Eppendorf centrifuge for 20 min. The pellet is rinsed with 80% ethanol, dried in a SpeedVac (Savant, Hicksville, NY), and resuspended in 15 μl of RNase-free deionized, distilled H_2O. The RNA is quantified by measurement of absorbance at 260 nm. The molecular weight of the standard RNA molecule (2.79 $\times$ 10^5 g/mol) is calculated using the molecular weight and number of residues of each type of ribonucleotide. The yield in this experiment was 2.52 μg of RNA, which means that 2.34 mol of poly(A)-containing standard RNA was recovered per input mole of plasmid DNA template. From this number, we calculate that 5.45 $\times$ 10^{12} molecules of Oncoquant standard RNA were recovered.

Preparation of Endogenous RNA

Total RNA is isolated from frozen tumor tissue or frozen cell pellets by a modified version of the method of Chomczynski and Sacchi.[33] From

---

[33] P. Chomczynski and N. Sacchi, *Anal. Biochem.* **162,** 156 (1987).

---

agarose gel followed by ethidium bromide staining. To generate enough product to see by staining, each reaction mix was subjected to an additional 20 PCR cycles before gel analysis. Following steps 5, 13, and 18, products from preparative agarose gels were purified by electroelution and used for the subsequent reaction. The purified fragment from step 18 (done in quintuplicate) was cloned into M13mp18 and the sequence determined using a chain termination procedure.[32] Following mutagenesis to repair point mutations introduced during synthesis, the *Hin*dIII–*Bam*HI fragment was subcloned into pSP64 poly(A). The first and last lanes of the gel contain ϕX174 *Hae*III digest as molecular weight markers.

0.1 to 0.5 g of frozen tissue (ground into a powder in liquid N_2 with a mortar and pestle) or 10^6 to 10^8 frozen cells are added to a 50-ml polypropylene centrifuge tube containing 5 ml of solution D [4 M guanidinium thiocyanate, 25 mM sodium citrate (pH 7.0), 0.5% (w/v) sarkosyl, 0.1 M 2-mercaptoethanol]. Then 0.5 ml of 2.0 M sodium acetate (pH 4), 5 ml of phenol equilibrated to pH 4 with 0.1 M sodium acetate (pH 4), and 1 ml of CHCl$_3$ are added sequentially. This mixture is subjected to approximately 1 min of simultaneous homogenization and sonication with a homogenizer (Brinkman Instruments, Westbury, NY) set at high speed. For "doping" experiments, known quantities of Oncoquant standard RNA were added to the mixture following homogenization. Following a 30-min incubation on ice, the homogenate is centrifuged at 2000 g for 20 min at 4°. The aqueous supernatant is removed from the organic phase and RNA is precipitated from the mixture with an equal volume of 2-propanol for a minimum of 1 hr at $-20°$. The total RNA precipitate is then recovered by centrifugation at 2000 g for 30 min at 4°. The RNA pellet is resuspended in 1 ml of solution D and transferred to a 2-ml Eppendorf tube. The mixture is then extracted with 1 ml of a 1 : 1 (v/v) mixture of phenol and chloroform, then spun for 5 min at 12,000 g at 4° in an Eppendorf centrifuge. The aqueous phase is transferred to a 2-ml Eppendorf tube and precipitated with 1 ml of 2-propanol for 20 min at $-20°$. The resulting RNA pellet is rinsed with 70%, then 100% ethanol, dried, and resuspended in RNase-free deionized, distilled H$_2$O. The final RNA concentration is quantified by measurement of the absorbance at 260 nm.

For RNA preparations from fewer than 10^6 cells, all manipulation are performed in 1.5 ml microfuge tubes. Cell pellets are resuspended with 200 μl of solution D and lysed by vigorous vortexing. Then 20 μl of 2.0 M sodium acetate, pH 4.0, 200 μl equilibrated phenol, and 40 μl equilibrated CHCl$_3$ are added sequentially, with vigorous vortexing between each addition. The aqueous phase is removed following centrifugation to a new tube containing 20 μg of RNase-free glycogen, and RNA precipitated by the addition of an equal volume of 2-propanol. Following centrifugation the RNA pellet is resuspended with 200 μl of solution D, 2-propanol precipitated again, and then washed with 70% ethanol. The dried pellet is resuspended with 20–100 μl deionized, distilled H$_2$O. Often it is helpful to heat the purified RNA to 65° for 10–15 min with intermittent vortexing to dissolve.

Synthesis of First-Strand cDNA

First-strand cDNAs are generated from mixtures containing 8.0 ng to 2.0 μg of total RNA and known numbers of Oncoquant standard RNA molecules. The endogenous RNA plus the standard RNA are mixed with an appropriate volume of RNase-free deionized, distilled H_2O (to a final volume of 14 μl) and heated at 65° for 2 min in a 1.5-ml Eppendorf tube. This mix is then centrifuged briefly to cool the tube and recover condensed liquid from the inner tube surfaces. Then 2 μl of 10× reaction buffer [1×: 50 mM Tris-HCl (pH 8.3 at 22°), 8 mM $MgCl_2$, 30 mM KCl, 10 mM DTT], 2 μl of 10× deoxyribonucleotides (1×: 0.25 mM of each dNTP), 1 μl of random DNA 6-mer primer [pd(N)$_6$, 1 μg/μl in deionized, distilled H_2O; Pharmacia], 8 U (0.5 μl) of FPLC-pure Moloney murine leukemia virus (M-MuLV) reverse transcriptase (Pharmacia), and 25 U (0.5 μl) human placental RNase inhibitor (Boehringer Mannheim) are added. This mixture is incubated at 37° for 1 hr, then stored at −20° until PCR analysis is performed. In some cases 80 μl of deionized, distilled H_2O is added to the completed cDNA reaction mix to facilitate subsequent analysis.

Standard Polymerase Chain Reactions

Polymerase chain reactions are performed essentially as follows. Mixes (50 μl) contain 0.5 or 1.0 μl of the cDNA reaction as template; 10 pmol of each oligonucleotide primer; dATP, dCTP, dGTP, and dTTP (0.1 mM each); 2 U *Taq* polymerase in 1× reaction buffer [10 mM Tris-HCl (pH 8.3) at 25°, 50 mM KCl, 1.5 mM $MgCl_2$, 0.01% (w/v) gelatin]. We have routinely used reagents and enzyme from Perkin-Elmer Cetus (Norwalk, CT); however, the technique should be compatible with PCR reagents from any source. The experiments described in this chapter have usually been performed on a Techne (Princeton, NJ) PHC-1 programmable dryblock apparatus with the following cycling parameters: denaturation at 94° for 0.6 min, annealing at 40° for 1.0 min, and extension at 72° for 1.5 min. The annealing temperature should be determined empirically. Conditions under which the maximum amount of specific product is generated with minimal amounts of nonspecific background products are optimal. For primers between 15 and 25 nucleotides in length with moderate G/C contents, optimal annealing temperatures range between 40 and 65°. More recent analysis has shown that, in general, all of these steps can be shortened as follows: denaturation for 0.4 min, annealing for 0.5 min, and extension for 0.5 min. We have also found that the addition of low concentrations of formamide (~2%) can greatly increase the signal-to-noise ratio with certain PCR primer sets (see also Sarkar *et al.*[34]).

For accurate quantification of PCR products, ^{32}P-labeled primers are used. Oligonucleotides (100 pmol) are labeled with 20 μCi of [γ-^{32}P]ATP (>5000 Ci/mmol) and 10 U of T4 polynucleotide kinase in 50 μl of kinase

---

[34] G. Sarkar, S. Kapelner, and S. S. Sommer, *Nucleic Acids Res.* **18**, 7465 (1990).

buffer [50 mM Tris-HCl (pH 7.6), 10 mM MgCl$_2$, 5 mM DTT, 0.1 mM spermidine, 0.1 mM EDTA].[4] Following a 90-min incubation at 37° the kinase is inactivated for 10 min at 65° and the labeled primer separated from unincorporated nucleotides on a P-6 spin column (Bio-Rad, Richmond, CA). For the PCR reactions 1–2 × 10^5 cpm of labeled primer (~1 pmol) is included in the standard reaction mix.

Gel Electrophoresis and Quantification

Following the PCR reaction, 10 μl of gel electrophoresis sample buffer [50% (v/v) glycerol, 5× TAE buffer, 0.25% (w/v) bromphenol blue, 0.25% (w/v) xylene cyanol] is added and a portion loaded into the slots of a standard agarose gel (1.2–2.0%) in 1× TAE buffer (40 mM Tris acetate, 2 mM EDTA). The gel is electrophoresed at ~80 V until the bromphenol blue is three-fourths of the way down the gel. The gel is then soaked in 0.4 N NaOH for 30 min, and placed in a one gel-volume pool of 0.4 N NaOH for semidry transfer to a nylon filter (Gene Screen Plus; New England Nuclear, Boston, MA) by capillary action. The nylon filter is prewetted in deionized, distilled H$_2$O and then placed on top of the gel. Two pieces of Whatman (Clifton, NJ) filter paper are placed on top of the nylon filter, and then a stack of paper towels. The towels are weighed down with a 1-kg weight, and the sandwich left overnight. The next day the paper towels and Whatman paper are carefully removed, and the nylon filter is wrapped in plastic wrap and exposed to autoradiography or phosphorimaging (see below). Although it is also possible to dry the agarose gel down using a vacuum gel dryer, we found that the bands are much sharper if the transfer procedure is followed. Alternatively, a vacuum blotter (Bio-Rad) can be used to transfer the labeled DNA to a filter; this system can save a considerable amount of time.

When radioactive primers are used, the quantification can be achieved by a number of procedures, including densitometric scanning of autoradiographic film or cutting out the labeled DNA band and scintillation counting. We have been using a phosphorimager system in combination with Imagequant software (Molecular Dynamics, Sunnyvale, CA), which we find superior to either of the other methods. This technique involves the use of a signal storage technology in which radioactive decay is detected and stored on a screen during the exposure time. The signal can then be released as photons following illumination with a laser emitting a certain wavelength of light. Using the Imagequant software the signal can then be directly quantified. The two main advantages to this system of quantification are that it is not labor intensive (as opposed to cutting out bands and scintillation counting), and that it gives a linear signal over a large range of radioactive concentrations (~10,000 fold versus ~50 fold for preflashed X-ray film).

Analysis and Considerations

Polymerase Chain Reaction Products from Different Primers and RNA Sources

We expect to find two PCR product bands following gel electrophoresis when PCR reactions are performed with primers specific for a particular gene, and cDNA from a mixture of standard RNA molecules and endogenous mRNA. The bands should be of the sizes predicted from the endogenous gene sequence and from the sequence of the Oncoquant construct. Figure 4 shows the results of this type of analysis. If cDNA is made from a mixture of 10^7 molecules of Oncoquant standard RNA and 1.0 μg of total RNA, in this case isolated from a pre-B cell tumor,[31] two PCR product bands can be seen using primers specific for β-actin, GAPDH, c-*myc*, and c-*myb* (Fig. 4A, lanes 2, 5, 8, and 11, respectively). In addition, the molecular weights observed match the molecular weights predicted. It should also be noted that these products are derived from aliquots of the same cDNA reaction mix. On the other hand, if only 10^5 molecules of the standard RNA are used, only a band from the endogenous RNA is seen (lanes 3, 6, 9, and 12, Fig. 4A). Because half of the PCR mix was loaded, and only 1/20 of the cDNA mix was used for the PCR reactions, one can estimate the limit of detection in ethidium bromide-stained gels to be between 2.5×10^3 and 2.5×10^5 molecules of Oncoquant standard, and the ability to detect these four different mRNAs to be from as little as 25 ng of total RNA.

Earlier we discussed that efficiency differences between reaction samples can cause problems for the accurate quantification of mRNA in PCR reactions without internal standards. In Fig. 4A efficiency differences can be observed for the same RNA molecules under different conditions. The upper PCR product bands (Fig. 4A) in lanes 1 and 2, and lanes 10 and 11, come from 10^7 standard RNA molecules. However, the addition of 1.0 μg of endogenous RNA to the cDNA reaction results in an increase in the amount of PCR product from the standard RNA with β-actin primers (lane 2, Fig. 4A), and a decrease with c-*myb* primers (lane 11, Fig. 4A). This result demonstrates that reaction conditions can alter reaction efficiencies, and emphasizes the importance of the internal standard to control for these changes. The problems of differences in reaction efficiencies will be addressed in more detail in a subsequent section.

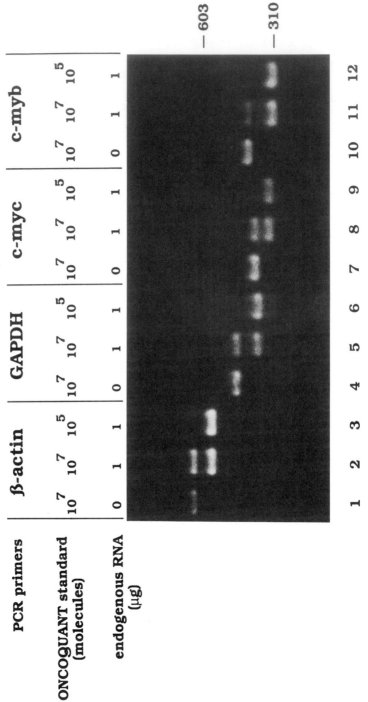

PCR primers	ß-actin		GAPDH		c-myc		c-myb	
ONCOQUANT standard (molecules)	10^7	10^7 10^5	10^7	10^7 10^5	10^7	10^7 10^5	10^7	10^7 10^5
endogenous RNA (μg)	0	1 1	0	1 1	0	1 1	0	1 1

—603

—310

1   2   3   4   5   6   7   8   9   10   11   12

Fig. 4. PCR products from endogenous and standard RNA mixtures. Ethidium bromide-stained agarose gels of PCR reactions using different primer pairs and different RNA sources are shown. (A) Total RNA was isolated from a pre-B cell tumor originating in a *c-myc* transgenic mouse.[31] cDNA was synthesized from mixtures of 0 or 1.0 μg of endogenous RNA and 10^5 or 10^7 molecules of standard RNA as indicated. PCR reactions contained 1/20 of the total cDNA mix and primers specific for *β-actin*, GAPDH, *c-myc*, and *c-myb*. Following 30 PCR cycles, agarose gel sample buffer was added and half of the sample loaded onto a 2% agarose gel.

462

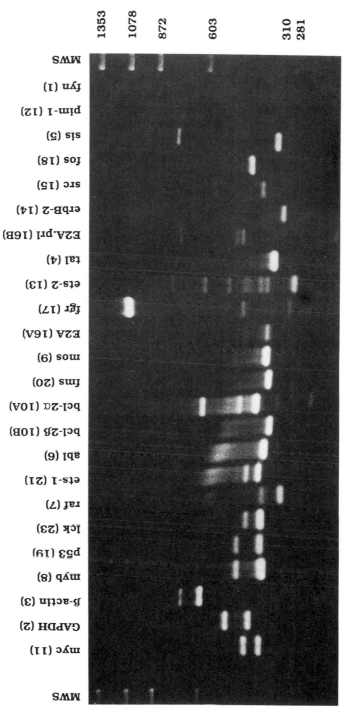

B

FIG. 4. (*continued*) (B) Poly(A)⁺ RNA was purified with oligo(dT)-cellulose from MOLT-4 (human T cell line) total RNA. cDNA was synthesized from a mixture of poly(A)⁺ RNA derived from the equivalent of 10⁶ cells and 5 × 10⁷ molecules of standard RNA. PCR reactions contained 1/100 of the total cDNA mix and PCR primer pairs specific for the genes indicated at the top of each lane. Following 30 PCR cycles, sample buffer was added and half of the sample loaded onto a 2% agarose gel.

463

The system was also tested using RNA from a human T cell line, MOLT4 (Fig. 4B). Each lane contains the product of PCR reactions with aliquots of the same cDNA mix as template and different PCR primer pairs. In this case the products are derived from 5×10^3 cells, and 2.5×10^5 molecules of Oncoquant standard. For the first eight primer pairs (*myc* through *ets-1*) two bands can be seen, derived from the endogenous mRNA and the standard RNA, at the appropriate sizes (see Table I). For *abl* and *mos*, strong standard bands and weak endogenous bands are present. In the case of *ets-2*, several DNA bands are visible in addition to the standard and predicted endogenous bands. Two explanations for this type of result are possible. The extra bands might be due to false priming by cross-reaction with RNA species derived from other genes, that is, artifacts or pseudogenes. The other possibility is that one or more of the extra bands derive from RNAs that are alternative splice products from the *ets-2* gene itself. These possibilities also hold for the *bcl-2α*, *fgr*, and *sis* primers, for which no bands are seen at the predicted molecular weight for the endogenous product, but strong bands are seen elsewhere. The possibility that the unexpected bands are really derived from the same gene can be determined with subsequent Southern blot analysis. A third possibility, that the band is derived from contaminating genomic DNA, can be ruled out based on the size of these unexpected products, and using control reactions containing genomic DNA as template.

In cases in which only products derived from the standard molecules are seen there are two possible explanations. Either the PCR primers do not work efficiently with the endogenous mRNA, possibly due to the presence of secondary structures, or these particular messages are not present at significant levels in these samples. For instance, one would not expect to detect the expression of platelet-derived growth factor (PDGF; product of the c-*sis* gene) or the receptor for epidermal growth factor (EGF; product of the c-*erbB-2* gene) in a T cell tumor. Finally there are primer pairs, like *pim-1* and *fyn*, that do not function with this standard molecule. Apparently there are problems with the efficiency of priming with these oligonucleotides.

To avoid the use of sequences that do not prime efficiently, or which give unexpected patterns following the PCR reactions, we recommend that primer pairs be tested with RNA preparations before the sequences are included in the design of the synthetic RNA and before its construction. We use RNA isolated from a panel of tissues in which we expect the gene in question to be expressed. Interestingly, we find that brain tissue is often a good source for this analysis because many oncogenes we have tested seem to be transcribed at some level in the brain, even when we have not detected their expression elsewhere.

Quantification of β-Actin Levels by Template Titration

To quantify the amount of product generated in our PCR reactions we use ^{32}P-labeled PCR primers and phosphorimaging analysis. The amount of a specific mRNA in a sample can be estimated by the titration of the endogenous and standard RNAs against each other in the cDNA mix. Conditions under which the same radioactive signal is achieved between the standard and endogenous products will then allow one to calculate the number of endogenous mRNA molecules in the RNA preparation. For these experiments, PCR is performed for small numbers of cycles to ensure that the reactions are in an exponential phase. In this section we will assume that the reaction efficiencies for the two templates, endogenous and standard, are the same.

Figure 5 demonstrates this analysis for β-actin mRNA abundance in a pre-B cell tumor. If cDNA mixes contain a constant amount of endogenous RNA and varying amounts of standard RNA, the composition is reflected in the intensities of the PCR product bands following gel electrophoresis (Fig. 5A). When duplicate PCR reactions are performed the band intensities are highly reproducible. Quantification of these bands reveals that although there is little change in the amount of endogenous product, the amount of standard product increases in proportion to its initial concentration (Fig. 5B). The point at which these two curves intersect indicates the number of mRNA molecules in this sample; in this case, $\sim 1.1 \times 10^8$ molecules/μg of total RNA. Conversely, a similar analysis can be done in which the amount of standard RNA is held constant and varying amounts of total RNA are included in the cDNA reaction mix (Fig. 5C); in this case the intersection point gives 10^8 molecules in 0.8 μg or 1.25×10^8 molecules/μg. So from this analysis we can estimate that there are $\sim 1.2 \times 10^8$ β-actin mRNA molecules per microgram of total RNA isolated from this pre-B cell tumor. It should also be noted that the amount of product from the constant amount of standard RNA decreases with increasing concentration of total RNA in the cDNA reaction. Once again this points to the importance of having an internal standard to control for efficiency differences under different reaction conditions.

RNA Purification Yield

In many cases it would be more informative to express the amount of mRNA for a particular gene in terms of the number of molecules per cell. If total RNA is isolated from counted numbers of cells, the cDNA synthesis can simply be done by adding the amount of RNA derived from a certain number of cells for the quantification. However, whereas quantification in terms of molecules per microgram is not affected by the yield of RNA

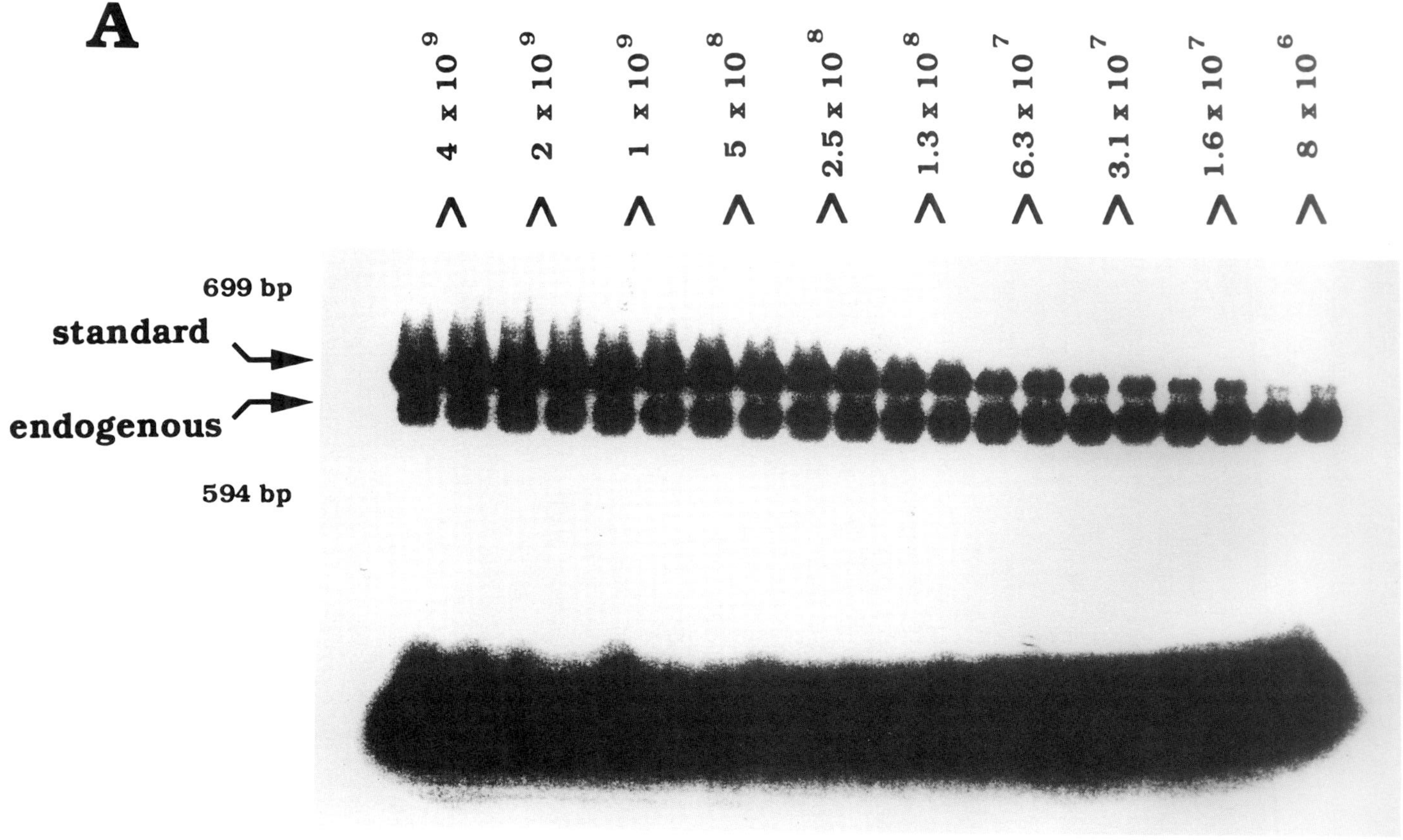

A

8 x 10^6 >

1.6 x 10^7 >

3.1 x 10^7 >

6.3 x 10^7 >

1.3 x 10^8 >

2.5 x 10^8 >

5 x 10^8 >

1 x 10^9 >

2 x 10^9 >

4 x 10^9 >

699 bp
standard

endogenous
594 bp

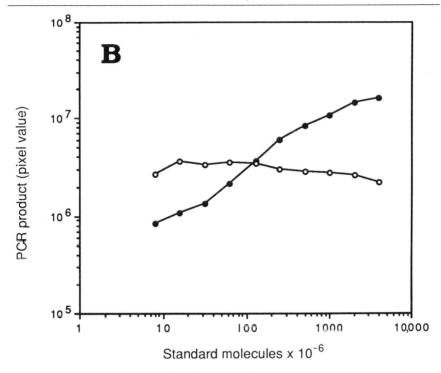

Standard molecules x 10^{-6}

FIG. 5. Quantification using radioactive PCR primers. PCR reactions were done in duplicate, and contained 100,000 cpm of ^{32}P end-labeled 3′ β-actin primer and 1/20 cDNA reaction mix. (A) cDNA was synthesized from a mixture containing the number of standard molecules indicated, and 1.0 μg of total endogenous RNA from a pre-B cell tumor derived from a c-*myc* transgenic mouse.[31] Following 26 PCR cycles, agarose gel sample buffer was added and one-third of the sample run on a 1.5% agarose gel at 80 V for 6 hr. The gel was processed as described in Materials and Methods, and exposed to phosphorimaging overnight. (B) The radioactive signal from the gel in (A) was quantified, and the amount of PCR product derived from standard RNA (●) and endogenous RNA (○) (pixel value is a measure of the number of photons released per unit area) plotted against the amount of standard RNA included in the cDNA reaction mix. Each point represents the average of duplicate reactions. (C) Reactions similar to (A) and (B) were performed except that the amount of standard RNA (●) in the cDNA reaction was held constant (10^8 molecules) and the amount of endogenous RNA (○) was varied.

during the isolation procedure, RNA yield is an important consideration when quantifying in terms of molecules per cell. We determined the yield of intact RNA during our purification procedure by "doping" the cell sample with a known amount of standard RNA and then determining how much was left after purification. Cell cultures were divided in half and standard RNA was added to one of the lysed cell preparations. RNA was prepared from both samples, with or without doping, by our standard

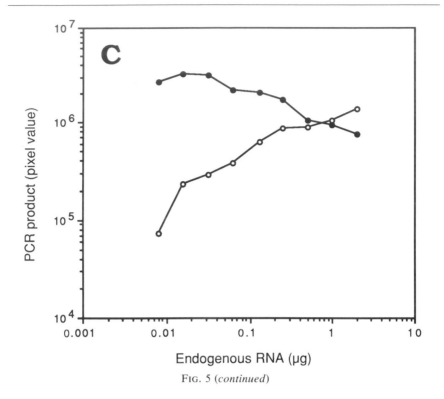

FIG. 5 (*continued*)

procedure. cDNA reactions were then set up with the two samples, and the equivalent amount of standard RNA was added at this point to the nondoped sample. A comparison of the PCR product signal using ^{32}P-labeled primers specific for β-actin showed that the standard products from the doped and nondoped samples were virtually identical, indicating that the isolation procedure we used resulted in a yield of >90% of the RNA. This result is especially important because it means that the standard RNA can be added after RNA preparation, directly before cDNA synthesis, and the results will still be accurate.

Addressing Problems of Efficiency Differences between Templates during β-Actin Quantification

Quantification of mRNA levels using an internal standard controls for efficiency differences from sample to sample and between different primer pairs. However, it is also possible that there would be different reaction efficiencies between the endogenous mRNA and the standard RNA templates because the internal sequences between the priming sites are not the same. There are two potential sources for these efficiency differences. First, there could be inherent differences in the efficiencies of PCR amplification for the standard and endogenous templates. These could arise from differences in secondary structure or length of the sequences located between the priming sites in the two molecules. In the design of our synthetic standard we tried to avoid these potential problems as much as possible. The other source of efficiency differences could result from reaction conditions that differentially affect one template more than the other. Contaminants carried over during RNA preparation or components of the cDNA reaction mix might generate these types of effects.

Problems of efficiency differences can be detected if PCR products are generated with varying numbers of PCR cycles. The amount of product (N_c) generated in the PCR reaction after c cycles is given by the equation

$$N_c = N_i(1 + f)^c \tag{1}$$

where N_i is the initial number of template molecules and f is the reaction efficiency. In a graph of log N_c vs c, the slope will equal log$(1 + f)$, and the y intercept will equal log N_i.

The amount of product generated following c cycles for the endogenous mRNA (E_c) and the standard RNA (S_c) would be

$$E_c = E_i(1 + f_e)^c \tag{2}$$
$$S_c = S_i(1 + f_s)^c \tag{3}$$

In this case the efficiencies for the two template molecules, f_e and f_s, may be different.

Equations (2) and (3) can be combined to give Eq. (4):

$$E_c/S_c = (E_i/S_i)[(1 + f_e)/(1 + f_s)]^c \tag{4}$$

Equation (4) can then be rearranged as

$$\log(E_c/S_c) = \log(E_i/S_i) + c \log[(1 + f_e)/(1 + f_s)] \tag{5}$$

In a graph of log(E_c/S_c) vs c, the slope will equal log$[(1 + f_e)/(1 + f_s)]$, and the y intercept will equal log(E_i/S_i). This is true whether the efficiencies for the two molecules are the same or different. Therefore, if PCR analysis is done by varying the cycle number, and E_c and S_c are measured, then the ratio of E_i/S_i can be calculated from the y intercept. Because we know what the input number of standard molecules is (S_i), the starting number of mRNA molecules (E_i) can be determined.

An example of this type of analysis is presented in Fig. 6. Products from both the endogenous and standard RNA molecules were analyzed by phosphorimaging following varying PCR cycle numbers of the same reaction mix containing RNA isolated from the B cell line WEHI279.

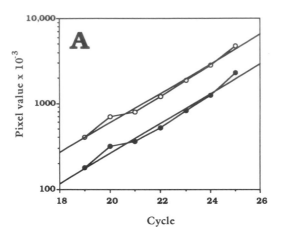

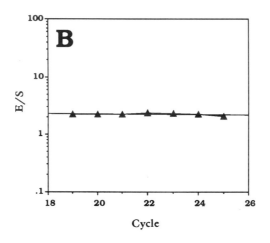

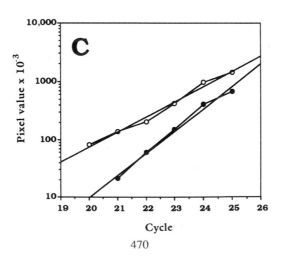

470

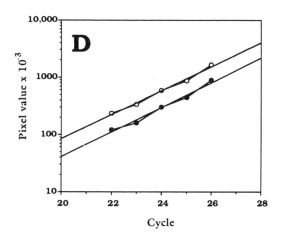

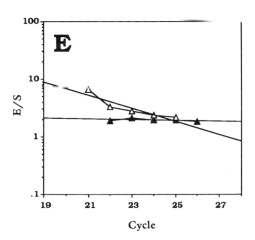

FIG. 6. Quantification to identify efficiency differences between the endogenous and standard molecules. Standard PCR reactions contained primers for β-actin, including 200,000 cpm of end-labeled 3' β-actin primer, and template derived from cDNA mixes containing RNA isolated from 5×10^5 cells. A stock mix was prepared containing all ingredients and aliquotted into individual reaction tubes. After the indicated cycle numbers, tubes were placed on ice and agarose gel sample buffer added. For these experiments, standard RNA was added to the lysed cells before RNA isolation (1000 standard molecules per cell), to avoid inaccuracies associated with RNA yield during purification. (A) RNA was isolated from the mouse mature B cell line WEHI279, and 1/20 of the cDNA mix (1 μl) was used in each PCR reaction. The amount of product derived from the endogenous RNA (○) and standard RNA (●) was measured by phosphorimaging. (B) The data from (A) was analyzed in terms of the E/S ratio to illustrate PCR reaction efficiency differences (see text for details). (C) The same as (A) except that RNA was isolated from the pre-B cell line 70Z/3. (D) Same as (C) except that 1/100 of the cDNA mix (1 μl of a one-fifth dilution in deionized, distilled H_2O) was used as the PCR template. (E) The data from (C) (△) and (D) (▲) are expressed in terms of the E/S ratio.

When the data are plotted on a semilog graph of PCR product (pixel value) versus cycle number, the slope of the curve is related to the reaction efficiency (Fig. 6A). In this case, the slopes of the curves for the endogenous and standard products are virtually identical, indicating that the reaction efficiencies for these two templates are the same. Because the data give straight lines it also demonstrates that the PCR reactions are still in the exponential phase of amplification, during which the reaction efficiencies are constant. When these data are expressed in terms of the ratio of endogenous product to standard product (E/S) versus cycle number the line is nearly horizontal (Fig. 6B). A horizontal line results when the efficiencies of the reactions with the two templates are the same ($f_e = f_s$). In contrast, in the plot of PCR product versus cycle number using RNA from the pre-B cell line 70Z/3, the slopes of the lines from the endogenous and standard templates were quite different (Fig. 6C). This indicates that under these conditions the reaction efficiencies for the two templates were different. However, in Fig. 6D a 1/5 dilution of the cDNA mix used in Fig. 6C was amplified with the same PCR parameters. The use of the cDNA dilution now gives parallel curves, indicating equal efficiencies. Whereas the data from Fig. 6C give an E/S curve with a significant slope, the data from Fig. 6D give a slope ≈ 0 (Fig. 6E).

If significant differences in the slopes of the standard and endogenous curves are observed, resulting in a nonhorizontal E/S curve, it is important to distinguish between the possibilities of effects inherent to the two templates and effects of contaminants. One should try to alter the reaction conditions to see what effect this has on the slopes of the product curves. In the analysis presented in Fig. 6 for 70Z/3, simply diluting the cDNA mix before amplification was sufficient to obtain parallel curves. In general, if dilution generates curves that are more parallel, it would suggest that a contaminant was present that adversely affected one of the templates. However, if the relationship between the slope of the endogenous product and the standard product is maintained following changes in reaction conditions (cDNA dilution, annealing temperature, Mg^{2+} concentration, etc.), it indicates that there is an inherent difference in the reaction efficiencies between the two templates. These inherent differences can be dealt with by extrapolating the E/S curve back to zero cycle numbers. Solving Eq. (5) for $c = 0$, the y intercept will equal the starting number of endogenous molecules (E_i) divided by the starting number of standard molecules (S_i). Because S_i is known, E_i can be calculated. In practice, the best results are obtained when the E/S curve is horizontal.

The number of β-actin mRNA molecules per cell in WEHI279 and 70Z/3 calculated from Eq. (5) and Fig. 6 (closed triangle data) is 2570 and 2750, respectively. The similarity in β-actin abundance between the two cell lines is in good agreement with analysis by other methods. Indeed, β-actin is often used as a control for standardization with these other techniques. β-Actin concentrations of 1000–2000 molecules/cell have previously been reported for chicken fibroblasts and undifferentiated myoblasts.[35–37]

Conclusion

We have described a method for the quantification of mRNA levels for specific genes using PCR and a synthetic RNA as an internal standard. The plasmid Oncoquant I exemplifies the design considerations important for the generation of such a standard RNA. There are several advantages to the use of this procedure for mRNA quantification. First, this method provides results in terms of molecules per cell or molecules per microgram. Quantification in absolute numbers allows easy comparison of results from different experiments or different research groups. Second, combining PCR cycle titration data with the mathematical analysis we have described allows one to detect problems relating to differential amplification efficiencies of the standard and endogenous templates. Without the assessment of PCR reaction efficiencies many potential measurement errors could go undetected. We have discussed the possible causes of efficiency differences and show how these problems can be alleviated. Finally, PCR gives the procedure exquisite sensitivity; we have detected as few as 10,000 standard molecules using β-actin primers by our standard procedure. Because β-actin is expressed at $\sim$2500 molecules/cell, message from as few as 4 cells ($\sim$10 pg total RNA) could be quantified. This kind of sensitivity is crucial for the quantification of message levels in rare cell types, for example, bone marrow cells at early stages of lymphoid development, or when the source of material is limited, for example, a human biopsy. Although we have developed the system for the quantification of a set of mouse and human oncogenes, it is clear that the technique can be applied to any problem in which the measurement of mRNA levels is required.

Acknowledgments

The work described in this chapter was completed while the authors were members of the Basel Institute for Immunology. We thank M. Wiles and J. Salomonsen for critical reading of the manuscript, and all of the members of our institute for helpful discussion. We would also especially like to thank M. Wiles and A. Traunecker for helpful suggestions about PCR conditions, H. R. Kiefer for synthesis of oligonucleotides, A. Schroepel and T. Uchida for the MOLT4 experiment, and C. Steinberg for help with the mathematical analysis. The Basel Institute for Immunology was founded and is supported by F. Hoffman-La Roche, Ltd. (Basel, Switzerland).

[35] R. J. Schwartz and K. N. Rothblum, *Biochemistry* **20,** 4122 (1981).
[36] J. B. Lawrence and R. H. Singer, *Nucleic Acids Res.* **13,** 1777 (1985).
[37] K. Taneja and R. H. Singer, *Anal. Biochem.* **166,** 389 (1987).

[34] Quantification of Polymerase Chain Reaction Products by Affinity-Based Collection

By Ann-Christine Syvänen and Hans Söderlund

In the polymerase chain reaction (PCR) a defined polynucleotide region can be amplified one-billionfold.[1] The PCR is a powerful tool in the analysis of nucleic acids, and it has won widespread use for many different applications. In many cases accurate quantification of the PCR product, as well as an estimate of the initial amount of template prior to the amplification, is desired. The latter question has been difficult to address.

To quantify the amount of specific PCR product it is usually not sufficient to measure the total amount of DNA synthesized in the reaction. Depending on the conditions and experimental design, the PCR can create various amounts of product that do not correspond to the desired template. Such erroneous products may arise by amplification of sequences by essentially random priming or by creation of primer–dimer artifacts.[2] Once a polymerization product has been formed, it becomes a true PCR template and is amplified in subsequent cycles. Consequently, a qualitative step, such as hybridization with a probe internal to the PCR primers, should be included in the analysis.

Here we describe the affinity-based hybrid collection procedure,[3] in which the correct PCR products are quantified using specific hybridization probes and selective isolation by immobilization on a solid matrix.

The efficiency of PCR is high only at low concentrations of template. At high initial template concentrations or at high cycle numbers when a significant amount of template has accumulated, the efficiency of PCR decreases. This effect is due to several factors, such as the increased rate of template reannealing, limiting concentrations of reactants, and inactivation of the DNA polymerase.[3] The decrease in efficiency of the amplification cannot be eliminated, and thus quantification of the initial concentration of template is not possible when too many PCR cycles are employed. Consequently, methods for detection of PCR products should

---

[1] K. B. Mullis and F. A. Faloona, this series, Vol. 155, p. 335.

[2] M. A. Innis and D. H. Gelfand, *in* "PCR Protocols: A Guide to Methods and Applications" (M. A. Innis, D. H. Gelfand, J. J. Sninsky, and T. J. White, eds.), p. 3. Academic Press, San Diego, 1990.

[3] A.-C. Syvänen, M. Bengtström, J. Tenhunen, and H. Söderlund, *Nucleic Acids Res.* **16**, 11327 (1988).

METHODS IN ENZYMOLOGY, VOL. 218

be relatively sensitive to allow quantification before the plateauing effect of PCR becomes dominant.

Internal standards aid in the determination of the initial concentration of template,[4] but it should be noted that the reannealing rate of the template molecules is concentration and sequence dependent, and the efficiency of the amplification depends on the size of the template. Thus the efficiency of amplification of unrelated sequences in the same reaction may vary significantly. Therefore the internal standard used should be as similar as possible to the analyte sequence.[5]

We have also applied the affinity-capture technology for quantification of mixtures of closely related sequences.[6] In this modification of the procedure the relative concentration of two sequences differing from each other, for example, by a point mutation is accurately determined. Both sequences behave identically in the amplification and one of them serves as an internal standard for quantification of the other sequence.

Principle of Affinity Capture

The basic idea of the affinity-capture technologies is to combine the advantageous reaction kinetics of reactions in free solution with the fractionating power and convenient format of solid-phase assays.[7] To facilitate affinity capture, one of the reactants is modified to contain a chemical residue with high affinity for another component. When the latter component is immobilized on a solid support, complexes carrying the affinity tag can easily be removed from the solution for final analysis. We have previously applied this approach in sandwich hybridization assays using a labeled detector probe and a capturing probe modified with biotin residues for subsequent isolation of the hybrids with the aid of an avidin-affinity matrix.[8] Similarly, homopolynucleotides[9] or haptens[8,10] have been used as capturing residues.

For quantification of PCR products applying the affinity-capture principle, one of the PCR primers is modified with a biotin residue at its 5' end.

[4] J. Chelly, J.-C. Kaplan, P. Maire, S. Gautron, and A. Kahn, Nature (London) 333, 858 (1988).

[5] A. M. Wang, M. V. Doyle, and D. F. Mark, Proc. Natl. Acad. Sci. U.S.A. 86, 9717 (1989).

[6] A.-C. Syvänen, K. Aalto-Setälä, L. Harju, K. Kontula, and H. Söderlund, Genomics 8, 684 (1990).

[7] H. Söderlund, Ann. Biol. Clin. 48, 489 (1990).

[8] A.-C. Syvänen, M. Laaksonen, and H. Söderlund, Nucleic Acids Res. 14, 5037 (1986).

[9] D. Gillespie, J. Thompson, and R. Solomon, Mol. Cell. Probes 3, 73 (1989).

[10] K. Mühlegger, H.-G. Batz, S. Böhm, H. van der Eltz, H.-J. Höltke, and C. Kessler, Nucleosides Nucleotides 8, 1161 (1989).

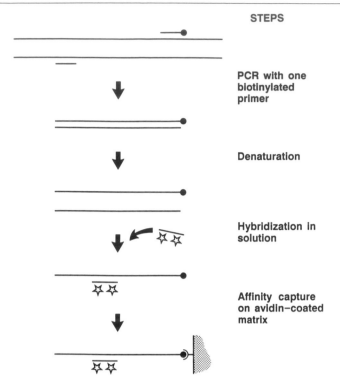

FIG. 1. Principle of the affinity-based hybrid collection method (method A).

As a consequence of PCR one of the strands of the product becomes 5'-biotinylated (Figs. 1 and 2). For further analysis of the PCR product alternative routes are available. In method A (Fig. 1) the PCR product is denatured and allowed to hybridize to a labeled probe complementary to the amplified sequence between the primers. The hybridization mixture is exposed to a matrix carrying avidin or streptavidin. The hybrids formed with the biotinylated strand of the PCR product and the labeled probe, and also the excess of free biotinylated primer, will bind to the matrix. After washing the matrix the amount of bound probe is determined. This approach is used for different purposes, including the detection of human immunodeficiency virus, cytomegalovirus, and papillomavirus.[11]

[11] L. Harju, P. Jänne, A. Kallio, M.-L. Laukkanen, I. Lautenschlager, S. Mattinen, A. Ranki, M. Ranki, V. R. X. Soares, H. Söderlund, and A.-C. Syvänen, *Mol. Cell. Probes* **4,** 223 (1990).

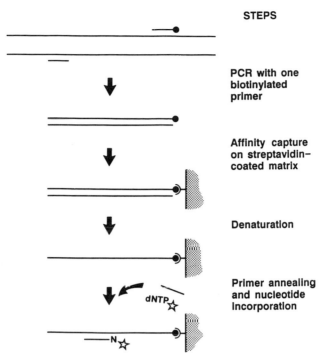

FIG. 2. Principle of the primer-guided nucleotide incorporation method (method B).

Alternatively, in method B (Fig. 2), the biotinylated PCR product is captured on the affinity matrix in double-stranded form, the components of the PCR mixture are removed by washing the matrix, and the unbiotinylated strand of the PCR product is released by alkaline treatment of the matrix. The immobilized DNA strand is now utilized as template for a primer-guided nucleotide incorporation assay, in which a single nucleotide is incorporated by a DNA polymerase. Using a detection step primer, which anneals immediately upstream of a variable site in the template, two forms of an essentially identical sequence can be quantified by measuring the amount of the respective nucleotides incorporated. We have applied this approach for genotyping[6,12,13] and also for quantification of the

[12] A.-C. Syvänen, E. Ikonen, T. Manninen, M. Bengtström, H. Söderlund, P. Aula, and L. Peltonen, *Genomics* **12**, 590 (1992).

[13] A. Jalanko, J. Kere, E. Savilahti, M. Schwartz, A.-C. Syvänen, M. Ranki, and H. Söderlund, *Clin. Chem.* **38**, 39 (1992).

proportion of cells carrying mutations in the N-*ras* gene in samples from patients with acute myeloid leukemia.[14]

Materials and Reagents

Target DNA

Any type of DNA sample, treated as is suitable for PCR amplification,[15] can be analyzed by the methods described here. In the experiments with method A purified cloned DNA served as the target. The human cytomegalovirus (HCMV) DNA template was a 12.2-kb *Hind*III L fragment of the long unique region of the CMV genome cloned into the pAT153 vector.[16] The human immunodeficiency virus type 1 (HIV-1) template was a 1.8-kb DNA segment from the *gag* region of the HIV-1 genome cloned into an expression vector.[17] The plasmids were linearized by restriction enzyme digestion before amplification.

For method B human leukocytic DNA was purified from ethylenediaminetetraacetic acid (EDTA) blood samples.[18] PA-1 cells, which harbor a heterozygous mutation in the second nucleotide of codon 12 of the N-*ras* gene, were obtained from the American Type Culture Collection (ATCC, Rockville, MD; CRL 1572). The DNA was extracted from the cells by standard methods.[19]

Oligonucleotides: Synthesis, Biotinylation, and Labeling

The oligonucleotides were synthesized on an Applied Biosystems (Foster City, CA) 381A DNA synthesizer using the β-cyanoethyl phosphoramidite method.[20] Table I gives the sequences of the oligonucleotides used as PCR primers, hybridization probes, and standards (method A), and detection step primers (method B).

[14] A.-C. Syvänen, H. Söderlund, E. Laaksonen, M. Bengtström, M. Turunen, and A. Palotie, *Int. J. Cancer* **50**, 713 (1992).

[15] R. Higuchi, *in* "PCR Technology: Principles and Applications for DNA Amplification" (H. A. Erlich, ed.), p. 31. Stockton Press, New York, 1989.

[16] J. D. Oram, R. G. Downing, A. Akrigg, A. A. Dollery, C. J. Duggleby, W. G. Wilkinson, and P. J. Greenaway, *J. Gen. Virol.* **59**, 111 (1982).

[17] A. Jalanko, A. Kallio, M. Ruohonen-Lehto, H. Söderlund, and I. Ulmanen, *Biochim. Biophys. Acta* **949**, 206 (1988).

[18] G. I. Bell, J. H. Karam, and W. J. Rutter, *Proc. Natl. Acad. Sci. U.S.A.* **78**, 5759 (1981).

[19] T. Maniatis, E. F. Fritsch, and J. Sambrook, "Molecular Cloning: A Laboratory Manual." Cold Spring Harbor Press, Cold Spring Harbor, New York, 1982.

[20] S. L. Beaucage and M. H. Caruthers. *Tetrahedron Lett.* **22**, 1859 (1982).

TABLE I

OLIGONUCLEOTIDES USED FOR QUANTIFICATION OF POLYMERASE CHAIN REACTION PRODUCTS BY AFFINITY-BASED COLLECTION

Target	Name[a]	Purpose	Size of PCR product (bp)	Sequence (5' to 3')
HCMV[b]	Bio-151	PCR	155	AGCTCTTTCCCGGCCTGGCT
	152	PCR		GCGCGAACATGTAGTCGGCC
	38P/Bio-38P	Probe/standard		GGTTGCTCTTGCTGAGCTGCATGAGCAGCGCGCCGCC
	20P	Probe primer		GGCGGCGGCGCTGCTCATGC
HIV-1[c]	Bio-SK38	PCR	115	ATAATCCACCTATCCCAGTAGGAGAAAT
	SK39	PCR		TTTGGTCCTTGTCTTATGTCCAGAATGC
	SK19/bio-SK-9	Probe/standard		ATCCTGGGATTAAATAAAATAGTAAGAATGTATAGCCCTAC
	20SK	Probe primer		GTAGGGCTATACATTCTTAC
ApoE[d]	Bio-P2	PCR	265	TCGCGGCCCCGGCCTGGTACA
	P3	PCR		GAACAACTGACCCGGGTGGCGG
	D112	Detection step		GCGCGGACATCGAGGACGTG
N-ras[e]	12/13A1	First PCR	109	GGAGCTTGAGGTTCTTGCTG
	12/13B1	First PCR		GCCTCACCTCTATGGTGGGA
	12/13A2	Second PCR		GACTGAGTACAAACTGGTGG
	Bio-12/13B2	Second PCR		CCTATGGTGGCGATCATATT
	D12:2	Detection step		ACTGGTGGTGGTTGGAGCAG

[a] Bio- denotes biotinylation.

[b] The sequence of the HCMV primers was modified from Ref. 3.

[c] The HIV-1 primers have been described in S. Kwok, D. H. Mack, K. B. Mullis, B. Poiesz, G. Erlich, D. Blair, A. Friedman-Klein, and J. Sninsky, *J. Virol.* **61**, 690 (1987).

[d] The apoE primers were designed on the basis of the sequence of the apolipoprotein E (*apoE*) gene published in Y.-K. Paik, D. J. Chang, C. A. Reardon, G. E. Davies, R. W. Mahley, and J. M. Taylor, *Proc. Natl. Acad. Sci. U.S.A.* **82**, 3445 (1985).

[e] The N-ras primers were designed on the basis of the sequence of the N-ras gene published in A. Hall and R. Brown, *Nucleic Acids Res.* **13**, 5255 (1985).

One of each set of PCR primers and the oligonucleotides used as hybridization standards were modified with a biotin residue at their 5′ ends. For biotinylation we use the following procedure: The oligonucleotides are first aminated at their 5′ ends during the synthesis, using the Aminolink 2 reagent (Applied Biosystems). The amino groups (500 μM oligonucleotide) are biotinylated with the water-soluble sulfo-NHS-biotin ester (Pierce Chemical, Rockford, IL) at 50 mM concentration in 100 μl of 0.1 M phosphate buffer, pH 7.5, for 2 h at 37°. The biotinylated oligonucleotide is purified on a reversed-phase C_{18} high-performance liquid chromatography (HPLC) column using a 20-min linear gradient of 5 to 20% acetonitrile in 0.1 M triethylammonium acetate, pH 6.9.[21] It is important to purify the biotinylated oligonucleotide from possible unbiotinylated oligonucleotide and from the excess of biotin present in the reaction. This can also be achieved by means other than HPLC, such as purification from an acrylamide gel.[22] The 5′-biotin residue can also be introduced during the oligonucleotide synthesis using biotinyl phosphoramidite derivatives.[23]

As hybridization probes we use 30 to 70-mer oligonucleotides complementary to the region between the PCR primers of the nonbiotinylated strand of the amplified fragment. The oligonucleotides are labeled by a primer extension reaction from a 20-mer primer complementary to the 5′ region of the probe oligonucleotide.[3] The labeled (^{32}P or ^{35}S) dNTP used is chosen on the basis of the sequence of each probe so that as many labeled dNTPs as possible will be incorporated. A typical labeling reaction consists of 3 pmol of probe oligonucleotide (38-mer for HCMV and 41-mer for HIV-1), 30 pmol of 20-mer probe primer, 66 pmol of [α-^{32}P]dCTP [>3000 Ci/mmol; PB 10205 (Amersham International, Amersham, UK)] or [α-^{35}S]dCTP (>1000 Ci/mmol; SJ 1305) for HCMV or [α-^{32}P]dTTP (>3000 Ci/mmol; PB 10207) or [α-^{35}S]dTTP (>600 Ci/mmol; SJ 307) for HIV-1, and the three other dNTPs at 200 μM concentration in 50 μl of 50 mM Tris-HCl, pH 8.0, 50 mM KCl, 10 mM MgCl$_2$ and 10 mM dithiothreitol (DTT). The reaction mixture is kept at 65° for 15 min and 4 units of *Escherichia coli* DNA polymerase I (the Klenow fragment; Boehringer GmbH, Mannheim, Germany) is added, and the reaction is allowed to proceed for 30 min at 20°. The labeled oligonucleotide is purified from excess labeled dNTP by a suitable method, such as gel filtration on a Sephadex G-25 spin column. Probes with specific activities of 2–8 × 10^7 cpm/pmol are obtained in this procedure. Alternatively, the oligonucleo-

---

[21] M. Bengtström, A. Jungell-Nortamo, and A.-C. Syvänen, *Nucleosides Nucleotides* **9**, 123 (1990).
[22] R. Wu, N.-H. Wu, Z. Hanna, F. Georges, and S. Narang, *in* "Oligonucleotide Synthesis: A Practical Approach" (M. J. Gait, ed.), p. 135. IRL Press, Oxford, 1984.
[23] K. Misiura, I. Durrant, M. R. Evans, and M. J. Gait, *Nucleic Acids Res.* **18**, 4345 (1990).

tide probes can be labeled using polynucleotide kinase or terminal transferase by standard methods.[19] In these cases the probe oligonucleotide is complementary to the biotinylated strand of the amplified DNA fragment.

Affinity Matrices

Avidin-coated polystyrene microparticles (0.77 μm; Fluoricon assay particles) are purchased from IDEXX Corp. (Portland, ME). The binding capacity of the microparticles is >2 nmol of biotinylated oligonucleotide per milligram of particles.[3] Streptavidin-coated magnetic polystyrene beads (Dynabeads M-280, streptavidin; biotin-binding capacity, 300 pmol/mg) are obtained from Dynal AS, Oslo, Norway). The microparticles and the magnetic beads are washed once with 20 mM sodium phosphate buffer (pH 7.5), 0.15 M NaCl, 0.1% (v/v) Tween 20 before use. Microtitration wells (Maxisorb; Nunc, Roskilde, Denmark) are coated with streptavidin (Porton Products, Salisbury, UK, or Sigma, St. Louis, MO) by passive absorption.[24] The biotin-binding capacity of the wells is about 5 pmol of biotin per well. Such plates are also commercially available, for instance, from Labsystems (Helsinki, Finland).

Other Reagents

[^{3}H]dNTPs are obtained from Amersham ([^{3}H]dATP, TRK 633; [^{3}H]dCTP, TRK 576; [^{3}H]dGTP, TRK 627; [^{3}H]dTTP, TRK 576). *Thermus aquaticus* (*Taq*) DNA polymerase is from Perkin-Elmer (Norwalk, CT) or Promega (Madison, WI) and modified T7 DNA polymerase (Sequenase) from United States Biochemical Corporation (Cleveland, OH).

Methods

Amplification with Biotinylated Primers

The amount of primers used in the PCR is determined by the biotin-binding capacity of the affinity matrix to be used for capturing the amplified DNA. With the avidin microparticles the primers are used at 1 μM concentration, with the streptavidin-coated magnetic beads at 0.5 μM concentration and with the microtitration wells at 0.1 μM concentration[11] or, alternatively, the unbiotinylated primer is used at 1 μM and the biotinylated primer at 0.1–0.2 μM concentration.[13] Varying amounts of target DNA can be used. For instance, from 30 to 3 × 10^7 molecules of plasmid DNA or 10–100 ng (corresponding to 3 × 10^3–3 × 10^4 molecules) of human

[24] T. Ternynck and S. Avrameas, this series, Vol. 183, p. 469.

genomic DNA is successfully amplified per reaction. The PCR mixture (100 μl) consists of dATP, dCTP, dGTP, and dTTP (0.2 mM each), 20 mM Tris-HCl (pH 8.8), 15 mM $(NH_4)_2SO_4$, 1.5 mM $MgCl_2$, 0.1% (v/v) Tween 20, 0.01% (w/v) gelatin, and 2.5 units of *Taq* DNA polymerase. The PCR cycles (20–30) are carried out for 1 min at 96°, 1 min at a primer annealing temperature, which is determined primarily on the basis of the sequence of the primer,[25] and 1 min at 72°. In some cases it is advantageous to improve the quality of the PCR product by using nested primers. In this case a first amplification is carried out with a pair of unbiotinylated primers at 1 μM concentration, and an aliquot (10 μl of a 1 : 300 dilution) of the first PCR product is amplified with a pair of nested primers, one of which is biotinylated, at 0.5 μM concentration. We used this approach when quantifying N-*ras* mutations by method B (see below).

Detection Method A: Quantitation Using Labeled Probe

Experimental Details

Hybridization in Solution. One-tenth (10 μl) of the PCR product is analyzed without any purification steps in parallel with a series of standards containing known amounts of biotinylated oligonucleotide. The PCR product and 0.01–0.04 pmol (2 × 10^5 cpm) of the labeled probe are denatured and added to the hybridization mixture to a final volume of 50 μl containing 0.6 M NaCl, 20 mM sodium-phosphate (pH 7.5), 1 mM EDTA, 0.1% (w/v) sodium dodecyl sulfate (SDS), and 0.02% (w/v) Ficoll, 0.02% (w/v) polyvinylpyrrolidone, 0.02% (w/v) bovine serum albumin (1× Denhardt's solution) in an Eppendorf tube. The hybridization reaction is allowed to proceed for 1–2 hr at 65°.

Affinity Capture on Avidin-Coated Polystyrene Microparticles. After hybridization 20 μl of a 5% (w/v) suspension of avidin-coated polystyrene particles is added to the reaction mixture, the tubes are vortexed briefly, and the capturing reaction is allowed to proceed for 30 min at 20°. Because of their small size (0.77 μm) the particles stay in suspension without mixing during the incubation. The particles are collected by centrifugation at 6000–13,000 g for 2 min and the supernatant is discarded. The particles are washed three times with 1 ml of 15 mM NaCl, 1.5 mM sodium citrate (0.1× SSC), 0.2% (w/v) SDS at 50°. The washing solution, preheated to 50°, is added and the particles are resuspended by vortexing the tubes vigorously for 15–30 sec, followed by incubation in a water bath for 5 min at 50°. The tubes are centrifuged as above and the washing solution is

[25] S. L. Thein and R. B. Wallace, *in* "Human Genetic Diseases: A Practical Approach" (K. E. Davis, ed.), p. 33, IRL Press, Herndon, Virginia, 1986.

discarded. The particles are easy to handle when the solution contains a detergent and the concentration of NaCl is not above 0.15 M. When the label is ^{32}P, the washed microparticles are directly measured by Cerenkov radiation in a liquid scintillation counter. Alternatively, with ^{32}P, and when ^{35}S is used, the labeled probe is released from the captured hybrid by treating the particles with 200 μl of 50 mM NaOH for 5 min at 20°, and the eluted radioactivity is measured. If a xylene-based scintillation fluid is used the polystyrene particles dissolve in it and the measurement can be done without elution. Nonradioactive detection can also be employed, measuring either bound or eluted probe.

Affinity Capture in Microtitration Wells. The hybridization mixture is transferred to streptavidin-coated microtitration wells and the capturing reaction is allowed to proceed for 2 hr at 37° by gently shaking the plates. The wells are washed three times with 200 μl of 0.1 × SSC, 0.2% (w/v) SDS for 5 min at 50°, and the hybridized probe is released by treating the wells for 5 min at 20° with 60 μl of 50 mM NaOH. The eluted radioactivity is measured in a liquid scintillation counter. The microtitration well format is well suited for measuring probes labeled directly or indirectly with enzymes or fluorescent groups.[8,26]

Results

Quantification of Polymerase Chain Reaction Products. The biotinylated PCR product is quantified after hybridization to a labeled probe by comparing the amount of radioactive hybrid captured on an affinity matrix to that captured in a series of standards with known amounts of biotinylated oligonucleotide. Figure 3 shows two standard curves obtained with varying amounts (10^7–10^{11} molecules) of the biotinylated SK 19 oligonucleotide hybridized to the ^{32}P-labeled SK 19 probe, applying the microparticle and microtitration well formats, respectively. A linear relationship between the amount of radioactivity collected and the amount of standard added to the hybridization reaction is observed over a 10,000-fold range.

Six samples containing varying amounts of HIV-1 DNA (from 30 to 3×10^5 molecules) were amplified for 20 PCR cycles using the primers Bio-SK38 and SK39. The amount of biotinylated PCR products obtained was determined by affinity-based hybrid collection on microparticles (Table II). The calculated mean efficiencies of the PCR cycles for the six samples show that the efficiency of PCR depends on the initial amount of DNA template, the process being more efficient at low inputs of template.

[26] P. Dahlen, A.-C. Syvänen, P. Hurskainen, M. Kwiatkowski, C. Sund, J. Ylikoski, H. Söderlund, and T. Löfgren, *Mol. Cell. Probes* **1**, 159 (1987).

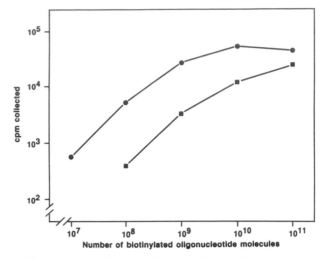

FIG. 3. Hybridization standard curves obtained with the affinity-based hybrid collection method using avidin-coated polystyrene microparticles (●) and streptavidin-coated microtitration wells (■) as matrix. Biotinylated SK 19 oligonucleotide was used as standard and ^{32}P-labeled SK 19 as probe. (Figure 3 is modified from Ref. 11.)

The efficiency at different stages of the PCR process was also determined. Varying amounts of HCMV DNA (300 to 3×10^7 molecules) were amplified for 10, 15, 20, and 25 cycles using the primers Bio-152 and 151. The amount of amplified DNA in each sample was quantified using the biotinylated 38P oligonucleotide as hybridization standard. The calculated

TABLE II
QUANTIFICATION OF AMPLIFIED HUMAN IMMUNODEFICIENCY VIRUS 1 DNA
BY AFFINITY-BASED HYBRID COLLECTION

Initial number of HIV plasmid molecules	Amount of amplified DNA		Mean efficiency[b] (%)
	Radioactivity collected (cpm)	Number of molecules[a]	
3×10^5	41,000	2.6×10^9	57
3×10^4	36,000	1.7×10^9	73
3,000	31,000	1.3×10^9	91
300	600	1.1×10^8	90
30	530	9.5×10^7	91
0	120	—	—

[a] Determined from the standard curve in Fig. 1. Data from Ref. 11.
[b] The efficiencies were calculated using the formula $\sqrt[n]{\text{amplification level}} - 1$; n = number of cycles; according to R. Saiki, S. Scharf, F. Faloona, K. B. Mullis, G. T. Horn, and H. A. Erlich, *Science* **230,** 1350 (1985).

TABLE III
EFFICIENCY OF POLYMERASE CHAIN REACTION AS FUNCTION OF CYCLE
NUMBER AND AMOUNT OF TEMPLATE

Initial number of CMV plasmid molecules	Mean efficiency (%) for cycle number				
	1–10	11–15	16–20	21–25	1–25
3×10^7	86	22	20	5.3	40
3×10^6	92	41	19	3.7	45
3×10^5	96	56	33	22	57
3×10^4	—	86[a]	57	26	66
3000	—	—	97[b]	28	81
300	—	—	106[b]	26	87

[a] Mean efficiency of cycles 1–15.
[b] Mean efficiency of cycles 1–20, data from Ref. 3. The efficiencies were
calculated as in Table II.

efficiencies at different stages of PCR show that the process is efficient during the first 10 cycles and at low initial template concentrations. With a high template input the efficiency of the last five cycles is low. Comparison of the data in Tables II and III shows that the efficiency of PCR is similar with the HIV-1- and HCMV-specific reagents. Thus it is clear that with high template concentrations, either in the original sample or as a result of the amplification, the efficiency of PCR decreases significantly. This leads to the well-known plateauing effect of PCR; the amount of PCR product is essentially independent of the initial template concentration if too many cycles are employed. With 20–25 cycles the amount of PCR product is related to the initial template concentration only over relatively narrow range of less than 10,000 input molecules.[3,11]

Detection Method B: Quantification by Primer-Guided Nucleotide Incorporation

Method B quantifies the proportion of two DNA sequences, which are present in a mixture and differ from each other by a single nucleotide. The ratio of the two sequences determined by the primer-guided nucleotide incorporation method reflects the initial proportion of the two sequences independently of variations in efficiency of the PCR amplification, depending on the initial amount of template and cycle number (cf. Table III).

Experimental Details

Affinity Capture in Microtitration Wells. Two 10-μl aliquots of each amplified sample and 40 μl of 0.15 M NaCl, 20 mM sodium phosphate buffer (pH 7.5), 0.1% (v/v) Tween 20 are transferred to streptavidin-coated microtitration wells. The biotinylated amplified DNA is allowed to bind to the wells for 1.5 hr at 37° with gentle shaking. The wells are washed three times with 200 μl of a solution containing 40 mM Tris-HCl (pH 8.8), 1 mM EDTA, 50 mM NaCl, and 0.1% (v/v) Tween 20 at 20°. The unbiotinylated strand of the captured DNA fragment is removed by treating the wells with 100 μl of 50 mM NaOH for 5 min at 20°, after which the wells are washed three times as described above.

Detection Step Primer Extension Reaction in Microtitration Wells. Fifty microliters of reaction mixture containing the detection step primer at 0.2 μM and the labeled dNTP to be incorporated at 0.05–0.1 μM concentration, for example, the primer D112 together with either [³H]dTTP or [³H]dCTP in two separate reactions for apolipoprotein E (apoE), and 0.05 unit of *Taq* DNA polymerase in 20 mM Tris-HCl (pH 8.8), 15 mM (NH$_4$)$_2$SO$_4$, 1.5 mM MgCl$_2$, 0.1% (v/v) Tween 20, 0.01% (w/v) gelatin is added to each well. The primer annealing and extension reactions are carried out simultaneously for 10 min at 50° and the wells are washed as described above. The elongated primer is released by treating the wells with 60 μl of 50 mM NaOH for 5 min at 20° and the eluted radioactivity is measured in a liquid scintillation counter. Here also nonradioactive detection using nucleoside triphosphates modified with haptens or fluorescent groups can be used.[6]

Affinity Capture on Streptavidin-Coated Magnetic Beads. Eighty microliters of PCR product and 20 μl of 0.75 M NaCl in 100 mM sodium phosphate buffer (pH 7.5), 0.5% (v/v) Tween 20 are added to 300 μg of streptavidin-coated magnetic beads in Eppendorf tubes. The beads are suspended by shaking the tubes briefly and the capturing reaction is allowed to proceed for 30 min at 20° with occasional mixing. To achieve efficient capture the reaction mixture should contain at least 0.1 M NaCl. The beads are separated from the reaction mixture in a magnetic test tube rack (MPC-E; Dynal AS), which takes less than 5 sec. The beads are washed three times with 1 ml of 0.15 M NaCl in 20 mM sodium phosphate buffer (pH 7.5), 0.1% (v/v) Tween 20. The washing solution is added, the tubes are vortexed briefly, the particles are collected with the aid of the magnet, and the solution is discarded. The captured DNA is denatured by treating the beads twice with 100 μl of 50 mM NaOH and the beads are washed as described above.

Detection Step Primer Extension Reaction on Magnetic Beads. The beads carrying the denatured amplified DNA fragment are suspended in 10 μl of 50 mM NaCl, 20 mM MgCl$_2$, 40 mM Tris-HCl (pH 7.5), containing the detection step primer (D12 : 2 in the case of N-*ras*) at 0.2 μM concentration. The primer is allowed to anneal to the DNA template for 10 min at

TABLE IV
GENOTYPING OF APOLIPOPROTEIN E BY DETERMINATION OF
RATIO OF $\varepsilon2$ AND $\varepsilon4$ ALLELES

Genotype of sample	Radioactivity incorporated (cpm)[a]		T_{cpm}/C_{cpm}
	T reaction	C reaction	
$\varepsilon2/\varepsilon2$	13,700	270	51
$\varepsilon2/\varepsilon4$	14,100	5,800	2.4
$\varepsilon4/\varepsilon4$	200	12,100	0.017

[a] L. Harju (unpublished data, 1990).

37°. One microliter of 0.1 M dithiothreitol, 0.1 unit of modified T7 DNA polymerase, and 3 pmol of either labeled dNTP ([^{3}H]dGTP or [^{3}H]dATP in the case of N-*ras*) are added to a final volume of 15 μl. The primer extension reaction is allowed to proceed for 5 min at 37° and the beads are washed six times as above. The elongated primer is released by treating the beads with 100 μl of 50 mM NaOH for 5 min at 20° and the eluted radioactivity is measured in a liquid scintillation counter.

Results

Genotyping by Determination of Allelic Ratios. Three DNA samples, known to represent three common genotypes ($\varepsilon2/\varepsilon2$, $\varepsilon2/\varepsilon4$, and $\varepsilon4/\varepsilon4$) of apoE[6] were analyzed for the nucleotide variation in the first nucleotide of codon 112 by the nucleotide incorporation method. The DNA was amplified with the primers Bio-P3 and P4 and the nucleotide at the variable site (T in $\varepsilon2$ or C in $\varepsilon4$) was determined using the primer D112 and applying the microtitration well format of the test. In samples from homozygous individuals, either ^{3}H-labeled T or C is incorporated, and in samples from heterozygous individuals both ^{3}H-labeled T and C are incorporated. The ratio of ^{3}H incorporated in the T reaction to ^{3}H incorporated in the C reaction unequivocally determines the genotype of each sample (Table IV).

Quantification of Minority Point Mutations. Mixtures of normal human DNA and varying amounts of DNA isolated from PA-1 cells, which carry a heterozygous mutation (G > A) in the second nucleotide of codon 12 of the N-*ras* gene, were analyzed by the primer-guided nucleotide incorporation assay. The DNA was amplified using the primers 12/13A2 and Bio-12/13B2 and analyzed applying the magnetic bead format of the test with the primer D12:2. The ratio of ^{3}H incorporated in the reaction with the

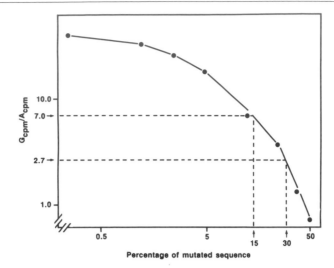

FIG. 4. Quantification of the amount of mutated N-*ras* DNA in two cell samples by the primer-guided nucleotide incorporation method with the aid of a standard curve constructed with known amounts of PA-1 cells mixed with normal human DNA. G_{cpm}/A_{cpm}, Ratio of ^{3}H incorporated in the reaction with dGTP to ^{3}H incorporated in the reaction with dATP. (A.-C. Syvänen, H. Söderlund, and A. Palotie, unpublished data, 1990.)

normal nucleotide (G) to the ^{3}H incorporated in the reaction with the mutated nucleotide (A) was determined. The G_{cpm}/A_{cpm} ratio is inversely proportional to the percentage of mutated N-*ras* DNA in the sample (Fig. 4). Comparison of the G_{cpm}/A_{cpm} ratio obtained in two samples known to carry a G > A mutation in the second nucleotide of codon 12 of the N-*ras* gene[14] allows quantification of the amount of mutated N-*ras* DNA present in the samples (Fig. 4).

Discussion

The choice of affinity matrix depends on the intended application of the method. The microtitration well offers the advantage of a convenient test format, in which existing semiautomatic laboratory equipment, such as multichannel pipettors and washers as well as fully automatized laboratory robots, can be used. This matrix is also well suited for nonradioactive detection with enzymes[6,8] or fluorescent groups[26] as markers because the test format is identical to that of enzyme immunoassays or fluoroimmunoassays. A drawback of the microtitration well as affinity matrix is its limited biotin-binding capacity, which is due to the small surface area (1 cm²) of the matrix and unfavorable reaction kinetics of the capturing reaction.

From our experience the polystyrene microparticles perform reliably in the affinity-based collection methods. Because of the extremely small size of the particles the binding area is large (70 cm²/mg of particles), resulting in high biotin-binding capacity and kinetics approaching those of reactions in solution. The major drawback of these particles is the requirement for separation by centrifugation. We have noticed some batch-to-batch variation with respect to the optimal g value during centrifugation and in the requirement for low NaCl concentration for easy handling of the particles.

The magnetic polystyrene particles are a compromise between the two above-mentioned formats with the advantage of rapid and easy separation, but with intermediate biotin-binding capacity and reaction kinetics.[27]

The affinity-based hybrid collection procedure (method A) gives a (semiquantitative) estimate of the presence or the absence of a specific DNA sequence in the analyzed sample. The PCR process proceeds exponentially only over a 1000- to 10,000-fold range when the initial amount of target is low. Consequently a sensitive detection method is required for analysis of the PCR product after a minimum number of PCR cycles and for measurement of the PCR product at stages of the process when the amount of product is still exponentially increasing. We therefore use relatively long (30 to 70-mer) oligonucleotide probes in the solution hybridization reaction. This gives the advantage of introducing more than one labeled dNTP into each probe molecule. Furthermore, the critical conditions and risk for strand displacement[28] typical for hybridization reactions with short oligonucleotides are avoided. For exact quantification of the PCR product by affinity-based hybrid collection relevant standards are required. The amount of hybridized probe does not directly reflect the amount of target present, because the target coverage of the probe is higher with low amounts than with high amounts of target, where reannealing of the strands becomes significant.

The primer-guided nucleotide incorporation method (method B) allows quantification of DNA sequences that contain single nucleotide variations, deletions, or insertions, and are present in the sample as a mixture with the normal sequence. Thus the method can easily be applied for genotyping of samples with defined alleles present in a 2 : 0 (homozygote) or 1 : 1

[27] A. Jungell-Nortamo, A.-C. Syvänen, P. Luoma, and H. Söderlund, *Mol. Cell. Probes* **2**, 281 (1988).

[28] C. Green and C. Tibbets, *Nucleic Acids Res.* **9**, 1905 (1981).

(heterozygote) ratio. Mutations present in a minority of the analyzed DNA molecules can also be quantitatively determined because the ratio of label incorporated in the reaction with the normal to that incorporated in the reaction with the mutated nucleotide is inversely proportional to the amount of mutated sequence present.[14,29]

A factor directly influencing the ratio between the incorporated labels is the specific activity of the labeled dNTP used. This factor is easily accounted for. More problematic is the amount of dNTP unspecifically incorporated by the DNA polymerase. *Taq* DNA polymerase and the modified T7 DNA polymerase perform satisfactorily with less than 2% misincorporation, whereas the *E. coli* DNA polymerase I (the Klenow fragment) yields significant misincorporation (20%) of labeled dNTP.[6] This is expected because the exonuclease activity of the Klenow enzyme will allow exchange of nucleotides within the primer.[30] The background misincorporation even using *Taq* or T7 DNA polymerase limits the sensitivity of detection. This background is partially due to other dNTPs present as impurities in the labeled dNTPs, and is thus sequence dependent. The specificity of the DNA polymerase may also be reduced by the chemical moiety by which the label is introduced into the dNTP. Consequently, standards with known amounts of mutated sequence mixed with the normal sequence are required for accurate quantification.

In the primer-guided nucleotide incorporation assay one sequence functions as an internal standard to its mutated counterpart. Any nucleic acid sequence can be quantified after addition of a known amount of the same DNA or RNA sequence carrying a nucleotide variation to the sample.[31] Thus sequence-specific quantification by affinity-based collection offers an accurate method for quantification of PCR products as well as of the original amount of template amplified by the PCR.

[29] A. Suomalainen, P. Kollman, J.-N. Octave, H. Söderlund, and A.-C. Syvänen, *Eur. J. Hum. Gen.* **1,** in press (1992).
[30] A. Kornberg, *Science* **131,** 1503 (1960).
[31] E. Ikonen, T. Manninen, L. Peltonen, and A.-C. Syvänen, *PCR Meth. Applic.* **1,** 234 (1992).

Section III

Methods for Detecting DNA–Protein Interaction

[35] Cross-Species Polymerase Chain Reaction: Cloning of
TATA Box-Binding Proteins

By MICHAEL GREGORY PETERSON and ROBERT TJIAN

Introduction

The method described here is an application of the polymerase chain reaction (PCR)[1] to isolate genomic or cDNA fragments of a gene from one species, given that the polypeptide or DNA sequence of the gene from a second species is known. The example used here is the TATA box-binding protein, a critical component for eukaryotic RNA polymerase II (Pol II) transcription initiation that is found in species as diverse as yeast and humans.

Reconstituted basal level transcription requires at least six protein fractions (TFIIA, B, D, E, F, and H) in addition to RNA Pol II (reviewed in Ref. 2). The TFIID fraction contains an activity that binds to the TATA box. The mammalian TATA-binding protein (TBP) has not been amenable to purification by conventional chromatography, or by DNA affinity chromatography. Thus there had been no obvious method available to clone mammalian TBP directly. An activity in yeast that can functionally substitute for the human TBP has been purified and cDNA clones isolated.[3,4] However, all attempts to use the yeast cDNA clone to obtain a human clone by cross-hybridization had failed. One possibility was that TBP in yeast and humans had diverged structurally to such an extent that this strategy becomes untenable. However, the strong conservation of function between yeast and humans suggested that the polypeptide sequences are similar. Thus, it seemed more reasonable that species-specific codon usage results in DNA sequences with an insufficient degree of homology for this approach to be successful. Although changing the entire coding region of the yeast TBP sequence to reflect mammalian codon usage could represent a legitimate strategy, this would be highly impractical. Using shorter oligonucleotide hybridization probes applying mammalian codon usage against a number of regions of the protein sequence may also be feasible; however, the problem of false positives due to nonspecific hybridization would

[1] R. K. Saiki, S. Scharf, F. Faloona, K. B. Mullis, G. T. Horn, H. A. Erlich, and N. Arnheim, *Science* **230,** 1350 (1985).
[2] A. G. Saltzman and R. Weinmann, *FASEB J.* **3,** 1723 (1989).
[3] S. Hahn, S. Buratowski, P. A. Sharp, and L. Guarente, *Cell* **58,** 1173 (1989).
[4] D. M. Eisenmann, C. Dollard, and F. Winston, *Cell* **58,** 1183 (1989).

probably be intolerable. Specificity can be improved dramatically by using the PCR and the time taken to analyze "positives" can be significantly reduced.

Principle of the Method

The procedure used for cross-species PCR amplification based on a known amino acid sequence is a derivative of the mixed oligonucleotide-primed amplification of cDNA (MOPAC) technique.[5] In MOPAC, one starts with a stretch of known amino acid sequence. Primer sites are selected based on polypeptide sequences encoded by minimally degenerate oligonucleotides. The region between the primers is then amplified by PCR. In the original report, PCR was performed with the large fragment of *Escherichia coli* DNA polymerase, necessitating the use of lower temperatures, which resulted in more nonspecific priming. Thus, typically a fragment of the correct size was not discernable and a probe against a region between the primers was needed to screen for the correct product after cloning.

The technique described here varies in a number of important respects. The first is that heat-stable *Taq* DNA polymerase[6] is used instead of *E. coli* DNA polymerase, allowing higher annealing and polymerization temperatures. This is important as it greatly increases the specificity of the primers to such a point that the PCR products can be directly visualized under ultraviolet (UV) on agarose gels. This circumvents the requirement for a probe located between the priming sites, which in many cases may not be available.

The second difference is that the exact polypeptide sequence (and therefore DNA sequence) of the gene or cDNA to be amplified is not known. All that is known is the sequence of the homologous gene from another closely or even distantly related species. The new strategy requires only that short stretches of the polypeptide sequence be identical or nearly identical between the two species. As these regions may not be known in advance, a "shotgun" approach is utilized, in which a number of primers are used in all possible combinations in order to maximize the probability of finding two short conserved stretches within the polypeptide sequence. If two such regions (six to seven amino acids each) do not exist within the protein, this method will not work. However, many proteins do

[5] C. C. Lee, X. Wu, R. A. Gibbs, R. G. Cook, D. M. Muzny, and C. T. Caskey, *Science* **239**, 1288 (1988).
[6] R. K. Saiki, D. H. Gelfand, S. Stoffel, S. J. Scharf, R. Higuchi, G. T. Horn, K. B. Mullis, and H. A. Erlich, *Science* **239**, 487 (1988).

show this (low) degree of sequence similarity. Where the protein sequence from more than one species is known, the most highly conserved regions of the polypeptide sequence can be used for cross-species PCR. However, in the example described here the sequence from only one species was known. Potentially important structural features likely to be conserved in amino acid sequence were therefore targeted as PCR priming sites.

Third, we have made use of "guessmer" primers and the nucleotide inosine to reduce the degeneracy of the primers. This may be important for generating sufficient amounts of fragment for direct visualization by gel electrophoresis and limiting the background of nonspecific priming. Although guessmer primers may result in a number of mismatches between the primer and template, as described here and in the original paper on MOPAC, a few mismatches can be tolerated. Finally, a modification of the PCR technique termed "touchdown PCR"[7] greatly increases the yield of specific PCR products versus the background artifactual products.

Materials and Reagents

Polymerase Chain Reaction Equipment and Primers

Polymerase chain reactions are performed in a programmable thermal controller (MJ Research, Inc.). Oligonucleotides are synthesized using a DNA synthesizer 380B (Applied Biosystems, Foster City, CA). No further purification of the oligonucleotides was found to be necessary for PCR. The sequences of the primers are shown in Table I.

Enzymes and Reagents

Taq DNA polymerase (AmpliTaq, 5 units/ul) is purchased from Perkin-Elmer Cetus (Norwalk, CT); T4 DNA polymerase, polynucleotide kinase, and T4 DNA ligase are purchased from New England BioLabs (Beverly, MA). Deoxynucleoside triphosphates (ultrapure) are purchased from Pharmacia (Piscataway, NJ). NuSieve GTG agarose is purchased from FMC (Rockland, ME).

Working Stocks

Primers: 0.25 μg/μl in 1× TE [10 mM Tris-HCl (pH 8.0), 1 mM ethylene-diaminetetraacetic acid (EDTA)]

PCR buffer (10×): 100 mM Tris-HCl (pH 8.4 at room temperature), 500 mM KCl, 15 mM MgCl$_2$

dNTP mix: dATP, dCTP, dGTP and dTTP (2.5 mM each in 1× TE)

---

[7] R. H. Don, P. T. Cox, B. J. Wainwright, K. Baker, and J. S. Mattick, *Nucleic Acids Res.* **19**, 4008 (1991).

TABLE I
OLIGONUCLEOTIDE PRIMERS USED IN THIS STUDY[a]

Number	Species	Sequence	Peptide	Orientation
1	D	5′-ATGCG(CT)ATCCG(CT)GAGCC(CT)AA	MRIREPK	Sense
2	D	5′-CG(CT)AAGTACGC(CT)CG(CT)ATCAT	RKYARII	Sense
3	D	5′-AAGATCGG(ACT)TTCGC(CT)GC(CT)AA	KIGFAAK	Sense
4	D	5′-TTCAAGAT(CT)CAGAACAT(CT)GT	FKIQNIY	Sense
5	D	5′-GA(CT)GT(CG)AAGTTCCC(CT)ATCCG	DVKFPIR	Sense
6	D	5′-GGGAA(CG)AGCTC(AGT)GGCTCGTA	YEPELFP	Antisense
7	D	5′-GGCTT(CG)ACCAT(AG)CGGTA(AG)AT	IYRMVKP	Antisense
8	D	5′-TA(AG)ATCTCCTC(AG)CGCTGCTT	KQREEIY	Antisense
9	D	5′-AG(CG)AC(AG)GGGTAGAT(AG)GCCTC	EAIYPVL	Antisense
10	D	5′-CTACTT(AGCGGAACTC(CG)GA(CG)A	LSEFRKM	Antisense
11	H	5′-TGTGATGTGAA(AG)TT(CT)CC(ACT)AT	CDVKFPI	Sense
12	H	5′-GATGTGAAGTTCCC(ACT)AT(CT)(AC)G	DVKFPIR	Sense
13	H	5′-GGGAACAGCTC(AGT)GG(CT)TC(AG)TA	YEPELFP	Antisense
14	H	5′-GGCTTCACCAT(CT)C(GT)(AG)TA(AG)AT	IYRMVKP	Antisense
15	H	5′-ACAATCTT(AGT)GG(CT)TT(CG)ACCAT	MVKPKIV	Antisense

[a] Parentheses indicate an equal mixture of bases at that position. D, *Drosophila;* H, human.

DNA template: 100 ng/μl in 1× TE

Paraffin oil: Mineral oil, light, white

DDW: Deionized and distilled water, autoclaved (any high-quality water)

Molecular weight markers: *Hae*III-digested pBluescript SK$^+$ [Stratagene (La Jolla, CA) plasmid], 100 ng/μl in 0.5× TBE (0.045 M Tris-borate, 0.001 M EDTA) gel loading buffer, use 5 μl (0.5 μg) per lane

Methods

Choosing Target DNA Template

Suitable DNA targets for cross-species PCR might be cDNA from poly(A)$^+$ RNA or total RNA known to contain the desired message, a cDNA library from mRNA known to contain the desired message, or genomic DNA. Potential problems may arise using all of these sources. Total RNA known to contain the desired message is probably the safest starting material. Using this source of template does not assume that the mRNA is polyadenylated. The total RNA can be reverse transcribed with random primers and potential secondary structure problems may be overcome using heat-stable reverse transcriptase. A shortcoming of using total RNA instead of poly(A)$^+$ RNA is that the desired message is effectively diluted by approximately 10-fold. If the message is rare, genomic DNA may indeed represent a more abundant source of the target gene, especially when using organisms with small genomes. A potential problem with genomic DNA is the presence of introns, the size and frequency of which is species dependent. Although cDNA libraries may not faithfully represent the RNA population of the starting material, it is useful to perform cross-species PCR on the cDNA library before screening to ascertain whether the clone is indeed in the library to be screened for potential clones.

Preparation of Drosophila Genomic DNA

Drosophila melanogaster (Oregon R strain) are homogenized gently in 0.1 M Tris-HCl (pH 9.0), 0.1 M EDTA, 1% (w/v) sodium dodecyl sulfate (SDS), and 1% (w/v) diethyl pyrocarbonate (DEPC) (10 μl/fly) and incubated at 70° for 30 min. Potassium acetate is added to 1 M and incubated for a further 30 min on ice. Debris is pelleted by centrifugation at 15,000 g for 15 min at 4°. DNA is precipitated from the supernatant by addition of 0.5 vol of 2-propanol and collected by centrifugation at 15,000 g for 5 min at room temperature.

Preparation of λ DNA from HeLa cDNA Library

A HeLa λgt10 cDNA library (500,000 pfu; original complexity, 10^7) is used to infect 0.5 ml of *E. coli* BNN102 plating cells and grown at 37° in 100 ml L broth, 10 mM MgCl$_2$ until lysis occurs (at 7 hr). λ DNA is then prepared as described.[8] This procedure gives approximately 100 μg of λ DNA. A simpler method that harvests the entire complexity of the library is to prepare λ DNA from a portion of an amplified library. For example, approximately 10 μg of λ DNA can be prepared by phenol–chloroform extraction of 50 μl of an amplified library supernatant. This is sufficient for approximately 100 PCR reactions as described below.

Designing Oligonucleotide Primers

When designing primers for cross-species PCR, a number of factors need to be considered. Knowledge of protein structure and function can play an important role in determining the location of the PCR primers. If no knowledge exists of the important structural regions of the molecule

---

[8] J. Sambrook, E. F. Fritsch, and T. Maniatis, "Molecular Cloning: A Laboratory Manual," 2nd Ed. Cold Spring Harbor Press, Cold Spring Harbor, New York, 1989.

(i.e., the sequence of only one species is known and it is not known which structural features are likely to be the most conserved) the positioning of the primers will be determined solely on the basis of the polypeptide sequences that are encoded by minimally degenerate oligonucleotides. The length of the oligonucleotides that have been successfully used varies from 17-mers to 23-mers, 20-mers being used in the example shown here. In the original report on MOPAC,[7] 17-mers were used, but in this case a probe was needed to identify the correct product. We have successfully used 17-mers with *Taq* polymerase, but in this case an internal primer was used in a second round of PCR to obtain a single product (discussed below). Thus peptides between six and eight residues in length encoded by minimally degenerate oligonucleotides are suitable. With more knowledge, such as the sequence from more than one species, or knowledge of structure–function relationships, the regions to be scanned for 6- to 8-mer peptides encoded by minimally degenerate oligonucleotides can be narrowed. In the example described here, the known conserved function of the TBP was its ability to bind to the TATA box. Thus potential DNA-binding domains were likely to be conserved. The only available sequence (from yeast) had three structural features that limited the search for potential priming site targets. The first is that the C-terminal 60% of the molecule is almost entirely composed of an imperfect direct repeat. This, combined with two other features (the observation that this region is highly basic in nature, and that it contains a short stretch showing some homology with the σ factor), leads to the idea that this region represents a novel DNA-binding motif. Thus the region showing weak similarity with the σ factor, and the surrounding basic regions, were scanned for 7-mer peptides (or greater) encoded by minimally degenerate oligonucleotides.

Finally, the degeneracy of the primers used can be limited by using either known codon biases, or by using inosine. Two related potential problems may arise if the oligonucleotide primers are too degenerate. The first is that insufficient product may be generated to observe directly by ethidium staining and UV exposure of an agarose gel. The second is that the level of nonspecific priming may be so great that it obscures the correct product. Strikingly, very degenerate primers can yield observable quantities of the correct product even though the theoretical maximum amount of product based on the molar amount of the correct oligonucleotide is below the level of visual detection. This is due to "correct" priming by oligonucleotides with multiple mismatches with the target sequence. However, we have always chosen to limit the degeneracy of primers by using some guesses based on codon usage or, more recently, by using

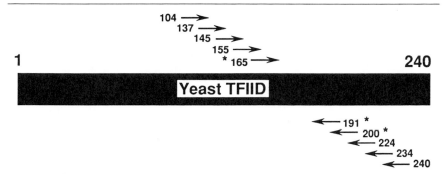

FIG. 1. Scheme of primers based on the yeast TBP amino acid sequence that were used for PCR on *Drosophila* DNA. Arrows indicate the position and direction of primers. Numbers indicate the first (top or sense arrows) or last (bottom or antisense arrows) amino acids of the primers. An asterisk denotes primers that yielded PCR products that were characterized and found to be correct.

inosine. We typically insert guesses near the 5' end of the oligonucleotide, as anecdotal evidence suggests that mismatches at these positions are more readily tolerated. Inosine can be usefully inserted in the third codon position in place of all nucleotides to limit the degeneracy of the oligonucleotide primer.

We first amplified *Drosophila* TBP by cross-species PCR as a "stepping stone" to the human TBP. Using *Drosophila* as an "intermediate" species was useful for two reasons. The first is that *Drosophila* has a much smaller genome and a stronger codon bias and thus is more amenable to this technique. The second is that the sequence of *Drosophila* TBP will provide valuable knowledge, allowing oligonucleotide primers to be designed against sequences absolutely conserved from yeast to *Drosophila* (and presumably humans). Ten degenerate oligonucleotide probes were designed for PCR of *Drosophila* TBP based on the yeast TBP polypeptide sequence (Fig. 1). Each 20-mer probe is between 4- and 12-fold degenerate, and contains 6 to 8 guesses (i.e., not every codon base is used at that position). *Drosophila* has a strong codon bias and this was utilized in designing the oligonucleotides. Where there was a very strong bias for one or two codons, these were used instead of all possible codons. For amino acids with a weak or no codon bias, all possible codons were utilized.

Choosing Cross-Species Polymerase Chain Reaction Conditions

The reaction conditions for cross-species PCR differ from those of conventional PCR in one important respect: the annealing temperature

A

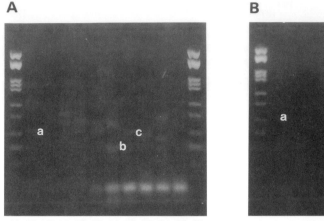

B

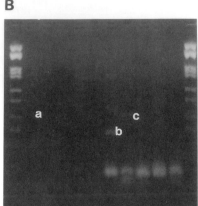

M 16 17 18 19 20 21 22 23 24 25 M   M 16 17 18 19 20 21 22 23 24 25 M

FIG. 2. Agarose gel analysis of PCR products from reactions 16 through 25. (A) Standard PCR conditions were used; (B) touchdown PCR conditions were used. The letters a, b, and c denote fragments that are of the predicted size based on the yeast TBP sequence. The molecular weight markers are *Hae*III-digested pBluescript SK⁺ [sizes (bp): 608, 468, 434, 290, 267, 254, 174, 143, 102, 80/79, 47, (18), (11)].

of the correct primer to the target is uncertain because (1) the oligonucleotides are degenerate and thus have significantly different calculated melting temperatures and (2) tolerating a small number of mismatches between the primer and "correct" target sequence may be desirable. Thus, standard PCR conditions can be used, except that a higher concentration of primer is included to compensate for degeneracies, and the annealing temperature needs to be determined experimentally. We first tried a low annealing temperature, 30° (for 30 sec). This temperature is 30° below the calculated melting temperature for a 20-mer oligonucleotide of 50% GC content [using the formula $T_m = 4(G + C) + 2(A + T)$]. An extension time of 1 min at 72° was chosen using the "rule of thumb" that *Taq* polymerase requires 1 min per kilobase. Subsequent reactions were then performed using annealing temperatures of 40, 45, and 50° (for 30 sec). Surprisingly, the effect of shifting the annealing temperature 20° was not dramatic, with perhaps only 30% of the fragments generated at 30° being absent at 50° (Fig. 2 shows the result at 45°). In addition, there was perhaps only a slight difference in the background smear. These same PCR reactions were repeated using a new modification to the PCR procedure termed "touchdown PCR."[7] In this procedure, the annealing temperature is decreased over the course of the PCR reaction. This is certainly the method of choice for cross-species PCR for two reasons. The first is that as the required annealing temperature is unknown, this effectively allows a range of annealing temperatures to be tested in the one reaction. The second is that, as described later, touchdown PCR selects the product with the most correct priming reactions, which will likely be the desired products. With touchdown PCR parameters that can be varied are the following:

1. Maximum annealing temperature: This might be the calculated T_m for a perfect match with the most GC-rich oligonucleotide in the mix.

2. Minimum annealing temperature: This might be below the calculated T_m of the most AT-rich oligonucleotide in the mix.

3. The rate of decrease of the annealing temperature per cycle: Theoretically, dropping the temperature 1° every second cycle will give a 1024-fold advantage to a correct primer over an incorrect primer with a 5° lower T_m.

Standard Cross-Species Polymerase Chain Reaction Conditions

In Fig. 2 A and B, two sets of PCR reactions are compared. The reagents in each set of reactions are identical. The difference is that the annealing temperature in Fig. 2A was held constant at 45° for 30 cycles, whereas in Fig. 2B the annealing temperature was lowered 1° every second cycle from 55 to 45°, followed by 10 cycles at 45°.

When performing cross-species PCR, the same care to avoid cross-contaminations must be used as for standard PCR. As described below, species of unknown origin can be amplified from minute amounts of contaminating material. For amplification of *Drosophila* TBP the 10 primers used were assayed in all combinations of sense and antisense pairs (5 sense and 5 antisense), giving a total of 25 reactions to perform. As the reaction constituents differ only by the primers, a mix can be made of all other components. Performing a contamination control for each set of primers requires an additional 25 reactions. If this is desired, a second mix containing all the constituents except the primers and template DNA can be set up. Typically 0.5-ml microcentrifuge tubes are used for PCR. Having labeled 25 microcentrifuge tubes, add 40 μl (2 drops) of paraffin oil to each tube. This covers the sample to stop evaporation. Then add 2 μl (0.5 μg) of the appropriate primer as described below:

Tube number	Primer number	Tube number	Primer number
1	1 and 6	14	3 and 9
2	1 and 7	15	3 and 10
3	1 and 8	16	4 and 6
4	1 and 9	17	4 and 7
5	1 and 10	18	4 and 8
6	2 and 6	19	4 and 9
7	2 and 7	20	4 and 10
8	2 and 8	21	5 and 6
9	2 and 9	22	5 and 7
10	2 and 10	23	5 and 8
11	3 and 6	24	5 and 9
12	3 and 7	25	5 and 10
13	3 and 8		

Make the following mix (26× for 25 reactions):

Constituent	1× (μl)	26× (μl)
DDW	37.5	975
Buffer (10×)	5	130
dNTPs	2	52
DNA	1	26
Taq polymerase	0.5	13
Total	46	1196

Add 46 μl of the mix to each microcentrifuge tube. Spin briefly in microcentrifuge to mix.

Polymerase Chain Reaction Programs

   Standard

Step 1: 94° for 80 sec
Step 2: 94° for 40 sec
Step 3: 45° for 30 sec
Step 4: 72° for 1 min
Step 5: Cycle to step 2, 29 more times
Step 6: 72° for 2 min
Step 7: 4° indefinite

After step 6 the tubes can be removed.

   Touchdown Polymerase Chain Reaction

Step 1: 94° for 80 sec
Step 2: 94° for 40 sec

Step 3: 55° for 30 sec

Step 4: 72° for 1 min

Step 5: Cycle to step 2, lower the annealing temperature 1° every second cycle for 20 cycles, that is, until an annealing temperature 45° is reached

Step 6: 10 further cycles (steps 2 to 4), using an annealing temperature of 45°

Workup and Further Analysis

Following PCR, the aqueous layer under the paraffin is transferred to a new 1.5-ml microcentrifuge tube. Then add 5 μl 3 M sodium acetate and 125 μl of ethanol, mix, and spin in a microcentrifuge for 5 min at 4° (this is a standard ethanol precipitation). A small pellet should be visible. Take off the supernatant, spin briefly in centrifuge, take off the last of the supernatant, vacuum dry briefly, dissolve in 8 μl of DDW, add 2 μl of 0.5× TBE loading dye, and load each sample into one well (2.5 mm wide) of a 4% (w/v) NuSieve/0.5× TBE agarose gel. The gels shown in Fig. 2 were run approximately 5 cm.

Figure 2 shows the products from reactions 16 to 25 only, for both standard PCR and touchdown PCR. Having determined the predicted sizes of the correct products based on the yeast polypeptide sequence, fragments of approximately the correct size were located. Three fragments in Fig. 2A are within a few base pairs of the calculated molecular size (marked with a, b, and c). As two of these share a common primer, we concentrated our efforts on these. Note that during the cloning of *Drosophila* TBP, only the standard PCR reactions were performed. Thus we did not notice that fragment b was of a much greater intensity than the other fragments. The first set of PCR reactions was performed using an annealing temperature of 30°. After having repeated the same set of reactions using annealing temperatures of 40 and 50° and finding that the same two (b and c) fragments were present at all temperatures, these two fragments were excised for further analysis. Because these fragments share a common primer (primer 5), fragment b should be contained within fragment c and thus, using PCR primers 5 and 6, the identity of these fragments can be confirmed. The excised fragment c was melted at 65° and 2 μl of the agarose solution was used in two PCR reactions under the conditions described above, the first containing primers 5 and 7 (a control) and the second primers 5 and 6. Controls lacking target DNA were also included. Fragments were amplified using 20 cycles of 94° for 40 sec, 50° for 30 sec, and 72° for 1 min. Both the predicted shorter and longer PCR products were obtained, confirming the identity of these fragments.

The PCR fragments were excised from the gel and the DNA recovered by phenol extraction of the melted gel slice, followed by ethanol precipitation. Fragments were prepared for cloning into M13 by creating flush ends with T4 DNA polymerase (*Taq* polymerase typically leaves a 1-bp 3′ overhang) followed by kinasing with polynucleotide kinase. Fragments were then ligated with *Sma*I-digested and dephosphorylated M13mp18, and transformed into *E. coli*. The sequence obtained by the chain termination method is shown in Fig. 3. The sequence is similar to the yeast TBP sequence in this region and encodes a polypeptide differing by just five amino acids. Across the region spanned by the internal PCR primer, we find that the amino acid sequences are identical. The primer successfully used in this case contains two mismatches with the target sequence. The sequence of a *Drosophila* cDNA later showed that primer 7 falls within a region differing by one amino acid between yeast and *Drosophila* (K to R), resulting in a 3-bp mismatch and a total of four mismatches (denoted by asterisks). Primer 5 contained one mismatch with the target. This indicates that a number of mismatches can be tolerated in a primer of 20 nucleotides, although the exact position and spacing of the mismatches may be important.

Having identified conserved amino acids between yeast and *Drosophila* TBP, five further oligonucleotide primers were designed for PCR of human TBP. Once again, guessmers based on mammalian codon usage were utilized. However, for these primers, "guesses" were restricted to the 5′ half of the oligonucleotide and degeneracies to the 3′ half of the oligonucleotide. This results in the primers containing a minimum of 11 consecutive matches at the "priming" end and a maximum of 3 mismatches at the 5′ end, assuming that the human polypeptide sequence is identical to the *Drosophila* and yeast sequences in these regions. The positions of the five "human" primers are shown in Fig. 3.

The conditions used for PCR of fragments of human TBP were similar to those used for *Drosophila* above. The six combinations of the five primers were performed on both λgt10 DNA from a HeLa cDNA library and human genomic DNA (100 ng), using the following program:

Step 1: 94° for 1 min and 20 sec
Step 2: 94° for 40 sec
Step 3: 40° for 30 sec
Step 4: 72° for 15 sec
Step 5: Cycle to step 2, 34 more times
Step 6: 4°

The combination of primers 12 and either 13, 14, or 15 produced specific PCR products of the predicted size from the cDNA template. Using the

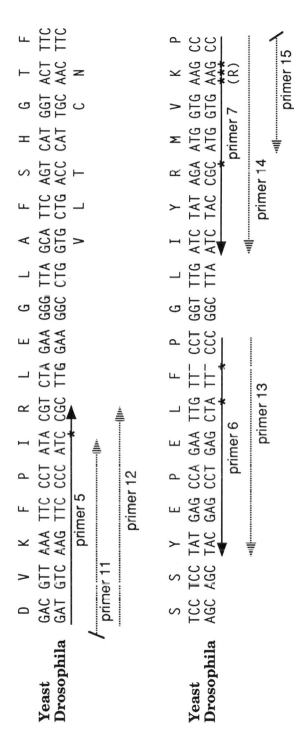

		D	V	K	F	P	I	R	L	E	G	L	A	F	S	H	G	T	F
Yeast		GAC	GTT	AAA	TTC	CCT	ATA	CGT	CTA	GAA	GGG	TTA	GCA	TTC	AGT	CAT	GGT	ACT	TTC
Drosophila		GAT	GTC	AAG	TTC	CCC	ATC	CGC	TTG	GAA	GGC	CTG	GTG	CTG	ACC	CAT	TGC	AAC	TTC
												V	L		T		C	N	

primer 5

primer 11

primer 12

		S	S	Y	E	P	E	L	F	P	G	L	I	Y	R	M	V	K	P
Yeast		TCC	TCC	TAT	GAG	CCA	GAA	TTG	TT⁻	CCT	GGT	TTG	ATC	TAT	AGA	ATG	GTG	AAG	CC
Drosophila		AGC	AGC	TAC	GAG	CCT	GAG	CTA	TT⁻	CCC	GGC	TTA	ATC	TAC	CGC	ATG	GTG	AAG	CC

primer 6

primer 13

primer 14

primer 7   (R)

primer 15

FIG. 3. Derived sequence of the *Drosophila* TBP PCR product compared with yeast TBP. The yeast sequence shown is translated into single amino acid code whereas the *Drosophila* translation is only shown where it differs from yeast. Primers used to amplify the *Drosophila* sequence are denoted by solid arrows, and differences between the primer sequence and the target *Drosophila* DNA sequence are denoted by an asterisk. The position and orientation of primers used to amplify the human TBP are denoted by dotted arrows. Primers 11 and 15 extend beyond the sequence shown here (see Table I for details).

conditions described above, these fragments were the most abundant products in the reactions. The combination of primers 12 and either 13 or 14 resulted in a number of discrete products from genomic DNA, two of which were of the predicted size. The largest fragment generated from HeLa cDNA (116 bp with primers 12 + 15) was reamplified to generate more product for cloning into M13 (as described above). Sequence analysis of 18 M13 subclones revealed that no sequences similar to the yeast and *Drosophila* TBP sequences had been cloned. This was apparently due to the cloning of nonspecifically amplified sequences that contaminated the correct PCR product. To regain specificity in the PCR reaction, the 116-bp product was reamplified using primer 12 and the internal primer 14. The correct size product was subcloned into M13 and sequenced as described above. This resulted in 7 out of 18 subclones bearing sequences similar to the yeast and *Drosophila* TBP sequences. Interestingly, not all of the sequences were the same. Evidently, the PCR reaction had been contaminated with DNA from yeast and as many as three other unknown sources. This at once points to the sensitivity of PCR and the need to avoid even the smallest amounts of contamination.

Discussion

As described above, the PCR technique is extremely sensitive, which means that the utmost care must be taken to avoid cross-contamination of samples, especially from micropipetting equipment. Physically separating work areas and equipment used for setting up the reactions and the subsequent analysis of the PCR products is perhaps the best way to avoid cross-contamination.

As an initial screen of potentially correct PCR products, PCR can be repeated on the fragment with the primers individually to ensure that the product is dependent on both primers. The product can be excised from the gel and a second round of PCR performed directly on the melted gel slice. The same strategy of melting the gel slice and repeating the PCR can be utilized to obtain large amounts of the product for cloning, and so on.

Finally, the modification of the PCR technique termed touchdown PCR[7] is extremely useful for any PCR technique involving the use of mixed oligonucleotide primers. Unlike some previously described procedures for using mixed oligonucleotides, this technique specifically selects for primers having the greatest similarity with the target sequence. This can dramatically improve the ratio of correct signal to incorrectly primed noise. This also allows the technique to be extended to situations in which the size of the correct PCR product is not known. One such situation occurs when PCR is being used to amplify a fragment of a cDNA based on short stretches of amino acid sequence obtained by microsequencing.

Conclusion

Having obtained the complete sequence of *Drosophila*[9] and human[10] TBPs, we are now in a position to determine the matches between the primers used in this study and the actual target sequence. This analysis shows that of the 10 regions of the yeast amino acid sequence chosen for designing PCR primers, 3 show perfect identity between the 2 species and 3 others show 1 amino acid sequence difference between the 2 species. Two of the three primers successfully used were regions showing perfect matches and the third successful primer was to a region containing one amino acid substitution. The other primer to a region of perfect identity that did not obviously give correct PCR products contained four mismatches with the target due to "bad" guesses. The use of inosine instead of a guess may overcome this problem. Inosine has been successfully used for a number of applications similar to that described here and appears to be the method of choice. Also, limiting "guesses" to the 5' end of the primers may be helpful. If we had chosen to use a fully degenerate primer instead of using a guessmer for this region, then the primer would have had 1728-fold degeneracy. In summary, the success of this technique appears to be closely tied to the likelihood of there being short, six- to eight-amino acid regions showing identity between the known sequence and the target sequence.

Acknowledgments

We thank Brian Dynlacht for the *Drosophila* genomic DNA. This work was supported in part by an NIH grant to R.T. M.G.P. is a Special Fellow of the Leukemia Society of America.

[9] T. Hoey, B. D. Dynlacht, M. G. Peterson, B. F. Pugh, and R. Tjian, *Cell* **61,** 1179 (1990).
[10] M. G. Peterson, N. Tanese, B. F. Pugh, and R. Tjian, *Science* **248,** 1625 (1990).

[36] Magnetic DNA Affinity Purification of Yeast Transcription Factor

By ODD S. GABRIELSEN and JANINE HUET

General Introduction

Studies of transcription factors have profited from the fact that many of these proteins are able to recognize a DNA target and bind to a specific DNA sequence much more strongly than to nonrelated DNA sequences. The ratio of association constants for specific to nonspecific DNA binding is often in the range of 10^3–10^5, meaning that it is possible to concentrate the protein of interest selectively on a solid support containing the specific recognition sequence and reduce the association of contaminating DNA-binding proteins by adding an excess of soluble nonspecific DNA. This is the principle of DNA affinity chromatography. Thus, use of immobilized plasmids[1] or oligonucleotides[2] harboring the specific sequence has allowed efficient purification of several sequence-specific DNA-binding proteins. This development represents a significant improvement over classical purification techniques and has made possible studies of transcription factors in highly purified form.[3,4]

The method presented here is a further development of this affinity principle.[5] The main improvement is that columns and chromatography techniques are replaced by the more rapid magnetic solid-phase technology, a separation principle now used in solid-phase DNA sequencing,[6–8] mRNA purification,[9–11] *in vitro* mutagenesis,[12] and in many immunological

[1] S. Camier, O. Gabrielsen, R. Baker, and A. Sentenac, *EMBO J.* **4**, 491 (1985).

[2] J. T. Kadonaga and R. Tjian, *Proc. Natl. Acad. Sci. U.S.A.* **83**, 5889 (1986).

[3] M. R. Briggs, J. T. Kadonaga, S. P. Bell, and R. Tjian, *Science* **234**, 47 (1986).

[4] M. R. Montminy and L. M. Bilezikjian, *Nature (London)* **328**, 175 (1987).

[5] O. S. Gabrielsen, E. Hornes, L. Korsnes, A. Ruet, and T. B. Øyen, *Nucleic Acids Res.* **17**, 6253 (1989).

[6] T. Hultman, S. Stahl, E. Hornes, and M. Uhlen, *Nucleic Acids Res.* **17**, 4937 (1989).

[7] M. Uhlen, *Nature (London)* **340**, 733 (1989).

[8] J. Wahlberg, J. Lundberg, T. Hultman, and M. Uhlen, *Proc. Natl. Acad. Sci. U.S.A.* **87**, 6569 (1990).

[9] K. S. Jakobsen, E. Breivold, and E. Hornes, *Nucleic Acids Res.* **18**, 3669 (1990).

[10] C. Albretsen, K. H. Kalland, B. I. Haukanes, L. S. Havarstein, and K. Kleppe, *Anal. Biochem.* **189**, 40 (1990).

[11] E. Hornes and L. Korsnes, *Genet. Anal. Techniques Applic.* **7**, 145 (1990).

[12] T. Hultman, M. Murby, S. Stahl, E. Hornes, and M. Uhlen, *Nucleic Acids Res.* **18**, 5107 (1990).

and cell-separation techniques.[13,14] What is gained is primarily speed of operations and the possibility of working close to conditions in solution. The latter implies that protein adsorption occurs with kinetics approaching binding to DNA in solution, which by itself contributes to the speed of the whole procedure. The binding occurs in a few minutes, magnetic separation takes seconds, and washes and elution take a few minutes. Speed can be a crucial parameter, especially when unstable gene regulatory proteins are to be purified extensively. As for the previous column methods, the magnetic DNA affinity variant was found to be a powerful purification technique.[5]

This chapter describes the preparation of magnetic DNA affinity beads and their use in the purification of sequence-specific DNA-binding proteins. The protein chosen for illustration of the method is yeast transcription factor τ (or TFIIIC), one of the most complex transcription factors characterized so far (reviewed in Geiduschek and Tocchini-Valentini[15] and in Gabrielsen and Sentenac[16]). We present an improved variant of the previously reported magnetic DNA affinity purification method.[5] Starting with a partially purified protein fraction (purity < 0.2%), essentially pure preparations of yeast transcription factor τ (TFIIIC) were obtained in less than 30 min with a single adsorption step. Improved purity and yields were obtained by saturating the DNA affinity beads with the specific DNA-binding protein and by including competitor DNA only during washing of the beads. With the same starting material, this 30-min procedure yielded higher purity than the normal method, which uses columns and requires several days. The procedure should work for any high-affinity sequence-specific DNA-binding protein after optimalization of binding and elution conditions.

Principle of Method

Figure 1 outlines the principle of the method. An isolated DNA fragment harboring a specific recognition sequence for the protein of interest is labeled at one end through a fill-in reaction incorporating a biotinylated deoxyribonucleotide. This allows the subsequent attachment of the fragment to Dynabeads coated with streptavidin (Dynabeads M-280 streptavidin; Dynal A/S, Oslo, Norway). The Dynabeads are uniform, superpara-

[13] S. Funderud, K. Nustad, T. Lea, F. Vartdal, G. Gaudernack, P. Stenstad, and J. Ugelstad, in "Lymphocytes: A Practical Approach" (G. G. B. Klaus, ed.), p. 55. IRL Press, Oxford, 1987.

[14] T. Lea, F. Vartdal, and K. Nustad, J. Mol. Recogn. 1, 9 (1988).

[15] E. P. Geiduschek and G. P. Tocchini-Valentini, Annu. Rev. Biochem. 57, 873 (1988).

[16] O. S. Gabrielsen and A. Sentenac, Trends Biochem. 16, 412 (1991).

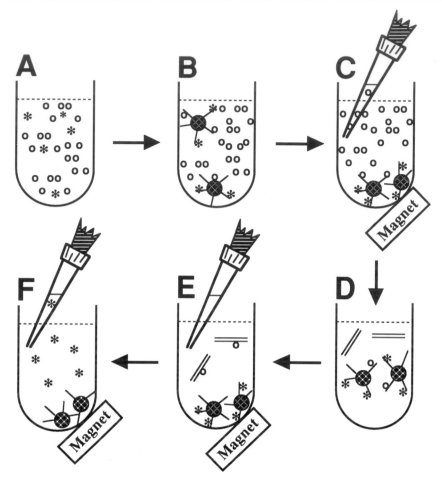

FIG. 1. Schematic representation of the principle of magnetic DNA affinity purification. (A) Starting material, a partially purified protein fraction containing a sequence-specific DNA-binding protein (*) and a large excess of other proteins, including nonspecific DNA-binding proteins (○). (B) A specific recognition sequence immobilized on magnetic DNA affinity beads bind preferentially to the specific DNA-binding protein. (C) A strong magnet separates in seconds the magnetic DNA affinity beads with the specific protein adsorbed from the vast majority of contaminating proteins. (D) The beads are incubated with excess nonspecific competitor DNA (=) to displace contaminating DNA-binding proteins from the affinity beads. (E) A magnet separates the specific protein bound to the affinity beads from nonspecific proteins bound to competitor DNA. (F) Resuspension in high salt buffer followed by magnetic separation liberates the specific protein in essentially pure form.

magnetic, polystyrene beads with chemically bound high-quality streptavidin. The beads have high stability, low particle–particle interaction, and high dispersibility. The biotin–streptavidin complex is extremely strong ($K_{ass} = 10^{15}\ M^{-1}$)[17] and resistant to high concentrations of salt and urea. The resulting magnetic DNA affinity beads have a high capacity for binding the sequence-specific DNA-binding protein in question. Because the affinity beads are nonporous, adsorption and desorption of proteins occurs on the surface of the beads, which is kinetically favorable.

For purification, a protein fraction containing the sequence-specific DNA-binding proteins (Fig. 1A) is incubated with DNA affinity beads using a ratio of specific protein to immobilized DNA that nearly saturates the affinity beads. During competition between specific and nonspecific DNA-binding proteins for limited amounts of sites, the specific protein will be favored if it has the strongest affinity for the recognition sequence immobilized on the beads (Fig. 1B). After a short incubation, a strong magnet is placed against the wall of the tube. In seconds this separates the magnetic beads, with the specific protein adsorbed, from the contaminating proteins remaining in solution (Fig. 1C). Some nonspecific proteins may still bind the affinity beads. The inclusion of excess competitor DNA in subsequent washes will, however, greatly improve purity (Fig. 1D). The distribution of proteins with general DNA affinity between solid and liquid phase will be determined by the relative concentrations of DNA in the two phases. High relative excess competitor will cause these proteins to be enriched in the liquid phase (Fig. 1E). After a few washes of the affinity beads, the specifically bound protein is eluted by resuspending the beads in a buffer of high ionic strength that dissociates the DNA-bound protein. The beads are removed by magnetic separation, leaving a protein fraction highly enriched for the sequence-specific DNA-binding protein (Fig. 1F). Essentially pure proteins can normally be obtained by proper loading and competitor treatment. The method is rapid enough to allow binding, repeated washes, and elution in less than one-half hour.

Materials and Reagents

Enzymes and Chemicals

Restriction enzymes: New England BioLabs (Beverly, MA)
DNA polymerase I (Klenow fragment): New England BioLabs
One of the following biotinylated deoxyribonucleotides: Bio-7-dATP or
    Bio-14-dATP (GIBCO-Bethesda Research Laboratories, Gaithers-

---

[17] M. Wilchek and E. A. Bayer, *Anal. Biochem.* **171**, 1 (1988).

burg, MD), Bio-4-dUTP or Bio-11-dUTP (Sigma, St. Louis, MO), Bio-16-dUTP (Boehringer Mannheim, Indianapolis, IN)
Streptavidin: Sigma

Buffers

TE buffer: 10 mM Tris-HCl, 1 mM ethylenediaminetetraacetic acid (EDTA)
TEN buffer: 10 mM Tris-HCl, 1 mM EDTA, and 100 mM NaCl
TGED buffer: 20 mM Tris-HCl (pH 8.0), 1 mM EDTA, 10% (v/v) glycerol, 1 mM dithiothreitol, and 0.01% (v/v) Triton X-100, with NaCl added to the indicated concentration

Special Equipment

Dynabeads M-280 streptavidin, 10 mg/ml: Dynal A/S (Oslo, Norway)
Neodymium–iron–boron permanent magnet: As a magnet, we have used a magnetic particle concentrator designed to hold six microcentrifuge tubes (Dynal A/S)

Methods

Preparation of DNA Affinity Beads

Comments on Strategy

DNA harboring the specific recognition sequence can be either in the form of a DNA fragment prepared by restriction enzyme digestion of a plasmid, in the form of a duplex oligonucleotide, preferentially polymerized,[2] or the product of a scaled-up polymerase chain reaction (PCR) using biotinylated oligonucleotide primers.[6,12] The protein we have studied, yeast transcription factor τ, interacts with a split intragenic promoter and protects more than eight helical turns of DNA.[1,15,16] We have therefore preferred to use affinity beads constructed from DNA fragments harboring entire tRNA genes. A convenient source of DNA fragment is a plasmid with the fragment cloned into a polylinker sequence. Because the preparation of DNA affinity beads for protein purification requires the use of milligram amounts of plasmid, a large-scale plasmid preparation is needed. The plasmid can be purified by any method[18] that allows efficient digestion by restriction enzymes. Because the coupling of DNA to the beads by itself represents a purification step that selects the biotin-labeled molecules, an

---

[18] J. Sambrook, E. F. Fritsch, and T. Maniatis, "Molecular Cloning: A Laboratory Manual." Cold Spring Harbor Laboratory Press, Cold Spring Harbor, New York, 1989.

extremely high plasmid purity is not required to obtain a clean preparation of affinity beads.

The fragment harboring the specific recognition sequence is excised with two restriction enzymes, one of which must generate a 5' extension that can be exclusively filled in by a biotinylated deoxyribonucleotide. This allows the fragment to attach to the beads only at one site, leaving the rest of the molecule free to interact with soluble proteins. However, the only deoxyribonucleotides available in biotinylated form are Bio-dATP (Bethesda Research Laboratories; will directly label ends generated by *Hind*III or *Eco*RI) and Bio-dUTP [Sigma, Boehringer Mannheim, and Clontech (Palo Alto, CA); compatible with ends generated by *Sal*I, *Xho*I, and *Nde*I]. However, efficient biotinylation can also be achieved when the biotinylated nucleotide is incorporated in the second position, which increases the choice of restriction enzymes. In the example given below, the inclusion of dGTP in the fill-in reaction permitted the use of Bio-dATP to label a *Bam*HI site quantitatively. For some purposes, especially for testing for the presence of nucleases in the protein extracts, it is convenient to be able to introduce a radioactive label on the free end of the DNA through a fill-in reaction. To have this possibility, any enzyme generating a 5' extension can be used for the other end of the fragment.

In contrast to small-scale purification of DNA fragments, classic methods for fragment purification based on agarose gel electrophoresis are less adapted to milligram-scale isolations. We have previously purified DNA fragments by fast protein liquid chromatography (FPLC) using a Mono Q column (Pharmacia LKB Biotechnology, Uppsala, Sweden) eluted by a salt gradient. This works adequately for some fragments, less so for others, probably because a Mono Q column separates DNA fragments not exclusively according to size. In our experience this step can be the most bothersome in the preparation of DNA affinity beads, and we have therefore tried to eliminate it.

Fragment purification can be dispensed with by using a proper strategy for restriction enzyme digestion. If the fragment is excised by two restriction enzymes (A for biotinylation and B at the other end) with no further purification, both the insert and the vector fragments will be biotinylated at their A sites and become attached to the beads, leaving a relatively high concentration of nonspecific binding sites on the beads. However, because the concentration of nonspecific binding sites is proportional to the amount of DNA (or concentration of nucleotides in DNA), it can be reduced to an unimportant level by making the biotinylated vector fragment very small. It suffices to cleave the vector sequence by a third enzyme close to the A site, leaving only a tiny fragment (preferentially <20 bp) of vector sequence biotinylated. Provided the DNA fragment is cloned into a polylinker sequence in the plasmid, this is often feasible. Even if equimolar amounts of insert and vector fragment now will be coupled to the streptavidin-coated beads, the vector fragment will represent only a minor concentration of nonspecific binding sites due to its small size. The rest of the DNA fragments in the digest will not become attached because they are not biotinylated. This strategy can be further improved if the plasmid contains multiple copies of the insert. In the example given below, we show the construction of DNA affinity beads without fragment purification using a plasmid pRB1 containing 12 copies of the tRNA^{Tyr} *Sup 4-o* gene.[1]

An even more elegant strategy, which totally eliminates vector sequences, is to use a vector that allows asymmetric labeling like the ones developed by Volckaert *et al.*[19] and improved by Eckert *et al.*[20] Provided that the DNA sequence of interest contains no *Tth*111I site (GACN ↓ NNGTC), the fragment can be cloned into the *Sma*I site of pSP64CS or pSP65CS vectors and excised with *Tth*111I. This generates four termini, each with a unique 5′ protruding nucleotide. Using Bio-dUTP the excised fragment can be labeled at one end, leaving the vector ends unlabeled. When the streptavidin-coated beads are added, only the excised insert becomes coupled.

Below we describe a procedure for the construction of magnetic DNA affinity beads using a DNA fragment generated by restriction enzyme digestion. Using the same basic principles, it should be easy to construct a protocol for alternative approaches, like the use of polymerized duplex oligonucleotides[2] or the use of a product from a scaled-up PCR reaction with biotinylated oligonucleotide primers.[6,12]

Step 1: Large-Scale Digestion of Plasmid DNA. Plasmid DNA (1–5 mg) containing the specific protein-binding site (in single or multiple copies) is digested with enough units of the appropriate restriction enzymes to give complete digestion overnight. Completeness of digestion is verified by agarose gel electrophoresis of an aliquot. More enzyme is added if needed. The final digest is extracted with phenol, precipitated with ethanol, and dissolved in TE buffer.

Step 2: Selective Biotinylation of DNA. The fragment containing the specific binding site is end labeled at one end using Bio-dATP or Bio-dUTP and the Klenow fragment of DNA polymerase I. A typical reaction mixture (200–600 μl) contains the DNA digest with 0.2–1.0 nmol of specific fragment, a slight excess of biotinylated deoxyribonucleotide triphosphate

[19] G. Volckaert, E. De Vleeschouwer, H. Blöcker, and R. Frank, *Gene Anal. Tech.* **1,** 52 (1984).
[20] R. L. Eckert, *Gene* **51,** 247 (1987).

(total, 4–5 nmol), a second deoxyribonucleotide triphosphate, if necessary, to fill in the first position (total, 20 nmol), 200 units DNA polymerase I Klenow fragment, all in 10 mM Tris-HCl, 10 mM MgCl$_2$, 50 mM NaCl, and 1 mM dithiothreitol, pH 8. Incubation is performed at 15° for 45 min. The reaction is stopped by the addition of EDTA to 15 mM final concentration, and heating for 15 min at 60°. The efficiency of biotinylation is assayed by agarose gel electrophoresis of two aliquots (1–2 μl), one loaded directly as a reference, the other loaded after mixing with 1 μl 10 mg/ml streptavidin. Streptavidin will bind to the biotinylated molecules and cause retarded migration of the modified bands. If the fragment of interest is quantitatively modified, the entire band should be shifted. Note that streptavidin is multivalent and can give rise to multiple shifted bands.

Unincorporated Bio-dATP is removed by passage through a small Sephadex G-50 desalting column (Pharmacia LKB Biotechnology), pre-equilibrated with TEN buffer. The fractions containing DNA are easily identified by a spot test. Aliquots of 2 μl of each fraction are mixed with 2 μl of 2 μg/ml ethidium bromide directly on the inner surface of an empty petri dish. The petri dish is inverted and placed on a standard ultraviolet (UV) transilluminator. Fractions containing DNA are identified as fluorescent drops.

Step 3: Coupling of DNA to Magnetic Beads. A suspension of 10 mg/ml Dynabeads M-280 streptavidin (1–5 ml; Dynal A/S) is washed three times by resuspension and magnetic separation in portions of 5 ml of TEN buffer. After removal of the last wash solution, the beads are resuspended in 0.2–0.5 ml TEN buffer and 0.5–2 ml of the desalted DNA solution is added. The suspension is gently agitated in a roller (just to prevent the beads from sedimenting) at room temperature for 30 min. The efficiency of coupling is visualized by agarose gel electrophoresis of DNA samples taken before and after coupling (see Fig. 2). Finally the beads are washed again three times by resuspension and magnetic separation in portions of 5 ml of TEN buffer.

Normally, the fragment of interest is quantitatively adsorbed to the beads. From the inputs of Dynabeads and specific DNA fragments, the DNA density of the beads (pmol DNA/mg beads) is calculated. We have used beads with densities in the range of 5–20 pmol/mg. Even higher capacities might be obtained due to the high biotin binding capacity of Dynabeads M-280 streptavidin (>300 pmol/mg). However, too much DNA on the beads is not practical to work with for several reasons. The DNA affinity beads can be stored at 4° for months without loss of protein-binding capacity. We have successfully used beads stored at 4° for more than 2 years.

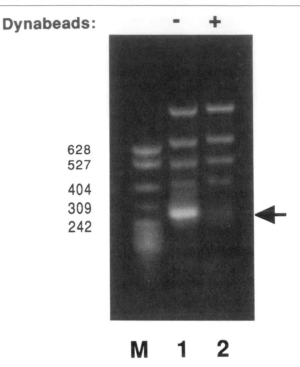

FIG. 2. Selective coupling of biotinylated DNA fragments to Dynabeads M-280 streptavidin. The plasmid pRB1 containing 12 copies of the tRNA^{Tyr} *Sup 4-o* gene[1] was digested to completion with *Bam*HI and *Alu*I and biotinylated at the *Bam*HI sites by a fill-in reaction using dGTP and Bio-14-dATP as described in text. The biotinylated fragments were adsorbed to Dynabeads M-280 streptavidin. Aliquots of the biotinylated digest before (lane 1) and after (lane 2) incubation with Dynabeads are shown. A molecular size marker, pBR322 digested with *Msp*I, is shown in lane M with fragment lengths (bp) indicated on the left. The 270-bp gene fragment is indicated by an arrow.

Purification of Transcription Factor

Partially Purified Protein Extract

For optimal results, the magnetic DNA affinity beads should not be used directly on a crude extract. Some partial purification is desirable. To what extent will depend on the protein, its source, abundance, and binding affinity, and on the amount and nature of contaminating proteins interacting with DNA (see Concluding Remarks, below), in particular on the presence of nucleases in the extract that can degrade the DNA on the beads. In the example given below, transcription factor τ is purified from a cell-free extract of the *Saccharomyces cerevisiae* strain 20B-12 (*trp1*, *pep4-3*) purified by phosphocellulose and heparin Ultrogel (IBF Biotechnics, Villeneuve-la-Garenne, France) as previously described.[1,21,22]

The presence of DNases in the protein fractions causes a problem that can be controlled to some extent by including EDTA in all buffers and by keeping incubation times short, and implies that it is easier to work with proteins that do not require divalent metal ions and bind DNA efficiently in the presence of EDTA. Nuclease activity is not only a problem for the stability of the affinity beads, but also for activity assays. Normally, a sequence-specific DNA-binding protein is assayed by the electrophoretic mobility shift assay.[23,24] Due to the high concentration of specific DNA on the affinity beads, even a moderate (in relative terms) liberation of DNA into the solution can cause serious problems in this assay, the liberated DNA acting as a specific unlabeled competitor that titrates out binding to the labeled probe. Nuclease activity can easily be assayed by end labeling an aliquot of the affinity beads through a fill-in reaction with DNA polymerase I and a ^{32}P-labeled deoxyribonucleotide compatible with the free end of the immobilized DNA. Unincorporated deoxynucleotide is removed by magnetic separation and washing of the beads. Labeled beads are then incubated with the protein fraction in question and nuclease activity is detected as radioactive label liberated to the supernatant.

Optimalization of Conditions for Binding

It is advantageous to have some ideas of the DNA-binding properties of the protein of interest. One critical parameter is the ionic strength. Too high a concentration of salt abolishes protein binding to DNA, and too low a salt concentration may hinder a rapid exchange of factor and nonspecific proteins between sites on the affinity beads and sites on the competitor DNA in solution. For binding of factor τ to a tRNA^{Glu} gene a narrow range of ionic strengths (150–200 mM NaCl) was found optimal, 100 mM being too low, 250 mM too high.[5]

Three different strategies are possible for protein purification with DNA affinity beads. The protein can be purified through cycles of adsorption and desorption as previously described.[5] The sequence-specific DNA-binding protein is adsorbed to the affinity beads in the presence of competitor DNA, washed, and eluted in a high salt buffer. Then the eluate is mixed with new competitor DNA and washed beads, and the ionic strength is

[21] A. Ruet, S. Camier, W. Smagowicz, A. Sentenac, and P. Fromageot, *EMBO J.* **3**, 343 (1984).

[22] O. S. Gabrielsen, N. Marzouki, A. Ruet, A. Sentenac, and P. Fromageot, *J. Biol. Chem.* **264**, 7505 (1989).

[23] M. G. Fried, *Electrophoresis* **10**, 366 (1989).

[24] R. Baker, O. Gabrielsen, and B. D. Hall, *J. Biol. Chem.* **261**, 5275 (1986).

adjusted to binding conditions. After binding, magnetic separation, and wash, a second eluate is obtained, and so on. This is analogous to successive DNA affinity column separations. Our experience is, however, that improved yields are obtained if the specific protein is kept bound to the affinity beads after the first adsorption. Purification is enhanced by equilibrations with competitor DNA in successive washes. Each wash with competitor DNA will preferentially dissociate nonspecific proteins. Finally the specific protein is eluted in high-salt buffer. A third strategy to reduce nonspecific protein binding is to overload the affinity beads with protein. In a situation of high loading and an excess of nonspecific DNA-binding proteins relative to available DNA-binding sites on the affinity beads, specific and nonspecific DNA-binding proteins will compete for a limiting amount of available sites. Normally, the specific protein has the highest affinity for the immobilized DNA and will be preferentially bound. Our experience is that the best results are obtained by a combination of the second and third approach, performing binding at near-saturating conditions and including competitor DNA in one or several subsequent washes, as illustrated in Fig. 1.

The amount of competitor needed will vary from case to case, depending on initial purity and concentrations of other DNA-binding proteins, and on the protein saturation of the affinity beads. The ratio of competitor to immobilized DNA will determine the distribution of nonspecific DNA-binding proteins. To illustrate this, let us assume that a protein X, with high affinity for the immobilized DNA, is to be purified from a partially purified extract containing 0.01% X and 50% proteins with general DNA affinity. Under nonsaturating conditions one cycle of adsorption and desorption in the presence of a 100-fold excess of competitor DNA will give a maximal theoretical purity of 2%. However, successive cycles of equilibrations with competitor can improve the purity significantly. A second equilibration with 100-fold excess competitor would increase the theoretical purity from 2 to 67%, a third wash to 99%. The corresponding figures for an initial purity of 0.1% and 10 times excess of competitor, is 2% purity after 1 competitor wash, 17% after 2, 67% after 3, and 95% after 4. Thus the required number of washes with competitor depends on the initial purity, and the increase in purity per wash, depends on the excess of competitor used. Under saturating conditions the situation will be quite different. Bound nonspecific proteins will be displaced by the stronger binding specific protein, giving very high initial purity. Consequently, only one or two equilibrations with competitor DNA will normally be required to obtain highly purified preparations. With a typical initial purity of 80%, a single wash with 10 times excess competitor DNA will theoretically increase the purity to 98%, 2 washes to 99.8%. It should be noted that all sequence-specific DNA-binding proteins will also have a general DNA-binding affinity. Therefore the use of a large excess of competitor DNA will decrease the yield of the purified protein.

Working under near-saturating conditions will also reduce the consumption of competitor DNA that otherwise can be considerable. Several incubations with a 100-fold excess of competitor DNA will easily amount to milligram quantities of DNA even for small-scale experiments. If saturating conditions are not obtainable, it should be noted that the competitor consumption can be reduced by doing repeated washes with a moderate excess of competitor DNA instead of few washes with a large excess. Assuming an experiment with beads containing 10 μg of specific DNA and an initial purity of protein X of 0.05% (otherwise as in the examples above), the theoretical consumption of competitor DNA to reach 90% purity would be 100 mg for 1 cycle, 2 mg for 2 cycles (100-fold excess each), 650 μg for 3 cycles (22-fold excess each), and 400 μg for 4 cycles (10-fold excess each).

The kinetics of adsorption is normally not a limiting factor. Our experience is that quantitative binding is achieved in a few minutes. In general, the kinetics will vary with the amount of magnetic affinity beads used, the DNA density on the bead surface, the concentration of competitor DNA, as well as with temperature and ionic strength. The minimal adsorption time necessary to achieve quantitative binding should be tested in each case. The kinetics of adsorption is easily monitored by the electrophoretic DNA-binding assay[23,24] on the supernatant after magnetic removal of the affinity beads. For factor τ, we found more than 90% adsorbed after 4 min under the conditions described.[5]

Step 1: Finding Optimal Concentrations of Salt for Protein Binding. Equilibrate 1.5 mg DNA affinity beads with TGED buffer containing 50 mM NaCl by successive resuspensions in buffer and magnetic separations using a magnetic particle concentrator. When the magnet is placed against the wall of the tube, the magnetic beads are rapidly collected in a firm pellet. The "supernatant" can then be pipetted off while keeping the magnet in place. After three washes in 0.5 ml buffer, resuspend in 300 μl buffer and divide between seven tubes (40 μl or 0.2 mg beads in each). After magnetic separation and removal of buffer, add a fixed, nonsaturating amount of protein fraction in TGED buffer adjusted to 50, 100, 150, 200, 250, 300, and 400 mM NaCl. Incubate for 5–10 min. The optimal temperature for specific DNA binding will depend on the binding properties and stability of the protein. We have carried out the technique on ice and at 25°, the main difference observed being that the kinetics of DNA binding is slightly slower on ice. Incubation times should thus be adjusted according to the temperature chosen. After magnetic separation, the "supernatants'' are transferred to new tubes and assayed by the electrophoretic mobility shift assay, using nontreated protein as reference. Quantitative adsorption is seen as disappearance of specific DNA-binding activity in the "supernatants." A typical example can be found in Ref. 5. When a functional range of salt concentrations is determined, it is recommended to choose working concentrations in the upper part of this concentration range.

Step 2: Finding Optimal Concentrations of Salt for Protein Elution. Equilibrate 1.5 mg DNA affinity beads with TGED buffer containing the optimal concentration of NaCl for specific DNA binding. Incubate with an appropriate amount of protein fraction in the same buffer to allow binding. After magnetic separation and washes in TGED buffer (still with the optimal NaCl concentration), divide a suspension of the loaded beads between seven tubes, giving 0.2 mg beads in each. After magnetic separation and removal of buffer, resuspend on ice in 20–30 μl cold TGED buffer containing 0.2, 0.4, 0.6, 0.8, 1.0, 1.5, and 2.0 M NaCl. To determine the salt concentration that results in quantitative elution, test the specific DNA-binding activity by the electrophoretic mobility shift assay, using small aliquots of the eluates (keeping final salt concentrations within acceptable limits for the binding assay).

Step 3: Determining Saturation Conditions. If the initial protein fraction contains a high salt concentration, a saturation experiment will require impractical dilutions at high loadings. The protein fraction is therefore preferably desalted to the optimal salt concentration on a small Sephadex G-25 column (column volume three to four times loaded volume) or by dialysis. To portions of 0.2–0.5 mg equilibrated beads prepared as above, add a range of desalted protein fraction (typically 5–1000 μl) at the chosen working temperature. The volumes might be adjusted by addition of buffer (optional). After 5–10 min of binding reaction, the affinity beads are washed once with buffer (with the optimal NaCl concentration), once with buffer containing a 10 times excess of competitor DNA {poly[d(I–C)] or any plasmid DNA without the actual binding site; excess competitor is on a microgram basis relative to the amount of DNA immobilized}; and once in buffer. Adsorbed protein is eluted on ice in a small volume (20–30 μl) of buffer containing a sufficiently high concentration of salt. To determine saturation ratios, the eluates are analyzed by sodium dodecyl sulfate (SDS)-polyacrylamide gel electrophoresis (see Fig. 3). Alternatively, saturation can be determined by following the loss of specific DNA binding activity in the "supernatants."

Step 4: Scaled-Up Experiment. If satisfactory results are obtained in step 3, a scaled-up purification can be performed. If not, further parameters should be analyzed such as the extent of competitor washing (see Fig. 3B), the need for further initial purification (see Concluding Remarks, below), kinetics of binding, temperature dependence of binding, effect of stabilizing components such as glycerol or bovine serum albumin, the presence of nuclease activity in the protein fraction, and the nonspecific adsorption to unmodified Dynabeads M-280 streptavidin. A moderately scaled-up purification can be performed with 10 mg affinity beads, and carried out otherwise as in step 3, that is, washing once with buffer, once with buffer containing a 10–20 times excess of competitor DNA, and once in buffer. Adsorbed protein is eluted on ice in a small volume of buffer containing a sufficiently high concentration of salt.

Used beads are regenerated by repeated washes in high salt buffer (TGED with 2 M NaCl, or TGED with 2 M NaCl and 6 M urea) and a few washes in TEN buffer, and are stored at 4° for further use.

Results of Experiment

Preparation of DNA Affinity Beads Containing the Yeast tRNA^{Tyr} Sup 4-o Gene

Here we describe an example of preparation of DNA affinity beads without fragment purification. One milligram of plasmid pRB1 containing 12 copies of the tRNA^{Tyr} *Sup 4-o* gene[1] was digested to completion with *Bam*HI and *Alu*I in 1 ml at 37° overnight. After phenol extraction and ethanol precipitation, the gene fragment (1.0 nmol) was end labeled in the *Bam*HI site using dGTP, Bio-14-dATP, and the Klenow fragment of DNA polymerase I as described above. Unincorporated Bio-14-dATP was removed by passage through a small Sephadex G-50 desalting column. A suspension of 50 mg Dynabeads M-280 streptavidin in 0.4 ml TEN buffer was mixed with 1.6 ml of the desalted DNA solution as described above. Figure 2 shows an assay of the efficiency of coupling. Before addition of the beads, the 270-bp tRNA^{Tyr} gene fragment is clearly distinguished due to its stoichiometric overrepresentation (arrow). After incubation with the beads this fragment is absent (the weak band remaining in that position is a vector band of similar size), showing quantitative adsorption. All main vector fragments remained in solution. Two small vector fragments of 345 and 311 bp were also attached (representing 17% of the immobilized DNA), as can be seen by careful inspection of the agarose gel shown. The beads now contained 20 pmol specific DNA fragment per milligram beads or 3.6 μg DNA/mg beads.

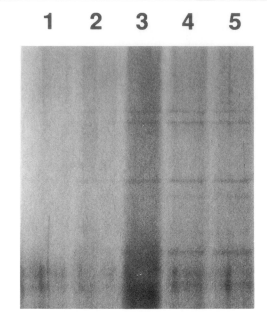

FIG. 3. Analytical purification of transcription factor τ by DNA affinity Dynabeads. (A) The saturation point for the DNA affinity beads was determined by incubation of 0.2 mg of beads with 5 μl (lane 1), 20 μl (lane 2), 80 μl (lane 3), 300 μl (lane 4), and 1 ml (lane 5) of a partially purified factor τ preparation (heparin fraction[23]) that had previously desalted on a Sephadex G-25 fine column equilibrated with TGED buffer containing 100 mM NaCl. After binding, the affinity beads were washed once in 0.5 ml TGED buffer containing 100 mM NaCl, once in the same buffer also containing poly[d(I–C)] competitor DNA (10-fold excess in micrograms), and once in TGED buffer containing 100 mM NaCl. Factor was eluted in 40 μl TGED buffer containing 1 M NaCl and analyzed by SDS-polyacrylamide gel electrophoresis. (B) The effect of various competitor treatments after saturation of the affinity beads with factor τ. The initial heparin fractions (3.5 μg protein, 10 μl) before (lane 1) and after (lane 2) incubation with affinity beads are shown. Lanes 3–5 show the polypeptides eluted from 1 mg of beads after saturation with 400 μl of heparin fraction and one (lane 3), two (lane 4), or three (lane 5) subsequent competitor treatments {5 min with a 10-fold excess of poly[d(I–C)]}. Migrations of size markers (lanes 6 and 7) are indicated to the right, and positions of factor subunits are indicated to the left.

Analytical Purification of Transcription Factor τ by DNA Affinity Dynabeads

Optimal conditions of binding and elution were determined from previous experiments with binding of transcription factor τ to the tRNA[Tyr] gene. Protein binding was performed at 100 mM NaCl at 25° and elution at 1 M NaCl on ice. Figure 3A shows the results of a saturation experiment. Using portions of 0.2 mg affinity beads, maximal staining of factor τ-specific

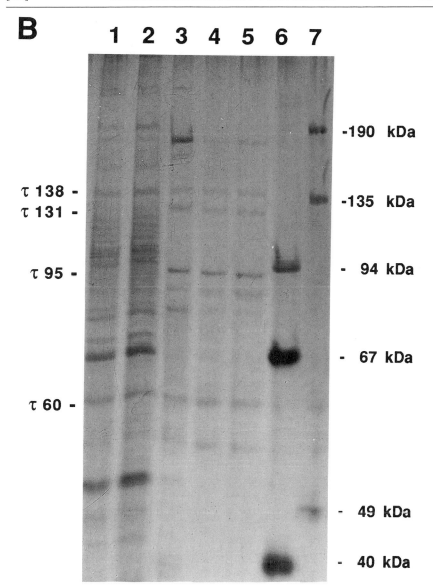

FIG. 3 (*continued*)

polypeptides was obtained with 80 μl partially purified protein fraction (saturation at 0.4 ml protein fraction per milligram beads). Figure 3B shows the results of various competitor treatments after saturation of the affinity beads with factor. The initial heparin fractions before and after incubation with affinity beads are shown in lanes 1 and 2 (note that 10 μl heparin fraction was loaded in lanes 1 and 2 whereas lanes 3–5 show material purified from about 400 μl heparin fraction). No difference in band pattern was evident, showing that the vast majority of the proteins remained in the supernatant. Lane 3 shows the polypeptides eluted after saturation and a single competitor treatment {5 min with a 10-fold excess of poly[d(I–C)]}. The degree of purification was already excellent. Only a few bands that clearly do not belong to factor τ, and thus decrease on further competitor treatment, are seen. The polypeptides remaining after a second competitor treatment are shown in lane 4. Except for a trace of a 180-kDa band, factor τ now seems to be essentially pure, containing the four characterized subunits of 138, 131, 95, and 60 kDa[25] that have previously been found in factor τ–tDNA complexes.[22,26] (Note that τ95 in the present gel system migrates more like a 92-kDa polypeptide. The cloned τ95 gene predicts a polypeptide of 73 kDA.[25]) A faint band at the 88-kDa region is also seen that might belong to factor τ.[16] The band below τ60 is probably a silver stain artifact. A third wash with a 10-fold excess of poly[d(I–C)] did not result in any significant further purification (lane 5). Note that no significant reduction in yield was observed between lanes 3 and 5. This contrasts with what was observed in a parallel experiment with three cycles of adsorption and elution resulting in a dramatic fall in recovered factor (not shown). The purified factor was active in transcription and mobility shift assays (not shown).

Concluding Remarks

We have described an improved protocol for the use of magnetic DNA affinity beads in the purification of sequence-specific DNA-binding proteins. Provided that sufficient amounts of partially purified protein fraction are available to allow operations near saturation, the method has the potential of rapidly yielding highly purified protein. Problems could be envisioned in which the protein of interest does not retain specific DNA binding in the presence of EDTA, thus making it difficult to limit DNase

[25] R. N. Swanson, C. Conesa, O. Lefebvre, C. Carles, A. Ruet, E. Quemeneur, J. Gagnon, and A. Sentenac, *Proc. Natl. Acad. Sci. U.S.A.* **88,** 4887 (1991).

[26] B. Bartholomew, G. A. Kassavetis, B. R. Braun, and E. P. Geiduschek, *EMBO J.* **9,** 2197 (1990).

activity in impure protein fractions. A low ratio of specific to nonspecific DNA-binding constants could also cause problems during equilibrations with competitor DNA. If other proteins are present in the fraction that have higher affinity for the immobilized sequence than the protein of interest, the saturation strategy will not work. For example, we were not able to isolate pure factor τ from a less purified fraction of τ (a heparin-Ultrogel fraction for which the previous phosphocellulose step was omitted) due to a high concentration in these fractions of the EF1α protein[27] that bound avidly to the affinity beads. End-binding proteins might also cause problems. We have observed retention of yeast RNA polymerases on the DNA beads, probably due to the strong affinity of these enzymes for ends of DNA. In an experiment in which a fraction containing RNA polymerase B was added to affinity beads loaded with factor τ, the enzyme was retained and eluted with factor τ at high salt concentration. When the starting material contained RNA polymerase A, this enzyme was also retained and remained adsorbed during elution of factor τ, but was eluted in a subsequent step using 6 M urea and 1 M NaCl. It is thus recommended that the magnetic DNA affinity approach be tested on protein fractions at different stages of purification to find the best starting material. However, in the majority of cases in which traditional DNA affinity columns work, magnetic DNA affinity purification should work as well or better.

A still unexploited possibility is to use magnetic DNA affinity beads as a template for the assembly of active transcription complexes. Such complexes should include transcription factors that are not by themselves sequence-specific DNA-binding proteins, but which become assembled through protein–protein interactions with the DNA-binding factors, like transcription factor TFIIIB.[28] Transcription factors in this category have been less well characterized due to difficulties with their purification. Work in progress shows that active tRNA gene transcription complexes can indeed be assembled on the affinity beads, giving accurate transcription when supplemented with RNA polymerase C and ribonucleotides. We are currently pursuing the purification of TFIIIB using this strategy.

Acknowledgments

We thank André Sentenac, Robert Swanson, and Tordis B. Øyen for critically reading the manuscript, and Anne Françoise Burnol for kindly providing partially purified factor τ. We also thank E. Hornes and J. Andreassen for the gift of Dynabeads. O.S.G. is supported by a fellowship from the Laboratoire d'Ingéniérie des Protéines, Direction des Sciences du Vivant.

[27] D. Thiele, P. Cottrelle, F. Iborra, J.-M. Buhler, A. Sentenac, and P. Fromageot, *J. Biol. Chem.* **260,** 3084 (1985).

[28] G. A. Kassavetis, B. R. Braun, L. H. Nguyen, and E. P. Geiduschek, *Cell* **60,** 235 (1990).

[37] Affinity Selection of Polymerase Chain Reaction Products by DNA-Binding Proteins

By ANDREW M. LEW, VIKKI M. MARSHALL, and DAVID J. KEMP

Anchoring DNA Amplified by Polymerase Chain Reaction with DNA-Binding Proteins

There are numerous DNA-binding proteins that bind specific sequences and some of these, including GCN4,[1,2] tyrR,[3] and the *lac* repressor,[4] have been used for affinity selection of polymerase chain reaction (PCR) products. If the binding site is incorporated into the oligonucleotide primer, the amplified DNA molecule can be anchored to a solid phase for subsequent manipulation. To date, affinity selection of PCR products on DNA-binding proteins has been applied to diagnostics,[1,3,4] clone isolation,[2] and DNA sequencing (see below). Potentially, it could be developed for any DNA manipulation. One advantage over other possible ligands is that binding is reversed and the DNA is released at 65° or with 0.1% sodium dodecyl sulfate (SDS).

To develop these procedures we have used GCN4, a yeast regulatory protein that has been studied in detail.[5,6] It binds a 9-base palindromic sequence with a central C or G (ATGAC/GTCAT) and has a higher affinity for double-stranded DNA (dsDNA) than for single-stranded DNA (ssDNA). This latter property is useful in that if an oligonucleotide containing the recognition site is used as a primer in a PCR, then GCN4 will bind the PCR product in preference to the oligonucleotide. Such a circumstance does not occur for affinity selection with avidin using biotinylated oligonucleotides.

[1] D. J. Kemp, D. B. Smith, S. J. Foote, N. Samaras, and M. G. Peterson, *Proc. Natl. Acad. Sci. U.S.A.* **86,** 2423 (1989).

[2] A. M. Lew and D. J. Kemp, *Nucleic Acids Res.* **17,** 5859 (1989).

[3] T. Triglia, V. P. Argyropoulos, B. E. Davidson, and D. J. Kemp, *Nucleic Acids Res.* **18,** 1080 (1990).

[4] J. Lundberg, J. Wahlberg, M. Holmberg, U. Pettersson, and M. Uhlen, *DNA Cell Biol.* **9,** 287 (1990).

[5] I. A. Hope and K. Struhl, *Cell* **46,** 885 (1986).

[6] A. R. Oliphant, C. J. Brandt, and K. Struhl, *Mol. Cell. Biol.* **9,** 2944 (1989).

METHODS IN ENZYMOLOGY, VOL. 218

Materials and Reagents

Production and Isolation of Recombinant GCN4 Fusion Protein

To obtain large quantities of purified GCN4 protein, a recombinant form of GCN4 was produced. This has been described in detail elsewhere (D. J. Kemp, this series, Volume 216 [12]). GCN4 belongs to a class of sequence-specific DNA-binding proteins that have a bipartite structural motif consisting of a highly charged (basic) DNA-binding domain and a dimerization domain called a leucine zipper.[7] A segment of GCN4 encoding both these regions was cloned into the bacterial expression vector, pGEX, to produce a fusion protein with glutathione S-transferase and GCN4 (GST–GCN4). The GST fusion partner allows the protein to be rapidly isolated on glutathione beads in one step by affinity chromatography.[1] These beads can be made[8] or purchased from Sigma (St. Louis, MO). This preparation contains 1% (w/v) SDS but usually is diluted 1 in 100 before use so the SDS concentration does not interfere with GCN4 binding to the tube (see below). The GST-GCN4 is now commercially available (AM-RAD Corporation, Ltd., Victoria, Australia).

GST–GCN4-Coated Tubes

Purified GST–GCN4 [100 μl of a 5-μg/ml solution in phosphate-buffered saline (PBS)] is allowed to absorb onto microcentrifuge tubes for 2 hr at room temperature or overnight at 4°. Before using, the excess GST–GCN4 is eliminated by three PBS washes. No centrifugation is required for these washes; the solution is simply removed by flicking.

Methods and Results

Cloning of Regions Adjacent to Regions of Known Sequence

There are many situations where it would be useful to obtain the DNA sequence of regions flanking a known segment, for example, completing the sequence of an incomplete cDNA clone. Several expedient strategies for doing this have been described.[9,10] We describe a method ("GCN4-PCR")[2] that can be used when the sequence of only a short segment is known, for example, when only the N-terminal sequence of a protein is

---

[7] M. G. Oakley and P. B. Dervan, *Science* **248,** 847 (1990).
[8] A. M. Lew, D. I. Beck, and L. M. Thomas, *J. Immunol. Methods* **136,** 211 (1991).
[9] T. Triglia, M. G. Peterson, and D. J. Kemp, *Nucleic Acids Res.* **16,** 8186 (1988).
[10] M. Kalman, E. T. Kalman, and M. Cashel, *Biochem. Biophys. Res. Commun.* **167,** 504 (1990).

Step 1. PRC, 20 cycles

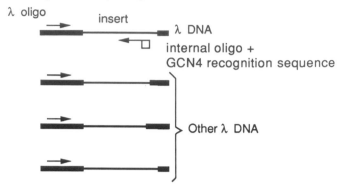

Step 2. Binding of specific dsDNA
to GCN4-coated tube

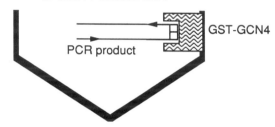

Step 3. Wash to clear other λ DNA

Step 4. Repeat PCR in the tube with
bound PCR product, 30 cycles

FIG. 1. Outline of strategy of using GCN4 enrichment to clone outside regions.

known. This method allows one to isolate DNA segments from phage libraries without the need for screening plaques by hybridization.

Figure 1 demonstrates the steps of a strategy devised to obtain the 5' end of a cDNA from a λgt10 cDNA library of the human malaria parasite, *Plasmodium falciparum*. We have found that using an internal oligonucleotide and a vector oligonucleotide (representing the sequence flanking the insert) to amplify the segments from a library[11] is inefficient, presumably because of competition for the vector oligonucleotide by the many irrele-

[11] R. Jansen, F. Kalousek, W. A. Fenton, C. E. Rosenberg, and F. D. Ledley, *Genomics* **4**, 198 (1989).

vant clones. Therefore, to enrich for the desired product, a GCN4 recognition sequence (GGATGACTCA) was added onto the 5' end of the 20 bases of the known internal sequence. Polymerase chain reaction was carried out with this oligonucleotide (2 μM), the vector oligonucleotide (2 μM), and 10–100 ng DNA from a λgt10 cDNA library in a final volume of 100 μl for 20 cycles at 95° for 60 sec, 55° for 60 sec, 70° for 90 sec.

The entire PCR mixture was transferred into the GST–GCN4-coated tube and the dsDNA containing the GCN4 recognition sequence was allowed to bind for 1 hr at room temperature. Washing three times with PBS eliminates the irrelevant DNA and the single-stranded oligonucleotides. A second set of PCR (using fresh reagents) was then carried out in this GST–GCN4-coated tube, this time for 30 cycles. As shown in Fig. 2, this produced a visible band on agarose electrophoresis and ethidium bromide staining. Gel purification of the band and direct sequencing with dimethyl sulfoxide[12] confirmed that the DNA was indeed the segment of interest (the full cDNA sequence was already known).

Above this predominant band, there was a smear of larger molecules, probably representing longer clones. Multiple clones of varying length should still be amenable to sequencing *en masse* from the internal oligonucleotide, although not from the vector oligonucleotide. Alternatively, these larger species could be cloned into a plasmid for subsequent sequencing.

Polymerase Chain Reaction for Direct Sequencing from 5 μl of Whole Blood

There are many situations when the sample of biological material from which DNA sequence is to be derived is precious or available only in minute amounts. We describe an expeditious method that requires only 5 μl of whole blood. The processing time from blood sampling to loading on the sequencing gel is approximately 10 hr. The method consists of (1) partial DNA purification from blood/guanidine sample using glass milk (Bio-Rad, Richmond, CA), (2) a primary round of PCR, (3) enrichment by GCN4 capture, (4) a second round of PCR using nested oligonucleotides (one of these contains the M13 universal primer sequence and is phosphorylated), (5) glass milk purification, (6) λ-exonuclease digestion to obtain single-stranded DNA (the strand containing the 5' phosphate is digested 10 times faster than the nonphosphorylated strand), and (7) heat denaturation of the exonuclease, 2-propanol precipitation, and DNA sequencing.

---

[12] P. R. Winship, *Nucleic Acids Res.* **17**, 1266 (1989).

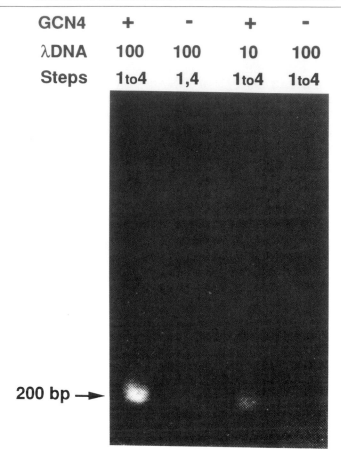

GCN4	+	–	+	–
λDNA	100	100	10	100
Steps	1to4	1,4	1to4	1to4

200 bp ➞

FIG. 2. PCR products obtained after the procedure shown in Fig. 1 using steps 1 to 4, or just steps 1 and 4, as indicated. As controls, the experiment was repeated with and without GST–GCN4 (+, –). The amount of initial substrate DNA is indicated (100 or 10 ng). The products were fractionated by agarose gel electrophoresis and stained with ethidium bromide.

The strategy is outlined in Fig. 3. This method requires no gel purification if the sequencing primers are labeled with fluorescent dyes (as for automated DNA sequencing) or end labeled with [32]P.

Preparation of Blood and GCN4 Enrichment

Five microliters of whole blood is collected into an Eppendorf tube containing 45 μl (9 vol) 8 M guanidine hydrochloride in 0.1 M sodium acetate, pH 5.3. This sample is partially purified by adding 20 μl glass

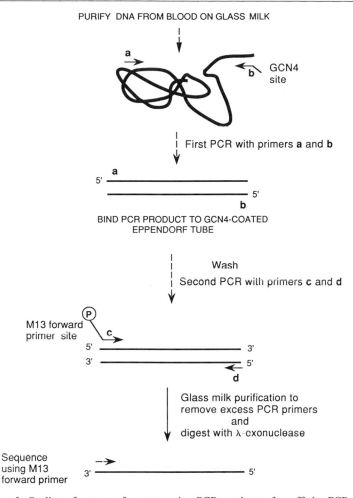

PURIFY DNA FROM BLOOD ON GLASS MILK

a

GCN4
site

b

First PCR with primers **a** and **b**

a

5'

5'

b

BIND PCR PRODUCT TO GCN4-COATED
EPPENDORF TUBE

Wash
Second PCR with primers **c** and **d**

(P)

M13 forward
primer site

c

5'   3'

3'   5'

d

Glass milk purification to
remove excess PCR primers
and
digest with λ-exonuclease

Sequence
using M13
forward primer

3'   5'

FIG. 3. Outline of strategy for sequencing PCR products after affinity PCR.

milk (Prep-A-Gene; Bio-Rad) and binding the DNA for 5 min at room temperature by shaking on an Eppendorf 5432 mixer (intermittent flicking is adequate if a shaker is not available).

The glass milk DNA is then washed three times with 500 μl ice-cold wash buffer [20 mM Tris, 2 mM (EDTA), 0.4 M NaCl, 50% ethanol, pH 7.4] and spun in an Eppendorf centrifuge (20-sec spin, room temperature at 1000 rpm). All remaining traces of buffer are pipetted from the last wash, and the genomic DNA (now sheared) eluted from the glass milk in 20 μl of purified H$_2$O.

Five microliters of this DNA template is used in a primary PCR containing 50 pmol each of oligonucleotides a and b, which represent DNA sequences at the N- and C-terminal conserved regions, respectively, of the *MSA-2* gene from *P. falciparum*.[13] Oligonucleotide b has the 10-bp GCN4 recognition sequence at its 5′ end (see Cloning of Regions Adjacent to Regions of Known Sequence, above).

The PCR is carried out for 35 cycles at 94° for 60 sec, 50° for 60 sec, 70° for 70 sec. Half of the PCR is transferred to GST–GCN4-coated tubes as described previously and allowed to bind at room temperature for 1 hr. During this incubation 1–5 μg oligonucleotide c, which contains the M13 forward primer sequence, is phosphorylated in a 50-μl volume for 1 hr at 37° in buffer containing 50 mM Tris-HCl (pH 7.5), 10 mM MgCl$_2$, 5 mM dithiothreitol (DTT), 1 mM ATP, and 20 U T4 polynucleotide kinase (T4 PNK).

Following GCN4 binding, unbound material is removed from the GCN4 tubes by three 1 ml PBS washes. A PCR mix (100 μl) containing 50 pmol each of internal oligonucleotides c and d is then added directly to the PCR-bound GCN4-coated tubes that have bound the previous PCR products. This second-round PCR is carried out for 38 cycles under the same conditions as for the primary PCR. Five percent of the total material from the first- and second-round PCR is loaded onto a 1% (w/v) agarose gel and stained with ethidium bromide. As shown in Fig. 4, material that is not detectable on an agarose gel after the first round of PCR is easily visualized after the second PCR, following GCN4 enrichment.

Using this method we obtained a visible band from blood samples of <0.002% parasitemia (<500 parasites). We have not yet determined the actual lower limit of detection. However, the sample in lane 6 gave a visible band following GCN4 enrichment, although no parasites were detected by light microscopy, indicating the different levels of sensitivity of the two techniques, that is, higher sensitivity of GCN4 enrichment.

Direct Sequencing of Polymerase Chain Reaction-Amplified DNA Using λ-Exonuclease

We have developed a method for the direct automated sequencing of DNA amplified from blood by PCR following the GCN4 enrichment protocol outlined above. This method has been modified from Higuchi and Ochman.[14] Following the second-round PCR, the entire 100-μl reaction is purified on glass milk (Prep-A-Gene; Bio-Rad); following the instructions

[13] J. A. Smythe, M. G. Peterson, R. L. Coppel, A. J. Saul, D. J. Kemp, and R. F. Anders, *Mol. Biochem. Parasitol.* **39,** 227 (1990).
[14] R. G. Higuchi and H. Ochman, *Nucleic Acids Res.* **17,** 5865 (1989).

SAMPLE   **SAMPLE**

M 1 2 3 4 5 6 7 8 9 N - +   M 1 2 3 4 5 6 7 8 9 - +

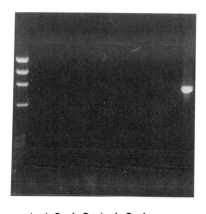   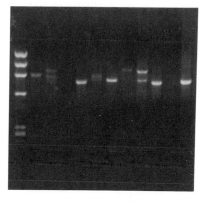

1 4 0 4 0 1 4 6 4   1 4 0 4 0 1 4 6 4

PARASITEMIA   **PARASITEMIA**

A   **B**

FIG. 4. Agarose gel electrophoresis of 5% of the total PCR reaction following the first (A) and second (B) round of PCR. M, ϕX174-*Hae*III size markers; N, sample from first-round PCR not carried through to second round; (−), negative control (no target DNA): (+), positive control (target is 100 ng DNA purified from cultured laboratory isolate of *P. falciparum*). The PCR products fractionated in tracks 1–9 were derived from blood samples from Papua New Guineans. The degree of *P. falciparum* parasitemia shown is as defined by Bruce-Chwatt.[15] The range of parasitemia for each classification is as follows: 0, no detectable parasites; 1, <0.002%; 4, 0.008–0.016%; 6, 0.032–0.064%. The presence of more than one band in some samples is due to mixed infections rather than nonspecific PCR products.

of the manufacturer. Glass milk binds short single-stranded DNA such as PCR primers, but does not elute them efficiently. Removal of excess PCR primers by this method is only approximately 90% efficient (data not shown) but is adequate, provided sequencing primers are fluorescently labeled or end labeled, because DNA sequences extended from residual PCR primers are not detected.

[15] L. J. Bruce-Chwatt, "Essential Malariology," 2nd Ed., p. 184. Heinemann, London, 1985.

The entire eluted PCR product is then digested with 8 U λ-exonuclease (Bethesda Research Laboratories, Gaithersburg, MD) in a 50-μl volume [67 mM glycine–KOH (pH 9.4), 2.5 mM MgCl$_2$, 50 μg/ml bovine serum albumin (BSA)] for 20 min at 37°. The λ-exonuclease is then heat killed at 65° for 10 min. The resulting single-stranded DNA is precipitated by adding an equal volume of fresh 4 M ammonium acetate and 2 vol 2-propanol. Following 10 min at room temperature the mixture is spun in an Eppendorf centrifuge for 15 min at room temperature at 13,000 rpm and washed in 70% ethanol. Half to all of the PCR product, now single stranded, is used in a sequencing reaction with fluorescent M13 universal primers (Applied Biosystems, Foster City, CA). We routinely obtain 450–500 bp of sequence using this protocol. For manual DNA sequencing where the radioisotope is incorporated into the extending chain, the PCR product must be gel purified prior to λ-exonuclease digestion.

Discussion

The number of DNA-binding proteins that have been sequenced is growing rapidly. Many of these have recognition sites encoded by DNA only 5–20 bases long (e.g., Rap1, C/EBP); these can therefore be relatively inexpensively added onto the 5′ end of a desired oligonucleotide. It is important to note that if the PCR product is abundant enough after the first PCR, binding to GCN4 achieves very little—an uncoated control tube will give similar results. However, if the amount of PCR product is limiting after the first round, the yield is considerably improved using the GCN4 enrichment. This can be particularly important for applications such as the sequencing procedure described above. The method described above for cloning outside regions also need not be limited to λ or pUC libraries. It should be suitable to any DNA source that has linkers long enough that the corresponding oligonucleotide will prime PCR.

[38] Protein-Blotting Procedures to Evaluate Interactions of Steroid Receptors with DNA

By Douglas B. Tully and John A. Cidlowski

Introduction

The Southwestern blotting protocol developed by Bowen *et al.*[1] and modified by Miskimins *et al.*[2] for transferring electrophoretically separated proteins onto nitrocellulose filters for subsequent binding of radioactively labeled DNA has proved to be a powerful tool for the analysis of DNA–protein interactions. We have further modified this technique to improve detection of selective interactions between glucocorticoid receptors (GR) and DNA.[3] These modifications have also been shown to enhance analysis of other DNA–protein interactions.[4–8] In this chapter, after a brief discussion of other techniques that have been employed for analysis of interactions between steroid receptors and DNA, we will give a detailed description of the Southwestern blotting protocol we have used for study of glucocorticoid receptor–DNA interactions.

The physiological effects of glucocorticoids and other steroid hormones (including mineralocorticoids, progestins, androgens, and estrogens) are mediated through interaction of the steroid molecules with specific receptor proteins present in hormonally responsive cells.[9,10] Steroid molecules are lipid soluble and are generally believed to diffuse freely across cell membranes. Inside target cells, the steroids are bound specifically and with high affinity by the cognate steroid hormone–receptor proteins. Following steroid binding, the hormone–receptor complex un-

[1] B. Bowen, J. Steinberg, U. K. Laemmli, and H. Weintraub, *Nucleic Acids Res.* **8**, 1 (1980).
[2] W. K. Miskimins, M. P. Roberts, A. McClelland, and F. H. Ruddle, *Proc. Natl. Acad. Sci. U.S.A.* **82**, 6741 (1985).
[3] C. M. Silva, D. B. Tully, L. A. Petch, C. M. Jewell, and J. A. Cidlowski, *Proc. Natl. Acad. Sci. U.S.A.* **84**, 1744 (1987).
[4] O. Andrisani and J. E. Dixon, *J. Biol. Chem.* **265**, 3212 (1990).
[5] W. Knepel, L. Jepeal, and J. F. Habener, *J. Biol. Chem.* **265**, 8725 (1990).
[6] S. Kitajima, T. Kawaguchi, Y. Yasukochi, and S. M. Weissman, *Proc. Natl. Acad. Sci. U.S.A.* **86**, 6106 (1989).
[7] J. Kwast-Welfeld, C.-J. Soong, M. L. Short, and R. A. Jungmann, *J. Biol. Chem.* **264**, 6941 (1989).
[8] M. Giacca, M. I. Gutierrez, F. Demarchi, S. Diviacco, G. Biamonti, S. Riva, and A. Falaschi, *Biochem. Biophys. Res. Commun.* **165**, 956 (1989).
[9] M. Beato, *Cell* **56**, 335 (1989).
[10] K. L. Burnstein and J. A. Cidlowski, *Annu. Rev. Physiol.* **51**, 683 (1989).

METHODS IN ENZYMOLOGY, VOL. 218

dergoes a process termed transformation or activation, which converts the receptor into a form that has high affinity for specific sites in DNA. Although still not completely understood, the activation process may involve both a conformational change in the receptor protein itself and dissociation of the receptor protein from a heteromeric complex with heat-shock proteins and/or other unidentified proteins.[11] For glucocorticoid receptors, which reside in the cytoplasmic compartment of the cell in the absence of hormone, activation leads to rapid translocation of the receptor into the nucleus. The other steroid receptors may already exist in the nucleus prior to steroid binding and activation. Inside cell nuclei, activated hormone–receptor complexes interact with specific sites in DNA either to enhance or repress the transcription of particular genes (for reviews see Refs. 9 and 10). These receptor-binding sites in DNA are small, 12 to 15-bp sequences[12] that have been shown, in recombinant DNA transfection experiments, to confer hormone-responsive transcriptional regulation onto linked heterologous genes and are thus known as hormone response elements (HREs).[13] Hormone-response elements can occur in various positions and in either left- or right-hand orientation with respect to the transcription start sites of regulated genes, and thus resemble enhancers in their mode of action. By binding to these sites to modulate transcription of associated genes, steroid receptors behave as ligand-dependent transcription factors.[9,10]

Selective binding of steroid hormone receptors to HRE-containing DNA fragments has been demonstrated by a variety of techniques for study of DNA–protein interaction, including DNA–cellulose competitive binding,[14,15] nitrocellulose filter binding,[16,17] gel retardation,[17,18] sucrose gradient shift,[19,20] DNA footprinting (nuclease protection),[12,17,21] and meth-

[11] L. C. Scherrer, F. C. Dalman, E. Massa, S. Meshinchi, and W. B. Pratt, *J. Biol. Chem.* **265,** 21397 (1990).

[12] F. Payvar, D. DeFranco, G. L. Firestone, B. Edgar, O. Wrange, S. Okret, J.-A. Gustafsson, and K. R. Yamamoto, *Cell* **35,** 381 (1983).

[13] V. L. Chandler, B. A. Maler, and K. R. Yamamoto, *Cell* **33,** 489 (1983).

[14] M. Pfahl, *Cell* **31,** 475 (1982).

[15] D. B. Tully and J. A. Cidlowski, *Biochem. Biophys. Res. Commun.* **144,** 1 (1987).

[16] S. Geisse, C. Scheidereit, H. M. Westphal, N. E. Hynes, B. Groner, and M. Beato, *EMBO J.* **1,** 1613 (1982).

[17] L. Hennighausen and H. Lubon, this series, Vol. 152, p. 721.

[18] T. Perlmann, P. Eriksson, and O. Wrange, *J. Biol. Chem.* **265,** 17222 (1990).

[19] D. B. Tully and J. A. Cidlowski, *Biochemistry* **28,** 1968 (1989).

[20] D. B. Tully and J. A. Cidlowski, *Biochemistry* **29,** 6662 (1990).

[21] C. Scheidereit and M. Beato, *Proc. Natl. Acad. Sci. U.S.A.* **81,** 3029 (1984).

ylation interference assays,[21] in addition to Southwestern blotting.[3,22,23] Each of these techniques has its own set of advantages and limitations (see also Ref. 17).

DNA–Cellulose Competitive Binding Assays. In DNA–cellulose competitive binding assays, activated ^{3}H-labeled receptor extracts are incubated with DNA–cellulose in the presence of various concentrations of a specific DNA fragment, after which the DNA–cellulose is pelleted.[14,15] Receptor bound to the free DNA fragment remains in suspension, thus decreasing the amount of ^{3}H-labeled receptor detected in the DNA–cellulose pellet. DNA–cellulose competitive binding assays are not particularly sensitive and require relatively large amounts of purified DNA fragment as competitor DNA.

Nitrocellulose Filter-Binding Assays. Filter-binding assays are based on the selective retention of DNA–protein complexes, but not free DNA, on nitrocellulose filters. These assays are simple, inexpensive, sensitive, and therefore widely used. The assays are not quantitative, however, because weak complexes may dissociate during the course of filtration and washing, and even strong complexes may not be quantitatively retained on the filters.[24,25] Additionally, the technique cannot discriminate among DNA molecules having one or multiple proteins bound.

Gel Retardation Assays. In gel retardation experiments, *in vitro*-formed DNA–protein complexes are separated from free DNA and free protein by electrophoresis in low ionic strength polyacrylamide gels.[17,18] Gel shift is a sensitive technique for detection of proteins that bind to a specific DNA oligonucleotide, but the assay is more difficult to accomplish with larger proteins (intact GR has a molecular weight of 97K) and it does not allow specific identification of the DNA-binding protein, except by supershift with specific antibody, which is also confusing if multiple bands are shifted.

Sucrose Gradient Shift (Zonal Sedimentation) Assays. In this technique, DNA–protein complexes are separated from free DNA and free proteins by centrifugation on sucrose density gradients.[17,19,20,26] Although this technique does not have the high resolution and sensitivity achieved by gel electrophoresis systems, it has nonetheless been useful in studies

[22] U. Hubscher, *Nucleic Acids Res.* **15**, 5486 (1987).
[23] J. M. Strawhecker, N. A. Betz, R. Y. Neades, W. Houser, and J. C. Pelling, *Oncogene* **4**, 1317 (1989).
[24] C. P. Woodbury, Jr. and P. H. von Hippel, *Biochemistry* **22**, 4730 (1983).
[25] A. D. Riggs, H. Suzuki, and S. Bourgeois, *J. Mol. Biol.* **48**, 67 (1970).
[26] M. Ptashne, *Nature (London)* **214**, 232 (1967).

of steroid hormone receptors since radioactive steroid ligands have become readily available for tracing the migration of the receptor proteins.

DNA Footprinting (Nuclease Protection). In DNA footprinting assays, a DNA fragment that is radioactively labeled at one end is incubated with a DNA-binding protein and subsequently subjected to controlled degradation by nucleases, such as DNase I[12,21,27] or exonuclease III,[28,29] or by low molecular weight chemical reagents, such as dimethyl sulfate[28,30] or methidiumpropyl-EDTA · Fe(II).[31] The DNA is then electrophoresed on high-resolution denaturing polyacrylamide gels of the type used for DNA sequencing. Regions of the DNA fragments that were protected by bound protein will lead to the absence of bands in corresponding regions of the sequencing gel.

Methylation Interference Assays. Methylation interference assays complement footprinting assays. End-labeled DNA fragments are methylated prior to incubation with a DNA-binding protein, and are then electrophoresed on a DNA sequencing gel in a lane adjacent to a standard footprinting lane to determine if methylation of C or G nucleotides in the footprint region interferes with binding of the protein of interest.[21]

Together, these assays provide the most precise identification of the DNA-binding sites of any technique currently available, but, as is the case for gel retardation, footprinting assays do not allow specific identification of the DNA-binding protein and thus require purified protein to allow definitive interpretation of results.

In contrast, Southwestern blot analysis can be performed with crude protein extracts and requires only small amounts of radiolabeled DNA.[1–3] Southwestern blot analysis further permits identification of DNA-binding proteins by their relative mobility in the polyacrylamide gel, by ^{3}H-labeled steroid binding, in the case of steroid receptors, and by immunodetection, if appropriate antibodies are available.

Principle of Method

Southwestern blot analysis provides a means for detecting selective interactions between DNA and steroid hormone receptors or other DNA-

[27] D. Galas and A. Schmitz, *Nucleic Acids Res.* **5,** 3157 (1978).
[28] U. Siebenlist, R. T. Simpson, and W. Gilbert, *Cell* **20,** 269 (1080).
[29] C. Wu, *Nature (London)* **317,** 84 (1985).
[30] D. Gidoni, W. S. Dynan, and R. Tjian, *Nature (London)* **312,** 409 (1984).
[31] M. W. Van Dyke, R. P. Hertzberg, and P. B. Dervan, *Proc. Natl. Acad. Sci. U.S.A.* **79,** 5470 (1982).

binding proteins after immobilization of the electrophoretically separated proteins on nitrocellulose filters.[1-3] The method involves the following general steps, as illustrated schematically in Fig. 1. (1) Protein extracts are electrophoretically separated on polyacrylamide gels, using any of a number of electrophoresis protocols.[32-34] (2) After electrophoresis, the polyacrylamide gels are incubated in a 4 M urea buffer to remove sodium dodecyl sulfate (SDS) and facilitate partial renaturation of the proteins in the gels. (3) Following the renaturation step, the proteins are electrophoretically transferred onto nitrocellulose filters. (4) The protein-containing nitrocellulose filters are pretreated with a buffer containing 5% (w/v) nonfat dry milk to block nonspecific DNA binding, and are subsequently incubated with [^{32}P]DNA fragments. (5) Protein bands exhibiting selective interactions with the labeled DNA fragment are detected by autoradiography.

Reagents and Buffers

Cell culture medium: Joklik's minimal essential medium (JMEM) containing 2 mM glutamine, 75 units of penicillin G per milliliter, 50 units of streptomycin sulfate per milliliter, and 3.0% (v/v) of a 1 : 1 (v/v) mixture of fetal calf serum and calf serum

Glucocorticoids: [^{3}H]Dexamethasone mesylate, a glucocorticoid affinity label[35] [48.9 Ci/mmol (1 Ci = 37 GBq); New England Nuclear, Boston, MA]; [^{3}H]dexamethasone, a high-affinity glucocorticoid analog (45.8 Ci/mmol; New England Nuclear)

Homogenization buffer: 10 mM Tris-HCl (pH 8.3) and 1 mM ethylenediaminetetraacetic acid (EDTA)

Dextran-coated charcoal: 1.0% (w/v) activated charcoal in 0.1% (w/v) dextran and 1.5 mM MgCl$_2$

Renaturation buffer: 50 mM NaCl, 10 mM Tris-HCl (pH 7.0), 20 mM EDTA, 0.1 mM dithiothreitol, and 4 M urea

Nitrocellulose filters: BA 85 (Schleicher & Schuell, Keene, NH)

Blotto, DNA binding buffer: 5% (w/v) nonfat dry milk (Carnation) in 50 mM NaCl, 10 mM Tris HCl (pH 7.4), and 1 mM EDTA.[36]

[32] G. Fairbanks, T. L. Steck, and D. H. F. Wallach, *Biochemistry* **10**, 2606 (1971).

[33] U. K. Laemmli, *Nature (London)* **227**, 680 (1970).

[34] P. H. O'Farrell, *J. Biol. Chem.* **250**, 4007 (1975).

[35] S. S. Simons and E. B. Thompson, *Proc. Natl. Acad. Sci. U.S.A.* **78**, 3541 (1981).

[36] D. A. Johnson, J. W. Gautsch, J. R. Sportsman, and J. H. Elder, *Gene Anal. Tech.* **1**, 3 (1984).

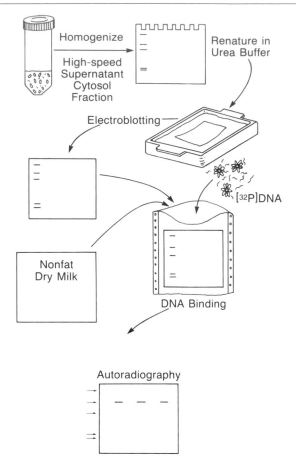

FIG. 1. Schematic flow diagram of Southwestern blot protocol. Crude or partially purified protein extracts are electrophoresed on polyacrylamide gels. Gels are incubated in a urea-containing buffer to remove SDS and permit renaturation of the proteins in the gels. Proteins are then electrophoretically transferred onto nitrocellulose filters. The nitrocellulose filters are incubated with a buffer containing 5% (w/v) nonfat dry milk (Carnation) to minimize nonspecific binding, and are subsequently incubated with a [^{32}P]DNA fragment to permit DNA binding. Protein bands that bind the labeled DNA are visualized by autoradiography.

Methods

Preparation of Glucocorticoid Receptor Extracts from HeLa S$_3$ Cells

HeLa S$_3$ cells are grown in suspension culture at 37° in Joklik's minimal essential medium (JMEM) containing 2 mM glutamine, 75 units of penicillin G per milliliter, 50 units of streptomycin sulfate per milliliter, and 3.0% (v/v) of a 1 : 1 (v/v) mixture of fetal calf serum and calf serum. Cells are harvested by centrifugation at 2500 g for 5 min at 4°, washed in cold unsupplemented JMEM to dilute endogenous steroids, and resuspended to a final cell density of 2–4 × 10^7 cells/ml. Typically approximately 2–4 × 10^8 cells are incubated with 10–20 nM [3H]-labeled steroid for 2 hr at 0°. The glucocorticoid affinity label, [3H]dexamethasone mesylate, is used in experiments for which we wish to link the [3H]-labeled steroid covalently to the glucocorticoid receptor in order to follow migration of the receptor in polyacrylamide gels. If covalent labeling is not necessary, 10–20 nM [3H]dexamethasone is used instead, with no apparent loss in subsequent DNA-binding activity, even though the steroid appears to dissociate from the receptor during electrophoresis. With either steroid, incubation of intact cells at 4° maintains the glucocorticoid receptor in a nonactivated state during steroid binding and reduces the potential for proteolytic degradation of receptor. Following steroid binding, the cells are pelleted and resuspended in an equal volume of ice-cold 10 mM Tris-HCl at pH 8.3, 1 mM EDTA and homogenized by using a prechilled Tekmar Ultra Turrax homogenizer (Tekmar Co., Cincinnati, OH). Homogenization under conditions of alkaline pH induces the formation of a covalent bond between [3H]dexamethasone mesylate and a cysteine residue of the receptor.[22] The homogenate is immediately centrifuged at 100,000 g for 1 hr at 0°. The high-speed supernatant obtained (cytosol) is collected and added to the pellet from an equal volume of dextran-coated charcoal [1.0% (w/v) activated charcoal in 0.1% (w/v) dextran, and 1.5 mM MgCl$_2$]. The cytosol/charcoal suspension is incubated on ice for 5 min to adsorb excess steroid. The charcoal is removed by centrifugation at 7000 g for 10 min at 0°, and the supernatant is collected. Aliquots of cytosol are mixed 1 : 1 (v/v) with the electrophoresis sample buffer appropriate for the gel system being used, and heated at 100° for 5 min. Samples may then be electrophoresed immediately or frozen at −70° for later use.

Polyacrylamide Gel Electrophoresis of Receptor Extracts

We have successfully performed Southwestern blot analysis after separation of protein extracts by a variety of polyacrylamide gel electrophoresis protocols, including one-dimensional polyacrylamide gel electrophoresis based on the procedures of either Fairbanks et al.[32] or Laemmli,[33] and two-dimensional polyacrylamide gel electrophoresis as described by O'Farrell.[34]

One-Dimensional Polyacrylamide Gel Electrophoresis. For electrophoresis on Fairbank's SDS-polyacrylamide gels,[32] cytosol protein extracts are mixed 1 : 1 (v/v) with Fairbank's sample buffer [2% SDS, 20 mM Tris, 2 mM Na$_2$EDTA, 10% (w/v) sucrose, and 20 μg of pyronin Y tracking dye per milliliter at pH 7.5], and heated at 100° for 5 min. Aliquots (2 ml) of the denatured cytosols are electrophoresed on preparative slab gels (16 × 16 × 0.3 cm) consisting of a 3.0% (w/v) acrylamide stacking gel and a 7.5% (w/v) acrylamide separating gel in a Hoeffer model SE 600 vertical gel electrophoresis unit until the tracking dye reaches the bottom of the gel.

Alternatively, samples are mixed 1 : 1 (v/v) with Laemmli sample buffer and heated for 5 min at 100° in preparation for electrophoresis in the Laemmli discontinuous buffer system.[25]

Two-Dimensional Polyacrylamide Gel Analysis. Samples of cytosol containing steroid receptor complexes are mixed 1 : 1 with lysis buffer [9.5 M urea, 2% (w/v) Nonidet P-40, 1.6% (w/v) ampholytes of pI 6–8, 0.4% (w/v) ampholytes of pI 3–10, and 2% (w/v) SDS] and stored at $-70°$ until use. Tube gels (115 × 3 mm) consisting of 10 M urea, 3.69% (w/v) acrylamide, 0.21% (w/v) methylenebisacrylamide, 2% (w/v) Nonidet P-40, 4% (w/v) ampholytes of pI 6–8, and 1% (w/v) ampholytes of pI 3–10 are prepared as described by O'Farrell.[34] After this treatment, samples that have been thawed and sonicated are layered onto tube gels, followed by the overlay solution and cathode buffer (20 mM NaOH), and are subjected to isoelectric focusing for a total of 5000–9000 V-hr. Anode buffer consists of 10 mM phosphoric acid. After electrophoresis, gels are extruded from the tubes, placed in SDS-PAGE sample buffer [10% (v/v) glycerol, 2.3% (w/v) SDS, 0.0625 M Tris-HCl, pH 6.8], frozen in a dry ice–ethanol bath, and stored at $-70°$ until use. The tube gels are electrophoresed in the second dimension on a 5–12% (w/v) polyacrylamide gradient gel with a 3% (w/v) acrylamide stacking gel (45–60 mA/gel). Gels are then transferred to nitrocellulose.[37] The nitrocellulose is dried and sprayed with EN^{3}HANCE (New England Nuclear) to visualize [^{3}H]dexamethasone mesylate. Alternatively, gels are incubated with renaturation buffer (see Blotting Procedure, below), transferred to nitrocellulose, and probed with nick-translated mouse mammary tumor virus (MMTV) DNA.

Blotting Procedure

Sodium Dodecyl Sulfate Removal and Protein Renaturation. After separation of proteins from crude cytosolic extracts by polyacrylamide gel electrophoresis, gels are incubated for two 1-hr washes in 200 ml of renaturation buffer (50 mM NaCl, 10 mM Tris-HCl at pH 7.0, 20 mM

---

[37] H. Towbin, T. Staehelin, and J. Gordon, *Proc. Natl. Acad. Sci. U.S.A.* **76,** 4350 (1979).

EDTA, 0.1 mM dithiothreitol, and 4 M urea)[1] with gentle agitation at room temperature. Although the precise degree to which proteins in the gel are renatured is not known, the 4 M urea washes apparently remove SDS from the proteins and permit functional restoration of the DNA-binding capacity of the glucocorticoid receptor.

Electroblotting. Following the urea wash steps, the proteins are electrophoretically transferred onto nitrocellulose filters by the method of Towbin *et al.*[37] Briefly, electrophoretic transfers are conducted in a 4° cold room, using a Hoeffer model TE-50 electrophoretic transfer apparatus with the gel/nitrocellulose filter sandwich immersed in 25 mM Tris, 192 mM glycine, 20% (v/v) methanol at pH 8.3. The gel/filter sandwiches are inserted into the transfer apparatus with the gels oriented toward the negative electrode and the filters toward the positive electrode. Transfers are begun at 0.5 A, 100 V and continued for approximately 1.5–2 hr, until the current rises to 1.0 A. *Note*: It is important to monitor the progress of transfers because the current will rise more rapidly as the temperature of the buffer increases, and the apparatus can seriously overheat if left unattended for too long. We prefer to use prestained protein molecular weight standards (GIBCO-Bethesda Research Laboratories, Gaithersburg, MD), which give a direct indication of the efficiency of transfer, even though they do not migrate absolutely true to their expected molecular weights. In our experience, transfers conducted under the conditions described are essentially quantitative for proteins with molecular weights less than 120 K. Transfer of much larger proteins, however, is clearly not complete. When transfers are stopped, the nitrocellulose filters are each marked with a black ball-point pen to indicate the position of the top of the gel after electroblotting. Filters are then lifted off the gels, placed onto Whatman (Clifton, NJ) 3MM filter paper moistened with transfer buffer, and stored in plastic bags at 4° for up to several days prior to use.

Isolation and Labeling of DNA Fragments. Plasmid DNAs are digested with the appropriate restriction enzymes, and the resulting fragments are electrophoresed on 6.5% (w/v) polyacrylamide gels containing 90 mM Tris–borate and 2 mM EDTA. Bands containing the desired DNA fragments are excised from the gels. The gel slices are crushed with a glass rod and extracted with 500 mM ammonium acetate, 10 mM magnesium acetate, 1 mM EDTA, and 0.1% (w/v) SDS[38] overnight in a 37° shaker–incubator. Samples are then centrifuged at 7000 g for 10 min at room temperature. The supernatants are collected, and the DNA fragments are concentrated by precipitation with ethanol. The isolated DNA fragments

---

[38] A. M. Maxam and W. Gilbert, this series, Vol. 65, p. 499.

are labeled with $[\alpha\text{-}^{32}P]dCTP$ by nick translation[39,40] (specific activity, $\sim 10^8$ cpm/μg of DNA). Alternatively, DNA fragments or synthetic oligonucleotides may be 5' end labeled by T4 polynucleotide kinase with $[\gamma\text{-}^{32}P]ATP$, or DNA fragments with recessed restriction ends may be labeled by using Klenow fragment of DNA polymerase and $[\alpha\text{-}^{32}P]dNTPs$ to fill in the recessed ends.

Filter Blocking and DNA Binding. Vertical strips are cut from the nitrocellulose filters, placed into heat-sealable plastic bags containing binding buffer [5% (w/v) nonfat dry milk in 50 mM NaCl, 10 mM Tris-HCl (pH 7.4), and 1 mM EDTA],[36] and pretreated for 2 hr with gentle agitation to block nonspecific DNA binding. The preincubation buffer is then removed and replaced with 5 ml of binding buffer containing 6×10^6 cpm of $[^{32}P]DNA$ per milliliter, and the filters are incubated at room temperature for 3 hr with gentle agitation. After DNA binding, the filter strips are washed for four 30-min washes in 50–100 ml of binding buffer, rinsed briefly in STE buffer (100 mM NaCl, 10 mM Tris-HCl at pH 7.6, and 1 mM EDTA), and air dried. The filter strips are placed in X-ray cassettes with Kodak X-ray film and exposed for 24 to 48 hr to detect protein bands that bound the $[^{32}P]DNA$.

Results

Southwestern blot analysis was used to study the interaction between human glucocorticoid receptors and DNA. We wished to determine if the DNA-binding domain of human glucocorticoid receptors retained the ability to interact selectively with glucocorticoid response element (GRE)-containing DNAs after immobilization of the receptors on nitrocellulose filters. To address this question, three distinct DNA fragments were chosen for analysis. One fragment chosen was a 326-bp *Hae*III–*Hpa*II DNA fragment obtained from the long terminal repeat of mouse mammary tumour virus (MMTV-LTR). This DNA fragment corresponds to nucleotides -222 to $+104$ relative to the primary transcription start site in the 5' long terminal repeat of MMTV. This region of the MMTV-LTR DNA has been shown to contain three regions of sequence protected to varying degrees in nuclease protection experiments with purified glucocorticoid receptor[12] and has further been shown to confer hormone-responsive transcriptional regulation onto linked heterologous genes in DNA transfection experiments.[13] The nuclease-protected regions in this DNA fragment, in combination with protected regions in other parts of the MMTV genome, were

[39] P. W. J. Rigby, M. Dieckmann, C. Rhodes, and P. Berg, *J. Mol. Biol.* **113,** 237 (1977).
[40] J. Meinkoth and G. M. Wahl, this series, Vol. 152, 91.

used to develop the first identification of a GRE DNA consensus sequence.[12] Two additional DNA fragments, similar in size to the MMTV-LTR DNA fragment, were chosen from the plasmid, pBR322. The pBR322, *Taq*I–E DNA fragment is a 312-bp DNA that contains no regions of sequence exhibiting detectable homology with the GRE DNA consensus sequence and the pBR322, *Taq*I–D DNA fragment is a 368-bp DNA fragment that contains a single octanucleotide matching the DNA consensus sequence for a GRE half-site.[12,15] Each of these three DNA fragments was labeled with ^{32}P by nick translation to a specific activity of approximately 10^8 cpm/μg DNA.

HeLa S$_3$ cells were incubated with 20 nM [^{3}H]dexamethasone mesylate for 2 hr and a cytosol was prepared as described in Methods. An aliquot of cytosol was subjected to electrophoresis on a Fairbank's SDS polyacrylamide gel. The gel was incubated in renaturation buffer and proteins from the gel were subsequently transferred onto a nitrocellulose filter. Identical adjacent lanes were cut from the resulting filter and incubated with each of the three [^{32}P]DNA fragments described above. Figure 2 shows the autoradiographs obtained after incubation of the nitrocellulose filter strips with these three DNA fragments. The filter strip incubated with the GRE-containing MMTV-LTR DNA fragment (lane 2) showed a prominent band of DNA-binding activity that migrated near the phosphorylase *b* marker at a position consistent with the expected relative mobility of the human glucocorticoid receptor. By contrast, in lane 1, the nitrocellulose filter-immobilized glucocorticoid receptor band showed barely detectable binding of the pBR322, *Taq*I–E DNA fragment, which contains no GRE sequences, but showed appreciable binding of the pBR322, *Taq*I–D DNA fragment containing the single GRE half-site (lane 3). The identity of the DNA-binding activity shown by an approximately 18K protein seen in all three lanes is unknown. Because the apparent specificity of binding of these three DNA fragments is similar to that of the glucocorticoid receptor, it is possible that this low molecular weight protein may be a proteolytic fragment of the receptor. We have not experimentally investigated this possibility.

We next sought to analyze the DNA-binding characteristics of human glucocorticoid receptors immobilized on nitrocellulose filters after two-dimensional polyacrylamide gel electrophoresis. Our previous analysis of the affinity-labeled glucocorticoid receptor on two-dimensional gels[41] had demonstrated the presence of distinct isoelectric point variants, which were indicative of variations in the physiological state of the cell. Two-dimensional gel electrophoresis[34] separates proteins by ionic charge in the

---

[41] J. A. Cidlowski and V. Richon, *Endocrinology* **115**, 1588 (1984).

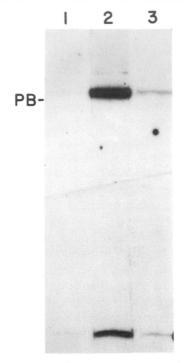

FIG. 2. Selective binding of isolated [^{32}P]DNA fragments by [^{3}H]dexamethasone mesylate affinity-labeled glucocorticoid receptor immobilized on nitrocellulose filters. HeLa S$_3$ cells were incubated with 20 nM [^{3}H]dexamethasone mesylate for 2 hr, a cytosol was prepared, and samples were treated as described for analysis on a Fairbank's SDS-polyacrylamide gel. The gel was incubated in renaturation buffer and proteins from the gel were subsequently transferred onto a nitrocellulose filter as described. Adjacent lanes were cut from the resulting filter and incubated with each of three distinct DNA fragments, labeled with ^{32}P to a specific activity of approximately 10^8 cpm/ug DNA. Lane 1 shows the autoradiograph of the filter strip incubated with 7.5 × 10^6 cpm of the 312-bp *Taq*I–E fragment of pBR322 DNA, which contains no detectable GREs. Lane 2 shows the autoradiograph of an adjacent filter strip incubated with 7.5 × 10^6 cpm of the 325-bp *Hae*III–*Hpa*II fragment derived from the long terminal repeat of mouse mammary tumor virus DNA, which has been shown in DNA transfection experiments to confer hormone responsiveness onto linked heterologous genes and which contains three regions of sequence protected by purified glucocorticoid receptor in DNA footprinting experiments. Lane 3 shows the autoradiograph of a filter strip incubated with 7.5 × 10^6 cpm of the 368-bp *Taq* I–D fragment of pBR322 DNA, which contains a single occurrence of an octanucleotide matching the DNA consensus sequence for a GRE half-site. The position of the 97K phosphorylase *b* protein marker is indicated by PB.

first dimension and by molecular weight in the second dimension, thus permitting discrimination among similarly sized, differently charged proteins. A cytosol was prepared from HeLa S_3 cells incubated with the glucocorticoid affinity label, [^{3}H]dexamethasone mesylate. Aliquots of the cytosol were subjected to two-dimensional electrophoresis on replicate gels. One of the gels (Fig. 3A) was stained with Coomassie blue and showed a complex pattern of staining, as would be expected from the large number of proteins presumed to be present in a crude cytosol extract. A second gel (Fig. 3B) was electrophoretically transferred onto a nitrocellulose filter, which was subsequently sprayed with EN^{3}HANCE (New England Nuclear) and fluorographed at −70° for 5 days. The fluorograph in Fig. 3B shows the saturable [^{3}H]dexamethasone mesylate binding pattern typically seen after two-dimensional gel electrophoresis of cytosol extracts. The glucocorticoid receptor focused as a series of charged species, with molecular weights of approximately 97K and isoelectric points ranging from 7.7 to 7.2. The third replicate gel (Fig. 3C) was incubated with the urea-containing renaturation buffer and then electrophoretically transferred onto a nitrocellulose filter. The resulting filter was incubated with MMTV-LTR [^{32}P]DNA. As seen in Fig. 3C, a series of spots of bound MMTV-LTR [^{32}P]DNA at approximately 97K comigrated with the bound [^{3}H]dexamethasone mesylate seen in Fig. 3B and also ranged in pI from approximately 7.7 to 7.2 (Fig. 3C). This comigration of the MMTV-LTR [^{32}P]DNA-binding activity with the [^{3}H]dexamethasone mesylate-labeled band on two-dimensional gels strongly suggested that the DNA binding observed was attributable to glucocorticoid receptor. The binding buffer used in the two-dimensional gel series of experiments did not contain 5% (w/v) nonfat dry milk as blocking agent, which may account for the greater extent of DNA binding observed in the lower molecular weight regions of the gel in Fig. 3C, as compared with the extent of binding of this DNA fragment observed in lane 2 of Fig. 2.

Discussion

In our initial use of the Southwestern blotting procedure to study glucocorticoid receptor–DNA interactions we employed a variety of blocking agents in an effort to minimize background binding of the [^{32}P]DNA fragments. Different blocking reagents, including bovine serum albumin (fraction V), sonicated *Escherichia coli* DNA, and poly(dI–dC), produced varying degrees of success as blocking agents, but none was as effective as 5% nonfat dry milk, implemented as suggested by Johnson *et al.*[36] An indication of the effectiveness of nonfat dry milk as a blocking reagent can be seen by comparing the extent of binding of the MMTV-

Protein Staining

pI   4.3   6.4   7.3   7.7

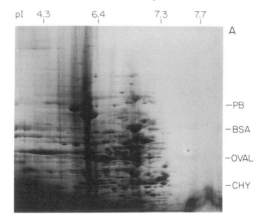

A

—PB
—BSA
—OVAL
—CHY

Steroid Binding

pI   4.3   6.4   7.3   7.7

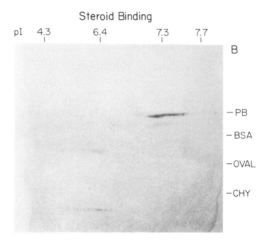

B

—PB
—BSA
—OVAL
—CHY

DNA Binding

pI   4.3   6.4   7.3   7.7

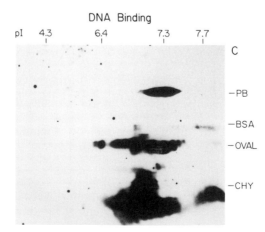

C

—PB
—BSA
—OVAL
—CHY

LTR [^{32}P]DNA fragment in Fig. 3 C, where 5% nonfat dry milk was not present in the binding buffer, with the binding of this same DNA fragment seen in lane 2 of Fig. 2, where 5% nonfat dry milk was used as the blocking reagent. Because nonfat dry milk is inexpensive, readily available, and at least as effective as any other blocking reagent we have tried, we have used it as the blocking reagent of choice in nearly all subsequent blotting experiments. We have routinely used Carnation brand nonfat dry milk because it is readily available and has proved to be of consistently good quality with regard to ease of solubility in aqueous buffers. We have not systematically investigated the composition of the nonfat dry milk. Presumably, it would be enriched in milk proteins and glycoproteins and in lactose and other carbohydrates. We have some preliminary indication[42] that it may also contain appreciable quantities of nucleic acids. This would not be surprising, because milk contains lymphocytes whose nucleic acids would not necessarily be removed by the defatting process employed in the production of nonfat dry milk.

In our original description of this procedure,[3] we presented several lines of evidence to satisfy ourselves and others that the DNA-binding activity we observed could, in fact, be attributed to glucocorticoid receptor. Observations that support the conclusion that the approximately 97K protein with which the GRE-containing DNA interacts is the glucocorticoid receptor include the following: (1) comigration of MMTV-LTR DNA-binding activity with the affinity-labeled glucocorticoid receptor on both one- and two-dimensional gels, (2) the detection of MMTV-LTR DNA-binding activity in subcellular fractions that parallel the subcellular distribution of glucocorticoid receptor under defined conditions (i.e., cytosolic localization prior to activation), (3) the detection of DNA-binding activity

[42] D. B. Tully and J. A. Cidlowski, unpublished results, 1987.

FIG. 3. Two dimensional polyacrylamide gel electrophoresis of cytoplasmic glucocorticoid receptor. HeLa S$_3$ cells were incubated with 20 nM [^{3}H]dexamethasone mesylate for 2 hr, a cytosol was prepared, and samples were treated as described. In the first dimension samples were subjected to denaturing isoelectric focusing, and in the second dimension samples were separated by SDS-polyacrylamide gel electrophoresis. Three replicate gels were prepared. (A) Gel A was stained with Coomassie blue. (B) Gel B was electrophoretically transferred to nitrocellulose and the filter was sprayed with EN^{3}HANCE and fluorographed to detect the locations of [^{3}H]dexamethasone mesylate-labeled proteins. (C) Gel C was treated in renaturation buffer, transferred to nitrocellulose, and the resulting filter probed with nick-translated MMTV-LTR [^{32}P]DNA and autoradiographed. Positions of prestained molecular weight markers electrophoresed as standards (PB, 97K phosphorylase b; BSA, 68K bovine serum albumin; OVAL, 43K ovalbumin; CHY, 25K chymotrypsinogen) are shown to the right of each panel.

in purified preparations of receptor immobilized on nitrocellulose, and (4) the absence of detectable DNA-binding activity in cytosol extracts from cells that have been depleted of glucocorticoid receptor. Based on these results, we concluded that the Southwestern blotting procedure could be beneficially employed in studies of glucocorticoid receptor interactions with DNA. Although we are not aware of specific cases in which Southwestern blotting has been employed to study interactions of other steroid receptors with DNA, given that other steroid receptors have comparable affinities for their cognate HREs and that there is a high degree of sequence homology among all the members of the steroid hormone receptor superfamily, we can see no reason to expect that the technique would not work.

Others have shown that the Southwestern blotting procedure, as implemented here or with minor modifications, can be used to study interactions between DNA and a wide variety of other DNA-binding proteins.[4–8] In particular, the technique has been applied to assist in the identification and isolation of new DNA-binding proteins that recognize known binding sites in DNA.[4–7] Southwestern blotting is also well adapted for qualitative comparison of glucocorticoid receptor binding of different DNA fragments, as shown, for example, in Fig. 2. In addition, we have used this procedure to compare the relative binding activity of nitrocellulose filter-immobilized glucocorticoid receptors for a series of point mutations of the GRE consensus sequence produced by doped synthesis of a GRE oligonucleotide and subsequent cloning into an M13 library.[43,44] The Southwestern blotting technique thus provides a versatile protocol that can be used to study the interaction of a known DNA-binding protein with various DNAs containing known or suspected binding sites, or alternatively can be used to identify proteins that bind with high affinity to a particular DNA sequence.

Southwestern blotting, unlike nuclease protection or methylation interference studies, does not require prior purification of the DNA-binding protein, and can, in fact, be used to study DNA–protein interactions in crude cellular extracts or in partially purified protein preparations at essentially any stage of purification. Southwestern blotting offers further advantages in that once a protein extract has been electrophoretically separated and finally immobilized on a nitrocellulose filter, proteins of interest can potentially be characterized by a number of criteria, including relative mobility on the SDS-polyacrylamide gel, isoelectric point on two-

[43] C. A. Hutchison, III, S. K. Nordeen, K. Vogt, and M. H. Edgell, *Proc. Natl. Acad. Sci. U.S.A.* **83,** 710 (1986).
[44] D. B. Tully, C. A. Hutchison, and J. A. Cidlowski, unpublished results, 1987.

dimensional gels, DNA-binding activity, ligand binding, or immunoreactivity if antibodies for the protein are available.

Acknowledgments

We thank Dr. Corinne Silva for permission to use the experimental data presented in Fig. 3. We thank Dr. Kerry Burnstein for critical review of this manuscript. This work was supported by National Institutes of Health Grants DK 32459 and 32460.

[39] Specific Recognition Site Probes for Isolating Genes Encoding DNA-Binding Proteins

By HARINDER SINGH

Introduction

Sequence-specific DNA-binding proteins play central roles in DNA-based trans-actions including transcription, replication, and site-specific recombination. In genetically tractable prokaryotes and eukaryotes, most sequence-specific DNA-binding proteins have been characterized as products of trans-acting regulatory loci. In higher eukaryotes a similar genetic approach to their identification has not been possible. Instead sensitive biochemical assays, in particular DNase I footprinting[1] and gel electrophoresis of protein–DNA complexes,[2,3] have led to the identification and characterization of numerous sequence-specific DNA-binding proteins. A majority of these proteins bind selectively to distinct transcriptional control elements and are thereby implicated in regulating the activity of genes containing functional binding sites.[4] Even though the purification of sequence-specific DNA-binding proteins has been greatly facilitated by the development of oligonucleotide affinity matrices,[5] the requirement for large amounts of starting material (cells or tissue) makes purification on a preparative scale difficult. The strategy described in this chapter obviates purification of a sequence-specific DNA-binding protein for the purpose of isolating its gene. It simply requires an appropriate recombinant DNA library constructed for expression in *Escherichia coli,* typically cDNAs in

[1] D. Galas and A. Schmitz, *Nucleic Acids Res.* **5,** 3157 (1978).
[2] M. Fried and D. Crothers, *Nucleic Acids Res.* **9,** 6505 (1981).
[3] M. Garner and A. Revzin, *Nucleic Acids Res.* **9,** 3047 (1981).
[4] T. Maniatis, S. Goodbourn, and J. A. Fischer, *Science* **236,** 1237 (1987).
[5] J. T. Kadonaga and R. Tijian, *Proc. Natl. Acad. Sci. U.S.A.* **83,** 5889 (1986).

λgt11, and a DNA recognition site probe. Therefore this strategy is ideally suited for isolating clones encoding rare DNA-binding proteins.[6,7]

Principle

Sequence-specific DNA-binding proteins are modular in design and contain functionally as well as structurally isolatable DNA-binding domains.[8] Furthermore, in many such proteins the DNA-binding domains are contained within relatively small protein segments, approximately 60–200 amino acids, that can be expressed in a functional form as fusion proteins in bacteria.[9-11] The cloning strategy depends on the functional expression in *E. coli* of high levels of the DNA-binding domain of a regulatory protein and a specific interaction between this domain and its recognition site. If these conditions are fulfilled, a recombinant clone encoding a sequence-specific DNA-binding protein can be detected by probing protein replica filters of an expression library with radioactive recognition site DNA. An outline of the steps involved in identifying and analyzing such a clone, using a recombinant library constructed in the expression vector λgt11, is depicted in Fig. 1. The initial phase involves the identification of a recombinant clone that is specifically detected with the binding site DNA probe (X) but not with a mutant probe (X*) or a DNA probe (Y) that lacks the given binding site. The clone that binds to probe X is then purified. Such a clone is then shown to encode a β-galactosidase fusion protein of the expected DNA-binding specificity. This strategy is derived from that developed for the isolation of genes using antibodies to screen recombinant expression libraries.[12-14] The experimental steps are similar, except that the nitrocellulose filters carrying immobilized proteins are screened with ^{32}P-labeled double-stranded DNA rather than with an antibody.

It has been demonstrated that DNA probes containing a single recognition site can be used to isolate the relevant DNA-binding protein clones (MBP-1, XBP, and CREB; see Table I). However, in many cases the

[6] H. Singh, J. H. LeBowitz, A. S. Baldwin, and P. A. Sharp, *Cell* **52,** 415 (1988).
[7] H. Singh, R. G. Clerc, and J. H. LeBowitz, **7,** 252 (1989).
[8] R. Schleif, *Science* **241,** 1182 (1988).
[9] I. A. Hope and K. Struhl, *Cell* **46,** 885 (1986).
[10] L. Keegan, G. Gill, and M. Ptashne, *Science* **231,** 699 (1986).
[11] D. R. Rawlins, G. Milman, S. D. Hayward, and G. S. Hayward, **42,** 859 (1985).
[12] S. Broome and W. Gilbert, *Proc. Natl. Acad. Sci. U.S.A.* **75,** 2746 (1978).
[13] D. M. Helfman, J. R. Feramisco, J. C. Fiddes, G. P. Thomas, and S. H. Hughes, *Proc. Natl. Acad. Sci. U.S.A.* **80,** 31 (1983).
[14] R. A. Young and R. W. Davis, *Science* **222,** 778 (1983).

λgt11 library

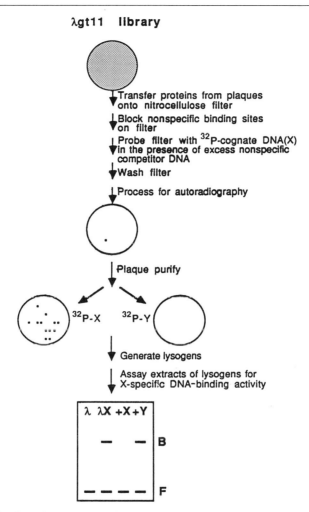

Fig. 1. Outline of the strategy for the molecular cloning of sequence-specific DNA-binding proteins using the expression vector λgt11. X is a recognition site DNA probe, whereas Y is a control DNA probe that lacks the given recognition site. The initial phase involves the identification of λgt11 recombinants that are specifically detected with DNA probe X (λX). After plaque purification, the gel electrophoresis DNA-binding assay is used to analyze extracts of λX and λgt11 (λ) lysogens. Radiolabeled X-DNA is used as a probe in the binding reactions. F and B refer to free and bound X-DNA, respectively. Reactions in lanes +X and +Y are carried out with the λX extracts and contain an excess of either unlabeled X-DNA or unlabeled Y-DNA as competitors. (Reproduced with permission from Ref. 7.)

TABLE I

CLONES ENCODING SEQUENCE-SPECIFIC DNA-BINDING PROTEINS ISOLATED BY
SCREENING OF λ LIBRARIES WITH RECOGNITION SITE DNA PROBES

Clone	Binding site/oligonucleotide probe	Organism	Refs.
MBP-1	GGGGATTCCCC	Human	(6, 15)
Oct-2	ATGCAAAT	Human	(16–18)
Oct-1	ATGCAAAT	Human	(19)
E12	GGCAGGTG	Human	(20)
XBP	GCTGGCAACTGTGTGACGTCATCACAAGA	Mouse	(21)
RF-X	CCCCCTAGCAACAG	Human	(22)
IRF-1	AAGTGA	Mouse	(23)
Pit-1	GATTACATGAATATTCATGA	Rat	(24)
CREB	TGACGTC	Human	(25)
CNBP	GTGCGGTG	Human	(26)
HBP-1	ACGTCA	Wheat	(27)
ITF-1, 2	AACACCTGCAGCAGCAGCTGGCAGG	Human	(28)
Isl-1	TTAATAATCTAATTA	Rat	(29)
PU.1	GAGGAA	Mouse	(30)
ICSBP	AGTTTCACTTCT	Mouse	(31)
Ets-1	AGCCACATCCTCTGGAA	Human	(32)
OBF1	GATCCAAGTGCCGTGCATAATGATGTGGG	Yeast	(33)
FRG-Y1, Y2	CTGATTGGCCAA	*Xenopus*	(34)
TEF-1	GTGGAATGT	Human	(35)
HTF4	CAGCTGG	Human	(36)
NF-E1	CCTCCATC	Human	(37)

signal was appreciably enhanced with concatenated DNA probes containing several copies (3–10) of the appropriate binding site (Oct-2, Oct-1, and E12; see Table I[15–37]). Enhancement of the signal with a multisite probe may be due to the fact that such a probe can simultaneously interact with two or more immobilized protein molecules, thereby increasing the overall

[15] A. S. Baldwin, Jr., K. P. LeClair, H. Singh, and P. A. Sharp, *Mol. Cell. Biol.* **10,** 1406 (1990).

[16] R. G. Clerc, L. M. Corcoran, J. H. LeBowitz, D. Baltimore, and P. A. Sharp, *Genes Dev.* **2,** 1570 (1988).

[17] M. M. Müller, S. Ruppert, W. Schaffner, and P. Matthias, *Nature (London)* **336,** 544 (1988).

[18] L. M. Staudt, R. G. Clerc, H. Singh, J. H. LeBowitz, P. A. Sharp, and D. Baltimore, *Science* **241,** 577 (1988).

[19] R. A. Sturm, G. Das, and W. Herr, *Genes Dev.* **2,** 1582 (1988).

[20] C. Murre, P. Schonleber-McCaw, and D. Baltimore, *Cell* **56,** 777 (1989).

[21] L. Hsiou-Chi, M. R. Boothby, and L. H. Glimcher, *Science* **242,** 69 (1988).

[22] W. Reith, E. Barrass, S. Satola, M. Kobr, D. Reinhart, C. H. Sanchez, and B. Mach, *Proc. Natl. Acad. Sci. U.S.A.* **86,** 4200 (1989).

stability of the protein–DNA complexes. This type of probe is particularly suitable for the isolation of clones encoding DNA-binding proteins with low affinity for their recognition sites. The equilibrium association constants of sequence-specific DNA binding proteins range over many orders of magnitude (10^8 to 10^{12} M^{-1}). Assuming that a protein has an equilibrium association binding constant of 10^{10} M^{-1} and an association rate constant of 10^7 M^{-1} sec^{-1},[38] the dissociation rate constant of the resulting protein–DNA complex will be 10^{-3} sec^{-1} and its half-life will be approximately 10 min.[6] Thus, only one-eighth of the complexes formed in the binding incubation will survive a 30-min wash. This type of calculation suggests that screening of expression libraries with single-site oligonucleotide probes is unlikely to be successful if the equilibrium association binding constant of the DNA-binding protein is less than 10^9 M^{-1}. This problem appears to be alleviated by using a concatenated probe that contains multiple binding sites.

Using a ^{32}P-labeled concatenated binding site probe (specific activity, $\geq 10^8$ cpm/μg), it is possible to detect 10^{-2} fmol of active protein in a

[23] M. Miyamoto, T. Fujita, Y. Kimura, M. Maruyama, H. Harada, Y. Sudo, T. Miyata, and T. Taniguchi, *Cell* **54**, 903 (1988).

[24] H. A. Ingraham, R. Chen, H. J. Mangalam, H. P. Eisholtz, S. E. Flynn, C. R. Lin, D. M. Simmons, L. Swanson, and M. G. Rosenfeld, *Cell* **55**, 519 (1988).

[25] J. P. Hoeffler, T. E. Meyer, Y. Yun, J. L. Jameson, and J. F. Habener, *Science* **242**, 1430 (1988).

[26] T. B. Rajavashisth, A. K. Taylor, A. Andalibi, K. L. Svenson, and A. J. Lusis, *Science* **245**, 640 (1989).

[27] T. Tabata, H. Takase, S. Takayama, K. Mikami, A. Nakatsuka, T. Kawata, T. Nakayama, and M. Iwabuchi, *Science* **245**, 965 (1989).

[28] P. Henthorn, M. Kiledjian, and T. Kadesch, *Science* **247**, 467 (1990).

[29] O. Karlsson, S. Thor, T. Norberg, H. Ohlsson, and T. Edlund, *Nature (London)* **344**, 879 (1990).

[30] M. J. Klemsz, S. R. McKercher, A. Celada, C. Van Beveren, and R. A. Maki, *Cell* **61**, 113 (1990).

[31] P. H. Driggers, D. L. Ennist, S. L. Gleason, W. Mak, M. S. Marks, B. Levi, J. R. Flanagan, E. Appella, and K. Ozato, *Proc. Natl. Acad. Sci. U.S.A.* **87**, 3743 (1990).

[32] I. Ho, N. K. Bhat, L. R. Gottschalk, T. Lindsten, C. B. Thompson, T. S. Papas, and J. M. Leiden, *Science* **250**, 814 (1990).

[33] E. E. Biswas, M. J. Stefanec, and S. B. Biswas, *Proc. Natl. Acad. Sci. U.S.A.* **87**, 6689 (1990).

[34] S. R. Tafuri and A. P. Wolffe, *Proc. Natl. Acad. Sci. U.S.A.* **87**, 9028 (1990).

[35] J. H. Xiao, I. Davidson, H. Matthes, J. Garnier, and P. Chambon, *Cell* **65**, 551 (1991).

[36] Y. Zhang, J. Babin, A. L. Feldhaus, H. Singh, P. A. Sharp, and M. Bina, *Nucleic Acids Res.* **19**, 4555 (1991).

[37] K. Park and M. L. Atchison, *Proc. Natl. Acad. Sci. U.S.A.* **88**, 9804 (1991).

[38] O. G. Berg, R. B. Winter, and P. H. von Hippel, *Trends Biochem. Sci.* **7**, 52 (1982).

plaque (assuming a 1 : 1 stoichiometry for the protein–DNA complex). This detection limit represents 1 pg of a β-galactosidase fusion protein ($\sim M_r$ 170,000), which is an amount that is well below the expected level of expression for such a protein in a plaque of the desired λgt11 phage. In fact, overexpression of the lacZ fusion gene should result in the accumulation of 100 pg of the fusion protein in a phage plaque, assuming that there are 10^5 infected cells/plaque and that the β-galactosidase fusion protein represents 1% of the total protein mass (0.1 pg) of an infected cell.[7]

Overexpression of proteins in E. coli leads to the formation of large, insoluble aggregates termed inclusion bodies. In these aggregates only a small proportion of the fusion protein produced by the bacteriophage-infected cells is likely to be folded into a conformation suitable for binding to its DNA ligand. For some DNA-binding proteins the fraction of properly folded, that is, active molecules can be increased by exposing the filter-immobilized protein briefly to the denaturing agent, guanidine hydrochloride.[39] During this process, aggregates of the fusion protein are solubilized and partially denatured. The denaturant is then removed in a stepwise manner, allowing an increased proportion of the fusion protein to refold into a conformationally active form. However, it is important to note that exposure to the denaturant may not necessarily be beneficial because some proteins are not readily renatured after treatment with guanidine hydrochloride. The best course of action, therefore, is to screen two sets of filters, only one of which has been processed by denaturation and renaturation steps.

The prospects for successfully isolating the clone for a given sequence-specific DNA-binding protein are greatly enhanced if the native protein can be detected after Western blotting using the multisite DNA probe.[17] In this technique a crude fraction of the native protein, for example, a nuclear extract, is resolved by sodium dodecyl sulfate-polyacrylamide gel electrophoresis (SDS-PAGE). After electrophoresis proteins are electroblotted onto a nitrocellulose membrane. The membrane is then screened with a radioactive binding site probe. As indicated above for protein replica filters generated by the plating of a recombinant cDNA library, the electroblotted proteins can be screened as such with labeled binding site DNA or first subjected to a denaturation/renaturation cycle with guanidine hydrochloride and then screened. If this procedure detects the relevant DNA-binding protein it proves that a single polypeptide is capable of site-specific recognition. Thus, unless the DNA-binding protein of interest

---

[39] C. R. Vinson, K. L. LaMarco, P. F. Johnson, W. H. Landschulz, and S. L. McKnight, Genes Dev. 2, 801 (1988).

is a heterodimer in which both subunits are essential for site-specific recognition, it should be possible to clone the DNA-binding subunit by the above strategy of ligand-based expression screening. Furthermore, if the denaturation/renaturation cycle enhances the detection signal this protocol should be used to screen the expression library.

A given DNA-binding site can be recognized by a family of structurally related proteins and in some cases unrelated proteins.[4] Thus additional structural as well as functional analyses are necessary to definitively relate the recombinant protein cloned by ligand-based expression screening to that identified by previous biochemical analysis.[16]

Materials and Reagents

Escherichia coli Y1089 and Y1090
λgt11 cDNA library
Minimal binding site duplex oligonucleotides (wild type and mutant pairs)
Recombinant pUC plasmids containing 3–10 tandem copies of wild-type/mutant binding site
Isopropyl-β-D-thiogalactopyranoside (IPTG), 1 *M*: Prepare with sterile water and store at −20°
Sonicated calf thymus DNA (~1 kb), 1 mg/ml
Nitrocellulose filters and membrane (Schleicher & Schuell, Keene, NH)
Elutip-d columns (Schleicher & Schuell)
Hybridization chambers (Altec Plastics, Boston, MA)

Methods

Construction of Expression Library

cDNA Synthesis and Cloning. Successful screening is critically dependent on the frequency with which functional recombinants (in-frame fusions of the DNA-binding domain with a bacterial protein segment) are represented in a given cDNA expression library. The cDNA library should be made from mRNA isolated from a cell or tissue source with the highest levels of the desired DNA-binding protein. First-strand cDNA synthesis should be carried out using random primers rather than oligo(dT), because the DNA-binding domain may be encoded in the amino-terminal part of the desired protein (5' end of the corresponding mRNA). Adaptors rather than linkers are preferred for ligating the cDNA inserts to the vector, because their use does not require digestion of the cDNA with a restriction enzyme.[40,41]

Expression Vectors. The phage λgt11 appears most suitable for expression screening.[14] It offers the advantages of high cloning efficiency, the expression of relatively stable β-galactosidase fusion proteins, and a simple means of preparing protein replica filters. An alternate bacteriophage λ expression vector (λZAP) can be used.[42] This vector obviates subcloning of cDNA inserts into plasmid vectors for their analysis. The presence of multiple cloning sites makes possible the use of "forced cloning" strategies for expression of cDNA inserts from its *lac* promoter. Unlike λgt11, λZAP expresses fusion proteins containing a small amino-terminal segment of β-galactosidase. Therefore, the stability of λZAP-encoded fusion proteins may be different from that of their counterparts encoded in λgt11.

Screening of Expression Library

Preparation of Protein Replica Filters. Protein replica filters suitable for screening with DNA recognition site probes are most easily prepared using a series of steps derived from the immunoscreening protocol.[43] This simple procedure has permitted the isolation of many clones encoding different DNA-binding proteins, for example, MBP-1, Oct-2, E12, XBP, IRF-1, and CREB (see Table I). Vinson *et al.*[39] have shown that processing dried nitrocellulose replica filters through a denaturation/renaturation cycle, using 6 M guanidine hydrochloride, significantly enhances the signal from a λgt11 recombinant encoding C/EBP. As indicated earlier, the denaturation/renaturation cycle may in some cases increase the detection signal by facilitating the correct folding of a larger fraction of the *E. coli*-expressed protein. It may also help to dissociate insoluble aggregates that form as a consequence of overexpression. This modified procedure has been successfully used to isolate clones encoding Oct-1, Pit-1, and RF-X (see Table I). Because the denaturation/renaturation cycle may not necessarily be beneficial a given expression library should be screened by both protocols detailed below.

1. Plate λgt11 library on host strain *E. coli* Y1090 (3–5 × 10⁴ pfu/150-mm plate).

[40] C. P. Bahl, R. Wu, R. Brousseau, A. K. Sood, H. M. Hsiung, and S. A. Narang, *Biochem. Biophys. Res. Commun.* **81,** 695 (1978).

[41] H. Haymerle, J. Herz, G. M. Bressan, R. Frank, and K. K. Stanley, *Nucleic Acids Res.* **21,** 8615 (1986).

[42] J. M. Short, J. M. Fernandez, J. A. Sorge, and W. D. Huse, *Nucleic Acids Res.* **16,** 7583 (1988).

[43] T. V. Huynh, R. A. Young, and R. W. Davis, *in* "DNA Cloning: A Practical Approach" (D. M. Glover, ed.), p. 49. IRL Press, Oxford, 1985.

2. Incubate the LB plates at 42° until tiny plaques are visible (~3 hr).

3. In the meantime soak 132-mm nitrocellulose filters in 10 mM IPTG for 30 min and then allow them to air dry.

4. Overlay each LB plate with an IPTG-impregnated filter. Avoid trapping air bubbles between the filter and the top agar.

5. Incubate the LB plates at 37° for 6 hr.

6. Cool LB plates at 4° for 10 min. Mark the position of the filter on each plate.

The protocols diverge at this point: proceed either to steps 7 and 8, or to steps 7a–11a.

7. Lift the nitrocellulose filters and immerse in a deep dish containing Blotto [5% (w/v) nonfat milk powder (Carnation), 50 mM Tris-HCl (pH 7.5), 50 mM NaCl, 1 mM ethylenediaminetetraacetic acid (EDTA), 1 mM dithiothreitol (DTT)]. Incubate for 60 min at room temperature with gentle swirling on an orbital platform shaker.

8. Transfer filters to a dish containing binding buffer [10 mM Tris-HCl (pH 7.5), 50 mM NaCl, 1 mM EDTA, 1 mM DTT] and incubate as for step 7 for 5 min. Repeat this wash step twice with fresh binding buffer. Filters can be stored in binding buffer at 4° for up to 24 hr. prior to screening.

7a. Lift the nitrocellulose filters and air dry them for 15 min at room temperature.

8a. Immerse the filters in HEPES binding buffer [25 mM N-2-hydroxy-ethylpiperazine-N'-2-ethanesulfonic acid (HEPES), pH 7.9, 25 mM NaCl, 5 mM MgCl$_2$, 0.5 mM DTT] supplemented with 6 M guanidine hydrochloride. Incubate with gentle shaking at 4° for 10 min. Repeat this step with fresh HEPES binding buffer containing 6 M guanidine hydrochloride.

9. Incubate the filters in HEPES binding buffer containing 3 M guanidine hydrochloride for 5 min at 4° (a 1 : 1 dilution of the 6 M guanidine hydrochloride from the previous step). Repeat this step four times. Each time use HEPES binding buffer that contains a twofold dilution of the guanidine hydrochloride from the previous step.

10a. Incubate the filters in HEPES binding buffer for 5 min at 4°. Repeat this step and then block the filters by incubating in Blotto (see step 7) at 4° for 30 min.

11a. Immerse the filters in HEPES binding buffer supplemented with 0.25% (w/v) nonfat milk powder (Carnation) for 1 min at 4°. Screen the filters as described below (see Binding and Wash Conditions).

Recognition Site DNA Probe. The highest affinity site among a set of related sequences should be chosen for the synthesis of an oligonucleotide probe. Multisite probes can then be generated by cloning multiple copies (3–10) of the binding site oligomer in a pUC vector. The DNA fragment containing tandem binding sites is then end labeled using [α-^{32}P]dNTP and Klenow polymerase.[44] The labeled binding site DNA fragment is purified by electrophoresis in a 6% (w/v) nondenaturing polyacrylamide gel. Probe DNA is eluted from the gel and purified on an Elutip-D disposable column (Schleicher & Schuell). Using 20 μg of a recombinant pUC plasmid DNA and 200 μCi of [α-^{32}P]dNTP (5000 Ci/mmol), the probe yield should be 10^8 to 2×10^8 cpm with a specific activity of 2×10^7 to 4×10^7 cpm/pmol. Multisite probes can also be prepared for screening simply by concatenation of a binding site oligonucleotide with DNA ligase, followed by labeling using nick translation.[39]

Nonspecific Competitor DNA. The addition of an excess of nonspecific competitor DNA in the probe solution reduces background as well as minimizes the detection of recombinant phage encoding nonsequence specific DNA-binding proteins.[7] Several different competitor DNAs have been used to screen expression libraries successfully, including poly(dI–dC) · poly(dI–dC) and denatured calf thymus DNA. The latter DNA is preferred because its inclusion in the probe solution reduces background as well as eliminates signal from recombinant phage encoding single-stranded DNA-binding proteins.

Binding and Wash Conditions. The association constants of DNA-binding proteins are dependent on ionic strength, temperature, and pH. Therefore these parameters can be manipulated in the binding and washing steps to optimize the detection of a relevant DNA-binding protein clone. If the transcription factor being cloned has an exogenous metal ion requirement (e.g., Mg^{2+}), the binding and wash buffers should be appropriately supplemented.

1. Incubate replica filters as a stack in binding buffer containing 2.5×10^6 cpm/ml of ^{32}P-labeled recognition site DNA and 5 μg/ml sonicated and heat-denatured calf thymus DNA. Plastic hybridization chambers (Altec Plastics) are convenient for this incubation. Shake the chamber containing the filters gently for 60 min at room temperature.

2. Wash the filters in batches, four times (7.5 min each wash, 30 min total) at room temperature with aliquots of the binding buffer. Filters are washed in large plastic containers. To minimize background, filters should be kept from adhering to one another during the wash steps.

3. Dry the filters on blotting paper and perform autoradiography with a tungstate intensifying screen at $-70°$ for 12 to 24 hr.

[44] J. Sambrook, E. F. Fritsch, and T. Maniatis, *in* "Molecular Cloning: A Laboratory Manual" 2nd Ed., p. 10.51. Cold Spring Harbor Press, Cold Spring Harbor, New York, 1989.

4. Identify the putative positive phage plaques by aligning the autoradiographs with the LB plates. To reduce the number of false positives, generate autoradiography exposures of varying times with the primary filters. Short versus long exposures help to distinguish spots with intense centers (likely to be artifacts) from those with a diffuse, halo-like appearance (likely to represent true positives).

5. Pick putative positives for secondary screening. True positives will show an enrichment of 10- to 100-fold.

6. Screen secondary platings of true positives with the wild-type recognition site probe as well as with control DNA probes that either lack the binding site or contain mutant versions.

7. Plaque purify phage that are specifically detected with the wild-type recognition site probe but not with control DNA probes.

Characterization of Recombinant DNA-Binding Proteins

After the isolation of a recombinant phage that is specifically detected with a given binding site probe, but not with control DNAs (see Fig. 2), it is necessary to demonstrate that this clone encodes a recombinant protein of the expected DNA-binding specificity. In the case of a λgt11 recombinant, this is simply achieved by isolating lysogenized E. coli clones[43] and assaying extracts of induced lysogens for a β-galactosidase fusion protein that specifically binds the recognition site probe used in the screen (see below and Fig. 3). Figure 3 shows an analysis of lysogen extracts using the gel electrophoresis DNA-binding assay. More rigorous characterization of the DNA-binding specificity of the recombinant protein is achieved by chemical and enzymatic footprinting along with the analysis of mutant binding sites. It is important to note that even though the above analyses allow the DNA-binding specificity of the recombinant protein to be thoroughly compared with the native protein, they do not prove that the cloned protein is the one identified by earlier biochemical analysis. Eventually, direct structural analyses are necessary to resolve this issue. Antibodies generated against the cloned protein permit the detection of shared antigenic determinants. Peptide mapping performed on analytical amounts of the native and cloned proteins constitutes a definitive structural comparison.[16]

1. Recombinant phage lysogens are isolated by infecting E. coli strain Y1089 and screening for temperature-sensitive clones.[43]

2. Grow 2-ml cultures of the recombinant lysogens at 32° with good aeration.

3. When cultures reach an $OD_{600} = 0.5$, shift them to a 44° incubator for 20 min.

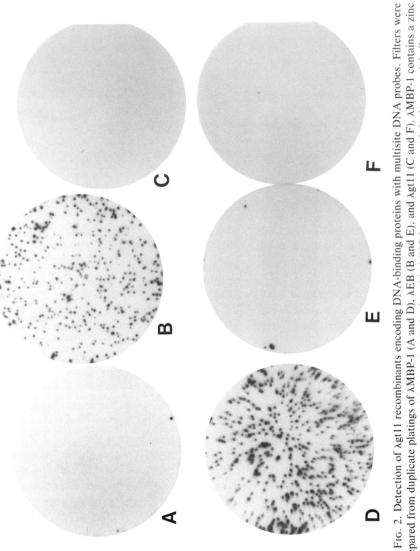

FIG. 2. Detection of λgt11 recombinants encoding DNA-binding proteins with multisite DNA probes. Filters were prepared from duplicate platings of λMBP-1 (A and D), λEB (B and E), and λgt11 (C and F). λMBP-1 contains a zinc finger domain fused to β-Gal that binds to the κB site found in MHC I gene promoters and the immunoglobulin κ gene enhancer.[6] λEB contains the DNA-binding domain of the EBV EBNA-1 protein fused to β-Gal.[7] Filters A, B, and C were screened with an EBNA-1-binding site probe (11 sites) whereas filters D, E, and F were screened with an MBP-1-binding site probe (9 sites).

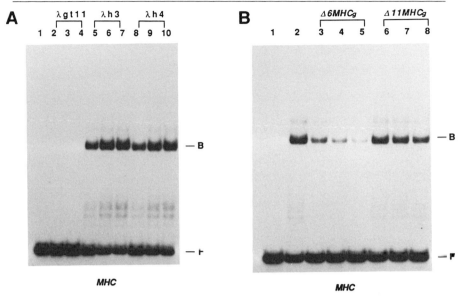

FIG. 3. Gel electrophoresis DNA-binding assays with extracts of λgt11 and λMBP-1 lysogens. λh3 and λh4 are identical independent isolates of MBP-1 recombinants. The DNA-binding domain of MBP-1 recognizes a palindromic element in MHC I gene promoters.[6] The MHC probe contains a single binding site for the β-Gal–MBP-1 fusion protein. Extracts prepared from induced cultures of λgt11, λh3, and λh4 were incubated with MHC DNA (25,000 cpm) and 4 μg poly(dI–dC) · poly(dI–dC) and the reactions resolved by electrophoresis in a nondenaturing polyacrylamide gel (4%, 30 : 1). Free and bound probes are indicated by F and B, respectively. (A) The total protein concentrations in lanes 2, 3, and 4 were 9, 18, and 27 μg, respectively; in lanes 5, 6, and 7 the concentrations were 6, 12, and 18 μg, respectively; and in lanes 8, 9, and 10 they were 6, 12, and 18 μg, respectively. (B) Aliquots (6 μg total protein) of an extract of the λh3 lysogen were incubated with MHC DNA as detailed in (A) in the absence or presence of varying amounts of specific competitor DNAs. The control binding reaction (no specific competitor DNA added) is shown in lane 2. The reactions in lanes 3, 4, and 5 contained a 10-, 20-, and 30-fold molar excess, respectively, of Δ6MHCg DNA. The reactions in lanes 6, 7, and 8 contained a 10-, 20-, and 30-fold molar excess, respectively, of Δ11MHCg DNA. Δ6MHCg contains an MBP-1 binding site whereas Δ11MHCg lacks an MBP-1-binding site. The free DNA probe was resolved in lane 1. (Reproduced with permission from Ref. 6.)

4. Add IPTG to 10 mM to induce expression of the β-galactosidase fusion protein and shift the cultures to a 37° incubator for 60 min.

5. Centrifuge 1-ml aliquots of induced cultures in a microfuge at 12,000 rpm for 1 min at room temperature.

6. Discard the supernatants and resuspend the pellets in 100 ml extract buffer [50 mM Tris-HCl (pH 7.5), 1 mM EDTA, 1 mM DTT, 1 mM phenylmethylsulfonyl fluoride (PMSF)].

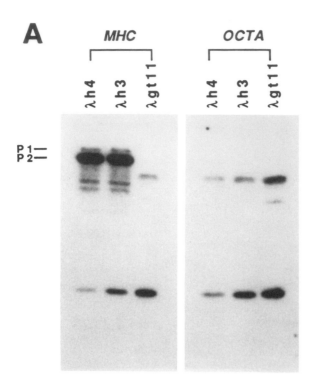

A

MHC   OCTA

λh4 λh3 λgt11   λh4 λh3 λgt11

P1 —
P2 —

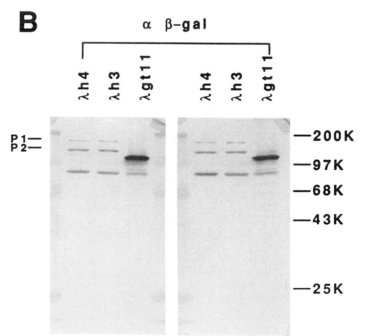

B

α β-gal

λh4 λh3 λgt11   λh4 λh3 λgt11

P1 —
P2 —

— 200K

— 97K

— 68K

— 43K

— 25K

7. Quickly freeze the resuspended cells in liquid nitrogen.

8. Thaw frozen cell suspensions, adjust to 0.5 mg/ml lysozyme, and incubate on ice for 15 min.

9. Adjust the cell suspensions to 1 M NaCl, mix thoroughly, and incubate on a rotator for 15 min at 4°.

10. Centrifuge the lysates in a microfuge at 12,000 rpm for 30 min at 4°.

11. Dialyze the supernatants on Millipore (Bedford, MA) filters (type VS, 0.025-μm pore size) against 100 ml extract buffer for 60 min at 4°. Millipore filters should be floated on dialysis buffer in a 150-mm petri dish before applying samples.

12. Freeze the dialyzed extracts immediately and store at −70° until needed.

13. The DNA-binding properties of the recombinant protein can be tested in various ways, including the gel mobility shift assay[2,3] (see Fig. 3) and DNase I footprinting.[1] If the insolubility of the recombinant protein does not permit detection of DNA binding in crude *E. coli* extracts, then the cDNA insert can be conveniently cloned into the vector pBS-ATG[15] for *in vitro* transcription and translation. This also permits analysis of the DNA-binding properties of the cloned protein in the absence of β-galactosidase.

Detection of Sequence-Specific DNA-Binding Proteins by Western Blotting

As previously indicated, the prospects for successfully isolating the clone for a given sequence-specific DNA-binding protein are significantly enhanced if it can be detected after Western blotting using the cognate multisite DNA probe. The same technique can also be used to demonstrate that a recombinant phage isolated by screening a library with a recognition site probe encodes a sequence-specific DNA-binding fusion protein (see Fig. 4).

FIG. 4. Western blot analysis of λMBP-1 recombinants. (A) Total proteins from induced cultures of λh4, λh3, and λgt11 lysogens were resolved by SDS-PAGE and transferred to nitrocellulose (see caption to Fig. 3 for additional details). Equivalent filters were probed with MHC and OCTA DNAs (each at 10^6 cpm/ml). Bands detected with both probes represent proteins unrelated to the recombinant phage because they are present in λgt11 lanes. Furthermore, these bands do not comigrate with native β-galactosidase or any of its major cleavage products. (B) The same filters analyzed in (A) were probed with anti-β-galactosidase antibodies. The two largest β-galactosidase fusion polypeptides detected in the λh3 and λh4 lanes (labeled P1 and P2) comigrated with the upper two of the four bands specifically detected with MHC probe. (Reproduced with permission from Ref. 6.)

1. Resolve 10–100 μg of the protein sample (nuclear extract or λ lysogen extract) by SDS-PAGE (10%).

2. Electroblot proteins onto nitrocellulose using 25 mM Tris-HCl (pH 8.3), 192 mM glycine, 20% (v/v) methanol buffer. The gel should be pre-equilibrated in transfer buffer (30–60 min) before electroblotting.

3. After transfer, immerse the filter in Blotto and incubate for 60 min at room temperature with gentle shaking.

4. Incubate the filter in denaturation buffer [50 mM Tris-HCl (pH 8), 7 M guanidine hydrochloride, 50 mM DTT, 2 mM EDTA, and 0.25% (v/v) Blotto] for 60 min at room temperature.

5. Rinse the filter twice with renaturation buffer [50 mM Tris-HCl (pH 8), 100 mM NaCl, 2 mM DTT, 2 mM EDTA, 0.1% (v/v) Nonidet P-40, and 0.25% (v/v) Blotto] and leave in renaturation buffer at 4° for 16–24 hr.

6. Rinse the filter twice with binding buffer [10 mM Tris-HCl (pH 7.5), 50 mM NaCl, 1 mM EDTA, and 1 mM DTT] and incubate in binding solution for 60 min at room temperature (see Screening of Expression Library, above). DNA probe is used at a concentration of 10^6 to 2.5×10^6 cpm/ml. Denatured calf thymus DNA can be included in the binding solution at a concentration of 5–10 μg/ml to reduce background as well as binding of probe to nonspecific DNA-binding proteins.

7. Wash the filter with binding buffer (three or four changes at room temperature, 5 min/wash step). Dry the filter on blotting paper and perform autoradiography with intensifying screen(s) at $-70°$ for 12–48 hr.

Discussion

A large number of sequence-specific DNA-binding proteins have been cloned using recognition sites as ligands to screen expression libraries. Table I shows a partial listing of the various clones. Most of these proteins are mammalian transcription factors. However, the technique is not restricted by the source or function of the DNA-binding protein.

Sequence-specific DNA-binding proteins generally contain one of the following structural motifs: the helix–turn–helix, zinc fingers, the leucine zipper, or the helix–loop–helix.[20,45–48] Multiple clones encoding proteins belonging to these various structural classes have been isolated by in situ screening with the relevant recognition site DNAs. The Oct-2 and Oct-1 cDNA clones (see Table I) encode proteins with a predicted helix–

[45] R. M. Evans and S. M. Hollenberg, *Cell* **52**, 1 (1988).
[46] W. J. Gehring, *Science* **236**, 1245 (1987).
[47] W. H. Landschulz, P. F. Johnson, E. Y. Adashi, B. J. Graves, and S. L. McKnight, *Genes Dev.* **2**, 786 (1988).
[48] C. O. Pabo and R. T. Sauer, *Annu. Rev. Biochem.* **53**, 293 (1984).

turn–helix motif. The protein encoded by the MBP-1 cDNA (see Table I) contains zinc finger motifs. C/EBP[47] and CREB[25] are leucine zipper proteins whereas E12 is the prototype of the helix–loop–helix family (see Table I). It should be noted that these latter examples represent proteins that bind DNA as obligate homo- or heterodimers. Therefore clones encoding proteins that bind DNA as dimers, using different dimerization domains, can be successfully screened as a consequence of their functional expression in *E. coli.*

The screening strategy, although a very powerful tool, has limitations. Because it relies on functional expression of a DNA-binding domain in *E. coli,* it is highly unlikely to enable the cloning of proteins that depend either on a cell-specific posttranslational modification or a second distinct subunit for high-affinity DNA binding.[49–52] Another limitation of this strategy is that initially a recombinant protein can be related to a previously identified native protein only by comparing the DNA-binding specificities of the two. However, in situations in which distinct DNA-binding proteins recognize the same sequence, this criterion is difficult to apply. Eventually, direct structural analyses are necessary to resolve this issue. A third limitation of this strategy is that its application can result in the isolation of phage whose β-galactosidase fusion proteins do not apparently bind the recognition site DNA with detectable affinity when assayed in soluble fractions.

The strategy of cloning a gene on the basis of detection of its functional recombinant product with a ligand probe has considerable potential. It has been successfully used with other ligand probes, such as apurinic DNA,[53] and protein subunits.[54]

Acknowledgments

This work was supported by the Howard Hughes Medical Institute. I would like to thank Charlie Eisenbeis and Kara Arvin for testing some of the parameters in the experimental protocols and Bethann Muehlhausen for patiently and carefully preparing the manuscript.

[49] L. A. Chodosh, A. S. Baldwin, R. W. Carthew, and P. A. Sharp, *Cell* **53,** 11 (1988).
[50] T. Curran and B. R. Franza, *Cell* **55,** 395 (1988).
[51] S. Hahn and L. Guarante, *Science* **240,** 317 (1988).
[52] T. D. Halazonetis, K. Georgopoulos, M. E. Greenberg, and P. Leder, *Cell* **55,** 917 (1988).
[53] J. Lenz, S. A. Okenquist, J. E. LoSardo, K. K. Hamilton, and P. W. Doetsch, *Proc. Natl. Acad. Sci. U.S.A.* **87,** 3396 (1990).
[54] E. Y. Skolnik, B. Margolis, M. Mohammadi, E. Lowenstein, R. Fischer, A. Drepps, A. Ullrich, and J. Schlessinger, *Cell* **65,** 83 (1991).

[40] Footprinting of DNA-Binding Proteins in Intact Cells

By PETER B. BECKER, FALK WEIH, and GÜNTHER SCHÜTZ

Introduction

The molecular analysis of protein factors that regulate important processes in the cell nucleus, such as transcription, replication, and recombination, has intensified over recent years. The mechanism of action of one class of protein factors (exemplified by transcription factors) requires their specific binding to short DNA sequence modules at regulatory loci in the genome.[1–3] Powerful techniques such as DNase I footprinting[4] have been developed to demonstrate specific binding of purified proteins or components of crude nuclear extracts to DNA in vitro. The DNA in these assays usually is of low complexity and easily accessible to the protein to be studied. Genomic DNA in the nucleus of eukaryotes, however, is of high sequence complexity and associated with histones and nonhistone proteins to form the hierarchical structures of chromatin.[5] The density of structural protein components associated with DNA in the nucleus and its resulting highly compacted state raise questions as to whether the proteins that bind to naked DNA in vitro will have access to their potential binding sites in chromatin. It is of particular interest in this context, whether chromatin structure contributes to the mechanisms of regulation by modulating the accessibility of factors to their binding sites in response to extracellular stimuli.[6]

The development of genomic footprinting enabled the direct visualization of specific protein–DNA interactions within the living cell.[7] The refinement of the indirect end-labeling technique[8,9] by Church and Gilbert,[10] along with an improved hybridization protocol and single-nucleotide resolution of stretches of genomic DNA, allows the mapping of highly complex genomic DNA. Although originally designed to obtain genomic sequence

[1] N. C. Jones, P. W. Rigby, and E. B. Ziff, Genes Dev. 2, 267 (1988).
[2] W. S. Dynan, Cell 58, 1 (1989).
[3] P. J. Mitchell and R. Tjian, Science 245, 371 (1989).
[4] D. J. Galas and A. Schmitz, Nucleic Acids Res. 5, 3157 (1978).
[5] D. S. Pederson and R. T. Simpson, ISI Atlas Sci.: Biochem. 1, 155 (1988).
[6] A. Wolffe, New Biol. 2, 211 (1990).
[7] A. Ephrussi, G. M. Church, S. Tonegawa, and W. Gilbert, Science 227, 134 (1985).
[8] C. Wu, Nature (London) 286, 854 (1980).
[9] S. A. Nedospasov and G. P. Georgiev, Biochem. Biophys. Res. Commun. 92, 532 (1980).
[10] G. M. Church and W. Gilbert, Proc. Natl. Acad. Sci. U.S.A. 81, 1991 (1984).

METHODS IN ENZYMOLOGY, VOL. 218

information and to map the occurrence of modified nucleotides,[10] the protocols were immediately applicable for footprinting experiments in the context of the entire genome; the direct labeling of a cloned DNA fragment used in an *in vitro* experiment is replaced by indirect end-labeled probing, which visualizes the sequence of interest selectively.[7,11]

The methods we will describe in detail are refinements of the original protocol[7,10] and have been found to reproducibly yield *in vivo* footprints of high quality from mammalian genomes. The importance of the genomic footprinting approach is highlighted in a study[12] of a series of ubiquitous proteins that interact with sequences upstream of the rat tyrosine amino-transferase (TAT) gene *in vitro*. While all factors were present in the nuclei of cells whether they did or did not express the TAT gene, productive interaction of the proteins with their potential binding sites was observed only in the TAT-expressing hepatoma cells.[12] Because all binding sites in the chromatin of the TAT-expressing cells, but not of the nonexpressing cells, are within regions that are hypersensitive to DNase I, these results may document an involvement of chromatin components in the determination of binding site accessibility. Occlusion of proteins from potential binding sites *in vivo* may also be due to CpG methylation in mammalian DNA. Binding of a number of transcription factors has been shown to be sensitive to modification of CpG residues in their recognition sites.[13,14] Genomic sequencing can be used to obtain important information about the methylation status of CpGs in the nucleus and—in combination with *in vivo* footprinting—their effects on protein interaction.[12,15] Genomic footprinting has yielded valuable information about the mechanisms of action of well-known transcriptional activator proteins. For example, although the interaction of steroid receptors with their response elements *in vivo* is dependent on the presence of the hormone,[16-18] purified glucocorticoid and progesterone receptors can interact with their binding sites *in vitro* in the absence of hormones.[19,20] Binding of a factor to the metal-responsive element of the metallothionein promoter was also observed

[11] P. B. Becker and G. Schütz, *Genet. Eng.* **10,** 1 (1988).

[12] P. B. Becker, S. Ruppert, and G. Schütz, *Cell* **51,** 435 (1987).

[13] M. Höller, G. Westin, J. Jirizny, and W. Schaffner, *Genes Dev.* **2,** 1127 (1988).

[14] F. Watt and P. L. Molloy, *Genes Dev.* **2,** 1136 (1988).

[15] H. P. Saluz, J. Jiricny, and J. P. Jost, *Proc. Natl. Acad. Sci. U.S.A.* **83,** 7167 (1986).

[16] P. B. Becker, B. Gloss, W. Schmid, U. Strähle, and G. Schütz, *Nature (London)* **324,** 686 (1986).

[17] J. N. J. Philipsen, B. C. Hennis, and G. AB, *Nucleic Acids Res.* **16,** 9663 (1988).

[18] J. Wijnholds, J. N. J. Philipsen, and G. AB, *EMBO J.* **7,** 2757 (1988).

[19] T. Willmann and M. Beato, *Nature (London)* **324,** 688 (1986).

[20] A. Bailly, C. lePage, M. Rauch, and E. Milgrom, *EMBO J.* **5,** 3235 (1986).

only after metal induction.[21,22] In contrast, a cAMP-responsive element (CRE) located within the tissue-specific enhancer of the TAT gene is contacted by protein even in uninduced hepatoma cells, reflecting the basal state of protein interaction at this CRE and the basal level of transcription. A detailed genomic footprinting analysis monitoring protein interaction at the CRE during cAMP stimulation revealed a transient modulation of binding activity correlating with transcriptional induction. This led to the conclusion that activation was brought about by a cAMP-dependent posttranslational modification of the factor that interacts with the cAMP response element.[23,24] Protein binding at the serum response element in the control region of the c-*fos* gene is observed before and after serum induction.[25] In this case protein–protein interactions or posttranslational modifications of the DNA-binding protein may be required for transcriptional activation.

In summary, genomic footprinting is the method of choice to study (1) the biological relevance of protein–DNA interactions previously established *in vitro,* (2) the binding sites of factors in order to identify potential sites for functional analysis, (3) the accessibility of target sites in chromatin, (4) the mechanisms of action of DNA-binding proteins, (5) the action of factors that are unstable during purification, and (6) the occurrence of CpG methylation at specific sites.

Principle

The genomic footprinting methodology can be subdivided into three distinct steps: (1) a chemical reaction performed with intact cells to randomly modify nuclear DNA under conditions in which DNA-binding proteins protect their binding sites from the modification; (2) purification and single nucleotide resolution of the genomic DNA; and (3) visualization of the actual footprint. We will focus in this section on the strategy that has been successfully applied in the studies mentioned above: the use of dimethyl sulfate (DMS) as a reagent to chemically modify DNA *in vivo* and the indirect end labeling by hybridization to visualize the footprint. An alternative detection method that employs polymerase chain reaction

[21] R. D. Andersen, S. J. Taplitz, S. Wong, G. Bristol, B. Larkin, and H. R. Herschman, *Mol. Cell. Biol.* **7,** 35 (1987).

[22] P. R. Müller, S. J. Salser, and B. Wold, *Genes Dev.* **2,** 412 (1988).

[23] F. Weih, A. F. Stewart, M. Boshart, D. Nitsch, and G. Schütz, *Genes Dev.* **4,** 14 (1990).

[24] M. Boshart, F. Weih, A. Schmidt, R. E. K. Fournier, and G. Schütz, *Cell* **61,** 905 (1990).

[25] R. E. Herrera, P. E. Shaw, and A. Nordheim, *Nature (London)* **340,** 68 (1989).

(PCR) amplification of the genomic sequences has been described[26-29] and will be discussed in the final section of this chapter. Alternative routes can also be followed to create the footprint on the genomic DNA. Dimethyl sulfate has been widely used as a modifying agent for DNA because of its small size and hydrophobic character, which enable it to diffuse through the cellular and nuclear membranes of intact cells. Various nucleases that are often used in *in vitro* footprinting experiments cannot enter intact cells and thus will work only on permeabilized cells or isolated nuclei. If leaking of proteins from nuclei is carefully minimized genomic footprints can also be obtained with nucleases.[30-33]

Figure 1 summarizes the technical steps (1)–(9) involved in the procedure detailed below. In step (1) of Fig. 1 a hypothetical protein–DNA interaction in a nucleus is schematized. The binding site for the protein contains a guanine (G) residue that is contacted by the protein. Protein binding also distorts the nearby DNA structure, symbolized by the adjacent *G* residue. Cells in suspension are treated with DMS, which diffuses into the nuclei and methylates the N-7 position of guanines as well as the N-3 position of adenines [Fig. 1, step (2)]. The conditions are chosen such that a partial methylation of guanine residues is achieved as in the G-specific sequencing reaction in the Maxam–Gilbert protocol.[34] The N-7 position of guanine is directed toward the major groove of the DNA and is frequently contacted by sequence-specific DNA-binding proteins. A guanine in the recognition sequence of a protein will be protected from methylation in the presence of bound protein and thus methylated at reduced frequency as compared with neighboring guanines. Protein binding to DNA can also result in hyperreactivity of guanines toward DMS. This may be caused either by a distortion of the DNA structure in the vicinity of a bound protein [see also Fig. 1, step (1)], or by a locally increased concentration of DMS due to its trapping in hydrophobic pockets of the protein.[35] To distinguish DMS reactivity differences of guanines caused by protein binding from those due to the microsequence environ-

[26] P. R. Mueller, and B. Wold, *Science* **246**, 780 (1989).
[27] G. P. Pfeiffer, S. Steigerwald, P. R. Mueller, B. Wold, and A. D. Riggs, *Science* **246**, 810 (1989).
[28] Reference deleted in proof.
[29] H. P. Saluz and J. P. Jost, *Proc. Natl. Acad. Sci. U.S.A.* **86**, 2602 (1989).
[30] K. Zinn and T. Maniatis, *Cell* **45**, 611 (1986).
[31] G. Albrecht, B. Devaux, and C. Kedinger, *Mol. Cell. Biol.* **8**, 1534 (1988).
[32] R. L. Tanguay, G. P. Pfeifer, and A. D. Riggs, *Nucleic Acids Res.* **18**, 5902 (1990).
[33] I. M. Huibregste and D. R. Engelke, *Mol. Cell. Biol.* **9**, 3244 (1989).
[34] A. M. Maxam and W. Gilbert, this series, Vol. 65, p. 499.
[35] R. T. Ogata and W. Gilbert, *Proc. Natl. Acad. Sci. U.S.A.* **75**, 5851 (1978).

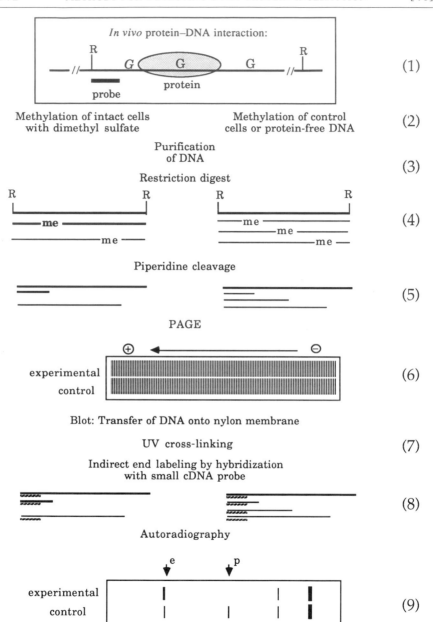

FIG. 1. A schematic outline of the procedures involved in the *in vivo* footprinting methodology. Steps (1) to (9) are referred to in text.

ment, control reactions are carried out in parallel on cells that serve as a control for the effect to be observed (e.g., cells that were not treated with an inducing agent or cells that do not express the gene under study). If such control cells are not available, protein-free genomic DNA can be methylated as a standard. This is, however, not the ideal control because it is conceivable that guanine residues in chromatin react differently with DMS from those in protein-free DNA *in vitro*.[16]

After the methylation reaction has been terminated, genomic DNA is isolated and then cleaved with a restriction enzyme to create the fragments needed for indirect end labeling analysis [Fig. 1, step (3)]. The DNA sequence to be analyzed is represented in the genomic DNA pool as a family of fragments [Fig. 1, step (4)]. Due to the partial methylation reaction, many fragments will carry only one methylated guanine. Most of the guanine residues in the nucleus will be similar in DMS reactivity and fragments with the corresponding methylated G's will be found at approximately equal frequency. If guanines were protected from modification, the fragment with a methylated G at the corresponding position will be missing from the pool. Similarly, a fragment with a DMS-hyperreactive guanine will be overrepresented in the pool as compared to the DNA from the control reaction [Fig. 1, step (4), compare left- and right-hand sides].

To map the positions and frequency of methylated guanines by indirect end labeling, the modification must be converted into a break in the DNA backbone [Fig. 1, step (5)]. This is achieved using piperidine, following standard sequencing procedures.[34] The complex mixture of DNA fragments resulting from restriction enzyme digest and piperidine cleavage at sites of methylated guanines is now separated with single-nucleotide resolution on an acrylamide gel [Fig. 1, step (6)]. The DNA is then transferred from the gel onto a nylon membrane to which it is covalently cross-linked with UV [Fig. 1, step (7)]. From among the vast excess of unrelated DNA sequences on the membrane, the fragments of interest are now visualized by hybridization of a small DNA probe that abuts the end set by the restriction enzyme [see also Fig. 1, step (1)]. The indirect end label thus highlights the family of fragments that share the end to which the probe hybridizes but differ in the other end, which is determined by the position of the methylated guanine [Fig. 1, step (8)]. The labeled bands are detected by autoradiography. Each band on the film corresponds to the position of a guanine (or a group of unresolved guanines) in the genomic DNA. The intensity of each band is correlated with the reactivity of the corresponding G residue toward methylation *in vivo*. Comparison of the intensity of each band in the experimental DNA with the one from the control reaction allows identification of the guanines within or close to DNA-binding sites of proteins in the cell. Both the reduced intensity of a

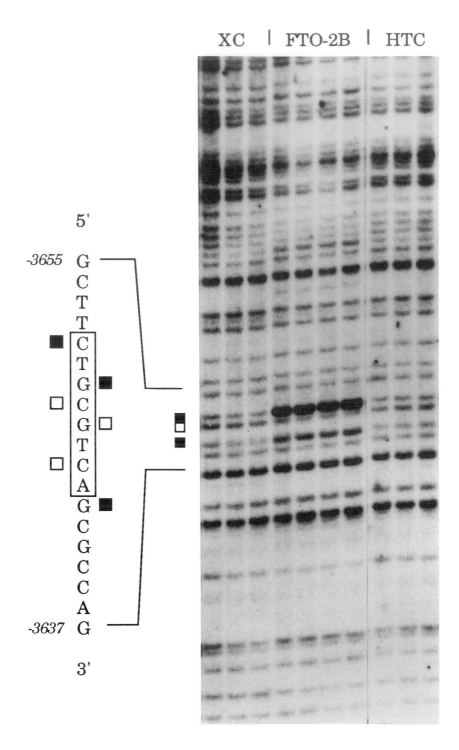

band [Fig. 1, step (9), protection] or an increased intensity [Fig. 1, step (9), enhancement] are indicative of protein–DNA interaction.

Figure 2 shows an example of a genomic footprinting experiment. A pattern of strong enhancements and weaker protection is caused by protein binding to a functional CRE within the cAMP-responsive enhancer of the TAT gene in FTO-2B hepatoma cells. CREB protein binding to this sequence is neither detected in fibroblast cells, in which the gene is not expressed, nor in hepatoma (HTC) cells, in which the gene can no longer be induced by cAMP (see Refs. 22 and 23 for further details).

Procedures

Each section begins with a list of reagents and materials needed, followed by a basic description of the procedure and finally a section with additional comments, hints, and background information. High-quality chemicals are used throughout the procedures.

In Vivo Methylation of Cells

Materials

Dimethyl sulfate (DMS): Fluka (Ronkonkoma, NY), puriss. p.A., stored in the dark at 4° under nitrogen; DMS is a potent carcinogenic agent and thus all solutions containing DMS as well as used plasticware should be inactivated with 5–10 M NaOH

Phosphate-buffered saline (PBS): 140 mM NaCl, 2.5 mM KCl, 8.1 mM NaH$_2$PO$_4$, 1.5 mM KH$_2$PO$_4$, pH 7.5

Nuclei buffer: 0.3 M sucrose, 60 mM KCl, 15 mM NaCl, 60 mM Tris-HCl (pH 8.2), 0.5 mM spermidine, 0.15 mM spermine, 0.5 mM ethylene glycol-bis (β-aminoethyl ether)-N,N,N',N'-tetraacetic acid (EGTA), 2 mM ethylenediaminetetraacetic acid (EDTA)

FIG. 2. A typical result of an *in vivo* footprinting experiment: Protein interaction at a cAMP response element within the tissue-specific enhancer of the tyrosine aminotransferase (TAT) gene (Ref. 22). The binding of a protein is inferred from prominent enhancements (closed boxes) and a weak protection (open box) that correlate with cAMP induction of TAT transcription. Control cells, such as a fibroblast cell line (XC) in which the gene is not expressed, or HTC hepatoma cells, in which the cAMP induction is impaired, do not show protein binding in chromatin. The changes on the upper strand, visible on the autoradiogram, are marked on the right-hand side of the sequence, enhancements and protections on the lower strand (not shown; see Ref. 22) are indicated to the left of the sequence. The TAT CRE sequence is boxed.

Nonidet P-40
Sarkosyl: N-Lauroylsarcosine (Sigma, St. Louis, MO), 20% (w/v)

Procedure

Cells are mildly trypsinized and resuspended in 1 ml of the original medium at 25° in a 14-ml polypropylene tube. To this 5 μl of DMS is added, and the reaction is mixed by swirling and incubated for 5 min at 25°. The reaction is stopped by the addition of ice-cold PBS to fill the tube and chilled on ice. The following steps are performed at 4° without delay. Cells are collected by centrifugation in a cooled centrifuge (5 min, 2500 rpm), resuspended in 10 ml of cold PBS, and pelleted as before. The cells are resuspended in 1.5 ml nuclei buffer and an equal volume of nuclei buffer containing 1% (v/v) Nonidet P-40 is added, thoroughly mixed, and incubated on ice for 5 min, during which cell lysis occurs. Nuclei are recovered by centrifugation (5 min, 3000 rpm), washed with 5 ml nuclei buffer without sucrose or Nonidet P-40, and lysed by suspension in 1 ml 0.5 M EDTA with vigorous mixing. Sarkosyl and RNase A are added to 0.5% (w/v) and 250 μg/ml, respectively. After 3 hr of incubation at 37° proteinase K is added to a final concentration of 250 μg/ml and the incubation is continued at 37° overnight.

The above protocol usually results in a partial methylation of 1 in 500 base pairs in the genomic DNA. The reaction is dependent on the DMS concentration in the nuclei, which in turn is determined by the poor solubility of the hydrophobic chemical in water. Increasing the amount of DMS or the incubation times does not influence the reaction significantly and thus suitable guanine ladders are obtained under a variety of conditions. Once the reaction is stopped the chemical must be removed as completely as possible by thorough washes of the cells and nuclei. Traces of DMS in the following overnight incubation at 37° will continue to react with the now protein-free DNA and obscure potential footprints.

Although the volume of the cell suspension should not be varied, the actual cell number is not critical to obtain the desired partial modification. Between 4×10^7 and 2×10^8 cells have been methylated with satisfactory results. Depending on the availability of the cells, it is convenient to process a larger number of cells at a time to obtain several hundred micrograms of methylated DNA. This DNA can be used for many experiments to visualize proteins interacting with chromatin at different sites. Thus the presence of a footprint at one specific site can serve as an "internal control" for the results of a new site obtained from the same methylated genomic DNA. If, for example, a clear effect of an inducing agent is seen at one particular genomic location when methylated DNAs from induced cells and control cells are compared, the absence of such effects at another site can be interpreted with confidence.

Preparing DNA Samples for Electrophoresis

Materials

Phenol: Contains 0.1% (w/v) 8-hydroxyquinoline; equilibrate with 100 mM Tris-Cl (pH 8.0), 10 mM EDTA

Phenol–chloroform (1 : 1): Equilibrate as above

TE buffer: 10 mM Tris-HCl (pH 8.0), 1 mM EDTA

Bovine serum albumin (BSA): Nucleic acid enzyme grade, 10 mg/ml

Piperidine [10 M stock, grade 1 (Sigma)]: Store at 4° in the dark; before use dilute freshly to 1 M with water

Loading buffer: 94% (w/v) deionized formamide (Fluka), 0.05% (w/v) xylene cyanole, 0.05% (w/v) bromphenol blue, 10 mM EDTA

Procedure

The clear and viscous DNA solution after proteinase K treatment (see above) is diluted by addition of 1 ml 0.5 M EDTA and extracted twice with phenol. The high EDTA concentration renders the aqueous phase more dense than the organic phase, allowing easy removal of organic phase together with the interphase. In the subsequent extraction with phenol–chloroform, the aqueous phase is the upper phase again. The DNA is dialyzed overnight against 3 liters of TE buffer at 4° (one change of buffer). To the dialysate add 1/10 vol of 3 M sodium acetate (pH 7), followed by 2.5 vol of ethanol. The precipitate is collected by centrifugation, washed with 80% ethanol, dried in a desiccator, and finally redissolved in 1 ml of TE. The DNA is stored frozen at −20°. The concentration of each sample should be determined by OD measurement. The DNA is analyzed by electrophoresis on a 0.6% (w/v) agarose gel.

Genomic DNA purified by this procedure is sufficiently clean to be easily digested with the restriction endonuclease [see Fig. 1, step (4) and below] to create the fixed end for the indirect end labeling. Thirty micrograms of each DNA sample is digested with 60 units of restriction enzyme overnight at 37° in a volume of 300 μl. Bovine serum albumin is added to 100 μg/ml for stabilization of the enzyme. The reaction is stopped by adding EDTA to 10 mM and sodium acetate (pH 7) to 0.3 M and the DNA is precipitated with 2.5 vol ethanol. After chilling, the DNA is collected by centrifugation (10 min, Eppendorf table centrifuge), washed with 80% ethanol, and dried in a vacuum concentrator (SpeedVac; Savant, Hicksville, NY). Pellets are dissolved in 100 μl 1 M piperidine and incubated at 85–90° for 30 min [Fig. 1, step (5)]. After chilling on ice, the solution is transferred into a fresh tube and precipitated as above. The DNA, which usually is spread over the tube wall as a film, is carefully dissolved in 100 μl of water and dried for 2 hr in a vacuum concentrator. The pellets are dissolved in 20 μl of water and dried again for at least 1 hr (or overnight). The resulting fluffy pellet is dissolved in 3 μl loading buffer.

Methylated DNA should not be heated above 37° prior to the piperidine reaction in order to avoid uncontrolled depurination in the neutral buffer conditions. DNA that appears to be larger than 10 kb when analyzed on a 0.8% agarose gel will be suitable for a genomic footprinting experiment.

Preparing Membranes for Hybridization

Materials

GeneScreen membrane (Du Pont/NEN, Boston, MA)

Electrophoresis buffer (TBE, pH 8.8) (10× stock solution): 162 g Tris base, 27.5 g boric acid, 9.5 g Na$_2$EDTA · 2H$_2$O/liter

Ashless hardened paper (Cat. No. 103 00187; Schleicher & Schull)

Blotting buffer (0.5× TBE, pH 8.4) (10× stock solution): 109 g Tris base, 55 g boric acid, 9.3 g Na$_2$EDTA · 2H$_2$O/liter. Due to the diffusion of urea from the gel into the blotting buffer, the buffer must be changed before every use

Procedure

The piperidine-treated genomic DNA is now separated on a denaturing polyacrylamide sequencing gel [Fig. 1, step (6)]. We routinely run wide gels (30 × 35 cm), which differ from common sequencing gels in their thickness (1 mm) and the cross-linking ratio (acrylamide : bisacrylamide, 39 : 1). Prior to pouring of the gel both glass plates must be freshly siliconized to facilitate the later blotting of the gel. Gel and electrophoresis buffer is TBE, pH 8.8. The gel is prerun at a constant 900 V for about 3 hr, until the current stabilizes [or, for convenience, overnight (~12 hr) at 220–250 V], followed by 30 min at 900 V. The DNA samples in loading buffer are denatured for 3 min at 85–90°, chilled on ice, and loaded on the gel. Electrophoresis is carried out at a constant 900 V (under those conditions the gel does not heat up) until the desired resolution is achieved. From time to time during the run, dye markers are loaded into the slots adjacent to the samples to mark the area to be blotted. After the run, the glass plates are carefully separated and the gel covered with a sheet of ashless hardened paper to which it sticks tightly and thus can be removed from the supporting plate [do not use Whatman (Clifton, NJ) 3MM paper].

The apparatus used for electroblotting is a replica of the Harvard Biological Laboratories model detailed in Ref. 10. Its size (38 × 46 × 20 cm, holding 20 liters of buffer) accommodates a wide sequencing gel. Two spirals of platinum wire at both the bottom and the lid constitute the two electrodes. A supportive plastic grid with feet is covered with a spongy pad. The gel supported on paper is placed directly on the dry pad and its surface wetted with blotting buffer. The GeneScreen membrane, prewetted in the same buffer, is placed on top of the gel and trapped bubbles are squeezed out carefully. The membrane should be marked so that identification of the side that is in contact with the gel is possible. A second layer of dry spongy pad is directly put on top of the membrane, and another plastic grid with feet completes the "sandwich," which is tied together with rubber tubings. The sandwich is flipped over (thus blotting will be downward) and *slowly* submerged into the blotting buffer in a slanted position. It is important to make sure that no air bubbles are trapped underneath the spongy pad, which would cause severe blotting artifacts. The electrodes are connected such that the anode is at the bottom of the tank (and itself covered by a spongy pad). Thus gas bubbles are not trapped in the sandwich. The transfer is performed in the cold room with precooled buffer. It is complete after 1 hr of blotting at constant voltage of 90–100 V (3–4 A). The membrane is now air dried on a filter paper and baked in a vacuum oven at 80° for 20 min. Next, the side of the membrane that was in contact with the gel is irradiated with UV light (254 nm) [Fig. 1, step (7)] to cross-link the DNA covalently to the nylon matrix. A variety of devices can be used for irradiation, some of which are commercially available. We use an inverted transilluminator from which the quartz glass panel is removed and irradiation comes from six bulbs at a distance of 20 cm for 20 sec (5000 μW/cm^2). The membrane is now ready for hybridization.

We advise that the conditions for UV cross-linking be carefully optimized for each individual device used. With GeneScreen membranes optimal hybridization signals are obtained under cross-linking conditions that result in the retention of 30–40% of the blotted DNA on the membrane after hybridization and wash. Tighter cross-linking adversely affects its base-pairing capacity to the DNA probe during hybridization.

The spongy pads of the blotting device should not be stored in the tank under blotting buffer, but rather separately, as it needs to be dry when the gel is placed on it for blotting. When the pads are wet, air bubbles cannot escape through the meshes when the sandwich is submerged in the buffer, thus causing blotting artifacts.

To identify the genomic sequence unambiguously, sequence standards produced with cloned DNA should be run alongside the genomic samples.

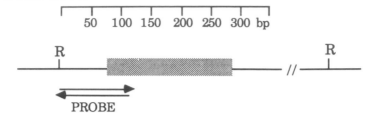

FIG. 3. Probing strategy. The probe abuts the strategic restriction site (R) and partially overlaps the genomic region that can be scanned for protein–DNA interaction with good resolution (shaded field). See text for details.

The plasmid standards must be diluted to a concentration that approaches the amount of specific sequence in a genomic sample to yield a signal of roughly equal intensity following hybridization. A plasmid containing the cloned sequence to be analyzed in the genome is cleaved with the restriction enzyme that is also used to cut the genomic DNA and then subjected to Maxam–Gilbert sequencing reactions.[34] The reaction products after piperidine cleavage are diluted into loading buffer (including 100 μg/ml sheared salmon sperm DNA) to genome equivalent abundance, taking into account the complexity of the genome (3×10^9 bp for mammals) and the absolute amount of processed genomic DNA to be loaded on the gel (30 μg). Thus for a plasmid of 3 kb size, 30 pg of DNA equals one genome equivalent. Thirty picograms is applied to the sequencing gel in 3 μl of loading buffer. Plasmid dilutions can be used to evaluate the sensitivity of the analysis and provide useful standards to monitor improvements during subsequent experiments.

Probing Strategy

To visualize interactions of protein at one particular site in the genome the indirect end labeling of genomic fragments, constituting the guanine sequence ladder, must be achieved. The probing strategy, that is, the selection of the "strategic" restriction site in the vicinity of the sequences under study as well as the DNA probe itself, deserves some consideration (see Fig. 3). Under the stringent conditions of UV cross-linking and hybridization described, fragments smaller than 70 nucleotides will not yield a signal. The polyacrylamide gel resolves genomic DNA with single-nucleotide resolution for about 200–250 bp beyond the point where the hybridization signal is picked up (shaded bar in Fig. 3). Thus the restriction site that will become the invariant end should be further than 90 bp away from the region of focus, the resulting fragment on cleavage longer than 400 bp. The "strategic" restriction endonuclease should cut genomic DNA reliably, be reasonably priced, and not be inhibited by possible methylation of its recognition sequence. Ideally, the probe fragment itself should be a single-copy sequence between 100 and 150 nucleotides long (i.e., short in comparison to the restriction fragment to be labeled, to minimize hybridization to fragments that do not have the invariant end), should be of moderate GC content (35–55% will work), and should abut the strategic restriction site.

Once a suitable restriction site has been selected it is wise to confirm its presence in the genome of the cells used by Southern blotting to rule out the existence of a restriction site polymorphism.

Probe Synthesis

To obtain a suitable hybridization probe as much ^{32}P as possible needs to be incorporated into a relative short stretch of nucleic acid. We favor single-stranded DNA probes over RNA probes produced by transcription with phage polymerases[36] because precursor dNTPs of higher specific activity than the corresponding NTPs can be obtained and because DNA polymerases appear to be more efficient in incorporating nucleotides at limiting concentrations. Originally[10] the single stranded DNA probe was synthesized by primer extension using single-stranded DNA of M13-derived plasmids into which the short probe sequence were cloned. We have described a protocol[11] that reproducibly yields high-quality probes following this strategy. A number of disadvantages with this procedure, such as the necessity to clone the probe fragment in both orientations into the M13 vector, the difficulty to confine radioactive labeling to insert sequences only, and the requirement for a gel purification step, have led to the development of a new strategy[37] that yields probes of high specific activity from either strand with minimal effort.

Materials

All solutions are prepared with diethyl pyrocarbonate-treated water.

T3/T7 transcription buffer (5× stock): 200 mM Tris-HCl (pH 8.0), 250 mM NaCl, 40 mM MgCl$_2$, 10 mM spermidine, 50 mM dithiothreitol (DTT)

NTP mix: ATP, GTP, CTP, and UTP (3.3 mM each); store in aliquots at −20°

DNase I buffer (10× stock): 500 mM Tris-HCl (pH 7.5), 50 mM MgCl$_2$, 10 mM DTT

---

[36] K. Zinn and T. Maniatis, *Cell* **45**, 611 (1986).
[37] F. Weih, A. F. Stewart, and G. Schütz, *Nucleic Acids Res.* **16**, 1628 (1988).

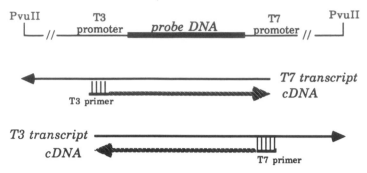

FIG. 4. Probe synthesis by reverse transcription of an RNA template. The upper line displays the *Pvu*II fragment from the probe plasmid (probe DNA inserted into the polylinker of pBluescript). The strategy makes use of the symmetric location of two bacteriophage promoters and two "universal" primers that correspond to sequences within the promoters. The procedure is described in detail in the text.

RNase inhibitor: 40 U/μl (Stratagene, La Jolla, CA)

RNase-free DNase I: 1 U/μl (e.g., Promega, Madison, WI)

RT buffer (5×): 250 mM Tris-HCl (pH 8.5; pH 8.3 at 42°), 50 mM MgCl$_2$, 50 mM DTT, 600 mM KCl

dGCT mix: 20 mM dGTP/dCTP/dTTP; store in aliquots at −20°

[α-^{32}P] dATP: 5000–6000 Ci/mmol

AMV reverse transcriptase: Avian myeloblastosis virus reverse transcriptase, 20 U/μl (Boehringer Mannheim, Indianapolis, IN)

Procedure

Figure 4 illustrates the principle of the probe synthesis. The small probe fragment is cloned into pBluescribe/pBluescript (Stratagene) or any other vector that contains a multiple cloning site flanked by phage promoters on either side. After digestion of the plasmid with *Pvu*II (which cuts on either side of the insert in the vector) a phage polymerase is used to synthesize a large quantity of RNA containing probe and vector sequences up to the *Pvu*II site. An aliquot of this RNA then serves as a template for cDNA synthesis. The RNA derived from transcription with T7 polymerase is annealed with the T3 primer and extended by reverse transcriptase in the presence of radioactive dATP to yield a highly labeled cDNA probe specific for one strand of the genomic DNA (see Fig. 4). Correspondingly, a probe for the opposite strand can be derived from the same plasmid construct simply by using the T3 polymerase and T7 primer. The probe synthesis by reverse transcription of an RNA template has a number of advantages over the methods that use M13-derived constructs.

First, by making use of the symmetric structure of vectors carrying two phage promoters, probe DNAs need only be cloned in one orientation. Second, when primers hybridizing close to the insert DNA are employed, the labeling of vector sequences is minimized. Third, large quantities of RNA can be synthesized with the phage polymerases. This RNA can be characterized, accurately quantitated, and then serve as template for multiple rounds of actual probe synthesis. Fourth, once the RNA template has been synthesized, probe preparation is performed in a short time and with minimal exposure to radioactivity. Finally, reverse transcriptase incorporates dNTPs at low concentrations efficiently and thus mostly full-length products are obtained, even under conditions in which only one of the dNTPs is provided as an isotope of the highest specific activity.[37]

The probe plasmid is cleaved with PvuII and the completeness of the digestion verified by agarose gel electrophoresis. Ten micrograms of this DNA (5 pmol of plasmid) is incubated with 20 μl 5 × transcription buffer, 10 μl NTP mix, 2.5 μl RNase inhibitor (100 U; Stratagene), 5 μl RNA polymerase (50 U T3 or T7) in a volume of 100 μl for 15 min at 37°. Thirteen microliters of 10 × DNase I buffer and 20 μl of RNase-free DNase I (20 U; Promega) are added and the reaction is incubated for further 15 min at 37° to degrade the DNA template after transcription. The reaction is terminated by phenol and phenol–chloroform (1 : 1) extraction. The following precipitations remove unincorporated nucleotides as well as short DNA fragments. A one-tenth volume of 7.5 M ammonium acetate and 2.5 vol of ethanol are added, mixed, and the reaction chilled for 5 min on ice. After a 5 min spin, the pellet is resuspended in 50 μl water, precipitated as above, and washed twice with 80% ethanol. After a short period of drying in a speed vacuum concentrator it is dissolved in 20 μl water. An aliquot can be checked on a RNase-free minigel. The RNA concentration is determined by optical density (OD) measurement of a 1/200 dilution at 260 nm and adjusted to 1 pmol/μl. Usually 10–20 pmol transcript is obtained per picomole plasmid DNA. The RNA is stored at −70° and can be used as template for multiple probe synthesis reactions.

The following conditions of reverse transcription attempt to achieve optimal incorporation of a relatively low amount of isotope into largely full-length cDNA. For probe synthesis 250–300 μCi [α-^{32}P] dATP (50 pmol, final concentration 5 μM) is dried down in a SpeedVac and redissolved in 4 μl of template RNA (4 pmol). The RNA is denatured by incubation at 70° for 5 min and chilled on ice. To the denatured RNA 2 μl of the appropriate (T7 or T3) primer (40 pmol; Stratagene), 2 μl 5 × RT mix, 0.5 μl dGTC mix (final concentration, 1 mM for each dNTP), 0.5 μl RNase inhibitor (Stratagene), and 1 μl reverse transcriptase are added and the reaction is incubated for 45 min at 42°. Synthesis is terminated by addition of 1 μl each of 0.5 M EDTA and 20% (w/v) sodium dodecyl sulfate (SDS). The template RNA is hydrolyzed by addition of 12 μl 0.4 M NaOH and incubation for 15 min at 70° followed by neutralization with 1 μl 1 M Tris-HCl (pH 7.5) and 12 μl 0.4 M HCl. After the addition of 20 μg carrier tRNA the single-stranded DNA probe is recovered by one selective precipitation, as above. The pellet is dried in a SpeedVac and dissolved in 200 μl of TE. Incorporation of isotope can be determined with chromatography on polyethylene imine cellulose; the length distribution of the probe DNA should be checked on a denaturing polyacrylamide gel. In general a reaction with a probe fragment (100–150 nucleotides) under the above conditions will result in an incorporation of 1–5 $\times$ 10^8 dpm.

It is advantageous to insert the probe fragment into the vector such that much of the polylinker is deleted, thus minimizing the contribution of vector sequences to the probe DNA. Should the probe DNA contain a PvuII site, PvuII cannot be employed to cleave the plasmid. Alternative enzymes should be chosen that produce either blunt or 5' overhang ends (such as RsaI or DdeI) in order to avoid unspecific initiation of the phage polymerases at 3' overhang ends.

Hybridization

The visualization of the G-specific pattern of a genomic sequence requires that the cDNA probe be able to detect femtogram amounts of specific sequences in the context of a 10^7-fold excess of unrelated DNA in a highly specific and sensitive hybridization [Fig. 1, step (8)]. The highest sensitivity is needed when working with mammalian genomes, requiring exposure times of at least 1 week. The success of an experiment is thus largely determined by the ratio of specific hybridization to membrane background. The hybridization procedure that was originally introduced by Church and Gilbert[10] is crucial for achieving this goal.

Materials

Na$_2$HPO$_4$ (0.5 M), pH 7.2 (solution is 1 M with respect to Na$^+$): Dissolve 89 g Na$_2$HPO$_4 \cdot$ 2H$_2$O in water, adjust the pH with 85% H$_3$PO$_4$, and make up to 1 liter with water

Hybridization buffer: 7% (w/v) SDS [Bio-Rad (Richmond, CA) electrophoresis purity reagent], 1% (w/v) BSA (A7906; Sigma), 1 mM EDTA, 0.25 M Na$_2$HPO$_4$ (0.5 M Na$^+$), pH 7.2

Wash buffer: 20 mM Na$_2$HPO$_4$ (40 mM Na$^+$), pH 7.2, 1 mM EDTA, 1% (w/v) SDS; 100 mM NaCl are added for each 5% of reduced GC contents of the probe DNA (from 50%)

Procedure

The most reliable results with a favorable signal-to-noise ratio have been obtained when using a hybridization incubator that contains rotating cylinders made of polypropylene or Plexiglas. The initial experiments in our laboratory were performed with an adapted bacterial incubator and polypropylene measuring cylinders with silicone stoppers. Sophisticated hybridization ovens are now commercially available from a number of companies.

Before each hybridization the cylinders must be cleaned thoroughly with detergent and 10 M NaOH, followed by extensive rinsing with water. The inner surface should not be cleaned mechanically (e.g., with a brush) to avoid scratching the surface. The dry GeneScreen membrane after UV irradiation is wetted by floating on blotting buffer, rolled up, and transferred into the hybridization cylinder. The membrane is manipulated with a thick pipette until it attaches smoothly to the walls without trapping air bubbles. Filling the cylinder with buffer facilitates these manipulations. Multiple layers of membrane do not adversely affect signal or background hybridization. Excess blotting buffer is poured off and 10–20 ml hybridization buffer is added. The cylinder is closed and rotated in the incubator at 65° for at least 30 min. Meanwhile the radioactive probe DNA (1–5 × 10⁰ dpm) is added to 10 ml of hybridization buffer and mixed well. After the prehybridization (the length of which is not critical) the hybridization mix is exchanged for the one containing the probe and the cylinder is again rotated at 65° for 16–24 hr. The solution is then poured off and the membrane is rinsed five times with 100 ml wash buffer at 65° while still in the cylinder. The following washes are performed at room temperature in large trays on a platform shaker. Hot wash buffer is poured into a tray and allowed to cool to 65° before the membrane is submerged. Each wash (1 liter of wash buffer) takes about 5 min, during which the buffer slowly cools down. The membrane is then transferred into a second tray again with 1 liter wash buffer at 65°. After eight washes of this kind the membranes are generally clean. The wet membranes are stretched between two layers of Saran wrap and exposed to Kodak (Rochester, NY) XAR film with Du Pont (Wilmington, DE) Cronex Lightning Plus screen at −70°, the DNA-bound side being in contact with the film. After a 20-hr exposure weak signals should be visible that usually allow complete evaluation of the experiment with regard to the quality of the DNA samples and possible footprinted regions. This first exposure also helps to estimate the time finally needed (7–14 days) to obtain publication-quality exposures.

After a satisfactory exposure is obtained, the membrane can be hybridized with the probe specific for the complementary strand. To strip the old probe, the membrane is submerged in 1 liter 0.2 M NaOH and agitated for 15 min at room temperature. After neutralization with two washes in 100 mM Na$_2$HPO$_4$ (pH 7.2), 1 mM EDTA the membrane is ready for rehybridization. The NaOH treatment of the nylon membrane often reduces background as well. With every stripping of the membrane, however, about 10% of the original signal is lost.

Under the given experimental conditions only fragments larger than 70 nucleotides are labeled. We have not determined whether smaller fragments are not retained on the membrane because they are not efficiently cross-linked or because the stringency of hybridization and washing does not allow smaller fragments to hybridize.

The extremely sensitive hybridization technique described above can be used to detect femtogram amounts of specific genomic sequence. However, it also detects equally small amounts of plasmid that may contaminate one or more of the solutions used to prepare the genomic DNA. To avoid such contaminations with, for example, the probe plasmid, we recommend a strict separation of solutions used for *in vivo* footprinting experiments (such as buffers, ethanol, and phenol).

Concluding Remarks

The procedures described in detail have been used with great success in our laboratory as well as in others to obtain information about a large number of different protein–DNA interactions. Among the factors studied using this method were proteins binding to GREs, CREs, CAAT boxes, CACC boxes, G-strings, octamer motifs, and a variety of other sequence modules that have yet to be characterized in more detail. Dimethyl sulfate has proved to be an extremely useful reagent, even though it presents a certain bias to the study of protein binding. Binding to sites that do not contain crucial guanines will go undetected and it is also conceivable that some proteins themselves may be methylated by the agent and dissociate from their binding site. Besides the agent used to footprint a protein *in vivo,* the success of the experiment will be largely determined by the occupancy of a given site. Thus it is of great importance that the cell population to be analyzed be homogeneous with regard to the protein interaction. If only a fraction of the binding sites in the cell pool is occupied by a factor because of asynchrony in cell cycle, variable responsiveness to environmental stimuli, mixed cell types, or simply because the factor does not bind tightly enough *in vivo,* no clear footprint will be obtained. To establish the significance of a specific effect, it is helpful to use control cells that are treated in parallel with the ones under study.

Even though the methods described yield reliable, satisfying results they should by no means be considered perfect. Without doubt, considerable improvements will be made in the near future as more investigators concentrate their efforts and creativity on further developing the technique. To overcome the need for exposure times in excess of 1 week, linear or polymerase chain reaction (PCR) amplification of genomic sequences has been used to create genomic sequence and footprints.[26-29] Although exposure times can be reduced using these procedures, we feel that the need to anneal or ligate different oligonucleotides to genomic DNA, with the possibilities of false priming and PCR amplification artifacts, are disadvantages. Direct genomic sequencing will clearly be the method of choice, when less complex genomes (such as *Drosophila* or yeast) are analyzed. We are confident that future improvements will also result in a considerable shortening of exposure times for mammalian genomic footprinting. The most exciting development in this regard will be the substitution of chemiluminescent probes for radioactive labeling.

Acknowledgments

We thank Dr. A. F. Stewart and A. Reik for stimulating discussions and Dr. A. F. Stewart for the initial suggestion of the probe synthesis strategy described in this chapter. We also thank Drs. C. DeVack and S. K. Rabindran for reading the manuscript.

[41] *In Situ* Detection of DNA-Metabolizing Enzymes following Polyacrylamide Gel Electrophoresis

By MATTHEW J. LONGLEY and DALE W. MOSBAUGH

Introduction

Assignment of catalytic activities to specific polypeptides or protein complexes is one fundamental objective of the enzymologist. Protein electrophoresis methods have proved extremely useful for correlating enzyme activity to a particular gene product. A powerful technique emerged for studying DNA-metabolizing enzymes with the development of "activity gels." These techniques permitted electrophoretically separated enzymes to act on DNA substrates that were embedded in a polyacrylamide gel. The location of reaction products within the gel identified the enzyme bands. Several laboratories have reported activity gel systems for the detection of various nucleic acid-metabolizing enzymes, including RNA polymerase,[1] RNase,[2] terminal deoxynucleotidyltransferase,[3] DNA methyltransferase,[4] polymerase,[5-7] primase,[8] endonuclease,[9] exonuclease,[5] and ligase.[10] These systems have generally utilized a different, heterogeneous substrate for the detection of each activity, which does not allow the simultaneous detection of more than one enzymatic activity.

We have extended the versatility of the activity gel technique by developing procedures to simultaneously detect multiple DNA-metabolizing enzymes from a single lane of an activity gel. Two elements have provided the improvement to this activity gel technique. First, defined ^{32}P-labeled oligonucleotides annealed to M13 DNA were used to act as dual substrates for various enzymes. Second, a unique approach was developed for resolving and characterizing [^{32}P]DNA reaction products from activity gels. This sensitive technique has proved extremely useful for identifying the catalytic subunits of numerous multimeric or multifunctional enzymes.[11-13] Modification of DNA substrates within the gel either before or after *in situ* enzyme reactions can create intermediate substrates. Thus, activities of a complete DNA metabolic pathway may be identified by the sequential modification of substrates or by casting intermediate substrates in a gel. Described below are the general procedures and conditions for detecting DNA polymerase, exonuclease, endonuclease, ligase, and uracil-DNA glycosylase with this novel activity gel system.

Principle of Method

This activity gel technique is based on the use of a two-dimensional gel system. In the first dimension proteins are resolved through either a nondenaturing or a denaturing polyacrylamide gel containing defined ^{32}P-

[1] A. Spanos and U. Hubscher, this series, Vol. 91, p. 263.

[2] A. Blank, R. H. Sugiyama, and C. A. Dekker, *Anal. Biochem.* **120,** 267 (1982).

[3] L. M. S. Chang, P. Plevani, and F. J. Bollum, *J. Biol. Chem.* **257,** 5700 (1982).

[4] U. Hubscher, G. Pedrali-Noy, B. Knust-Kron, W. Doerfler, and S. Spadari, *Anal. Biochem.* **150,** 442 (1985).

[5] A. Spanos, S. G. Sedgwick, G. T. Yarranton, U. Hubscher, and G. R. Banks, *Nucleic Acids Res.* **9,** 1825 (1981).

[6] A. Blank, J. R. Silber, M. P. Thelen, and C. A. Dekker, *Anal. Biochem.* **135,** 423 (1983).

[7] N. F. Insdorf and D. F. Bogenhagen, *J. Biol. Chem.* **264,** 21491 (1989).

[8] U. Hubscher, *EMBO J.* **2,** 133 (1983).

[9] S. A. Lacks and S. S. Springhorn, *J. Biol. Chem.* **255,** 7467 (1980).

[10] M. Mezzina, J.-M. Rossignol, M. Philippe, R. Izzo, U. Bertazzoni, and A. Sarasin, *Eur. J. Biochem.* **162,** 325 (1987).

[11] M. J. Longley and D. W. Mosbaugh, *Methods Mol. Cell. Biol.* **1,** 79 (1989).

[12] M. J. Longley, S. E. Bennett, and D. W. Mosbaugh, *Nucleic Acids Res.* **18,** 7317 (1990).

[13] M. J. Longley and D. W. Mosbaugh, *Biochemistry* **30,** 2655 (1991).

labeled DNA substrates. Following electrophoresis, enzymes catalyze *in situ* reactions that alter (elongate, degrade, or modify) their DNA substrates. The [^{32}P]DNA substrates and products are resolved according to size in a second dimension of electrophoresis through a denaturing DNA sequencing gel. The position and size of the DNA products permits identification and characterization of the enzymes.

The basic procedure can be described in five steps, as illustrated in Fig. 1. The simultaneous detection of DNA polymerase and 3' → 5'-exonuclease activities by *Escherichia coli* DNA polymerase I (large fragment) will be used as a prototypic example.

1. Substrates are designed and constructed as either 3' or 5' end-labeled oligonucleotides annealed to M13 DNA. For DNA polymerase I large fragment (LF), two synthetic 5' end ^{32}P-labeled oligonucleotides were designed and separately annealed to M13mp2 DNA. The sequence of each oligonucleotide is given in Table I. A 15-nucleotide oligomer (15-mer) was fully complementary to positions 106–120 of the *lacZα* sequence of M13mp2 DNA, and it served as a primer for DNA synthesis. A 24-mer was complementary to positions 107–129 and noncomplementary at position 106, forming a 3'-terminally mispaired (T · C) substrate for the 3' → 5'-exonuclease.

2. The [^{32}P]DNA substrates are then cast within a polyacrylamide gel and standard protein electrophoresis of DNA polymerase I (or other nucleic acid-metabolizing enzymes) is performed. Either native or sodium dodecyl sulfate (SDS)-polyacrylamide gels may be used in this "analytical activity gel" format. This gel format utilizes a single, wide sample well (1.3 × 1.3 cm), a 1-cm high stacking gel and a 5-cm resolving gel. For this example, equal mixtures of ^{32}P-labeled 15-mer/M13mp2 and ^{32}P-labeled 24-mer/M13mp2 DNA are cast into separate nondenaturing and denaturing activity gel matrices, samples of DNA polymerase I (LF) are applied, and electrophoresis is performed.

3. *In situ* enzyme reactions are performed following electrophoresis. In the case of SDS-polyacrylamide gels, the SDS is extracted from the gel and enzymes are allowed to renature prior to assaying for *in situ* activity. Nondenaturing gels are processed without these manipulations. For both types of gels, the sample lane is sliced vertically into several narrow gel strips (six to eight identical sublanes). Each slice is placed in a test tube and incubated with reaction components. To detect the DNA polymerase and 3' → 5'-exonuclease activities of DNA polymerase I (LF), reactions are initiated by the addition of 7 mM MgCl$_2$ and 100 μM ddTTP. Following incubation at 25°, *in situ* reactions are terminated by the addition of buffer containing 10 mM ethylenediaminetetraacetic acid (EDTA).

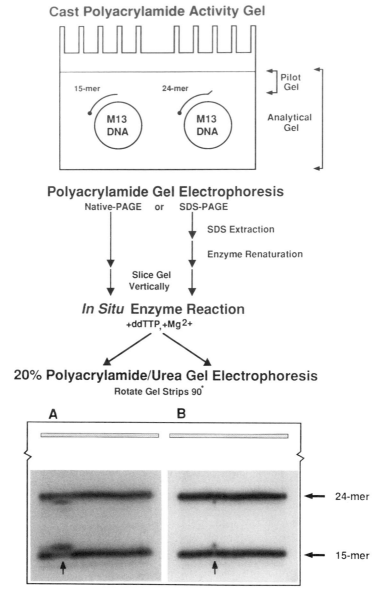

FIG. 1. Overview of the essential steps for the *in situ* detection of DNA-metabolizing enzymes. (1) Defined ^{32}P-labeled oligonucleotides are hybridized to M13 DNA and cast within either nondenaturing or SDS-polyacrylamide gels. (2) Standard protein gel electrophoresis of DNA-metabolizing enzymes is performed. For denaturing gels, electrophoresis is followed by SDS extraction and enzyme renaturation steps. (3) The resolving gel containing [^{32}P]DNA and enzymes is vertically subdivided into thin strips. *In situ* enzymatic reactions are initiated

4. A second dimension of electrophoresis through a denaturing DNA sequencing gel resolves [^{32}P]DNA reaction products from the first gel. Each gel strip is rotated 90° and cast within a 20% (w/v) polyacrylamide/ 8.3 M urea DNA sequencing gel. Reorientation of analytical gels permits separation of DNA-metabolizing enzymes in the first dimension of electrophoresis and a size analysis of the ^{32}P-labeled products in the second dimension.

5. Autoradiography is performed to locate the substrates and reaction products. For DNA polymerase I (LF), unaltered ^{32}P-labeled 15-mer and ^{32}P-labeled 24-mer migrate as uniform bands, whereas ^{32}P-labeled oligonucleotides of different sizes are observed at the position corresponding to the enzyme. The 3′ → 5′-exonuclease activity degrades the mismatched 24-mer, almost exclusivley generating a matched 23-mer product. On the other hand, DNA polymerase catalyzes the incorporation of a single ddTMP residue onto the matched 3′ terminus of the 15-mer, producing a 16-mer. Incorporation does not occur on the 3′ ends of the 24-mer or the 23-mer because ddTTP is a noncomplementary nucleoside triphosphate for this reaction. The observation of the two activities in the same band under both nondenaturing (Fig. 1A) and SDS-polyacrylamide (Fig. 1B) gel electrophoresis confirmed that DNA polymerase and 3′ → 5′-exonuclease activities reside in the same polypeptide.[14]

Materials and Reagents

Bacteria

Bacteriophage M13mp2, M13mp2(−C106), and M13mp19 DNAs are grown in *E. coli* strain JM107 and are isolated as described in this series by Kunkel *et al.*[15]

Chemicals

Radioisotopes {[α-^{32}P]dGTP (3000 Ci/mmol) and [γ-^{32}P]ATP (6000 Ci/ mmol)} are obtained from New England Nuclear (Boston, MA). Unlabeled

[14] P. Setlow, D. Brutlag, and A. Kornberg, *J. Biol. Chem.* **247**, 224 (1972).
[15] T. A. Kunkel, J. D. Roberts, and R. A. Zakour, this series, Vol. 154, p. 367.

---

by immersing each strip into buffers containing diffusible cofactors and precursors. (4) Gel strips are reoriented by a 90° rotation and cast within a denaturing DNA sequencing gel. [^{32}P]DNA products are resolved by size in a second dimension of electrophoresis. (5) *In situ* reaction products are analyzed by autoradiography. Shown are examples of native (A) and denaturing (B) activity gel analysis of *E. coli* DNA polymerase I (large fragment). The position of the enzyme in each gel system is indicated by a vertical arrow.

TABLE I
DNA Substrates and in Situ Reaction Conditions for Various DNA-Metabolizing Enzymes

Oligonucleotide[a]

	Sequence
12-mer	5'-TAACGCCAGGGT-3'
15-mer	5'-GGCGATTAAGTTGGG-3'
24-mer	5'-GTGCTGCAAGGCGATTAAGTTGGT-3'
U-24-mer	5'-GTGCTGCAAGGCGAUTAAGTTGGT-3'
50-mer	5'-CGGCCAGTGAATTCGAGCTCGGTACCCGGGGATCCTCTAGAGTCGACCTG-3'

DNA substrate	Enzyme source	Reaction initiation[b]	Product
I. DNA polymerase			
5' ^{32}P-Labeled 15-mer/M13mp2 DNA	E. coli Pol I (LF) T4 DNA polymerase	+ 7 mM MgCl$_2$ + 100 μM dNTP	≤625 Nucleotides
	E. coli Pol I (LF) E. coli Pol III (Holo) Novikoff Pol β Taq polymerase	+ Mg^{2+} ^{c} + 100 μM ddTTP	16-mer
IIa. 3' → 5'-Exonuclease			
5' ^{32}P-Labeled 15-mer/M13mp2 DNA	E. coli Exo III	+ 5 mM MgCl$_2$	<15-mer
5' ^{32}P-Labeled 24-mer/M13mp2 DNA	E. coli Exo III	+ 5 mM MgCl$_2$	<24-mer
IIb. 3' → 5'-Exonuclease (Proofreading)			
5' ^{32}P-Labeled 24-mer/M13mp2 DNA	E. coli Pol I (LF)	+ 7 mM MgCl$_2$	23-mer
	E. coli Pol III (Holo)	+ 15 mM Magnesium acetate	23-mer
	E. coli ε subunit (DNA Pol III)	(− 50 mM NaCl)[d] + 4 mM MgCl$_2$	23-mer

Substrate[a]	Enzyme	Reaction components[b,c]	Product
III. 5′ → 3′-Exonuclease			
3′ ^{32}P-Labeled 24-mer/M13mp2 DNA	Taq DNA Pol	+ 10 mM MgCl$_2$	<24-mer
5′ ^{32}P-Labeled 24-mer/M13mp2 DNA	λ Exonuclease	+ 2.5 mM MgCl$_2$	Unlabeled oligomer
IV. DNA Ligase			
5′ ^{32}P-Labeled (12-mer + 15-mer)/M13mp2 DNA	E. coli ligase	+ 4.6 mM MgCl$_2$ + 10 mM (NH$_4$)$_2$SO$_4$ + 150 µM NAD$^+$ (− 50 mM KCl)[d]	27-mer
	T4 ligase	+ 10 mM MgCl$_2$ + 1 mM ATP	27-mer
V. Uracil-DNA glycosylase			
5′ ^{32}P-Labeled U-24-mer/M13mp2 DNAt-C106)	E. coli uracil-DNA glycosylase	No addition	AP 24-mer
5′ ^{32}P-Labeled AP-24-mer/M13mp2 DNA(-C105)		+ 1 mM Sodium cacodylate (pH 6.5) + 0.1 mM EDTA + 1 mM NaCl + 100 µM Lys-Trp-Lys	
VI. Restriction endonuclease			
5′ ^{32}P-Labeled 50-mer/M13mp19 DNA	EcoRI	(− 25 mM NaCl)[d] + 8 mM MgCl$_2$	9-mer
	KpnI		25-mer
	XmaI		25-mer
	SmaI		27-mer
	XbaI		36-mer

[a] The U-24-mer contains a single, site-specific uracil residue at position 15 nucleotides from the 5′ end. Removal of the free base (uracil) generates an oligomer with an apyrimidinic (AP) site, designated AP 24-mer. The hexameric recognition sequences of EcoRI, KpnI, SmaI, XmaI, and XbaI are highlighted on the 50-mer. SmaI and XmaI cleave at different positions within the same recognition sequence.

[b] Reaction buffer components are the same as renaturation buffer components (Table IV) with the additions and deletions noted here.

[c] The divalent metal cation concentrations for E. coli DNA polymerase I (large fragment), E. coli DNA polymerase III (holoenzyme), Novikoff hepatoma DNA polymerase β, and Taq DNA polymerase reactions were 7 mM MgCl$_2$, 15 mM magnesium acetate, 7 mM MgCl$_2$, and 10 mM MgCl$_2$, respectively. The final salt concentration in Novikoff hepatoma DNA polymerase β reactions was 50 mM NaCl.

[d] As indicated, final salt concentrations are reduced from the amount used in renaturation buffers (Table IV).

deoxyribonucleoside triphosphates and bovine fibrinogen are purchased from Sigma (St. Louis, MO). Pharmacia (Piscataway, NJ) is the source of 2',3'-dideoxythymidine 5'-triphosphate. 2-Propanol is obtained from J. T. Baker (Phillipsburg, NJ). Sodium dodecyl sulfate (specially purified for biochemical work—SDS-44215) is from Gallard-Schlessinger (BDH, Poole, England). Prestained protein molecular weight markers are from Sigma (SDS-7B) and Bio-Rad (Richmond, CA; low range).

Apparatus

Gel electrophoresis of proteins is performed on a Mini-Protean II (Bio-Rad) apparatus. DNA sequencing gel electrophoresis is done on a model S2 apparatus (Bethesda Research Laboratories, Gaithersburg, MD).

Enzymes

Escherichia coli DNA polymerase I large fragment (LF), T4 DNA polymerase, *E. coli* DNA ligase, T4 DNA ligase, and restriction endonucleases *Eco*RI, *Kpn*I, *Sma*I, *Xma*I, and *Xba*I are from New England BioLabs (Beverly, MA). *Thermus aquaticus* DNA polymerase, *E. coli* exonuclease III, λ-exonuclease, and T4 polynucleotide kinase are from Bethesda Research Laboratories. Homogeneous Novikoff hepatoma DNA polymerase β (fraction VI)[16] and *E. coli* uracil-DNA glycosylase (fraction V)[17,18] are purified as previously described. *Escherichia coli* DNA polymerase III holoenzyme (fractions IV and V), prepared as described,[19] is a gift of C. S. McHenry (University of Colorado, Denver). The ε subunit of *E. coli* DNA polymerase III, purified as previously described,[20] is a gift of F. W. Perrino and L. A. Loeb (University of Washington).

Stock Solutions

Kinase buffer (10×): 500 mM Tris-HCl (pH 7.6), 100 mM MgCl$_2$, 50 mM dithiothreitol (DTT), 1 mM EDTA

SSC (20×): 3 M NaCl, 0.3 M sodium citrate

Solution A: 36.3 g Tris base, 48 ml 1 M HCl; dissolve in distilled H$_2$O, adjust volume to 100 ml

[16] D. M. Stalker, D. W. Mosbaugh, and R. R. Meyer, *Biochemistry* **15,** 3114 (1976).
[17] T. Lindahl, S. Ljungquist, W. Siegert, B. Nyberg, and B. Sperens, *J. Biol. Chem.* **252,** 3286 (1977).
[18] J. D. Domena, R. T. Timmer, S. A. Dicharry, and D. W. Mosbaugh, *Biochemistry* **27,** 6742 (1988).
[19] R. Oberfelder and C. S. McHenry, *J. Biol. Chem.* **262,** 4190 (1987).
[20] R. H. Scheuermann and H. Echols, *Proc. Natl. Acad. Sci. U.S.A.* **81,** 7747 (1984).

Solution B: 20 g acrylamide, 0.74 g N,N'-methylenebisacrylamide; dissolve in distilled H_2O, adjust volume to 100 ml

Solution C: 3 g Tris base, 24 ml 1 M HCl; dissolve in distilled H_2O, adjust volume to 50 ml

Solution D: 5 g acrylamide, 1.25 g N,N'-methylenebisacrylamide; dissolve in distilled H_2O, adjust volume to 50 ml

PAGE running buffer (5×): 15 g Tris base, 72 g glycine; dissolve in distilled H_2O, adjust volume to 1000 ml. Immediately prior to use dilute fivefold in distilled H_2O. For denaturing polyacrylamide gel electrophoresis, add 0.5 g SDS per 500 ml of dilute solution

Cracking buffer (3×): 26 μl 1.25 M Tris-HCl (pH 6.8), 10 μl 0.1 M EDTA, 5 μl 2-mercaptoethanol, 50 μl 75% (w/v) glycerol, 50 μl 10% (w/v) SDS, 50 μl 0.4% (w/v) bromphenol blue, 9 μl distilled H_2O

SDS extraction buffer: 10 mM Tris-HCl (pH 7.5), 5 mM 2-mercaptoethanol, 25% (v/v) 2-propanol

Enzyme reaction buffers: See Table I

Enzyme renaturation buffers: See Table IV

TBE buffer (10×); 162 g Tris base, 27.5 g boric acid, 9.3 g EDTA; dissolve in distilled H_2O, adjust volume to 1000 ml

40% Acrylamide solution (29 : 1): 193 g acrylamide, 6.7 g N,N'-methylenebisacrylamide; dissolve in distilled H_2O, adjust volume to 500 ml, and filter through Whatman (Clifton, NJ) #1 filter paper

20% Polyacrylamide DNA sequencing gel solution: 75 ml 40% acrylamide solution, 15 ml 10× TBE buffer, 75 g urea; adjust volume to 150 ml with distilled H_2O, filter through a 0.45-μm pore size filter, and degas solution. Immediately prior to pouring the gel, 1050 μl of 10% (w/v) ammonium persulfate and 16.5 μl of N,N,N',N'-tetramethylethylenediamine (TEMED) are carefully mixed with the gel solution

Formamide/tracking dye: 95% (w/v) deionized formamide, 10 mM EDTA, 0.1% (w/v) bromphenol blue, 0.1% (w/v) xylene cyanol

Methods

Procedure 1: Preparation of Oligonucleotide Substrates

1. Design, chemically synthesize, deblock, and deprotect[21] the oligonucleotide preparation (0.2 μmol).

2. Resuspend the lyophilized oligonucleotide in 1 ml of 10 mM triethylamine–bicarbonate (TEA) buffer (pH 7.0).

[21] M. H. Caruthers, A. D. Barone, S. L. Beaucage, D. R. Dodds, E. F. Fisher, L. J. McBride, M. Matteucci, Z. Stabinsky, and J.-Y. Tang, this series, Vol. 154, p. 287.

3. Apply the oligonucleotide sample to a Sephadex G-50 column (1.8 cm² × 3.1 cm) equilibrated with 10 mM TEA buffer (pH 7.0) at room temperature. Collect 15 fractions (1 ml each) while washing the column with equilibration buffer.

4. Identify oligonucleotide-containing fractions by absorbance at 260 nm and combine peak fractions (~3–4 ml). Remove the equilibration buffer by vacuum evaporation in a SpeedVac centrifuge (model SVC100H; Savant, Hicksville, NY) operating at room temperature.

5. Resuspend the dry pellet in 300 μl of distilled water and measure the optical density (260 nm) of the solution.

6. Adjust to 0.1 units A_{260} and distribute 50-μl aliquots into 1.5-ml Eppendorf tubes.

7. Vacuum evaporate the solution to dryness using a SpeedVac and store samples at $-70°$ until needed.

Procedure 2: 5' End ^{32}P Labeling of DNA Substrates

1. Resuspend an oligonucleotide sample in 50 μl of distilled water ($OD_{260} = 0.1$).

2. Phosphorylation reactions are carried out in a mixture (75 μl) containing 1× kinase buffer, 30 units of T4 polynucleotide kinase, 0.4 μM [γ-^{32}P]ATP (6000 Ci/mmol), and 30 μl of the oligonucleotide sample (final $OD_{260} = 0.04$ or ~83 ng).

3. After incubation for 60 min at 37° the reaction is terminated by the addition of 9 μl of 0.1 M EDTA and by a heat treatment at 70° for 5 min.

4. Reaction products can be analyzed for purity by 20% polyacrylamide DNA sequencing gel electrophoresis. In general, oligonucleotide preparations should be >95% pure and have a specific activity of 3.0–3.5 × 10⁶ cpm/pmol of 5' ends. When necessary, further purification of the unlabeled oligonucleotide can be carried out by electrophoresis through a 20% polyacrylamide/8.3 M urea DNA sequencing gel followed by concentration on a NENSORB-20 (Du Pont), Wilmington, DE) column,[18] and the phosphorylation reaction can then be performed.

5. Defined DNA substrates are constructed by annealing ^{32}P-labeled oligonucleotides to various single-stranded M13 DNA molecules (Table I). Hybridization reaction mixtures (800 μl) contained 1× SSC, 63 μg/ml of M13 DNA, and 80 μl of the ^{32}P-labeled oligonucleotide preparation. Hybridization is performed by placing the mixture in an Eppendorf tube suspended in a 500-ml beaker of water at 70° and allowing the water to cool slowly to room temperature (2–4 hr).

6. ^{32}P-Labeled DNA substrates are stored at $-20°$ and used within 3 weeks.

Procedure 3: 3' End ^{32}P Labeling of DNA Substrates

1. An oligonucleotide (24-mer; see Table I) is prepared and phosphorylated at the 5' end as described in procedure 2, except 1 mM ATP is substituted for the radioactive precursor and EDTA is omitted from the termination reaction.

2. The 24-mer is hybridized to M13mp2 DNA (0.2 pmol of 3' ends per microgram DNA at 63 μg/ml) as described in procedure 2.

3. Labeling reactions (106 μl) contain 53 μg of 24-mer/M13 DNA, 21 pmol [α-^{32}P]dGTP (3000 Ci/mmol), and 5 units of *E. coli* DNA polymerase I (LF).

4. After DNA synthesis at 37° for 40 min, reactions are terminated by adding 10.6 μl of 0.1 M EDTA and by heating at 70° for 5 min. Analysis of purity and/or further purification may be performed as before (procedure 2, step 4). Labeled oligonucleotides have a specific activity of 1.8–3.1 $\times$ 10^6 cpm/pmol of 3' ends.

5. Add 42 μl of 20× SSC and 681 μl of distilled H$_2$O. Rehybridize the oligonucleotide to the existing M13 DNA by slowly cooling the 70° mixture to room temperature (see procedure 2, step 5).

Procedure 4: Nondenaturing Polyacrylamide Activity
  Gel Electrophoresis

Nondenaturing analytical activity gels are cast in two sections (Fig. 1), using the resolving and stacking gel solutions described in Table II.

1. The analytical resolving gel solution is poured between the glass plates to a height of approximately 5.3 cm and overlaid with ~0.5 cm of distilled water. Polymerization should occur within 20 to 30 min. The polymerized gel has dimensions of 8.5 cm (*w*) × ~5 cm (*h*) and is 0.075 cm thick.

2. Following polymerization, the water layer is removed with a Kimwipe by inverting the gel apparatus. Caution should be exercised because this solution generally contains ^{32}P radioactivity.

3. Finally, the stacking gel solution is poured to fill most of the remaining space and a Teflon comb (10 well, 0.5 × 1.3 cm) is inserted. Polymerization should occur within 20 min after placing the gel about 10 cm from a Sylvania (F15T8/D) Daylight fluorescent bulb. During this period, the resolving gel should be shielded from light to prevent possible DNA damage.

4. Following polymerization, the comb is removed and one stacking gel finger between two adjacent wells is removed with a bent 22-gauge needle (Fig. 1), forming a wide sample well (1.3 × 1.3 cm).

TABLE II
COMPONENTS OF NONDENATURING 10% POLYACRYLAMIDE ACTIVITY GELS[a]

Stock solution	Analytical gel format		Pilot gel format	
	Stacking	Resolving	Support	Resolving
Solution A	—	500 μl	500 μl	125 μl
Solution B	—	2000 μl	2000 μl	500 μl
Solution C	313 μl	—	—	—
Solution D	625 μl	—	—	—
Sucrose (40%)	1250 μl	—	—	—
EDTA (0.1 M)	50 μl	80 μl	80 μl	20 μl
Fibrinogen (1 mg/ml)	—	200 μl	200 μl	50 μl
Riboflavin (0.01%)	125 μl	—	—	—
^{32}P-Labeled N-mer/M13 DNA	—	120 μl	—	30 μl
Distilled H_2O	135.6 μl	1070.8 μl	1190.8 μl	267.7 μl
Ammonium persulfate (10%)	—	28 μl	28 μl	7 μl
TEMED	1.4 μl	1.2 μl	1.2 μl	0.3 μl
Gel concentration	2.5%	10%	10%	10%
Volume	2.5 ml	4 ml	4 ml	1 ml
Dimensions (cm, $w \times h$)	8.5 × 1	8.5 × 5	8.5 × 4	8.5 × 1

[a] Gel composition is based on that described by B. J. Davis, *Ann. N.Y. Acad. Sci.* **121,** 404 (1964), as modified by M. J. Longley and D. W. Mosbaugh, *Biochemistry* **30,** 2655 (1991).

5. An enzyme sample (20–30 μl) containing a final concentration of 30% (w/v) glycerol and 0.008% (w/v) bromphenol blue is loaded in each wide lane. Electrophoresis is carried out at 4° in polyacrylamide gel electrophoresis (PAGE) running buffer (1×), using 100 V, until the bromphenol blue dye has migrated through the stacking gel (~1.0 cm). After stacking occurs, the potential is increased to 200 V and migration is continued until the dye approaches the bottom of the resolving gel.

Procedure 5: Sodium Dodecyl Sulfate-Polyacrylamide Activity Gel Electrophoresis

Denaturing analytical activity gels are cast in two sections (Fig. 1), using the resolving and stacking gel solutions described in Table III.

1. The analytical resolving gel and stacking gel are formed following the procedure for native gels (procedure 4, steps 1–4), except that ultraviolet (UV) polymerization of the stacking gel is unnecessary.

2. Enzyme samples (60 μl) are prepared by mixing 2 parts (40 μl) of an enzyme preparation with 1 part (20 μl) of 3× cracking buffer, followed by heating at 37° for 3 min.

TABLE III
COMPONENTS OF DENATURING 10% POLYACRYLAMIDE ACTIVITY GELS[a]

Stock solution	Analytical gel format		Pilot gel format	
	Stacking	Resolving	Support	Resolving
Tris-HCl (1.5 M), pH 8.8	—	750 μl	750 μl	250 μl
Tris-HCl (1.25 M), pH 6.8	250 μl	—	—	—
Acrylamide (30%)	375 μl	1000 μl	1000 μl	333 μl
N,N'-Methylenebisacrylamide (2%)	250 μl	400 μl	400 μl	133 μl
Sodium dodecyl sulfate (10%)	25 μl	30 μl	30 μl	10 μl
EDTA (0.1 M)	50 μl	60 μl	60 μl	20 μl
Fibrinogen (1 mg/ml)	—	150 μl	150 μl	50 μl
[32]P-Labeled N-mer/M13 DNA	—	90 μl	—	30 μl
Distilled H_2O	1531 μl	500 μl	590 μl	167 μl
Ammonium persulfate (10%)	17.5 μl	18 μl	18 μl	6 μl
TEMED	1.5 μl	2 μl	2 μl	1 μl
Gel concentration	4.5%	10%	10%	10%
Volume	2.5 ml	3 ml	3 ml	1 ml
Dimensions (cm, $w \times h$)	8.5 × 1	8.5 × 5	8.5 × 4	8.5 × 1

[a] Gel composition is based on that described by A.Blank, J. R. Silber, M. P. Thelen, and C. A. Dekker, *Anal. Biochem.* **135,** 423 (1983), as modified by M. T. Longley and D. W. Mosbaugh, *Methods Mol. Cell. Biol.* **1,** 79 (1989).

3. Gel samples (50 μl) are immediately loaded into wide wells (1.3 × 1.3 cm; see procedure 4, step 4), and prestained molecular weight protein markers are applied to both adjacent lanes.

4. Electrophoresis [100 V (stacking gel) and 200 V (resolving gel)] is performed as described (see procedure 4, step 5) at 4° until the dye approaches the bottom of the resolving gel.

5. Record the distance migrated through the resolving gel by the tracking dye and by each prestained protein standard. It is most convenient to measure these mobilities *prior* to separation of the glass plates. This avoids inaccurate measurements due to band/dye diffusion and physical distortion of the gel, while it also minimizes manipulation of the gel containing ^{32}P radioactivity and fragile enzymes.

6. Prior to the removal of the stacking gel, the resolving gel should be nicked with a razor blade so that the left and right edges of the lanes containing samples are readily identifiable following SDS extraction and enzyme renaturation.

7. The resolving gel is immersed in 33 gel volumes (100 ml/analytical resolving gel) of SDS extraction buffer and gently agitated on a gyratory shaker (60 rpm) at 25°.

TABLE IV
COMPOSITION OF RENATURATION BUFFERS

Enzyme	Renaturation buffer
Novikoff hepatoma DNA polymerase β	25 mM Tris-HCl (pH 8.4), 5 mM 2-mercaptoethanol, 0.5 mM EDTA, 15% (w/v) glycerol, 400 μg/ml BSA, 500 mM NaCl
E. coli DNA polymerase I (large fragment)	50 mM Tris-HCl (pH 7.5), 5 mM 2-mercaptoethanol, 16% (w/v) glycerol, 400 μg/ml BSA, 50 mM KCl
E. coli DNA polymerase III (holoenzyme)	35 mM HEPES-KOH (pH 7.6), 7 mM DTT, 15% (w/v) glycerol, 400 μg/ml BSA, 50 mM KCl
E. coli ε subunit (DNA polymerase III)	20 mM Tris-HCl (pH 8.0), 8 mM DTT, 15% (w/v) glycerol, 400 μg/ml BSA, 50 mM NaCl
Taq DNA polymerase	50 mM Tris-HCl (pH 9.0), 15% (w/v) glycerol, 400 μg/ml BSA, 50 mM NaCl
T4 DNA polymerase	67 mM Tris-HCl (pH 8.8), 10 mM 2-mercaptoethanol, 6.7 $\mu$$M$ EDTA, 16.6 mM (NH$_4$)$_2$SO$_4$, 15% (w/v) glycerol, 400 μg/ml BSA
E. coli exonuclease III	50 mM Tris-HCl (pH 8.0), 5 mM 2-mercaptoethanol, 15% (w/v) glycerol, 400 μg/ml BSA, 50 mM KCl
λ-Exonuclease	67 mM Glycine-KOH (pH 9.4), 5 mM 2-mercaptoethanol, 15% (w/v) glycerol, 400 μg/ml BSA
E. coli uracil-DNA glycosylase	70 mM HEPES-KOH (pH 7.5), 5 mM 2-mercaptoethanol, 1 mM EDTA, 15% (w/v) glycerol, 400 μg/ml BSA
E. coli or T4 DNA ligase	40 mM Tris-HCl (pH 8.0), 20 mM DTT, 5% (w/v) glycerol, 400 μg/ml BSA, 50 mM KCl
Restriction endonucleases EcoRI, SmaI, KpnI, XbaI, or XmaI	20 mM Tris-HCl (pH 7.5), 5 mM 2-mercaptoethanol, 5% (w/v) glycerol, 400 μg/ml BSA, 50 mM NaCl

8. After 30 min, the buffer is discarded and replaced with the same volume of fresh SDS extraction buffer. Extract for an additional 30 min.

9. The SDS extraction buffer is completely removed and 10 ml of the appropriate enzyme renaturation buffer (Table IV) is added to wash the extraction buffer from the resolving gel.

10. After brief (30 sec) agitation, this solution is replaced with 27 gel volumes (80 ml) of enzyme renaturation buffer and the gel is incubated with gentle agitation for 18–25 hr at 4°.

Procedure 6: In Situ Enzymatic Reactions

Following nondenaturing polyacrylamide gel electrophoresis or after renaturation of enzymes within SDS-polyacrylamide gels, *in situ* enzyme reactions are initiated by immersing resolving gels into reaction buffers.

In situ reaction buffers consist of renaturation buffers (Table IV) supplemented with diffusible cofactors, as described in Table I. In some cases, renaturation buffers contain inhibitory concentrations of NaCl, which are reduced for initiation of these *in situ* reactions (Table I).

1. Before carrying out reactions, each lane of an analytical activity gel is longitudinally cut into six (0.2 × 5 cm) slices. Then individual gel strips are placed into separate 16 × 100 mm test tubes containing 5 ml of reaction buffer (Table I). *In situ* reactions are performed in stoppered (Parafilm) tubes that are placed on their sides and gently shaken.

2. Generally, incubation occurs at 25° for various times (0–60 min) while shaking (60 rpm).

3. Reactions are terminated by substituting reaction buffer with an equal volume of ice-cold buffer containing 10 mM EDTA but lacking diffusible cofactors (e.g., MgCl$_2$, ATP, NAD$^+$, dNTPs). Then tubes are gently mixed at 4° for 30 min.

Procedure 7: Electrophoretic Separation of [^{32}P]DNA Products in a Second Dimension

Following *in situ* enzyme reactions, analytical activity gels containing [^{32}P]DNA products are cast within a 20% polyacrylamide/8.3 M urea DNA sequencing gel. The second dimension of electrophoresis resolves DNA substrates and products, allowing the identification and characterization of numerous DNA metabolizing enzymes.

1. Wash and allow to air dry one set of sequencing gel plates (30 × 40 cm) for every four activity gel strips to be analyzed. Temporarily assemble the plates without spacers and draw a line on the back side of the large plate at a depth of 1.5 cm below the top edge of the small plate. Disassemble the sandwich and place the large plate horizontally on a laboratory bench.

2. Arrange individual activity gel strips along the line on the large plate, taking care to remove any small bubbles that might have adhered to the gel strips. Position the spacers (0.075 cm) at the sides and bottom of the gel. Carefully lower the smaller gel plate into its final position, clamp the gel plates together, and seal the edges with agarose.

3. Extreme care should be exercised when pouring the 20% polyacrylamide/8.3 M urea sequencing gel, because small bubbles cannot easily be removed from beneath the gel strips without loss of the sample. We have found the easiest way to avoid bubbles is to hold the gel sandwich at a compound angle while slowly pouring the gel solution into one corner through a syringe. In this orientation the gel strips are not horizontal, and bubbles are swept away as the rising acrylamide solution passes. Gels

should be overfilled and placed horizontally. Polymerization of the sequencing gel around the recessed gel strips is complete within 45 to 60 min. The uppermost edge of the sequencing gel is in contact with air and often has an irregular appearance after polymerization. This is no cause for alarm as long as the gel strips are completely embedded in the polymerized sequencing gel.

4. Following polymerization, the gel sandwich is mounted on the electrophoresis apparatus, and the reservoirs are filled with 1× TBE. Then, 200 μl of formamide/tracking dye is underlaid onto the surface of the sequencing gel. Electrophoresis is performed at 1200 V (~45 mA/gel) until the bromphenol blue has migrated approximately 20 cm.

5. After electrophoresis the gel is removed from the bottom plate by soaking in water. Agitation under water should be gentle so as not to tear the gels at the discontinuity between the sample gel strips and the sequencing gel. The gel is transferred to filter paper, dried, and autoradiographed.

General Comments

Molecular Weight Determination

Sodium dodecyl sulfate-polyacrylamide *in situ* activity gels can be used to correlate enzyme activity definitively with a particular protein band and to determine polypeptide molecular weight. The approach for molecular weight determination is almost identical to standard SDS-polyacrylamide gel electrophoresis.[22] A standard curve (log M_r versus R_f) is constructed from the relative electrophoretic mobilities of prestained marker polypeptides. The relative mobility of a protein possessing *in situ* activity indicates its molecular weight. As an example, Novikoff hepatoma DNA polymerase β was separated on denaturing 7.5, 10, and 12.5% polyacrylamide activity gels. Following electrophoresis, SDS extraction, and enzyme renaturation, *in situ* DNA polymerase reactions (Table I) in the presence of all four dNTPs (A, C, G, and T) were performed on a vertical slice from each gel. Resolution of [^{32}P] DNA reaction products in a second dimension of electrophoresis was followed by autoradiography (Fig. 2A–C). The position of the extended primers within the sample gel strip indicates the location of the polymerase. Because bovine fibrinogen was uniformly cast in the protein gel to facilitate the recovery of enzyme activity,[6,23] protein staining of the gels became uninformative and the use of prestained molec-

[22] U. K. Laemmli, *Nature (London)* **227**, 680 (1970).
[23] E. Karawya, J. A. Swack, and S. H. Wilson, *Anal. Biochem.* **135**, 318 (1983).

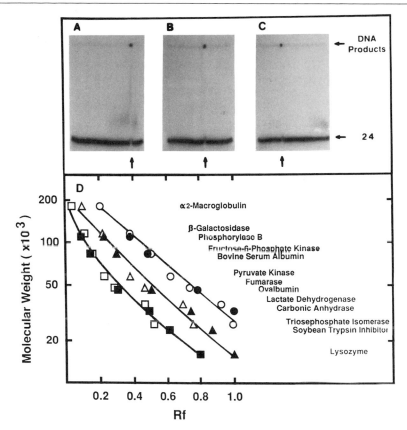

FIG. 2. Molecular weight determination by activity gels. Novikoff hepatoma DNA poly merase β was analyzed on separate denaturing 7.5% (A), 10% (B), and 12.5% (C) polyacryl-amide analytical activity gels. Following electrophoresis, SDS extraction, and enzyme renaturation, DNA synthesis reactions were conducted in the presence of all four dNTPs. Unutilized [32P]DNA primers migrated as uniform bands [arrow, 24 (24-mer)]. The location of highly extended primers (arrow, DNA products) indicated the position of the renatured polymerase. (D) The apparent molecular weight of the polymerase was determined from a standard curve (log M_r versus R_f) of the relative mobilities of prestained protein molecular weight markers (Sigma SDS-7B, open symbols; Bio-Rad low range, closed symbols) in 7.5% (circles), 10% (triangles), and 12.5% (squares) polyacrylamide activity gels.

ular weight markers became necessary. Standard curves for DNA polymerase β (Fig. 2D) indicate apparent molecular weights of 34,000, 38,000, and 39,000 as analyzed on 7.5, 10, and 12.5% polyacrylamide gels, respectively. Clearly, the presence of DNA in these gels does not affect these determinations, because they confirm the molecular weight (32,000) determined by standard SDS-polyacrylamide gel electrophoresis.[16]

Substrate Design

Several factors should be considered in the design of the oligonucleotide substrates, including (1) length of the oligonucleotide, (2) position of the radiolabel, (3) temperature of the *in situ* reaction, and (4) the size of the expected DNA products. We have successfully used oligonucleotides ranging in length from 12 to 50 nucleotides. An ideal oligonucleotide substrate would remain bound to M13 DNA during protein electrophoresis and *in situ* reactions, but it would freely dissociate on electrophoresis in denaturing DNA sequencing gels. Oligonucleotide denaturation is influenced by length, base composition, exposure to urea, and the temperature of the sequencing gel. Unlike conventional DNA sequencing gels, these gels cannot be warmed by preelectrophoresis to facilitate oligonucleotide denaturation. Thus, longer oligonucleotides ($\geq$24-mer) may remain partially hybridized to M13 DNA until the temperature increases during electrophoresis. Some smearing can result, but this effect is readily alleviated by prewarming the mounted gel and the electrophoresis apparatus with a hand-held hairdryer. However, oligonucleotides as large as a 50-mer are not observed to enter a 20% polyacrylamide/8.3 M urea DNA sequencing gel despite this prewarming procedure.

The position of the radioactive label in the substrate depends on the nature of the activity to be detected. For example, a 3'-terminal ^{32}P label is best for detecting 5' $\rightarrow$ 3'-exonuclease activity and a 5'-terminal ^{32}P label is preferred for 3' $\rightarrow$ 5'-exonucleases. However, 5' $\rightarrow$ 3'-exonuclease activity can be detected using a 5' end-labeled oligonucleotide. Under this condition, an unlabeled gap in the substrate band is observed on the autoradiogram. A variety of DNA substrates and reactions (Table I) have proved useful in the detection of many DNA-metabolizing enzymes. With appropriate modification of substrate design and labeling, this technique could be adapted to detect other enzymes.

In situ reactions have customarily been performed at room temperature to reduce oligonucleotide denaturation. In some cases reactions have successfully been carried out at 37°. Detection of *Thermus aquaticus* DNA polymerase and 5' $\rightarrow$ 3'-exonuclease activities was done at 50°, but the ^{32}P signal was greatly reduced during extended reaction times at this elevated temperature. This was presumably due to thermal denaturation and diffusion of the oligonucleotide into the reaction mixture. We have combatted these problems by using longer oligonucleotides and shorter reaction times. Reactions as short as 5 min or as long as 3 hr have been conducted. These results suggest only a brief lag time exists before initiation of reactions by diffusion of cofactors and precursors (e.g., Mg^{2+}, dNTPs, NAD^+, and ATP) into the gel.

We have found the use of mixed oligonucleotide substrates (15-mer and 24-mer) permits the simultaneous detection of multiple activities. For example, the use of 5' end-labeled oligonucleotides of different lengths has permitted both DNA polymerase and 3' → 5'-exonuclease activities to be detected in the same *in situ* activity gel (Figs. 1 and 3A). Similarly, the use of a 5' ^{32}P-labeled 15-mer and a 3' ^{32}P-labeled 24-mer hybridized to M13mp2 DNA has established that *Taq* DNA polymerase contains a 5' → 3'-exonuclease activity (Fig. 3B). Examples described above utilize two overlapping sequences of oligonucleotides individually hybridized to separate M13 DNA molecules. Alternatively, nicked or gapped substrates can be created by hybridizing multiple oligonucleotides to the same M13 DNA molecule. We have detected both *E. coli*[13] and T4 DNA ligase (Fig. 3C) using a 5' ^{32}P-labeled 12-mer and 5' ^{32}P-labeled 15-mer hybridized to M13mp2 DNA. These two oligonucleotides are juxtaposed and create a 27-mer when joined by DNA ligase. In some cases a single oligonucleotide served as a simultaneous substrate for different enzymes. A 5' end-labeled oligonucleotide (5' ^{32}P-labeled 50-mer) was designed and annealed to the multiple cloning region of M13mp19 DNA. This substrate contained double-stranded recognition sites for *Eco*RI, *Kpn*I, *Sma*I, *Xba*I, and *Xma*I (Table I). *In situ* activity gel reactions with these restriction endonucleases generate fragments of different lengths. Therefore, one or more of the enzymes can be identified by determining the specific size of the restricted product on a DNA sequencing gel.[13] For instance, *Kpn*I cleaves at the GGTACC sequence, producing a ^{32}P-labeled 25-mer (Fig. 3D). Finally, DNA substrates within polyacrylamide gels can be modified by exogenous reagents or enzymes either before or after *in situ* detection of enzyme activities. To detect uracil-DNA glycosylase in an SDS-polyacrylamide gel, a 5' ^{32}P-labeled U-24-mer containing a site-specific uracil residue was hybridized to M13mp2(−C106) DNA and cast within a gel (Table I). *In situ* uracil-DNA glycosylase activity would be expected to remove uracil and produce an apyrimidinic (AP) site, 15 residues from the 5' end. The tripeptide Lys-Trp-Lys was then diffused into the gel to facilitate cleavage of the AP site.[24,25] Following this β-elimination reaction, a 5' ^{32}P-labeled 15-mer containing a 3'-terminal AP site was generated. The location of this oligonucleotide identified *E. coli* uracil-DNA glycosylase (Fig. 3E). In addition to modifying DNA with Lys-Trp-Lys, we have previously shown that several enzymes (restriction endonucleases, polynucleotide kinase, and alkaline phospha-

[24] T. Behmoaras, J.-J. Toulme, and C. Helene, *Nature (London)* **292**, 858 (1981).
[25] B. Weiss and L. Grossman, *Adv. Enzymol.* **60**, 1 (1987).

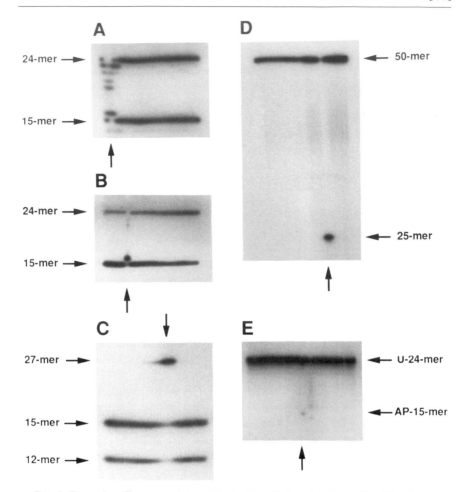

FIG. 3. Examples of enzymes detected by *in situ* activity gel analysis. Specific substrates, reaction conditions, and products are given in Table I. (A) The comigration of the DNA polymerase and $3' \rightarrow 5'$-exonuclease activities of *E. coli* DNA polymerase III (holoenzyme) were observed by nondenaturing activity gel analysis. (B) The *Taq* DNA polymerase polypeptide was shown to possess a $5' \rightarrow 3'$-exonuclease activity by denaturing activity gel analysis. (C) Renatured T4 DNA ligase formed a ^{32}P-labeled 27-mer by joining adjacent ^{32}P-labeled 15-mer and ^{32}P-labeled 12-mer. (D) The restriction endonuclease *Kpn*I cleaved a ^{32}P-labeled 50-mer. The location of the ^{32}P-labeled 25-mer product indicated the mobility of the enzyme. (E) *Escherichia coli* uracil-DNA glycosylase was shown to act on a ^{32}P-labeled 24-mer containing a site-specific uracil residue. Subsequent treatment with Lys-Trp-Lys generated a ^{32}P-labeled 15-mer with a $3'$-terminal AP site.

tase) can penetrate activity gels and modify the immobilized DNA substrates.[13] This feature allows additional flexibility in detecting various DNA metabolizing enzymes.

Troubleshooting

Unsuccessful experiments are usually attributed to one of two types of problems. The first concerns the difficulties of manipulating and processing activity gels. The second relates to determining optimal renaturation and/ or reaction conditions for individual enzymes. Helpful hints and systematic methods for overcoming these types of problems are discussed below.

The procedure for longitudinally subdividing activity gel lanes into strips prior to *in situ* reactions can be tedious and requires some practice. This process can be simplified by first aligning the nicks that mark the borders of the sample lane(s). Then the entire length of the resolving gel should be sliced in a single motion to avoid generating a jagged edge. A device composed of two double-edged razor blades clamped between microscope slides works well for this purpose. We generally divide a lane (13 mm wide) into six to eight vertical strips. To avoid decreased resolution in the second dimension of electrophoresis, care should be taken to make strips ≤2 mm in width. Individual gel strips are somewhat fragile and should be carefully transferred to *in situ* reaction test tubes using the edge of a scalpel blade. Prewetting the test tubes with reaction buffer will prevent the sticking and tearing of the delicate gel strips. The mixing and free movement of the gel strip during *in situ* reactions is important, but this often results in inversion of the strip in the test tube. To eliminate ambiguity concerning the top and bottom of the gel, we have found it convenient to cut a "fish tail" in the end nearest the dye front. This aids in the alignment of the wet gel strips on the sequencing gel plate. Care should be taken to prevent the accumulation of small bubbles below the gel strips once they are positioned on the sequencing gel plate. Slightly wetting the gel plate will reduce bubble formation and will facilitate manipulation of the gel strips.

When examining a new enzyme preparation with this activity gel technique, the most frequently encountered problem concerns the quantity of enzyme needed to obtain an acceptable signal. If too much enzyme is used, the signal is broad and resolution is low. Also, excess enzyme can diffuse out of the gel and react at a uniform, low level across the entire length of the gel strip. On the other hand, if not enough enzyme is applied, the signal may be too weak to detect. Within reasonable limits, weak signals can be intensified by lengthening reaction times, by loading more enzyme, or by increasing the autoradiographic exposure time. However, optimization of renaturation and/or reaction conditions may also produce rewarding results. As an initial approximation, the renaturation buffer conditions should be the standard *in vitro* reaction conditions, with the omission of diffusible cofactors and precursors. We have compiled an extensive list of renaturation conditions for various enzymes (Table IV). Although these buffers have proved sufficient, not all have been optimized for recovery of maximal activity. In most cases individual *in situ* reactions are initiated by introducing the required cofactors and precursors (Table I).

We have developed another type of activity gel system for the optimization of renaturation and reaction conditions. "Pilot activity gels" permit the rapid determination of the optimal (1) renaturation buffer conditions, (2) reaction components, (3) *in situ* reaction temperature, (4) reaction time, and (5) quantity of enzyme needed to obtain an acceptable signal. Pilot gels are ideal for evaluating these parameters for DNA polymerases, but they may be adapted for use with other enzymes. Gels in the pilot format are cast in three sections. From the bottom to the top of a Mini Protean II (Bio-Rad) gel apparatus, the gel is composed of a support gel lacking DNA [8.5 cm (w) × 4 cm (h)], an activity resolving gel (Fig. 1) containing the [^{32}P]DNA substrate [8.5 cm (w) × 1 cm (h)], and an ~1 cm stacking gel. The composition of each gel solution is given in Table II (native) and Table III (denaturing), and gels are cast as previously described (procedures 4 and 5). Identical enzyme samples are applied to each lane (0.5 cm wide) of the pilot gel, and electrophoresis is performed as before until the dye approaches the bottom of the short resolving gel containing the [^{32}P]DNA. Following electrophoresis, SDS extraction, and enzyme renaturation, *individual* lanes are incubated under different renaturation and/or reaction conditions to test various parameters. For DNA polymerases, *in situ* reactions are done with all four deoxyribonucleoside triphosphates to permit extended DNA synthesis. Then pilot gels are reassembled and cast within a 20% polyacrylamide denaturing DNA sequencing gel (procedure 7). Unlike analytical gels, reassembled pilot gels are not reoriented by a 90° rotation on the sequencing gel plate. Highly extended [^{32}P]DNA products are retained in the pilot gel, whereas unutilized [^{32}P]DNA primers migrate into the DNA sequencing gel, as in Fig. 2. Because equal amounts of enzyme are applied to each sample well, renaturation and reaction efficiencies are directly comparable as variable autoradiographic intensities. The pilot gel system has proved extremely useful for the characterization and systematic optimization of conditions for *in situ* DNA polymerase activity.[11,13]

Summary

We have presented several protocols for producing an *in situ* activity gel that allows detection of various DNA-metabolizing enzymes. Both

nondenaturing polyacrylamide and SDS-polyacrylamide activity gel electrophoresis procedures were detailed. Combining the use of defined
[^{32}P]DNA substrates with product analysis, these procedures detected a
wide spectrum of enzymatic activities. The ability to detect 7 different
catalytic activities of 15 different enzymes provides encouragement for
expanded applications. It is hoped that others will find this technique
applicable for detecting these enzymes and other activities in different
biological systems. The modification of DNA *in situ* and the creation of
intermediate substrates within activity gels should prove extremely useful
for dissecting the enzymatic steps of DNA replication, repair, recombination, and restriction, as well as the metabolic pathways of other nucleic
acids.

Acknowledgments

This work was supported by National Institutes of Health Grant GM32823 and National
Institute of Environmental Health Sciences Grant ES00210. Technical Report Number 9532
is from the Oregon Agricultural Experiment Station.

[42] Simultaneous Characterization of DNA-Binding Proteins and Their Specific Genomic DNA Target Sites

By Jean-Claude Lelong

Growth, development, and differentiation of the cell involve important
cellular processes such as DNA replication, recombination, and mRNA
transcription, which are controlled by specific interactions between nuclear proteins and defined DNA regulatory sequences. For instance, most
transcription factors that regulate gene activity, either positively or negatively, are cell-specific or ubiquitous proteins that recognize and bind to
their own promoter DNA motif, thereby influencing the function of the
RNA polymerase at a variable distance in cis. Thus, a complex interplay
between multiple DNA-binding proteins and their specific binding sites
located within the promoter regions of eukaryotic genes is a fundamental
determinant of both temporal regulation and tissue specificity of transcription.

Practically, it has proved difficult to resolve one specific protein–DNA
complex from the many found in cells. The most successful strategy to
date has been to purify a DNA sequence and to use it to select from a
mixture any protein that binds to it. The reverse approach is to use purified

protein to select a specific DNA sequence. In both cases, prior purification of the DNA or protein used as the selective agent is necessary. The stepwise combination of protein blotting and selective DNA binding allows one to circumvent these requirements, and, by associating the two above strategies, permits the identification, among a natural or reconstituted mixture of promoter DNA elements, of those that are specifically bound to individual tissue-specific or ubiquitous DNA-binding proteins detected by Southwestern blotting from different cell crude extracts.

Principle

The method involves four steps summarized in Fig. 1: separation of proteins by gel electrophoresis, transfer of the separated proteins to a nitrocellulose filter, detection of DNA-binding proteins by incubating the filter with a mixture of radioactively labeled DNA restriction fragments, and identification of the DNA fragment(s) bound to individual proteins by elution and gel electrophoresis.

Total, cytoplasmic, or nuclear cell extracts are loaded without prior heating on a sodium dodecyl sulfate (SDS)-polyacrylamide slab gel and electrophoresed along with ^{14}C-labeled molecular weight marker proteins. The fractionated proteins are allowed to renature by removing the SDS from the gel in a buffer containing appropriate concentrations of salt, urea, and dithiothreitol. Proteins are then transferred bidirectionally to nitrocellulose sheets by a diffusion sandwich method. A pair of identical protein blots is obtained. One is stained by China ink, the other is probed with an equimolar mixture of end-labeled DNA fragments of different sizes covering a putative regulatory region of genomic DNA, in the presence of an excess of unspecific competitor poly(dI–dC) · poly(dC–dI). Characterization of the bound DNA fragment(s) is achieved by excising from the nitrocellulose sheet the radioactive spots of interest.

The retained DNA is eluted from the filter with SDS, ethanol precipitated, run out on a gel alongside a sample of the probe, and visualized by autoradiography. Binding specificity, defined as a relative enrichment in one or more of the restriction fragments over the others, is reached at an appropriate DNA/protein ratio that varies from one protein to another, therefore requiring the use of serial dilutions on the nuclear extracts.

Materials and Reagents

All solutions should be autoclaved as $10\times$ stock solutions, omitting dithiothreitol, phenylmethylsulfonylfluoride (PMSF), and glycerol, which are added just before use.

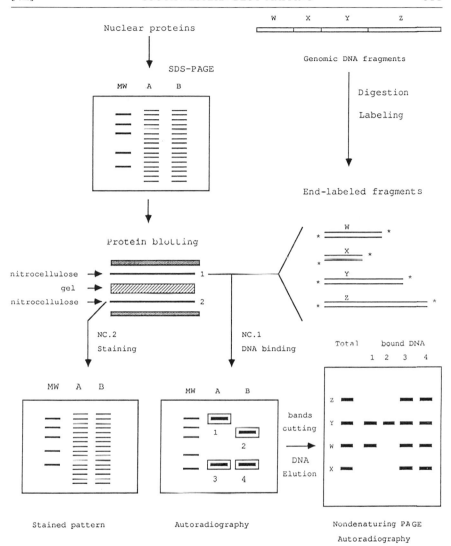

Fig. 1. Summary of the South Western blot mapping procedure. A detailed protocol is described in text.

Solutions for Preparation of Nuclear Proteins

Buffer D-I: 10 mM N-2-hydroxy ethylpiperazine-N'-2-ethanesulfonic acid (HEPES) (pH 8), 2 mM MgCl$_2$, 3 mM CaCl$_2$, 0.2% (w/v) Nonidet P-40 (NP-40), 0.5 mM dithiothreitol (DTT), 0.5 mM PMSF

Buffer D-II: 2 M sucrose in buffer D-I

Buffer D-III: 10 mM HEPES (pH 8), 4 mM MgCl$_2$, 0.1 mM ethylenedi-aminetetraacetic acid (Na-EDTA), 0.33 M sucrose, 0.5 mM dithio-threitol, 0.5 mM PMSF

Buffer D-IV: 20 mM HEPES (pH 8), 1 mM Na-EDTA, 0.42 M NaCl, 15% (v/v) glycerol, 1 mM dithiothreitol, 1 mM PMSF

Buffer D-V: 20 mM HEPES (pH 8), 6 mM MgCl$_2$, 0.1 M KCl, 0.2 mM Na-EDTA, 15% (v/v) glycerol, 0.5 mM dithiothreitol, 0.5 mM PMSF

Buffer G-I: 10 mM Tris-HCl (pH 7.9), 1 mM Na-EDTA, 5 mM dithio-threitol

Buffer G-II: 50 mM Tris-HCl (pH 7.9), 10 mM MgCl$_2$, 2 mM dithiothrei-tol, 25% (w/v) sucrose, 50% (v/v) glycerol

Buffer G-III: 50 mM Tris-HCl (pH 7.9), 6 mM MgCl$_2$, 40 mM (NH$_4$)$_2$SO$_4$, 0.2 mM Na-EDTA, 1 mM dithiothreitol, 15% (v/v) glycerol

Solutions for Sodium Dodecyl Sulfate Polyacrylamide Gel Electrophoresis

Lower gel buffer (4×): 1.5 M Tris-HCl (pH 8.8), 0.4% (w/v) SDS
Upper gel buffer (4×): 0.5 M Tris-HCl (pH 6.8), 0.4% (w/v) SDS
Acrylamide (30%, w/v), ultrapure grade, filtered
Bisacrylamide (1%, w/v), ultrapure grade, filtered
Running buffer (10×): 250 mM Tris, 1.92 M glycine (pH 8.8), 1% SDS
Sodium Dodecyl Sulfate-polyacrylamide slab gels (0.9 mm thick)

Lower gel: 12.5% (w/v) acrylamide, 0.1% (w/v) bisacrylamide, 0.027% (w/v) N,N,N',N'-tetramethylethylenediamine (TEMED), 0.027% (w/v) ammonium persulfate, 375 mM Tris-HCl (pH 8.8), 0.1% (w/v) SDS

Upper gel: 5% (w/v) acrylamide, 0.13% (w/v) bisacrylamide, 0.05% (w/v) TEMED, 0.03% (w/v) ammonium sulfate, 125 mM Tris-HCl (pH 6.8), 0.1% (w/v) SDS

Gel renaturation buffer: 4 M urea, 50 mM NaCl, 2 mM Na-EDTA, 0.1 mM dithiothreitol; 0.1 mM PMSF, 10 mM Tris-HCl (pH 7.5)

Transfer buffer: 50 mM NaCl, 2 mM Na-EDTA, 0.1 mM dithiothreitol, 0.1 mM PMSF, 10 mM Tris-HCl (pH 7.5)

Nitrocellulose membranes: BA 85 (Schleicher & Schuel, Keene, NH)

Binding buffer: 100 mM NaCl, 1 mM Na-EDTA, 0.02% (w/v) Ficoll 400, 0.02% (w/v) polyvinylpyrrolidone, 0.02% (w/v) bovine serum albumin (BSA) fraction V, 10 mM Tris-HCl (pH 7.2). The use of 5% (w/v) nonfat dry milk instead of Denhardt's solution[1] has been reported[2]

---

[1] D. T. Denhardt, *Biochem. Biophys. Res. Commun.* **23,** 641 (1966).
[2] W. K. Miskimins, M. P. Roberts, A. McClelland, and F. H. Ruddle, *Proc. Natl. Acad. Sci. U.S.A.* **82,** 6741 (1985).

DNA extraction buffer: 10 mM Tris-acetate (pH 8), 1 mM EDTA, 0.4 M NaCl, 0.1% (w/v) SDS

Methods

Preparation of Nuclear Proteins

Cell growth is arrested by washing the plates once with cold phosphate-buffered saline (PBS) solution containing 10 mM MgCl$_2$. Crude extracts for fractionation of DNA-binding proteins can be made either from purified nuclei by the method of Dignam[3] (procedure 1, below), or from total cells by the method of Manley and Gefter[4] (procedure 2, below). All manipulations, unless otherwise stated, are carried out at 0–4°.

Procedure 1. The cells are harvested by scraping and centrifuged for 10 min at 1500 rpm, resuspended in a small volume of buffer D-1 to determine the packed cell volume (PCV), and lysed for 10 min in a final volume of 10 ml/ml PCV of the same buffer with 20 strokes of a loose fitting Dounce homogenizer (Wheaton, Millville, NJ). The extract is brought to 0.33 M sucrose by adding 1/5 vol of buffer D-II and 1 vol of buffer D-III, then centrifuged through a 10-ml cushion of buffer D-II in an SW27 or SW28 Beckman (Fullerton, CA) rotor, 60 min at 15,000 rpm. The pellet is then resuspended by gentle agitation in buffer D-IV for 60 min. The nuclear suspension is centrifuged for 30 min at 30,000 g. The supernatant extract thus obtained is carefully removed, dialyzed against three changes of buffer D-V, and stored in aliquots in liquid nitrogen.

Procedure 2. The cells are suspended in 4 PCV of buffer G-I, and lysed by incubation for 20 min in buffer G-II, followed by homogenization with 8 strokes of a Dounce homogenizer. To the lysate, 4 PCV of buffer G-II is immediately added. After gentle stirring for 10 min, 1 PCV of an (NH$_4$)$_2$SO$_4$-saturated solution is added dropwise and the resulting viscous suspension is further stirred for an additional 60 min. After centrifugation (3 hr at 50,000 rpm, in a 60Ti Beckman rotor) the clear supernatant solution is precipitated overnight by (NH$_4$)$_2$SO$_4$ (0.33 g/ml).

The proteins are collected by centrifugation (30,000 g, 30 min), suspended in 1/10 vol of the original supernatant in buffer G-III, dialyzed against three changes of the same buffer, clarified by centrifugation (10,000 g, 15 min), and stored in aliquots in liquid nitrogen.

[3] J. D. Dignam, R. M. Lebovitz, and R. G. Roeder, *Nucleic Acids Res.* **11**, 1475 (1983).
[4] J. L. Manley, A. Fire, A. Cano, P. Sharp, and M. L. Gefter, *Proc. Natl. Acad. Sci. U.S.A.* **77**, 3855 (1980).

Gel Electrophoresis, Renaturation, and Protein Blotting

The subsequent steps are carried out at room temperature. The procedure is based on the original method of Bowen *et al.*,[5] and modified. Protein samples are prepared in running buffer containing 5% (v/v) 2-mercaptoethanol, 8% (w/v) sucrose, and 0.001% (w/v) bromphenol blue, without heating, and coelectrophoresed along with [14]G-labeled molecular marker proteins serving also as controls for protein transfer efficiency. Sodium dodecyl sulfate gels are without urea. Electrophoresis is performed overnight at 8 mA/gel (0.9 mm thick). The fractionated proteins are renatured by gentle agitation of the gel in three changes of renaturation buffer, but this step is reduced to 90 min to avoid a progressive wash-out of proteins. Although the extent of renaturation is not known, these washes in 4 M urea remove SDS from the proteins and permit functional recognition of DNA-binding domains such as of the *lac* repressor,[5] the glucocorticoid receptor,[6] and the *SNF1* yeast protein.[7] Omitting it increases the yield and number of blotted proteins, but results in a complete loss of DNA sequence specificity of the DNA-binding proteins. The transfer of proteins from the gel to prewetted nitrocellulose filters occurs by diffusion in a "sandwich" apparatus[5] submerged in two changes of transfer buffer over a period of 36–48 hr. Varying the pH of transfer buffer from 3.7 to 8.9 allows one to select the acidic or basic nature of the DNA-binding proteins being transferred. A pH of 7.5 is routinely used. The extent of protein transfer can be determined by thoroughly washing one filter replicate in phosphate-buffered saline containing 0.2% (w/v) Tween 20, overnight staining in 0.1% (v/v) China ink (17 Black, Pelican), and final washing in water.

Alternative Blotting and Renaturation Procedures. Electrophoretic transfer[8] has been used with some success by several authors, in which case methanol is omitted from the buffer and proteins are renatured either in the gel before transfer[6] or on the blot after transfer.[7,9] Renaturation of blotted proteins is carried out by incubating the filters "back to back" into a plastic box containing renaturation buffer: The choice of the later (particularly the use of additional ions such as $MgCl_2$ or $ZnSO_4$) is critical and will largely depend on each protein. However, a general procedure is described[7] that involves two steps: denaturation with 7 M guanidine-HCl

[5] B. Bowen, J. Steinberg, U. K. Laemmli, and H. Weintraub, *Nucleic Acids Res.* **8,** 1 (1980).
[6] C. M. Silva, D. B. Tully, L. A. Petch, C. M. Jewell, and J. A. Cidlowski, *Proc. Natl. Sci. U.S.A.* **78,** 1744 (1987).
[7] J. L. Celenza and M. Carlson, *Science* **233,** 1175 (1986).
[8] H. Towbin, T. Staehlin, and J. Gordon, *Proc. Natl. Acad. Sci. U.S.A.* **76,** 4350 (1979).
[9] U. Hüscher, *Nucleic Acids Res.* **15,** 5486 (1987).

in 50 mM Tris-HCl (pH 8.3), 50 mM dithiothreitol, 2 mM Na-EDTA, 0.25% (w/v) nonfat dry milk for 1 hr at 25° and renaturation in 50 mM Tris-HCl (pH 7.5), 100 mM NaCl, 2 mM dithiothreitol, 2 mM Na-EDTA, 0.1% (w/v) Nonidet P-40, 0.25% (w/v) nonfat dry milk for 16 hr at 4°.

DNA-Binding Assay

Preparation of Labeled DNA Probe. Specific protein-binding sites usually require the integrity of both DNA strands, whereas nicked or single-stranded DNAs often bind proteins unspecifically. Therefore end labeling by nucleotide filling is more suitable than nick translation or random priming, because high specific activities are usually not required. The genomic DNA region of interest is digested by appropriate restriction enzymes so as to yield DNA fragments of unequal sizes, ranging from 300 to 700 bp, well characterized by native gel electrophoresis. After phenol–chloroform extraction and ethanol precipitation, the restriction enzyme-digested DNA is end labeled with the four [^{32}P]dXTPs using the Klenow large fragment of DNA polymerase I, as described.[10] The reaction mixture is passed through an Elutip-d column (Schleicher & Schuell), and the eluted material (specific activity, 1–5 × 10^7 cpm/μg) is prefiltered through a nitrocellulose filter and used as radioactive DNA probe without further purification.

Identification of DNA-Binding Proteins. All incubations are performed with gentle shaking at room temperature. The nitrocellulose blots are immersed for 30 min in binding buffer containing 30 μg/ml poly(I)–poly(C) (Boehringer GmbH, Mannheim, Germany) to block high-affinity unspecific DNA binding. The filters are then washed for 30 min in three changes of binding buffer without poly(I)–poly(C). They are subsequently incubated, using a sealed plastic bag or a plastic box, in a minimum volume of the same buffer containing the labeled specific DNA fragments in equimolar amounts (2–3 × 10^5 cpm/ml, about 10^{-11} M) with an appropriate molar excess (50 to 200-fold) of poly(dI–dC) · poly(dI dC) (Pharmacia, Piscataway, NJ), usually 1–2 μg/ml, as unspecific competitor, for 60 min. The filters are exhaustively washed until no radioactivity is detectable in the wash and then autoradiographed without drying.

Mapping of DNA Fragments Bound to Individual Proteins. Radioactive spots of interest are excised from the blots and extracted with phenol–chloroform in a minimum volume of DNA extraction buffer. The purified DNA fragments eluted from each band are ethanol precipitated in the presence of 4 μg yeast tRNA, dissolved in a minimum TAE buffer,[10]

---

[10] T. Maniatis, E. F. Fritsch, and J. Sambrook, *in* "Molecular Cloning: A Laboratory Manual." Cold Spring Harbor Press, Cold Spring Harbor, New York, 1982.

FIG. 2. Map of the LTR-MMTV DNA *Sau*3A restriction fragments used as probe. P, *Pst*1; S, *Sau*3A. Fragments are numbered 1 to 4 from 5' to 3' on coding strand. Distances are indicated in base pairs from the transcription initiation site. Thick bars represent the known regulatory elements: HRE, hormone responsive element; NFI, nuclear factor 1; TATA, TATA box.

and run out on a nondenaturing 5% (w/v) polyacrylamide gel[10] alongside a sample of the labeled probe mixture as a standard. Specific binding would result in preferential retention of one fragment rather than others. Therefore, the DNA fragment(s) bound to each separated protein is revealed by autoradiography and characterized by reference to the coelectrophoresed DNA probing mixture used.

Characterization of Cell-Specific LTR-MMTV DNA–Protein Complexes

The long terminal repeat (LTR) of the mouse mammary tumor provirus (MMTV) contains promoter DNA sequences that control positively or negatively both the hormonal induction and the tissue specificity of its transcription. Among them, the glucocorticoid hormone-responsive elements (GRE) and other proximal promoter elements are well documented as well as their interactions with several regulatory DNA-binding proteins (hormone receptors, NFI).[11] However, little is known about the multiple factors that might interact with the upstream promoter region of the LTR-MMTV DNA, which, according to functional studies by *in vivo*[12] or *in vitro*[13] gene transfer, are implicated in the tissue specificity of transcription. Therefore, nuclear proteins extracted from the GR mouse mammary epithelial cell line, expressing *in vivo* low basal or hormone-induced high levels of MMTV RNA, and from the nonexpressing 10T1/2 mouse fibroblast cell line, were probed, according to the above-described procedure, with a radioactive mixture of four sequential MMTV-LTR DNA restriction fragments (Fig. 2). The high-affinity DNA-binding proteins detected by this technique from either GR or 10T1/2 cell nuclear extracts are shown

[11] H. Ponta, W. H. Gunzburg, B. Salmons, B. Groner, and P. Herrlich, *J. Gen. Virol.* **66,** 931 (1985).
[12] T. A. Stewart, P. G. Hollingshead, and S. L. Pitts, *Mol. Cell. Biol.* **8,** 473 (1988).
[13] S. Mink, H. Ponta, and C. B. Cato, *Nucleic Acids Res.* **18,** 2217 (1990).

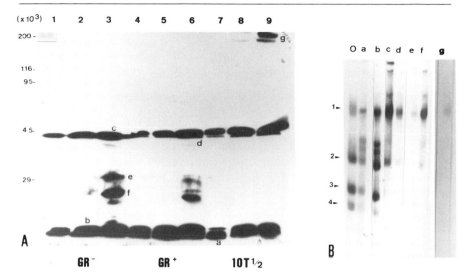

FIG. 3. Characterization of LTR-MMTV DNA–protein complexes by Southwestern blot. (A) Autoradiography of a nitrocellulose blot probed in 30 ml of binding buffer with 300 ng (10^7 cpm) of equimolecular amounts of the four LTR-MMTV *Sau*3A subfragments in the presence of 1 μg of poly(I)–poly(C). Lower-case letters indicate the radioactive bands from which the DNA was eluted and analyzed in (B). (B) Electrophoretic analysis on 5% polyacrylamide gel of the LTR-MMTV DNA subfragments bound to the protein–DNA complexes in (A), showing the selective affinity of some tissue-specific or ubiquitous DNA-binding proteins for different LTR-MMTV DNA regions. The letter at the top of each lane indicates the protein in (A) from which the corresponding bound DNA fragments were eluted. Lane 0 was loaded with the equimolecular mixture of the four end-labeled fragments used as a probe.

in Fig. 3A: Some of them are present in both cell types (proteins a, b, c, and d) whereas others appear restricted to GR cells (proteins e, f, and g). The DNA fragments bound to these proteins are shown in Fig. 3B: By reference to lane 0, containing the original probing mixture, it is clear that the ubiquitous DNA-binding proteins a and b (M_r 12,000–15,000) exhibit no preferential affinity for the three promoter DNA fragments 1, 2, and 3. However, the proteins c (43 kDa) and d (40 kDa), albeit apparently ubiquitous, bind either selectively to both DNA fragments 1 and 2 (protein c) or highly preferentially to DNA fragment 1 (protein d). Interestingly, as expected from their cell specificity, proteins e (30 kDa), f (25 kDa), and g (200 kDa) also select the only DNA fragment 1 that includes the far upstream promoter MMTV DNA sequences. Note that DNA fragment 4, which maps downstream from the transcription start, has almost no apparent affinity for any of the DNA-binding proteins detected and therefore can be taken as a useful internal control of the sequence specificity of the above DNA–protein interactions.

Limitations, Advantages, and Extension of Method

The described technique is suitable only for proteins that satisfy the following conditions: (1) efficient transfer to nitrocellulose filters, (2) efficient renaturation of the DNA-binding domain, (3) high affinity and sequence specificity for DNA, and (4) active monomeric DNA-binding domain. Therefore except for the last requirement, which is difficult to circumvent without using native gels, optimal conditions must be worked out for each case. Perhaps the most critical point is to minimize unspecific interactions by optimizing the ratio between specific DNA, competitor synthetic polydesoxynucleotide, and protein concentration.

Nevertheless, the main advantage of the method is obviously its high sensitivity and resolving power, allowing the direct screening of crude nuclear extracts. An attractive extension would be the protein-mediated cloning[14] of regulatory sequences in a given phenotype. Finally, perhaps the most flagrant evidence for the validity of the method[15] is provided by Huet et al.,[16] who demonstrated by a two-step "blot and footprint" procedure that the DNA retained by a filter-bound yeast TUF protein was protected from DNase I digestion exactly as in solution. Thus, the South Western blot mapping procedure described here might include direct footprinting of each individual blotted specific DNA-binding protein. However, it is not obvious that the binding sites we pick up in the in vitro assays are also binding sites in vivo.

[14] R. M. Gronostajski, K. Nagata, and J. Hurwitz, Proc. Natl. Acad. Sci. U.S.A. **81,** 4013 (1984).
[15] J. C. Lelong, G. Prevost, K. I. Lee, and M. Crepin, Anal. Biochem. **179,** 299 (1989).
[16] J. Huet and A. Sentenac, Proc. Natl. Acad. Sci. U.S.A. **84,** 3648 (1987).

Section IV

Other Methods

[43] Preparation and Storage of Competent *Escherichia coli* Cells

By CATHIE T. CHUNG and ROGER H. MILLER

Introduction

The uptake of foreign DNA by *Escherichia coli* can be induced either through electroporation, which involves discharging an electrical voltage across bacterial cell membranes, or by making bacteria competent via chemical methods. Two widely used chemical methods involve treating bacteria with calcium chloride[1] or hexamminecobalt[2] and subjecting the cells to a heat shock. Transformation efficiencies of these methods are on the order of 10^7 to 10^8 transformants/μg plasmid DNA. Moreover, storage of competent cells in both methods is possible by freezing cells after the addition of extra reagents. We have developed a method to prepare competent *E. coli* that is more rapid and convenient than existing methods. Our technique is especially useful in routine cloning experiments[3] and has been used to transform specific strains of *E. coli* that are found not to be transformable by conventional methods.[4]

Principle of Method

The method we developed to prepare competent bacterial cells is basically a one-step procedure in which bacteria are grown to the early exponential phase and then treated with a transformation and storage solution (TSS) containing polyethylene glycol (PEG), dimethyl sulfoxide (DMSO), and magnesium. Subsequently, competent cells are mixed with plasmid DNA, kept on ice for 5–30 min, incubated at 37° to allow expression of an antibiotic resistance gene, and plated. Using this method, we routinely obtain transformation efficiencies of 10^7 to 10^8 transformants/μg plasmid DNA.

Unlike other methods to prepare competent bacteria, our procedure does not involve multiple washing or incubation steps and requires a

---

[1] M. Dagert and S. D. Ehrlich, *Gene* **6**, 23 (1979).
[2] D. Hanahan, *J. Mol. Biol.* **166**, 557 (1983).
[3] J. P. Huff, B. J. Grant, C. A. Penning, and K. F. Sullivan, *BioTechniques* **9**, 570 (1990).
[4] A. Z. Ge, R. M. Pfister, and D. H. Dean, *Gene* **93**, 49 (1990).

minimal number of chemical components that can be added as one reagent rather than requiring multiple additions. Furthermore, we found that incubation of bacteria in TSS for as little as 5 min is sufficient to obtain a transformation efficiency of 5×10^7 transformants/μg DNA. Thus, unlike the $CaCl_2$ method, prolonged treatment of bacteria with TSS is not necessary. Another characteristic of our method is that it does not require the administration of a heat shock. In fact, we found that subjecting cells to heat shocks at various temperatures lowered transformation efficiencies. Finally, bacteria treated with TSS can be frozen directly and stored for future use, which makes this procedure convenient.

Materials and Reagents

Chemicals

PEG: M_r 3350 or 8000 (Sigma Chemical Company, St. Louis, MO)
DMSO (Fluka Chemical Corporation, Ronkonkoma, NY)
Bacto-tryptone and Bacto-yeast extract (Difco Laboratories, Detroit, MI)

Bacteria

Escherichia coli strain JM109 (Stratagene, La Jolla, CA) was used in the development of this assay

Method

Preparation of Stock Solutions and Reagents

Luria–Bertani (LB) broth (per liter): 10.0 g Bacto-tryptone, 5.0 g Bacto-yeast extract, 5.0 g NaCl; autoclave to sterilize
Mg^{2+} (2.0 M): 1.0 M $MgCl_2$ plus 1.0 M $MgSO_4$ or 2.0 M concentration of either solution alone; prepare in deionized, distilled water and sterile filter
Glucose (2.0 M): Prepare in deionized, distilled water and sterile filter

Preparation of TSS (1×)

1. Add solid PEG (M_r 3350 or 8000) to LB broth to make a 10% (w/v) solution.
2. Add an aliquot of the 2.0 M Mg^{2+} stock solution to achieve a final concentration of 20–50 mM.
3. Adjust the pH of the solution to 6.5–6.8.

4. Sterilize by filtration through a disposable cellulose nitrate filter (0.45-μm pore size).

5. Add an aliquot of DMSO to the filtered solution to achieve a final concentration of 5% (v/v) and keep the solution cold. Alternatively, PEG, Mg^{2+}, and DMSO can be added to LB broth and sterile filtered through a chemically resistant nylon filter (available from Nalge Company, Rochester, NY) after adjusting the pH to 6.5–6.8.

Transformation Procedure

LB broth containing the appropriate antibiotic is inoculated with an overnight culture of bacteria, making a 1 : 100 dilution of the culture. Bacteria are grown at 37°, with shaking (225 rpm), to the early exponential phase (OD_{600} 0.3–0.4). Subsequently, bacteria are pelleted by centrifugation at 1000 g for 10 min at 4° and the cell pellet is resuspended with ice-cold TSS in one-tenth of the original volume, until the cell suspension is homogeneous. Alternatively, the centrifugation step can be omitted by diluting cells directly with an equal volume of ice-cold 2× TSS. TSS-treated cells are incubated on ice for 5–15 min and then pipetted into sterile, prechilled polypropylene tubes in 100-μl aliquots. For transformation, 1 μl (100 pg–1 ng) of plasmid DNA is added per tube, followed by gentle mixing and incubation of the bacteria/DNA on ice for 5–60 min. To obtain maximum transformation efficiencies, a 30- to 60-min incubation on ice is required. A negative control in which no DNA is added to cells should also be included. After the incubation period on ice, 0.9 ml of LB broth (or TSS) to which an aliquot of the 2.0 M glucose stock solution has been added to yield a final concentration of 20 mM, is pipetted into each tube and the tubes are incubated at 37°, with shaking (225 rpm), for 1 hr. Cells are then diluted, if necessary, and plated on agar plates prepared with LB broth and the appropriate antibiotic, according to standard methods. Plates are incubated at 37° for 17–20 hr. Transformation efficiency can be expressed by calculating the number of transformants per microgram DNA.

Excess TSS-treated cells can be prepared for storage after the incubation period on ice by pipetting aliquots into sterile tubes and immediately freezing the tubes in a dry ice–ethanol bath. Frozen bacteria should be stored at −70° and used immediately on thawing in ice.

Development of Procedure

The effect of the chemical components in TSS on inducing uptake of DNA by bacterial cells is shown in Table I. We previously examined the effects of varying concentrations of other divalent cations (e.g., Ca^{2+},

TABLE I
EFFECT OF CHEMICAL COMPONENTS ON TRANSFORMATION EFFICIENCY[a]

Components	Transformation efficiency,[b] (number of transformants/μg DNA)
LB alone	$<10^1$
LB + DMSO	$<10^1$
LB + PEG	$<10^1$
LB + Mg^{2+}	$<10^1$
LB + DMSO + PEG	$<10^1$
LB + DMSO + Mg^{2+}	$<10^1$
LB + PEG + Mg^{2+}	$1.38 \pm 0.11 \times 10^7$
LB + DMSO + PEG + Mg^{2+}	$4.05 \pm 0.26 \times 10^{7c}$

[a] Assays were performed as described in the text. TSS-treated JM109 cells (100 μl) were mixed with 100 pg of pUC19 DNA and incubated on ice for 30 min. Concentrations of components in TSS were 10% (w/v) PEG (M_r 8000), 5% DMSO (v/v), and 20 mM Mg^{2+}.

[b] Values represent mean $\pm$ SEM of triplicate plates. The experiment was repeated twice with similar results. A value of $<10^1$ indicates that no transformants were obtained when 0.1 ml of bacteria (undiluted) was plated.

[c] $P < 0.001$, compared to LB + PEG + Mg^{2+} (two-way analysis of variance).

Mn^{2+}, and Zn^{2+}) in our assay system; however, concentrations of Mg^{2+}, ranging from 20 to 50 mM, yielded the highest transformation efficiencies. The effectiveness of PEG of different molecular weights was also measured and these results showed that PEG 3350 and 8000 preparations were equivalent and gave optimal results. In addition, we varied concentrations of PEG, DMSO, and input DNA in transformation mixtures and harvested bacteria at different stages of cell growth, in order to determine the optimal conditions for obtaining maximal numbers of transformants.[5] These data showed that under the currently described assay conditions, we could consistently achieve transformation efficiencies of 10^7 to 10^8 transformants/μg DNA. Moreover, as shown in Table II, a number of commonly used strains of *E. coli* were found to be transformable with this method.

Conclusion

The protocol we developed for the preparation of competent *E. coli* was designed to be rapid, convenient, and inexpensive, yet suitable to achieve transformation efficiencies sufficient for routine cloning experiments. As shown in Table III, the number of steps in the procedure can

[5] C. T. Chung, S. L. Niemela, and R. H. Miller, *Proc. Natl. Acad. Sci. U.S.A.* **86**, 2172 (1989).

TABLE II
Bacterial Strains Transformed with Method

D1210[a]	JM105[b]	MM294[c]	SG21166[d]
DH1[c]	JM107[e]	MV1190[f]	SMH50[g]
DH5α[c]	JM109[c]	N4830-1[h]	TG1[i]
GM161[j]	K802[k]	RA2021[l]	TP610A[m]
HB101[c]	LE392[c]	RB79li[l,n]	TP2010[m]
JA221[o]	LK111[p]	RR1[c]	TP2339[m]
JC7623[q]	MC1061[r]	RZ1032[j]	UT5600[s]
JE5513[t]	MC4100[s]	SCS-1[c]	VJS697[u]
JM83[v]	MJM413[w]	SG1117[d]	XL-1[c]

[a] S. Bolland, M. Llosa, P. Avila, and F. de la Cruz, *J. Bacteriol.* **172**, 5795 (1990).

[b] R. L. Haining and B. A. McFadden, *J. Biol. Chem.* **265**, 5434 (1990).

[c] C. T. Chung, S. L. Niemela, and R. H. Miller, *Proc. Natl. Acad. Sci. U.S.A.* **86**, 2172 (1989).

[d] A. Z. Ge, R. M. Pfister, and D. H. Dean, *Gene* **93**, 49 (1990).

[e] T. Tao, J. C. Bourne, and R. M. Blumenthal, *J. Bacteriol.* **173**, 1367 (1991).

[f] S. Y. Yang, X. Y. H. Yang, G. Healy-Louie, H. Schulz, and M. Elzinga, *J. Biol. Chem.* **265**, 10424 (1990).

[g] D. B. Olsen, G. Kotzorek, J. R. Sayers, and F. Eckstein, *J. Biol. Chem.* **265**, 14389 (1990).

[h] T. Unge, H. Ahola, R. Bhikhabhai, K. Bäckbro, S. Lövgren, E. M. Fenyö, A. Honigman, A. Panet, J. S. Gronowitz, and B. Strandberg, *AIDS Res. Hum. Retroviruses* **6**, 1297 (1990).

[i] D. B. Olsen and F. Eckstein, *Proc. Natl. Acad. Sci. U.S.A.* **87**, 1451 (1990).

[j] P. Liljeström and H. Garoff, *J. Virol.* **65**, 147 (1991).

[k] L. M. Hoffman and J. Jendrisak, *Gene* **88**, 97 (1990).

[l] A. DeChavigny, P. N. Heacock, and W. Dowhan, *J. Biol. Chem.* **266**, 5323 (1991).

[m] A. Beuve, B. Boesten, M. Crasnier, A. Danchin, and F. O'Gara, *J. Bacteriol.* **172**, 2614 (1990).

[n] C. M. Skoglund, H. O. Smith, and S. Chandrasegaran, *Gene* **88**, 1 (1990).

[o] G. S. Dahler, F. Barras, and N. T. Keen, *J. Bacteriol.* **172**, 5803 (1990).

[p] J. T. Burger, R. J. Brand, and E. P. Rybicki, *J. Gen. Virol.* **71**, 2527 (1990).

[q] D. S. Waugh and N. R. Pace, *J. Bacteriol.* **172**, 6316 (1990).

[r] M. El Hassouni, M. Chippaux, and F. Barras, *J. Bacteriol.* **172**, 6261 (1990).

[s] D. Cavard and C. Lazdunski, *J. Bacteriol.* **172**, 648 (1990).

[t] Y. Asai, Y. Katayose, C. Hikita, A. Ohta, and I. Shibuya, *J. Bacteriol.* **171**, 6867 (1989).

[u] J. V. Platko, D. A. Willins, and J. M. Calvo, *J. Bacteriol.* **172**, 4563 (1990).

[v] J. R. Sayers and F. Eckstein, *J. Biol. Chem.* **265**, 18311 (1990).

[w] A. Burkovski, G. Deckers-Hebestreit, and K. Altendorf, *FEBS Lett.* **271**, 227 (1990).

TABLE III
TRANSFORMATION PROCEDURE[a]

1. A fresh overnight culture of bacteria is diluted 1 : 100 into prewarmed LB broth and the cells are incubated at 37° with shaking (225 rpm) to an OD_{600} of 0.3–0.4.
2. An equal volume of ice-cold 2× TSS is added and the cell suspension is mixed gently and incubated on ice for 5–15 min. [TSS is sterile filtered and contains LB broth with 10% PEG (M_r 3350 or 8000), 5% DMSO, and 20–50 mM Mg^{2+} (MgCl$_2$ or MgSO$_4$), at a final pH of 6.5–6.8.]
3a. For long-term storage, cells are frozen immediately in a dry ice–ethanol bath and stored at −70°.
3b. For transformation, a 0.1-ml aliquot of cells is pipetted into a cold polypropylene tube and 1 μl (100 pg–1 ng) of plasmid DNA is added followed by gentle mixing. (When frozen cells are used, cells are thawed slowly on ice and used immediately.)
4. The cell–DNA mixture is incubated on ice for 5–60 min.
5. A 0.9-ml aliquot of LB broth (or TSS) plus 20 mM glucose is added and the cells are incubated at 37°, with shaking (225 rpm), for 1 hr to allow expression of the antibiotic resistance gene.
6. Transformants are selected by standard methods.

[a] Steps should be performed aseptically.

be minimized without affecting the yield of transformants. Because our protocol does not involve any complicated manipulations or steps which require exact timing, there are few places for sources of error. The correct preparation of TSS is, however, critical. TSS should be made with relatively fresh, ultrapure DMSO and should be kept cold. Although we have not found it necessary to use tissue culture-grade water or PEG in our laboratory, these can be tried if poor results are obtained with the quality of existing reagents. For convenience, it is possible to store TSS at 4° for several days without loss in activity. Most importantly, TSS should *not* be autoclaved.

Many strains of *E. coli* have been successfully transformed by our method (Table II). Transformation efficiencies among the strains we tested[5] were at least 5 × 10^6 transformants/μg DNA. When testing new strains of *E. coli* or other bacteria, we suggest that parallel experiments be done in which pUC19 or pBR322 is used to transform a previously tested strain of *E. coli* (e.g., JM109, HB101, and DH5α) as well as the particular strain of bacteria of interest. Although transformation efficiencies obtained with our protocol are comparable to those of other commonly used methods to prepare competent bacteria, the technique is advantageous in that it is simple and can be completed in a short time once the bacteria are grown. Furthermore, although we have specified conditions necessary to achieve maximum numbers of transformants, there is flexibility in the assay. For instance, transformation efficiencies of $\geq 10^5$ transformants/μg DNA are possible even when bacteria are harvested as late in the growth cycle as the stationary phase or treated with TSS and stored at 4° for 4 days. Last, the capability of using the same solution to both transform and freeze cells for storage is a convenient feature.

[44] Storage of Unamplified Phage Libraries on Nylon Filters

By Paul A. Whittaker

Introduction

The construction and screening of libraries made using bacteriophage λ vectors has been, and continues to be, an essential step in the analysis of the structure and function of many genes and gene regions.[1-4] Over the years the accepted way to store these libraries has been as an amplified liquid stock.[5,6] Aliquots of this stock are then plated out and screened using the plaque screening protocol of Benton and Davis.[7] The advantage of library amplification is that an almost unlimited supply of the library is generated that is stable for many years. The disadvantage is that the complexity of the library is altered due to variations in the growth rates of different recombinant phages such that some clones become overrepresented and some clones become underrepresented, or even lost. In the case of the former this means that repeat isolates of the same clone will be characterized and, in the case of the latter, several library equivalents will have to be screened if a positive phage is to be detected. Therefore, the most efficient way to obtain clones from a λ library is to screen the

[1] D. C. Page, R. Mosher, E. M. Simpson, E. M. C. Fisher, G. Mordon, J. Pollack, B. McGillivray, A. de la Chappelle, and L. G. Brown, *Cell* **51**, 1091 (1987).

[2] J. M. Rommens, M. C. Ianuzzi, B. S. Kerem, M. L. Drumm, G. Melmer, M. Dean, R. Rozmohel, J. L. Cole, D. Kennedy, N. Hidaka, M. Zsiga, M. Buchwald, J. R. Riordan, L. C. Tsui, and F. S. Collins, *Science* **245**, 1059 (1989).

[3] E. R. Fearon, K. R. Cho, J. M. Nigro, S. E. Kern, J. W. Simons, J. W. Ruppert, S. R. Hamilton, A. C. Presinger, G. Thomas, K. W. Kinzler, and B. Vogelstein, *Science* **247**, 49 (1990).

[4] N. E. Murray, this series, Vol. 204, p. 280.

[5] A.-M. Frischauf, this series, Vol. 152, p. 190.

[6] J. Sambrook, E. F. Fritsch, and T. Maniatis, "Molecular Cloning: A Laboratory Manual." Cold Spring Harbor Press, Cold Spring Harbor, New York, 1989.

[7] W. D. Benton and R. W. Davis, *Science* **196**, 180 (1977).

METHODS IN ENZYMOLOGY, VOL. 218

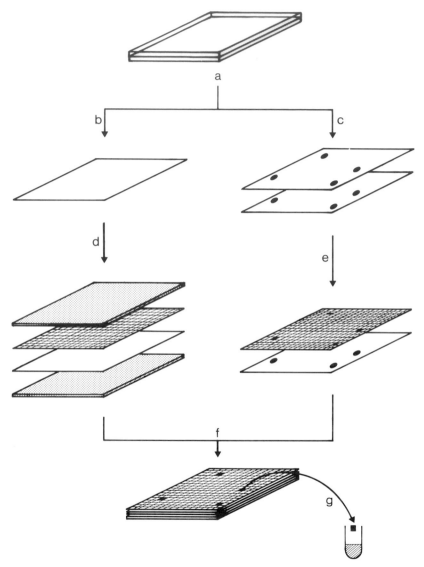

FIG. 1. Cryopreservation of phage λ libraries on nylon filters: storage and screening. (a) Recombinant phage are plated out at high density on 245 × 245 mm agar plates in top agarose containing 30% (v/v) glycerol as a cryopreservative.[8] After phage growth three plaque lifts are prepared using nylon filters. (b) The first lift (master filter) is stored at −70°. (c) The second and third lifts are screened using standard plaque hybridization protocols.[6] (d) For the accurate identification of phage positives a photocopied acetate of a piece of graph paper is placed onto the master filter before sandwiching between two plastic plates and storing at −70°. (e) A replica acetate is used to obtain the exact coordinates of positives on the

unamplified library directly. This is particularly advantageous in chromosome walking experiments, in which screening of the same primary library at all steps of the walk means that backward walking steps can be avoided because clones already positive at the preceding steps are easily identified. Unfortunately, the longevity of plated phage rarely exceeds several weeks and library plates soon become overgrown with molds, and so on. A useful method for the long-term storage of primary phage cDNA libraries in a form that allows the subsequent identification and recovery of relevant clones has been described by Klinman and Cohen.[8] However, the method does require the storage and handling of plates of frozen agar. Therefore, their protocol has been modified such that the phage are stored on nylon filters at −70°.[9] This makes storage and handling easier and makes it possible to prepare both an unamplified and amplified version of the same library. This chapter updates and expands the earlier report.[9]

Principle of Method

Phage λ particles grown in top agarose supplemented with glycerol and transferred to nylon filters using the technique of Benton and Davis[7] remain viable during long-term storage at −70°.[9] This finding is the basis for the λ library storage and screening protocol described in the following sections and diagrammed in Fig. 1. The technique makes repeated screening of the same primary library over a long period very straightforward. In addition, because the library is stored on filters, the top agarose from the primary plates can be scraped off after the lifts have been prepared and the remaining phage eluted to give an amplified library.

Materials and Reagents

Growth Media

pH is adjusted to 7.5 with NaOH prior to autoclaving for 20 min at 15 lb/in^2.

[8] D. M. Klinman and D. I. Cohen, Anal Biochem. **161**, 85 (1987).
[9] P. A. Whittaker and F. L. Lavender, Nucleic Acids Res. **17**, 4406 (1989).

---

autoradiographs of the hybridized filters. (f) Phage positives are recovered for secondary screening by placing the master filter/photocopied acetate sandwich on dry ice and overlaying with the replica acetate containing the position of the phage positives. The appropriate area is cut out, using a scalpel, and the master filter is replaced at −70°. (g) Phage are eluted from the excised filter piece by placing in phage buffer. Dilutions of the eluted phage are then plated out for secondary screening.

T broth: 1% (w/v) Difco (Detroit, MI) tryptone, 1% (w/v) NaCl
CY broth: 1% (w/v) Difco casamino acids, 0.5% (w/v) Difco yeast
    extract, 0.3% (w/v) NaCl, 0.2% (w/v) KCl, 0.2% (w/v) $MgCl_2 \cdot 6H_2O$
L broth: 1% (w/v) Difco tryptone, 0.5% (w/v) Difco yeast extract, 1%
    (w/v) NaCl, 0.2% (w/v) $MgCl_2 \cdot 6H_2O$
Bottom agar: Prepare liquid medium as above. Just before autoclaving
    add Difco Bacto-agar to 1.5% (w/v)
Top agarose: Prepare liquid medium as above. Just before autoclaving
    add electrophoresis-grade agarose (e.g., type II; Sigma, St. Louis,
    MO) to 0.7% (w/v). Add prewarmed sterile glycerol (Analar, BDH)
    to 30% (v/v) before use

Solutions

Phage buffer: 50 mM Tris-HCl (pH 7.5), 100 mM NaCl, 10 mM MgSO$_4$,
    0.01% (w/v) gelatin; autoclave
SSC (20×): 3.0 M NaCl, 0.3 M sodium citrate; adjust pH to 7.0
Denhardt's (50×): 1% (w/v) Ficoll, 1% (w/v) polyvinylpyrrolidone, 1%
    (w/v) bovine serum albumin; sterilize by filtration through a 0.45-μm
    filter and store at $-20°$

General

Nunc bioassay dishes (245 × 245 mm): GIBCO Bethesda Research
    Laboratories, (Gaithersburg, MD)
Hybond-N nylon filters: Amersham (Arlington Heights, IL)

Method

Preparation of Plating Cells

A culture of the appropriate host bacteria is grown in T broth supplemented with 0.4% (w/v) maltose[10] at 37° overnight with vigorous aeration.
The next day the cells are concentrated twofold by centrifugation (2500 g,
10 min) and resuspension in sterile 10 mM MgSO$_4$. The plating cells can

---

[10] T broth is preferred to L broth, as this helps prevent the culture from reaching stationary
phase and so giving cultures of reduced plating efficiency. Certain bacterial strains [e.g.,
NM514; W. Arber, L. Enquist, B. Hohn, N. E. Murray, and K. Murray, *in* "Lambda II"
(R. W. Hendrix, J. W. Roberts, F. Stahl, and R. A. Weisberg, eds.), p. 433. Cold Spring
Harbor Press, Cold Spring Harbor, New York, 1983] do not grow well in T broth, so in
these cases L broth may be substituted. Maltose is added from a filter-sterilized 20%
(w/v) stock. Bacteria grown in the presence of maltose adsorb phage λ more efficiently
because the maltose operon, which contains the gene (*lamB*) that codes for the phage λ
receptor, is induced.

be stored at 4° for up to 2 weeks, but fresh stocks are recommended for the highest plating efficiencies.

Preparation of Plates for Library Plating

Because the identification of the appropriate clone in a mammalian genomic DNA or cDNA library may involve the screening of 10^6 recombinants or more, the use of Nunc bioassay dishes (245 × 245 mm) reduces the number of filters that must be prepared, as up to 2×10^5 phage can be accommodated on one plate. The plastic plates are reusable and can be adequately resterilized by rinsing with ethanol and treating with ultraviolet (UV; 10 min under a germicidal lamp). Liquid bottom agar[11] (250 ml, cooled to 45–50°) is poured into each plate and allowed to set. Bubbles on the surface of the agar are removed by quickly passing a Bunsen flame over the surface before the agar hardens. Before use, plates are dried[12] either by leaving them uncovered under a laminar flow hood with the air blower on for 1–2 hr, or by leaving them with their lids on at room temperature for 2 days.

Plating of Phage Libraries

Forty-five milliliters of CY or L top agarose[13] (depending on the cloning vector used; see Ref. 11) is equilibrated at 45–50° in a sterile 50-ml screw-

---

[11] The use of CY bottom agar or other similarly less rich medium [e.g., BBL trypticase agar; K. Kaiser and N. E. Murray, in "DNA Cloning" (D. M. Glover, ed.), Vol. 1, p. 1. IRL Press, Oxford, 1985] is essential if plaques of an acceptable size (0.5–1 mm in diameter) are to be obtained when growing $red^- gam^-$ recombinants (the red/gam status for recombinants made using various λ vectors can be found on pages 2.19–2.55 of Ref. 6). This is because $red^- gam^-$ phage produce small plaques, and as a major determinant of plaque size is the rate at which the bacterial lawn grows, the use of "weak" media increases plaque size by slowing down lawn growth. Growth of $red^- gam^-$ recombinants on richer medium (e.g., L broth agar) results in minute plaques (~0.1 mm) and a reduction in plating efficiency. In contrast, $red^+ gam^+$ phage [e.g., λgt10 recombinants (J. Jendrisak, R. A. Young, and J. D. Engel, this series, Vol. 152, p. 359)] give plaques of 0.5–1 mm diameter on L broth agar and 1- to 2-mm diameter on CY agar. For high-density plating, however, the smaller plaques obtained on L broth agar are preferable.

[12] The excessive moisture in freshly poured plates will tend to form in pools on the agar surface. If these pools are not removed, they will encourage phage to spread and "run" across the agarose surface during plaque growth. Care should be taken not to overdry the plates as this will restrict phage diffusion and so reduce plaque size. As a guide, properly dried plates show a slight puckering of the agar surface.

[13] Inclusion of glycerol in the top agarose can result in turbid plaque morphology. For example, recombinants made using the vector EMBL3cos [P. A. Whittaker, A. J. B. Campbell, E. M. Southern, and N. E. Murray, *Nucleic Acids Res* **16**, 6725 (1988)] give clear plaques when plated in top agarose lacking glycerol, but turbid plaques when glycerol is present. The reason for this is unclear, but control experiments have shown that the yield of phage from the turbid plaques is identical to clear plaques. In contrast, λgt10 recombinants give clear plaques irrespective of the presence or absence of glycerol.

capped polypropylene tube (e.g., Falcon 50-ml conical centrifuge tube from Becton Dickinson, Oxnard, CA). In a separate 50-ml tube an infection is performed using 1 ml of $MgSO_4$ plating cells and 2×10^5 recombinant phage in a volume of 1 ml or less. The plating cells and phage are incubated for 20 min at 37°. Top agarose is added to the infected bacteria and the mixture is poured onto the surface of a plate of CY or L bottom agar (prewarmed to 37°). The agarose is quickly spread across the surface of the bottom agar by tilting the plate. After the agarose has solidified, the plates are incubated inverted overnight at 37°.

Plaque Lifts and Filter Storage

Before making filter replicas the plates are cooled to 4° for 60 min to harden the top agarose. To ease the placement and removal of nylon filters during plaque lift preparation, 22×22 cm nylon filters are cut down to 21×22 cm. Three filters are needed per plate and these are used directly from the box with no pretreatment (i.e., no prewetting or sterilization). The first lift (master filter) is made by placing a nylon filter onto the surface of the top agarose.[14] The filter is keyed to the plate by stabbing through the filter and into the agar beneath using a needle that has been dipped in drawing ink (remember to make an asymmetric pattern of spots). After 1 min on the plate the replica filter is removed for storage at $-70°$ (see below). The second and third replicas (for hybridization screening) are made in the same way with the filters being left on the plate for 2 and 3 min, respectively.

For the accurate identification of phage positives the technique described by Herrmann et al.[15] for cosmid libraries is used. The master filter is placed plaque side up on a plastic plate (the lids of Nunc bioassay dishes with the edges removed work well). A photocopied acetate of a piece of graph paper is placed on top of the filter (do not move the acetate once it has contacted the filter). The keying marks on the filter are copied onto the acetate with a waterproof marker pen and these are copied in turn onto a second acetate grid placed on top of the first grid. This replica acetate grid is used to identify the area containing the positive phage after hybridization of the replica filters and it is kept on file. The master filters from all the plates are treated identically. The stack of plastic plates/nylon filter/acetate grid sandwiches is then clamped together using bulldog clips and stored at $-70°$.

---

[14] Air bubbles between the filter and top agarose can be avoided by bending the filter into a U shape so that the middle of the filter contacts the center of the plate first. The rest of the filter can then be gently rolled into position.

[15] B. G. Herrmann, D. P. Barlow, and H. Lehrach, *Cell* **48,** 813 (1987).

Filter Screening

The second and third replicas are placed plaque side up on Whatman (Clifton, NJ) 3MM sheets soaked in 0.5 *M* NaOH, 1.5 *M* NaCl for 5 min, then transferred to 0.5 *M* Tris-HCl (pH 7.0), 1.5 *M* NaCl for another 5 min. The filters are rinsed in 2× SSC for 5 min before drying at room temperature (alternatively, 1 min at full power in a microwave oven speeds up the process). The DNA can be fixed to the filters either by UV cross-linking (2–5 min on a UV transilluminator) or by baking at 80° for 1–2 hours. Both replicas are hybridized to avoid effort on false positives (only rarely is a positive that comes up on one filter a true positive). To simplify the hybridization procedure all the replica filters (10 or more in a typical library screen) are hybridized together in a 24 × 24 cm polyethylene freezer box in a water bath at 65°. The prehybridization and hybridization mix is 5× SSC, 5× Denhardts, 0.1% (w/v) sodium dodecylsulfate (SDS), 0.1% (w/v) sodium pyrophosphate.[16] Filters are prehybridized in 150 ml of mix for 1–4 hr before replacing with 100 ml of new mix and radiolabeled probe (50 ng labeled using the oligoprime method[17,18]). The filters are hybridized overnight without shaking and then washed in 2× SSC, 0.1% SDS, 0.1% sodium pyrophosphate (15 min at 65°) and 0.1× SSC, 0.1% SDS, 0.1% sodium pyrophosphate (twice for 15 min at 65°).[19] The filters are sealed into plastic bags while still damp and then autoradiographed with intensification at −70°. Exposure for 1–3 days is adequate, with positive signals normally being detected after an overnight exposure. To facilitate alignment of the filters with the autoradiographs, backing paper (Whatman Benchkote works well) marked with several asymmetrically spaced dots of radioactive ink should be placed in the X-ray cassettes used for the autoradiography.

Recovery of Positive Phage

After exposure and development of the autoradiograph, the filters are matched to it, using the radioactive spots on the backing paper, and the keying marks on the filters are transferred from the filters onto the autoradiograph. The keying marks on the autoradiograph are matched with the markings on the replica acetate grid on file and the positions of

[16] Dextran sulfate is not needed to enhance signal strength. Denatured sonicated salmon sperm DNA can be added to 100 μg/ml if background is a problem; however, this is not usually necessary with Hybond-N.

[17] A. P. Feinberg and B. Vogelstein, *Anal. Biochem.* **132**, 6 (1983).

[18] A. P. Feinberg and B. Vogelstein, *Anal. Biochem.* **137**, 266 (1984).

[19] These washing conditions are for perfectly matched single-copy probes. Less stringent washing conditions may have to be used for probes of divergent sequence.

the positives are copied onto the grid. The appropriate master filter/acetate grid sandwich is taken from the $-70°$ freezer and left on its plastic plate on a bed of dry ice pellets in a polystyrene box lid. The replica acetate grid containing the position of the positives is placed on top of the acetate grid overlaying the master filter and clamped in place using two bulldog clips. The appropriate area containing the phage positive is cut out using a fresh scalpel blade and the filter piece is placed in 1 ml of phage buffer. The stack of master filters is then reassembled and placed at $-70°$. Phage are eluted for at least 60 min at room temperature and then plated out for secondary screening using standard methods.[6] Another one or two screenings are normally sufficient to give pure positives. The results of a typical screening experiment are given in Table I.

Preparation of Amplified Library Stocks

Even after three plaque lifts have been made, there are still large numbers of viable phage left on the library plates that can be used to prepare an amplified library for distribution. The top agarose is scraped off the plates into phage buffer in a sterile conical flask (use 50 ml of buffer per plate) and chloroform is added to 5% (v/v). Phage are left to elute at room temperature for at least 60 min (or overnight at 4°) with gentle mixing before the agarose and debris are removed by centrifugation (2500 g for 10 min in 50-ml screw-capped Falcon polypropylene centrifuge tubes). The supernatant is then transferred to a fresh tube and stored at 4° with 2% (v/v) chloroform.[20–24]

Discussion

The filter protocol has been successfully used for the storage of both genomic DNA and cDNA libraries.[25] Positive phage are recovered from the filter at the same efficiency as agarose plugs taken from the same area

[20] The lysate does tend to look a bit murky even after clarification, but this is not a problem.

[21] EMBL3cosW is a modified version of the fast mapping vector EMBL3cos[13] made by inserting oligonucleotide cassettes containing SP6 and T7 RNA polymerase promoters directly adjacent to and flanking the replaceable stuffer fragment so that end-specific RNA probes can be generated for fast chromosome walking [P. A. Whittaker, *Clin. Biotechnol.* **3**, 67 (1991)].

[22] L. M. Kunkel, A. P. Monaco, W. Middlesworth, H. D. Ochs, and S. A. Latt, *Proc. Natl. Acad. Sci. U.S.A.* **82**, 4778 (1985).

[23] A. P. Monaco, C. J. Bertelson, C. Colletti-Feener, and L. M. Kunkel, *Hum. Genet.* **75**, 221 (1987).

[24] P. N. Ray, B. Belfall, C. Duff, C. Logan, V. Kean, M. W. Thompson, J. E. Sylvester, J. L. Gorski, R. D. Schmickel, and R. G. Worton, *Nature (London)* **318**, 672 (1985).

[25] Vectors EMBL3cos[13] and EMBL3cosW[21] were used for the construction of the genomic libraries and λgt10[11] for the construction of the cDNA libraries. Recombinant phage were propagated on *E. coli* strains NM621[13] and NM514,[10] respectively.

TABLE I
Efficiency of Recovery of Phage Positives from Nylon Filters and Plugs Taken from Primary Plates[a]

Parameter	Number of plates screened	Average pfu[b] per plate[c]	Average number of positives for three probes	Percentage of plated phage showing positive hybridization	Number of plates containing zero phage positives
Agarose plugs	22	800[d]	6[e]	0.75[e]	5
Filter areas	22	1800[f]	15[g]	0.83[g]	3

[a] A library of 10[6] recombinants made using the phage vector EMBL3cosW[21] and 48XXXX cell line DNA was plated out on five Nunc bioassay dishes containing CY bottom agar and plaque lifts for storage and screening were prepared exactly as described in text. The second and third plaque lifts were hybridized with a mixture of 50 ng each of radiolabeled probes pERT84,[22] J47,[23] and pXJ1.1.[24] Twenty-two positive phage in duplicate were identified on the resulting autoradiographs. Areas (0.25 cm[2]) containing the positive phage were recovered from the master filters and the primary plates and placed in 1 ml (filters) and 5 ml (plugs) of phage buffer. After elution of phage for 4 hr at room temperature 10 μl of each suspension was removed and diluted into 1 ml of phage buffer. Ten microliters of each diluted phage suspension was mixed with 50 μl of NM621[13] MgSO₄ cells and incubated at 37° for 20 min. The infected cells were then added to 3 ml of CY top agarose and plated out on 9-cm petri dishes. Plates were incubated overnight at 37°. Single plaque lifts from each plate were prepared using Schleicher & Schuell (Keene, NH) BA85 nitrocellulose filters, hybridized with the three probes used for the primary screen, washed in 0.1 × SSC, 0.1% (w/v) SDS, 0.1% (w/v) sodium pyrophosphate at 65°, and autoradiographed at −70° for 24 hr.

[b] Plaque-forming units (pfu).

[c] For 22 plates.

[d] Range, 400–1300.

[e] Figure based on the 17 plates containing positive phage.

[f] Range, 1200–2500.

[g] Figure based on the 19 plates containing positive phage.

of the primary plates (Table I) and phage show only a modest reduction in titer on prolonged storage at $-70°$. For example, over a 22-month period the average number of phage recovered from 0.25-cm^2 filter areas of the EMBL3cosW library (Table I) decreased from 1.5×10^8 plaque-forming units (pfu) to 0.95×10^8 pfu. Because, on average, 0.8% of recovered phage are positive on secondary screening (Table I) the yield of phage is still orders of magnitude over that needed experimentally. Therefore the stored libraries are expected to be useful for several years.

The stability of the amplified libraries made by scraping off the top agarose from the plates after plaque lifts have been made is also very good. The titer of an amplified EMBL3cosW library prepared in this way showed a reduction in titer from 4×10^8 pfu/ml to 1.15×10^8 pfu/ml after 22 months at 4°.

The ratio of positive to negative phage on secondary screening could be increased by cutting out smaller filter areas (e.g., 1 mm^2 as opposed to 25 mm^2), but this would probably increase the number of failures as it is still possible to miss recovering phage when cutting out larger areas (Table I). This is because of shrinkage and distortion of the filter replicas during probe hybridization.

We have not tried filters other than Hybond-N for library storage and screening, although they should work equally well. However, the hybridization conditions described here may well have to be altered if other makes of filter are used (follow the recommendations of the manufacturers; see also Khandjian[26]). A potential limitation of the protocol is the number of times the replica filters can be screened before loss of signal is apparent. Stripping of probe signal from the filters before rehybridizing with other probes is not recommended as positive signals from previous screens can be easily identified and so disregarded. Obviously, for chromosome walking experiments the ability to see positive phage from previous screens is advantageous as it avoids wasted effort on characterizing clones already isolated at the previous walking step.

Finally, the ability to make both an amplified and unamplified version of the same plated library removes the dilemma that has faced many workers in the past: whether to screen the primary library directly or to amplify it first and run the risk of missing an underrepresented clone.

Acknowledgments

This work was carried out with support from the Nuffield Foundation, Southampton University Research Fund and the Medical Research Council, as part of the U.K. Human Genome Mapping Project. I would like to thank Louise Lavender for trying the method in the first instance and Diane Brown for typing.

[26] E. W. Khandjian, *Bio/Technology* **5**, 165 (1987).

[45] Magnetic Affinity Cell Sorting to Isolate Transiently Transfected Cells, Multidrug-Resistant Cells, Somatic Cell Hybrids, and Virally Infected Cells

By Raji Padmanabhan, R. Padmanabhan, Tazuko Howard,
Michael M. Gottesman, and Bruce H. Howard

Introduction

In recent years, many techniques have been developed to introduce foreign DNA into eukaryotic cells for stable expression of genes, transient analysis of gene products, and gene regulation. Several methods of transfection have been used such as (1) calcium phosphate coprecipitation,[1] (2) DEAE-dextran,[2] (3) protoplast fusion,[3] (4) microinjection,[4] (5) electroporation,[5,6] and (6) lipofection.[7] With the exception of DEAE-dextran transfection, which can be used only for transient assays, other techniques are employed to study both transiently and stably transfected cell populations.

The expression of readily assayable reporter gene products such as chloramphenicol acetyltransferase (CAT),[8] firefly luciferase,[9] or *Escherichia coli* β-galactosidase[10] is often used in optimizing the efficiency of transient expression by transfection. The expression of the CAT and luciferase genes by DNA-mediated gene transfer can be easily quantitated, because mammalian cells are devoid of any endogenous CAT or luciferase activity.

Isolation of stably transfected cells generally requires the expression of a dominant selectable marker gene, which often confers a drug resistance phenotype. *Ecogpt*[11] and *neo*[12] are commonly used for stable transfections.

[1] F. L. Graham and A. Van Der Eb, *Virology* **52,** 456 (1973).
[2] J. H. McCutchan and J. S. Pagano, *J. Natl. Cancer Inst.* **41,** 351 (1968).
[3] W. Schaffner, *Proc. Natl. Acad. Sci U.S.A.* **77,** 2163 (1980).
[4] M. R. Capecchi, *Cell* **22,** 479 (1980).
[5] E. Neumann, M. Schaefer-Ridder, Y. Wang, and P. H. Hofschneider, *EMBO J,* **1,** 841 (1982).
[6] G. Chu, H. Hayakawa, and P. Berg, *Nucleic Acids Res.* **15,** 1311 (1987).
[7] C. Nicolau and C. Sene, *Biochim. Biophys. Acta* **721,** 185 (1982).
[8] C. M. Gorman, L. F. Moffat, and B. H. Howard, *Mol. Cell. Biol.* **2,** 1044 (1982).
[9] J. R. Dewet, K. V. Wood, M. DeLuca, D. R. Helinski, and S. Subramani, *Mol. Cell. Biol.* **7,** 725 (1987).
[10] C. V. Hall, P. E. Jacob, G. M. Ringold, and F. Lee, *J. Mol. Appl. Genet.* **2,** 101 (1983).
[11] R. C. Mulligan and P. Berg, *Proc. Natl. Acad. Sci. U.S.A.* **78,** 2072 (1981).
[12] P. J. Southern and P. Berg, *J. Mol. Appl. Genet.* **1,** 327 (1982).

However, this approach has several drawbacks. Selection of stable trans-fectants is time consuming and at times may also result in selection of cells with rearrangement(s) or insertional mutagenesis of cellular genes. In addition, the chromatin structure at the integration sites may influence the expression of the gene of interest, so such studies may not be suitable for analysis of gene expression.

In studies involving regulation of cellular growth control by a gene of interest, selection of stable transfectants may be further complicated if that gene negatively regulates growth, and thus impedes formation of identifiable colonies. In this regard, transient expression studies[13] have a definite advantage over those involving stably transformed cells, because such studies can be performed within a relatively short period of time after transfection (within 24–72 hr). However, standard methods of transient gene expression are generally not useful in studies involving the elucida-tion of the functional role of a gene product in growth control, because the inefficiency of transfection procedures in most cell types gives rise to a high background from the large fraction of untransfected cells.

In view of these problems, there is clearly a need for a method to isolate a small population of cells having a specific genetic trait or a uniquely differentiated state away from a large population of cells having different characteristics. Our experimental goal was to develop a method for the isolation of cells that are homogeneous with respect to the expres-sion of the gene of interest without using any drug selection. We developed a novel methodology[14] that involves the transient expression of the gene of interest along with a marker gene that encodes a cell surface protein. The latter would render the cells amenable to selection by the magnetic affinity cell sorting (MACS) technique.[15] Magnetic affinity cell sorting allows the rapid separation of a small population of cells transiently ex-pressing the surface protein and the gene of interest away from a large fraction of nontransfected cells.

Early methods used in cell fractionation of heterogeneous populations, in order to identify a specific cell type, relied mostly on the cell size, density, or surface charge, which often lacked specificity. Subsequently developed immunomagnetic techniques used surface proteins to distin-guish specific cell types. This technique allows the separation of a cell type expressing a surface protein away from those lacking the marker. In this procedure, antibodies against the surface protein, which are attached

[13] C. M. Gorman, R. Padmanabhan, and B. H. Howard, *Science* **221**, 551 (1983).
[14] R. Padmanabhan, C. D. Corsico, T. H. Howard, W. Holter, C. M. Fordis, M. Willingham, and B. H. Howard, *Anal. Biochem.* **170**, 341 (1988).
[15] R. S. Molday, S. P. S. Yen, and A. Rembaum, *Nature (London)* **268**, 437 (1977).

to a solid (magnetic) matrix, specifically react with the surface marker of the target cells, and immobilize them, while the nontargeted cells remain unaffected. Used in this manner, the immunomagnetic separation is limited to special cell types expressing a unique surface protein. To develop a more general procedure applicable to all cell types, we made use of the DNA-mediated transient expression system in which a cell surface marker and the gene of interest were coexpressed in a specific cell type. Cells transiently expressing the transfected genes were then isolated by an immunoaffinity procedure. An important consideration for the choice of the surface marker is that its expression should not interfere with the normal growth of the cell type subjected to this fractionation procedure, or with the coexpression of the gene of interest.

Principle of Method

The MACS technique relies on the coexpression of the gene product of interest and a surface marker for which a specific antibody is readily available. Initially we chose cell surface markers quite foreign to normal mammalian cells, or expressed only on specialized cells, such as the vesicular stomatitis virus glycoprotein (VSV-G),[16] and the Tac subunit of interleukin 2 (IL-2) receptor[17] (IL-2R). Highly specific monoclonal antibodies are readily available for these proteins. In later applications, we have made use of an endogenous surface marker in order to select a specific cell type, such as cells expressing the multiple drug resistance gene (*mdr*), the expression of which is activated in cells in response to treatment with antitumor drugs.[18,19] The selection of the *mdr*-expressing cells was made possible due to the availability of a highly specific monoclonal antibody (MRK-16) against an external epitope of the *mdr* gene product, P-glycoprotein (P170).[20]

Two types of magnetic beads have been used routinely in our laboratory. The first is the BioMag M4100 (Advanced Magnetics, Cambridge, MA), which contains a magnetic ferric oxide core coupled to primary amino groups to which antibodies (or any other proteins) can be covalently attached using glutaraldehyde. The second commercial magnetic beads are Dynabeads M-450 (Dynal, Great Neck, NY), which are precoated with

[16] H. Riedel, C. Kondor-Koch, and H. Garoff, *EMBO J.* **3**, 1477 (1984).
[17] S. Z. Salahuddin, P. O. Markham, F. Wong-Staal, G. Franchini, V. S. Kalyanaraman, and R. C. Gallo, *Virology* **129**, 51 (1983).
[18] M. M. Gottesman and I. Pastan, *J. Biol. Chem.* **263**, 12163 (1988).
[19] V. Ling, *Annu. Rev. Biochem.* **58**, 137 (1989).
[20] H. Hamada and T. Tsuruo, *Proc. Natl. Acad. Sci. U.S.A.* **83**, 7785 (1986).

goat anti-mouse IgG to which a monoclonal antibody against a surface marker can be noncovalently attached.

The surface protein IL-2R is expressed from the plasmid pRSVIL2R, which contains the coding sequence for the Tac subunit of the IL-2R under the control of Rous sarcoma virus long terminal repeat (RSV-LTR) promoter. This promoter drives the expression of the Tac subunit of IL-2R at fairly high levels in many actively dividing cell types and, more importantly, the protein itself does not interfere with the growth of any of the cell types studied so far.[21]

Reagents and Methods

Reagents

The sources of the chemicals, media, and other reagents used in the MACS protocol are as follows: bovine serum albumin, [14]C- and ^{3}H-labeled chloramphenicol, and thymidine are from NEN-DuPont, (Wilmington, DE); phosphate-buffered saline (PBS), PBS without Ca^{2+} or Mg^{2+}, 1 M N-2-hydroxyethylpiperazine-N'-2-ethanesulfonic acid (HEPES), fetal bovine serum, Dulbecco's modified Eagle's medium (DMEM), and RPMI are from GIBCO-Bethesda Research Laboratories (Gaithersburg, MD). The MEM for suspension cultures and DMEM are also purchased from Quality Biologicals (Gaithersburg, MD). Chondroitin sulfate, gelatin, ethylenediaminetetraacetic acid (EDTA), $MgCl_2$, and $MgSO_4$ are from Sigma Chemical (St. Louis, MO). The nonfat dry milk is from Carnation (Los Angeles, CA). Glutaraldehyde is purchased from EM Sciences (Gibbstown, NJ). Chromatographically purified rabbit IgG is from Copper Medical (Malvern, PA). The two types of magnetic beads, BioMag M4100 and Dynabeads M-450 (Dynabeads) are from Advanced Magnetics (Cambridge, MA) and Dynal (Great Neck, NY; distributed by P&S Biochemicals, Gaithersburg, MD), respectively. The drugs Adriamycin, vinblastine, and colchicine, and soybean trypsin inhibitor, are from Sigma. Polyethylene glycol used for cell fusion studies is purchased from the American Type Culture Collection (ATCC; Rockville, MD). The sources of the antibodies used in these studies are as follows: polyclonal antibodies against VSV-G protein[22] from Dr. R. Lazzarini [National Institutes of Health (NIH)]; polyclonal antibodies against the epidermal growth factor (EGF) receptor from Dr. G. Merlino (NIH); the monoclonal antibody

[21] W. C. Greene, R. J. Robb, P. B. Svetlik, C. M. Rusk, J. M. Depper, and W. J. Leonard, *J. Exp. Med.* **162**, 363 (1985).

[22] H. Arnheiter, M. Dubois-Dalcq, and R. A. Lazzarini, *Cell* **39**, 99 (1984).

[23] T. Uchiyama, S. Broder, and T. A. Waldmann, *J. Immunol.* **126**, 1393 (1981).

against IL-2R[23] from Dr. T. Waldmann (NIH); the monoclonal antibody against P-glycoprotein[20] from Dr. T. Tsuruo (Institute of Applied Microbiology, Univ. of Tokyo, Japan).

Preparation of Plasmids

Plasmid DNAs used for transfections are purified by CsCl gradient centrifugation, following standard procedures.[24] The details of the plasmid constructs encoding IL-2R, VSV-G, and CAT, such as pRSVIL2R, pMSVG, and pRSVCAT, respectively, have been described elsewhere.[13,14]

Cell Lines

African green monkey kidney cells and HeLa S3 cells are from the ATCC, and are grown in DMEM containing 10% (v/v) fetal bovine serum. Human lymphocyte cell line C-8166, constitutively expressing IL-2R,[17] and KB-3-1, KB-V1, and NIH 3T3 cells expressing IL-2R are established cell lines described elsewhere.[25] They are grown in DMEM containing 10% (v/v) fetal bovine serum. Lymphocytes infected with a retrovirus expressing the human *MDR-1* gene, and selected in colchicine, and A431 cells are obtained from S. Goldenberg and I. Pastan, respectively (National Cancer Institute). DNA-mediated transfections are performed as described.[13] KB-V1 cells are transfected with pRSVIL2R using electroporation.

Preparation of Magnetic Beads Coated with Monoclonal Antibodies

BioMag M4100 particles are precoated to provide primary amino groups to which antibodies can be covalently attached using the glutaraldehyde protocol recommended by the manufacturer. A wide variety of proteins, such as monoclonal antibodies, albumins, lectins, and enzymes, and both single-[26] and double-stranded[27] DNA, can be attached to these beads. Attachment of avidin to these beads would allow binding of a biotinylated moiety of DNA or RNA. The derivatized particles are stored either at 4° (short term) or at −20° (long term) in a storage buffer containing 50% (w/v) glycerol and sodium azide (0.02%, w/v).

[24] T. Maniatis, E. F. Fritsch, and J. Sambrook, "Molecular Cloning: A Laboratory Manual." Cold Spring Harbor Press, Cold Spring Harbor, New York, 1982.
[25] R. Padmanabhan, T. Tsuruo, S. E. Kane, M. C. Willingham, B. H. Howard, M. M. Gottesman, and I. Pastan, *J. Natl. Cancer Inst.* **83,** 565 (1991).
[26] V. Lund, R. Schmid, D. Rickwood, and E. Hornes, *Nucleic Acids Res.* **16,** 10861 (1988).
[27] O. S. Gabrielsen, E. Hornes, L. Korsnes, A. Ruet, and T. B. Oyen, *Nucleic Acids Res.* **17,** 6253 (1989).

The binding of primary monoclonal antibodies to the Dynabeads is noncovalent because the commercially available beads are precoated with goat anti-mouse IgG. For attachment of rabbit polyclonal antibodies to these beads, use of Dynabeads precoated with goat anti-rabbit IgG is recommended when these become available. Dynabeads have OH groups on their surface to allow antibody or the antibody-binding molecules to be covalently linked to the surface of the beads.[28]

Reagents for Binding Antibodies to Magnetic Particles

1. Pyridine hydrochloride (10 mM) is prepared by diluting 0.4 ml of pyridine to 400 ml with water, and the pH is adjusted to 6.0 with 10 mM HCl; the volume is then brought to 500 ml with water.

2. Phosphate coupling buffer (10 mM K_2HPO_4) is prepared by dissolving 1.72 g of K_2HPO_4 in 1 liter of water, and adjusting the pH to 7.0 with HCl.

3. Glycine quenching solution (1 M glycine) is prepared by dissolving 75 g of glycine in water (1 liter), and adjusting the pH to 7.0 with HCl.

Note: The solutions made in steps 1–3 are sterilized by filtration through a 0.45 μm (pore size) filter.

4. Azide buffer [10 mM K_2HPO_4, 0.15 M NaCl, 0.1% (w/v) bovine serum albumin (BSA), 0.1% (w/v) sodium azide] is prepared by dissolving 1.745 g of K_2HPO_4, 8.7 g of NaCl, 1.0 g of BSA, and 1.0 g of sodium azide in 1 liter of water, and adjusting the pH to 7.4.

Treatment of BioMag Particles with Glutaraldehyde

1. Between 0.2 and 1.0 g BioMag (low settling) particles (10^8–10^9) are mixed with 1.0 ml of pyridine coupling buffer in a 1.5-ml Eppendorf tube.

2. The particles are gently mixed, magnetically separated, and the supernatant is discarded. This washing procedure is repeated three more times.

3. The magnetically separated particles after the third wash are resuspended in 0.4 ml of 5% (v/v) glutaraldehyde solution prepared in pyridine coupling buffer. The tube containing the suspension is vortexed and placed in a rotary mixer. It is allowed to rotate for 3 hr at room temperature with intermittent vortexing.

4. The particles are washed four times with 1.0 ml each of pyridine buffer and magnetically separated to remove excess glutaraldehyde, and resuspended in 1.0 ml of pyridine buffer.

[28] J. T. Kemshead and J. Ugelstad, *Mol. Cell. Biochem.* **67,** 11 (1985).

Coupling of Antibodies to BioMag Particles

1. Purified antibody (polyclonal antibodies containing 1.0 mg of IgG, or monoclonal antibody from tissue culture supernatants or ascites fluid containing 200 μg of IgG) is diluted to 200–500 μl with phosphate coupling buffer containing 8 mg/ml of crystalline BSA. An aliquot (200 μl) of the antibody solution is added to the derivatized magnetic particles (1 ml suspension from the previous section), and the mixture is gently mixed in a rotary mixer at room temperature for 10–12 hr.

2. At the end of the incubation period, the particles are magnetically sorted to remove any unbound antibodies. Glycine quenching buffer (1 ml) is added, the particles are mixed in a rotary mixer for 10 min, then sorted by a magnet to remove the glycine buffer and any unbound antibody.

3. The particles are washed twice with 1.0 ml of azide buffer, and stored at −20° in azide buffer containing 50% (v/v) glycerol.

Coupling of Antibodies to Dynabeads

1. Solutions needed include PBS, pH 7.4 (0.15 M NaCl in 10 mM sodium phosphate), PBS/BSA solution [prepared by adding 0.1% (w/v) BSA to PBS, and sterilized by filtration through a 0.45 μm (pore size) filter], and PBS/BSA/0.02% sodium azide solution [prepared by adding sodium azide to 0.02% (w/v) to PBS/BSA solution].

2. Dynabeads (1.0 ml) coated with goat anti-mouse IgG are washed with PBS/BSA solution four times. Then an aliquot of a suitable monoclonal antibody containing 100–200 μg IgG purified from a hybridoma culture supernatant, or ascites fluid, is added. The mixture is gently mixed and allowed to bind to the beads by placing the tube in a rotary mixer for about 24 hr. On the following day the beads coated with the primary antibody are washed four to five times with PBS/BSA to remove any unbound antibody. Finally, the beads sorted are resuspended in a solution containing 0.02% (w/v) azide in PBS/BSA.

Figure 1 shows the efficiency of coating the Dynabeads with the monoclonal MRK-16 antibody. The commercial magnetic beads precoated with goat anti-mouse IgG are visualized as such (Fig. 1A), or subsequent to staining with the rhodamine-labeled goat anti-mouse IgG (Fig. 1B). Only the background staining is seen in each case. Next, the magnetic beads are first treated with the primary monoclonal MRK-16 antibody, and then a small aliquot (20 μl) is treated with the rhodamine-labeled second antibody. It may be seen that the second antibody binds to the magnetic beads almost quantitatively (seen as bright white dots in Fig. 1C), indicating that

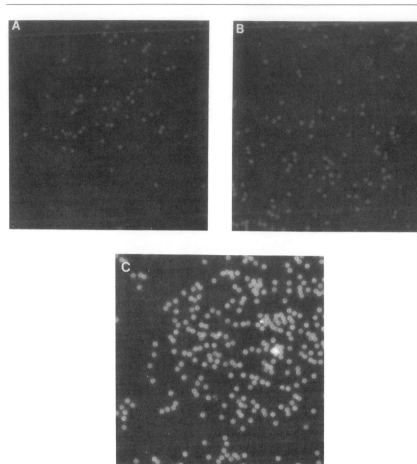

Fig. 1. Efficiency of coating magnetic beads with the monoclonal MRK-16 antibody assayed by immunofluorescence. (A) Dynal M-450 magnetic beads coated with goat anti-mouse IgG. Commercial Dynabeads were visualized by immunofluorescence prior to coating with the monoclonal antibody. (B) Dynabeads treated with rhodamine-labeled, affinity-purified goat anti-mouse IgG. The background staining was found to be minimal. (C) Dynabeads were first treated with the MRK-16 antibody (directed against the *mdr* gene product, P-glycoprotein),[20] followed by rhodamine-labeled, affinity-purified goat anti-mouse IgG. Note that all the beads, visualized as strong white dots due to binding of the rhodamine-labeled second antibody, were uniformly and efficiently bound to the monoclonal antibody in the first treatment.

the coating of the beads with the monoclonal antibody, MRK-16, is very efficient.

Magnetic Affinity Cell Sorting Protocol

Solutions Required for Sorting

Medium MS: Spinner MEM containing 40 mM EDTA, 10 mM HEPES, pH 7.3

Medium C: Spinner MEM containing 10% (v/v) fetal bovine serum, 20 mM HEPES (pH 7.3), 100 μg/ml chondroitin sulfate

Sorting medium: PBS containing 4 mM EGTA, 1 mM MgCl$_2$, 10 mM HEPES (pH 8.0), 100 μg/ml chondroitin sulfate, 1 mg/ml gelatin, 8 mg/ml nonfat dry milk, 10 μg/ml BSA

Sorting Protocol. At 48 hr after transfection of CV-1 cells or HeLa cells with the plasmid encoding a cell surface marker, the monolayer of transfected cells is incubated with MS medium (5 ml for 10 min at 37°) to dislodge the cells. Then medium C is added (25 ml) to the cell suspension, and incubated until the cells are dislodged to >90%. The detached cells are pelleted by centrifugation and resuspended in sorting medium (5 ml) in a 25-cm² flask, and then mixed with magnetic beads coated with antibody at a bead-to-cell ratio of 10:1 for 15–30 min with gentle agitation. The particle-bound cells are isolated by using Biomagnets (Advanced Magnetics) and resuspended in sorting medium (10 ml). The sorting procedure is repeated twice to remove any unbound cells.

Alternate Protocol Using Trypsin.[25] During the course of the work, we found an alternate protocol that is more effective and easier to use than the one described above. It should be applicable when surface antigen–antibody interactions, which are presumably protected by beads, become resistant to degradation by trypsin. The monolayer of cells to be sorted is treated with the magnetic particles (Dynabeads or BioMag) coated with the antibody in medium S or PBS/BSA (5 ml/25-cm² flask). It is incubated for 30 min with gentle intermittent mixing at 37°. Then at the end of the incubation the unattached beads are aspirated, leaving the monolayer with the attached beads. The monolayer is rinsed twice with PBS alone, and 2 ml of 0.25% (w/v) trypsin containing 2 mM EDTA is then added to cells. Incubation of cells is continued until they start to dislodge from the flask. Surprisingly, the cells attached to the magnetic beads via antigen–antibody interaction are resistant to trypsin digestion and remain attached to the beads. The cells are then immediately resuspended and diluted to 25 ml in sorting medium or PBS/BSA. Alternatively, addition of soybean trypsin inhibitor (2 mg) is effective to stop trypsin digestion prior to dilution. The cells are then sorted, using Biomagnets, by sandwiching the flask between

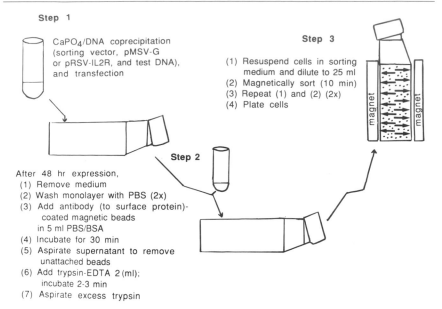

Step 1

CaPO₄/DNA coprecipitation
(sorting vector, pMSV-G
or pRSV-IL2R, and test DNA),
and transfection

Step 3

(1) Resuspend cells in sorting
    medium and dilute to 25 ml
(2) Magnetically sort (10 min)
(3) Repeat (1) and (2) (2x)
(4) Plate cells

Step 2

After 48 hr expression,
(1) Remove medium
(2) Wash monolayer with PBS (2x)
(3) Add antibody (to surface protein)-
    coated magnetic beads
    in 5 ml PBS/BSA
(4) Incubate for 30 min
(5) Aspirate supernatant to remove
    unattached beads
(6) Add trypsin-EDTA 2 (ml);
    incubate 2-3 min
(7) Aspirate excess trypsin

FIG. 2. Magnetic affinity cell sorting protocol. Step 1: Plasmid DNAs encoding a surface protein and the gene of interest are mixed with calcium phosphate (CaPO₄) to form a coprecipitate before transfection. Other methods of DNA-mediated gene transfer are also suitable. Step 2: After 48 hr of expression, the monolayer of cells is treated directly with magnetic beads coated with a specific antibody against the surface protein, and subsequently trypsinized and sorted as outlined here. Alternatively, the cells are detached to form a suspension with minimal aggregation by treatment with a medium containing EDTA, chondroitin sulfate, gelatin, and BSA as decribed in text. Step 3: The particle-bound cells are sorted magnetically away from the unbound cells, which are then plated.

two such magnets as illustrated in Fig. 2, step 3. The cells attached to beads are separated from free cells, which are then removed by aspiration. The sorted cells are washed twice with PBS and BSA with magnets still placed outside the flask, and finally the cells are resuspended in 10 ml of growth medium. The cells are counted and plated. To establish stable colonies of transfectants, cells with magnetic beads still attached can be plated in a suitable medium and selective conditions can be applied after 24 hr.

Results

Preparation of Cell Suspensions for Magnetic Affinity Cell Sorting

In general, DNA-mediated gene transfer protocols such as calcium phosphate–DNA coprecipitation, or liposome-mediated gene transfer,[7] are tailored for cells growing in monolayers. However, for the application of MACS, a suspension of cells free of aggregates is required. Initially, a medium containing a high EDTA concentration in the presence of chondroitin sulfate was used to reduce nonspecific aggregation.[29] However, high EDTA concentration and/or chondroitin sulfate was found to be toxic to cells, and caused some cell lysis. This drawback was partially corrected by the inclusion of gelatin and BSA. The addition of BSA to reduce cell lysis caused some nonspecific cell–particle interactions and particle aggregation. Addition of nonfat dry milk, gelatin, and low concentration of BSA gave fairly reproducible cell sorting without much aggregation.[14]

In later experiments, we found that the cells expressing a surface marker protein can be directly subjected to MACS in some cases without the necessity of dislodging them by treatment with solutions containing a high EDTA concentration and chondroitin sulfate. For example, cells that express P-glycoprotein on their surface can be directly treated with the magnetic beads attached to the monoclonal antibody, MRK-16,[20] directed against the P-glycoprotein. The particle-bound monolayer cells are then treated with trypsin–EDTA to dislodge them from the tissue culture flask, and the suspended cells are subjected to MACS to select the population of cells expressing the P-glycoprotein.[25] It has been shown that both the MRK-16 antibody and P-glycoprotein are resistant to trypsin. This unique advantage may be restricted to only certain antibody–cell surface marker interactions, but one should first test whether trypsin treatment can be carried out to dislodge the cells prior to MACS protocol for each test system. For certain cell types, it may be necessary to avoid treatment with trypsin and, in those cases, sorting medium containing varying concentrations of EDTA, chondroitin sulfate, gelatin, and BSA would be required.

Application of Magnetic Affinity Cell Sorting

Monitoring Efficiency and Specificity. To demonstrate that the MACS technique is applicable to select a small population of cells expressing the gene of interest away from a large population of negative cells, we used cotransfection of pRSVCAT and pMSVG into CV-1 cells. Due to the availability of a sensitive assay[8] for the CAT enzyme, it is possible to quantitate the efficiency and specificity of the MACS technique. pMSVG directs the expression of vesicular stomatitis virus G (VSV-G), a surface marker protein, under the control of the Moloney sarcoma virus LTR.[16] When the cells were cotransfected with pMSVG and pRSVCAT, and subjected to MACS protocol, about 94% of CAT activity was recovered

[29] G. J. Cole, D. Schubert, and L. Glaser, *J. Cell Biol.* **100**, 1192 (1985).

in the magnetic particle-bound fraction. When the cells were individually transfected with either of the two plasmids, and then mixed before MACS, only 0.4% of the CAT activity was in the particle-bound fraction.[14] The CAT activity was recovered almost quantitatively in the unbound fraction in the latter experiment. Thus this technique is highly specific and efficient in enriching cells expressing a surface marker, and it can be applied for both positive and negative selection of a cell population for further study.

Application to Select Multidrug-Resistant Cells. One of the major problems of cancer chemotherapy is the acquisition of multidrug resistance by tumor cells due the expression of a 170-kDa glycoprotein, which functions as an efflux pump.[18,19] To study the mechanism of drug resistance mediated by overexpression of P-glycoprotein in these cells, it is important to establish conditions for the separation of these cells from normal cells. The MACS technique is ideally suited to achieve this aim, as these cells are endowed with a surface marker, and a highly specific monoclonal antibody is available.[20] This antibody is directed against an extracellular domain of human P-glycoprotein.[20] The sorting protocol previously described[14] was simplified, because both the P-glycoprotein and the antibody MRK-16 were found to be resistant to trypsin digestion.

Bead-to-cell ratios of 3 : 1 for lymphoid cells, or 10 : 1 for other cells, were used. The sorting data showed that less than 1% of the drug-sensitive parental human KB-3-1 adenocarcinoma cells, human FEM-X melanoma cells, mouse L5178Y lymphoma cells, mouse NIH 3T3 cells, or NIH 3T3-mdr cells expressing an endogenous mouse *mdr* gene were selected by this procedure. Thus, a high degree of specificity for human P-glycoprotein was associated with this selection technique. The sorting efficiency was also found to be proportional to the levels of expression of the P-glycoprotein, which varied significantly depending on the cell line tested.[25]

To determine whether P-glycoprotein-positive cells could be sorted from a population of negative cells, two different mixing experiments were carried out.[25] In the first, KB-3-1 (drug-sensitive parent) cells were mixed with varying amounts of KB-8-5 (drug-resistant) cells. KB-8-5 cells were efficiently recovered from such mixtures even when they represented only 10% of the total population. In the second mixing experiment, mouse L5178Y cells expressing the human *MDR-1* gene were mixed with normal human bone marrow cells. Again the *MDR-1*-expressing cells were efficiently recovered from the bone marrow cells. The baseline expression of P-glycoprotein in normal bone marrow cells was too low to cause any interference in this selection protocol. The fact that the MACS protocol did not affect the inherent biological property of the cells was shown by the retention of the same drug resistance by the sorted cells at 24 hr postselection period, as the resistant population from which they were originally selected. However, when drug resistance was examined immediately after sorting (while the cells were still bound to the beads), it was found that these cells were about twofold more sensitive to the drug vinblastine or colchicine than the parent cell population. If the magnetic beads were bound to these drug-resistant cells via another cell surface marker such as the IL-2R, no drug sensitivity was observed, suggesting that the binding of the magnetic beads per se did not cause this effect. The drug sensitivity of these cells immediately after sorting could be explained by the interference of the efflux of the drug by the MRK-16 antibody binding to the P-glycoprotein.[20] Thus, the MACS technique should be applicable to select P-glycoprotein-positive cells from a heterogeneous population of cells normally found in a human tumor. We also found that human A431 cells expressing high levels of epidermal growth factor (EGF) receptor at the cell surface can be selected by MACS using anti-EGF receptor antibody-bound beads.

Application to Select Somatic Cell Hybrids. Somatic cell hybridization is a useful method to analyze whether a phenotype is dominant or recessive in hybrids formed, based on selectable phenotypic markers expressed by the parental cell types. There are potential problems with this form of selection in that the parental cell type must exhibit a mutant phenotype such as TK$^-$ or HPRT$^-$. To circumvent this problem, a hybrid cell selection strategy was used based on transfection of two exogenous DNAs, one of which codes for a cell surface marker such as IL-2R, and the other for the selectable marker, the *neo*^{R} gene. When both exogenous DNAs. were transfected into a single cell type, it was shown that the IL-2R-expressing cells selected by MACS technique could be directly subjected to selection for neomycin resistance.[30] However, in the strategy for hybrid cell selection, one of the parental cell lines, a spontaneously transformed rat liver epithelial (RLE) cell line,[31] was transferred with the *neo*^{R} gene conferring G418 resistance to this cell type. The second one, which was a nontransformed RLE cell line,[31] was transfected with the plasmid encoding the Tac subunit of the IL-2R under the control of cytomegalovirus (CMV) promoter.[32] Figure 3 shows the attachment of magnetic beads to the IL-2R-expressing RLE cells. When these two cell lines were fused, the resulting hybrids were initially selected by MACS technology using IL-2R as the handle, and then subjected to selection for G418 resistance. The cells that are IL-2R$^+$ and neo$^+$ are often contaminated with hybrid

[30] T. Giordano, T. Howard, and B. H. Howard, *Nucleic Acids Res.* **17**, 7540 (1989).
[31] J. B. McMahon, W. L. Richards, A. A. del Campo, M.-K. Song, and S. S. Thorgeirsson, *Cancer Res.* **46**, 4665 (1986).
[32] Raji Padmanabhan, T. Tan, T. Giordano, and S. S. Thorgeirsson, in preparation.

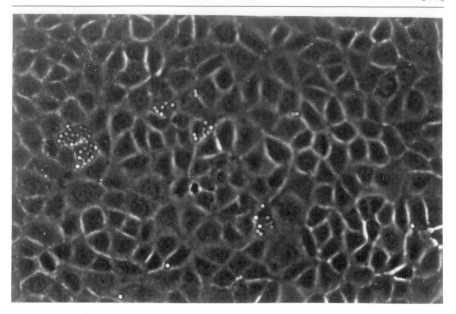

FIG. 3. Selection of parental rat liver epithelial (RLE) cells[29] by MACS. The RLE cells were transfected with the plasmid pCMVIL2R by the $CaPO_4$ coprecipitation method.[1,13] Forty-eight hours after transfection the cells treated with the magnetic beads attached to anti-IL-2R antibody (1 to 5 under step 2 in Fig. 2). Note the specific association of beads with the IL-2R-expressing cells.

cells having only neo[+] characteristic. To remove these contaminating cells, the hybrid cells were subjected to another round of the MACS protocol. Subsequent to this step, all selected hybrid cells were neo[+]/IL-2R[+], representing the proper fusion of each parental cell type. This allows one to carry out cell fusions of any cell type, independent of its normal phenotype.

Application to Viral Diagnostics. Viruses cause a number of diseases in humans, and correct diagnosis of the viral pathogen in a timely fashion is a major challenge in the field of medicine today. Many of the viruses encode proteins that are expressed on the surface of the virions, or on the infected cells of the host, which often become the targets for the immune response of the host.These surface antigens could serve as the basis for application of MACS technology as a sensitive method of detection.

For example, the MACS technology is applicable for the selection of cells infected by dengue virus, a member of the flavivirus family.[33] The

[33] E. G. Westaway, M. A. Brinton, S. Y. Gaidamovich, M. C. Horzinek, A. Igarashi, L. Kaariainen, D. K. Lvov, J. S. Porterfield, P. K. Russell, and D. W. Trent, *Intervirology* **24,** 183 (1985).

nonstructural glycoprotein NS1 of flaviviruses exists in multiple forms, including cell surface and extracellular (secreted) forms.[34,35] Flavivirus infections produce an antibody response to NS1 in the infected host, and these antibodies have complement fixing activity. Magnetic beads coated with a monoclonal antibody against NS1 (3E9) bound specifically to the surface of infected mosquito (C6/36) cells (data not shown), suggesting that it should be possible to separate infected monocytes from uninfected-cells in patients for diagnostic studies.

[34] E. G. Westaway, *Adv. Virus Res.* **33,** 45 (1987).
[35] T. J. Chambers, C. S. Hahn, R. Galler, and C. M. Rice, *Annu. Rev. Microbiol.* **44,** 649 (1990).

[46] Construction of Restriction Fragment Maps of 50- to 100-Kilobase DNA

By MING YI, NORIO ICHIKAWA, and PAUL O. P. Ts'o

Introduction

Construction of the restriction fragment maps obtained from plasmid, cosmid, and YACs (yeast artificial chromosomes) is a key step in many molecular biological experiments, particularly involving sequencing, such as in the genome sequencing process. To obtain the nucleotide sequence, a genome consisting of 10^8–10^9 base pairs (bp) is usually broken down to the level of cosmid or YACs with the size range of 30–300 kilobases (kb). The DNA restriction fragments from the cosmids or YACs 1–5 kb in size represent an obligatory intermediate stage for the nucleotide sequencing procedure.

This chapter describes a procedure for a simultaneous construction of two restriction fragment maps. This procedure consists of two-enzyme digestion, two-dimensional electrophoresis, as well as the construction of a sequence matrix based on the data.[1] λ phage DNA is used as an example to illustrate this approach. This approach is applicable to both linear DNA and circular DNA.

Principle

Using a two-enzyme digestion, two-dimensional (2-D) electrophoresis procedure, all the restriction fragments in 50- to 100-kb DNA can be

[1] N. Ichikawa, M. Yi, and P. O. P. Ts'o, *Anal. Biochem.* **201,** 158 (1992).

METHODS IN ENZYMOLOGY, VOL. 218

individually resolved and displayed on a 2-D plane.[2] This 2-D gel pattern, with appropriate markers, provides a fixed set of x, y coordinates for each fragment obtained from the single and double digestion as well as the relationship between all the doubly digested restriction fragments (obtained after in gel digestion) and their corresponding parental restriction fragments (obtained after the first digestion in solution).

Under ideal conditions, which are demonstrated in this chapter with λ phage DNA, each and every restriction fragment obtained from the cleavage by the first enzyme (i.e., *Eco*RI) contains one restriction site for the second enzyme (i.e., *Bam*HI). Conversely, each and every fragment obtained from the treatment with the second enzyme (*Bam*HI) contains one restriction site for the first enzyme (*Eco*RI). Thus, the same set of doubly digested fragments (i.e., *Eco*RI + *Bam*HI fragments) can be obtained either by a forward order of treatment (i.e., *Eco*RI first, followed by *Bam*HI) or by a backward order (i.e., *Bam*HI first, followed by *Eco*RI). However, there will be one set of singly digested fragments from *Eco*RI, and another set of singly digested fragments from *Bam*HI. Through the two-enzyme–2-D analyses described above, the relationship between the singly digested fragments and the doubly digested fragments of both forward and backward order can be experimentally and unambiguously determined. The connection of one doubly digested fragment to one singly digested fragment from the first enzyme (*Eco*RI), as well as to the other singly digested fragment from the second enzyme (*Bam*HI), provides the unambiguous overlapping relationship that is the cornerstone of the mapping process.

Based on the above reasoning, a sequence matrix is constructed from the one-dimensional electrophoresis pattern and two 2-D electrophoresis patterns. Each doubly digested fragment in the sequence matrix defines the overlap of two adjacent fragments produced after a two-enzyme digestion. In the restriction map construction, each doubly digested fragment contains the information for defining the overlap of the parental, singly digested fragment obtained from the *first* enzyme as well as the parental, singly digested fragment obtained from the *second* enzyme. Therefore a continuous overlapping can be produced when all the doubly digested fragments are placed in proper order. By tracing in the sequence matrix following the rules described in this section, two DNA restriction fragment maps of the same 50- to 100-kb DNA can be simultaneously constructed in a self-reconfirming manner (Fig. 1).

[2] M. Yi, L. C. Au, N. Ichikawa, and P. O. P. Ts'o, *Proc. Natl. Acad. Sci. U.S.A.* **87**, 3919 (1990).

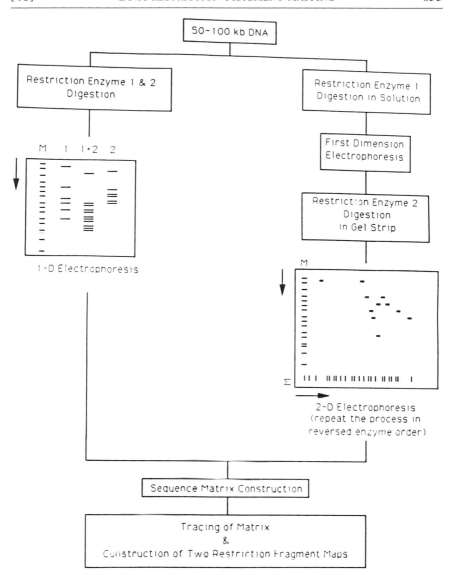

FIG. 1. The general scheme for the construction of two DNA restriction maps based on 1-D electrophores, two two-enzyme digestions, two 2-D electrophoresis, and sequence matrix. M, DNA fragments as size markers; arrows denote the direction of electrophoresis. The choice of enzyme pair has been made already (see Discussion).

Methods

Restriction Enzyme Digestion in Solution

Choice of Restriction Enzyme Pairs. The enzyme pairs chosen for this demonstration are *Eco*RI/*Bam*HI, *Eco*RI/*Hin*dIII.

Reagents

*Eco*RI reaction buffer: 100 mM Tris (pH 7.75), 50 mM NaCl, 10 mM MgCl$_2$

*Bam*HI reaction buffer: 100 mM Tris (pH 7.75), 100 mM NaCl, 10 mM MgCl$_2$

*Hin*dIII reaction buffer: 50 mM Tris (pH 8.0), 50 mM NaCl, 10 mM MgCl$_2$

TE buffer: 10 mM Tris, 1 mM ethylenediaminetetraacetic acid (EDTA) (pH 8.0)

Procedure

1. Digest λ phage DNA (Bethesda Research Laboratories, Gaithersburg, MD) with *Eco*RI and *Bam*HI or *Hin*dIII (each with 2.5 units/μg DNA) separately in their respective reaction buffer at 37°, overnight.

2. Digest 1.5 μg of λ phage DNA with *Eco*RI together with *Bam*HI (each with 2.5 units/μg DNA).

3. Extract the DNA solutions with the same volume of phenol once and the same volume of chloroform twice.

4. Add 1/9 vol of 3 M sodium acetate, pH 7.0.

5. Precipitate DNA fragments by adding 2.5 vol of 95% (v/v) ethanol overnight at $-20°$.

6. Centrifuge the DNA precipitates at 14,000 rpm at 4° for 10 min.

7. Rinse the DNA pellets with 70% (v/v) ethanol twice, 95% (v/v) ethanol once.

8. Dry the DNA pellets *in vacuo*.

9. Redissolve the DNA pellets in TE buffer.

All the following procedures are described using the *Eco*RI/*Bam*HI enzyme pair as an example.

One-Dimensional Electrophoresis

Reagent

TPE: 0.08 M Tris–phosphate, 0.008 M EDTA (pH 7.7)

Procedure

1. Load 1.0 μg of *Eco*RI fragments of λ phage DNA, 1.0 μg of *Bam*HI fragments of λ phage DNA, and 1.5 μg of *Eco*RI/*Bam*HI doubly digested fragments of λ phage DNA separately onto different lanes in a horizontal 25 × 20 × 0.46 cm 1% (w/v) agarose gel slab.

2. Load the mixture of 0.5 μg of λ/*Hin*dIII fragments (Bethesda Research Laboratories) and 1.6 μg of 1-kb DNA ladder fragments (Bethesda Research Laboratories) onto the gel slab at the first lane as size markers.

3. Perform electrophoresis at 40 V for 20 hr in TPE buffer at room temperature.

4. Stain the gel with ethidium bromide.

5. Make a photograph (Fig. 2a).

6. Determine the molecular sizes of all the fragments by comparison with the size markers (Table I).

Two-Dimensional Electrophoresis: First Dimension

Reagents

Agarose: The agarose used should be highly purified and devoid of sulfated polysaccharides

Procedure

1. Load 1.5 μg of *Eco*RI fragments of λ phage DNA onto a horizontal 25 × 20 × 0.46 cm gel slab containing 1% (w/v) ultrapure agarose (Bio-Rad, Richmond, CA). The sample well is 5 mm wide and 1 mm thick.

2. Load 0.5 μg of λ/*Hin*dIII fragments and 1.6 μg of 1-kb DNA ladder fragments as size marker.

3. Perform electrophoresis in TPE buffer at 40 V for 20 hr at room temperature.

Second enzyme in situ digestion in gel

1. Cut the DNA-containing portion of the gel into a 16 × 0.8 cm strip.

2. Equilibrate the gel strip with *Bam*HI reaction buffer by shaking at 100 rpm at room temperature for 1 hr.

3. Put the gel strip into a 21 × 1.25 cm tube (total volume, 26 ml) containing *Bam*HI reaction buffer, 325 μg/ml of molecular biology-grade bovine serum albumin (BSA) (Sigma, St. Louis, MO)[2] and a 4 m*M* concentration of molecular biology-grade spermidine (Sigma).[3]

4. Rotate the tube gently at 37° for 1 hr.

5. Add 2000 units of *Bam*HI to the tube.

[3] J. P. Bouche, *Anal. Biochem.* **115,** 42 (1981).

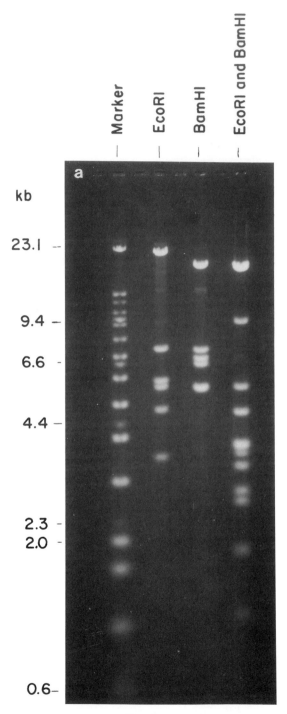

FIG. 2a

656

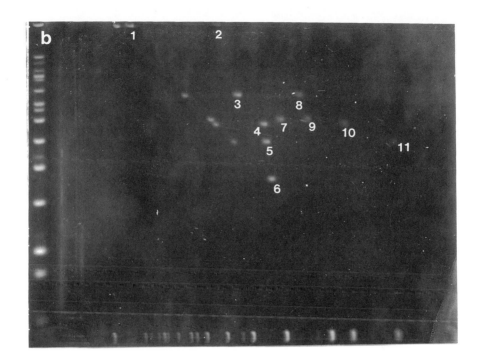

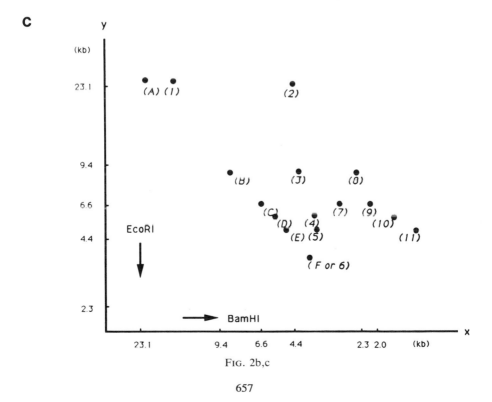

FIG. 2b,c

657

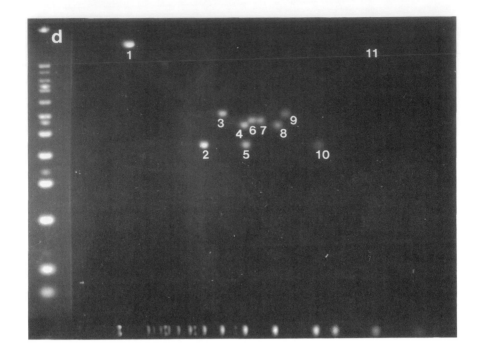

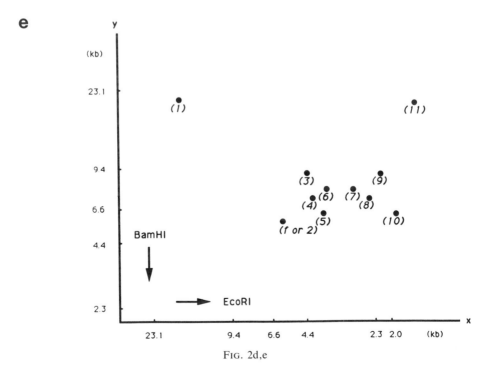

Fig. 2d,e

6. Rotate the tube gently at 37° for 20 hr.

7. Equilibrate the gel strip with TPE buffer by shaking at room temperature for 1 hr.

Two-Dimensional Electrophoresis: Second Dimension

1. Fuse the gel strip with a 1% (w/v) agarose gel slab by adding melted agarose to fill the gap.

2. Load the size markers (same as those used in first-dimension electrophoresis) in a well located at the end of the fused gel strip.

3. Perform the second-dimensional electrophoresis, with a direction perpendicular to the first one, at 35 V for 20 hr, in TPE buffer at room temperature.

4. Stain the gel slab with ethidium bromide.

5. Make a photograph (Fig. 2b).

6. Determine the molecular sizes of all the fragments by comparison with the size markers (Fig. 2c).

7. In the case in which BamHI is used as first restriction enzyme and EcoRI is used as second enzyme, the procedure is the same as described above except the enzyme order is reversed, together with their corresponding reaction buffers (Fig. 2d and e).

8. The same procedure can be used for the EcoRI/HindIII pair with the corresponding buffers (Fig. 3).

Construction of Matrix

1. Set the vertical column on the very left side of the matrix (Fig. 4a): the upper section contains EcoRI singly digested fragments, and the lower section contains BamHI singly digested fragments listed in Table I. The

---

FIG. 2. (a) First-dimensional electrophoresis pattern of λ phage DNA. One microgram of DNA is used in each single digestion (EcoRI or BamHI) and 1.5 μg of DNA in each double digestion (EcoRI and BamHI). (b) Two-dimensional pattern of λ phage DNA restriction fragments (1.5 μg) digested by EcoRI → BamHI. The spots in the diagonal line represent either the partial digestion of the singly digested fragments (A) to (E) or fragment (F), which does not have a cutting site for the second enzyme BamHI. (c) Diagrammatic interpretation of (b). (d) Two-dimensional pattern of λ phage DNA restriction fragments (1.5 μg) digested by BamHI → EcoRI. The left and bottom edges of the gel are lined with λ/HindIII and 1-kb DNA ladder as markers. (e) Diagrammatic interpretation of (d). The vertical arrow denotes the direction of the first dimension, whereas the horizontal arrow denotes the direction of the second-dimension electrophoresis. See Table I for designation.

TABLE I
MOLECULAR SIZES OF SINGLY AND DOUBLY
DIGESTED RESTRICTION FRAGMENTS
FROM λ PHAGE DNA[a]

EcoRI	BamHI	EcoRI and BamHI[b]
21.30 (A)	17.80 (a)	16.00 (1)
7.50 (B)	7.40 (b)	5.70 (2)
6.00 (C)	6.95 (c)	4.75 (3)
5.75 (D)	6.70 (d)	3.85 (4)[c]
4.80 (E)	5.75 (e)[d]	3.85 (5)[c]
3.65 (F)	5.65 (f)[d]	3.70 (6)[e]
		3.37 (7)
		2.88 (8)
		2.68 (9)
		1.90 (10)
		1.14 (11)

[a] As analyzed by 1-D electrophoresis. Data from Fig. 2a.

[b] The 9.2-kb band in Fig. 2a is not listed in this table because its existence is due to the "sticky ends" of λ DNA and this band disappears in 2-D electrophoresis (Fig. 2b and c).

[c] These two fragments (4) and (5) have nearly the same molecular size and cannot be separated in 1-D electrophoresis (Fig. 2a) but can be clearly distinguished in 2-D electrophoresis (Fig. 2b–e).

[d] This band in the 1-D electrophoretic pattern of the BamHI fragments is found to consist of two bands with close molecular sizes. These two bands can be distinguished in 1-D or 2-D electrophoresis after the EcoRI digestion. One band [5.75 kb, (e)] is digested by EcoRI to produce a 3.85-kb fragment (5) and a 1.90-kb fragment (10). The 5.65-kb fragment (f) contains no cleavage site for EcoRI (see Fig. 2c and e).

[e] Fragment (F) from EcoRI digestion does not contain a cleavage site for BamHI.

order of arrangement is always from the smallest to the largest fragments, from top to bottom.

2. Set the dividing horizontal row between the upper and lower sections with EcoRI and BamHI doubly digested fragments listed in Table I. The order of arrangement is always from the smallest to the largest fragments, from left to right.

3. Define the locations of E (for *Eco*RI fragments overlapping with *Bam*HI fragments in the doubly digested fragments), B (for *Bam*HI fragments overlapping with *Eco*RI fragments in doubly digested fragments), and S (for fragments that are not digested by the second enzyme or located at the end). S fragments arising from the end do not exist in the digestion of circular DNA.

Figure 5 is a diagrammatic interpretation of the data in Fig. 2. The 1-D pattern comes from Fig. 2a and the data is listed in Table I, which gives the molecular sizes of all the singly digested and doubly digested DNA fragments and their designation. The evaluation of the molecular size of the restriction fragments based on comparison with the markers is estimated to have a range of ±0.5 kb in accuracy. The solid lines indicate the relationship between the singly digested DNA fragments and doubly digested DNA fragments that are obtained from the two 2-D patterns shown in Fig. 2b and d. Figures 2c and e are the graphic interpretation of Fig. 2b and d, based on the designation listed in Table I. All the information needed for the construction of the matrix is summarized in Fig. 5.

For example, results in Fig. 2b and c show that the 7.5-kb *Eco*RI fragment, designated (B), is cleaved into a 4.75-kb fragment (3) and a 2.88-kb fragment (8) by the *Bam*HI in gel digestion. This is readily seen because spots (B), (3), and (8) are on the same horizontal line. From this data, these two E fragments are marked in the matrix (Fig. 4a) as follows: The location of the first E, the E3 fragment (4.75 kb), is defined by the coordinate of the X axis with fragment (B) (left column) and the Y axis with fragment (3) (horizontal row), and this spot is placed on Fig. 4a; the second E, the E4 fragment (2.88 kb), is marked by the X axis again with fragment (B) (left column) and the Y axis with fragment (8) (horizontal row), and this spot is placed on the figure. This construction shows that both E3 and E4 fragments originate from an *Eco*RI digestion of the 7.50-kb fragment (B) and it is cleaved by *Bam*HI into two fragments, a 4.75-kb fragment (3) and a 2.88-kb fragment (8).

In a similar process based on the data shown in Fig. 2a (1-D analysis) and Fig. 2d and e (2-D analyses), one can identify that the same 2.88-kb fragment (8) originates from the singly digested *Bam*HI fragment [6.7 kb, (d)] and the same 4.75-kb fragment (3) originates from the singly digested *Bam*HI fragment [7.40 kb, (b)]. The origination of these two doubly digested fragments (8) and (3) from the one singly digested *Eco*RI fragment (B) and the two singly digested *Bam*HI fragments (d and b) is clearly shown in Fig. 5. Figure 5 provides the experimental basis for the connection between the upper half of the matrix (*Eco*RI digestion) and the lower half of the matrix (*Bam*HI digestion). The ordering of the overlapping

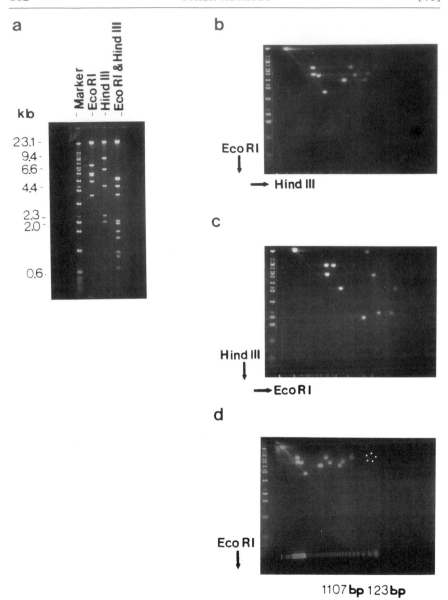

FIG. 3. (a) First-dimensional electrophoresis pattern of λ DNA. One microgram of DNA is used in each single digestion (*Eco*RI or *Hin*dIII) and 1.5 μg of DNA in each double digestion (*Eco*RI and *Hin*dIII). (b) Two-dimensional pattern of λ phage DNA restriction fragments digested by *Eco*RI → *Hin*dIII. (c) Two-dimensional pattern of λ phage DNA

fragments shown in the upper half in connection with the lower half in the matrix is the basic concept and step in the simultaneous construction of two linear restriction fragment maps.

Construction of the Restriction Fragment Maps

1. Connect E, B, and S locations throughout the matrix (Fig. 4a) by starting the trace with the smallest fragment at the top of the left column (e.g., 3.65 kb): after reaching S_1, go down vertically (passing the 3.70-kb fragment in the medial row) to meet B (B_1) in the lower section, move horizontally to another B (B_2) and then vertically upward (passing the 3.37-kb fragment in the medial row) to E (E_1, then E_2) and so on. The tracing is carried out alternately between a horizontal movement followed by a vertical one and vice versa. In the meantime, when a fragment is reached, the order of this fragment in its section is determined according to the numerical order in the tracing. For example, the first B reached by tracing in the lower section of the matrix is ordered as B_1, the second B reached is ordered as B_2. When tracing goes through the first turn in the upper section, it determines the order of E_1 and E_2. The next tracing goes to the lower section again, giving the order of B_3 and B_4, and so on.

2. Construct two restriction fragment maps simultaneously from the sequence matrix. In Fig. 6a, the top section of the map is for the ordering of the λ DNA EcoRI restriction fragments. The bottom section is for the ordering of BamHI fragments. The construction of these two maps is based on the tracing in Fig. 4a. Following the tracing procedure described above, starting from the 3.65-kb fragment (F) at the top of the left column in the upper section, the tracing reaches S_1 first. The 3.65-kb fragment (F), which is the singly digested parental fragment of S_1, is determined as the first fragment in the EcoRI map. Tracing goes down vertically in the matrix to reach B_1 and B_2 in the lower section. The singly digested 6.95-kb fragment (c) shown in the left column, which is the parental fragment of B_1 and B_2, is now determined as the first fragment in the BamHI map. Then the tracing goes up vertically to reach E_1 and E_2, which originate from the 6.0-kb fragment (C). Thus this fragment (C) is now the second fragment in the EcoRI map. Continuing the tracing, the parental fragment of B_3 and B_4, the 7.40-kb (b) fragment, will be denoted as the second

restriction fragments digested by HindIII → EcoRI. (d) A 0.130-kb fragment can be detected on the 2-D map obtained in a 1.5% agarose gel slab (10 μg of DNA and 10,000 units of HindIII). A 123-bp DNA ladder is used as DNA size markers during second-dimension electrophoresis. This small fragment is encircled.

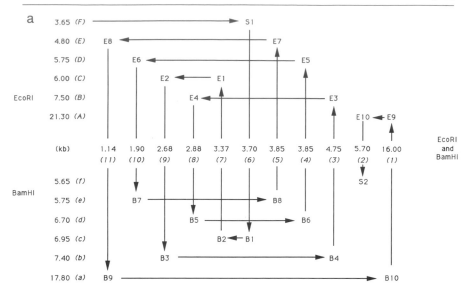

FIG. 4. Sequence matrices for construction of the linear restriction fragment map of λ phage DNA. (a) The vertical column at the left shows the size of all the *Eco*RI-digested fragments in the upper half and all the *Bam*HI-digested fragments in the lower half as listed in Table I (top to bottom, from small to large). The horizontal row shows all the sizes of the doubly digested fragments as listed in Table I (left to right, from small to large). "E" and "B" denote *Eco*RI- and *Bam*HI-digested fragments, respectively. "S" denotes a fragment that is not digested by the second enzyme, that is, no cutting site is present for the second enzyme or located at the end. The numerical subscripts denote the linear order for the restriction fragments in the map. (b) The matrix has the same pattern as that in (a), with *Hin*dIII substituting for *Bam*HI. The 0.13-kb fragment in the lower part of the matrix can be determined as a singlet by the electrophoresis pattern in a 1.5% agarose gel slab and 10 μg of DNA (data not shown). The tracing procedure for the construction of the matrices and the linear maps of the λ phage DNA are described in text.

fragment in the *Bam*HI map. The next tracing will add 7.50 kb (B) to the *Eco*RI map as the third fragment. The tracing in the sequence matrix represents the order and the overlap of the fragments.

As shown in the constructed λ DNA restriction maps, each fragment from enzyme 1 is bound by two restriction sites of enzyme 1 (except the end fragments S) and is ordered by the two overlapping fragments generated by the second enzyme. For example, the order of the 6.0-kb E fragment (for the *Eco*RI mapping) is defined by E_1 and E_2 overlapping B_2 and B_3 generated by *Bam*HI. A similar procedure of tracings leads to the construction of the two maps of *Eco*RI/*Hin*dIII in Fig. 6b and c.

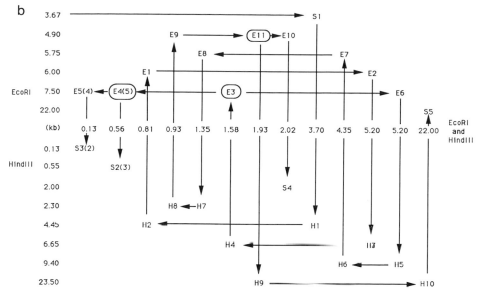

FIG. 4 (*continued*)

The numerical determinants in Figs. 4 and 6 (such as E_{1-10}, B_{1-10}, and S_{1-2}) are the overlapping orders of the fragments in the map. These orders are not defined by the polarity (5' end and 3' end) of the DNA strands in the duplex. Therefore, these orders can be reverted depending on the choice of polarity.

Results

The construction of two definitive restriction fragment maps for a given restriction enzyme pair requires performing the procedure involving forward and backward ordering of sequential enzyme treatment. For example, the forward treatment is done with *Eco*RI digestion in solution first, followed by *Bam*HI digestion in gel; and the backward treatment is done with *Bam*HI digestion in solution first, followed by *Eco*RI digestion in gel.

Figure 2 and Table I show all the experimental data necessary for the construction of the linear restriction maps. The 1-D electrophoretic patterns for the singly digested and doubly digested λ DNA fragments obtained from the chosen pair of restriction enzymes together with the DNA size markers are shown in Fig. 2a. The number and molecular sizes of all the fragments obtained have been noted and accounted for (Table

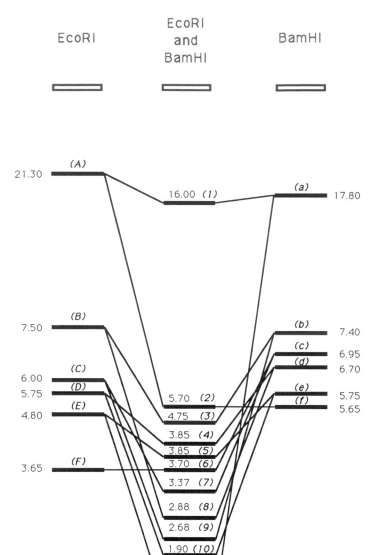

FIG. 5. A diagrammatic interpretation of the relationship between singly digested and doubly digested λ phage DNA fragments by *Eco*RI and *Bam*HI, based on the data shown in Fig. 2 and Table I. The singly digested fragments are listed along two sides together with their molecular sizes (in kilobases) and their designation. The doubly digested fragments are listed in the middle together with their molecular sizes and their designation. The solid lines (—) represent the relationship between the singly digested fragments and doubly digested fragments.

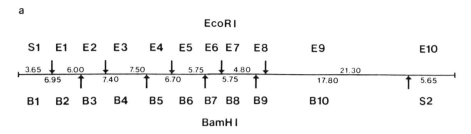

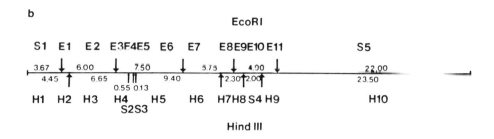

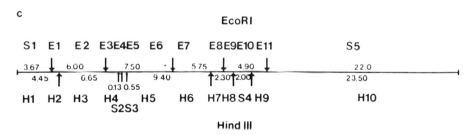

FIG. 6. (a) The linear map of λ phage DNA constructed for *Eco*RI and *Bam*HI restriction fragments. (b and c) The linear maps of λ phage DNA constructed for *Eco*RI and *Hind*III restriction fragments, with the uncertainty of the order of sequences S₂ and S₃ in the *Hind*III map.

I). This is the essential step to screen and choose the appropriate pair of restriction enzymes (see Discussion, below). Figure 2b and d indicate the 2-D patterns from the forward treatment (*Eco*RI → *Bam*HI) and backward treatment (*Bam*HI → *Eco*RI). From these two 2-D patterns, the molecular sizes of all the fragments as well as the relationship of the origination of all the doubly digested restriction fragments with the parental, singly digested restriction fragments are unambiguously provided by the defined *x, y* coordinate values of each fragment (Fig. 5). Based on Fig. 5, a matrix is constructed in Fig. 4a, the vertical column showing all the *Eco*RI fragments (denoted by their sizes) in the upper section and all the *Bam*HI fragments (denoted by their sizes) in the lower section. The medial horizontal row shows all the fragments obtained from the double digestion by both *Eco*RI and *Bam*HI (as denoted by their individual sizes). The locations of E, B, and S are defined by the singly digested fragment sizes (column) and doubly digested fragment sizes (row). The arrows in the matrix reveal the connections among E fragments or B fragments and the overlaps between E and B through the fragment in the medial horizontal row, which is shared by the corresponding E and B in the double digestion. For example, the 4.75-kb fragment [designated as (3)] in the medial horizontal row is shared by the 7.50-kb *Eco*RI fragment [designated as (B)] in the vertical column as well as E_3 and E_4 in the matrix and by the 7.40-kb *Bam*HI fragment [designated as (b)] in vertical column as well as B_3 and B_4 in the matrix in the double digestion. It should be noted that the lines connecting the fragments in Fig. 5 correspond to the same tracing lines in Fig. 4a.

Using the sequence matrix in Fig. 4a, two linear restriction fragment maps of λ phage DNA from *Eco*RI and *Bam*HI digestion are constructed (Fig. 6a). A tracing method in matrix has been described in Methods (above). However, the tracing may be started from any fragment listed not only from the end fragments, particularly for the circular DNA. In the case of starting with a fragment in the middle of the matrix, the tracing will go in two directions. For example, starting with the 7.50-kb fragment in Fig. 4a, as E_4, one direction is to continue going downward, through point B_5, B_6, and so on, until S_2 is reached, and the other direction is to go horizontally, through point E_3, B_4, and so on, until S_1 is reached.

The *Eco*RI/*Bam*HI enzyme pair is ideal for the restriction fragment map construction of λ phage DNA, because each and every singly digested fragment contains sites for the second digestion except for the end fragments. The linear maps for both *Eco*RI and *Bam*HI fragments are unambiguous. In the second example, utilizing the enzyme pair *Eco*RI and *Hin*dIII (Fig. 3), two difficulties arise during the linear map construction. One problem is the presence of a small fragment (i.e., 0.130 kb—S_3 in Fig. 6b) that may escape detection. This small fragment may migrate out of the gel because of its high mobility; or it may not be detectable. To overcome these difficulties, the following three steps can be adopted: (1) electrophoresis is conducted in a gel with higher agarose concentration (1.5%); (2) a larger amount of initial DNA sample of 10 μg is used (Fig. 3d); and (3) a more sensitive detection technique is used [(3' end labeling with ^{32}P-labeled nucleoside triphosphate *in situ* with Sequenase (United States Biochem. Corp., Cleveland, OH; data not shown)].

The second problem arises from the fact that not all the restriction fragments obtained from the first enzyme digestion contain cleavage sites for the second enzyme. In this case, the *Hin*dIII fragments S_2 and S_3 are not cleaved by *Eco*RI. In such a situation, the order or the orientation of these S_2 and S_3 fragments in the *Hin*dIII map can *not* be explicitly determined between the two choices, as indicated in Fig. 6b and c. However, as indicated, the *Eco*RI linear restriction map remains to be unique. This example demonstrates that the success of this approach depends heavily on the choice of the enzyme pairs that are selected on the basis of the 1-D experiments. In the case of Fig. 4b, the construction of the matrix and the way of tracing are the same as in Fig. 4a. When the fork is reached, represented by the encircled marker, tracing always goes to the singlet S first, and then in the other direction.

Discussion

The requirements of this method are as follow.

1. All the restriction fragments obtained from the first/single digestion must be completely retained and resolved in the 1-D electrophoresis and all the doubly digested fragments must be retained and resolved within the 2-D gel.

2. The correct choice of a restriction enzyme pair is crucial. To get the appropriate enzyme pair, the DNA sample should be singly digested by 8 to 10 enzymes. From the 1-D electrophoresis pattern, the enzymes that produce 6 to 10 fragments in a single digestion should be chosen. If too few fragments are produced, the obtained map of the restriction sites would not be useful. If too many fragments are produced, an overcrowding in the 2-D gel pattern may be observed, leading to overlapped and unresolved DNA spots. Also, tracing the matrix would be excessively difficult and the mapping process would be complicated. For an ideal enzyme pair, the set of restriction fragments produced by the first enzyme would contain one or two cutting sites for the second enzyme in each fragment. If only two restriction sites exist for the second enzyme in a restriction fragment obtained from the first enzyme, no ambiguity can arise as to the arrangement of the singlet fragment in the middle, because this singlet is always bound by two neighboring fragments that can be ordered. Clearly, then, the presence of multiple singlets within a fragment would cause ambiguity in mapping the order of these singlets for that restriction map.

3. Some of the restriction enzymes either exhibit poor activity in the gel or are rather expensive. Currently, the preferred list of acceptable enzymes includes *Eco*RI, *Eco*RV, *Hin*dIII, *Bst*EII, *Sal*I, *Sty*I, and *Xho*I. *Bam*HI usually digests DNA incompletely in the gel, leaving around 20% of the fragments uncut. *Xba*I and *Ban*I also fail to digest to completion in our experience.

In comparison to the previously published approaches[4,5] the current procedure has the following advantages.

1. After 1-D electrophoresis, the second restriction enzyme in gel digestion of the first restriction enzyme products establishes the definite relationship between the doubly digested fragments and the parental, singly digested fragments. By a symmetric approach of using the forward order digestion and the backward order digestion with a pair of enzymes, the relationship of all the overlapping fragments in this mapping procedure is also unambiguously established.

2. The results are reliable. This approach not only produces two restriction maps at one time, but is also a self-validating process. Not only are the two maps established in a self-validating manner, but the maps also require arithmetic reconfirmation for proper summation of all restriction fragments.

3. The process is comparatively short and simple once the proper choice of restriction enzyme pair is made. The analytical and map construction processes are well defined and easy to follow.

4. All the fragments in the 2-D gel are properly indexed and can be stored indefinitely, awaiting future nucleotide-level sequence analysis.

Finally, using a larger gel (50 × 40 cm; data not shown), restriction fragments from a large DNA size sample (200–300 kb) can be readily analyzed and displayed by this 2-D electrophoresis procedure. Also, to define the molecular sizes of all the DNA fragments precisely, for proper indexing and comparison, three gel slabs could be run simultaneously in the same electrophoresis tank. The first gel displays the map from the forward order of the two enzyme digestions, the second gel displays the map from the backward order of the two enzyme digestions, and the third gel contains all the DNA markers.[1]

This procedure is demonstrated for a linear DNA but is also applicable to a circular DNA, such as cosmid or plasmid.

[4] W. M. Fitch, T. F. Smith, and W. W. Ralph, *Gene* **22**, 19 (1983).
[5] W. Bautsch, *Nucleic Acids Res.* **16**, 11461 (1988).

Acknowledgments

We would like to express our appreciation to Gerard F. Leblond, Ph.D., and Hideaki Nakashima, M.D., Ph.D., for their assistance and enthusiastic suggestions.
This work was supported by the Department of Energy, Office of Health and Environmental Research (Washington, D.C.) Grant #DE-FG02-88-ER6036-03. Reprint requests should be sent to P.O.P.T.

[47] Tissue-Print Hybridization on Membrane for Localization of mRNA in Plant Tissue

By Yan-Ru Song, Zheng-Hua Ye, and Joseph E. Varner

When a freshly cut section of plant tissue is pressed on a nitrocellulose or nylon membrane, the soluble contents of the cut cells are transferred onto the membrane. The cell contents on the membrane make a latent print that can then be visualized with appropriate probes. The rigid cell walls (especially xylem and phloem fibers) leave a physical print that makes the anatomy of the tissue visible without any further treatment The tissue print technique allows rapid and convenient tissue-level localization of proteins, enzymes, mRNAs, viral DNA, and metabolites. The tissue print technique has been used to localize cell wall extensin in soybean seeds and stems[1,2]; polyphenol oxidase, peroxidase, glycosidases, dehydrogenase, and phosphatase in bean leaf petioles and oil palm fruit[3]; cysteine rich proteins in barley seed, potato tuber parenchyma, and soybean stem[4]; cellulase in bean abscission zones and in the stem[5]; and cell wall glycine-rich proteins (GRPs) in bean and soybean tissues.[2,6] The tissue print technique was also developed to localize extensin mRNA in soybean pods and α-subunit mRNA of β-conglycinin in different developmental stages of soybean seeds using ^{32}P-labeled DNA probes.[7] Furthermore, it was used to study the tissue-specific expression of auxin-regulated genes in elongating hypocotyl regions of etiolated soybean seedlings and the rapid turnover of RNAs encoded by these genes during gravistimulation

[1] G. I. Cassab and J. E. Varner, *J. Cell Biol.* **105**, 2581 (1987).
[2] Z.-H. Ye and J. E. Varner, *Plant Cell* **3**, 23 (1991).
[3] J. Spruce, A. M. Mayer, and D. J. Osborne, *Phytochemistry* **26**, 2901 (1987).
[4] R. F. Pont-Lezica and J. E. Varner, *Anal. Biochem.* **182**, 334 (1989).
[5] E. del Campillo, P. D. Reid, R. Sexton, and L. N. Lewis, *Plant Cell* **2**, 245 (1990).
[6] B. Keller, N. Sauer, and C. J. Lamb, *EMBO J.* **7**, 3625 (1988).
[7] J. E. Varner, Y. R. Song, L.-S. Lin, and H. Yuen, *UCLA Symp. Mol. Cell. Biol.* No. 92, p. 161 (1989).

by using ^{35}S-labeled antisense RNA probes,[8] and to localize cell wall hydroxyproline-rich glycoprotein (HRGP) and GRP mRNAs in developing soybean tissues.[2] Although the tissue print technique in its present state of development has a lower resolution than the traditional cytochemical localization for proteins and *in situ* hybridization localization for mRNAs, it provides a quick and convenient way to screen a large quantity of tissues from different developmental stages or from different plants at the same time on the same membrane; it allows one to study the spatial distribution of specific proteins or mRNAs in large organs such as tomato fruit or whole seedlings; it provides a unique way to study insolubilization of cell wall structural proteins because the detection of cell wall proteins depends on the transfer of soluble proteins onto the membrane; and it gives tissue-level localization of proteins and mRNAs. In the following sections, we describe the application of the tissue print technique to localize mRNAs in plant tissues by using ^{32}P-labeled DNA probes and ^{35}S-labeled antisense RNA probes.

Tissue Print

Both nitrocellulose membranes and nylon membranes can be used for tissue prints without any pretreatment. The principle for retaining transferred mRNAs on the membrane is the same as that for RNA gel blot or RNA dot blot. We have used nitrocellulose membrane (Schleicher & Schuell, Keene, NH), Nytran (Schleicher & Schuell), GeneScreen (New England Nuclear Research Products, Boston, MA), and Zeta-Probe blotting membrane (Bio-Rad, Richmond, CA) for tissue print and hybridization. All of these membranes show essentially the same resolution. Generally nylon membrane is preferred because of its easy handling.

One advantage for the tissue print technique is that the fresh tissue can be cut and the freshly cut surface can be pressed on the membrane immediately without any treatment. The cellular mRNAs are immobilized on the membrane and are then detected by specific probes. The specific cellular mRNAs show little lateral diffusion during or after printing on the membrane. The precise localization of the mRNAs can then be determined by comparison with the anatomy of corresponding stained sections or by comparison with the physical print on the membrane. Another advantage of the tissue print technique is that large organs or whole seedlings can be printed on the same membrane, and

[8] B. A. McClure and T. J. Guilfoyle, *Plant Mol. Biol.* **12**, 517 (1989).

the spatial distribution of specific mRNAs can then be analyzed. We successfully localized extensin mRNA in the seed coat regions of intact tomato fruit by tissue print technique, that is, we cut the tomato fruit in half, pressed the cut surface on the nylon membrane, then probed with ^{35}S-labeled extensin antisense RNA probe (Y.-R. Song and J. E. Varner, unpublished results, 1992). The distribution of SAURs (i.e., small auxin up-regulated RNAs) accumulated in gravistimulated soybean seedlings was demonstrated simply by cutting longitudinally the whole seedling, pressing the cut surface on the membrane, then probing with specific probes.[8] Normally many prints can be done on the same membrane. This tool may therefore be used to screen for organ and tissue-specific gene expression in different plants and different developmental stages at the same time on the membrane. Because all the prints are on the same membrane and they are treated with the same hybridization, washing, and exposure conditions, the specific mRNAs from different developmental stages can then be compared qualitatively by comparing the signal intensities of different prints on the same membrane.

The purpose of the tissue print localization of mRNAs is different from that of the RNA gel blot. For the tissue print, we not only want to know the quantitative changes of specific mRNAs, but also to study the spatial distribution of specific mRNAs on the prints. So it is important to achieve the highest resolution possible. The key determinant of resolution for *in situ* hybridization is the choice of radioisotope, for example, ^{3}H-labeled probes provide the best resolution of any isotope, ^{35}S-labeled probes next, and ^{32}P-labeled probes least. The same situation applies for tissue print hybridization. However, for tissue print hybridization, we cut the section first, then press the cut surface on the membrane. To obtain good prints we suggest the following: (1) use each razor blade edge once, because the surface of the blade is contaminated after cutting; (2) use forceps when transferring sections; (3) press sections on the membrane carefully and evenly (do not press so hard as to crush the sections); (4) if the tissue is juicy, or quantities of fluid come out from vascular bundles on the section, adsorb the surface on Kimwipes gently. Generally it is wise to print several sections, then the signals of these prints can be compared to see whether they show the same pattern. It is also convenient to see whether an even physical print is obtained or whether there is the same distribution of chlorophyll as the section (if there is chlorophyll) just by examining with a hand-held lens after printing. The printed membrane can be stored at 4° up to several months without loss of signal.

As in all procedures involving RNA, precautions to prevent RNase contamination should be followed. Wear gloves when making prints. Endogenous RNase in our experience is not a problem.

Reagents

Nitrocellulose membrane (Schleicher & Schuell)
Nylon membranes [e.g., Nytran (Schleicher & Schuell), GeneScreen (New England Nuclear Research Products), Zeta-Probe blotting membrane (Bio-Rad)]

Procedures

1. Put six layers of Whatman (Clifton, NJ) No. 1 paper on a plastic plate. On the Whatman paper, put one sheet of photocopy paper, then lay the nitrocellulose or nylon membrane on it.

2. Use a double-edge razor blade to cut a thick section free-hand (the thickness of sections can be varied because only the cellular contents on the cut surface can be transferred), then carefully transfer the freshly cut section to the membrane with forceps. Do not move the section after it is transferred onto the membrane.

3. Put four layers of Kimwipes on the section; press the section gently and evenly for 15–20 seconds with one finger.

4. Remove the Kimwipes and section carefully by forceps, keeping the section or cut thin section for anatomical comparison. The section can be stained with toluidine blue for observation of anatomy.

5. Repeat steps 2–4 for the next prints.

6. Bake at 80° for 2 hr. The printed membrane is then ready for hybridization.

Preparation of Radiolabeled Probes

Both ^{32}P-labeled DNA probes and ^{35}S-labeled RNA probes have been used for tissue print hybridization.[7,8] The choice of ^{32}P-labeled or ^{35}S-labeled probes depends on the tissue size used and the resolution desired. ^{32}P-Labeled probes give very high specific activity. Such probes can then be used conveniently in conjunction with X-ray film autoradiography to identify organs or regions of organs containing individual mRNAs. However, the resolution is usually insufficient for precise localization of specific labeled cells. As for *in situ* hybridization, ^{35}S-labeled probes are preferred for tissue print hybridization due to the high specific activity attainable, high autoradiographic efficiency, reasonable half-life, and safety. The resolution is usually adequate to see tissue-level localization of specific mRNAs, for example, in a 2-mm cross-section print of soybean stem it is easy to distinguish GRP mRNA hybridization signal in primary xylem from HRGP mRNA hybridization signal in the cambial region. DNA probes radiolabeled to high specific activity can be prepared by random oligonucleotide-primed synthesis.[9] RNA probes can be prepared by cloning the corresponding DNA fragment into a vector containing RNA polymerase promoters (e.g., pBluescript vector from Stratagene, La Jolla, CA) and synthesis of strand-specific RNA by corresponding RNA polymerase.[9] For making strand-specific RNA probes plasmid templates should be linearized with an appropriate restriction endonuclease. Generally, single-stranded RNA probes are preferred over DNA probes because the higher thermal stability of RNA–RNA duplexes allows more stringent wash conditions, thus resulting in lower nonspecific hybridization background. Partial hydrolysis of RNA probes seems unnecessary for tissue print hybridization.

Reagents

$[\alpha\text{-}^{32}\text{P}]$dCTP, 10 mCi/ml (NEG-013A; New England Nuclear)
$[\alpha\text{-}^{35}\text{S}]$UTP, 12.5 mCi/ml (NEG-039H; New England Nuclear)
Klenow fragment of *Escherichia coli* DNA polymerase I (600071; Stratagene)
Sephadex G-50–150
Dithiothreitol (DTT)
ATP, CTP, GTP, and UTP
RNase block II (300153; Stratagene)
RNA polymerase: T7 (600123; Stratagene); T3 (600111; Stratagene)
DNase I, RNase free (600031; Stratagene)
Stock buffer (4×):
    200 mM Tris-HCl (pH 8.0), 20 mM MgCl$_2$, 20 mM DTT, 80 μM dATP, dGTP, dTTP, 800 mM N-2-hydroxyethylpiperazine-N'-2-ethanesulfonic acid (HEPES), 21.6 units/ml hexamer
Nick translation stop solution: 20 mg/ml Blue Dextran 2000, 0.1% (w/v) sodium dodecyl sulfate (SDS), 50 mM ethylenediaminetetraacetic acid (EDTA)
Transcription buffer (10×): 400 mM Tris-HCl (pH 7.5), 60 mM MgCl$_2$, 20 mM spermidine hydrochloride, 50 mM NaCl

[9] J. Sambrook, E. F. Fritsch, and T. Maniatis, "Molecular Cloning: A Laboratory Manual," 2nd Ed. Cold Spring Harbor Press, Cold Spring Harbor, New York, 1989.

Procedures

Preparation of ^{32}P-Labeled DNA Probes

1. Denature template DNA by incubating in boiling water for 5 min, then quickly immersing in an ice bath.

2. Assemble reaction mixture in an ice bath as follows:

Denatured DNA	0.2 μg
Stock buffer (4×)	7.5 μl
Bovine serum albumin (BSA)	4 μg
Klenow fragment of *E. coli* DNA polymerase I	3 units
[α-^{32}P]dCTP (10 mCi/ml)	3 μl

Bring the final volume to 30 μl by addition of water.

3. Incubate at room temperature for 2–4 hr.

4. Add nick translation stop solution to the reaction mixture. Labeled probes can be purified by Sephadex G-50–150 column chromatography.

Preparation of ^{35}S-Labeled RNA Probes

1. Assemble transcription reaction at room temperature as follows:

Linearized DNA template	0.5 μg
DTT (100 mM)	1 μl
Transcription buffer (10×)	2.5 μl
ATP (10 mM)	1 μl
CTP (10 mM)	1 μl
GTP (10 mM)	1 μl
UTP (1 mM)	1 μl
RNase block II	10 units
[α-^{35}S]rUTP	50 μCi
Bacteriophage DNA-dependent RNA polymerase	10 units

Add H$_2$O to a final volume of 25 μl.

2. Incubate at 37° for 1–2 hr.

3. Add 1 μl of RNase-free DNase I (1 mg/ml). Incubate the reaction for 15 min at 37°.

4. Add 50 μl H$_2$O into reaction, and purify the RNA by Sephadex G-50–150 column chromatography or ethanol precipitation.

Hybridization

Reagents

Hybridization solution (for DNA probes): 40% (v/v) formamide, 5× SSC, 5× Denhardt's solution, 0.1% (w/v) SDS, 0.1 mg/ml salmon sperm DNA

Hybridization solution (for RNA probes): 2× SSC, 1% (w/v) SDS, 5×
    Denhardt's solution, 0.1 mg/ml salmon sperm DNA, 10 mM DTT
Washing solution I: 0.2× SSC, 1% (w/v) SDS
Washing solution II: 2× SSC, 0.1% (w/v) SDS
Washing solution III: 0.2× SSC, 0.1% (w/v) SDS

Procedure

1. Wash the membrane in washing solution I at 65° for 4 hr.
2. Prehybridize in hybridization solution at 68° (for RNA probes) or at
42° (for DNA probes) for 4 hr.
3. Hybridize by adding radiolabeled probes at 0.5–1 × 10^7 cpm/ml at
68° (for RNA probes) or at 42° (for DNA probes) for 20 hr.
4. Wash the membrane in washing solution II three times at 42° for 20
min each.
5. Wash the membrane in washing solution III at 65° for a period of
time by monitoring the amount of radioactivity on the membrane, using a
hand-held minimonitor. After washing, the membrane hybridized with
sense-RNA probe should not emit signal, whereas the membrane hybridized with antisense-RNA probe generally emits detectable signal with
intensity depending on the abundance of the interested mRNA.
6. Briefly wash the membrane with 0.2× SSC at room temperature.
Air dry the membrane.

Autoradiography and Data Analysis

Either X-ray film or Tmax film (Eastman Kodak, Rochester, NY) can
be used for autoradiography.[7,8] The choice of the film depends on the
resolution desired and the tissue sizes to be used. X-ray film normally
gives lower resolution, but the film can be automatically developed and it
is sufficient to identify organs, or regions of organs, containing individual
mRNAs. Tmax film is generally preferred for precise localization of specific labeled cells. Tmax film is a single-side emulsion film. The membrane
must be exposed to the emulsion side of the Tmax film. There is an easy
way to recognize the emulsion side in darkness, that is, there is a notch
on one corner of the Tmax film. When film is held so that the notch is on
the lower right corner, then the emulsion side is underneath. Develop the
Tmax film in total darkness. Tmax film is developed in Tmax developer.
Exposure time depends on the intensity of signal. According to our
experience, it will take as short a time as 4 hr and as long as 4 days.
mRNA localization on the film can be observed with a hand-held lens or a
dissection microscope. The disadvantage with the tissue prints is that we
cannot see anatomical details on the film. It is therefore important to

compare the image on the film with the anatomy of the corresponding stained section to identify with certainty the tissue-level localization of the signal.

Reagents

X-Ray film (XAR5 1651454; Eastman Kodak)
Kodak Tmax 400 film (4 × 5 in.)
Kodak Tmax developer (1402767)
Kodak stop bath
Kodak rapid fixer
Kodak Technical Pan 2415 black-and-white film
Dissection microscope

Procedures

1. Expose the membrane to X-ray film or Tmax 400 film. X-Ray film is developed in an X-ray film processor. Tmax 400 film is developed as follows:

10–15 min in Kodak Tmax developer
30 sec in Kodak stop bath
5 min in Kodak rapid fixer
10 min in H_2O, then air dry

2. Observe autoradiographic results on the film under the dissection microscope. Photographically record the results on Kodak Technical Pan 2415 black-and-white film.

Examples of Tissue Print Localization of mRNAs in Plant Tissue

We have developed the tissue print method to study gene expression of the α subunit of β-conglycinin during soybean seed development by using ^{32}P-labeled β-conglycinin DNA probes.[7] As seen from Fig. 1, there is a marked change in the abundance of the α subunit mRNA of β-conglycinin in the soybean seeds at different developmental stages. We have also used the tissue print method to study gene expression of HRGPs and GRPs in developing soybean tissue by using ^{35}S-labeled antisense HRGP and GRP RNA probes.[2] The results shown in Fig. 2 demonstrate that the resolution is sufficient to tell that HRGP mRNAs are localized in the cambial region and in a few layers of cortex cells surrounding primary phloem, whereas GRP mRNAs are found mainly in primary xylem region. Control hybridization of RNA probe synthesized from pBluescript vector does not show any signal.[2]

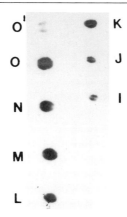

FIG. 1. Localization of the α subunit mRNA of β-conglycinin at different developmental stages of soybean seeds by tissue print hybridization on nitrocellulose membrane. Hybridization was performed with the α-subunit gene of β-conglycinin, which is a 12-kb EcoRI fragment of soybean genomic DNA (Gmg 17.1; obtained from Dr. R. Beachy). Probe was labeled with [α-^{32}P]dCTP (1–2 × 10^7 cpm/ml). Soybean seeds (*Glycine max* cv. Provar) in the different developmental stages were used. 0′, Mature yellow seed; 0, mature green, more than 28 days after anthesis; N, 24–28 days after anthesis; M, 21–23 days after anthesis; L, 20–22 days after anthesis; K, 19–21 days after anthesis; J, 18–20 days after anthesis; I, 17–19 days after anthesis. Hybridization was performed at 42° for 20 hr and washing done in 2× SSC and 0.1% SDS at 42° for 1 hr and in 0.5× SSC and 0.1% SDS at 65° for 1 hr. The filter was exposed to X-ray film for 6 hr. (From Varner *et al.*[7])

Controls

Several kinds of controls can be used to check for specific hybridization on tissue print; the following are examples.

1. Sense RNA probes: Sense RNA probes can normally be used as control of background for most mRNA localization, but this is not suitable for localization of some plant cell wall structural protein mRNAs. We found that sense HRGP RNA probes hybridized with GRP mRNAs, and sense GRP RNA probes hybridized with HRGP mRNAs, even under high-stringency washing conditions (30 min at 70° in 0.2× SSC, 0.1% SDS after normal wash). The reason for this cross-hybridization is that the codon for proline is CCX, and the codon for glycine is GGX; HRGPs and GRPs contain more than 40 and 60% of proline and glycine, respectively. With strand-specific probes, each can hybridize with complementary HRGP or GRP mRNA.[2]

2. Heterologous probes: Examples are probes synthesized from vectors that can be used for control of background.

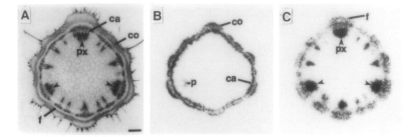

FIG. 2. Localization of HRGP mRNAs and GRP mRNAs in developing soybean stem by tissue print hybridization on nylon membrane. (A) Soybean stem free-hand section stained by toluidine blue; (B) localization of HRGP mRNAs in the cambial region and in a few layers of cortex cells around primary phloem; (C) localization of GRP mRNAs in primary xylem, in primary phloem, and in cambial region. Carrot HRGP genomic DNA (pDC5A1)[10] and bean GRP1.8 genomic DNA (a kind gift from Drs. B. Keller and C. J. Lamb)[6] cloned into pBluescript vector (Stratagene) were used to synthesize ³⁵S-labeled antisense RNA probes by using T7 or T3 RNA polymerase. Hybridization of probes to the membrane was carried out at 68° for 16 hr under the conditions described in the text. The membrane was washed in 2× SSC, 0.1% SDS at 42° three times for 20 min each, and in 0.2× SSC, 0.1% SDS at 65° twice for 30 min each. The membrane was exposed to Kodak Tmax 400 film for 12 hr at room temperature. The photographs were taken under a Zeiss stereomicroscope using Kodak Technical Pan 2415. Bar: 300 μm. ca, Cambium; co, cortex; f, primary phloem; p, parenchyma; px, primary xylem. (From Ye and Varner.[2])

3. A probe synthesized from a housekeeping gene or from another gene in which the spatial distribution of mRNA is known can be used for positive control.

4. Treatment of the membrane or probes with RNase before hybridization can be a control for background.

Sensitivity

When the cut section is printed on the membrane, the contents of the cut cells are transferred onto the membrane. The transferred mRNAs on the membranes are easily accessible to the probes. This is the advantage over *in situ* hybridization, in which mRNA accessibility could be dramatically different depending on the different fixation conditions, and extent of proteinase digestion. In our experiments, the HRGP mRNAs in the cambial region of soybean stem can be detected by tissue print hybridization after a 12-hr exposure on Tmax 400 film, whereas with a probe of the same specific activity, concentration, and fragment length, the HRGP

[10] J. Chen and J. E. Varner, *EMBO J.* **4,** 2145 (1985).

mRNAs from this tissue are detected by *in situ* hybridization only after a 2-week exposure on Kodak NTB-2 emulsion.

Resolution

In situ hybridization shows both the signal and anatomical information. Therefore it is easy to determine the specific mRNA localization at the cellular level. Tissue print hybridization gives signal information on the film without precise anatomical information. However, by comparing the spatial distribution of the signal with the anatomy of the corresponding physical print and the corresponding stained section, the localization of the signal can still be conveniently interpreted at the tissue level; for example, as seen in Fig. 2, localization of GRP mRNAs in primary xylem region can be easily determined. For the spatial localization of mRNAs that cannot be interpreted with certainty, we recommend the use of *in situ* hybridization for comparison. If the antibody against the gene product of interest is available, immunohistochemical data on localization of the protein can also be useful for comparison. The resolution also depends on the size of material used. In our experience, the resolution is sufficient to distinguish the signal in xylem from that in phloem when using tissues in which the diameter is larger than 1.5 mm. However, with those tissues in which the diameter is less than 1 mm, the spatial distribution of the signal generally could not be determined accurately.

With continuing use of tissue print hybridization, new methodologies will surely be developed to improve the resolution. It might be possible to use nonradioactively labeled probes to detect the mRNA of interest directly on the membrane. For example, one could use digoxigenin-labeled probes, visualized by alkaline phosphatase-conjugated antibody against digoxigenin.[11] This would produce a much sharper signal and the signal and physical print could be visualized on the membrane at the same time.

Acknowledgments

We thank Beat Keller and Christopher J. Lamb for providing the GRP gene and Roger N. Beachy for providing the α-subunit gene of β-conglycinin. This work was supported by grants from the U.S. Department of Energy (DE-FG02-84ER-13255) and the National Science Foundation (DMB 86-08166) to J. E. Varner. Y.-R. Song and Z.-H. Ye were supported by a Grant-in-Aid from the Monsanto Company.

[11] D. Tautz and C. Pfeifle, *Chromosome* **98,** 81 (1989).

[48] Localization of Cell Wall Proteins Using Tissue-Print Western Blot Techniques

By Gladys I. Cassab

Introduction

The presence of walls is one of the outstanding characteristics distinguishing the cells of plants from those of animals. The cell wall is a complex entity with unique characteristics related to the developmental stage of a given plant cell type. The cell wall is not, like the mitochondrion, an organelle with relatively constant duties, but rather a cell component subject to the continuous developmental processes that govern cell size, cell division, cell shape, and function.[1]

The cell walls are composed of cellulose, hemicellulose, pectic compounds, lignin, suberin, proteins, and water. Cell walls may contain structural proteins as well as enzymes. The best characterized and perhaps the most abundant structural protein of dicot cell walls are the extensins, a member of a class of hydroxyproline-rich glycoproteins present in a wide variety of plants and algae.[1,2] Extensins are basic proteins with a high content of hydroxyproline (36–42%) that typically are glycosylated with one to four arabinosyl residues. They contain repeating amino acid sequences and assume polyproline II helical structure. Until recently, it was thought that extensins were the major protein component of the primary wall of all plant cells and that they play a role in cell wall architecture.[3] However, when we studied the possible function of extensin in plant cells by using a cell biology approach, we demonstrated that extensin is most abundantly localized in cells that belong to the sclerenchyma tissue.[4] The sclerenchyma cells act as the skeletal elements of the plant body. These cells enable the plant body to withstand various strains, such as stretching, bending, compression, and tension.[5,6] Thus, the presence of extensin in the sclerenchyma cell walls together with other wall components, may determine the unique characteristics of these cells. The developing of the

---

[1] G. I. Cassab and J. E. Varner, *Annu. Rev. Plant Physiol. Plant Mol. Biol.* **39,** 321 (1988).
[2] J. E. Varner and L.-S. Lin, *Cell* **56,** 231 (1989).
[3] D. T. A. Lamport, *in* "The Biochemistry of Plants" (P. K. Stumpf and E. E. Conn, eds.), Vol. 3, p. 639. Academic Press, New York, 1980.
[4] G. I. Cassab and J. E. Varner, *J. Cell Biol.* **105,** 2581 (1987).
[5] K. Esau, "Plant Anatomy." Wiley, New York, 1965.
[6] G. Haberlandt, "Physiological Plant Anatomy." Today & Tomorrow's Book Agency, New Delhi, 1914.

tissue printing technique on nitrocellulose paper by Cassab and Varner[4] for the study of extensin localization in plant cells using anti-soybean extensin antibodies has been useful toward understanding the role of extensin in plants.[7,8] Extensin is a difficult protein to isolate, because a higher proportion of it becomes insolubilized in the wall compartment. The *in muro* interactions of extensin with other cell wall components is not known, and the mechanism of insolubilization has not yet been elucidated.[1] Thus, the use of the tissue printing technique on nitrocellulose paper offers a new approach to analyze the distribution of extensin and other cell wall proteins in different plant cells. This technique is a simple immunolocalization procedure that should be of general use for studying cell wall proteins.

To date, little is known of the cell wall proteins of monocots. However, an extensin-like molecule has been characterized in the walls of maize cells grown in suspension culture,[9] and in the developing maize pericarp.[10] Extensin is not the only structural protein of the cell wall. Cell wall proteins unrelated to extensin have been shown to accumulate in response to a variety of developmental and stress signals.[11–16] Some plant cells are rich in glycine.[17] A petunia gene encoding a protein that is 67% glycine has been characterized.[18] Two similar genes occur in *Phaseolus vulgaris;* one of the encoded proteins is associated with the walls of cells in the vascular tissues.[19,20] In fact, Keller *et al.,*[19] using the tissue-print Western blot technique with specific antibodies against glycine-rich protein (GRP 1.8), showed that this protein was distributed in a regular pattern in the vascular system of the bean young hypocotyl and developing pod. Recently, GRP 1.8 has also been localized in the unlignified phloem tissue of hypocotyls

[7] D. J. Meyer, C. L. Afonso, and D. W. Galbraith, *J. Cell Biol.* **107**, 163 (1988).

[8] Z.-H. Ye and J. E. Varner, *Plant Cell* **3**, 23 (1991).

[9] M. Kieliezweski and D. T. A. Lamport, *Plant Physiol.* **85**, 823 (1987).

[10] E. E. Hood, Q. X. Shen, and J. E. Varner, *Plant Physiol.* **87**, 138 (1988).

[11] C. Bozart and J. S. Boyer, *Plant Physiol.* **85**, 261 (1987).

[12] H. J. Franssen, J. P. Nap, T. Gloudemans, W. Stiekema, H. Van Dam, F. Govers, J. Louwerse, A. van Kammen, and T. Bisseling, *Proc. Natl. Acad. Sci. U.S.A.* **84**, 4495 (1987).

[13] M. L. Tierny, J. Weichert, and D. Pluymers, *Mol. Gen. Genet.* **88**, 61 (1988).

[14] K. Datta, A. Schmidt, and A. Marcus, *Plant Cell* **1**, 945 (1989).

[15] J. C. Hong, R. T. Nagao, and J. L. Key, *Plant Cell* **1**, 937 (1989).

[16] B. Keller and C. J. Lamb, *Genes Dev.* **3**, 1639 (1989).

[17] J. E. Varner and G. I. Cassab, *Nature (London)* **323**, 110 (1986).

[18] C. Condit and R. B. Meagher, *Mol. Cell. Biol.* **7**, 4273 (1987).

[19] B. Keller, N. Sauer, and C. J. Lamb, *EMBO J.* **7**, 3625 (1988).

[20] B. Keller, M. D. Templeton, and C. J. Lamb, *Proc. Natl. Acad. Sci. U.S.A.* **86**, 1529 (1989).

by using this technique.[21] Since the tissue-print Western blot technique was developed,[4] it has been successfully utilized to localize cellular and extracellular proteins.[22-24]

The arabinogalactan proteins (AGP), members of a class of hydroxy-proline-rich glycoproteins, are widely distributed in plants. They are primarily localized in the extracellular matrix and in gums and exudates of plants.[25] The AGP portion associated with the cell walls is freely soluble in water and is thought to be involved in cell–cell recognition rather than having a structural role. The styles of flowers frequently contain arabinogalactan, possibly associated with protein,[25] and the medulla of soybean root nodules is rich in arabinogalactan protein.[26] The characterization of monoclonal antibodies to AGPs[27,28] has been useful for studying the possible role of AGPs in plants. The expression of the AGP epitopes in plant cells shows strict developmental regulation.[28-31] The precise function of these glycoproteins is unknown but the observed patterns of expression, their location at the plasma membrane, and the known ability of AGPs to react with Yariv antigen[25] may indicate a role involving molecular recognition and cell–cell interaction in relation to cell identity or position.[31] The tissue-printing technique has also been utilized in screening for cell surface carbohydrate mutants in *Arabidopsis thaliana*.[32] The method depends on the use of monoclonal antibodies that recognize a variety of carbohydrate epitopes at the plant cell surface.

In addition, several enzymes are associated with cell walls. These include peroxidases, phosphatases, β-1,3-glucanases, β-1,4-glucanases, polygalacturonase, pectin methylesterases, malate dehydrogenases, β-glucuronidases, β-xylosidases, proteases, and ascorbic acid oxidase.[1,2] The localization of peroxidase activity by tissue printing in the cell walls of the vascular bundles of pea epicotyls has been reported.[33]

[21] U. Ryser and B. Keller, *Plant Cell.* **4,** 773 (1992).

[22] P. D. Reid, E. del Campillo, and L. N. Lewis, *Plant Physiol.* **93,** 160 (1990).

[23] E. del Campillo, P. D. Reid, R. Sexton, and L. N. Lewis, *Plant Cell.* **2,** 245 (1990).

[24] Z.-H. Ye and J. E. Varner, *Plant Cell.* **3,** 23 (1991).

[25] G. B. Fincher, B. A. Stone, and A. E. Clarke, *Ann. Rev. Plant Physiol.* **34,** 47 (1983).

[26] G. I. Cassab, *Planta* **168,** 441 (1986).

[27] R. I. Pennel, J. P. Knox, G. N. Scofield, R. R. Selvedran, and K. Roberts, *J. Cell Biol.* **108,** 196 (1989).

[28] J. P. Knox, S. Day, and K. Roberts, *Development* **106,** 47 (1989).

[29] R. I. Pennel and K. Roberts, *Nature (London)* **344,** 547 (1990).

[30] N. J. Stacey, K. Roberts, and J. P. Knox, *Planta* **180,** 285 (1990).

[31] J. P. Knox, *J. Cell Sci.* **96,** 557 (1990).

[32] N. Stacey, J. P. Knox, and K. Roberts, *in* "Tissue Printing" (P. D. Reid *et al.,* eds.), p. 35. Academic Press, San Diego, 1992.

[33] G. I. Cassab, J.-J. Lin, L.-S. Lin, and J. E. Varner, *Plant Physiol.* **88,** 522 (1988).

Principle of Method

Nitrocellulose membrane adsorbs relatively large quantities of protein that are tightly bound,[34] whereas salts, many small molecules, and RNA are usually not retained. Nitrocellulose paper has been successfully used for transferring proteins subjected to polyacrylamide gel electrophoresis in Western blot analysis.[35] The principle of the tissue-print Western blot is based on the observation that blotting tissue sections onto nitrocellulose paper leaves a stable, faithful image of the cut surface of the original tissue. Because extensins are usually extracted with a high salt concentration solution,[36] it was assumed that cell wall proteins would transfer to the nitrocellulose paper if the sheets were previously soaked with 0.2 M $CaCl_2$; this transfer does in fact occur. Using the following procedure,[4] the immunolocalization of cell wall proteins can be performed simply and the results correlate with conventional immunolocalization methods for light microscopy (Fig. 1).

Materials

Nitrocellulose (type BA85; Schleicher & Schuell, Keene, NH)
Anti-rabbit IgG (Fc), alkaline phosphatase conjugated, NBT (nitroblue tetrazolium), and BCIP (5-bromo-4-chloroindoxyl phosphate) (Pro mega Biotec, Madison, WI)
$CaCl_2 \cdot 2H_2O$ (0.2 M)
Tris-buffered saline (TBS): 0.9% (w/v) NaCl in 20 mM Tris-HCl (pH 7.4) plus 0.3% (v/v) Tween 20 and 0.05% (w/v) NaN_3
Antibody incubation solution: 0.25% (w/v) Bovine serum albumin (BSA), 0.25% (w/v) gelatin, 0.3% (v/v) Tween 20 in TBS
Alkaline phosphatase (AP) buffer: 50 mM Tris-HCl (pH 9.8) plus 1 mM $MgCl_2$
India ink (Pelikan, Hannover, Germany)

Procedure

1. The nitrocellulose paper is soaked in 0.2 M $CaCl_2$ for 30 min, and dried on Whatman (Clifton, NJ) 3MM paper. Once dried, place nitrocellulose and 3MM Whatman paper on top of a plastic plate.

2. Fresh tissue is cut in sections of about 300 μm to 3 mm thickness with a new razor blade, washed in distilled water for 3 sec, and dried on

[34] H. Kuno and H. K. Kihara, *Nature (London)* **215**, 974 (1967).

[35] H. Towbin, T. Staehlin, and J. Gordon, *Proc. Natl. Acad. Sci. U.S.A.* **76**, 4350 (1979).

[36] G. I. Cassab, J. Nieto-Sotelo, J. B. Cooper, G. J. Van Holst, and J. E. Varner, *Plant Physiol.* **77**, 532 (1985).

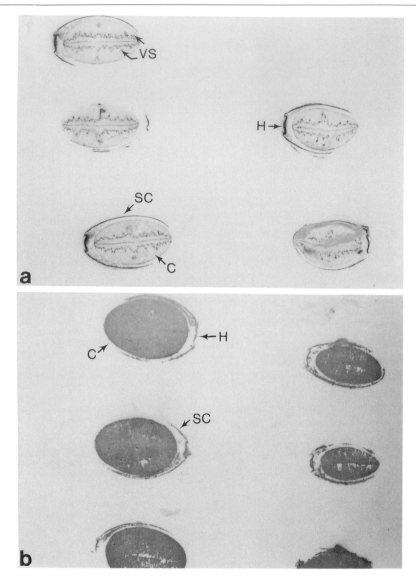

FIG. 1. Developing soybean seed prints on nitrocellulose membrane. Cross-sections of seeds were 3 mm thick. (a) The prints were reacted with polyclonal antibodies against soybean seed coat extensin diluted 1 : 15,000, and detected with alkaline phosphatase-conjugated anti-rabbit immunoglobulin antibodies. (b) The prints were stained with India ink. C, Cotyledon; H, hilum; SC, seed coat; VS, vascular supply of the seed. [Reproduced from *J. Cell Biol.* **105,** 2581 (1988), by copyright permission of the Rockefeller University Press.]

Kimwipes. Then the section is taken with Millipore (Bedford, MA) forceps and placed carefully on the nitrocellulose paper. The tissue is pressed onto the nitrocellulose paper for 15–30 sec, using a gloved fingertip. Finally, the tissue section is carefully removed with the aid of forceps, and the tissue print is immediately dried with warm air.

3. The nitrocellulose paper is then treated for the detection of alkaline phosphatase-conjugated second antibody as described for protein Western blots by Blake et al.[37] First, the nitrocellulose paper is blocked with the antibody incubation solution for 1–3 hr at room temperature with constant shaking.

4. The primary antibody is added to the desired dilution in the antibody incubation solution, and the nitrocellulose paper is then incubated for 1–3 hr as above.

5. After the antibody incubation, the nitrocellulose paper is washed three times (30 min each) in TBS with agitation.

6. Then the nitrocellulose paper is soaked with the alkaline phosphatase-conjugated anti-IgG antibody diluted 1 : 20,000 in the antibody incubation solution for 1–3 hr with agitation.

7. The nitrocellulose paper is then washed three times (30 min each) in TBS with agitation.

8. Before adding the substrate for visualizing AP, wash briefly the nitrocellulose paper with AP buffer, and then add the AP substrate solution containing 66 μl NBT plus 33 μl BCIP in 10 ml of AP buffer.

9. Develop the color until signal appears in the tissue print, and stop the reaction by washing the nitrocellulose paper quickly in distilled water.

10. Dry the nitrocellulose paper on Whatman 3MM paper.

11. Photographs of tissue prints are taken with Kodak (Rochester, NY) Technical Pan 2415 set at Asa 50.

For determining the total protein distribution pattern of the tissue section, follow the procedure for tissue printing as described above. Then the nitrocellulose paper is incubated 15 min with 1 μl/ml India ink in TBS with constant shaking as described by Hancock and Tsang.[38] Afterward, the nitrocellulose paper is washed three times (5 min each) in TBS, and the filter is air-dried on Whatman 3MM paper.

Notes and Precautions

1. All precautions noted on making the tissue print should be observed because double images are readily obtainable if the tissue section is not

[37] M. S. Blake, K. H. Johnston, G. J. Russell-Jones, and E. C. Gotschlich, Anal. Biochem. 136, 175 (1984).
[38] K. Hancock and V. C. W. Tsang, Anal. Biochem. 133, 157 (1983).

carefully blotted or removed from the nitrocellulose filter. Special attention must be taken when drying tissue sections that have high water content, such as developing shoots. This consideration will eliminate the transfer of soluble material from the tissue section outside the print made on the nitrocellulose membrane.

2. The detection of alkaline phosphatase-conjugated second antibody on tissue prints was selected over the peroxidase-conjugated second antibody procedure because the substrates used for detecting the peroxidase, such as o-phenylenediamine and H_2O_2, are capable of detecting endogenous peroxidase activity in plant tissue sections.[33] Thus the endogenous peroxidase activity in the tissue plant will mask the immunoblotting reaction.

Conclusions

As new structural cell wall proteins are discovered, it is likely that preparation of specific antibodies for these proteins will continue to provide useful information on how different cell walls are constructed. The use of the tissue-print Western blot technique with a new set of cell wall antibodies will be useful for screening many plant tissues and plant species because it is a simple immunolocalization procedure.

Acknowledgment

I am grateful to Dr. Joseph E. Varner for support and for reviewing the manuscript.

[49] Tissue-Print Hybridization for Detecting RNA Directly

By Tom J. Guilfoyle, Bruce A. McClure, Melissa A. Gee, and Gretchen Hagen

Varner et al.[1] first described tissue-print hybridization with plant organs by employing a modification of an immunological tissue-printing procedure developed by Cassab and Varner.[2] McClure and Guilfoyle[3,4] modified the hybridization procedure for detecting moderately abundant

[1] J. E. Varner, Y.-R. Song, L.-S. Lin, and H. Yuen, in "The Molecular Basis of Plant Development" (R. Goldberg, ed.), p. 161. Alan R. Liss, New York, 1989.
[2] G. I. Cassab and J. E. Varner, J. Cell Biol. 105, 2581 (1989).
[3] B. A. McClure and T. J. Guilfoyle, Science 243, 91 (1989).
[4] B. A. McClure and T. J. Guilfoyle, Plant Mol. Biol. 12, 517 (1989).

mRNA transcripts with ^{35}S-labeled antisense RNA probes. Subsequently, other modifications of these procedures have been reported by Mansky *et al.*[5]

Varner *et al.*[1] originally used tissue prints and ^{32}P-labeled cDNA probes to detect extensin mRNA in soybean pods and β-conglycinin α-subunit mRNA in developing soybean seeds. Tissue-print hybridization with ^{35}S-labeled antisense RNA probes was used to detect the organ and tissue distribution of auxin-responsive mRNAs in whole seedlings and organ sections[3,4,6] and in hypocotyl and epicotyl sections undergoing gravitropic curvature.[3] Ye and Varner[7] have localized the expression of mRNAs that encode hydroxyproline-rich glycoproteins (HRGPs) and glycine-rich proteins (GRPs) in developing soybean tissues using tissue print and *in situ* hybridization with ^{35}S-labeled antisense RNA probes.

Principle

Tissue printing is a simple method for detecting macromolecules blotted directly from the surfaces of severed organs onto nylon or nitrocellulose membranes. The blotting procedure produces an image of the cut surface of the tissues on the membrane, and macromolecules such as proteins, complex carbohydrates, and nucleic acids are fixed to the membrane. The retention of nucleic acids on the membrane allows the detection of RNAs by hybridization with either DNA or antisense RNA probes.

Materials

Reagents

Nylon membranes (Zeta-Probe; Bio-Rad, Richmond, CA)
Whatman 3MM paper (Fisher Scientific, St. Louis, MO)
Kimwipes (Kimberly-Clark Corp., Roswell, GA)
Ultraviolet (UV) light sources (260 and 300–320 nm)
Kapak/Scotchpak pouches (Kapak Corporation, Bloomington, MN)
Phenol : chloroform : isoamyl alcohol (25 . 24 : 1)
T7 RNA polymerase and vectors (GIBCO-Bethesda Research Laboratories, Gaithersburg, MD) or any one of a number of kits available commercially for synthesizing antisense or sense RNA probes (Promega, Madison, WI; Stratagene, La Jolla, CA)

[5] L. M. Mansky, R. E. Andrews, Jr., D. P. Durand, and J. H. Hill, *Plant Mol. Biol. Rep.* **8,** 13 (1990).
[6] A. R. Franco, M. A. Gee, and T. J. Guilfoyle, *J. Biol. Chem.* **265,** 15845 (1990).
[7] Z.-H. Ye and J. E. Varner, *Plant Cell.* **3,** 23 (1991).

Placental ribonuclease inhibitor and RNase-free DNase (Promega)
ATP, CTP, GTP, dithiothreitol (DTT), sodium dodecyl sulfate (SDS), salmon sperm DNA, poly(A), yeast tRNA, polyvinylpyrrolidone, bovine serum albumin, Ficoll, and formamide (Sigma Chemical Company, St. Louis, MO): The formamide is deionized prior to use by stirring for 30 min in AG501-X8 ion-exchange resin (Bio-Rad) as described by Sambrook et al.[8]
[^{35}S]Thio-UTP (>1200 Ci/mmol) (NEG039; New England Nuclear, Boston, MA)
Sephadex G-50 (Pharmacia, LKB Biotechnology, Inc., Piscataway, NJ)
India ink (Higgins No. 4415; Faber-Castell Corporation, Newark, NJ)
Kodak XRP-5 X-ray film, Tmax 400, Tech Pan 2415, and Tech Pan 4415 photographic films, Kodak Tmax and HC-110 developer, stop bath, and rapid fix (Eastman Kodak, Rochester, NY)

Solutions

SSC (10×): 1.5 M NaCl, 0.15 M sodium citrate
SSPE (10×): 1.8 M NaCl, 0.1 M sodium phosphate (pH 7.4), 0.01 M ethylenediaminetetraacetic acid (EDTA)
Denhardt's solution (50×): 5 g Ficoll, 5 g polyvinylpyrrolidone, and 5 g bovine serum albumin (BSA) brought to 500 ml with H$_2$O

Methods

Membranes and Printing Technique

For tissue printing, a dry nylon membrane is placed over a single layer of dry Whatman 3MM paper or some other absorbent paper. Vinyl medical gloves should be worn when handling the nylon membranes and when blotting the tissue sections to the membranes to prevent the transfer of fingerprints to the membrane, which can prevent proper wetting of the membrane.

Organs or organ sections are prepared for printing onto membranes by sectioning through the organ with a single- or double-edged razor blade.[9] The freshly cut surfaces are pressed immediately to the nylon membrane or lightly blotted with Kimwipes prior to blotting to the membrane (depending on the moisture content of the tissue). Tissue printing is performed by using firm pressure with the index finger above the sectioned organ for

---

[8] J. Sambrook, E. F. Fritsch, and T. Maniatis, "Molecular Cloning: A Laboratory Manual." Cold Spring Harbor Press, Cold Spring Harbor, New York, 1989.
[9] B. A. McClure and T. J. Guilfoyle, *Plant Mol. Biol.* **9**, 611 (1987).

30 to 120 sec. In some cases, we have used a thin piece of flexible cardboard to cover the tissue sections and, by applying a relatively uniform pressure above the cardboard, multiple or larger tissue sections can be printed at the same time. The amount of pressure applied should not be excessive, but should be sufficient to imprint an image of the section on the membrane.

After printing, the organ sections can be removed from the membrane with the aid of a forceps or spatula. The nylon membrane is then allowed to dry at room temperature.

We evaluate the quality of the tissue prints by examining the printed nylon membrane under a 300- to 320-nm UV light source. Under UV light, it is possible to observe whether any organ sections were crushed or distorted during blotting. Uneven blots may result from too much or too little finger pressure. Distortions of prints may result from uneven pressure over the section or movement of the section on the membrane during printing. Each type of organ or tissue has a characteristic consistency, turgidity, and cellular architecture and, therefore, the amount of pressure required to obtain even, consistent prints must be experimentally determined. In our experience, large organs of firm consistency such as cotyledons, stems, and petioles are much easier to tissue print than small or less firm organs such as roots, leaves, or floral parts.

We have kept dried prints at room temperature for several weeks, but it is best to use the prints immediately or store them at 4° in sealed Kapak/Scotchpak pouches until the prints are used for hybridization.

Preparation of Antisense and Sense RNA Probes

A number of suitable vectors with T7, T3, or SP6 promoters that flank multiple cloning sites for inserting full-length or partial-length cDNAs are commercially available (GIBCO-Bethesda Research Laboratories, Promega, Stratagene). To generate RNA probes, full-length or partial-length cDNAs are cloned into transcription vectors, and RNA is synthesized *in vitro* with the appropriate RNA polymerase and transcription buffer.

Prior to carrying out the RNA polymerase reaction, the DNA template is linearized by restriction enzyme cleavage. The linearization of the DNA template should be checked by agarose gel electrophoresis.[8] After confirmation of linearity, the DNA is extracted with 1 vol of phenol : chloroform : isoamyl alcohol (25 : 24 : 1), and precipitated from the aqueous phase with 2 vol of 95% (v/v) ethanol at −80° for 1 hr. The ethanol-precipitated DNA is recovered by centrifugation in a microfuge at top speed for 15 min. After removal of the supernatant, the DNA is dried *in vacuo*. The dried DNA pellet is suspended in sterile, deionized water. The concentration of the DNA can be determined by absorbance at 260 nm with a UV spectrophotometer.

The antisense or sense RNA is synthesized according to the instructions provided by the vendor that supplies the DNA template vector and RNA polymerase. We have used the following protocol for synthesis of RNA probes. Reactions are carried out in a 10-μl mixture containing T7 RNA polymerase buffer [40 mM Tris-HCl (pH 7.9), 6 mM MgCl$_2$, 10 mM dithiothreitol, 2 mM spermidine], 0.3 to 0.7 μg of linearized DNA template, 10 units (U) or 1 μl of T7 RNA polymerase, 30–40 U of placental ribonuclease inhibitor, ATP, CTP, and GTP (1 mM each), and 0.1 to 0.5 mCi of [^{35}S]thio-UTP. We incubate reaction mixtures at 37° for 40–60 min, and then add a second aliquot of T7 RNA polymerase (10 U) and incubation is continued for an additional 40–60 min. After the incubation period, the DNA template is removed by adding RNase-free DNase I (1 U) to the reaction mixture and incubating for an additional 15 min at 37°. Unincorporated nucleotides are removed from the RNA transcript by passing the reaction mixture through a 0.5-ml spun column of Sephadex G-50.[8]

Although we routinely use [^{35}S]thio-UTP, [α-^{32}P]UTP or any other ^{32}P-labeled ribonucleoside triphosphate can be used to synthesize the antisense or sense RNA probe. The ^{32}P-labeled probes allow shorter exposure times and are less expensive to synthesize, but provide less autoradiographic resolution compared to the ^{35}S-labeled probes. Although we generally use probes of high specific activity, the amount and specific activity of ^{32}P- or ^{35}S-labeled ribonucleotide used to synthesize the RNA probe can be altered depending on how rapidly one wants to detect a hybridization signal and how abundant the mRNA is within the tissue section.

Hybridization of RNA to Tissue Print

Prior to hybridization, dried tissue prints should be washed for 4–12 hr in 0.1–0.2× SSC containing 1% (w/v) SDS at 65°. After this washing step, prehybridization and hybridization are carried out in 1.5× SSPE, 1% (w/v) SDS, 1% (w/v) nonfat powdered milk, 0.5 mg/ml denatured, sonicated salmon DNA, and 100 mM dithiothreitol at 68°. We generally carry out prehybridization of the tissue print membranes for 12–16 hr, and follow this by hybridization with the antisense or sense RNA probes at 5 × 10^7 counts per minute (cpm)/ml in fresh buffer for 12–24 hr.

After hybridization, tissue prints should be rinsed briefly in 2× SSC, 1% SDS, and 10 mM dithiothreitol, and then washed two more times in fresh changes of 2× SSC, 1% SDS, and 10 mM dithiothreitol for 30 min each at 42° with gentle shaking. Two additional washes should be carried out in 0.2× SSC, 1% SDS, and 1 mM dithiothreitol at 65° for 30 min each. The membrane can then be dried and analyzed by autoradiography.

As an alternative to the prehybridization and hybridization buffer described above, we have substituted a buffer containing 50% (v/v) formamide, 5× Denhardt's solution,[8] 6× SSC, 2 mM EDTA, 0.1% SDS, 200 μg/ml poly(A), 100 μg/ml yeast tRNA, and 70 mM dithiothreitol. With this buffer system, prehybridization and hybridization should be carried out at 42°. Filters should be washed as described above following hybridization.

Staining Procedure

Tissue prints can be stained with India ink[2] or other dyes before autoradiography. Before staining, tissue prints should be briefly rinsed in ice-cold water and then immersed in ice-cold India ink (Higgins No. 4415; Faber-Castell Corporation) for 1–10 min. Tissue prints can be destained by briefly rinsing in ice water and then by rinsing several times in 0.2× SSC and 1% SDS.

Although we have found that the tissue prints are not always uniformly stained with India ink, the stained images, nevertheless, provide a useful comparison to the autoradiographic images. The ink-stained images reveal anatomical detail that is not obvious in the autoradiograms, and provide an image for better interpretation of localized mRNA expression patterns on the autoradiograms. The ink-stained images are also useful for interpreting autoradiograms in terms of ineffective, incomplete, or distorted blotting of the sections onto the membrane.

Autoradiography Procedure

We use Kodak XRP-5 film exposed for 24–48 hr at −70° to evaluate initially the quality and quantity of hybridization to the tissue prints. Although exposure of the prints on XRP-5 is suitable for some tissue prints, autoradiograms of higher resolution can be obtained by exposing the tissue prints to the photographic film, Tmax 400. Exposure times on Tmax 400 are about five times longer than tissue prints exposed on XRP-5, but a higher quality image is obtained with the Tmax 400. With SAUR (small auxin up-regulated RNA)[4] probes shown in Fig. 1, the autoradiograms on Tmax 400 were exposed for about 10 days at −70°. We develop the film for 6–11 min at 24° in Kodak Tmax developer. This is followed by Kodak stop bath for 30 sec and Kodak rapid fix for 5 min. The film is then washed in running tap water for 5 min, rinsed in Kodak, Photo-Flo 200, and air dried.

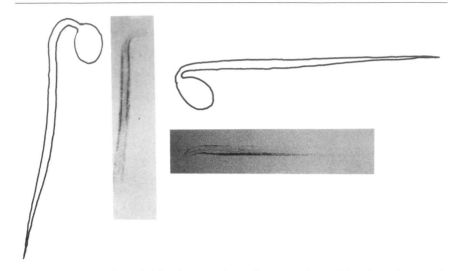

FIG. 1. Tissue-print hybridization reveals gravity-responsive mRNAs in soybean seedlings. Soybean seedlings were grown in the normal, vertical orientation or placed in a horizontal orientation for 20 min prior to tissue planting. The reorientation of the soybean seedlings from the vertical to the horizontal position results in negative gravitropic curvature 20–45 min after the seedlings are placed in the horizontal position. Auxin-responsive mRNAs are detected in the epidermis and cortex tissues of the elongating region of the vertical hypocotyl with an SAUR antisense RNA hybridization probe labeled with [35S]thio-UTP.[3] Tissue-print hybridization shows that the SAUR mRNAs disappear from the top of the horizontal hypocotyl prior to gravitropic bending. Schematic representations of the soybean seedlings are shown to the left or above the autoradiograms.

We have tested a number of other films for exposures of tissue-print hybridizations. Kodak Tri-X Pan does not perform as well as Tmax 400, and exposure time for Kodak Tmax 3200 is not substantially faster than Tmax 400. We have obtained the highest quality images with Kodak Tech Pan 2415 (35-mm format) or Kodak Tech Pan 4415 (4 × 5 in. format). The film speeds of Tech Pan 2415 and 4415 film are considerably slower than Tmax 400. Tech Pan films have a fine grain and produce high-quality autoradiograms of tissue prints, but the slow speed of these films limits their application to tissue prints that have strong hybridization signals.

Comments

Tissue-print hybridization provides an alternative method to *in situ* hybridization with fixed tissue sections. Tissue-print hybridization provides a reliable method to detect organ-specific and tissue-specific gene expression because the patterns of hybridization observed are similar, if not identical, to those observed with *in situ* hybridization.[3,5,6,10] Tissue-print hybridization is much less time consuming and less expensive than *in situ* hybridization, and requires a minimal amount of technical expertise and equipment. Tissue-print hybridization also has the advantage that numerous tissue treatments or manipulations can be examined with a minimal amount of effort. For example, all of the tissue sections can be printed on a single piece of nylon membrane and, once the tissue prints have been made for each treatment or manipulation, staining, prehybridization, hybridization, and autoradiography can be uniformly carried out on that single nylon membrane.

[10] M. A. Gee, G. Hagen, and T. J. Guilfoyle, *Plant Cell* **3**, 419 (1991).

[50] Recovery and Cloning of Genomic DNA Fragments from Dried Agarose Gels

By MICHAEL W. MATHER, J. ANDREW KEIGHTLEY, and JAMES A. FEE

We describe here a method for the cloning of bacterial genes that circumvents the need to prepare and maintain genomic DNA libraries. Genomic DNA samples, highly enriched in specific restriction fragments, are isolated from dried agarose gels and used directly for cloning.[1] The method is rapid and technically simple, and reduces the possibility that a DNA sequence of interest might be missed due to underrepresentation in a library.

Principle of the Method

An analytical in-gel hybridization is employed to detect those genomic DNA restriction fragments that contain the sequence(s) of interest. An enriched fraction is subsequently isolated from a preparative-scale dried agarose gel, ligated into an appropriate vector, and propagated in *Escherichia coli*. *Escherichia coli* clones containing the fragment of interest are identified by a low-density colony hybridization screen.

Hybridization in dried agarose gels was first described by Shinnick *et al.*[2] The technique is rapid, versatile, and sensitive. With radiolabeled

[1] M. W. Mather, *BioTechniques* **6**, 444 (1988).
[2] T. M. Shinnick, E. Lund, O. Smithies, and F. R. Blattner, *Nucleic Acids Res.* **2**, 1911 (1975).

oligonucleotide probes, the sensitivity is approximately fivefold greater than that obtained with an equivalent Southern blot.[3] Biotinylated oligonucleotides can be used,[4] and larger DNA probes can also be used when labeled by nick translation or random hexamer-primed DNA synthesis. Hybridizations can be carried out in standard hybridization solutions, and it is usually not necessary to prehybridize[3,5] or add carrier DNA/RNA.[4] In-gel hybridizations can also be carried out in tetraalkylammonium salt solutions,[1] which increase the hybridization specificity, especially in the case of mixed or AT-rich probes.[6-8] Dried agarose gels can be reprobed.[3]

Vogelstein and Gillespie reported[9] that agarose gels can be dissolved by high concentrations of sodium iodide, and the DNA from the dissolved gel isolated by binding to a glass matrix. Modifications of this procedure have been widely disseminated (see Appendix). Chaotropic salts can be used to dissolve dried agarose gels as well,[1] although a somewhat longer incubation period is required. Isolation of the DNA fragment of interest from the dried gel is advantageous because the thin nature of the dried gel allows a fairly precise excision of the region(s) containing the hybridizing fragment(s). In addition, the DNA in dried gels can be stored indefinitely in the dehydrated state.

Materials

Equipment. A full-size, horizontal gel electrophoresis apparatus is required. A variable-temperature vacuum gel drier is recommended for dehydrating agarose gels, and a constant-temperature shaker bath is needed for hybridizations.

Reagents. Ultrapure or genomic-grade agarose is recommended for use in gels from which DNA is to be purified. Tetraethylammonium chloride (Sigma, St. Louis, MO) is made up as a ~4.4 M stock solution. The grade sold by Sigma is generally pure enough to use without further purification.

[3] R. B. Wallace and C. G. Miyada, this series, Vol. 152, p. 432.
[4] N. F. Gontijo, J. C. C. Ribeiro, and S. D. J. Pena, *Focus (BRL)* **12**, 55 (1990).
[5] G. Dalbadie-McFarland, L. W. Cohen, A. D. Riggs, C. Morin, K. Itakura, and J. H. Richards, *Proc. Natl. Acad. Sci. U.S.A.* **79**, 6409 (1982).
[6] A. G. DiLella and S. L. C. Woo, this series, Vol. 152, p. 447.
[7] W. I. Wood, J. Gitschier, L. A. Lasky, and R. M. Lawn, *Proc. Natl. Acad. Sci. U.S.A.* **82**, 1585 (1985).
[8] K. A. Jacobs, R. Rudersdorf, S. D. Neill, J. P. Dougherty, E. L. Brown, and E. F. Fritsch, *Nucleic Acids Res.* **16**, 4637 (1988).
[9] B. Vogelstein and D. Gillespie, *Proc. Natl. Acad. Sci. U.S.A.* **76**, 615 (1979).

The exact molar concentration (c) is determined[10] by measuring the refractive index (n) at 25° and using the formula $c = (n_{25} - 1.325)30.30$. Rapid DNA purification kits were provided by Bio 101 (La Jolla, CA) and Bio-Rad (Richmond, CA).

Solutions

Denhardt's solution: 0.02% (w/v) bovine serum albumin (BSA), 0.02% (w/v) Ficoll 400, 0.02% (w/v) polyvinylpyrrolidone

NET: 0.15 M NaCl, 1 mM ethylenediaminetetraacetic acid (EDTA), 15 mM Tris · HCl (pH 7.5)

TAE running buffer: 40 mM Tris–acetate, 1 mM EDTA (pH 8.0)

TE: 10 mM Tris · HCl, 0.1 mM EDTA (pH 7.8)

TEA hybridization buffer[11]: 2.4 M tetraethylammonium chloride, 25 mM Tris, 2 mM EDTA, 0.2% (w/v) sodium dodecyl sulfate (pH 7.5)

Methods

Identification of Appropriate Genomic DNA Fragments for Cloning

Before the actual cloning steps, an analytical in-gel hybridization is carried out to identify which restriction endonucleases produce a hybridizing genomic fragment suitable for cloning. Generally, a fragment of between 1 and 10 kb is recommended. We have cloned fragments up to 13 kb.

Preparing and Labeling the Oligonucleotide Probe(s). The optimal design of oligonucleotide probes deduced from amino acid sequence data has been previously discussed.[3,12,13] Label the oligonucleotide to high specific activity using one of the standard methods.[14] We have had good success by end labeling with T4 DNA kinase using [γ-32P]ATP with a specific activity of 3000 Ci/mmol or higher.

---

[10] C.-T. Chang, T. C. Hain, J. R. Hutton, and J. G. Wetmur, *Biopolymers* **13,** 1847 (1974).

[11] Dried agarose gels can be hybridized in TEA buffer without any additives such as 5×
Denhardt's solution. The stringency is primarily determined by the hybridization temperature; thus, the washing steps can be carried out in a standard buffer such as NET, if desired. Tetraethylammonium is used so that the stringent temperature will not exceed the temperature limit for dried agarose gels (about 70°), regardless of probe length (3 M tetramethylammonium chloride is preferred for filter hybridizations; See Ref. 8). To a good approximation, the dissociation temperature, T_d, depends only on the length of the probe, and can be estimated with the formula: $T_d = 64.6° - (494.1/L)$, where L is the number of bases in the oligonucleotide probe. See Ref. 16 for examples of in-gel hybridizations using TEA.

[12] R. Lathe, *J. Mol. Biol.* **183,** 1 (1985).

[13] W. I. Wood, this series, Vol. 152, p. 443.

[14] F. Cobianchi and S. H. Wilson, this series, Vol. 152, p. 94.

Digesting Genomic DNA. Preparation of genomic DNA will vary depending on the source and is not discussed here (see, e.g., Refs. 15–18). Digest aliquots of the genomic DNA to completion with several individual restriction endonucleases (and/or prepare dual digests with two endonucleases). Use the same amount of DNA, relative to the slot size, as will be used in the preparative gel (see below) to ensure that the DNA fragments will electrophorese similarly. Prior to the digestion of high molecular weight genomic DNA, precautions should be taken to minimize shearing, including the use of wide-bore pipettes for transfer. Complete digestion of the DNA samples can usually be achieved by using a DNA concentration of about 50 ng/μl and 1.5 to 2 units of enzyme per microgram of DNA. For example, 5 μg of bacterial DNA is digested in 100 μl, concentrated to 35 μl by ethanol precipitation, and loaded in a 10 $\times$ 1 mm well (5-mm deep gel).

Preparing Radiolabeled DNA Marker Fragments. Labeled marker fragments are used to define the positions of the hybridizing bands. This eliminates the need for the second hybridization step previously used.[1] ^{35}S-Labeled λ DNA fragments are convenient in-gel markers. Digestion of λ DNA with *Eco*RI and *Hin*dIII yields DNA fragments from 0.125 to 21.2 kb. Other DNA fragments can be added, if desired, to fill the gaps between 2.0 and 3.5 kb and between 5.1 and 21.2 kb (see Fig. 1). Before labeling, disrupt the cohesive termini by heating the digested DNA to 68° for 10 min and cool on ice. The marker fragments are then labeled by filling in the ends with Klenow[14] or with Sequenase (United States Biochemical, Cleveland, OH). To fill in the ends of up to 5 μg of marker DNA with Sequenase, perform the initial labeling portion of a sequencing reaction (without addition of primer) according to the instructions of the manufacturer, using 1 μl of [α-^{35}S]dATP (about 10 pmol of 1000 Ci/mmol or greater activity) at 37°. Stop the reaction by addition of 4 μl of 0.25 M EDTA (pH 8.0) or an appropriate volume of agarose gel loading buffer. No purification is normally needed. Store at -20°.

Analytical Electrophoresis. Load about 1 μg of labeled marker DNA in flanking lanes surrounding the genomic DNA lanes in an agarose gel (0.7–1%, w/v). Electrophorese the gel on a *level surface* with TAE running buffer at 1 V/cm. Low voltage and level conditions are important for consistent mobility of the fragments across the gel (a temperature-controlled gel apparatus may allow the use of higher voltages). After

[15] T. M. Silhavy, M. L. Berman, and L. W. Entquist, "Experiments with Gene Fusions," p. 137. Cold Spring Harbor Press, Cold Spring Harbor, New York, 1984.
[16] M. W. Mather and J. A. Fee, *Plasmid* **24**, 45 (1990).
[17] B. G. Herrmann and A. M. Frischauf, this series, Vol. 152, p. 180.
[18] G. M. Wahl, S. L. Berger, and A. L. Kimmel, this series, Vol. 152, p. 399.

A

B

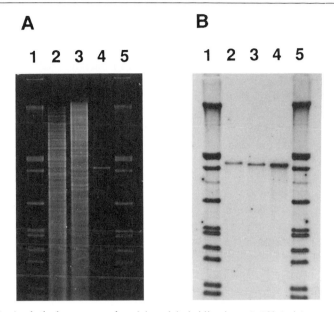

FIG. 1. Analytical agarose gel and in-gel hybridization. A 1% (w/v) agarose gel was prepared and run as described in text. A photograph of the ethidium bromide-stained gel (A) and an autoradiogram of the hybridized gel (B) are shown. About 1.5 μg/lane marker DNA (lanes 1 and 5), 5 μg *Thermus thermophilus* HB8 DNA restricted with *Bam*HI (lane 2) and *Hin*dIII (lane 3), and 100 ng of *Hin*dIII fragment isolated and cloned by the methods described in the text (lane 4) were electrophoresed at 1 V/cm. The radioactive marker is λ DNA digested with *Eco*RI and *Hin*dIII plus a linear 2.86-kb fragment (pTZ18U) added prior to labeling (see text). Hybridization was carried out at 48° for 16 hr in 6× NET, 5× Denhardt's solution, 1% (w/v) sodium dodecyl sulfate, 0.5% (w/v) sodium pyrophosphate, 250 μg/ml sonicated salmon sperm DNA, with ~300 ng of a 21-base, eightfold degenerate probe end labeled with [32]P. Stringent wash: 1× NET, 60°, 3 hr. *Note:* The full-length *Bam*HI fragment has resisted cloning.

electrophoresis the gel can be stained with ethidium bromide for photography, if desired.

Drying the Gel. The drying procedure described by Wallace and Miyada[3] gives good results. The DNA in the gel is alkali denatured, neutralized, and dried onto Whatman (Clifton, NJ) 3MM paper. After drying, the gel is a thin, nonpliant coating on the 3MM paper.

Hybridization/autoradiography. The dried gel is removed from the filter paper backing and rehydrated by soaking briefly in water. The rehydrated gel is a thin, flexible membrane. Seal the gel membrane into a Seal-a-Meal (Dazey, Industrial Airport, KS) plastic bag. Hybridization solution including the probe (>2 × 10⁶ cpm/ml) is added through a cut corner of the bag, and the bag resealed with a minimum of bubbles. Incubate at a temperature and time period appropriate for the probe used.[3,8,18] After hybridization, remove the gel from the bag, rinse twice with 6× NET, and proceed with appropriate stringent washing. In the case of bacterial DNA, when the apparent activity in the gel has been reduced to about 1000–4000 cpm/lane of genomic DNA (measured with a hand-held survey meter about 1 cm above the gel), blot the excess solution from the gel and wrap the gel in plastic wrap. Prepare a preliminary autoradiogram by exposing X-ray film [such as Eastman Kodak (Rochester, NY) XAR-5] to the gel in a cassette for a few hours. An appropriate exposure time can then be estimated (if necessary) from the initial autoradiogram. Multiple bands in each lane of genomic DNA may indicate that the hybridization temperature should be increased or that the probe is not specific. Additional stringent washes can be performed as necessary to improve the signal-to-noise level. An analytical gel and autoradiogram, based on the cloning of a bacterial oxidase gene, are shown in Fig. 1.

Cloning DNA from Dried Agarose Gel

Selection of Target Fragment and Preparative Gel Electrophoresis. Suitable target fragments are selected by examination of the analytical autoradiograms. The position of these fragments relative to the radioactive marker bands is carefully measured. Digest larger quantities of the genomic DNA with the appropriate restriction enzyme(s). Load a proportional amount (relative to the analytical gel) of digested DNA in a preparative slot. For example, load 40 μg of bacterial DNA into an 80 × 1 mm slot (5-mm deep gel) in 280 μl. Load the radioactive marker DNA in flanking lanes and electrophorese under conditions identical to those used for the analytical gel. After electrophoresis, dry the gel[3] (do not denature). Make orientation marks on the Whatman paper backing near three of the corners of the gel, using a pen containing radioactive ink (^{14}C or ^{35}S). Place the gel (still on the paper) in a cassette with X-ray film for autoradiography. After developing the film, the orientation marks and marker bands will be visible in the autoradiogram. Measuring from the marker bands, cut out a narrow slot in the autoradiogram corresponding to the position of the target DNA. As shown in Fig. 2, this film is now a template for excising the target DNA from the dried gel. Place the autoradiogram on top of the dried preparative gel, using the orientation marks to line it up, secure with tape, and cut out the strip of dried gel containing the target DNA with a scalpel or razor blade.

Purification of Target Fragments from Dried Agarose. Estimate the original volume of the excised gel strip. For example, the volume for an

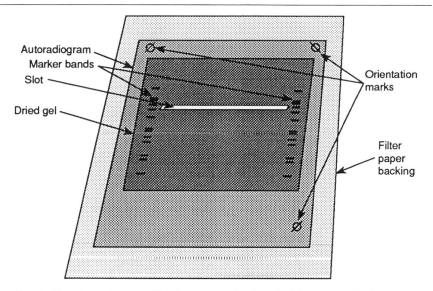

FIG. 2. Excision of target DNA fragments. As described in text, a slot is cut in the autoradiogram at the position of the target DNA as determined in the analytical procedure. The autoradiogram is aligned on the dried gel, using the radioactive pen markings, and taped down. A scalpel or razor blade is used to excise the portion of the gel exposed by the template.

80 × 2 mm gel strip from a 5-mm deep gel would be 800 μl. Multiply that volume by 4 to estimate the total volume required for the purification procedure and set out a sufficient number of microcentrifuge tubes to accommodate that volume. Cut the excised gel strip into small pieces to facilitate dissolution and distribute them evenly into the microcentrifuge tubes. To a volume of TE buffer equal to the original gel slice volume, add 3 vol of the sodium iodide or sodium perchlorate solution from a DNA purification kit (see Appendix), mix, and distribute evenly to the microcentrifuge tubes containing the gel pieces. Incubate at 50 to 55° with occasional mixing until the gel is dissolved (15–30 min) and proceed with the silica/silicate matrix DNA isolation (see Appendix). Extract the DNA solution with phenol and chloroform, precipitate with ethanol, dry, and resuspend in a small volume of TE. Estimate the yield and verify the size of the isolated fragments by electrophoresis; run 1/10 of the DNA in an agarose minigel with (unlabeled) marker. It is also advisable to dry and hybridize this minigel to confirm the presence of the target DNA before proceeding.

Cloning and Screening Transformants. The purified DNA fragments are joined to an appropriate, dephosphorylated[14] vector using T4 DNA ligase under standard conditions as recommended by the supplier. Following the ligation, transform competent *E. coli* (use a restriction and recombination deficient strain) by a high-efficiency procedure, such as that of Hanahan.[19] The transformants are then screened by standard colony hybridization techniques.[20-22] For bacterial systems, usually only 100 or so colonies need to be screened for each genomic fragment to be cloned. The microtiter tray replica method[22] is a convenient procedure for this low-density screening.

Discussion

We have found cloning genomic DNA from dried agarose gels to be a relatively simple and rapid method for cloning protein-encoding genes from bacteria.[23] The use of the autoradiogram with labeled marker DNA fragments as a template for cutting out the region of the preparative gel containing the target DNA allows precise excision of the gel band containing the DNA of interest. This minimizes the amount of screening subsequently required by reducing the carryover of nontarget genomic DNA. We have generally obtained 2–4 positive clones per 100 colonies in the screening step, using relatively wide excision slots (4–5 mm). This compares favorably to the many thousands of colonies one must screen in the

---

[19] D. Hanahan, *in* "DNA Cloning: A Practical Approach" (D. M. Glover, ed.), Vol. 1, p. 109. IRL Press, Oxford, 1985.

[20] M. Grunstein and J. Wallis, this series, Vol. 68, p. 379.

[21] G. Vogeli and P. S. Kaytes, this series, Vol. 152, p. 407.

[22] P. J. Mason and J. G. Williams, *in* "Nucleic Acid Hybridization: A Practical Approach" (B. D. Hames and S. J. Higgins, eds.), p. 113. IRL Press, Oxford, 1985.

[23] Because hybridization in dried agarose gels can be used to detect single-copy genes in mammalian genomic DNA (see Ref. 6) it seems likely that this method could be extended to more complex genomes. Mammalian genomes are about 3 orders of magnitude larger than bacterial genomes; thus to obtain each clone of a portion of a mammalian gene by this method, one would expect to have to prepare and screen about 25,000 or more *E. coli* transformants. This would still be a considerable reduction over the screening required to isolate clones from genomic libraries. Because up to 50 μg of digested mammalian DNA can be loaded in a 1-cm slot [see J. Sambrook, E. F. Fritsch, and T. Maniatis, "Molecular Cloning: A Laboratory Manual," p. 6.11. Cold Spring Harbor Press, Cold Spring Harbor, New York, 1989], there would be no theoretical impediment preventing the recovery of sufficient enriched DNA (micrograms) to yield the necessary number of transformants for screening. If the initial quantity of genomic DNA were limited, an adapter-mediated polymerase chain reaction (PCR) of the digested DNA might be used to amplify the amount of DNA available [see A. Akowitz and L. Manuelidis, *Gene* **81,** 295 (1989)]. Of course, this would limit the size of the DNA fragments that could be cloned to the range efficiently amplified by PCR. Similarly, after purification from the dried gel, the genomic DNA fraction containing the fragment of interest could be amplified to allow the generation of a larger number of transformants.

case of a bacterial genomic library, but can undoubtedly be improved on by using narrower excision slots.

Appendix: Preparation and Use of Silica Matrix for DNA Purification

Kits using binding matrices for the rapid purification of DNA are available from commercial sources, such as Bio 101 (Geneclean II) and Bio-Rad (Prep-A-Gene).[24] The yield of DNA recovered from dried gels varies somewhat from experiment to experiment (~55–85%), but is similar using kits from either source (or using the homemade matrix described below). The following recipes (modified from an unpublished procedure in circulation) are provided for those who wish to prepare their own materials.

NaI Solution (6 M). Dissolve 90.3 g NaI in distilled water and adjust the final volume to 100 ml. Add 1 g Na_2SO_3 (may not dissolve completely). Filter through a 0.45- or 0.2-μm (pore size) filter and add 0.5 g Na_2SO_3 (will not dissolve; saturates the solution). Store in the dark at 4°.

Washing Solution. Prepare a solution of 20 mM Tris · HCl (pH 7.5), 2 mM EDTA, and 0.4 M NaCl in distilled water. Mix 45 parts of this solution with 55 parts absolute ethanol. Filter through a 0.45- or 0.2-μm filter. Store at −20°.

Silica Fines. Suspend 250 ml of silica powder (325 mesh; Sargent Welch) in a total volume of 500 ml of distilled water. Stir for 1 hr. Let heavier particles settle for 1 hr. Pellet the fines in the supernatant by centrifugation in an SS-34 rotor at 4000 rpm for 5 min. Resuspend the pellet in 150 ml distilled water and add 150 ml nitric acid, heat to just below boiling in a fume hood, and let cool to room temperature. Wash four times with distilled water by suspending fines in water, pelleting at 3000 rpm for 5 min, discarding the supernatant (including any ultrafine particles that do not pellet), and resuspending in water. Resuspend in water, transfer to a glass bottle (half-full only), and sterilize by autoclaving for 30 min. Swirl to resuspend and transfer to sterile graduated centrifuge tubes (such as 50-ml conical polypropylene tubes). Measure the volume of the fines by pelleting in a swinging bucket rotor or allowing the fines to settle for 48 hr, discard the supernatant, and add an equal volume of sterile TE. Resuspend and aliquot a few 0.5-ml portions to sterile screwcap microcentrifuge tubes for daily use. Store at 4°.

---

[24] The Bio-Rad kit, containing a silicate binding matrix, is perhaps the easiest to use, due to more facile resuspension of the matrix during the washing steps. This may also be important, to minimize shearing, when trying to recover DNA fragments larger than 15 kb. However, this kit is somewhat more expensive per unit of DNA-binding capacity.

Use. For an aqueous DNA sample in a microcentrifuge tube:

1. Add 3 vol of NaI solution to the DNA sample. (If the DNA is in agarose, incubate at 50–55° until the agarose is completely dissolved.)
2. Resuspend the fines in a stock vial of silica fines, add 2 μl fines to the sample for each 1 μg DNA estimated to be in the sample, mix, and incubate on ice for 5 to 10 min with occasional brief mixing.
3. Pellet the fines for 30 sec in a microcentrifuge and remove as much of the supernatant as possible without disturbing the pellet.
4. For agarose samples only: Add 500 μl of 3 : 1 (v/v) NaI solution/TE buffer, resuspend the fines (with the bound DNA), incubate at 50–55° for 4–5 min, pellet the fines (30 sec), and remove the supernatant.
5. Add 30 pellet volumes or more of cold washing solution, resuspend the fines (with bound DNA), pellet the fines (20 sec), and discard the supernatant. Repeat the wash two more times. After the third wash, remove as much of the wash solution as possible.
6. Add two or three pellet volumes of TE, resuspend, and incubate at 50° for 5 min. Centrifuge for 60 sec and remove the supernatant containing the DNA to a new tube. Repeat the DNA elution once or twice more, combining the supernatants.

Acknowledgments

We thank the Bio 101 and Bio-Rad companies for providing materials. J.A.K. holds an Association of Western Universities graduate student fellowship from the Department of Energy and is a member of the Department of Biochemistry, University of New Mexico (Albuquerque). The authors' research is supported by National Institutes of Health Grant GM35342 to J.A.F. and performed under the auspices of the U.S. Department of Energy.

[51] Genetic Analysis Using Random Amplified Polymorphic DNA Markers

By John G. K. Williams, Michael K. Hanafey, J. Antoni Rafalski, and Scott V. Tingey

General Introduction

The use of DNA markers in genetic mapping, genetic diagnostics, molecular taxonomy, and evolutionary studies has been well estab-

TABLE I
COMPARISON OF RAPDs AND RFLPs

Characteristic	RAPD	RFLP
Dominance	Dominant	Codominant
Assay	DNA amplification	DNA blot hybridization
Detection	Fluorescence	Radioactive isotopes
Amount of DNA required	25 ng	2–10 μg

lished.[1,2] The most commonly used DNA markers are restriction fragment length polymorphisms (RFLPs). Detection of RFLPs by DNA blot hybridization is laborious and incompatible with applications requiring high throughput. Genetic tests based on the polymerase chain reaction (PCR) are simple to perform,[3] but target DNA sequence information is required to design specific primers.

We[4,5] and others[6,7] have described a novel type of genetic marker that is based on DNA amplification, but requires no knowledge of target DNA sequence. These markers, called RAPD markers (for random amplified polymorphic DNA), are generated by the amplification of random DNA segments with single primers of arbitrary nucleotide sequence. RAPD markers can be used for genetic mapping applications, as well as for genetic diagnostics. The assay is nonradioactive, requires only nanogram quantities of DNA, and is applicable to a broad range of species. Table I compares RFLP and RAPD polymorphism assays. In this chapter we present detailed experimental protocols for RAPD assays and applications, emphasizing their use for genetic analysis in plants.

Principle of the Method

To perform a RAPD assay, a single oligonucleotide of an arbitrary DNA sequence is mixed with genomic DNA in the presence of a thermosta-

[1] S. D. Tanksley, N. D. Young, A. H. Paterson, and M. W. Bonierbale, *Bio/Technology* **7**, 257 (1989).

[2] P. C. Watkins, *BioTechniques* **6**, 310 (1988).

[3] M. A. Innis, D. H. Gelfand, J. J. Sninsky, and T. J. White, eds., "PCR Protocols: A Guide to Methods and Applications." Academic Press, San Diego, 1990.

[4] J. G. K. Williams, A. R. Kubelik, K. J. Livak, J. A. Rafalski, and S. V. Tingey, *Nucleic Acids Res.* **18**, 6531 (1990).

[5] J. G. K. Williams, A. R. Kubelik, J. A. Rafalski, and S. V. Tingey, in "More Gene Manipulatons in Fungi" (J. W. Bennett and L. L. Lasure, eds.), p. 431. Academic Press, San Diego, 1991.

[6] J. Welsh and M. McClelland, *Nucleic Acids Res.* **18**, 7213 (1990).

[7] J. Welsh, C. Petersen, and M. McClelland. *Nucleic Acids Res.* **19**, 303 (1991).

ble DNA polymerase and a suitable buffer, and then is subjected to temperature cycling conditions typical of the polymerase chain reaction. The products of the reaction depend on the sequence and length of the oligonucleotide, as well as the reaction conditions. At an appropriate annealing temperature during the thermal cycle, the single primer binds to sites on opposite strands of the genomic DNA that are within an amplifiable distance of each other (e.g., within a few thousand nucleotides), and a discrete DNA segment is produced. The presence or absence of this specific product, although amplified with an arbitrary primer, will be diagnostic for the oligonucleotide-binding sites on the genomic DNA. In practice, the DNA amplification reaction is repeated on a set of DNA samples with several different primers, under conditions that result in several amplified bands from each primer. Polymorphic bands are noted, for example, between parents of a cross, and the polymorphisms can be mapped in a segregating population. Often a single primer can be used to identify several polymorphisms, each of which maps to a different locus.

Materials and Reagents

Thermostable DNA Polymerases

Thermus aquaticus DNA polymerase AmpliTaq is purchased from Perkin-Elmer Cetus (Norwalk, CT). *Thermus flavus* DNA polymerase is purchased from New England Nuclear (Boston, MA).

Sources of DNA

Soybean lines and F_2 individuals segregating from a cross of the inbred cultivars *Glycine max* variety Bonus and *Glycine soja* accession PI81762, and other species of the genus *Glycine,* are obtained from Dr. T. Hymowitz, University of Illinois. The *Zea mays* (corn) lines CM37 and T232 are obtained from Dr. B. Burr, Brookhaven National Laboratory. Human DNA samples are obtained from Drs. J Gilbert and A. Roses, Duke University. *Silene alba* DNA is from Herbicide Development, Inc. (Whatton in the Vale, Nottingham, England). *Neurospora crassa* DNA strains Oak Ridge FGSC 4488 and Mauriceville FGSC 2225 are obtained from Dr. R. L. Metzenberg, University of Wisconsin. Bacterial DNA samples are obtained from Dr. J. Webster, Du Pont (Wilmington, DE). Cyanobacterial DNA from strain PCC 6803 is purified as described in Williams.[8] DNA molecular size standards are purchased from Bethesda Research Labora-

---

[8] J. G. K. Williams, this series, Vol. 162, p. 766.

tories (Gaithersburg, MD) (1-kb DNA ladder, or ϕ-X174 RF DNA *Hae*III fragments).

Plant DNA Isolation

Plant DNA, extracted by the CTAB method of Bernatzky and Tanksley[9] from corn, soybean, rice, and *Brassica* species, is used as a substrate for the DNA amplification reaction. Briefly, 0.1 to 0.5 g of fresh leaf tissue (or a corresponding amount of lyophilized tissue) is ground in liquid nitrogen and mixed with 1 ml of CTAB extraction buffer[10] and 0.4 ml of chloroform. The sample is heated at 55° for 10 min, centrifuged for 5 min, and the supernatant is recovered and mixed with 1.2 vol of 2-propanol. The nucleic acid precipitate is recovered by centrifugation, washed with 1 ml of 70% (v/v) ethanol, dried, and dissolved in 10 mM Tris-HCl (pH 7.5), 0.1 mM ethylenediaminetetraacetic acid (EDTA).

Oligonucleotide Primers

The nucleotide sequence of each primer is generated randomly, within the following limits. Our laboratory generally uses oligonucleotides of 10 bases comprising 50–70% G + C. Palindromes greater than 6 bases in length, and complementary at the 3' end, were avoided.

Primers are synthesized using phosphoramidite chemistry. After deprotection, the samples are dried under vacuum, dissolved in 200 μl of water, and purified by gel filtration on Sephadex G-25 (NAP-5 disposable columns; Pharmacia, Inc., Piscataway, NJ). Sets of RAPD primers are now commercially available from Operon Technologies (Alameda, CA).

Methods

Standard Amplification Conditions

Amplification reactions are performed in 25-μl volumes containing 10 mM Tris-Cl (pH 8.3), 50 mM KCl, 2 mM MgCl$_2$, 0.001% (w/v) gelatin (Cat. No. G2500; Sigma, St. Louis, MO), 100 μM each of dATP, dCTP, dGTP, and TTP (Pharmacia), 0.2 μM primer, 25 ng of genomic DNA, and 0.5 unit of AmpliTaq polymerase, overlaid with 1 drop of mineral oil (Cat. No. M3516; Sigma). Tris buffer, KCl, MgCl$_2$, and gelatin are prepared at 10× concentration and autoclaved before use; the desired volume is restored by addition of a suitable amount of water after autoclaving. Amplifications

[9] R. Bernatzky and S. D. Tanksley, *Theor. Appl. Genet.* **72,** 314 (1986).
[10] M. W. Lassner, P. Peterson, and J. I. Yoder, *Plant Mol. Biol. Rep.* **7,** 116 (1989).

are performed in a Perkin-Elmer Cetus DNA thermal cycler programmed for 45 cycles of 1 min at 94°, 1 min at 36°, and 2 min at 72°, using the fastest available transitions between each temperature. Although we have typically amplified for 45 cycles, 35 cycles is sufficient in most experiments. Amplification products are analyzed by electrophoresis in 1.4% (w/v) agarose gels and detected by staining with ethidium bromide.

Reamplification of RAPD Bands for Use as Hybridization Probes

Bands of interest are visualized with a low-intensity ultraviolet (UV) light and excised from the gel by punching with a glass capillary. The agarose plugs containing the DNA are placed in 100 μl of 10 mM Tris-Cl (pH 7.5), 0.1 mM EDTA. The sample is heated at 94° to dissolve the agarose, and 1-μl aliquots of several dilutions are reamplified under standard conditions with the same primer that was originally employed to generate the band. The products from each sample dilution are analyzed in an agarose gel to identify pure samples containing only the desired band (the dilution series is employed to reduce reamplification of contaminating bands). Unincorporated deoxynucleotide triphosphates are removed by gel-filtration chromatography on Sephadex G-50 (NICK disposable column; Pharmacia). The product is collected in a 400-μl volume as described by the manufacturer, and 5-μl aliquots are used to label the probe by the random primer method.[11]

Documentation of Results

The RAPD gels are conveniently photographed under the standard conditions used for ethidium bromide-stained gels.[12] Alternatively, the gels can be imaged with a video camera and stored electronically.

Analysis of Genetic Data

An Apple Macintosh computer version of the Mapmaker program[13] is used for the calculation of genetic maps.

[11] A. P. Feinberg and B. Vogelstein, *Anal. Biochem.* **132,** 6 (1983).
[12] T. Maniatis, E. F. Fritsch, and J. Sambrook, "Molecular Cloning: A Laboratory Manual." Cold Spring Harbor Press, Cold Spring Harbor, New York, 1982.
[13] E. S. Lander, P. Green, J. Abrahamson, A. Barlow, M. J. Daly, S. E. Lincoln, and L. Newburg, *Genomics* **1,** 174 (1987). A version of Mapmaker for the Apple Macintosh computer is available from Scott Tingey (Du Pont). The Macintosh version was used to analyze the data of Fig. 14, while the UNIX version of Mapmaker v1.9 was used to analyze the simulated data of Figs. 16 and 17.

Results and Discussion

Reaction Conditions

In this section, we summarize our experiences with the variables of the RAPD assay. It is usually advisable to begin with the standard conditions (see Methods). In our experience, one of the most important variables is the concentration of genomic DNA. Because different DNA extraction methods produce DNA of widely different purity, it may be necessary to optimize the amount of DNA used in the RAPD assay to achieve reproducibility and a strong signal. Too much genomic DNA may result in smears or in a lack of clearly defined bands in the gel. Too little DNA gives unreproducible patterns. For soybean DNA (haploid DNA content, 2×10^9 bp), at least 850 haploid genome equivalents per 25-μl reaction volume are required for reproducibility (Fig. 1). At 285 genome equivalents and below (Fig. 1, lanes 4) some bands are lost, other bands appear, and the reaction appears to be chaotic. Because all portions of the genome are expected to be represented manyfold in the reaction mixture at 285 haploid equivalents, this result suggests that either a substantial proportion of the DNA was not available as a template, for example due to the presence of single strand breaks, or that several copies of the template are necessary to achieve detectable levels of amplification. In general, at a genomic DNA concentration of 1 μg/ml, reproducible amplification is achieved with DNA from a wide variety of organisms (Fig. 2; see also Ref. 6). Amplification is reproducible for genomes as small as 1.5 mb (*Haemophilus influenzae*), but is unreliable for smaller genomes such as bacteriophage λ (48.5 kb; data not shown). Loss of reliability in small genomes may result from poor complementarity to arbitrary primer sequences in these genomes. For example, to amplify five bands, the expected average number of complementary bases in the primer–target hybrid is 7.3 bp for a 1.5-Mb genome, and 6.1 bp for a 50-kb genome [see below; Eq. (1)].

For reproducibility, it is important to note that both the magnesium ion concentration and the annealing temperature affect the relative intensity of amplified bands. As the magnesium concentration increases, some DNA segments are amplified more efficiently while others are amplified less efficiently (Fig. 3; e.g., bands a and b). We typically use an approximately 1.5 mM excess of magnesium over the total nucleotide concentration. The annealing temperature in the thermal cycle profile is typically set at 36° in the Perkin-Elmer Cetus DNA thermal cycler; 40° is too high to obtain good amplification with many 10-base primers. Annealing temperature may affect the relative amounts of some amplified bands (Fig. 4). With the Perkin-Elmer Cetus thermal cycler, the fastest temperature transition from

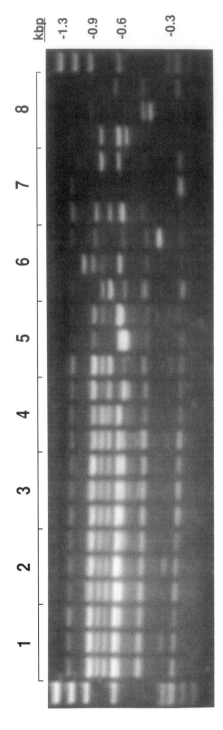

FIG. 1. Dilution of genomic DNA. Different amounts of soybean (*G. max* cv. Bonus) DNA were amplified under standard conditions with primer 5′-CTGATACGGA. Each sample was amplified in triplicate to assess the reproducibility of the reaction. Successive samples (lanes 1–8) contained genomic DNA at one-third the concentration of the previous sample. The starting amount of genomic DNA (lanes 1) was 15.2 ng/25 μl reaction, corresponding to about 7700 haploid genomes. The final genomic DNA amount was 7 pg or about 3.5 genomes per reaction (lanes 8). Molecular weight markers (kbp) are as indicated.

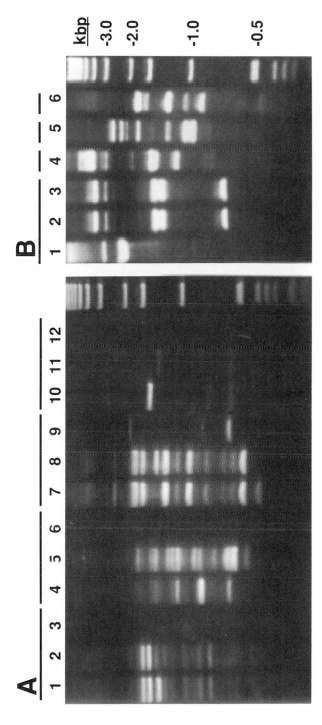

Fig. 2. Amplification of DNA from different species. (A) Amplification of eukaryotic DNA. DNA was amplified from a variety of species, using primers of arbitrary nucleotide sequence (Methods). Lanes 1 and 2, human DNA Hu2 and Hu3, respectively, amplified with primer 5'-ACGGTACACT. Lanes 4 and 5, corn CM37 and T232, respectively, amplified with 5'-GCAAGTAGCT. Lanes 7 and 8, soybean G. max and G. soja, respectively, amplified with 5'-CGGCCCCTGT. Lanes 10 and 11, N. crassa Oakridge and Mauriceville, respectively, amplified with 5'-CACATGCTTC. Genomic DNA was omitted in control reactions (lanes 3, 6, 9, and 12) to determine whether any of the bands seen with genomic DNA were primer artifacts; such an artifact is present in lane 9, but these bands were not seen when genomic DNA was present in the reaction mixture (lanes 7 and 8). Molecular weight markers (kbp) are as indicated. (B) Amplification of prokaryotic DNA. Lanes 1–3, Escherichia coli (strains 037, 641, and 642, respectively); lane 4, Listeria monocytogenes (strain 681); lane 5, Staphylococcus aureus (strain 684); lane 6, Salmonella typhimurium (strain 706). All genomic DNA samples were amplified with the primer 5'-TCACGATGCA. Molecular weight markers (kbp) are as indicated.

711

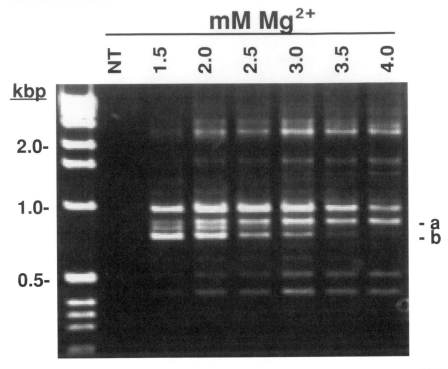

FIG. 3. Effect of magnesium concentration. Soybean DNA (*G. soja* PI81762) was amplified with the primer 5'-AGCACTGTCA under standard conditions (Methods), except that the genomic DNA concentration was 2 μg/ml, each dNTP was present at 200 μM, and the total MgCl$_2$ concentration varied from 1.5 to 3.5 mM, as indicated above each lane. Lane NT is a control without genomic DNA template (3.0 mM MgCl$_2$). This experiment shows that as the Mg^{2+} concentration increases, amplification of band a increases, while that of band b decreases.

36 to 72° is about 1 min; extending this transition to longer times had no significant effect on amplification results (data not shown).

Primer concentrations between about 0.1 and 2.0 μM are optimal. At lower concentrations it becomes difficult to detect amplification products in a stained agarose gel, and at higher concentrations smearing of the bands becomes evident. No effect of primer concentration on the relative intensities of bands has been noted (data not shown).

A deoxynucleotide triphosphate concentration of 100 μM for each of the four bases is adequate for generating RAPDs. At lower concentrations the intensity of stained bands in the gel becomes progressively weaker. No effect of deoxynucleotide triphosphate concentration on the relative intensities of the amplified bands has been noted (data not shown).

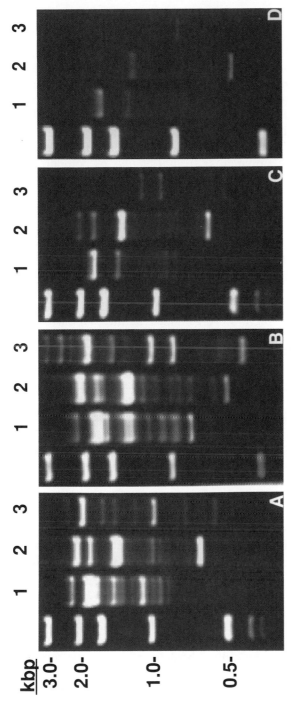

Fig. 4. Effect of annealing temperature. Soybean DNA (*G. soja* PI81762) was amplified in a Biocycler oven (Bios, Inc., New Haven, CT) using four different annealing temperatures in the thermal profile. (A) 20°; (B) 35°; (C) 36°; D) 37°. Standard conditions were used, except that the genomic DNA concentration was 2 μg/ml, MgCl₂ was 2.5 m*M*, each dNTP was 200 μ*M*, AmpliTaq was 25 units/ml, and the final volume was 50 μl. Three different primers were used: lanes 1, 5′-CTGATGCTAC; lanes 2, 5′-GCAA3TAGCT; lanes 3, 5′-TGGTCACTGA. Molecular weight markers (kbp) are as indicated. Samples were placed in a PVC 96-well plate (Falcon 3911), overlaid with 1 drop of mineral oil, and were amplified for 35 cycles of 93° for 1 min, *T*° for 2 min, 72° for 2 min in a Biocycler oven, where the annealing temperature *T*° was as indicated above. Although different results might be expected using a different thermal cycling instrument, this experiment illustrates qualitatively the dependence of amplification on annealing temperature.

The recommended concentration of *T. aquaticus* AmpliTaq is 20 units/ml (see Methods). Although this works well for the genomic DNA for most species, it may be worthwhile to optimize this parameter for some applications. For example, a concentration of 40 units/ml is generally adequate for amplifying soybean DNA, but was too high for our preparations of *N. crassa* DNA (not shown). The outcome of the amplification reaction is determined not only by the primer sequence employed, but also by the DNA polymerase. Different patterns of amplified bands are obtained with the DNA polymerase from *T. flavus* as compared to that from *T. aquaticus* (AmpliTaq) (Fig. 5).

Primer Sequence

To support DNA amplification under standard conditions (see Methods), a 10-base oligonucleotide primer should contain at least 4 G + C bases (Fig. 6). For primers containing 5 G + C bases, a length of 9 bases is the minimum that will support efficient amplification as detected by staining with ethidium bromide (Fig. 7). Although these results on primer base composition and length were obtained at the standard annealing temperature of 36°, the same results were seen at an annealing temperature of 15° (data not shown). Oligonucleotides shorter than 9 bases may be used, but smaller amounts of amplified products are obtained and staining methods of greater sensitivity are required to detect the products.[14] To determine the contribution of individual nucleotides in the primer to the specificity of the amplification reaction, a set of 11 related 10-base oligonucleotides was prepared. Each primer differed from the sequence 5'-TGGT-CACTGA by substitution of a single base at a successive position in the sequence. The G + C composition of each primer was maintained at 50%. In this experiment, nucleotide changes in the nine positions at the 3'-end caused nearly complete changes in the banding pattern, as compared to the original primer (Fig. 8). A change in the nucleotide at the 5' end had a smaller effect. We believe that nucleotide changes in the template site will have the same effect on specificity as those observed for changes in the primer, and that the RAPD assay may therefore be used to detect single base changes in the template. The presence of a single RAPD band would then be diagnostic for a sequence totaling 18 bp in the target genome (9 bases at each end of the genomic DNA segment that is amplified). An average of five amplified bands per primer would mean that each primer is diagnostic for 5 × 18 = 90 bp in the template. This is in contrast to an RFLP, which is diagnostic for only 12 bp per probe–enzyme combination.

[14] G. Caetano-Anolles, B. J. Bassam, and P. M. Gresshoff, *Bio/Technology* **9**, 553 (1991).

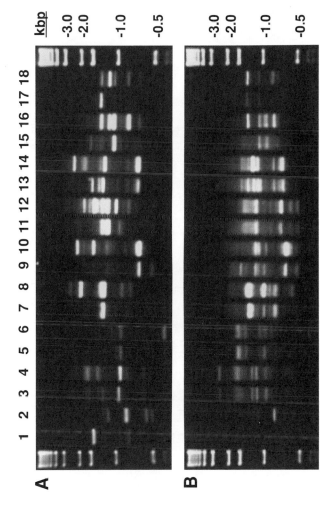

Fig. 5. Comparison of DNA polymerases. DNA samples from corn lines CM37 (odd-numbered lanes) and T232 (even-numbered lanes) were amplified with nine different primers using AmpliTaq from *T. aquaticus* (A) or a thermostable DNA polymerase from *T. flavus* (B). Primers are as follow: lanes 1 and 2 (5'-CGTAGCCAA), lanes 3 and 4 (5'-TGACGATGCA), lanes 5 and 6 (5'-TGGACACTGA), lanes 7 and 8 (5'-GCAAGTAGTG), lanes 9 and 10 (5'-CGGTCACTGT), lanes 11 and 12 (5'-CGGCCACTGT), lanes 13 and 14 (5'-CGGCCCCTGT), lanes 15 and 16 (5'-ATTGCGTCCA), lanes 17 and 18 (5'-TTGCGTCCA). Molecular weight markers (kbp) are as indicated. Standard conditions were used, except that the *T. flavus* enzyme was used in the buffer recommended by the manufacturer [50 mM Tris-Cl (pH 9.0) and 20 mM NH₄SO₄ are substituted for 10 mM Tris-Cl (pH 8.3) and KCl in the standard buffer]. AmpliTaq was used in the standard buffer (Methods).

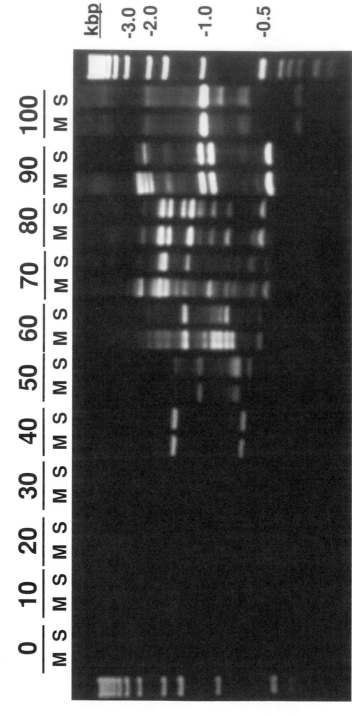

FIG. 6. Effect of primer G + C content. Primers of increasing G + C content were used to amplify genomic DNA from soybean *G. max* cv. Bonus (M) and *G. soja* PI81762 (S). The nucleotide sequence of each primer, with the percentage G + C, is 0% (5′-TAATTATTAT), 10% (5′-TAATTATTGT), 20% (5′-TAATTACTGT), 30% (5′-TAATCACTGT), 40% (5′-TAGTCACTGT), 50% (5′-TGGTCACTGT), 60% (5′-CGGTCACTGT), 70% (5′-CGGC-CACTGT), 80% (5′-CGGCCACTGT), 90% (5′-CGGCCCCTGT), 100% (5′-CGGCCCCGGC). Molecular weight markers (kbp) are as indicated.

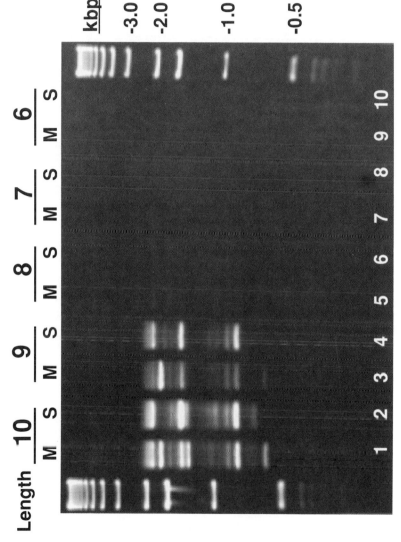

FIG. 7. Effect of primer length. Primers of decreasing length were used to amplify genomic DNA from soybean *G. max* cv. Bonus (M) and *G. soja* PI81762 (S). The nucleotide sequence of each primer, and number of nucleotides, is 10 (5′-ATTGCGTCCA), 9 (5′-TTGCGTCCA), 8 (5′-TGCGTCCA), 7 (5′-GCGTCCA), 6 (5′-CGCCCA). Molecular weight markers (kbp) are as indicated

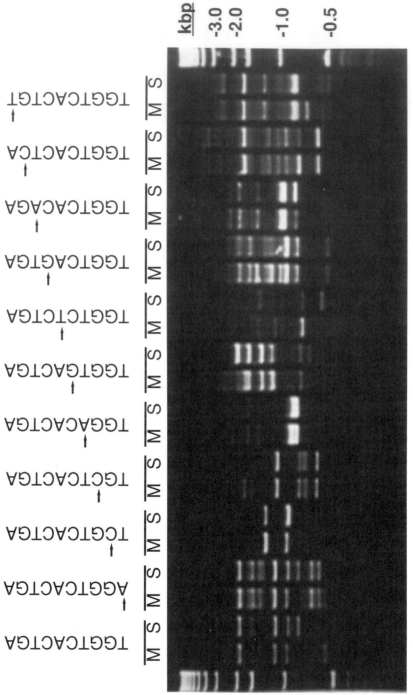

FIG. 8. Effect of nucleotide substitutions in the primer. Genomic DNA from soybean *G. max* cv. Bonus (M) and *G. soja* PI81762 (S) was amplified with the indicated primers. Arrows indicate nucleotide substitutions relative to the original primer (5′-TGGTCACTGA in lanes 1 and 2 at left of figure). Molecular weight markers (kbp) are as indicated.

Nature of Target Sites

The molecular nature of the polymorphisms detected in the RAPD assay has not been described. This would require DNA sequencing of several genomic primer-binding sites from a number of polymorphic parents. Amplification is probably initiated at many sites, which may often be imperfectly complementary to the primer, and which form short inverted repeats separated by up to several thousand nucleotides. On theoretical grounds, such sites would be expected to occur in a genome of random nucleotide sequence at a frequency that is related to genome complexity (see below). Experimental observations indicate that these sites are distributed throughout a genome, as we have seen no clear bias in the distribution of RAPD sites in the *Arabidopsis* and *Neurospora* genomes (R. Reiter and P. Scolnik; J. Williams and A. Kubelik; respectively, unpublished results, 1991). The amplified sequences between the sites belong to all abundance classes, from low copy to highly repetitive (see below). It has been suggested (H. Wu, personal communication, 1992) that some RAPD bands could result from self-priming events due to a hairpin loop formation at the 3' terminus of the first amplified strand. Such products would amplify as large inverted repeats. We have not tested this hypothesis.

Competition in Amplification

When genomic DNA samples from two individuals of the same species are mixed in different ratios, most polymorphic bands are amplified in proportion to the amount of their respective genomic DNA template (Fig. 9A). These bands may be amplified in proportion because they present equally good matches to the primer at their respective genomic target sites. Some polymorphic bands, however, may be poorly amplified and are detected only when their respective genome is present in severalfold excess over a competing genome of the same species (our unpublished data; Michelmore *et al.*;[15] G. Martin and S. Tanksley, personal communication, 1992).

When DNA from a soybean genome (high complexity) and a cyanobacterial genome (low complexity) were mixed in the same reaction, all of the detectable amplification products were from soybean, even when the smaller genome was present at up to a 460-fold molar excess (Fig. 9B). This mixing experiment and others (e.g., experiments with mixed primers; see below) suggest that the outcome of an amplification reaction is determined in part by a competition for priming sites in the genome. DNA

[15] R. W. Michelmore, I. Paran, and R. V. Kesseli, *Proc. Natl. Acad. Sci. U.S.A.* **88,** 9828 (1991).

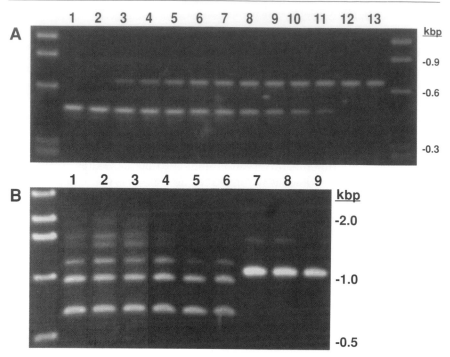

Fig. 9. Mixing of genomic DNA samples. (A) Genomes of the same complexity. *Silene alba* DNA from individuals 17 and 22 was amplified under standard conditions, using primer 5'-TCGTCACTGA. Reactions 1 through 13 contained 100, 99, 90, 80, 70, 60, 50, 40, 30, 20, 10, 1, and 0% DNA from individual 22, and 0, 1, 10, 20, 30, 40, 50, 60, 70, 80, 90, 99, and 100% DNA from individual 17, respectively. Molecular weight markers (kbp) are as indicated. (B) Genomes of different complexity. Mixtures of DNA from the soybean *G. soja* PI81762 and the cyanobacterium PCC 6803 were amplified under standard conditions with primer 5'-TCACGATGCA (Methods). Lanes 1–3, soybean DNA alone, at concentrations of 1.0, 0.2, and 0.05 μg/ml, respectively; lanes 7–9, cyanobacterial DNA alone, at 1.0, 0.2, and 0.05 μg/ml, respectively; lanes 4–6, soybean DNA at 1.0 μg/ml (all three lanes), mixed with cyanobacterial DNA at 1.0, 0.2, and 0.05 μg/ml, respectively. In lanes 4–6, the molar ratios of cyanobacteria : soybean haploid genomes are 460 : 1, 92 : 1, and 23 : 1, respectively. Molecular weight markers (kbp) are as indicated.

segments that compete poorly in the amplification reaction are expected to be unreliable as genetic markers.

One factor in the competition is likely to be the stability of the hybrid formed between the primer and its genomic target sites. A genome of high complexity should have more target sites with better complementarity to a primer, as compared to a genome of low complexity. Consider a genome comprising a random sequence of an equimolar mixture of the four nucleotides. Sequences complementary to an arbitrary oligonucleotide primer of length n bases will occur with probability $p = 4^{-n}$; two copies of the sequence arranged in inverted orientation at a particular distance suitable for amplification will occur with probability $p = 4^{-2n}$. Many distances between primer binding are suitable for amplification of the intervening sequence. For distances between 500 and 2500 bp, the probability of obtaining at least 1 of the possible 2000 distances is $p = 2000 \times 4^{-2n}$. For a given genome of complexity C (bp), the expected number of occurrences, b, is the following:

$$b = (2000 \times 4^{-2n})C \tag{1}$$

For example, for a primer–target hybrid of $n = 10$ complementary base pairs we would expect to see $b = 1.27$ bands per primer from the soybean genome $(C = 7 \times 10^8$ bp)[16] and $b = 0.007$ bands per primer from the cyanobacterial genome $(C = 4 \times 10^6$ bp). However, we generally observe an average of about five bands, in both the soybean and cyanobacterial genomes. To amplify five bands detectably, the average number of complementary bases in the primer–target hybrid is calculated from Eq. (1) to be 9.5 bp in soybean, and 7.6 bp in the smaller prokaryotic genome $(C = 4 \times 10^6$ bp). The availability of better target sites may account for the predominant amplification of soybean DNA in the mixing experiment of Fig. 9B.

Reamplification of RAPD Bands

The individual RAPD bands may be excised from the gel and reamplified for use as hybridization probes. The reamplified products usually appear as single bands of the predicted size, but background DNA picked up when excising the band may also be amplified. The purity of the reamplified probe is usually adequate for RFLP mapping or for confirmation of band identity on DNA blots of RAPD gels. For critical applications it may be necessary to subclone the band and confirm its identity, for example by following the ethidium bromide-stained RAPD segregation pattern with a DNA blot of the same gel probed with the cloned DNA.

Hybridizations

Gels containing RAPD bands can be blotted and hybridized with probes derived from selected RAPD bands. This procedure can be used to confirm the equivalence of bands, to confirm the identity of cloned RAPD bands, or to establish sequence similarity of two segregating RAPD bands of

[16] R. B. Goldberg, *Biochem. Genet.* **16,** 45 (1978).

different molecular weights, which would permit this pair of bands to be scored as a single codominant marker. Total genomic DNA or selected repetitive sequence clones can be used (see below).

RAPD Fingerprinting

RAPD markers are useful for fingerprinting genomes.[6] It is possible to reveal both a RAPD pattern visualized by ethidium bromide staining, and a second pattern visualized by blotting the RAPD gel and hybridizing it to a labeled probe containing repetitive DNA sequences (Fig. 10). Such a probe will preferentially visualize RAPD bands containing the corresponding repetitive sequence, even if they are too weak to be visible on ethidium bromide-stained agarose gels. In fact, many more bands are present among the amplification products than are detected by ethidium bromide staining. Caetano-Annoles et al.[14] were able to dtect over 100 bands amplified with a single random primer, by resolving the reaction products on a polyacrylamide gel and staining with silver.

Simplification of the RAPD Pattern by Prior DNA Cleavage with Restriction Enzymes

Digesting genomic DNA with frequently cutting restriction enzymes prior to the amplification reaction simplifies the pattern of amplified products and changes the relative intensities of the bands (Fig. 11). In some cases this approach may permit interpretation of a pattern that otherwise would be too complex. In addition, cutting the genomic DNA may reveal a restriction site polymorphism located between the primer-binding sites in an otherwise monomorphic band.

Use of Paired Primers

It should be possible to obtain additional amplified bands by using combinations of primers, as compared to single primers. This would permit more information to be obtained from a collection of primers. On purely stochastic grounds, four times as many bands would be expected from a single amplification reaction employing two primers, as compared to two separate reactions using each primer individually.[17] In fact, fewer additional bands are obtained with primer combinations than expected; many bands amplified by individual primers are no longer amplified when their respective primers are used in combination (Fig. 12). This observation is consistent with the competitive nature of the amplification reaction (see

[17] P. H. Dear and P. R. Cook, Nucleic Acids Res. 17, 6795 (1989).

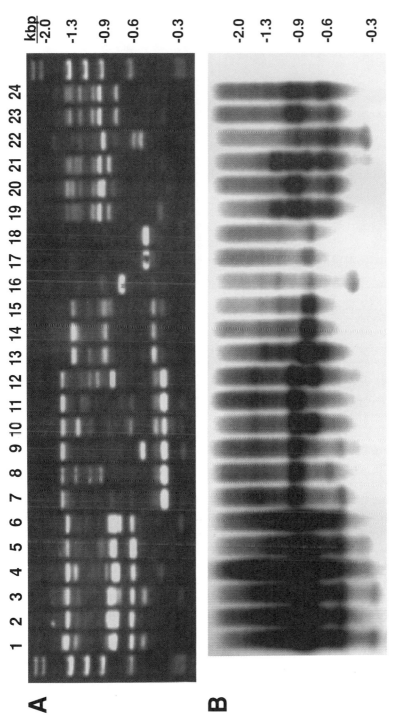

Fɪɢ. 10. RAPD fingerprinting. (A) DNA from male and female *Silene alba* plants was amplified with different primers. Lanes 1–6 (5′-CTGTAG-CATC), lanes 7–12 (5′-CTGATACGGA), lanes 13–18 (5′-TCGTCACTGA), lanes 19–24 (5′-TGCTCACTGA). Lanes 1, 7, 13, and 19, male 3; lanes 2, 8, 14, and 20, male 4; lanes 3, 9, 15, and 21, male 9; lanes 4, 10, 16, and 22, female 17; lanes 5, 11, 17, and 23, female 22; lanes 6, 12, 18, and 24, female 29. Molecular weight markers (kbp) are as indicated. (B) The gel shown in (A) was blotted onto a nylon membrane and probed with ^{32}P-labeled genomic DNA from a male *Silene* individual. The fingerprint obtained by probing with a total *Silene* DNA probe reflects the relative genomic abundance of the RAPD amplification products.

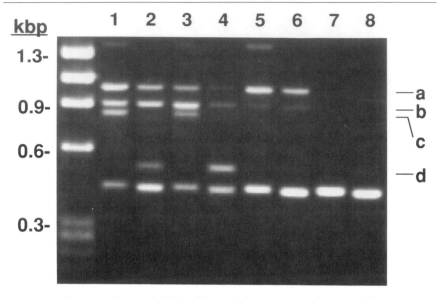

FIG. 11. Cleavage of genomic DNA with restriction enzymes prior to amplification. DNA from *G. soja* PI81762 (lanes 1, 3, 5, and 7) and *G. max* cv. Bonus (lanes 2, 4, 6, and 8) was amplified with the primer 5'-GCAAGTAGTG. DNA in lanes 1 and 2 was not cleaved; lanes 3 and 4, predigestion with *Alu*1; lanes 5 and 6, predigestion with *Hae*III; lanes 7 and 8, predigestion with *Sau*3A. Molecular weight markers (kbp) are as indicated. Predigestion of genomic DNA results in simplification of the RAPD pattern. Amplification of bands c and d was prevented by predigestion with *Hae*III, and of bands a to d by predigestion with *Sau*3A.

above). To characterize further the use of paired primers, we used a set of 7 primers individually and in all possible 21 pairwise combinations to amplify DNA segments from soybean. The number of detectable bands was counted for each reaction, and the results were averaged over the number of tested primers or primer combinations. Each primer when used individually amplified an average of 5.3 bands, whereas each pair of primers amplified only 4.4 bands. Of the average 4.4 bands seen with paired primers, half (2.2 bands) could be identified as "old" bands amplified by one of the individual primers comprising a pair, and half (2.2) were "new" bands dependent on both primers. In summary, 7 reactions using individual primers amplified 37 bands (7 reactions × 5.3 bands per reaction), and 21 additional reactions using the same set of primers in pairs amplified 46 "new" bands (21 reactions × 2.2 new bands per reaction). Thus, primers should be used individually to maximize the number of unique bands generated *per reaction;* and they should be used in pairs to maximize the number of unique bands obtained *per primer.*

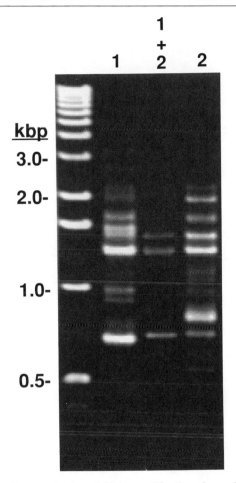

FIG. 12. Addition of a second primer inhibits amplification of some DNA segments. DNA from the soybean cultivar *G. soja* PI81762 was amplified with primer 4 (lane 1), primer 7 (lane 2), or a combination of both primers (lane 1 + 2). Molecular weight markers (kbp) are as indicated. Nucleotide sequences of the primers are as follow: 4, 5'-CTGAAGCTAC; 7, 5'-ACGGTACACT.

Applications

Genetic Mapping

When a RAPD marker is detected as a DNA segment amplified from one parent in a genetic cross but not from the other parent, the marker can be followed in the segregating progeny and can be assigned to a locus

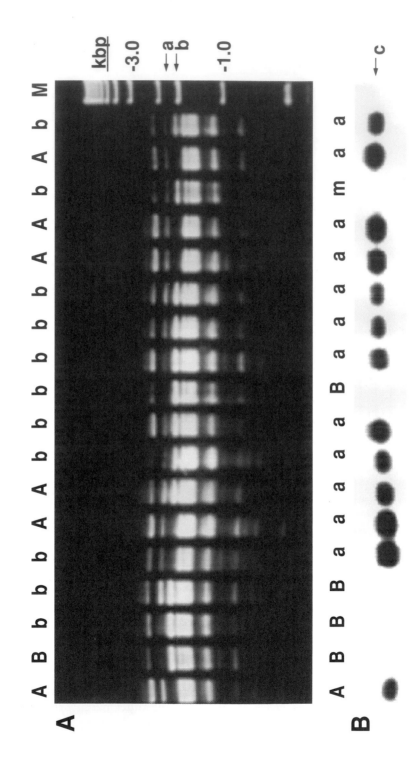

in a genetic map. For example, the RAPD marker AP11a was scored as a dominant marker among 66 soybean F_2 progeny (Fig. 13A), and its segregation correlated with the segregation of 430 RFLP markers[18] in the same population. Analysis of the data indicated that AP11a maps to linkage group 5 at the position shown in Fig. 14A. This result was confirmed by reamplifyng the AP11a RAPD band, using it as a radiolabeled hybridization probe to detect an RFLP, and showing cosegregation of this RFLP with its respective RAPD (Figs. 13B and 14B). In repeating this experiment with several different RAPD markers, it was noted that some of the amplified polymorphic DNA segments were not suitable as RFLP probes because of hybridization to repetitive DNA. Of 11 amplified probes tested, 6 hybridized to single-copy DNA (Fig. 15A), 3 hybridized to middle-repetitive DNA (Fig. 15B) and 2 hybridized to highly repetitive DNA in the soybean genome (Fig. 15C). This indicates that RAPDs can provide DNA markers in genomic regions that are not accessible to RFLP analysis due to the presence of repetitive DNA sequences.

RAPD markers are dominant markers, because the presence of a given RAPD band does not distinguish whether its respective locus is homozygous or heterozygous. If it is necessary to identify heterozygous regions, two closely linked RAPD markers, each amplified from a different parent, may be used as a pair. For example, amplification of both markers of the pair is diagnostic for a heterozygous genomic region, with an uncertainty equal to the recombination distance between the two markers. This use

---

[18] S. V. Tingey, J. A. Rafalski, J. G. K. Williams, and S. Sebastian. *Proc. NATO Adv. Study Inst. 6th, Plant Mol. Biol., Schloss Elmau, Ger., 1990* in press (1991).

FIG. 13. Genetic mapping with RAPD markers: (A) Segregation of a RAPD marker in soybean. Genomic DNA samples from 66 F_2 progeny were amplified with the primer AP11a (5'-ACCTCGAGCACTGTCT). Amplification products from the inbred (homozygous) parents *G. max* cv. Bonus, *G. soja* PI81762 (lanes 1 and 2, respectively), and 16 F_2 progeny are shown. Segregation scores are listed above each lane, and lane M contains molecular weight markers. Scores (for both Figs. 13A and B) are interpreted in the following way: A, homozygous genotype of parent A (lane 1); B, homozygous genotype of parent B (lane 2); a, ambiguous for either homozygote A or a heterozygote; b, ambiguous for either B or a heterozygote; m, missing data. Arrow b at the right side of the figure points to a RAPD band, AP11a, that can be scored with confidence. Arrow a points to a band that appears to be polymorphic in the parents, but cannot be scored with confidence among the progeny, and thus is not useful as a genetic marker. (B) Segregation of an RFLP marker detected by a RAPD DNA probe. The AP11a RAPD [(A), lane 2, at position noted by arrow b] amplified in *G. soja* was used as a hybridization probe to detect a *Bcl*II RFLP. Shown here is the hybridization of this probe to a DNA blot of *Bcl*II-digested genomic DNA from the same individuals shown in (A). Because only one parental allele could be scored with confidence on the autoradiogram (arrow c; and see Fig. 14B), it was scored as a dominant marker.

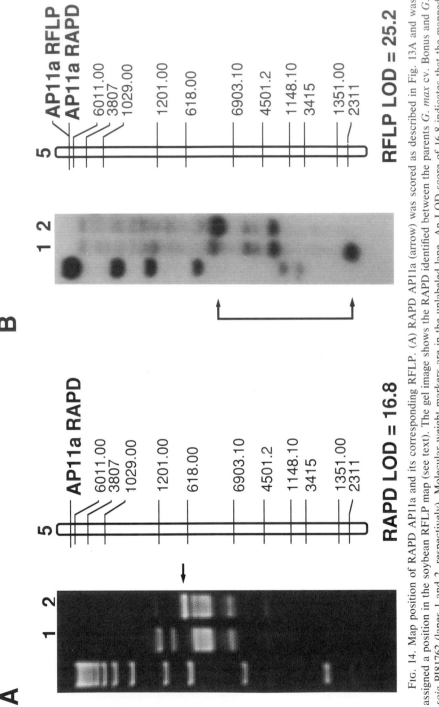

FIG. 14. Map position of RAPD AP11a and its corresponding RFLP. (A) RAPD AP11a and its corresponding RFLP. (A) RAPD AP11a (arrow) was scored as described in Fig. 13A and was assigned a position in the soybean RFLP map (see text). The gel image shows the RAPD identified between the parents *G. max* cv. Bonus and *G. soja* PI81762 (lanes 1 and 2, respectively). Molecular weight markers are in the unlabeled lane. An LOD score of 16.8 indicates that the mapped position of this RAPD is 10¹⁶·⁸ more likely than the probability of it being unlinked to this group of RFLP markers. (B) RAPD AP11a was used as a hybridization probe to detect the *Bc*/II RFLP AP11a (see Fig. 13B, and arrow in this figure). The RFLP maps to the same position as its corresponding RAPD probe, which confirms the equivalence of the two types of DNA marker.

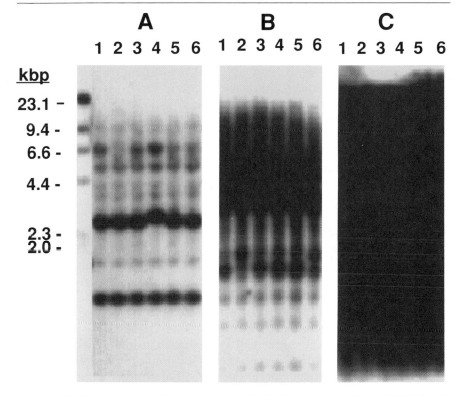

FIG. 15. Classes of genomic sequences amplified with arbitrary primers. RAPD bands were used as hybridization probes on Southern blots of *Eco*RI-digested genomic DNA from the soybean varieties *G. max* cv. Bonus (lane 1), PI81762 (lane 2), PI416937 (lane 3), N85-2176 (lane 4), PI153293 (lane 5), and PI230970 (lane 6). (A) The hybridization of a RAPD band amplified by the primer 5'-TGGTCACTGT to single-copy DNA. (B) The hybridization of a RAPD band amplified by the primer 5'-TCACGATGCA to middle repetitive DNA. (C) The hybridization of a RAPD band amplified by the primer 5' ATTGCGTCCA to highly repetitive DNA. Molecular weight markers (kbp) are as indicated.

of RAPD markers in pairs requires twice as many markers as compared to codominant markers (RFLPs). Alternatively, RAPD bands may be excised and used as probes to detect codominant RFLPs as described above. It is also possible to quantitate the intensity of RAPD bands using standard densitometric techniques. A RAPD band derived from a heterozygous region will have half the intensity of the band derived from a homozygous region (unpublished observations, 1990). However, the reliability of the quantitation remains to be determined.

Statistical Aspects of Genetic Mapping with RAPD Markers

In this section we discuss the relative information content of dominant and codominant markers in different mapping populations. Less information is obtained from dominant markers (RAPDs) than from codominant markers (RFLPs). Allard[19] provides equations for calculating error in estimating the recombination interval between markers for two-point mapping, but not for multipoint mapping. To estimate error in a multipoint map of dominant RAPD markers, we calculated multipoint maps from simulated data for several different marker densities.[20] The standard deviation for multipoint analysis was calculated from the difference between the "actual" percentage recombination (defined in the simulation model) and the computed percentage recombination (from analysis of the simulated data by Mapmaker[13]) over all intervals in the map. The standard deviation for two-point analysis was calculated using the equations provided by Allard[19] (for F_2 and backcross populations) and by Reiter *et al.*[21] (for recombinant inbred populations). The mean marker-to-marker interval for each simulated map was plotted against the respective multipoint standard deviation (Fig. 16, symbols and broken lines; see caption) and the predicted two-point standard deviation (solid lines). The results of this comparison indicate that in F_2 populations, the multipoint standard deviation is about two times greater for dominant markers than for codominant markers (Fig. 16A; compare curves F_2 dom-Mpt and F_2 codom). Dominant and codominant markers are equally informative, however, in recombinant inbred populations (because nearly all loci are homozygous) and backcross populations (because all alleles from the recurrent parent are homozygous, all alleles from the donor parent are heterozygous). When using dominant RAPD markers, recombination intervals are most accurately calculated in recombinant inbred and backcross populations (the best population depends on the recombination interval; Fig. 16A, curves RI and BC), and are least accurate in F_2 populations (Fig. 16A, curve F_2 dom-Mpt). Whereas populations of 800 individuals were used to calculate the standard deviations plotted in Fig. 16, standard deviations for other population sizes, n, may be calculated by multiplying the values of Fig. 16 by $(800/n)^{1/2}$.

We next discuss in detail the contributions of flanking markers to the estimation of recombination intervals in multipoint mapping. The data in

[19] R. W. Allard, *Hilgardia* **24**, 235 (1956).
[20] Simulations were performed by the UNIX computer program "Genotype," developed at the Du Pont Company by Mike K. Hanafey.
[21] R. S. Reiter, K. A. Feldman, J. G. K. Williams, J. A. Rafalski, S. V. Tingey, and P. A. Scolnik. *Proc. Natl. Acad. Sci. U.S.A.* **89**, 1477 (1992).

Fig. 16A show that there is very little difference in error between two-point and multipoint maps, except in the case of an F_2 population with dominant markers. In this case, the simulated multipoint maps have a significantly smaller error than the two-point predictions, especially for closely linked markers (Fig. 16A, curves F_2 dom-2pt and F_2 dom-Mpt). To examine the source of this difference in error, the results for the simulated F_2 dom-Mpt case were split into two categories of marker pairs according to the phase of the markers comprising the pair (Fig. 16B; see caption for explanation of marker phase). From this analysis, it is clear that there is no difference between two-point and multipoint error for markers linked in coupling (Fig. 16B, curve F_2 coupled-dom; the two-point prediction is superimposed on the simulated data points); all of the difference comes from the pairs of markers linked in repulsion (Fig. 16B, curves F_2 repelled-dom-2pt and F_2 repelled-dom-Mpt). Once the markers have been mapped, the relative phase of adjacent markers is known and a different level of confidence on distance can be given depending on phase. For coupled dominant markers, the two-point equation gives a good estimate of error, but for repelled dominant markers the error must be estimated from Fig. 16B (F_2 repelled-dom-Mpt).

The fact that the two-point equation closely describes the error for coupled dominant markers implies that the state of flanking markers is relatively unimportant. In contrast, for a pair of repelled dominant markers, flanking markers play a significant part in determining the multipoint map error. To examine in more detail the effect of the flanking marker configuration on multipoint mapping, an F_2 population of 500 individuals was simulated with 4 markers separated by intervals of 10 cM. The likelihood for a range of recombination distances between the second and third markers was calculated with Mapmaker. The results are plotted in Fig. 17 (see caption for an explanation of symbols). The steeper the likelihood drops off, the greater the information content of the markers. Without missing data or incorrect scoring, two-point and multipoint mapping of codominant markers is virtually identical (curves 1 and 2, Fig. 17). This follows from the information content of closely spaced codominant markers being almost equal to fully classified genotypes. The advantage of multipoint mapping of coupled-dominant markers is also very small (curves 3 and 4, Fig. 17). The relatively low information content of two-point repelled dominant markers is reflected in the less steep drop-off from the maximum in curve 5 (Fig. 17), and flanking coupled dominant markers add little information (curve 6, Fig. 17). The maximum likelihood distance (20 or 22 cM) is significantly displaced from the correct solution (10 cM), which is a reflection of the greater variance in this case. In contrast, a group of three pairs of repelled dominant markers has almost the same

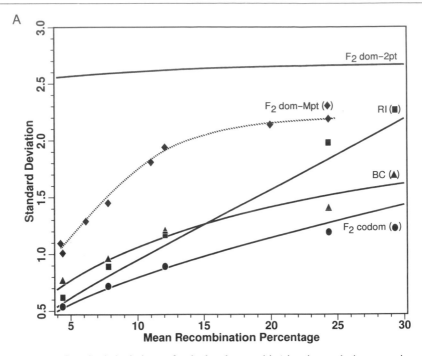

FIG. 16. Standard deviations of calculated recombination intervals in two-point and multipoint maps. (A) Two-point error was calculated from published equations, and multipoint error was calculated using simulated data for F_2, backcross, and recombinant inbred populations (labeled F_2, BC, and RI, respectively) of 800 individuals each.[20] The mean recombination percentage over all marker pairs (x axis) was calculated for each simulated map. Standard deviations (y axis) were calculated as described below. To calculate standard deviations for a different population size, n, the standard deviations in this figure should be multiplied by $(800/n)^{1/2}$. In an F_2 population, dominant and codominant markers have different information content, so the F_2 curves are distinguished as either F_2 dom or F_2 codom, respectively; in recombinant inbred and backcross populations marker type is not distinguished in the figure because both types have equal information content (see below). In most cases, the two-point solutions (solid lines) provide a good fit to the simulated data (symbols); the fit is poor, however, for dominant markers in an F_2 population, so the two-point prediction (F_2 dom-2pt, solid line) is drawn separately from the multipoint result (F_2 dom-Mpt, broken line).

Two-point standard deviation: For two-point analysis (solid lines in the figure), the standard deviation in recombination percentage for a pair of markers separated by the indicated distance was calculated as $100/(800I)^{1/2}$, with I defined for the different kinds of markers and populations according to Allard[19] and Reiter et al.[21]:

$$F_2 \text{ codom} = \frac{2(1 - 3\theta + 3\theta^2)}{\theta(1 - \theta)(1 - 2\theta + 2\theta^2)}$$

$$F_2 \text{ dom-2pt} = \frac{1}{\dfrac{(2 + \theta^2)(1 - \theta^2)}{4(1 + 2\theta^2)} + \dfrac{[2 + (1 - \theta)^2][1 - (1 - \theta)^2]}{4[1 + 2(1 - \theta)^2]}}$$

B

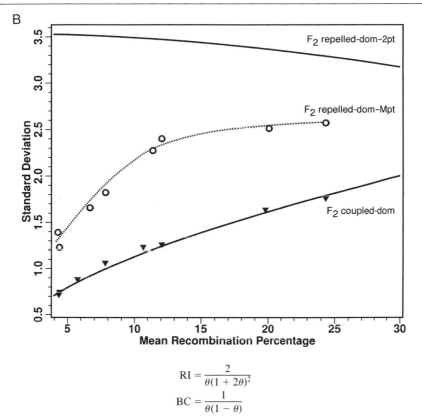

$$RI = \frac{2}{\theta(1 + 2\theta)^2}$$

$$BC = \frac{1}{\theta(1 - \theta)}$$

A two-point equation for i is not explicitly provided by Allard[19] for the case F_2 dom-2pt. However, because variance is additive, and because in a random map the expected frequencies of coupled and repelled marker pairs are each 0.5, the standard deviation for a random dominant two-point map is obtained by averaging the variances of both marker phases, $[(c^2 + r^2)/2]^{1/2}$, where c is the two-point standard deviation of coupled pairs, and r is the two-point standard deviation of repelled pairs.

Simulation and multipoint standard deviation: For multipoint maps, populations of 800 individuals were simulated for 400 randomly distributed marker pairs.[20] To obtain a total of 400 marker pairs for each average distance indicated in the figure, the simulation results for several independent maps (2000 cM, 10 equal-length linkage groups) were combined as required. Alternatively, and essentially equivalently, the desired 400 marker pairs were obtained by simulating larger genomes (e.g., 20,000 cM) of 10 equal-length linkage groups. The mean marker distance for each case was computed from the distribution of marker intervals in the respective simulated genomes. Standard deviations were computed from the difference between the recombination percentage as calculated by Mapmaker (multipoint) and the interval as defined in the simulation model

$$\sqrt{\frac{1}{N} \sum_{i=1}^{N} (calc_i - actual_i)^2}$$

information content of coupled dominant markers in a multipoint map (curve 7, Fig. 17). With one flanking coupled dominant marker, and one flanking repelled dominant marker, intermediate curves result (curves 8 and 9, Fig. 17). Analyses of several simulations indicate that although the maximum likelihood solution for each simulation may vary (the source of variance in estimating the interval), the shapes of the curves remain the same as those plotted in Fig. 17.

Near-Isogenic Lines and F_2 Pool Selection

Near-isogenic lines can be used with RAPD markers to identify the genomic regions derived from the trait donor parent, as described.[22,23] Several laboratories are currently using this approach for the identification of RAPD markers linked to disease resistance.

Recently, Michelmore[15] described a novel method for linking traits with DNA markers, named bulked segregant analysis (BSA). The BSA

[22] G. B. Martin, J. G. K. Williams, and S. D. Tanksley, *Proc. Natl. Acad. Sci. U.S.A.* **88,** 2336 (1991).

[23] I. Paran, R. Kessli, and R. Michelmore, *Genome* **34,** 1021 (1991).

Scoring of markers in simulated populations: In the F_2 populations, the presence of a dominant marker was scored as ambiguous (could be homozygous or heterozygous); and the absence of a dominant marker was scored as homozygous for the undetected allele. Recombinant inbred populations were simulated by a cross between the parents followed by eight self-crosses; because 99.6% of loci were homozygous in these populations, dominant markers scored as homozygous loci are nearly as informative as codominant markers; however, codominant markers were used in this analysis. Recombinant inbred map distances were computed by Mapmaker[13] as for an F_2, and the resulting apparent recombination percentages, R, were converted using the equation $\theta = R/2(1 - R)$. In backcross populations, the presence of a dominant marker was scored as heterozygous, the absence of a marker was scored as homozygous for the undetected allele, and map distances were computed for backcross populations by Mapmaker.[13]

(B) The marker pairs used to compute curve F_2 dom-Mpt (A) were split into two categories according to marker phase, and the standard deviations for the coupled and repelled phases were computed separately for the approximately 200 marker pairs in each category. For two-point analysis (solid lines in the figure), the standard deviation in recombination percentage for a pair of markers separated by the indicated distance was calculated as $100/(800i)^{1/2}$, with i defined for coupled and repelled markers according to Allard[19]:

$$F_2 \text{ coupled-dom} = \frac{2(1 + 2\Theta^2)}{(2 + \Theta^2)(1 - \Theta^2)}, \qquad \Theta = 1 - \theta$$

$$F_2 \text{ repelled-dom} = \frac{2(1 + 2\theta^2)}{(2 + \theta^2)(1 - \theta^2)}$$

Other calculations and symbols are as in (A).

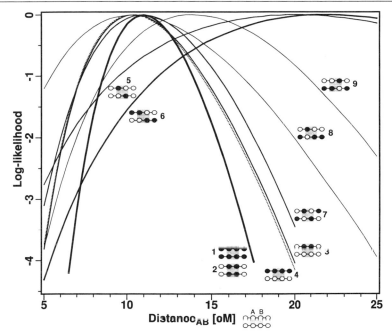

FIG. 17. Results of a single simulation showing log-likelihood vs recombination interval. An F_2 population of 500 individuals was simulated with four marked loci spaced at 10-cM intervals. Log-likelihood was calculated with Mapmaker for a range of distances between the second and third loci (shaded boxes), with the remaining intervals fixed at 10 cM. Nine different states of the four-locus set were analyzed (curves 1–9). Each cluster of eight circles represents the eight alleles of the four markers in the F_1 parent; one homolog of the linkage group is represented by the top line of four circles, and the other homolog is represented by the bottom line. Filled circles indicate alleles that can be detected; open circles are undetectable alleles. For example: —●— is a codominant marker at a locus because both alleles can be detected; —●— is a dominant marker because only one allele can be detected; and —○— means that the alleles cannot be distinguished, so this locus contributes no information to the map. Multipoint analysis is equivalent to two-point analysis for marker states 2, 3, and 5, where no flanking markers are available.

method can be used with either RFLP or RAPD markers. In a population segregating for a trait of interest, individuals are sorted into two groups based on the expression of that trait. The two groups will differ at loci linked to the trait, but will be randomly segregating for unlinked regions. DNA samples from each group are pooled separately, and the two DNA pools are screened for RAPD or RFLP polymorphisms. Markers polymorphic between the two pools are expected to be linked to the trait of interest. Three RAPD markers linked to *Dm5/8*, a downy mildew resistance locus for which no near isogenic lines were available, have been identified in

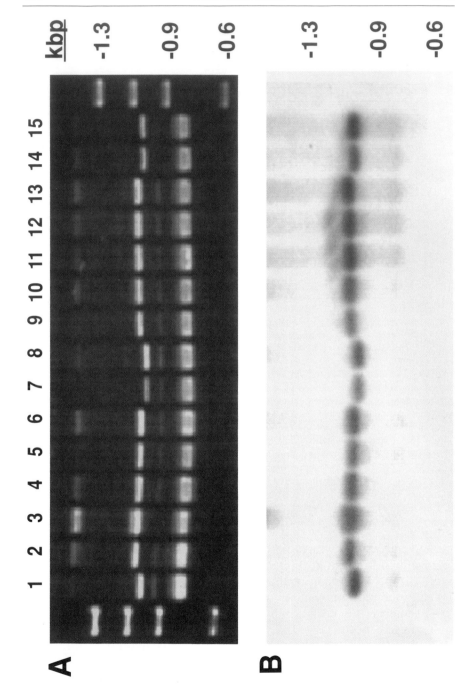

lettuce by screening with 100 RAPD primers.[15] The BSA method is applicable to a variety of simply inherited traits, making construction of near-isogenic lines or genetic maps unnecessary to identify linked DNA markers.

F_2 pools may also be selected from a genotyped population according to the genotype at a locus of interest.[24] This is an efficient strategy for identifying molecular markers that map specifically to a given locus, by screening markers on only two pooled DNA samples.

Use of Mapped RAPD Markers in Different Genetic Crosses

A RAPD marker mapped in one cross may be nonpolymorphic in another cross involving different parents. To use the same marker in the second cross, it may be necessary to find a polymorphism at the locus using another procedure. For example, if the RAPD band can be amplified from both parents, the DNA could be cleaved with restriction enzymes, either before or after amplification, to identify a restriction site polymorphism lying between the priming sites. Sequence-based differences in secondary structure could be detected on a gradient denaturing gel or by the single-stranded conformation polymorphism (SSCP) method.[25] Alternatively, the previously mapped band could be used as a probe to detect an RFLP, or could be sequenced to discover a sequence polymorphism that may then be detected using a PCR-based assay or other method.[26,27]

[24] D. A. Patton, L. H. Franzmann, and D. W. Meinke, *Mol. Gen. Genet.* **227,** 337 (1991).

[25] M. Orita, Y. Suzuki, T. Sekiya, and K. Hayashi, *Genomics* **5,** 874 (1989).

[26] U. Landegren, R. Kaiser, J. Sanders, and L. Hood, *Science* **241,** 1077 (1988).

[27] A.-C. Syvanen, K. Aalto-Setala, L. Harju, K. Kontula, and H. Soderlund, *Genomics* **8,** 684 (1990).

FIG. 18. RAPD amplification of DNA from different soybean cultivars with the primer 5'-CTCTTGCTAC. (A) Lane 1, *G. max* cv. Bonus; lane 2, *G. soja* PI81762; lane 3, cv. Williams 82; lane 4, cv. A4595; lane 5, cv. A3205, lane 6, cv. A3307; lane 7, cv. A5474; lane 8, cv. Essex; lane 9, cv. A3966; lane 10, cv. A1937; lane 11, cv. A3127; lane 12, cv. A4271; lane 13, cv. A4997; lane 14, cv. Mukden; lane 15, *G. max* PI54610. Two polymorphic bands are seen in the 1-kbp region. Molecular weight markers (kbp) are as indicated. (B) To determine whether the two polymorphic bands seen in (A) were amplified from the same locus, the gel was blotted onto a nylon membrane and probed with a ^{32}P-labeled, reamplified, 1-kb RAPD band excised from *G. max* cv. Bonus [e.g., see 1-kb band in (A), lane 1]. Hybridization of the probe to both polymorphic bands suggests that the two alleles seen in (A) are of the same locus. This particular RAPD is unusual in producing an amplified band for each allele, because most RAPD alleles are revealed as the presence or absence of a band (see text).

Population Genetics

DNA markers can be used to measure similarity among individuals within natural or artificial (through breeding) populations within a species. To compare the relative efficiencies of RAPD vs RFLP markers, consider that a single RAPD primer usually will amplify several independent genetic loci, but can be used to identify the presence of only one allele. Only the presence or absence of a RAPD band can be detected. In contrast, RFLP probes can usually be used to assay only one or a few loci of complementary nucleotide sequences, but can be used to detect multiple alleles at each locus due to the variations in the sizes of the RFLP fragments. In corn as many as eight alleles were distinguished at a single locus by an RFLP probe, whereas in soybean most RFLP probes detect no more than two alleles at each of two homologous loci.[18,28] Thus, it may be advisable to use RFLP probes for studies of genetically diverse outbreeding populations such as corn, whereas RAPD markers would be suitable for analysis of more uniform, inbred species such as soybean (Fig. 18). Diversity within populations and species is influenced by a multiplicity of factors including life cycle, generation time, outcrosser or inbreeder, means of pollen dispersal, geographic range, and ecological niche. These biological traits should be recognized in developing strategies for quantifying diversity with DNA markers.

Molecular Taxonomy

Amplification with RAPD primers is extremely sensitive to single-base changes in the primer–target site (Fig. 8). This feature suggests that RAPDs should be highly useful for phylogenetic analysis among closely related individuals (e.g., Fig. 18), but less useful for analysis of genetically diverse individuals. To test this, 10 different species of the genus *Glycine* were assayed with RAPD primers to identify shared characters within and between species. Five RAPD primers identified 61 different characters within the 10 species. Of 61 different characters identified among the tested species, 17 were shared between at least 2 different species, and 44 were shared only within a single species. As can be seen in Fig. 19, RAPD primers were useful for distinguishing species. They were also particularly useful for discriminating between different isolates of the same species (e.g., compare CAN 405, CAN 434, and CAN 666). The identification of shared characters in this analysis is limited to the resolution of the agarose gel. Figure 19 (panels 1, 2, and 3) illustrate the challenges to scoring identity for a particular character. Nucleic acid hybridizations were used

[28] P. Keim, R. C. Shoemaker, and R. G. Palmer, *Theor. Appl. Genet.* **77,** 786 (1989).

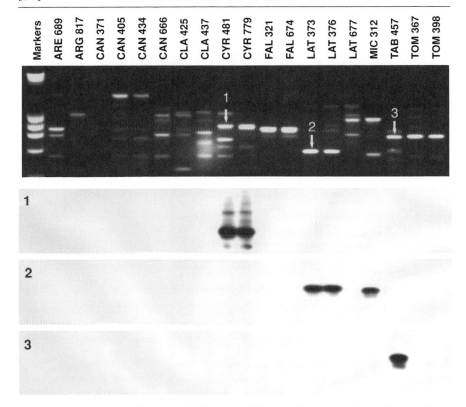

FIG. 19. RAPD amplification of DNA from different *Glycine* species with one primer. The RAPD primer 5'-TGGTCTCTGA was used to amplify DNA from several different isolates representing different species of the genus *Glycine*. These species were *G. arenaria* (ARE), *G. argyrea* (ARG), *G. canescens* (CAN), *G. clandestina* (CLA), *G. cyrtoloba* (CYR), *G. falcata* (FAL), *G. latifolia* (LAT), *G. microphylla* (MIC), *G. tabacina (TAB), and G. tomentella* (TOM). The tested species and individuals are indicated above each lane (e.g., ARE 689 is species ARE, isolate 689). Bands that comigrate on the gel may be identical, or they may represent different loci and their comigration may simply be fortuitous. To establish band identity, three different bands were excised from a gel (bands labeled 1, 2, and 3), labeled with [32]P (Methods), and hybridized individually to blots of the gel shown here. The results indicate that band 1 of CYR 481 is unique to the CYR individuals (panel 1), that band 2 of LAT 373 is present in only two of the three LAT individuals and is shared with the MIC individual (panel 2), and that band 3 of TAB 457 is unique to this individual (panel 3). Although band 3 comigrates with bands in both TOM individuals, they actually represent different loci.

to confirm the identity of characters 1 and 2, but revealed a potential misclassification for character 3 between *G. tabaccina* and *G. tomentella*. When the scores for nine different characters were confirmed by hybridization, only one character (Fig. 19, panel 3) would have been misclassified on the basis of gel resolution alone. This particular error could have been avoided simply by running the analytical gel longer to provide better resolution (data not shown).

Concluding Remarks

RAPD markers provide the geneticist with a new tool to explore the genetics of sexually reproducing organisms, with applications in gene mapping, population genetics, molecular systematics, and marker-assisted selection in plant and animal breeding. In most cases, data can be generated faster and with less labor than by previous methods. The process can be set up in a small laboratory and there is no need to use radioactive isotopes, making it accessible to a broad range of biologists.

Acknowledgments

The authors would like to thank Richard Michelmore, Mike Arnold, Robert Reiter, and Pablo Scolnik for communication of unpublished data; Steve Kresovich for helpful comments on utility of RAPDs in population genetics and systemics; Barbara Mazur for carefully reviewing the manuscript; and Karlene Butler, Al Ciuffetelli, Terri Grier, Will Krespan, and JoAnne Lynch for their technical assistance.

Author Index

Numbers in parentheses are footnote reference numbers and indicate that an author's work is referred to although the name is not cited in the text.

R

Rabinovitch, P. S., 314
Rafalski, J. A., 256, 704, 705, 727, 730, 732(21), 738(18)
Raff, M., 123, 127(3), 141(3)
Ragsdale, C. W., 353
Rajavashisth, T. B., 554(26), 555
Rajewsky, K., 80
Ralph, W. W., 670
Ramachandran, K. L., 189, 221(21)
Ramirez-Solis, R., 256, 314
Ranki, A., 476, 484(11), 485(11)
Ranki, M., 477, 481(13), 484(11), 485(11)
Rao, C. D., 449, 453(25)
Rapp, U. R., 449, 452(15)
Rappolee, D. A., 420, 421, 421(6)
Rauch, M., 569
Rawlins, D. R., 552
Rawlinson, W. D., 173
Ray, P. N., 634, 635(24)
Read, C. A., 176, 442
Reardon, C. A., 479
Recknor, M., 123, 127(3), 141(3)
Reddy, E. P., 449, 452(14)
Reddy, K. J., 237
Redmond, S., 449, 453(27)
Reece, K. S., 449, 452(10)
Reed, R. R., 281, 284, 285, 292(21), 295(21), 296(21)
Reeve, M. A., 176
Reid, P. D., 671, 684
Reinhart, D., 554
Reis, R.J.S., 352
Reiss, J., 334
Reiter, R. S., 730, 732(21)
Reith, W., 554
Rembaum, A., 638
Renz, M., 190
Revzin, A., 551, 565(3)
Reznikoff, W. S., 294
Rhoads, R. E., 85, 142, 379
Rhodes, C., 544
Ribero, J.C.C., 696
Rice, C. M., 651
Rich, J. J., 313
Richards, J. H., 696
Richards, W. L., 649
Richardson, C. C., 50, 51(21), 108, 113, 113(26), 128, 204, 211(39), 234, 247, 269

Richardson, T. C., 190, 207(30)
Richon, V., 545
Richterich, P., 26, 187, 188, 189, 189(11), 190(11), 191(11), 194(11), 195(11), 199(11), 204(11), 205(11), 221
Ricken, G., 335
Rickwood, D., 641
Riedel, H., 639, 647(16)
Rigby, P. W., 568
Rigby, P.W.J., 544
Riggs, A. D., 313, 421, 537, 542(25), 571, 696
Riley, J., 313
Ringold, G. M., 637
Riordan, J. R., 11, 627
Riva, S., 535, 550(8)
Robb, R. J., 640
Robbins, K. C., 449, 452(9), 453(25)
Roberg, K., 50
Roberts, J. D., 431, 591
Roberts, K., 684
Roberts, L., 380
Roberts, M. P., 535, 538(2), 539(2), 612
Roberts, R. A., 170
Roberts, T. M., 359
Robertson, C. W., 123, 153(4), 189, 242
Robinson, E. A., 280, 304(2)
Robinson, M. A., 335
Robinson, M. O., 421, 435(14)
Roe, A., 15
Roe, B., 108
Roe, S., 190, 207(30)
Roeder, R. G., 613
Rogers, S. G., 90, 91(30)
Roman-Roman, S., 335
Rommens, J. M., 11, 627
Ron, D., 403, 409(11)
Rosen, N. L., 338
Rosenberg, C. E., 528
Rosenberg, L. E., 27, 232
Rosenberg, W. M. C., 335
Rosenblatt, J. D., 421
Rosenfeld, M. G., 554(24), 555
Rosenthal, A., 93, 315
Rosenzweig, S., 235, 237(7)
Roskey, M., 294
Rosler, U., 334
Rossi, J. J., 447
Rossignol, J.-M., 588
Roth, J., 80
Rothblum, K. N., 473

Subject Index

A

from dried agarose
  excision of, 701
  purification of, 700–701
flanking IS*30*, nucleotide sequence of,
  318
identification of, for cloning, 697–700
  analytical electrophoresis for, 698–
    699
  digesting genomic DNA for, 698
  drying gel for, 699
  hybridization/autoradiography for,
    699–700
  preparing and labeling oligonucleotide
    probe for, 697
  preparing radiolabeled marker frag-
    ments for, 698
isolation and labeling of, 543–544
long, subcloning of
  procedures for, 47–48
  by reverse cloning, 47–58
purification of
  based on agarose gel electrophoresis,
    513
  by FPLC with Mono Q column, 513
DNA ligase. *See* T4 DNA ligase
DNA markers, RAPD, 704–740
DNA polymerase, 247. *See also Taq* DNA
  polymerase; T7 DNA polymerase
comparison of, 715
Hot Tub, 274
thermostable, 706
Vent, long-distance direct PCR amplifications by, products of, 274
DNA–protein complexes, long terminal
  repeat (LTR)-MMTV, cell-specific,
  characterization of, 616–618
  by Southwestern blot, 617
DNA substrates
3′ end ³²P labeling of, 597
5′ end ³²P labeling of, 596
for *in situ* detection of DNA-metabolizing enzymes, 592–593
Dot-blot typing, 370
  reverse, 369–381
Double-stranded DNA, dideoxy sequencing of, protocols for, 17
Double-stranded template, 5
  enzymatic conversion to single-stranded
    template, with T7 gene 6 exonu-
    clease, 105, 116, 118

preparation of, 17
purification of, by PEG precipitation,
  105, 116–117
sequencing of, with *Taq* DNA polymerase, 108
Drosophila
cross-species PCR on DNA from
  preparation of DNA for, 497
  yeast TBP amino acid sequence
    primers for, 499
  TBP PCR product, derived sequence of,
    compared with yeast TBP, 505
drPATTY. *See* Polymerization chain
  reaction (PCR)-aided transcript titration assay, differential restriction
dsDNA. *See* Double-stranded DNA
dsPATTY. *See* Polymerization chain
  reaction (PCR)-aided transcript titration assay, differential-size
DTE. *See* Direct transfer electrophoresis
Dye-labeled oligonucleotide
  reversed-phase HPLC of, 149
  synthesis of, 143
Dye-labeled primer
  fluorescent sequencing with, 100–102
  specific primer-directed sequencing with,
    122–153
  spectral characteristics of, 150
Dye-labeled terminators, fluorescent sequencing with, 99–100
Dye–primer sets
  purification of, 152
  synthesis of, 142–143, 152
Dye-terminator chemistry, 153
Dynabeads, 509–511
  analytical purification of transcription
    factor τ by, 522–524
  coupling of antibodies to, 643–645
  M-450, 639–640
  M-280 streptavidin, selective coupling of
    biotinylated DNA fragments to, 516

E

*Eco*RI/*Rsa*I 517-bp restriction fragment,
  from pBR322, comparison of G,
  G + A, and A reactions on, 224–225
Electroblotting, 543
Electrophoresis. *See also* Gel electropho-
  resis

Exometh sequencing, with transposon, 305

Exonuclease

choice of, for direct sequencing of PCR products, factors influencing, 90–92

sequencing reactions mediated by, 89–90

T7 gene 6, conversion of double-stranded template to single-stranded template with, 105, 116, 118

λ-Exonuclease

direct sequencing of PCR-amplified DNA with, 532–534

single-stranded DNA generated by, 7

Expression library

construction of, 557–558

cDNA synthesis and cloning for, 557–558

vectors for, 558

screening of, 558–561

binding and wash conditions for, 560–561

with nonspecific competitor DNA, 560

preparation of protein replica filters for, 558–559

protocols for, 558–559

with recognition site DNA probe, 559–560

Extensins, localization of, 682–683

F

Factor IX gene

Alu 4a polymorphism in, PAMSA of, 399

human, amino acid 397 of, oligonucleotides specific for transition at, 393

Fast protein liquid chromatography (FPLC), with Mono Q column, purification of DNA fragments by, 513

Fingerprinting

direct, 244–245

future of, 256–258

large-scale, 256–257

random, 257

RAPD, 722–723

by sampled sequencing, 241–258

Flexible primer strategy, 23–25

Fluorescent sequencing, 173

automated

of PCR products, 104–121

materials and reagents for, 110–111

troubleshooting, 118–121

specific primer-directed, 122–153

deoxy- and dideoxynucleotides for use in, composition of stock solutions, 166

developments in, 153

with dye-labeled primers, 100–102

with dye-labeled terminators, 99–100

gel handling for, 183–185

linear amplification, 179–181

reverse-primer, 181–182

Fluorochrome-labeled oligonucleotides

purification of, 147–148

apparatus for, 148

procedure, 148

reagents for, 148

synthesis of, 144, 146–147

procedure, 147

reagents for, 146–147

Footprinting, 538

detection methods for, 570–571

of DNA-binding protein

applications, 586

improvements, 587

in intact cells, 568–587

principle of, 570–575

procedures, 575–586

hybridization, 584–586

preparing DNA samples for electrophoresis, 577–578

preparing membranes for hybridization, 578–580

probe synthesis, 581–584

probing strategy, 580–581

in vivo methylation of cells, 575–577

in vivo

applications of, 569–570

procedures in, 572

results of, typical, 574–575

Forced cloning, 558

Frog, skeletal muscle α-tropomyosin, cDNA for, direct sequencing of, 37